# Mass Spectrometry

## Volume 10

A Specialist Periodical Report

# Mass Spectrometry

Volume 10

A Review of the Recent Literature Published between July 1986 and June 1988

Senior Reporter
**M. E. Rose**, *Department of Chemistry, The Open University*

Reporters
**J. H. Bowie**, *University of Adelaide, Adelaide, Australia*
**P. C. Burgers**, *PFW (Nederland) B.V., Amersfoort, The Netherlands*
**D. A. Catlow**, *Imperial Chemical Industries plc, Macclesfield*
**J. R. Chapman**, *Kratos Analytical Instruments, Manchester*
**J. Charalambous**, *Polytechnic of North London*
**R. P. Evershed**, *University of Liverpool*
**D. J. Harvey**, *University of Oxford*
**J. W. Hastie**, *National Institute of Standards and Technology, Gaithersburg, USA*
**C. Lifshitz**, *Hebrew University of Jerusalem, Jerusalem, Israel*
**F. A. Mellon**, *AFRC Institute of Food Research, Norwich*
**R. A. J. O'Hair**, *University of Adelaide, Adelaide, Australia*
**E. R. Plante**, *National Institute of Standards and Technology, Gaithersburg, USA*
**J. K. Terlouw**, *Utrecht University, Utrecht, The Netherlands*
**A. Viari**, *Institut Curie et Université Paris, VI, Paris, France*
**P. Vigny**, *Institut Curie et Université Paris, VI, Paris, France*
**K. W. P. White**, *Polytechnic of North London*

ROYAL SOCIETY OF CHEMISTRY

ISBN 0-85186-348-5
ISSN 0305-9987

© The Royal Society of Chemistry 1989

*All Rights Reserved*
*No part of this book may be reproduced or transmitted in any form or by any means—graphic, electronic, including photocopying, recording, taping, or information storage and retrieval systems—without written permission from The Royal Society of Chemistry*

Published by The Royal Society of Chemistry
Thomas Graham House, Science Park, Cambridge CB4 4WF

Printed in Great Britain by
Whitstable Litho Printers Ltd., Whitstable, Kent

# Preface

Can you spot the connection between the following chemists: J.H. Bowie, C.J.W. Brooks, M.I. Bruce, R.G. Cooks, I. Howe, G.S. Johnson, S.D. Ward, D.H. Williams, and J.M. Wilson? If you can, you have probably been reading these volumes for twenty years, because they were the contributors to the first volume in the *Specialist Periodical Reports: Mass Spectrometry* series, covering the literature at the end of the 1960s. In the spirit of looking only to the present and future, I will avoid retrospection in this tenth issue of the series. I will note, though, that most of the aforementioned chemists are still practising mass spectrometry, and that John Bowie deserves a medal for perseverance. He is the only mass spectrometrist to contribute a review to each and every one of the ten volumes! As for myself in the present and future, I have resolved not to include any more tongue-in-cheek remarks in these Forewords in case playfulness is again misconstrued as maliciousness.

As always in preparing these volumes I am indebted to a helpful editorial team at the RSC and to Alan Cubitt in particular. Despite relocation from London to new premises in Cambridge, the team has produced this book with little apparent disruption. Regarding the mass spectrometric content, Chava Lifshitz again covers the enormous area of ionization processes and ion dynamics in the first chapter. Because of ill-health, she was unable to cover uni- and bi-molecular reaction dynamics in this volume. I am sure that we all wish her a speedy recovery. Mike Baldwin's many commitments in mass spectrometry have forced him to relinquish his hold on Chapter 2 on structures and reactions of gas-phase positive ions. I thank him for his past contributions and welcome Hans Terlouw and Peter Burgers who

have written an admirable review on this key topic. Over
the years, Fred Mellon has contributed reviews on GC/MS and
LC/MS, and computers in mass spectrometry to *Specialist
Periodical Reports*. Now he returns to cover, in Chapter 3,
yet another topic: instrumentation across organic and
inorganic mass spectrometry. John Chapman and John Bowie
again serve us well with Reports on computers in Chapter 4,
and negative ions in Chapter 5, respectively. The latter is
longer than usual because I asked John Bowie to emphasize
the burgeoning area of anion/molecule reactions. To
facilitate this expanded brief, he enlisted the help of a
co-author, Richard O'Hair.

The analysis of mixtures is again covered in two
separate chapters: GC/MS in the sixth chapter (Richard
Evershed) and the complementary methods of LC/MS, SFC/MS,
CZE/MS, TLC/MS and analytical MS/MS in the following chapter
(David Catlow and myself). David Harvey contributes his
usual and excellent Report on drug metabolism and related
topics. Metal-containing compounds are again reviewed by
John Charalambous, along with his colleague, Keith White.
The two remaining reviews concern natural products and
high-temperature studies of inorganic systems. As I
explained in Volume 9 of this series, if the recurrent
chapter on natural products is to be an in-depth discussion,
the coverage of each contribution has to be restricted to
one branch of natural product chemistry. This is best done
by approaching different specialists to cover the different
areas and this time I approached Paul Vigny regarding the
important topic of nucleosides, nucleotides and nucleic
acids. I am grateful that he and his colleague, Alain
Viari, were able to oblige. In the previous volume, I
increased the proportion of pages devoted to inorganic mass
spectrometry by eliciting a one-off contribution on
quantitative analysis of metals. I have used the same
tactic with this volume, but changed the topic to
high-temperature studies, which have received scant coverage
in this series to date. Because of this, the references
cited in this Report are drawn from a wider time period than
the specified review period, but the review concentrates on
developments in 1986 - 1988. I thank John Hastie and Ernest

## Preface

Plante for this review which forms the last chapter of the book.

Now that the camera-ready copy (CRC) system is used to publish these texts, each Reporter has to produce not only a scientifically accurate but also a grammatically and visually pleasing typescript. There is little opportunity for editing. I thank the authors for undertaking this extra effort without complaint. It is an advantage of the CRC system that it again allows me to include a subject index which, I believe, makes the information in the book much more accessible. No doubt, the efforts of all of us in preparing Volume 10 will be judged in due course by our readers and reviewers.

MALCOLM ROSE

# Contents

| Chapter | 1 | Ionization Processess and Ion Dynamics<br>By C. Lifshitz | 1 |
|---|---|---|---|
| | 1 | Introduction | 1 |
| | 2 | Ionization Processes<br>Molecular Photoionization<br>Multi-photon Ionization<br>Electron Collision and Other<br>Ionization Phenomena<br>Doubly and Multiply Charged Ions<br>Cluster Ions | 1<br>1<br>6<br><br>8<br>13<br>17 |
| | 3 | Spectroscopy and Structure of Ions | 19 |
| | | References | 24 |

| Chapter | 2 | Structures and Reactions of Gas-phase Organic Ions<br>By P.C. Burgers and J.K. Terlouw | 35 |
|---|---|---|---|
| | 1 | Introduction | 35 |
| | 2 | Methodology<br>Ion and Neutral Thermochemistry<br>Photoelectron-Photoion Coincidence (PEPICO) Spectroscopy<br>Dissociation Characteristics<br>Reactivity: Ion/Molecule Reactons<br>Neutralization Reionization Mass Spectrometry<br>NRMS of Organic Ions<br>CIDI (Collision-induced Dissociative Ionization) Spectra of Neutral Species N in the Dissociation of Metastable Ions $m_1^+ \rightarrow m_2^+ + N$<br>Pyrolysis Experiments | 36<br>36<br><br>39<br>40<br>42<br>44<br>45<br><br><br><br>49<br>50 |
| | 3 | Ion Structures<br>Distonic Ions<br>Hydrogen-bridged Radical Cations<br>'Destabilized' Carbenium Ions<br>Multiply Charged Cations<br>Stability Reversal by (Multiple) Ionization | 51<br>51<br>52<br>54<br>54<br>55 |
| | 4 | Reaction Mechanisms<br>The [1,n]-Hydroxycarbene-Radical Migration<br>The [1,2]-Enol-Olefin Shift | 56<br>56<br>59 |

|  |  | Consecutive Enol-Olefin and Hydroxycarbene-Radical Migrations | 60 |
|---|---|---|---|
|  |  | The [1,2]-Amine-Radical Shift | 62 |
|  |  | Intermediate Ion-Neutral Complexes | 63 |
|  |  | The Dissociation Behaviour of Ion-Dipole Complexes | 66 |
|  |  | Methylnitrite | 67 |
|  |  | Remote Charge Site Fragmentations | 68 |
|  |  | References | 69 |

Chapter 3 **Developments and Trends in Instrumentation** 75
*By F.A. Mellon*

1 Introduction 75

2 Ionization Methods 76
   Lasers 76
   Secondary Ion Sources 81
   Plasma Desorption 83
   Rare Gas Plasma Sources 84
   Glow Discharge 86
   Atmospheric Pressure Ionization (API) 87
   Chromatography/Mass Spectrometry 87
   Miscellaneous 88

3 Mass Analysers 89
   Magnetic Sector Instuments 89
   Quadrupoles and Quadrupole Ion Storage Traps 92
   Fourier-transform Ion Cyclotron Resonance 97
   Time of Flight 99
   Accelerator Mass Spectrometry 101
   Ion Imaging 102
   Miscellaneous 103

4 Detectors 103

5 Sample Introduction 105

6 Integrated Techniques 105

  References 107

Chapter 4 **Application of Computers and Microprocessors in Mass Spectrometry** 118
*By J.R. Chapman*

1 Introduction 118

2 Instrumentation (Instrument Control and Data Acquisition) 119
   Computer Instrumentation 119
   Expert Systems and Robotics 120
   Tandem Mass Spectrometry (MS/MS) 121
   Inorganic Mass Spectrometry 122
   Isotopic Analysis 124

## Contents

|   |   |   |
|---|---|---|
| | Thermal Desorption | 124 |
| | Time-of-Flight Mass Spectrometry | 125 |
| | Fourier Transform Mass Spectrometry (FTMS) | 126 |
| | Process Control | 126 |
| | Miscellaneous Instruments | 127 |
| 3 | Data Analysis | 127 |
| | GC/MS | 127 |
| | Library Search and Related Techniques | 130 |
| | Spectrum Interpretation: Use of Other Spectroscopic Data | 131 |
| | Interpretation of MS/MS Data | 133 |
| | Pattern Recognition | 133 |
| | Biopolymer Sequencing | 134 |
| | Miscellaneous Organic Analysis | 135 |
| | Pyrolysis-Mass Spectrometry | 136 |
| | Inorganic Analysis | 137 |
| 4 | Other Software | 138 |
| | Instrument Design | 138 |
| | Miscellaneous | 138 |
| | References | 139 |

Chapter 5 **Organic Negative Ions: Structure, Reactivity, and Mechanism** 145
*By R.A.J. O'Hair and J.H. Bowie*

| | | |
|---|---|---|
| 1 | Introduction | 145 |
| 2 | Negative Ions Formed by Electron Capture (or Dissociative Electron Capture): Experimental and Theoretical | 145 |
| 3 | Negative Ion Chemical Ionization Mass Spectrometry | 148 |
| 4 | Negative Ion Fast Atom Bombardment Mass Spectrometry | 151 |
| 5 | Other Ionization Techniques | 153 |
| 6 | Ion-Molecule Reactions and Related Topics | 155 |
| 7 | Conclusion | 163 |
| | References | 163 |

Chapter 6 **Analysis of Mixtures by Mass Spectrometry Part I: Developments in Gas Chromatography/Mass Spectrometry** 181
*By R.P. Evershed*

| | | |
|---|---|---|
| 1 | General Considerations | 181 |
| | Introduction | 181 |
| | Instrumentation | 181 |
| | The Role of Data Systems | 184 |

|   |   |   |   |
|---|---|---|---|
|   |   | Quantification | 185 |
|   |   | Sampling | 185 |
|   |   | Chromotographic Aspects | 186 |
|   |   | Derivatization | 187 |
|   |   | Stereo- and Positional Isomeric Assignments | 190 |
|   | 2 | Applications | 192 |
|   |   | Long-chain Compounds | 192 |
|   |   | Prostaglandins and Related Eicosanoids | 192 |
|   |   | Steroids | 193 |
|   |   | Carbohydrates | 195 |
|   |   | Amines | 196 |
|   |   | Amino Acids and Peptides | 198 |
|   |   | Clinical and Metabolic Studies | 199 |
|   |   | Food and Agricultural Chemistry | 201 |
|   |   | Environmental Science and Technology | 202 |
|   |   | Organic Geochemistry and Fuel | 209 |
|   |   | Pyrolysis-GC/MS | 210 |
|   |   | References | 212 |

Chapter 7 **Analysis of Mixtures by Mass Spectrometry Part II: Techniques Other than Gas Chromatography/Mass Spectrometry** 222
*By D.A. Catlow and M.E. Rose*

|   |   |   |
|---|---|---|
| 1 | Introduction | 222 |
| 2 | Combinations with Capillary Zone Electrophoresis and Thin Layer Chromatography | 222 |
| 3 | Supercritical Fluid Chromatography/ Mass Spectrometry |   |
|   | Overview | 224 |
|   | Interfaces | 225 |
|   | Applications | 228 |
| 4 | High-performance Liquid Chromatography/ Mass Spectrometry |   |
|   | Overview | 229 |
|   | Thermospray LC/MS | 230 |
|   | Transport Devices | 234 |
|   | Direct Liquid Introduction Methods | 236 |
|   | Summary | 238 |
| 5 | Tandem Mass Spectrometry |   |
|   | Overview | 239 |
|   | Applications | 239 |
|   | References | 245 |

Contents xiii

| Chapter | 8 | Mass Spectrometry Applied to Natural Products: Nucleosides, Nucleotides and Nucleic Acids<br>*By P. Vigny and A. Viari* | 253 |
|---|---|---|---|
| | 1 | Introduction | 253 |
| | 2 | Nucleosides | 254 |
| | 3 | Nucleotides | 259 |
| | 4 | Dinucleotides | 261 |
| | 5 | Oligonucleotides | 266 |
| | | References | 270 |

| Chapter | 9 | The Use of Mass Spectrometry in Studies of Drug Metabolism and Pharmacokinetics<br>*By D.J. Harvey* | 273 |
|---|---|---|---|
| | 1 | Introduction | 273 |
| | | General | 273 |
| | | Books and Reviews | 273 |
| | 2 | General Comments | 273 |
| | | Current Trends | 273 |
| | | Artifacts and Contamination Problems | 275 |
| | 3 | Quantitative Studies | 276 |
| | 4 | Metabolic Studies | 281 |
| | | Model Compounds | 281 |
| | | Drug Metabolism Studies | 281 |
| | 5 | Anticancer Drugs | 281 |
| | 6 | Antimicrobial Drugs | 290 |
| | 7 | Drugs of Abuse | 292 |
| | | Drug Screening | 292 |
| | | Cannabinoids | 293 |
| | | Nicotine | 294 |
| | | Opiates | 294 |
| | | Phencyclidine | 295 |
| | | Amphetamines | 295 |
| | | Cocaine | 295 |
| | 8 | Cardiovascular Drugs | 295 |
| | | Beta Blockers | 296 |
| | | Dihydropyridine Calcium Antagonists | 297 |
| | | Cardiac Glycosides | 297 |
| | | Angiotensin-converting Enzyme Inhibitors | 297 |

| | 9 | Drugs Affecting Central Function | 298 |
|---|---|---|---|
| | | Tricyclic Antidepressants | 298 |
| | | Phenothiazines | 298 |
| | | Benzodiazepines | 299 |
| | | Barbiturates | 299 |
| | | Valproic Acid | 299 |
| | | Other Centrally Active Drugs | 300 |
| | 10 | Steroids | 301 |
| | 11 | Drugs Used to Treat Pain and Inflammation | 302 |
| | | Non-steroid Anti-inflammatory Agents | 302 |
| | | Paracetamol and Phenacetin | 302 |
| | | Local Anaesthetics | 302 |
| | | Analgesics | 303 |
| | | Antihistamines | 303 |
| | 12 | Other Miscellaneous Studies | 303 |
| | | References | 304 |

| Chapter | 10 | **Metal-containing and Inorganic Compounds Investigated by Mass Spectrometry** *By J. Charalambous and K.W.P. White* | 323 |
|---|---|---|---|
| | 1 | Introduction | 323 |
| | 2 | Main-group Organometallic Compounds | 323 |
| | 3 | Transition-metal Organometallic Compounds | 326 |
| | | Carbonyl and Related Compounds | 326 |
| | | Complexes Containing Hydrocarbon Ligands | 329 |
| | | Transition-metal Cluster Compounds | 336 |
| | 4 | Chelate, Macrocyclic, and Other Complexes | 338 |
| | | Neutral Chelates | 338 |
| | | Anionic Complexes | 341 |
| | | Cationic Complexes | 342 |
| | | Carboxylate and Related Complexes | 343 |
| | | Macrocycles | 344 |
| | 5 | Miscellaneous Inorganic Compounds | 346 |
| | 6 | Reactions of Gaseous Metal and Metal-containing Ions with Organic Compounds | 349 |
| | | References | 350 |

*Contents*  xv

| Chapter | 11 | High-temperature Mass Spectrometric Studies of Inorganic Systems<br>By E.R. Plante and J.W. Hastie | 357 |
|---|---|---|---|
| | 1 | Introduction | 357 |
| | 2 | Classical Method | 358 |
| | 3 | Electron Impact Ionization Cross Sections | 361 |
| | 4 | Recent Literature Data | 364 |
| | | Ion-Molecule Measurements | 364 |
| | | Gaseous Oxide Measurements | 365 |
| | | Activity of Oxide Systems | 365 |
| | | Simple Halide Systems | 365 |
| | | Mixed Halide Systems | 366 |
| | | Alloy Systems | 366 |
| | | Oxy-salts | 366 |
| | | Group VB and VIB | 366 |
| | | Appendix: Tables | 367 |
| | | References | 374 |

**Subject Index**  379

**Author Index**  388

# 1
# Ionization Processes and Ion Dynamics

BY C. LIFSHITZ

## 1. Introduction

Covering the literature concerning ionization processes and ion dynamics over a period of two years and doing justice to the field has become a formidable task, in view of the enormous expansion in these areas. It would perhaps be advisable to divide in the future the topic into two sub topics, one dealing with basic aspects of ionization processes (including dynamic aspects) and the other dealing with basic aspects of ion fragmentations and ion-molecule reactions. I have concentrated in this chapter on ionization processes. My task has been, as in the last review[1] to present work related to fundamental aspects of the behavior of relatively simple molecular ions. Other reviews of the mass spectrometry literature[2] are concentrating on the more applicative nature of ionization processes, unimolecular fragmentations and ion-molecule reactions with emphasis on the general trend in analytical mass spectrometry today, namely towards larger and larger molecular ions of biological importance.

## 2. Ionization Processes

**2.1 Molecular Photoionization.** - The emphasis in recent years has been on photoionization dynamics. Great progress has been made in experiments due to the development of synchrotron radiation giving continuously variable photon energies and in theory, since theoreticians are now able to perform calculations of the electronic continuum in the molecular field.[3] Several review articles have appeared on molecular photoionization.[3-6] Recent research has dealt with resonances which are quasi discrete states embedded in the ionization continuum and characterized by a finite lifetime. These resonances lead to peaks or dips in the ionization cross section. Three major categories have been recognized as explained previously:[1] Shape resonances, autoionization resonances and Cooper minima. Recently, each of these phenomena has been discussed separately.[7-9] Molecular shape resonances in diatomic molecules, which have no counterpart in atoms, are due to the existence of the $\sigma^*$ antibonding molecular orbital. Two diabatic states, the $R_\sigma$ Rydberg and a $\sigma^*$ valence state mix and give rise to two adiabatic states. One of these adiabatic states, which has $\sigma^*$ character at low internuclear distance and becomes a Rydberg orbital at large distance, constitutes the shape resonance,[7] reached by ionization from a bound $\sigma$ orbital. Shape resonances have been studied recently experimentally and theoretically for several diatomic[7,10-14] and polyatomic[15-19] molecules. The investigation of the valence shells of benzene[19] has allowed a comprehensive evaluation of the ability to predict the photoelectron dynamic properties of a moderately complex polyatomic system. The energy of the maximum in the ionization cross section due to a shape resonance is strongly dependent on the internuclear distance.[7] A search for a quantitative relationship between shape-resonance energies and bond lengths has been made[20] with the aim of developing a new analytical method competitive with EXAFS (extended X-ray absorption fine structure).

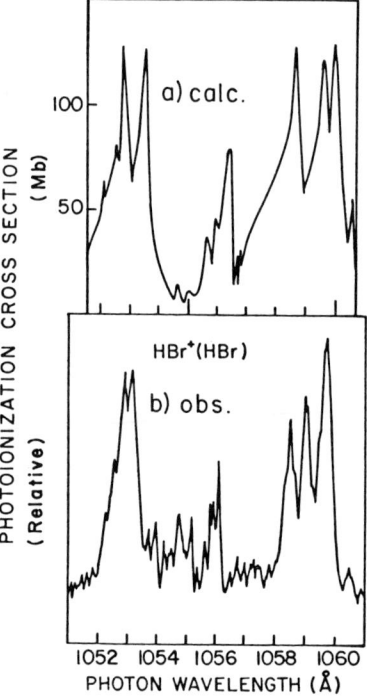

**Figure 1:** (a) Calculated photoionization cross section for HBr; (b) Experimental results
*(Reproduced with permission from "Molecules in Physics, Chemistry and Biology", Vol. II, J. Maruani (ed.), D. Reidel Publishing Co., 1988).*

# 1: Ionization Processes and Ion Dynamics

Autoionization resonances are due to couplings between the continuum and resonant Rydberg states. For light molecules electrostatic coupling terms are important while for heavy molecules it is the spin-orbit interaction which is important.[3] Electrostatic autoionization has been studied in detail[8,21,22] for CO and other diatomic molecules.[23,24] Spin-orbit autoionization has been studied for HBr[25] and HCl[26]. Both types of autoionization have been treated very successfully by multichannel quantum defect theory (MQDT) which treats the autoionization process on the basis of a breakdown of the adiabatic Born-Oppenheimer approximation.[23] A comparison between the calculated photoionization cross section for HBr and the experimental results[3,25] is reproduced in Figure 1. Twenty-four autoionizing resonances appear in the energy region studied, between the $^2\Pi_{3/2}$ and $^2\Pi_{1/2}$ ionic limits, and their wavelength positions, cross sections, widths and tentative assignments have been determined.[25] Fluorescence has been employed as a probe of molecular autoionization in $N_2O$,[27] $CS_2$[28] and HCl.[29] When predissociative levels mediate the autoionization process this leads to significant deviations from the vibrational autoionization "propensity rule"[29], which states that the change in the vibrational quantum number is $\Delta v = -1$ or that it is minimal. The fate of core excited molecules, $SiH_4$,[30] HI,[31] $CH_3I$,[31] HBr[31] and $CH_3Br$,[31,32] has been studied in detail recently. Two-step decay processes have been proposed for valence resonances – a fast dissociation followed by the autoionization of the excited fragment. In general, the dynamics of vibronic autoionization of Rydberg states of polyatomic molecules, including Rydberg-valence vibronic coupling via nontotally symmetric modes is now coming under theoretical scrutiny,[33] as more refined high resolution photoionization and threshold photoelectron spectra become available experimentally for polyatomic molecules.[34-36]

Many-body effects lead to a spreading of the spectral intensity of the inner-valence states over several lines in photoelectron spectra. This breakdown of the molecular orbital model has been explained in some detail in the previous review.[1] It has been studied further recently, employing high-energy-resolution synchrotron radiation on $H_2S$,[37] $NH_3$[38a,b], $HF$[38c] and $CS_2$.[39] Many valence satellites, some which have not been seen previously by binary (e,2e) spectroscopy and others which have not been predicted by theory, have been observed. Valence ionization spectra have been calculated for quite complex molecules such as $TiCl_4$,[40] o-Benzyne,[41] cyano-derivatives of organic molecules,[42] first and second row transition metal diatomics,[43] $S_2N_2$[44] and $S_3$,[45] quinone and benzene like molecules,[46] p-nitroaniline[47] by an *ab initio* Green's function formalism[48] that takes the effects of electron correlation and relaxation into account. In some cases satellite lines appear at very low energies.[47] Ionization energies are calculated by the extended two-particle-hole Tamm-Dancoff approximation (extended

2ph-TDA) which describes the satellite lines in a photoelectron spectrum by including the mixing of single hole with two hole one particle (2h1p) configurations.

The method called (e,2e) spectroscopy has been reviewed previously[1] together with molecular photoionization. Its lower energy resolution is giving way to using tunable synchrotron radiation as the method of choice for studying configuration interaction states discussed above. It has, however, a unique feature not available to photoionization, which makes it a powerful emerging technique for the study of molecular orbitals and the laboratory investigation of molecular wavefunctions and chemical bonding. The technique is now being called electron momentum spectroscopy (EMS) and has been reviewed recently.[49] EMS measurements provide information on orbital imaging and have thus given an empirical aspect to the orbital pattern concept, which was, until the advent of this method, of a somewhat unreal nature.[50] EMS is based upon the high energy binary (e,2e) reaction. It employs high energy electron impact ionization and measures all the necessary energies and angles of the particles involved, in coincidence. Energy conservation yields information on the binding (ionization) energy of the target and leads to information analogous to photoelectron spectra.[1] Momentum conservation yields information on the momentum p of the bound electron (which is ionized in the collision).[49] The measured (e,2e) cross section is directly proportional to the momentum density $\rho(\mathbf{p})$. Let $\psi(\mathbf{r})$ be a wave function; the quantity $|\psi(\mathbf{r})|^2$ has the interpretation of electron charge (probability) density in position (r) space. EMS measures the momentum probability density,

$$\rho(\mathbf{p}) = |\psi(\mathbf{p})|^2 \qquad \text{I}$$

and the momentum space wave function $\psi(\mathbf{p})$ is related to $\psi(\mathbf{r})$ by the Fourier transform[49],

$$\psi(\mathbf{p}) = (2\pi)^{-3/2} \int d\vec{r} \exp(-i\mathbf{p} \cdot \mathbf{r}) \psi(\mathbf{r}) \qquad \text{II}$$

To make comparison with experimental results, one calculates orbital densities by squaring the Fourier transform of wave functions obtained from SCF-MO calculations. EMS is a sensitive probe of phase space corresponding to small p, and therefore to the region of larger r which is particularly important for chemical bonding and reactivity. It has provided information on the methyl inductive effect by indicating appreciable delocalization of electron density away from the nitrogen in methyl amines.[50] It provides a powerful experimental tool in quantum chemistry for the testing, evaluation and design of molecular wavefunctions.[51] Quite a number of diatomic, triatomic and polyatomic molecules have been studied by EMS, including $H_2$,[49] NO,[49]

$Cl_2$,[52] $Br_2$,[53] $CO$,[54] $H_2O$,[49,55] $D_2O$,[55] $CO_2$, $COS$ and $CS_2$,[49] $H_2S$,[51] $NH_3$,[56,57] $NF_3$,[50,57] methyl amines,[50,57] $CF_4$,[49] $CH_3F$ and $CH_3Cl$.[58] Figure 2 reproduces a comparison between experimental and calculated momentum distributions for the $2b_1$ orbital of $H_2S$.[51] Calculations were carried out at various levels of theory and inclusion of correlation was found to have a minimal effect on the momentum distributions. Many of these (e,2e) studies measured binding energy spectra which demonstrate satellite lines. For example, in $H_2S$,[51] extensive many-body states arise from the $4a_1^{-1}$ hole state.

Absolute total absorption, photoionization and dissociative photoionization cross sections of ammonia have been measured from 80 to 1120 Å [59].

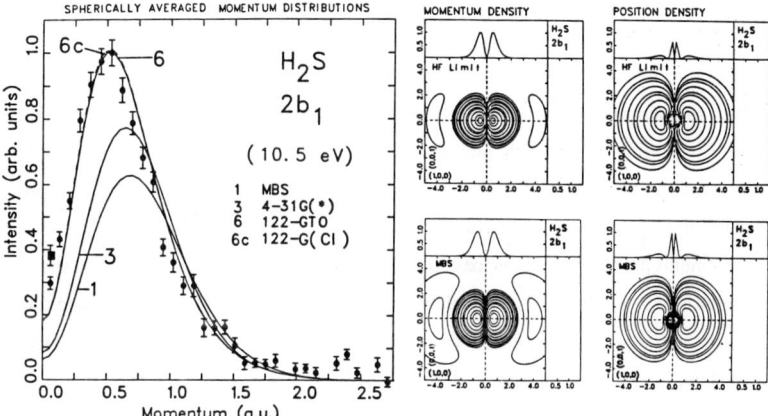

**Figure 2:** Comparison of the experimental momentum profile of the $2b_1$, orbital of $H_2S$ with calculated spherically averaged momentum distributions; minimal basis set wavefunction - MBS; Gaussian 76 calculation using the 4-31G* basis - 4-31G(*); Hartree Fock limit - 122-GTO; CI overlap - 122- G(CI). Also shown are orbital density maps in position and momentum space generated from MBS (bottom panels) and 122-GTO (top panels) wavefunctions. All dimensions are in atomic units. Contours are at 0.02, 0.05, 0.08, 0.2, 0.5, 0.8, 2, 5, 8, 20, 50 and 80% of the 122-GTO maximum intensity. The side panels of the density maps show density profiles along the dashed lines of the density maps.

(Reproduced with permission from Chem. Phys., 1988, 122, 247).

**2.2 Multi-photon Ionization.** - Molecular dynamic photoelectron spectroscopy using resonant multiphoton ionization (REMPI-PES) provides opportunities for the study of the dynamics of excited state photoionization and autoionization processes, the observation of neutral and ionic states that are symmetry forbidden in single photon excitation and the study of photodissociation and intramolecular relaxation. Several new review articles on MPI-PES have appeared recently.[60-62] Detailed experimental studies of two diatomic molecules, $H_2$[60-66] and NO[60-62,67-69] permit a direct comparison with theory. Photoelectron angular distributions have been compared with available theoretical calculations both for MPI[63] and for single photon ionization[70] of $H_2$. The need for substantial progress in understanding the photo-ionization dynamics of even the simplest excited molecular states has been pointed out.[63] Major contributions have come through the study by REMPI-PES of rotational and vibrational branching ratios of excited molecular states.[61,63] As expected on the basis of Franck-Condon arguments, photoionization via a particular vibrational level v' of the $C^1\Pi_u$ state of $H_2$ leads to a photoelectron spectrum strongly peaked at the ionic vibrational level $v^+ = v'$. Rotational propensity rules in photoionization were studied for $H_2$[61,63] and NO,[68] since specific rovibronic levels of different electronic states of the neutrals could be excited. In single photon ionization, such experiments are not possible since resolution is inadequate to resolve rotational structure and ionization takes place from a thermal distribution of rotational levels of the ground state.[68] The electronic excitation of the ion core is retained following photoionization of a molecular Rydberg state;[61] as a result ionization of $N_2$ via the $o_3$ $^1\Pi_u$ state leads exclusively to the $A^2\Pi_u$ state of the ion. If the rotational level of the resonant intermediate state is properly chosen, the angular momentum selection and propensity rules for single photon ionization of the excited neutral intermediate indicate that the ionic state might be produced in a single rotational level.[61] Thus, it is possible in the case of $N_2$ to produce an electronically excited state of the ion in a single vibrational and rotational state. The rotational distribution of $N_2^+$ formed by REMPI has been recently measured by laser induced fluorescence (LIF) and rotational propensity in the photoionization from the c' $^1\Sigma_u^+$ and $c^1\Pi_u$ states was further studied.[71] The autoionization dynamics of individual rotational levels has been studied for NO in the $9d\sigma\pi$, v = 2 band.[61,67] The branching ratio into the NO$^+$ $v^+ = 1$ state increases significantly on the autoionizing resonances, in accord with the $\Delta v = -1$ propensity rule for vibrational autoionization.

An added advantage of REMPI-PES lies in the ability to determine the ionic energy levels of species that are present in small concentrations in the presence of a species with an interfering PES. REMPI is used to selectively ionize the species of interest while leaving the interfering species unexcited. This has been applied to rare gas dimers such as $Xe_2$ in the presence of a large excess of their monomers, (Figure 3).[61,72]

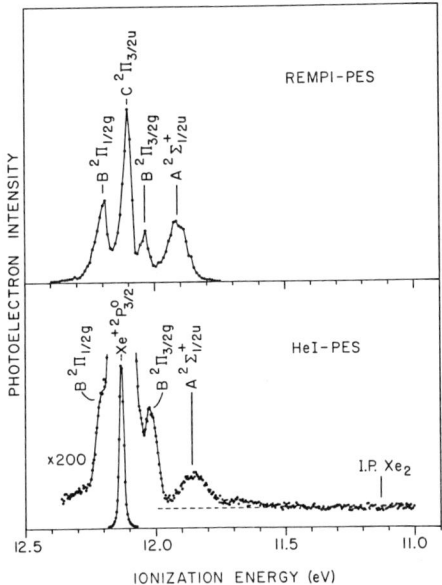

**Figure 3:** Illustrative example of photoelectron spectra of $Xe_2$ obtained by using HeI ionization (lower trace) and resonantly enhanced multiphoton ionization (upper trace) in the region of the $Xe^+$ $^2P_{3/2}$ ionization limit. Note that, in the HeI-PES, atomic ionization is more than 100 times as intense as the molecular ionization, but that no atomic ionization is observed in the PEMPI-PES.

*(Reproduced with permission from J. Phys. Chem., 1987, 91, 2593).*

MPI ion-current spectra have been obtained as a function of laser wavelength in the U.V./visible region[60] either on their own or in conjunction with MPI photoelectron spectra. Ion-current spectra were obtained for $H_2$,[64,66] $D_2$,[65] NO,[60,67,69,73] $O_2$,[74] $Xe_2$,[75] $C_2H_2$,[76,77] cyclic ketones,[78,79] and other molecules, yielding information about the structure and dynamics of the neutral intermediate states. A theory has been presented[80] to reduce 1+1 resonance enhanced multiphoton ionization spectra to accurate rovibrational state population distributions and alignment factors. The degree of alignment of desorbing NO molecules, which is a sensitive measure of desorption dynamics, was probed via 1+1 REMPI.[81] MPI ion-current spectra in combination with PES have detected the UV multiphoton dissociation of iron complexes.[60,82,83] Other studies of large polyatomic molecules involved the autoionization in diazabicyclooctane (DABCO)[84] and ionization thresholds of phenol-$(NH_3)_n$ clusters.[85]

**2.2.1. Mass-spectrometric Studies.** - MPI-MS and unimolecular ion decay have been reviewed recently[86]. Both experimental and theoretical aspects were covered in depth. The major breakthrough in this area has been the ability to produce state- and internal energy-selected molecular ions[86,87], which led to very significant contributions to the understanding of unimolecular fragmentations of polyatomic ions. Species selectivity, isomer specificity, soft ionization and hard fragmentations are now well understood on the basis of the unique excitation mechanism in multi-photon mass spectrometry[86]. Several of these aspects such as soft ionization[88], isomer specificity[89], wavelength-dependent fragmentation[90,91] and intensity dependent fragmentation[92] were discussed in several recent publications. The high relative abundance of $C^+$ in MPI-MS of benzene, particularly at higher laser intensities[86,92] has attracted considerable attention over the years. Recent experiments employing photoelectron spectroscopy[92] have unambiguously identified the presence of $^1D$ carbon atoms. While the contribution to the $C^+$ signal from fragmentation of larger ions is greater than from ionization of neutral carbon atoms[92], future theoretical interpretations of MPI-MS should take into account ionization of neutral fragments.

MPI-MS has been employed to probe non-fluorescing systems[93]. Electronically excited Ã states of ammonia clusters have been revealed by two-photon ionization mass spectrometry[94]. Picosecond MPI-MS has been used to monitor the dissociation of neutral methyl iodide excited to its A continuum[93]. A pulsed laser-pulsed extraction field technique has been employed[95] to differentiate between a statistical ladder and a nonstatistical ladder-switching mechanism in the 266-nm picosecond and nanosecond laser multiphoton absorption process by methyl iodide. A time-lag delay has been shown to be an important experimental parameter in the measurement of fragmentation patterns in MPI-MS[96].

**2.3 Electron Collision and Other Ionization Phenomena.** - Electron momentum spectroscopy (EMS), or (e, 2e) spectroscopy has been discussed in section 2.1 on molecular photoionization, since this is done traditionally in view of the analogy between photoionization and high energy (keV) electroionization. This section will deal mainly with low energy electroionization processes. Electron impact ionization has been reviewed in depth recently[97]. The review covered (1) the basics of the interaction between an ionizing electron and the target species,

# 1: Ionization Processes and Ion Dynamics

(2) the types of ions produced, (3) ionization efficiency and partial electron impact ionization cross sections. Electron impact ionization cross sections have been covered in additional monographs by Märk and coworkers[98-100]. Of special interest recently has been the ionization of clusters[97,100,101]. The chemistry of gas-phase ion clusters has been reviewed in volume 9 of this series[102] but is not covered in a separate chapter of this volume. Cluster ions will be discussed briefly in a separate section of this chapter, section 2.5.

Total ionization cross sections have been determined experimentally for $H_2O$ and $D_2O$[103]. Electron impact ionization cross sections have been calculated by utilizing empirically modified collision theories[99]. Corrections were made for deviations from the additivity of atomic ionization cross sections in fluorine containing compounds ($CF_4$, $SF_6$, $UF_6$, etc.) and calculated values were compared with available experimental results[99]. Ion and electron impact ionization of $SiF_4$ was studied via UV emission[104].

Electroionization spectra were measured for ethylene[105,106] and ethane[107,108]. Structures observed in the ionization efficiency curves were attributed to negative ion states (Feshbach resonances) and/or Rydberg autoionizing states. These studies involved monoenergetic electron impact and were done with very high signal-to-noise ratios. This reviewer has suggested long ago[109], on the basis of rather crude low-energy-resolution electron impact ionization experiments, that competitive decomposition processes influence ionization efficiency curves in ethylene. Lifshitz and Long[109] claimed that electron ionization of molecules shows competition between various final ions for their share of the total cross section and that the parent ethylene ion cross section becomes depleted at energies sufficient to produce a fragment ion. This idea came under sharp criticism from Rapp[110], who was unable to discover a 'pronounced decrease' in our published ionization efficiency curve for $C_2H_4^+$ at the onset of fragmentation. The new high quality data of Plessis and Marmet[106] clearly demonstrate competition between $C_2H_4^+$ production and $C_2H_3^+$ and $C_2H_2^+$ production in their respective ionization efficiency curves (Figure 4). It was clear to us several years ago[111] that direct ionization cannot be the cause for the observed competition. The recent results[106] indicate that features occurring at the same energy in several of the curves are due to a common progenitor, the $C_2H_4^{-*}$ complex. It is thus also obvious why this phenomenon is not apparent in photoionization of ethylene[111]. It has been suggested[106] that the $C_2H_4^{-*}$ complex is formed during impact with low-energy electrons because the interaction time between the molecule and the electron is relatively long, much longer than for photons that travel at the speed of light. The old idea of 'competition' in electron ionization processes[109] has thus been revived[106]. Another molecule for which a sharp drop in the parent electroionization efficiency curve, at the onset of fragmentation, has been claimed, is ammonia[109]. This has now clearly been demonstrated also in photoionization experiments[59] and thus cannot be the result of a negative ion complex intermediary.

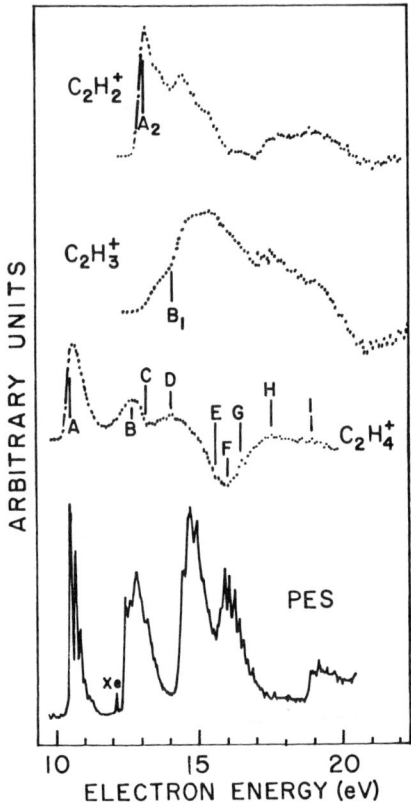

**Figure 4:** Filtered electron ionization efficiency (EIE) curves which bring out high frequency (narrow) structures of the original EIE curves. The EIE curves are for $C_2H_2^+$ (top), $C_2H_3^+$ (middle), $C_2H_4^+$ (bottom) ions from $C_2H_4$; PES - photoelectron spectrum for $C_2H_4$. The energy of peak $A_2$ (13.14 ± 0.03 eV) at the onset of the $C_2H_2^+$ curve coincides precisely with the energy of structure C (13.15 eV) of the $C_2H_4^+$ curve. The energy of peak $B_1$ of the $C_2H_3^+$ curve coincides with the energy of structure D of the $C_2H_4^+$ curve.

*(Adapted with permission from Can. J. Phys., 1987, 65, 803).*

## 1: Ionization Processes and Ion Dynamics

Rotational and vibrational distributions of $H_2^+$ formed in electroionization experiments have been topics of several recent studies. The rotational distribution of $H_2^+$ ions produced by 90 eV electron impact has been determined for the first time[112]. Rotational temperatures were found to decrease from 127 K for the vibrational level v = 7 of the ion to 110K for v = 11. A model that conserves angular momentum N during the ionization process was invoked to explain the results[112]. Ionization of high velocity neutral beams has been achieved in recent years through collision with various targets in neutralization-reionization mass spectrometry[113,114], which will be discussed in greater detail in the chapter on structures and reactions of gas-phase ions. Electron ionization[115] and field ionization[116] of fast neutrals has also been achieved. The latter method has been found particularly suitable for the ionization of high-Rydberg (HR) states[116-118]. Collisions between high-velocity (keV energies) diatomic ions ($H_2^+$, $D_2^+$, $N_2^+$, $O_2^+$) and noble gas target atoms (Xe, He) produced high-Rydberg molecules, ions and atoms. The neutral HR species were field-ionized and kinetic energy spectra of the resulting ions were obtained. Vibrational energy distributions of $H_2^+$ ion beams originating from 70 eV electron ionization of $H_2$, $H_2O$, $CH_4$ and $H_2S$ were characterized by this technique[118].

Penning ionization by $He(2^3S)$ and $He(2^1S)$ of $Cl_2$[119], $Br_2$ and $I_2$[120,121], ICl and IBr[121] has centered around 'transition state spectroscopy with electrons' a term used by analogy with 'optical transition state spectroscopy' which is a method employed for the investigation of the transition state region in the potential surfaces of chemical reactions. The Penning ionization reactions proceed predominantly by formation of an intermediate ion pair complex, ($He^+$ + $X_2^-$), where X is a halogen atom; these are autoionizing collision complexes. The experimental studies involve determinations of electron energy spectra and electron-ion-coincidence spectra which, with the aid of classical trajectory calculations, yield information on the reaction dynamics.

**2.3.1. Negative Ions.** - Electron attachment and detachment processes in electronegative gases have been reviewed recently[122]. Processes which were discussed include attachment to vibrationally/rotationally and electronically excited molecules, autodetachment, photodetachment and collisional detachment. The electron energy dependence and temperature dependence of the electron attachment rate constant have been measured for $F_2$[123], $n-C_4F_{10}$[124] and $c-C_4F_8$[125]. The cross section for dissociative attachment generally increases with temperature, T, while that for nondissociative attachment generally decreases with $T$[122-125]. Knowledge concerning interactions between electrons and excited molecules is of importance in connection with the control of electrical conduction and insulation properties of gaseous matter[122]. It has been demonstrated[126] that laser irradiation can be employed to produce large numbers of long-lived, electronically excited molecules which attach electrons with cross sections $10^5$-$10^6$ times larger than the respective unexcited species. This is achieved by strong absorption to an allowed singlet electronic state followed by fast internal conversion to a long-lived triplet state.

Sulfur hexafluoride, which has an electron capture cross section close to the maximum theoretically allowed value, remains a molecule of great interest for fundamental reasons as well as for its applicative qualities. Cross sections for parent and fragment negative ions have been determined over a wide energy range for $SF_6$[122,127], and $SF_5Cl$[127]. Of special interest over the years have been the autodetachment (or 'autoneutralization') lifetimes observed for $SF_6^-$. Recent measurements[128,129] yielded much more refined autoneutralization lifetime values through direct measurements of neutral $SF_6$ intensities. Autoneutralizing $SF_6^-$ ions have been developed into high-energy neutral molecular beams[130-132], which are applicable for secondary ion mass spectrometry (SIMS) of electrical insulators.

Fragmentation dynamics and energy partitioning in dissociative attachment of small molecules have been reviewed recently[133]. Systems included were $H^-/H_2O$ and $H^-/NH_3$, dissociations which take place on saddle point potential surfaces, $O^-/CO_2$, $O^-/SO_2$ and $S^-/CS_2$ which are dissociations on potential well surfaces, $Cl^-/HgCl_2$ and $Br^-/HgBr_2$ which are dissociations on purely repulsive surfaces. Ion kinetic energy analysis and angular distributions have given detailed information on the fragmentation dynamics.

Dissociative electron attachment, electron transmission spectroscopy and electron energy loss spectroscopy have been employed to study shape resonances and Feshbach resonances in $CS_2$[134], $C_2H_2$[135] and acetaldehyde[136]. Resonant vibrational excitation via shape and Feshach resonances has been studied with the use of a trochoidal electron spectrometer[137,138]. Resonant autodetachment of Feshach resonances in polyatomic molecules, for example in acetaldehyde, has been demonstrated through energy spectra of the detached electrons, to lead to highly excited vibrational levels of the neutral molecules. This has been termed "horizontal" decay[137] (as opposed to vertical near-Franck-Condon detachment of shape resonances in $N_2$ and $H_2$) and is due to very efficient coupling of the electronic and nuclear motions. "Horizontal" decay is phenomenologically related to resonant autoionization, which is usually revealed by threshold photoelectron spectroscopy populating final states of the same energy as the resonance, which often lies in the "Franck-Condon gaps" of the molecule[137].

A determination of the symmetry of $I_2^-$ resonant states, responsible for $I^-$ ion plus

*1: Ionization Processes and Ion Dynamics* 13

I($^2P_{3/2}$, $^2P_{1/2}$) atom formation in iodine molecules, has been reported[139]. The analysis relied on kinetic energy and angular momentum measurements of I$^-$ ions.

Several experiments with high energy ion beams have been performed in which negative ions were formed from positive ions by some type of charge reversal. The first of these[140] involves production of H$^-$ ions by double capture (double charge transfer) in proton-H$_2$ collisions, H$^+$ + H$_2$ → H$^-$ + H$_2^{2+}$. The energy distribution of H$^-$ served to show that the double ionization of H$_2$ follows a Franck-Condon type behavior. Collision-induced production of H$^-$ from H$_2^+$ on H$_2$ is a second experiment, for which several mechanisms have been proposed[141]. H$_3$O$^-$, H$_2$O$^-$ and NH$_4^-$ were topics of several recent studies on charge inversion and collision induced dissociations[142-145]. Finally, the involvement of neutral species in the NO$^-$ → O$^+$ · charge inversion has been studied[146].

**2.4 Doubly and Multiply Charged Ions.** - Doubly and multiply charged ions have attracted considerable attention in recent years. Of special interest are the single-photon double ionization experiments which have become possible with Synchrotron radiation[147]. Added to the arsenal of experimental coincidence methods has been the method of photoelectron photoelectron coincidence (PEPECO) in which the energies of both photoelectrons are measured, so that the energy transfer is completely determined[147]. Double charge transfer to bare protons (discussed briefly in the previous section) populates in the case of ground state, closed-shell species solely singlet states of the doubly charged ions[148]. On the other hand, double photoionization of closed-shell molecules is now known to populate both triplet and singlet states of doubly charged ions[148], but also to be subject to selection or propensity rules which are slowly coming to light[149].

Of particular significance is the discovery of emission[150] in NO$^{2+}$ which raises the question regarding the existence of certain quasibound excited states of doubly charged ions[151]. The emission was discovered[150] by using the PIFCO (photoion-photon of fluorescence coincidence) method. The fluorescence efficiency was measured as a function of the photon energy, indicating that the emitting state, whose onset was at 42.5 eV, is a stable or slowly predissociating state of NO$^{2+}$. The state was identified[150] as B$^2\Sigma^+$ and the emission was ascribed to the transition B$^2\Sigma^+$ → X$^2\Sigma^+$. Additional emission, observed more recently, was ascribed to the A$^2\Pi$ → X$^2\Sigma^+$ transition[151]. These results were discussed[150,151] in the light of *ab initio* calculations of potential energy curves[152,153] of NO$^{2+}$ and it was concluded that the three lowest electronic states of NO$^{2+}$ are quasibound. Emission has also been observed by PIFCO[154] from the quasibound $D^1\Sigma_u^+$ state of $N_2^{2+}$ and additional quasibound excited electronic states of other doubly charged ions are predicted to exist[151]. *Ab initio* configuration interaction calculations[155] on $N_2^{2+}$ have demonstrated the existence of eighteen quasibound states, one of which is $D^1\Sigma_u^+$ [154]. It has been concluded[154] that the picture according to which a doubly charged diatomic species $XY^{2+}$ is considered as two singly charged atoms interacting along a Coulomb repulsive curve is too simple. A perturbation by $X-Y^{2+}$ bonding states as a result of charge polarization forces has to be taken into account and gives rise to a local minimum in the potential curves. Additional characterization of $NO^{2+}$ states has come from double charge transfer experiments[156]. Six rapidly dissociating electronic states of $N_2^{2+}$ were observed by the PIPICO (photoion-photoion coincidence) method[157]. Other methods employed to determine energies of dissociative $N_2^{2+}$ states, namely differential photoion spectroscopy[158] and Auger electron-ion coincidence,[159] possess energy resolutions which are too low for the kind of state-to-state study of $N_2^{2+}$ fragmentations, which has been achieved by PIPICO[157].

Double- and multiple-ionization by a single photon occur by virtue of electron-electron interactions. Double photoionization of $H_2$ was studied by PIPICO in order to test electronic-correlation models[160]. Double-photoionization cross sections of hydrogen were determined experimentally as a function of the photon energy and compared[160] with *ab initio* calculations[161]. Various coincidence techniques available to the study of double and multiple ionization by photon impact were reviewed[162]. Triple ionization of a molecule (OCS) followed by fragmentation into three cations was demonstrated for the first time[162]. This was achieved through a modified PIPICO experiment demonstrating $C^+$-$O^+$-$S^+$ coincidences[162]. PIPICO was also applied to dissociative double and triple ionization of $CO^{163,164}$, to double ionization of $O_2^{149}$, $H_2O^{165}$, $N_2O^{148}$, $NH_3^{166}$, $C_2H_2^{149}$, $CH_3Br^{6,167}$, $SiH_4^6$ and $SiF_4^{17}$. Some of these studies involved valence shell ionizations[165] and others involved core-excited states[164]. Some processes which occur in water below the lowest vertical energy of formation of $H_2O^{2+}$ are considered to be two-step double ionization events[165]. A satellite state of $H_2O^+$ is first populated followed by threshold dissociative autoionization leading to a low kinetic energy electron and to $OH^+$ + $H^+$. The ejection of an Auger electron causes a sharp increase in the double photoionization cross section above core ionization edges[167]. Six features were observed in the photoelectron spectrum of $CS_2$ at binding energies above 28 eV, which were ascribed to direct double ionization continua[39]. At 10 eV or more above threshold, double photoionization is a one step process in which one electron leaves the ionic core with almost all the excess energy and the other is ejected with near zero kinetic energy[39].

A photoelectron-ion-ion triple coincidence technique termed photoelectron-photoion-photoion coincidence spectroscopy, PEPIPICO, has been developed in conjunction with ionization by He(II) light[168,169]. In the related PIPICO technique[1], only the time *difference* between correlated ions can be measured. In PEPIPICO, electrons are not energy analyzed, but are used as a time reference for measurement of the absolute flight times of the ions[162,168]. Advantages of the new technique are the unambiguous identification of ion pairs and a direct characterization of the dynamics of the charge separation reaction[168]. It has thus been demonstrated[168] that in $SO_2^{2+}$ the equal mass fragmentation produces predominantly $O_2^+ + S^+$ and not $O^+ + O^+$ and that the formation of $C^+ + S^+$ from $CS_2^{2+}$ can be modeled as a sequential, rather than a simultaneous explosion. The dynamics of three-body dissociations, i.e., fragmentations of a dication into two charged and one neutral particle, has been studied further by PEPIPICO in OCS, $CS_2$, $NO_2$, $SO_2$, $CH_3I$ and $SF_6$[169]. Sequential reactions involving initial charge separation, for example, $OCS^{2+} \rightarrow CO^+ + S^+$ followed by $CO^+ \rightarrow C^+ + O$, seem to be the most common dissociation route, although in $SF_6^{2+}$ a sequential reaction involving deferred charge separation, i.e., $SF_6^{2+} \rightarrow SF_4^{2+} + F_2$ followed by $SF_4^{2+} \rightarrow SF_3^+ + F^+$, is in better agreement with experimental results.

Double charge transfer spectroscopy has been employed to locate singlet states of $Cl_2^{2+}$ [170], $HCl^{2+}$ [171], $NO_2^{2+}$ [172] and $N_2O^{2+}$ [148]. Several of these experimental studies have been combined with theoretical calculations[170-172]. Other calculations involved $HCl^{2+}$ as well[173] and $CH^{2+}$ [174]. Calculations of the electronic states of $SO_2^{2+}$ were compared[175] with experimental results obtained previously by double-photoionization, double-charge-transfer and charge stripping.

Charge stripping of high kinetic energy singly charged ions by collisions with neutral targets, which has been demonstrated as an effective means of producing doubly charged ions, has been reviewed recently[176]. Inert atoms containing doubly charged ions, $AY^{2+}$, where A = Ne, Ar, Kr or Xe and Y = C, N, O or Cl have been produced by charge stripping and have been further characterized[177,178]. Helium containing multiply charged ions, for example, $CHe_4^{4+}$, have been predicted on the basis of *ab initio* calculations[179,180] to have remarkable stabilities. Other elusive interesting species, for example $CCl_4^{2+}$ and other chloromethanes, have been studied by *ab initio* calculations[181], charge stripping[181] and double charge transfer[182,183].

In charge stripping experiments the reactant ion beam is singly charged and the product doubly charged ion is usually in its ground state. In translational energy spectroscopy (TES), the ion beam is doubly charged and its energy loss spectra, upon inelastic collisions with a target gas, give information on its excited states. This information is in turn compared with previously discussed results from double-photoionization and double-charge transfer experiments and from *ab initio* calculations. A very interesting case at hand is again $NO^{2+}$ [184]. A novel high-resolution spectrometer yielded energy loss spectra[184] in excellent agreement with the calculated potential energy curves of Cooper[153] (see Figure 5). This lends further

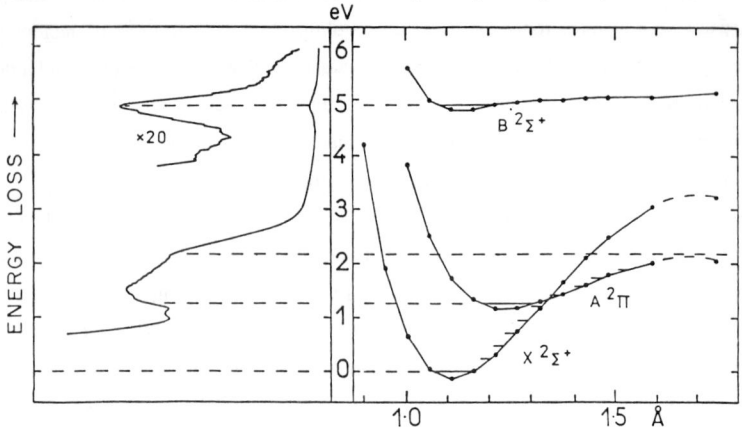

**Figure 5:** Correlation diagram between the experimental energy loss spectrum (left hand side) and Cooper's potential energy surfaces[153] for the $X^2\Sigma^+$, $A^2\Pi$ and $B^2\Sigma^+$ states of $NO^{2+}$.
*(Reproduced with permission from Chem. Phys. Lett., 1987, 141, 511).*

support in favor of a weakly bound $B^2\Sigma^+$ state[153] observed earlier through emission by PIFCO[150]. Other species studied by TES are $CH^{2+}$ [174,185], $CO_2^{2+}$, $OCS^{2+}$ and $CS_2^{2+}$ [186]. Translational energy spectra were studied also for electron capture collisions. Non-dissociative single electron capture occurs predominantly from the ground state doubly charged or triply charged ions into ground state product ions with occasional target excitation[187-189]. The situation is different in the case of $CO^{2+}$ [190]. Capture from $CO^{2+}(^3\Pi)$ into doublet states of $CO^+$ dominates. However, there are firm indications for the involvement of electronically excited $CO^{2+}$ in the projectile beam, and also for the formation of quartet states of $CO^+$ [190].

Various dissociative ion-beam techniques have been employed. A new method to study electron-capture collisions is electron capture-induced decomposition (ECID)[191,192]. TES of dissociating doubly charged ions is complimentary to PIPICO[193]; it does not have the capability for state-to-state studies but the kinetic energy analysis of the fragments is conducted much more precisely. The dissociation reactions of doubly charged polyatomic ions such as $CH_3I^{2+}$ have been demonstrated[193] to be highly non-statistical, proceeding via essentially isolated states. Charge separation reactions of $CF^{2+}$ and $CCl^{2+}$ were studied and the observed kinetic energy releases discussed on the basis of semiempirical potential energy curves[194]. Delayed predissociations were observed for these ions.

Double ionization and dissociation of $NH_3$ have been studied recently[195] through direct electroionization efficiency curves and the results have been compared with the PIPICO data[166] and various other data from previous measurements.

Double core vacancies[196] and outer valence double ionization transitions[197] in the benzene molecule have been computed using Green's functions. Double core ionization probes the bonding properties much more sensitively than single core ionization[196]. Double valence ionizations are characterized by strong correlation effects and the appearance of a very large number of satellite states[197]. Computational results were compared[197] with the experimental Auger spectrum of benzene. 226 computed dicationic states were analyzed[197].

**2.5 Cluster Ions.** - Cluster ions are covered in this review chapter, in spite of the fact that quite large molecular systems are sometimes involved. We feel that this is appropriate, since research on cluster ions contributes to the understanding of basic aspects of ionization processes and ion dynamics and is not covered in other chapters of this volume. Cluster ions, their formation and properties have been reviewed[101,198]. In this section we will concentrate briefly on ionization processes and stabilities. Other aspects, such as photodissociation will be discussed in coming sections. Combined laser/mass spectrometry techniques have become very efficient methods for the study of cluster ions[199]. Cluster ion production has been achieved through laser vaporization of solids, entrainment of the vaporized species in a molecular beam leading to condensation reactions followed by photoionization of the neutral clusters[199]. For example, carbon clusters were vaporized from a graphite rod by a Nd:YAG laser and photoionized by an excimer laser either by single photon ionization or, in the case of $C_{60}$ at 6.42 eV laser energy, by near-resonant two photon ionization[200]. Alternatively, cluster ions can be produced by direct laser vaporization[199]. Laser vaporization has been employed to prepare carbon cluster ions[201], as well as negative and positive cluster ions of gold, silver and copper[202,203]. Cluster ions of transition metals (Al and Fe) were formed by vaporization with a copper vapor laser and cooled in a He flow[204]. Alternative methods for ion cluster production have been reviewed[101,198]. A major method involves ionization of neutral van der Waals (vdW) clusters produced by adiabatic expansion of up to 10 bar of neat or mixed gas through a nozzle into vacuum[205]. When electron impact is employed, one gets negative ion clusters by electron attachment and positive ones by electron ionization[205]. Several ionization mechanisms have been proposed[101]. According to the model for electron ionization of a rare gas cluster, the positive charge, which is initially delocalized, gets localized at $\sim 10^{-12}$s after ionization on a dimer ion within the cluster[101,102]. Absolute ionization cross sections are lacking and no simple additivity rule exists for large clusters[206].

The special stability of certain cluster ion sizes (so-called 'magic numbers') is a topic of great interest. In the case of carbon clusters there is the remarkable stability of $C_{60}$[200], which shows up in the positive as well as in the negative ion spectra[207]. The production and stability of neon cluster ions has been studied by supersonic nozzle expansion[208]. Isotope enrichment in $^{22}$Ne was observed in Ne clusters[209].

The balance between Coulomb repulsion and surface energy as a function of cluster size n has been used to rationalize that multiply charged cluster cations smaller than a critical size limit $n_c$ do not exist[210]. For doubly charged argon cluster ions this critical size is $n_c = 91$[210]. For triply charged argon clusters, it is $n_c = 226$ [211]. It has been found[211] that $IE(Ar_n^{2+}) \approx 2\ IE(Ar^+)$ and $IE(Ar_n^{3+}) \approx 3\ IE(Ar^+)$ (where IE = ionization energy). This demonstrates unequivocally that multiple ionization of a rare gas cluster proceeds via successive single ionization collisions of one incoming electron within the cluster. Similar ionization mechanisms were proposed[212,213] for doubly charged $N_2$ and $O_2$ clusters. Critical sizes for doubly and triply charged clusters of ammonia are 51 and 121, respectively[214]. Unstable triply charged ions with sizes below the critical size live long enough to allow detection of the "Coulomb explosion" in which the doubly charged fragments have a size of about 90% of the precursor ion size[214]. Double ionization of mercury clusters, by photon as well as electron impact ionization, is a two-step process[215].

Electron impact ionization and electron attachment have been studied in $SF_6$ and $SF_6$/Ar clusters[216]. A strong zero energy resonance in the attachment cross section leading to the formation of $(SF_6)_n^-$ was discovered. The dissociative channel into $SF_5^-$ present in the monomer is efficiently quenched in the case of the cluster[216]. A similar result has been reported for $SO_2$ clusters[217]. Negative cluster ion production via free electron attachment is at least as efficient as collisional electron transfer from high Rydberg atoms[216,218,219]. Dissociative attachment from condensed $O_2$[220], $Cl_2$[221] and CO[222] has been compared with dissociative attachment in clusters; transitions which are forbidden in single gas phase molecules, by symmetry selection rules, are observed. Selection rules are violated because of the presence of neighboring molecules in the clusters or in the condensed phase.

Van der Waals and hydrogen bonded clusters have been ionized by single photo ionization[223-227] and multiphoton ionization[228]. Ionization efficiency curves, binding energies and mass spectra were determined. Evidence in favor of intramolecular Penning ionization of benzene-Ar and fluorobenzene-Ar van der Waals molecules has been presented[225]. Ionic clusters were grown from isolated $NO^+$ ion seeds formed by REMPI[229]. Appearance energies for water cluster ions $(H_2O)_2^+$, $(H_2O)_3^+$, $(H_2O)_2H^+$ and $(H_2O)_3H^+$ have been determined[230]. A high-pressure, temperature-variable ion source with coaxial electron beam, ion exist geometry has been employed to generate ionic hydrogen clusters. Even clusters, $H_4^+$, $H_6^+$, $H_8^+$ and $H_{10}^+$ were observed for the first time and characterized[231]. The hydrogen cluster spectrum is dominated by odd clusters[231,232]. The stabilities of odd hydrogen ion clusters have been determined from gas-phase ion equilibria[233]. Thermochemistry for gas-phase ion-molecule clustering has been studied in a new hybrid drift tube/ion source with coaxial electron beam and ion exit aperture[234].

Cluster stabilities and dissociation energies were determined for $Al_2^+$ - $Al_7^+$ [235] and $B_2^+$ - $B_{13}^+$ [236,237] by low energy collision-induced dissociation (CID) in a guided ion beam instrument. $B_5^+$ is a "magic" fragment in CID of larger Boron clusters.

Carbon clusters, $C_n$, have been classified as valence clusters since they contain conventional chemical bonds[101]. Most of mass spectrometry has been devoted to carbon containing ions. It is thus interesting to speculate whether $C_n^+$ ions formed by dissociative ionization of organic compounds have the same structures as the ones formed by laser vaporization of graphite. $C_{10}^+$ and $C_7^+$ carbon cluster ions were formed from overcrowded perchloro conjugated hydrocarbons and were found to undergo identical reactions to ions produced by laser vaporization, with $C_3$ elimination being the 'magic' reaction channel[238]. It is thus obvious that valence cluster ions such as $C_n^+$ can be formed by suitable dissociative ionization reactions, in this case exhaustive chlorine elimination.

### 3. Spectroscopy and Structure of Ions

Several aspects of ion spectroscopy have been discussed in previous sections, particularly in connection with emissions observed from doubly charged ions.

Vibrational and rotational absorption spectroscopy has developed rapidly through velocity-modulation laser spectroscopy[1] which has been reviewed by Saykally recently[239]. The alternative technique for high-resolution infrared spectroscopy of molecular ions, modulated hollow-cathode discharge, has been reviewed as well[240]. An earlier review[241] cited 41 ions and 12 more positive and negative ions were added to the compilation by Amano[240]. Oka, in a review on the infrared spectroscopy of carbo-ions[242], expressed the view that we are experiencing a period of a burst of activity in molecular-ion spectroscopy. The lasers employed are either difference-frequency laser systems[240,242] or tunable diode lasers[239]. Velocity modulation spectroscopy detects ionic absorption signals in the presence of a large background from the more abundant neutral molecules. The principle of the method has been explained in the last review chapter[1] and in the more recent review article[239]. Accurate spectroscopic constants, Born-Oppenheimer potential functions, abundance ratios of isomeric ions, comparisons of ion structures with theoretical predictions and correlations between laboratory data and their astrophysical implications have become possible. The following are some examples of ions studied by high resolution infrared spectroscopy in the last two years: $CCl^+$ [243], $NH^+$ [244], $H_2O^+$ [245], $H_3O^+$ [246], $CH_3^+$ [247], $SH_3^+$ [248,249], $C_2H_2^+$ [250], $HCNH^+$ [251,252], $ND_4^+$ [253], $HOCS^+$ [240], $HON_2^+$ [240], $C_2H_3^+$ [242], $HC_3NH^+$ [254], $OD^-$ [255], $SH^-$ [256], $C_2^-$ [242], $N_3^-$ [257], $NCO^-$ [258], $NCS^-$ [259] and $C_2H^-$[260]. Of particular interest for this series of reviews is the $CH_3^+$ ion[242,247], since $CH_3^+$ is one of the most ubiquitous carbonium ions in mass spectrometry. Its spectrum was obtained accidentally. Velocity modulation played a particularly important role in this case since many polymeric hydrocarbons are produced in the discharge which makes $CH_3^+$ and there are many strong neutral absorption lines. Over 200 absorption lines have been assigned to $CH_3^+$ and the isotopic species $^{13}CH_3^+$. The spectral pattern clearly indicates the plane symmetry of the $CH_3^+$ ion and settles an earlier controversy regarding the possibility of a pyramidal $CH_3^+$ with the carbon lying 0.2Å above the hydrogen plane[247]. The spectrum of $C_2H_3^+$ has been demonstrated to be dominated by the non-classical hydrogen bridged structure[242] in agreement with experiments by the Coulomb explosion method[261], whose potential as a method for determining ion structures has been pointed out in the previous review chapter[1].

Tunable diode lasers have been applied for the first time[263] to study high J rotational transitions of $ArH^+$, $NeH^+$, $HeH^+$, $OH^+$, $H_2O^+$ and $OH^-$.

Stimulated emission pumping is a two-laser pump-dump method, which gives information on highly vibrationally excited levels of ground electronic states, without recourse to infrared sources, has been applied to the diacetylene cation[264].

Visible photoabsorption spectra of ions have been measured by ion beam laser photodissociation spectroscopy. Examples for spectra obtained are for ions of quite varied nature such as $Ar_3^+$ [265], $MgCl^+$ [266], $Sr^+$ $(H_2O)$[267], and $C_6H_6^+$ isomers[268]. Most of these spectra are broad, that of $Sr^+$ $(H_2O)$ demonstrating some structure ascribable to sequence bands in the $Sr^+$-O-H stretch[267]. Laser photodissociation spectroscopy has been reviewed[269] and some high resolution structured spectra obtained in the past several years (for $Cl_2^+$, $Br_2^+$, $I_2^+$, $O_2^+$, $SO_2^+$ and $CH_3I^+$) have been discussed. Photodissociation processes combine the phenomena of spectroscopy and dynamics. Photodissociation spectra and dynamics of small cluster ions have been studied by Bowers and coworkers. Absolute cross sections for photodissociation of $CO_2$-$O_2^+$ were determined by comparison with $(CO_2)_2^+$ [270]. Theoretical studies of potential energy surfaces have been carried out in connection with the photodissociation spectra of $(CO)_2^+$ [271] and $MgCl^+$ [272]. Ionic cluster fragments of transition-metal carbonyls have been studied by laser-ion beam photodissociation[273]. The photodissociation of carbon cluster ions[274,275], including buckminsterfullerene ($C_{60}^+$) is of great interest in connection with the proposed structures for these ions. Small $C_n^+$ clusters (n < 32) fragment by loss of $C_3$; even numbered clusters with n > 32 fragment by loss of $C_2$. The photofragmentation cross sections for both 351 and 248 nm light show a significant change as a function of cluster size; there is a dramatic change[274] near n = 10 which may be related with a change in structure from an open chain configuration to a ring structure.

Kinetic energy spectra of $N_2^+$ photofragments in the laser photodissociation of $N_2O^+$ and of $S_2^+$ from $CS_2$ served to identify the states involved in the absorption and dissociation processes[276].

Vibrationally excited polyatomic ions in fast beams have been photodissociated by single photons in the infrared[277]. The infrared wavelength dependence of the process shows well-defined peaking, which can be interpreted as absorption at the normal-mode frequencies of the ion.

Maier[278] has reviewed his group's earlier work done in emission spectroscopy and laser excitation spectroscopy of open-shell polyatomic cations. Emission spectra obtained from effusive and supersonic sources, respectively, were discussed for hexafluorobenzene and chloroacetylene; resolutions of 0.015 nm have been achieved and vibrational frequencies in the ground and excited electronic states have been obtained for a series of substituted acetylenes and fluorobenzenes[278]. Emission spectra of several unstable molecules such as $HBS^+$ (thioborine cation) were reported[278]; rotational structure became apparent at resolutions of 0.006 nm. High resolution (0.0007 nm) laser excitation spectra were achieved and one of these spectra was presented in the previous review chapter[1]. Additional review articles by Maier have appeared[279-282] stressing the ability of the spectroscopic methods employed in vibrational and rotational characterization of ions by means of their electronic transitions. Molecular cations studied over the last two years by emission and laser excitation spectroscopy include:

diacetylene[283], bromocyanoacetylene[284], ClCN+ [285], X–C≡C–X+ (X = Cl, Br, I, a review paper)[286], $CH_3$– C≡C–X+ (X = Cl, Br)[287], FBS+ and ClBS+ [288], HCP+ and DCP+ [289], ClCN+ [290], BrCN+ [291]. Of particular interest is the $\tilde{B}\ {}^4\Sigma_u^-$ - $\tilde{X}\ {}^4\Sigma_g^-$ electronic spectrum of $C_2^+$ studied by laser excitation spectroscopy[292,293]. $C_2^+$ is a building block in the clustering reactions in graphite plasmas and an important constituent in extraterrestrial environments[293]. The spectroscopic constants $T_e$, $B_e$, $\alpha_e$, $\omega_e$, $\omega_e x_e$ were obtained for $^{12}C_2^+$ and $^{13}C_2^+$ in the $\tilde{X}\ {}^4\Sigma_g^-$ and $\tilde{B}\ {}^4\Sigma_u^-$ states. The observation and analysis of the $\tilde{B}\ {}^4\Sigma_u^-$ - $\tilde{X}\ {}^4\Sigma_g^-$ transition make it possible to monitor $C_2^+$ optically in comet tails, diffuse interstellar clouds, plasmas and flames[293]. Spectroscopic constants have been determined for the $A^2\Pi_i$ and $X^2\Pi_r$ states of SO+ [294,295]. Optical emission has been observed for $SiF_4^+$ [104] and $GeF_4^+$ [296] by high energy ion impact.

Emission spectra have been induced by Penning ionization[297-300] and by charge transfer[301-305] and nascent rovibrational distributions were determined. Absorption spectra have been measured by special techniques for cases in which the excited states are bound. Laser induced charge transfer (LICT) was applied to the study of $N_2^+$ Meinel bands and CO+ comet tail bands[306]. The charge transfer medium was Ar, which does not react with $N_2^+$ and CO+ in their ground states, but undergoes a strong reaction with their excited states. Two photon absorption spectroscopy has been applied to cases in which the intermediate state is bound and the upper state is predissociative so that its fragments may be detected. $N_2O^+$ and $CS_2^+$ have been studied by this method[307]. It was demonstrated that spectroscopic information can be obtained on the nondissociating states and in favorable cases on the predissociating levels[307].

The electronic spectroscopy and relaxation of some molecular cations (CS+, $CS_2^+$, SH+, $H_2S^+$, $CO_2^+$) which fluoresce and others whose fluorescence quantum yields $\phi_F$ are extremely small ($SO_2^+$, $CH_3CN^+$, $NH_3^+$, $C_2H_2^+$, $C_2N_2^+$) have been reviewed[308] because of their cometary interest. Fluorescence lifetimes of $N_2O^+$ [309] and SH+ [310] were determined experimentally by a high frequency deflection technique, the results compared with theoretical calculations and the degree of predissociation of the various vibronic levels was assessed. Emission spectra of NeAl+ and ArAl+ in the VUV region were calculated by *ab initio* methods[311] and radiative lifetimes of the excited excimer type states were estimated.

Photoelectron spectra (PES) were determined for molecules of different complexity. Chlorine was studied using synchrotron radiation[312]. Due to the employment of a variable wavelength source, ionization occurred via intermediate autoionizing states. This led to altered vibrational distributions of the final ionic state enabling accurate determinations of the vibrational constants $\omega_e$ and $\omega_e x_e$ and of the spin-orbit splitting for the $X^2\Pi g$ state[312]. The photoelectron spectrum of $SO_2$ was investigated using a supersonic molecular beam[313]. High resolution and rotational cooling led to the observation of new features and to resolution of the vibrational structure in the first six electronic states of the $SO_2^+$ ion[313]. A part of the first photoelectron band of the phenyl radical has been recorded[314]. While the ground state of the phenyl ion is $^1A_1$, the spectrum obtained has been interpreted in terms of the ionization $C_6H_5^+$ ($a^3B_1$) ← $C_6H_5$ ($X^2A_1$). A particularly original approach to photoelectron spectroscopy has bridged the gap between spectroscopy and music[315] - the photoelectron spectrum of phosphabenzene and matching model calculations were transcribed into music.

Photoelectron spectroscopy of negative ions has been pursued very actively in recent years. Information is obtained on the structure and bonding in the ions, electron affinities and singlet-triplet splittings of the corresponding neutrals and spectroscopic constants of the negative ions and corresponding neutrals. The species studied include: $C_3^-$, $C_3H^-$, $C_3H_2^-$, $C_3O^-$, $C_3O_2^-$, $C_4O^-$ and $CS_2^-$ [316], o-benzyne[317], $CH_2S^-$ [318,319], $S_2^-$, $HS_2^-$, $CH_3S_2^-$, $CH_3CH_2S^-$, $CH_3SCH_2^-$, and $CH_3S^-$ and deuterated analogues[319], $SiH_3^-$ and $SiD_3^-$ [320], $BH_3^-$ [321], alkali halides[322], $MgCl^-$ [323], $CrH^-$, $CoH^-$ and $NiH^-$ [324], $Si_2^-$ [325], $Fe_2^-$ and $Co_2^-$ [326], $Re_2^-$ [327]. Photoelectron spectroscopy of negative cluster ions included small clusters such as $O_4^-$ [328] and $N_2O_2^-$ [329] as well as large ones - $(CO_2)_n^-$ [330] and $(Cu)_n^-$ [331,332]. Vertical electron binding energies and adiabatic electron affinities were measured as a function of cluster size. Sharp discontinuities were observed for n = 2 and n = 6 in the case of $(CO_2)_n^-$ and this led to the conclusion that the dimer ion is the core of clusters 2 < n < 5 while the monomer ion forms the core for n ≥ 7 [330]. Electron affinities of the copper series increase with cluster size[331] and clusters 8, 20 and 40 which correspond to spherical shell closings are found to have unusually low electron affinities and large HOMO-LUMO gaps[332]. The HOMO-LUMO gap was extracted directly from the UPS spectrum of each even-numbered cluster as the energy difference between the lowest energy peak in the binding energy spectrum and the onset of the next feature. Subshell closings at 14 and 34 were also observed. The results were in agreement with detailed predictions of the ellipsoidal shell model for monovalent metal clusters[332].

The photoelectron spectra of the cyanomethide $CH_2CN^-$ and isocyanomethide $CH_2NC^-$ ions were studied in detail[333,334]. Cyanomethide and other dipolar anions, $CH_2CHO^-$ (acetaldehyde enolate anion) and $CH_2COF^-$ (acetyl fluoride enolate anion) were topics of experimental studies by high-resolution autodetachment spectroscopy[335-337] and theoretical studies[338]. In autodetachment spectroscopy the electron signal is measured as function of frequency of the laser photon absorbed. The upper level in autodetachment spectroscopy can be a vibrationally excited level of the same electronic state as the lower level; vibrational autodetachment spectroscopy has been used to obtain the vibration-rotation spectrum of $NH^-$ [335]. The upper level can also belong to an excited electronic state. In $CH_2CN^-$, the ground electronic state is a normal valence state while the outermost electron in the excited state is bound by the dipole moment of the neutral radical. Autodetachment occurs from excited rotational levels of this dipole bound state, giving some 5000 sharp features near the photodetachment threshold[336]. All of these features were assigned and spectroscopic constants for both states reported. The ion $PtN^-$ was observed to decay by both vibrational and rotational autodetachment[335,339].

Transition state spectroscopy is now a major goal of reaction dynamics. The transition state of a chemical reaction is at the saddle point of the potential surface. Due to the instability of the transition state and its short lifetime, experimental determination of its spectroscopic constants poses a unique challenge. Recently, photoelectron spectroscopy of negative ions has been employed. While $ClHCl^-$ is stable and resides in a potential well, ClHCl is the transition state for the Cl + HCl reaction. Photoelectron spectra of $ClHCl^-$ yielded vibrational structure which is a progression in the asymmetric stretch in the ClHCl complex[340]. This vibrational structure can be compared to calculated results on a model surface for the Cl + HCl reaction. This type of approach to transition state spectroscopy looks very promising and will most probably be extended and refined in the future.

### References

1. C. Lifshitz, 'Specialist Periodical Reports: Mass Spectrometry', 1987, **9**, 1.
2. A.L. Burlingame, D. Maltby, D.H. Russell and P.T. Holland, *Anal. Chem.*, 1988, **60**, 294R.
3. H. Lefèbvre-Brion, in 'Molecules in Physics, Chemistry and Biology', ed. J. Maruani, Vol. II, Reidel Publishing Co., Holland, 1988, p. 257.
4. J.L. Dehmer, A.C. Parr and S.H. Southworth, in 'Handbook on Synchrotron Radiation', ed. G.V. Marr, Vol. 2, Chapter 5, Elsevier Science Publishers, North-Holland, 1987, p. 241.
5. I. Nenner and J.A. Beswick, in 'Handbook on Synchrotron Radiation', ed. G.V. Marr, Vol. 2, Chapter 6, Elsevier Science Publishers, North-Holland, 1987, p. 355.
6. I. Nenner, in 'Electronic and Atomic Collisions', H.B. Gilbody, W.R. Newell, F.H. Read, A.C.H. Smith (eds.), Elsevier Science Publishers, 1988, p. 517.
7. F. Keller and H. Lefebvre-Brion, *Z. Phys. D-Atoms, Molecules and Clusters*, 1986, **4**, 15.
8. G. Raşeev, B. Leyh and H. Lefebvre-Brion, *Z. Phys. D-Atoms, Molecules and Clusters*, 1986, **2**, 319.
9. T.A. Carlson, M.O. Krause, W.A. Svensson, P. Gerard, F.A. Grimm, T.A. Whitley and B.P. Pullen, *Z. Phys. D-Atoms, Molecules and Clusters*, 1986, **2**, 309.
10. B. Basden and R.R. Lucchese, *Phys. Rev.* 1986, **34A**, 5158.

11  B. Basden and R.R. Lucchese, *Phys. Rev.* 1988, **37A**, 89.
12  S.H. Southworth, A.C. Parr, J.E. Hardis and J.L. Dehmer, *J. Chem. Phys.*, 1987, **87**, 5125.
13  H. Lefebvre-Brion, in 'Giant Resonances in Atoms, Molecules and Solids', J.P. Connerade, J.M. Esteva and R.G. Karnatak (eds.), NATO ASI Vol. 151, Plenum Publishing Co., 1987, p. 301.
14  G. Raşeev, F. Keller and H. Lefebvre-Brion, *Phys. Rev.*, 1987, **36A**, 4759.
15  E.D. Poliakoff, M.-H. Ho, M.G. White and G.E. Leroi, *Chem. Phys. Lett.*, 1986, **130**, 91.
16  S. Bodeur, I. Nenner and P. Millie, *Phys. Rev.*, 1986, **34A**, 2986.
17  I. Nenner, in 'Giant Resonances in Atoms, Molecules and Solids', J.P. Connerade, J.M. Esteva and R.G. Karnatak (eds.), NATO ASI Vol. 151, Plenum Publishing Co., 1987, p. 259.
18  T.A. Ferrett, M.N. Piancastelli, D.W. Lindle, P.A. Heimann, L.J. Medhurst, S.H. Liu and D.A. Shirley, *Chem. Phys. Lett.*, 1987, **134**, 146.
19  T.A. Carlson, P. Gerard, M.O. Krause, F.A. Grimm and B.P. Pullen, *J. Chem. Phys.*, 1987, **86**, 6918.
20  M.N. Piancastelli, D.W. Lindle, T.A. Ferrett and D.A. Shirley, *J. Chem. Phys.*, 1987, **86**, 2765; ibid. id. 1987, **87**, 3255.
21  B. Leyh and G. Raşeev, *Phys. Rev.*, 1986, **34A**, 2920.
22  B. Leyh, J. Delwiche, M.-J. Hubin-Franskin and I. Nenner, *Chem. Phys.*, 1987, **115**, 243.
23  A.L. Sobolewski and W. Domcke, *J. Chem. Phys.*, 1987, **86**, 176.
24  A.L. Sobolewski, *J. Chem. Phys.*, 1987, **87**, 331.
25  H. Lefebvre-Brion, P.M. Dehmer and W.A. Chupka, *J. Chem. Phys.*, 1986, **85**, 45.
26  H. Lefebvre-Brion, P.M. Dehmer and W.A Chupka, *J. Chem. Phys.*, 1988, **88**, 811.
27  E.D. Poliakoff, M.-H. Ho, G.E. Leroi and M.G. White, *J. Chem. Phys.*, 1986, **85**, 5529.
28  E.D. Poliakoff, J.L. Dehmer, A.C. Parr and G.E. Leroi, *J. Chem. Phys.*, 1987, **86**, 2557.
29  M.G. White, G.E. Leroi, M.-H. Ho and E.D. Poliakoff, *J. Chem. Phys.*, 1987, **87**, 6553.
30  G.G.B. de Souza, P. Morin and I. Nenner, *Phys. Rev.* 1986, **34A**, 4770.
31  P. Morin and I. Nenner, *Physica Scripta*, 1987, **T17**, 171.
32  I. Nenner, P. Morin, M. Simon, P. Lablanquie and G.G.B. de Souza, in 'DIET III', M. Knotek and R.H. Stulen (eds.), Springer, Berlin (1987).
33  A.L. Sobolewski and W. Domcke, *J. Chem. Phys.*, 1988, **88**, 5571.
34  T. Baer and P.M. Guyon, *J. Chem. Phys.*, 1986, **85**, 4765.
35  K. Börlin, M. Jungen, L. Karlsson and R. Maripuu, *Chem. Phys.*, 1987, **113**, 309.
36  R.H. Page, R.J. Larkin, Y.R. Shen and Y.T. Lee, *J. Chem. Phys.*, 1988, **88**, 2249.
37  M.Y. Adam, C. Cauletti and M.N. Piancastelli, *J. Electron Spectrosc. Relat. Phenom.*, 1987, **42**, 1.
38  a) M.N. Piancastelli, C. Cauletti and M.-Y. Adam, *J. Chem. Phys.*, 1987, **87**, 1982.
    b) M.S. Banna, H. Kossman and V. Schmidt, *Chem. Phys.*, 1987, **114**, 157.
    c) M.S. Banna, R. Malutzki and V. Schmidt, *J. Chem. Phys.*, 1987, **87**, 1582.
39  P. Roy, I. Nenner, P. Millie, P. Morin and D. Roy, *J. Chem. Phys.*, 1987, **87**, 2536.
40  W. von Niessen, *Inorg. Chem.*, 1987, **26**, 567.
41  I.H. Hillier, M.A. Vincent, M.F. Guest and W. von Niessen, *Chem. Phys. Lett.*, 1987, **134**, 403.
42  R. Cambi and W. von Niessen, *J. Electron Spectrosc. Relat. Phenom.*, 1987, **42**, 245.

43 W. von Niessen, *J. Chem. Phys.*, 1986, **85**, 337.
44 W. von Niessen, *J. Chem. Soc. Faraday Trans 2*, 1986, **82**, 1489.
45 W. von Niessen and P. Tomasello, *J. Chem. Phys.*, 1987, **87**, 5333.
46 W. von Niessen, L.S. Cederbaum and J. Schirmer, *J. Electron Spectrosc. Relat. Phenom.*, 1986, **41**, 235.
47 W. von Niessen, *J. Phys. Chem.*, 1988, **92**, 1035.
48 W. von Niessen, P. Tomasello, J. Schirmer and L.S. Cederbaum, *Aust. J. Phys.*, 1986, **39**, 687.
49 C.E. Brion, *Int. J. Quantum Chem.*, 1986, **29**, 1397.
50 A.O. Bawagan and C.E. Brion, *Chem. Phys.*, 1988, **123**, 51.
51 C.L. French, C.E. Brion and E.R. Davidson, *Chem. Phys.*, 1988, **122**, 247.
52 L. Frost, A.M. Grisogono, I.E. McCarthy, E. Weigold, C.E. Brion, A.O. Bawagan, P.K. Mukherjee, W. von Niessen, M. Rosi and A. Sgamelloti, *Chem. Phys.*, 1987, **113**, 1.
53 L. Frost, A.M. Grisogono, E. Weigold, C.E. Brion, A.O. Bawagan, P. Tomasello and W. von Niessen, *Chem. Phys.*, 1988, **119**, 253.
54 C.L. French, C.E. Brion, A.O. Bawagan, P.S. Bagus and E.R. Davidson, *Chem. Phys.*, 1988, **121**, 315.
55 A.O. Bawagan, C.E. Brion, E.R. Davidson and D. Feller, *Chem. Phys.*, 1987, **113**, 19.
56 A.O. Bawagan, R. Müller-Fiedler, C.E. Brion, E.R. Davidson and C. Boyle, *Chem. Phys.*, 1988, **120**, 573.
57 A.O. Bawgan and C.E. Brion, *Chem. Phys. Lett.*, 1987, **137**, 573.
58 A. Minchinton, J.P.D. Cook, E. Weigold and W. von Niessen, *Chem. Phys.*, 1987, **113**, 251.
59 J.A.R. Samson, G.N. Haddad and L.D. Kilcoyne, *J. Chem. Phys.*, 1987, **87**, 6416.
60 K. Kimura, *Int. Rev. Phys. Chem.*, 1987, **6**, 195.
61 P.M. Dehmer, J.L. Dehmer and S.T. Pratt, *Comments At. Mol. Phys.*, 1987, **19**, 205.
62 R.N. Compton and J.C. Miller, "Multiphoton Ionization Photoelectron Spectroscopy: MPI-PES", in 'Laser Applications in Physical Chemistry', D.K. Evans (Ed.), Marcel Dekker, Inc., in press.
63 S.T. Pratt, P.M. Dehmer and J.L. Dehmer, *J. Chem. Phys.*, 1986, **85**, 3379.
64 S.T. Patt, P.M. Dehmer and J.L. Dehmer, *J. Chem. Phys.*, 1987, **86**, 1727.
65 S.T. Pratt, P.M. Dehmer and J.L. Dehmer, *J. Chem. Phys.*, 1987, **87**, 4423.
66 M.A. O'Halloran, S.T. Pratt, F.S. Tomkins, J.L. Dehmer and P.M. Dehmer, *Chem. Phys. Lett.*, 1988, **146**, 291.
67 S.T. Pratt, P.M. Dehmer and J.L. Dehmer, *J. Chem. Phys.*, 1986, **85**, 5535.
68 K.S. Viswanathan, E. Sekreta, E.R. Davidson and J.P. Reilly, *J. Phys. Chem.*, 1986, **90**, 5078.
69 K.S. Viswanathan, E. Sekreta and J.P. Reilly, *J. Phys. Chem.*, 1986, **90**, 5658.
70 A.C. Parr, J.E. Hardis, S.H. Southworth, C.S. Feigerle, T.A. Ferrett, D.M.P. Holland, F.M. Quinn, B.R. Dobson, J.B. West, G.V. Marr and J.L. Dehmer, *Phys. Rev.*, 1988, **37A**, 437.
71 A. Fujii, T. Ebata and M. Ito, *J. Chem. Phys.*, 1988, **88**, 5307.
72 P.M. Dehmer, S.T. Pratt and J.L. Dehmer, *J. Phys. Chem.*, 1987, **91**, 2593.
73 D.C. Jabobs, R.J. Madix and R.N. Zare, *J. Chem. Phys.*, 1986, **85**, 5469.

74   S. Katsumata, K. Sato, Y. Achiba and K. Kimura, *J. Electron Spectrosc. Relat. Phen.*, 1986, **41**, 325.
75   P.M. Dehmer, S.T. Pratt, and J.L. Dehmer, *J. Chem. Phys.*, 1986, **85**, 13.
76   T.M. Orlando, S.L. Anderson, J.R. Appling and M.G. White, *J. Chem. Phys.*, 1987, **87**, 852.
77   M.N.R. Ashfold, B. Tutcher, B. Yang, Z.K. Jin and S.L. Anderson, *J. Chem. Phys.*, 1987, **87**, 5105.
78   T.J. Cornish and T. Baer, *J. Am. Chem. Soc.*, 1987, **109**, 6915.
79   T.J. Cornish and T. Baer, *J. Am. Chem. Soc.*, 1988, **110**, 3099.
80   D.C. Jacobs and R.N. Zare, *J. Chem. Phys.*, 1986, **85**, 5457.
81   D.C. Jacobs, K.W. Kolasinski, R.J. Madix and R.N. Zare, *J. Chem. Phys.*, 1987, **87**, 5038.
82   Y. Nagano, Y. Achiba and K. Kimura, *J. Phys. Chem.*, 1986, **90**, 615.
83   Y. Nagano, Y. Achiba and K. Kimura, *J. Phys. Chem.*, 1986, **90**, 1288.
84   M. Fujii, K. Sato and K. Kimura, *J. Phys. Chem.*, 1987, **91**, 6507.
85   D. Solgadi, C. Jouvet and A. Tramer, *J. Phys. Chem.*, 1988, **92**, 3313.
86   H.J. Neusser, *Int. J. Mass Spectrom. Ion Processes*, 1987, **79**, 141.
87   H. Kühlewind, A. Kiermeier and H.J. Neusser, *J. Chem. Phys.*, 1986, **85**, 4427.
88   J.B. Morris and M.V. Johnston, *Int. J. Mass Spectrom. Ion Processes*, 1986, **73**, 175.
89   T.-C. Chang and M.V. Johnston, *J. Phys. Chem.*, 1987, **91**, 884.
90   G.R. Kinsel, K.R. Segar and M.V. Johnston, *Org. Mass Spectrom.*, 1987, **22**, 627.
91   D.M. Szaflarski, E.L. Chronister and M.A. El-Sayed, *J. Phys. Chem.*, 1987,**91**, 3259.
92   E. Sekreta, K.G. Owens and J.P. Reilly, *Chem. Phys. Lett.*, 1986, **132**, 450.
93   L.R. Khundkar and A.H. Zewail, *Chem. Phys. Lett.*, 1987, **142**, 426.
94   H. Shinohara, K. Sato, Y. Achiba, N. Nishi and K. Kimura, *Chem. Phys. Lett.*, 1986, **130**, 231.
95   D.M. Szaflarski and M.A. El-Sayed, *J. Phys. Chem.* 1988,**92**, 2234.
96   W.B. Martin and R.M. O'Malley, *Int. J. Mass Spectrom. Ion Processes*, 1987, **77**, 203.
97   T.D. Märk, in 'Gaseous Ion Chemistry and Mass Spectrometry', Ed., J.H. Futrell, Chapter 3, John Wiley and Sons, New York, 1986, p. 61.
98   T.D. Märk, in 'The Physics of Ionized Gases', J.   Purić and D. Belić (eds.), World Scientific, Singapore, 1987, p. 145.
99   H. Deutsch, P. Scheier and T.D. Märk, *Int. J. Mass Spectrom. Ion Processes*, 1986, **74**, 81.
100  T.D. Märk, in 'Proceedings IAEA Advisory Group Meeting on Atomic and Molecular Data for Radiotherapy', IAEA-TECDOC, Wien, 1988, in press.
101  T.D. Märk, *Int. J. Mass Spectrom. Ion Proc.*, 1987, **79**, 1.
102  A.J. Stace, 'Specialist Periodical Reports: Mass Spectrometry', 1987, **9**, 96.
103  N.Lj. Djurić, I.M. Cadez and M.V. Kurepa, *Int. J. Mass Spectrom. Ion Processes*, 1988, **83**, R7.
104  J.F.M. Aarts, *Chem. Phys.*, 1986, **101**, 105.
105  P. Plessis and P. Marmet, *Can. J. Phys.*, 1987, **65**, 165.
106  P. Plessis and P. Marmet, *Can. J. Phys.*, 1987, **65**, 803.
107  P. Plessis and P. Marmet, *Can. J. Phys.*, 1987, **65**, 1424.
108  P. Plessis and P. Marmet, *Can. J. Phys.*, 1987, **65**, 2004.
109  C. Lifshitz and F.A. Long, *J. Chem. Phys.*, 1964, **41**, 2468.

110  D. Rapp, *J. Chem. Phys.*, 1971, **55**, 4154.
111  C. Lifshitz, *J. Chem. Phys.*, 1971,**55**, 4155.
112  W. Koot, W.J. Van der Zande and D.P. de Bruijn, *Chem. Phys.*, 1987, **115**, 297.
113  C. Wesdemiotis and F.W. McLafferty, *Chem. Rev.*, 1987, **87**, 485.
114  J.K. Terlouw and H. Schwarz, *Angew. Chem. Int. Ed. Engl.*, 1987, **26**, 805.
115  M.C. Blanchette, J.L. Holmes, C.E.C.A. Hop and A.A. Mommers, *Org. Mass Spectrom.*, 1988, **23**, 495.
116  J. Bordas-Nagy, J.L. Holmes and A.A. Mommers, *Org. Mass Spectrom.*, 1986, **21**, 629.
117  J. Bordas-Nagy and J.L. Holmes, *Chem. Phys. Lett.*, 1986, **132**, 200.
118  J. Bordas-Nagy and J.L. Holmes, *Int. J. Mass Spectrom. Ion Processes*, 1988, **82**, 81.
119  A. Benz and H. Morgner, *Mol. Phys.*, 1986, **57**, 319.
120  A. Benz and H. Morgner, *Mol. Phys.*, 1986,**58**, 223.
121  K. Beckmann, O. Leisin and H. Morgner, *Mol. Phys.*, 1986,**59**, 829.
122  L.G. Christophorou, *Contrib. Plasma Phys.*, 1987, **27**, 237.
123  D.L. McCorkle, L.G. Christophorou, A.A. Christodoulides and L. Pichiarella, *J. Chem. Phys.*, 1986, **85**, 1966.
124  P.G. Datskos and L.G. Christophorou, *J. Chem. Phys.*, 1987, **86**, 1982.
125  A.A. Christodoulides, L.G. Christophorou and D.L. McCorkle, *Chem. Phys. Lett.*, 1987, **139**, 350.
126  L.G. Christophorou, S.R. Hunter, L.A. Pinnaduwage, J.G. Carter, A.A. Christodoulides and S.M. Spyrou, *Phys. Rev. Lett.*, 1987, **58**, 1316.
127  M. Fenzlaff, R. Gerhard and E. Illenberger, *J. Chem. Phys.*, 1988, **88**, 149.
128  J.E. Delmore and A.D. Appelhans, *J. Chem. Phys.*, 1986, **84**, 6238.
129  A.D. Appelhans and J.E. Delmore, *J. Chem. Phys.*, 1988, **88**, 5561.
130  J.E. Delmore and A.D. Appelhans, *Int. J. Mass Spectrom. Ion Processes*, 1986, **68**, 327.
131  J.E. Delmore, A.D. Appelhans, R.E. Shomo II and D.A. Dahl, *Biomed. Environ. Mass Spectrom.*, 1988, **15**, in press.
132  A.D. Appelhans, J.E. Delmore and D.A. Dahl, *Anal. Chem.*, 1987, **59**, 1685.
133  M. Tronc, in 'Swarm Studies and Inelastic Electron-Molecule Collisions', L.C. Pitchford, B.V. McKoy, A. Chutjian and S. Trajmar (Eds.), Springer-Verlag, New York, 1987, p. 287.
134  R. Dressler, M. Allan and M. Tronc, *J. Phys. B: At. Mol. Phys.*, 1987, **20**, 393.
135  R. Dressler and M. Allan, *J. Chem. Phys.*, 1987, **87**, 4510.
136  R. Dressler and M. Allan, *J. Electron Spectrosc. Relat. Phenom.*, 1986, **41**, 275.
137  M. Allan, *Int. J. Quant. Chem.*, 1987, **31**, 161.
138  M. Allan, in 'Electronic and Atomic Collisions', H.B. Gilbody, W.R. Newell, F.H. Read and A.C.H. Smith, (eds.), Elsevier Science Publishers, 1988, p. 93.
139  R. Azria, R. Abouaf and D. Teillet-Billy, *J. Phys. B: At. Mol. Opt. Phys.*, 1988, **21**, L213.
140  P.G. Fournier, H. Aouchiche, V. Lorent and J. Baudon, *Phys. Rev.*, 1986, **34A**, 3743.
141  A.R. Lee, P. Jonathan, A.G. Brenton and J.H. Beynon, *Int. J. Mass Spectrom. Ion Processes*, 1987, **75**, 329.
142  W. deLange and N.M.M. Nibbering, *Int. J. Mass Spectrom. Ion Processes*, 1987, **80**, 201.
143  W.J. Griffiths and F.M. Harris, *Int. J. Mass Spectrom. Ion Processes*, 1987, **77**, R7.
144  W.J. Griffiths and F.M. Harris, *Org. Mass Spectrom.*, 1987, **22**, 559.

145 W.J. Griffiths and F.M. Harris, *Org. Mass Spectrom.*, 1987,22, 812.
146 R.G. Kingston, A.G. Brenton, M. Guilhaus and J.H. Beynon, *Org. Mass Spectrom.*, 1986, 21, 697.
147 P. Lablanquie, J.H.D. Eland, I. Nenner, P. Morin, J. Delwiche and M.-J. Hubin-Franskin, *Phys. Rev. Lett.*, 1987, 58, 992.
148 S.D. Price, J.H.D. Eland, P.G. Fournier, J. Fournier and P. Millie, *J. Chem. Phys.*, 1988, 88, 1511.
149 J.H.D. Eland, S.D. Price, J.C. Cheney, P. Lablanquie, I. Nenner and P.G. Fournier, *Phil. Trans. R. Soc. Lond.*, 1988, 324A, 247.
150 M.J. Besnard, L. Hellner, Y. Malinovich and G. Dujardin, *J. Chem. Phys.*, 1986, 85, 1316.
151 G. Dujardin, L. Hellner and M.J. Besnard, in 'Electronic and Atomic Collisions', H.B. Gilbody, W.R. Newell, F.H. Read and A.C.H. Smith (eds.), Elsevier Science Publishers, 1988, p. 471.
152 R.W. Wetmore and R.K. Boyd, *J. Phys. Chem.*, 1986, 90, 6091.
153 D. Cooper, *Chem. Phys. Lett.*, 1986, 132, 377.
154 L. Hellner, M.J. Besnard, G. Dujardin and Y. Malinovich, *Chem. Phys.*, 1988, 119, 391.
155 R.W. Wetmore and R.K. Boyd, *J. Phys. Chem.*, 1986, 90, 5540.
156 P.G. Fournier and R.E. March, *Chem. Phys. Lett.*, 1987, 137, 596.
157 M.J. Besnard, L. Hellner, G. Dujardin and D. Winkoun, *J. Chem. Phys.*, 1988, 88, 1732.
158 T. Masuoka and H. Fujikawa, *J. Chem. Phys.*, 1986, 84, 3771.
159 W. Eberhardt, E.W. Plummer, I.W. Lyo, R. Carr and W.K. Ford, *Phys. Rev. Lett.*, 1987, 58, 207.
160 G. Dujardin, M.J. Besnard, L. Hellner and Y. Malinovich, *Phys. Rev.*, 1987, 35A, 5012.
161 H. LeRouzo, *J. Phys. B.*, 1986,19, L677.
162 J.H.D. Eland, F.S. Wort, P. Lablanquie and I. Nenner, *Z. Phys. D. - Atoms, Molecules and Clusters*, 1986, 4, 31.
163 P. Lablanquie, J. Delwiche, M.-J. Franskin-Hubin, I. Nenner, J.H.D. Eland and K. Ito, *J. Mol. Structure*, in press.
164 A.P. Hitchcock, P. Lablanquie, P. Morin, E. Lizon, A. Lugrin, M. Simon, P. Thiry and I. Nenner, *Phys. Rev.*, 1988,37A, 2448.
165 D. Winkoun, G. Dujardin, L. Hellner and M.J. Besnard, *J. Phys. B.: At. Mol. Opt. Phys.*, 1988, 21, 1385.
166 D. Winkoun and G. Dujardin, *Z. Phys. D.- Atoms, Molecules and Clusters*, 1986, 4, 57.
167 I. Nenner, J.H.D. Eland, P. Lablanquie, J. Delwiche, M.J. Hubin-Franskin, P. Roy, P. Morin and A. Hitchcock, in 'Physics of Atoms and Molecules', P.G. Burke and J.B. West (eds.), Plenum, in press.
168 J.H.D. Eland, F.S. Wort and R.N. Royds, *J. Electron Spectrosc. Relat. Phenom.*, 1986, 41, 297.
169 J.H.D. Eland, *Mol. Phys.*, 1987, 61, 725.
170 P.G. Fournier, J. Fournier, F. Salama, D. Stark, S.D. Peyerimhoff and J.H.D. Eland, *Phys. Rev.*, 34A, 1657.
171 P.G. Fournier, M. Mousselmal, S.D. Peyerimhoff, A. Banichevich, M.Y. Adam and T.J. Morgan, *Phys. Rev.*, 1987, 36A, 2594.
172 P.G. Fournier, J.H.D. Eland, P. Millie, S. Svensson, S.D. Price, J. Fournier, G. Comtet, B.

Wannberg, P. Baltzer, A. Kaddouri and U. Gelius, *J. Chem. Phys.*, in press.
173 B.J. Olsson and M. Larsson, *J. Phys. B.*, 1987, **20**, L137.
174 D. Mathur and C. Badrinathan, *J. Phys. B.*, 1987, **20**, 1517.
175 D. Winkoun, D. Solgadi and J.P. Flament, *Chem. Phys. Lett.*, 1987, **139**, 546.
176 T. Ast, *Adv. Mass Spectrom.*, **10A**, J.F.J. Todd (ed.), John Wiley & Sons (1986), p. 471.
177 P. Jonathan, R.K. Boyd, A.G. Brenton and J.H. Beynon, *Chem. Phys.*, 1986, **110**, 239.
178 P. Johnathan, A.G. Brenton, J.H. Beynon and R.K. Boyd, *Int. J. Mass Spectrom. Ion Processes*, 1987, **76**, 319.
179 M.W. Wong, R.H. Nobes and L. Radom, *Rapid Commun. Mass Spectrom.*, 1987, **1**, 3.
180 L. Radom, P.M.W. Gill, M.W. Wong and R.H. Nobes, *Pure & Appl. Chem.*, 1988, **60**, 183.
181 T. Drewello, W. Koch, T. Weiske, H. Schwarz and D. Stahl, *Int. J. Mass Spectrom. Ion Processes*, 1986, **72**, 313.
182 W.J. Griffiths and F.M. Harris, *Rapid Commun. Mass Spectrom.*, 1988, **2**, 28.
183 W.J. Griffiths and F.M. Harris, *Rapid Commun. Mass Spectrom.*, 1988, **2**, 91.
184 P. Johnathan, Z. Herman, M. Hamdan and A.G. Brenton, *Chem. Phys. Lett.*, 1987, **141**, 511.
185 M. Hamdan, A.G. Brenton and D. Mathur, *Chem. Phys. Lett.*, 1988, **144**, 387.
186 P. Jonathan, M. Hamdan, A.G. Brenton and G.D. Willett, *Chem. Phys.*, 1988, **119**, 159.
187 D. Mathur, R.G. Kingston, F.M. Harris and J.H. Beynon, *J. Phys. B.: At. Mol. Phys.*, 1986, **19**, L575.
188 D. Mathur, R.G. Kingston, F.M. Harris, A.G. Brenton and J.H. Beynon, *J. Phys. B.: At. Mol. Phys.*, 1987, **20**, 1811.
189 D. Mathur, C.J. Reid and F.M. Harris, *J. Phys. B.: At. Mol. Phys.*, 1987,**20**, L577.
190 Z. Herman, P. Jonathan, A.G. Brenton and J.H. Beynon, *Chem. Phys. Lett.*, 1987, **141**, 433.
191 K. Vékey, A.G. Brenton and J.H. Beynon, *Int. J. Mass Spectrom. Ion Processes*, 1986, **70**, 277.
192 K. Vékey, A.G. Brenton and J.H. Beynon, *J. Phys. Chem.* 1986, **90**, 3569.
193 J.M. Curtis, A.G. Brenton, J.H. Beynon and R.K. Boyd, *Chem. Phys.*, 1987, **117**, 325.
194 J.M. Curtis, A.G. Brenton and R.K. Boyd, *Chem. Phys.*, 1987, **116**, 241.
195 R. Locht and J. Momigny, *Chem. Phys. Lett.*, 1987, **138**, 391.
196 L.S. Cederbaum, F. Tarantelli, A. Sgamellotti and J. Schirmer, *J. Chem. Phys.*, 1987, **86**, 2168.
197 F. Tarantelli, A Sgamellotti, L.S. Cederbaum and J. Schirmer, *J. Chem. Phys.*, 1987, **86**, 2201.
198 A.W. Castleman, Jr. and T.D. Márk, in 'Gaseous Ion Chemistry and Mass Spectrometry', Ed. J.H. Futrell, Chapter 12, John Wiley and Sons, New York, 1986, p. 259.
199 S.W. McElvany, M.M. Ross and A.P. Baronavski, *Anal. Inst.*, 1988, **17**, 23.
200 D.M. Cox, K.C. Reichmann and A. Kaldor, *J. Chem. Phys.*, 1988, **88**, 1588.
201 S.W. McElvany, H.H. Nelson, A.P. Baronavski, C.H. Watson and J.R. Eyler, *Chem. Phys. Lett.*, 1987, **134**, 214.
202 M. Moini and J.R. Eyler, *Chem. Phys. Lett.*, 1987, **137**, 311.
203 M. Moini and J.R. Eyler, *J. Chem. Phys.*, 1988, **88**, 5512.
204 S.K. Loh, D.A. Hales and P.B. Armentrout, *Chem. Phys. Lett.*, 1986, **129**, 527.

205 T.D. Márk and A. Stamatovic, in 'The Physics of Ionized Gases', J. Purić and D. Belić (eds.), World Scientific, Singapore, 1987, p. 161.
206 T.D. Márk, in 'Electronic and Atomic Collisions', H.B. Gilbody, W.R. Newell, F.H. Read and A.C.H. Smith (eds.), Elsevier Science Publishers, 1988, p. 705.
207 Y. Liu, S.C. O'Brien, Q. Zhang, J.R. Heath, F.K. Tittel, R.F. Curl, H.W. Kroto and R.E. Smalley, *Chem. Phys. Lett.*, 1986, **126**, 215.
208 T.D. Márk and P. Scheier, *Chem. Phys. Lett.*, 1987, **137**, 245.
209 P. Scheier and T.D. Márk, *J. Chem. Phys.*, 1987, **87**, 5238.
210 P. Scheier and T.D. Márk, *J. Chem. Phys.*, 1987, **86**, 3056.
211 P. Scheier and T.D. Márk, *Chem. Phys. Lett.*, 1987, **136**, 423.
212 P. Scheier, A. Stamatovic and T.D. Márk, *J. Chem. Phys.*, 1988, **88**, 4289.
213 P. Scheier, A. Stamatovic and T.D. Márk, *Chem. Phys. Lett.*, 1988, **144**, 119.
214 D. Kreisle, K. Leiter, O. Echt and T.D. Márk, *Z. Phys. D. Atoms, Molecules and Clusters*, 1988, **3**, 319.
215 C. Bréchignac, M. Broyer, Ph. Cahvzac, G. Delacretaz, P. Labastie and L. Wöste, *Chem. Phys. Lett.*, 1987, **133**, 45.
216 A. Stamatovic, P. Scheier and T.D. Márk, *J. Chem. Phys.*, 1988, **88**, 6884.
217 T.D. Márk, P. Scheier and A. Stamatovic, *Chem. Phys. Lett.*, 1987, **136**, 177.
218 K. Mitsuke, T. Kondow and K. Kuchitsu, *J. Phys. Chem.*, 1986, **90**, 1552.
219 T. Kondow, *J. Phys. Chem.*, 1987, **91**, 1307.
220 R. Azria, L. Parenteau and L. Sanche, *Phys. Rev. Lett.*, 1987, **59**, 638.
221 R. Azria, L. Parenteau and L. Sanche, *J. Chem. Phys.*, 1987, **87**, 2292.
222 R. Azria, L. Parenteau and L. Sanche, *J. Chem. Phys.*, 1988, **88**, 5166.
223 E.A. Walters, J.R. Grover and M.G. White, *Z. Phys. D - Atoms, Molecules and Clusters*, 1986, **4**, 103.
224 J.R. Grover, E.A. Walters and E.T. Hui, *J. Phys. Chem.*, 1987, **91**, 3233.
225 E. Rühl, P. Bisling, B. Brutschy, K. Beckmann, O. Leisin and H. Morgner, *Chem. Phys. Lett.*, 1986, **128**, 512.
226 P.G.F. Bisling, E. Rühl, B. Brutschy and H. Baumgärtel, *J. Phys. Chem.*, 1987, **91**, 4310.
227 P. Bisling, E. Rühl, B. Brutschy and H. Baumgärtel, in 'Structure and Dynamics of Weakly Bound Molecular Complexes', A. Weber (Ed.), D. Reidel Publishing Co., 1987, p. 303.
228 A. Kiermier, B. Ernstberger, H.J. Neusser and E.W. Schlag, *J. Phys. Chem.*, 1988, **92**, 3785.
229 C.-Y. Kung, R.A. Kennedy, D.A. Dolson and T.A. Miller, *Chem. Phys. Lett.*, 1988, **145**, 455.
230 H. Shiromaru, H. Shinohara, N. Washida, H.-S. Yoo and K. Kimura, *Chem. Phys. Lett.*, 1987, **141**, 7.
231 N.J. Kirchner and M.T. Bowers, *J. Chem. Phys.*, 1987, **86**, 1301.
232 N.J. Kirchner and M.T. Bowers, *J. Phys. Chem.*, 1987, **91**, 2573.
233 K. Hiraoka, *J. Chem. Phys.*, 1987, **87**, 4048.
234 A.J. Illies, *J. Phys. Chem.*, 1988, **92**, 2889.
235 L. Hanley, S.A. Ruatta and S.L. Anderson, *J. Chem. Phys.*, 1987, **87**, 260.
236 L. Hanley and S.L. Anderson, *J. Phys. Chem.*, 1987, **91**, 5161.

237 L. Hanley, J.L. Whitten and S.L. Anderson, *J. Phys. Chem.*, in press.
238 C. Lifshitz, T. Peres, S. Kababia and I. Agranat, *Int. J. Mass Spectrom. Ion Processes,* 1988, **82**, 193.
239 R.J. Saykally, *Science*, 1988, **239**, 157.
240 T. Amano, *Phil. Trans. R. Soc. Lond.*, 1988, **324A**, 163.
241 T.J. Sears, *J. Chem. Soc. Faraday Trans. 2*, 1987, **83**, 111.
242 T. Oka, *Phil. Trans. R. Soc. Lond.*, 1988, **324A**, 81.
243 M. Gruebele, M. Polak, G.A. Blake and R.J. Saykally, *J. Chem. Phys.*, 1986, **85**, 6276.
244 K. Kawaguchi and T. Amano, *J. Chem. Phys.*, 1988, **88**, 4584.
245 B.M. Dinelli, M.W. Crofton and T. Oka, *J. Mol. Spectrosc.*, 1988, **127**, 1.
246 M. Gruebele, M. Polak and R.J. Saykally, *J. Chem. Phys.*, 1987, **87**, 3347.
247 M.W. Crofton, M.-F. Jagod, B.D. Rehfuss, W.A. Kreiner and T. Oka, *J. Chem. Phys.*, 1988, **88**, 666.
248 T. Nakanaga and T. Amano, *Chem. Phys. Lett.*, 1987, **134**, 195.
249 T. Amano, K. Kawaguchi and E. Hirota, *J. Mol. Spectrosc.*, 1987, **126**, 177.
250 M.W. Crofton, M.-F. Jagod, B.D. Rehfuss and T. Oka, *J. Chem. Phys.*, 1987, **86**, 3755.
251 T. Amano and K. Tanaka, *J. Mol. Spectrosc.*, 1986, **116**, 112.
252 W.-C. Ho, C.E. Blom, D.-J. Liu and T. Oka, *J. Mol. Spectrosc.*, 1987, **123**, 251.
253 M.W. Crofton and T. Oka, *J. Chem. Phys.*, 1987, **86**, 5983.
254 S.K. Lee and T. Amano, *Astrophys. J.*, 1987, **323**, L145.
255 B.D. Rehfuss, M.W. Crofton and T. Oka, *J. Chem. Phys.*, 1986, **85**, 1785.
256 M. Gruebele, M. Polak and R.J. Saykally, *J. Chem. Phys.*, 1987, **86**, 1698.
257 M. Polak, M. Gruebele and R.J. Saykally, *J. Am. Chem. Soc.*, 1987, **109**, 2884.
258 M. Gruebele, M. Polak and R.J. Saykally, *J. Chem. Phys.*, 1987, **86**, 6631.
259 M. Polak, M. Gruebele and R.J. Saykally, *J. Chem. Phys.*, 1987, **87**, 3352.
260 M. Gruebele, M. Polak and R.J. Saykally, *J. Chem. Phys.*, 1987, **87**, 1448.
261 E.P. Kanter, Z. Vager, G. Both and D. Zajfman, *J. Chem. Phys.*, 1986, **85**, 7487.
262 T. Amano and T. Nakanaga, *Astrophys. J.*, 1988, **328**, 373.
263 D.-J. Liu, W.-C. Ho and T. Oka, *J. Chem. Phys.*, 1987, **87**, 2442.
264 F.G. Celii, J.P. Maier and M. Ochsner, *J. Chem. Phys.* 1986, **85**, 6230.
265 N.E. Levinger, D. Ray, K.K. Murray, A.S. Mullin, C.P. Schulz and W.C. Lineberger, *J. Chem. Phys.*, 1988, **89**, 71.
266 M. Larsson, S. Mannervik, R.T. Short, P. Sigray and D. Sonnek, *Chem. Phys. Lett.*, 1988, **146**, 507.
267 M.H. Shen, J.W. Winniczek and J.M. Farrar, *J. Phys. Chem.*, 1987, **91**, 6447.
268 W.J. van der Hart, L.J. de Koning, N.M.M. Nibbering and M.L. Gross, *Int. J. Mass Spectrom. Ion Processes*, 1986, **72**, 99.
269 J.D. Morrison, in 'Gaseous Ion Chemistry and Mass Spectrometry', J.H. Futrell (ed.), John Wiley and Sons, New York (1986), chapter 9, p. 179.
270 H.-S. Kim, C.-H. Kuo and M.T. Bowers, *J. Chem. Phys.*, 1987, **87**, 2667.
271 J.T. Blair, J.C. Weisshaar, J.E. Carpenter and F. Weinhold, *J. Chem. Phys.*, 1987, **87**, 392.
272 H. Åkeby and L.G.M. Pettersson, *Chem. Phys. Lett.*, 1988, **146**, 511.
273 R.E. Tecklenburg, Jr., and D.H. Russell, *J. Am. Chem. Soc.*, 1987,**109**, 7654.
274 M.E. Geusic, M.F. Jarrold, T.J. McIlrath, R.R. Freeman and W.L. Brown, *J. Chem. Phys.*,

1987, **86**, 3862.
275 S.C. O'Brien, J.R. Heath, R.F. Curl and R.E. Smalley, *J. Chem. Phys.*, 1988, **88**, 220.
276 M. Hamdan, F.M. Harris, I.W. Griffiths and J.H. Beynon, *Chem. Phys. Lett.*, 1987, **135**, 511; *Int. J. Mass Spectrom. Ion Processes*, 1986, **74**, 303.
277 M.J. Coggiola, P.C. Cosby, H. Helm, J.R. Peterson and R.C. Dunbar, *J. Phys. Chem.*, 1987, **91**, 2796.
278 J.P. Maier, *J. Electron Spectrosc.*, 1986, **40**, 203.
279 J.P. Maier, in 'Jahrbuch der Akademie der Wissenschaften in Göttingen', Vandenhoeck & Ruprecht in Göttingen, 1986.
280 J.P. Maier, *J. Chem. Soc., Faraday Trans. 2*, 1987, **83**, 49.
281 J.P. Maier, *Phil. Trans. R. Soc. Lond.*, 1988, **324A**, 209.
282 J.P. Maier, *Chem. Soc. Rev.*, 1988, **17**, 45.
283 R. Kuhn, J.P. Maier and M. Ochsner, *Mol. Phys.*, 1986, **59**, 441.
284 R. Kuhn, J.P. Maier, L. Misev and T. Wyttenbach, *J. Electron Spectrosc.*, 1986, **41**, 265.
285 F.G. Celii, J. Fulara, J.P. Maier and M. Rösslein, *Chem. Phys. Lett.*, 1986, **131**, 325.
286 D. Klapstein and J.P. Maier, in 'Proc. 17th Int. Symp. on Free Radicals', K.M. Evenson (Ed.), N.B.S. special publication 716, 1986.
287 D. Klapstein, R. Kuhn, J.P. Maier, M. Ochsner and T. Wyttenbach, *Chem. Phys.*, 1986, **101**, 133.
288 M.A. King, R. Kuhn and J.P. Maier, *J. Phys. Chem.*, 1986, **90**, 6460.
289 M.A. King, R. Kuhn and J.P. Maier, *Mol. Phys.*, 1987, **60**, 867; *ibid. id.* 1987, **62**, 1503.
290 F.G. Celii, M. Rösslein, M.A. Hanratty and J.P. Maier, *Mol. Phys.*, 1987, **62**, 1435.
291 M.A. Hanratti, M. Rösslein, F.G. Celii, T. Wyttenbach and J.P. Maier, *Mol. Phys.*, 1988, **64**, 865.
292 M. Rösslein, M. Wyttenbach and J.P. Maier, *J. Chem. Phys.*, 1987, **87**, 6770.
293 J.P. Maier and M. Rösslein, *J. Chem. Phys.*, 1988, **88**, 4614.
294 I.W. Milkman, J.C. Choi, J.L. Hardwick and J.T. Moseley, *J. Chem. Phys.*, 1987, **86**, 1679.
295 I.W. Milkman, J.C. Choi, J.L. Hardwick and J.T. Moseley, *J. Mol. Spectrosc.*, 1988, **130**, 20
296 H. Van Lonkhuyzen and J.F.M. Aarts, *Chem. Phys. Lett.*, 1987, **140**, 434.
297 H. Obase, M. Tsuji and Y. Nishimura, *J. Chem .Phys.*, 1987, **87**, 2695.
298 M. Tsuji, J.P. Maier, H. Obase and Y. Nishimura, *Chem. Phys.*, 1986, **110**, 17.
299 H. Sekiya, M. Tsuji and Y. Nishimura, *Chem. Lett., The Chemical Society of Japan*, 1986, p. 1997.
300 M. Tsuji, J.P. Maier, H. Obase and Y. Nishimura, *Chem. Phys. Lett.*, 1988, **147**, 619.
301 S. Yamaguchi, M. Tsuji, H. Obase, H. Sekiya and Y. Nishimura, *J. Chem. Phys.*, 1987, **86**, 4952.
302 M. Tsuji, J.P. Maier, H. Obase, H. Sekiya and Y. Nishimura, *Chem. Phys. Lett.*, 1987, **137**, 421.
303 S. Yamaguchi, M. Tsuji and Y. Nishimura, *J. Chem. Phys.*, 1987, **87**, 1637.
304 H. Obase, M. Tsuji and Y. Nishimura, *Chem. Phys. Lett.*, 1987, **141**, 133.
305 S. Yamaguchi, M. Tsuji and Y. Nishimura, *J. Chem. Phys.*, 1988, **88**, 3111.
306 C.-H. Kuo, I.W. Milkman, T.C. Steimle and J.T. Moseley, *J. Chem. Phys.*, 1986, **85**, 4269.
307 P.O. Danis, T. Wyttenbach and J.P. Maier, *J. Chem. Phys.*, 1988, **88**, 3451.

308 S. Leach, *Astron. Astrophys.*, 1987, **187**, 195.
309 G. Kindvall, M. Larsson, B. Olsson and P. Sigray, *Phys. Scripta*, 1986, **33**, 412.
310 O. Gustafsson, M. Larsson and P. Sigray, *Z. Phys. D. Atoms, Molecules and Clusters*, 1988, **7**, 373.
311 N. Sato, S. Nanbu and S. Iwata, *Chem. Phys. Lett.*, 1988, **146**, 275.
312 T. Reddish, A.A. Cafolla and J. Comer, *Chem. Phys.*, 1988, **120**, 149.
313 L. Wang, Y.T. Lee and D.A. Shirley, *J. Chem. Phys.*, 1987, **87**, 2489.
314 V. Butcher, M.L. Costa, J.M. Dyke, A.R. Ellis and A. Morris, *Chem. Phys.*, 1987, **115**, 261.
325 S. Leach and G.G. Englert, *J. Electron Spectrosc.*, 1986, **41**, 181.
316 J.M. Oakes and G.B. Ellison, *Tetrahedron*, 1986, **42**, 6263.
317 D.G. Leopold, A.E.S. Miller and W.C. Lineberger, *J. Am. Chem. Soc.*, 1986, **108**, 1379.
318 S. Moran and G.B. Ellison, *Int.J. Mass Spectrom. Ion Spectrosc.*, 1987, **80**, 83.
319 S. Moran and G.B. Ellison, *J. Phys. Chem.*, 1988, **92**, 1794.
320 M.R. Nimlos and G.B. Ellison, *J. Am. Chem. Soc.*, 1986, **108**, 6522.
321 C.T. Wickham-Jones, S. Moran and G.B. Ellison, *J. Chem. Phys.*, 1988, in press.
322 T.M. Miller, D.G. Leopold, K.K. Murray and W.C. Lineberger, *J. Chem. Phys.*, 1986, **85**, 2368.
323 T.M. Miller and W.C. Lineberger, *Chem. Phys. Lett.*, 1988, **146**, 364.
324 A.E. Stevens Miller, C.S. Feigerle and W.C. Lineberger, *J. Chem. Phys.*, 1987, **87**, 1549.
325 M.A. Nimios, L.B. Harding and G.B. Ellison, *J. Chem. Phys.*, 1987, **87**, 5116.
326 D.G. Leopold and W.C. Lineberger, *J. Chem. Phys.*, 1986. **85**, 51.
327 D.G. Leopold, T.M. Miller and W.C. Lineberger, *J. Am. Chem. Soc.*, 1986, **108**, 178.
328 L.A. Posey, M.J. DuLuca and M.A. Johnson, *Chem. Phys. Lett.*, 1986, **131**, 170.
329 L.A. Posey and M.A. Johnson, *J. Chem. Phys.*, 1988, **88**, 5383.
330 M.J. DeLuca, B. Niu and M.A. Johnson, *J. Chem. Phys.*, 1988, **88**, 5857.
331 D.G. Leopold, J. Ho and W.C. Lineberger, *J. Chem. Phys.*, 1987, **86**, 1715.
332 C.L. Pettiette, S.H. Yang, M. Craycraft, J. Conceicao, R.T. Laaksonen, O. Cheshnovsky and R.E. Smalley, *J. Chem. Phys.*, 1988, **88**, 5377.
333 S. Moran, H.B. Ellis, Jr., D.J. DeFrees, A.D. McLean and G.B. Ellison, *J. Am. Chem. Soc.*, 1987, **109**, 5996.
334 S. Moran, H.B. Ellis, Jr., D.J. DeFrees, A.D. McLean, S.E. Paulson and G.B. Ellison, *J. Am. Chem. Soc.*, 1987, **109**, 6004.
335 K.R. Lykke, K.K. Murray, D.M. Neumark and W.C. Lineberger, *Phil. Trans. R. Soc. Lond.*, 1988, **324A**, 179.
336 K.R. Lykke, D.M. Neumark, T. Andersen, V.J. Trapa and W.C. Lineberger, *J. Chem. Phys.*, 1987, **87**, 6842.
337 J. Marks, J.I. Brauman, R.D. Mead, K.R. Lykke and W.C. Lineberger, *J. Chem. Phys.* 1988, **88**, 6785.
338 D.C. Clary, *J. Phys. Chem.*, 1988, **92**, 3173.
339 K.K. Murray, K.R. Lykke and W.C. Lineberger, *Phys. Rev.*, 1987, **36A**, 699.
340 R.B. Metz, T. Kitsopoulos, A. Weaver and D.M. Neumark, *J. Chem. Phys.*, 1988, **88**, 1463.

# 2
# Structures and Reactions of Gas-phase Organic Ions

BY P. C. BURGERS AND J. K. TERLOUW

## 1 INTRODUCTION

In preparing for this task, we have collected the impressive number of c. 550 papers which are immediately relevant to our subject. The majority of these would deserve to be referred to in this chapter but in the space available an encyclopaedic review is out of the question. Fortunately several excellent review papers have appeared [1-21], some of which will be discussed below, and this has aided us in our decision to be strictly selective in deciding which areas of the subject deserve special attention. Topics which will be treated are :

i) developments in the methodology for obtaining information on the structures and reaction mechanisms of positive ions in the gas-phase. This is followed by a discussion of Neutralization Reionization Mass Spectrometry (NRMS), a rapidly growing branch of CID spectrometry (more than 10% of the papers collected) which, apart from its application to mechanistic studies, can successfully be used for the study of **neutral** species.

ii) classes of ions, including distonic ions, hydrogen-bridged radical cations, destabilized carbenium ions and dications, all of which have been studied extensively in the review period.

iii) dissociation mechanisms, in particular those where distonic ions and ion-dipole complexes seem to be key intermediates. The *detailed* study of reaction mechanisms is at a particularly interesting stage of development. Many mechanistic proposals are still at a crude and debatable stage but the current level of interest from both experimentalists and theoreticians will ensure further progress. In particular, state of the art ab initio molecular orbital theory calculations are increasingly being recognized and used as an indispensable auxiliary tool. In addition to calculations of the relative energies of isomeric ion structures, an established procedure, success is being realized in the much more difficult task of calculating energy barriers for isomerization and dissociation.

An area of rapidly growing research activity which falls within the scope of our subject is (transition) metal ion reactions with organic molecules. For reasons that reflect personal bias , the recent literature in this field has been reviewed in more than usual detail in the section "ion/molecule reactions" of the latest Analytical Chemistry review of mass spectrometry [8b] and will therefore not be treated here. The ion chemistry of (large) clusters has also been the subject of much research. However this field is less related to our topic and since two books [17,18] and two reviews [19,20] have appeared in 1987 with yet another one forthcoming, only some results on small clusters posing mechanistic and/or structural questions will be referred to. Similarly no specific attention will be paid to advances in flow reactor techniques for the study of gas-phase ion chemistry for which an up to date review by Graul and Squires [21] has just appeared.

## 2 METHODOLOGY

A review by Holmes of the techniques (and their limitations) currently available to assign ion structures has been discussed in Chapter 2 of the previous volume of this series. A new, highly sophisticated, but very promising method is the "Coulomb Explosion" technique, for which the first results, e.g. on the structure of the $C_2H_3^+$ ion, have been reported [22]. We will discuss recent developments in the areas of (a) ion and neutral thermochemistry, (b) photoelectron-photoion coincidence spectroscopy, (c) dissociation characteristics and (d) ion/molecule reactions.

### Ion and neutral thermochemistry

Reliable heats of formation for positive ions form a cornerstone in studies of ion structures and reaction mechanisms especially in the increasing number of studies where results of ab initio calculations are involved. Recently Lias et al. have finished their monumental task and the successor of the classical 1977 Rosenstock compilation "Energetics of Gaseous Ions" has now appeared. The new book " Gas-Phase Ion and Neutral Thermochemistry" [11] is a true treasure house of thermochemical data with up to date (literature coverage is up to 1987) information on the "best" available experimentally determined values for positive and negative ions. The positive ion tables not only cover data derived from IE and AE measurements but also a great many heats of formation for even-electron ions derived from proton affinities. Also included, and this we find very useful, are $\Delta H_f$ values of the corresponding neutral species. As stated in the introduction to the compilation - the reading of which is strongly recommended if only to prevent over-confidence in all the data presented - many $\Delta H_f$ neutral values are estimates, quite a few of which were made specifically for this publication. Regrettably no indication is given as to how these estimates, which, depending on the method which has been used and the supporting data available, may vary from very good to poor, were obtained. In this respect it should be recalled that there is a serious lack of data for many simple well known organic compounds [12]. For instance molecules such as $CH_3OCH=CH_2$, $CH_3OC\equiv CH$, $CH_3COCH=CH_2$, $CH_3COC\equiv CH$ and $HC\equiv CCOOH$, key compounds for an additivity scheme, are all missing. The estimates quoted in the book for these compounds are -100, 75, -138, 65 and -117 kJ/mol respectively whereas additivity values which we would favour, based on a consideration of the accumulated data in ref. 12, are : -107, 67, -121, 70 and -144 kJ/mol. The compilation does not provide literature references to the source of the IE/AE data (nor does it define the technique used) for the large number of data taken from the earlier NBS compilations. Unfortunately no indication is given from which of these compilations a result comes and this sometimes makes difficult the tracing of the primary literature source.

Another approach to obtain heats of formation of both ionic and neutral species involves the use of quantum mechanical calculations at varying levels of sophistication. In the above compilation only data from experimental determinations are included with the exception of some neutrals.

The use of the semiempirical MINDO/3 and MNDO methods for the determination of heats of formation has been critically examined for a great number of closed- and open-shell cations [23] and it was concluded that with proper consideration of failures in specific areas both methods can be used for the thermodynamics of carbocations containing C,H,N and O. The potential of the newer AM1 semiempirical method has been evaluated for proton affinities and deprotonation enthalpies [24] and the study of hydrogen bonding in the gas-phase hydration of (protonated) diamines [25]. Further, a comparison of AM1, MNDO and MNDOC methods has been reported for the calculation of the heat of formation and the IE of radicals [26]. Unfortunately comparison between theory and experiment is sometimes unreliable because of the authors' poor choice of experimental values.

Although the usefulness of these inexpensive methods cannot be denied, especially for larger systems not amenable to sophisticated ab initio calculations, they cannot be used reliably to establish heats of formation of organic ions and neutrals for which experimental values are unavailable or ambiguous. However, in the hands of skilled theoreticians ab initio molecular orbital theory based computations increasingly produce results which allow a critical assessment of experimental results. For instance, in a series of papers titled theoretical thermochemistry [27/28], Pople and Curtiss have shown that for small molecules, ionization energies, proton affinities and heats of formation of their cations can be obtained with an accuracy of ±8 kJ/mol. This was achieved via calculations involving Møller-Plesset theory to the fourth order and a series of extended basis sets and considering isogyric reactions (i.e processes which leave the number of unpaired electrons unchanged). The overall agreement with experimental values is excellent and the discrepancies between theory and experiment which emerge , such as that between the recommended experimental values for PA $CH_4$ (and hence $\Delta H_f$ $CH_5^+$), may well lead to a revision of the experimental values. A similar situation might obtain for the $HOC^+$ ion whose "best" experimental heat of formation [11] is 28 kJ/mol below that derived from the calculated proton affinity for oxygen protonation of CO [29]. On the other hand $\Delta H_f$ $HOC^+$ derived from the (ab initio) calculated energy difference between $HCO^+$ and $HOC^+$ [30] and $\Delta H_f$ (expt.) $HCO^+$ [11] is very close to the above experimentally derived $\Delta H_f$ $HOC^+$. The above approach remains confined to small systems because of the computational expenses involved. For larger systems, the less expensive isodesmic substitution procedure [31] has been used to calculate heats of formation [with an estimated accuracy of ±13 kJ/mol] for a series of neutral organic compounds containing the substituents $CH_3$, $CF_3$, $NH_2$, $NF_2$, $NO_2$, OH and F . Most of these species are at present unknown and so experimental values cannot directly be obtained.

For ions, the use of the isodesmic approach has been critically examined in a study of the stabilities of α-oxy ($CH_2OH^+$, $CH_2OCH_3^+$) and α-thio ($CH_2SH^+$, $CH_2SCH_3^+$) carbenium ions [32].

One of the isodesmic processes considered is the hydride transfer reaction $CH_2OH^+$ + $CH_3SH \rightarrow CH_3OH + CH_2SH^+$, which was calculated to be *endothermic* by 9.6 kJ/mol [the most elaborate calculation gives 5.4 kJ/mol [33] ].This result is quoted to be in excellent agreement with experiment; however it appears that the experimental values used for the comparison are obsolete and $\Delta H_f$ data from the new compilation show the reaction to be *exothermic*, by 20 kJ/mol.

When dealing with a set of isomeric ions it follows from many papers (in particular those from the Radom group [34-37]) that ab initio calculations involving geometry optimization at the Hartree Fock level and treatment of the electron correlation problem via Møller-Plesset perturbation theory can provide accurate *relative* energies for the isomers if the appropriate level of theory is used. If a well characterized experimental heat of formation for one (or more) of the isomers is available, this allows the confident assignment of $\Delta H_f$ values to the other species. The ylid ion $CH_2NH_3^{+\cdot}$ is an example of an ion whose computed heat of formation [36] is probably more reliable than the experimentally derived value [11]. However not every system, even if it is relatively small, can be satisfactorily treated by the above approach. There may be several reasons for a failure of the standard treatment, ranging from deficiencies in the basis set or in the Hartree Fock (HF) model to deceptive convergence in the Møller-Plesset perturbation expansion [38] or size consistency problems in Configuration Interaction calculations. Apart from the case of Unrestricted HF calculations, where the spin contamination may be used as a guideline for judging the adequacy of the single determinant representation of the wave function, it is usually difficult to assess the accuracy of a computational result from first principles.

The performance of the basis set may depend on the structure of the molecule, particularly if a classical cation is compared with an ion-dipole complex. Since most basis sets have energy-optimized exponents, other molecular properties such as the dipole moment and the polarizability are not necessarily accurate and this may affect the calculated relative energy of a complex which is mainly held together by electrostatic forces. There is also the question of the Basis Set Superposition Error, which may be rather large if a split-valence basis such as the standard 4-31 G basis is used [39].

Another problem lies within the HF model. A case in point is the $CH_3NO_2^{+\cdot}$ (nitromethane) system studied in detail by McKee [40] whose MP4SDQ/6-31G results yield energy differences between $CH_3NO_2^{+\cdot}$, $CH_3ONO^{+\cdot}$ and their various dissociation products which are in only moderate agreement with the well-defined experimental values taken from ref. 11. The results are shown in Table 1. As pointed out by McKee the assumption that each electronic state is dominated by one configuration, a prerequisite for the "standard" ab initio calculations which are based on single configuration wave functions, may not be valid for this system.

Table 1. Energy differences, ΔE, [kJ/mol] between $CH_3NO_2^{+\bullet}$, $CH_3ONO^{+\bullet}$ and some dissociation products from experimental values [EXP] and ab initio calculations [THEORY].

|  | ΔE(EXP) | ΔE (THEORY) |
|---|---|---|
| $CH_3NO_2^{+\bullet}$ | 0 | 0 |
| $CH_3ONO^{+\bullet}$ | -54 | -21 |
| $NO_2^+ + CH_3^{\bullet}$ | +131 | +144 |
| $NO_2^{\bullet} + CH_3^+$ | +138 | +79 |
| $NO^+ + CH_3O^{\bullet}$ | +13 | -38 |
| $NO^{\bullet} + CH_2OH^+$ | -193 | -199 |

HF calculations may also lead to poor geometries for saddle points or even to predicting false minima which turn out to correpond to transition states when the electronic correlation is taken into account. An example is the $C_2H_3O_2^+$ system [41], where the open structure $\overset{+}{C}H_2COOH$ corresponds to a minimum in the HF potential energy surface. If a multiconfigurational wave function is used in the geometry optimization this structure appears to be the saddle point separating the ring structure $\overline{CH_2\text{-}O\text{-}\overset{+}{C}OH}$ and the dissociation products $CH_2OH^+ + CO$. This situation is not always easily recognized. However, it is bound to occur if the open shell character of the transition state substantially differs from that of the isomers or from the reaction products, as in the torsional motion of ethene. Here breaking the double bond implies the need for a 2-configuration wave function in order to obtain a qualitatively correct description of the transition state for the rotational isomerization. This problem may very well arise if molecules or cations containing carbonyl groups are involved in the reaction, e.g. the dissociation reaction : $\overset{+}{C}H_2OH \rightarrow H\text{-}\overset{+}{C}=O + H_2$ [30].

Computational procedures based on multiconfiguration wavefunctions are available, but apart from the formidable expenses involved in their use, the interpretation of the results produced therefrom requires great care and skill.

### Photoelectron-photoion coincidence (PEPICO) spectroscopy.

This elegant but experimentally difficult and time-consuming technique has considerably contributed to our understanding of gas-phase ion reactions. Apart from the evaluation of the functions $k(E)$ and $T(E)$ ($k$= rate constant, T = kinetic energy release), obviously of fundamental importance, the technique yields accurate thermochemical information. Not only can an accurate AE be obtained but, when rearrangement preceeds fragmentation, the $\Delta H_f$ of the most stable isomerized form, a valuable piece of information, is also realized.

As an extension of PEPICO Baer and co-workers have developed the sophisticated technique of PEPICO-PD [42/43], whereby a (single) PEPICO selected ion is photodissociated (PD) by a laser. This elegant technique makes possible i) ion preparation in excited states and the recording of their absorption spectra; ii) the evaluation of $k(E)$ for fast unimolecular reactions and, perhaps most important for ion chemists; iii) the assessment of the height of the interconversion barriers among isomeric forms [42/43/44]. Thus earlier PEPICO work had indicated that [1-butyne]$^{+\bullet}$, [2-butyne]$^{+\bullet}$ and [1,3-butadiene]$^{+\bullet}$ fragment at common rates, i.e. through a common intermediate ( [ 1,3-butadiene]$^{+\bullet}$), but later PEPICO-PD [42] indicated that the isomerization of [1-butyne]$^{+\bullet}$ and [2-butyne]$^{+\bullet}$ occurs through a common transition state. The photodissociation experiments by Bunn and Bowers [45], also on *stable* $C_4H_6^{+\bullet}$ ions are in full agreement with the PEPICO-PD data of Bunn and Baer [42]; the surprising results of Preuninger and Farrar [ref. 11 in 45], namely that the KER distributions of decomposing $C_4H_6^{+\bullet}$ ions are much larger than predicted by statistical theory, most likely result from difficulties in accurately subtracting the intense metastable signal from the total signal [45].

## Dissociation characteristics

Spontaneous and collisionally induced (CID) reactions are often used to determine the structure of ions. The important difference in these techniques is that ions having different internal energies are being sampled; thus ions may be indistinguishable from their metastable ion (MI) characteristics but clearly identifiable from CID experiments.

Analysis of metastable peak shapes can lead to an accurate assessment of the average kinetic energy release (KER) . This is so because metastable ions have a velocity **v** much in excess of the relative velocities $\Delta v$ of their products, resulting in a magnification **A** of the KER : $A = 4v/\Delta v$. This well known and important property was stressed recently [43]. Very small kinetic energy releases can be accurately measured and they have been linked to ion-dipole fragmentations [46], a topic of current interest. A new method of obtaining accurate translational energy release distributions from the metastable peak shape has recently been described [47]. The method takes into account the more important instrumental factors affecting the peak shape. It was shown that the older method of Holmes and Osborne gives distributions ( for large KER) which are biased towards high energies, partly a result of the rectangular basis functions (i.e. the peak shapes for a given discrete energy release) used in the older method. The present work produces results in good agreement with those obtained by the Beynon group ([47], refs. 7 and 8 cited therein]).

The use of the KER as a parameter to differentiate among isomeric steroid type ions has been extensively used by Zaretskii et al. [48,49].

According to RRKM/QET, the most important factor which determines whether or not an ion will be metastable is the activation energy and not the frequency factor . If the activation energy is small and the ion does not isomerize to a thermodynamically more stable form, the ion will not be metastable. It has increasingly been found that dissociations having small measured activation energies [44] nevertheless produce intense metastable peaks.

It appears from the recent literature that a variety of mechanisms may be operative which slow a reaction sufficiently for it to generate an intense metastable peak. These are i) Isomerization into a thermodynamically more stable form, whence for example even exothermic reactions with small (measured) activation energies can produce intense metastable peaks [50,51]; ii) Fragmentation via ion-dipole complexes; iii) Dissociation via tunneling and iv) Non-adiabatic reactions. Specific examples of each mechanism are discussed in the appropriate sections below.

One problem associated with MI and CID experiments in terms of structure assignment is that it is often not possible to assign unique structures to ions whose interconversion barriers are relatively small. In such cases higher energy activation is necessary to access ions from deep within their potential wells. Dissociative charge stripping (DCS) requires energy depositions of c. 20 eV as opposed to a few eV for CID. DCS spectra have been successfully used to assign structures to ions where CID failed. However, the doubly charged ion signals are often very weak and/or hidden by those from singly charged products [52]. Beynon and co-workers [52,53] , following earlier work, have demonstrated that their technique of dissociative charge stripping/electron capture - using a $B/E_1/E_2$ instrument - allows the doubly charged product ions to be detected entirely free of interference from singly charged fragments (resolved E/2 mass spectra [53]). Briefly, after (dissociative) charge stripping a chosen doubly charged ion $A^{++}$, as well as the interfering singly charged ion, are passed through $E_1$. $A^{++}$ is then reduced by charge exchange to $A^+$ and these singly charged ions are subsequently selected by $E_2$ and detected. Resolved E/2 spectra are obtained by a linked scan of $E_1$ and $E_2$. The reduction step may also yield singly charged dissociation products but they are not detected by such a linked scan [52].

Cooks and co-workers [54-58] have developed the technique of surface-induced decomposition (SID) as an alternative to CID. A mass selected beam is decelerated to 20 - 150 eV before colliding with a stainless steel surface; the reflected beam is mass analyzed by a quadrupole. Depending on the system and the surface, association reactions may take place and the resulting fragment ions may be isomer specific [55-57].

The potential of both techniques is well exemplified by the results on $C_5H_8^{+\bullet}$, $C_6H_6^{+\bullet}$ [52,53] and $C_6H_6^{+\bullet}$, $C_2H_4O^{+\bullet}$ [55,57] for resolved E/2 and SID spectrometry, respectively. A very important feature of these methods is that, in contrast to ordinary CID, *all* ions *must* collide (or interact) before they are detected and so fragmentation efficiencies may be used to distinguish isomeric ions (relative abundance of the recovered parent ion). Agreement between the two techniques is excellent and these methods may well solve as yet intractible problems - for example the $C_4H_8^{+\bullet}$ structure formed by loss of $H_2O$ from ionized isobutanol.

In an interesting study on the *low* -energy CID of $CH_3OCH_2^+$ and $CH_3CHOH^+$ to $CH_3^+$ + ($H_2,C,O$), Kinter and Bursey [59] interpret the much higher threshold for the dissociation of $CH_3CHOH^+$ in terms of the formation of the neutral hydroxycarbene $H\ddot{C}OH$ instead of formaldehyde, $CH_2=O$, which is formed from $CH_3OCH_2^+$. Also, at threshold, both reactions proceed with approximately 100% conversion of translational energy into internal energy and this result strongly indicates that formation of $CH_3^+$ from both ions results from a simple bond breaking process, as proposed above. Further studies include the reaction of $CH_3OCH_2^+$ [59, 60] and $C_6H_5CO^+$ [61] with ammonia. The protonation of $NH_3$ at very low reactant ion axial kinetic energy has been proposed to involve two different and non-competitive mechanisms [62].

### Reactivity: ion/molecule reactions

Protonation of $\alpha,\beta$-unsaturated ethers occurs at the $\beta$-carbon atom [63,64], but protonation of $\alpha,\beta$-unsaturated ketones, aldehydes and acids takes place at the (carbonyl) oxygen atom [65]. Protonation of $\beta,\gamma$ unsaturated ethers occurs at the oxygen atom [63,64]. It was shown [65] that the $\Delta H_f$ of these protonated species obey the generic equation of Holmes and co-workers well ($\Delta H_f = a - b \log n$, n = number of atoms).

Ion/molecule reactions associated with acetone have been the subject of several investigations [66-71]. Methylation of acetone, $CH_3COCH_3$, using $CH_3I$ [71] or $CH_3N_2^+$ [70] produces methylated acetone; this species is also generated by protonation of 2-methoxypropene, $CH_3C(OCH_3)=CH_2$, at high pressures [71]. Interestingly "self chemical-ionization" of 2-methoxypropene at low pressures brings about - in addition to protonation at carbon - proton attack at oxygen generating the ion $CH_2=C(CH_3)-\overset{+}{O}(H)CH_3$ [71]. This species predominantly dissociates to $CH_3OH_2^+ + C_3H_4$ with a small KER via, it was suggested, a proton-bound methanol/propyne complex.

The rate-constant for the association reaction of protonated acetone with neutral acetone has been measured by using selective photoionization in conjunction with low-energy CID [67]. The structure of the protonated dimer of acetone is distinct from that of protonated diacetone alcohol as evidenced by infrared multiphoton dissociation [68] and CID [67] experiments. The methylated acetone-dimethyl ether complex $(CH_3)_2C(OCH_3)-\overset{+}{O}(CH_3)_2$ fragments predominantly to methylated dimethyl ether and this reaction proceeds via a 1,3-$CH_3$ shift with a small kinetic energy release.This should be contrasted with the 1,2-$CH_3$ shift in "destabilized" carbenium ions (see below) for which large reverse energy terms are involved. Following earlier work Garvey and Bernstein [72] observed intramolecular ion/molecule reactions within ionized hetero clusters of chloromethane-acetone. The fragmentation of protonated and methylated dimethylether has been studied at medium pressure [73].

The reactivity of methyl halide ions ($CH_3X^{+\bullet}$) with various nucleophilic compounds has been investigated [74]. Protonation , reaction **a** (see below), is less exothermic than alkylation, reaction **b**, but nevertheless protonation and not alkylation takes place (for X = F, Cl, Br).

$$B + H_3CX^{+\bullet} \rightarrow B\cdots HCH_2X^{+\bullet} \quad -TS_1 \rightarrow BH\cdots CH_2X^{+\bullet} \rightarrow BH^+ + CH_2X^{\bullet} \quad (a)$$
$$B + CH_3X^{+\bullet} \rightarrow B\cdots CH_3X^{+\bullet} \quad -TS_2 \rightarrow BCH_3\cdots X^{+\bullet} \rightarrow BCH_3^+ + X^{\bullet} \quad (b)$$

As mentioned by Houriet et al. [74] previous work had indicated that i) the efficiency of an ion/molecule reaction is related to the stability of the transition structure within the complex and not necessarily to the reaction's exothermicity and ii) that in nucleophilic substitution reactions proton bridged complexes are more stable than ion-dipole complexes. By analogy with (ii) , $TS_1$ lies lower in energy than $TS_2$ and following (i) reaction **a** will be faster than **b**. The reactions of $CH_3F$ and $CH_2F$ cations with ethene and propene [75] and with alkenes and fluoroalkanes have been studied [76]. Two microscopic reaction mechanisms in the reaction $CH_3Cl^{+\bullet} + CH_3Cl \rightarrow CH_4Cl^+ + CH_2Cl^\bullet$ have been separated by state selected ion/molecule reactions [77].

In very interesting studies on gas-phase proton transfer equilibria between chloro- and fluoro-toluene mixtures Jennings and co-workers [78,79] have observed that both the proton affinity and the entropy of binding are high only for the ortho and para forms. These authors conclude that although the proton is very firmly bound to the aromatic ring, it is able to move around the ring, essentially freely above 300 K , effectively generating a "dynamic" structure. The proton transfer from $CH_5^+$ to fluorobenzene was observed to have a dramatic temperature dependence; it was concluded that at lower temperatures the proton is transferred directly to the energetically least favoured site, the F atom [80].

The aromatic nitration of ten monosubstituted benzenes by $CH_3O(H)NO_2^+$ has been studied by a combined approach of Ion Cyclotron Resonance (ICR) spectroscopy, Chemical ionization (CI) and CID spectrometry [81]. The reaction is a typical well-behaved electrophilic substitution. The translational energy dependence of ion/molecule collisions has been studied by Fourier transform ICR (FTICR) [82] and two dimensional Fourier transform spectroscopy has been applied to the ICR technique [83] to obtain direct evidence for mass transfer due to ion/molecule collisions. FTICR has also been used to characterize for the first time a species which contains a two-center, three-electron sulphur-sulphur bond, $(C_3H_7)_2S\bullet\!:\!S(C_3H_7)_2^{+\bullet}$, [84]. The proton affinities of diacetylene, $HC\equiv C-C\equiv CH$, cyanoacetylene, $HC\equiv C-CN$, cyanogen, NC-CN, [85], isocyanides [86], methylcyanide and methylisocyanide, [87] and azoles [88] have been measured and in some cases compared with values derived from theoretical calculations [85,88]. In a pulsed ICR study of the alkylation of cyanides [89] it was proposed that a mixture of $HCN/CH_3OH$ generates , among other products, $CH_3\overset{+}{O}(H)CH_3$. Considering the high $\Delta H_f$ of protonated dimethylperoxide this is a remarkable observation.

Ion/molecule reactions of carbon chain molecules with diacetylene and its cation [90], of the vinyl cation with $H_2$ and $CH_4$ [91], of $C_3H_3^+$ with acetylene and diacetylene [92] have been studied. Diolefin dimer radical cations in the gas-phase were found to be doubly and singly linked, whereas in solution they are either singly or doubly linked [93]. The photodissociation and ion/molecule reactions of $C_6H_6^{+\bullet}$ have been investigated [94,95]. The reaction of $C_6H_6^{+\bullet}$ with $CH_2=CH-CH=CH_2$ in the gas-phase is a two-step cycloaddition which proceeds via distonic intermediate ions [95]. Oxirene was found to transfer a methylene group to benzene [96].

Ion/molecule reactions have been used to differentiate between enantiomeric menthols [97]. Rate constants for electron transfer between neutral and ionized tetra-alkylhydrazines, $(CH_3)_2N-N(CH_3)_2$ and analogues, have been measured by high pressure mass spectrometry and were found to be low [98]. It was concluded that a large neutral to radical cation geometry change is involved in this reaction. The chemistry involved in the generation of the $CH_3O_2^+$ ion in the reaction of $O_2^{+\bullet}$ with $CH_4$ has been investigated [99]. A pulsed high pressure mass spectrometry study of mixed clusters of water and methanol indicates that in such clusters the favoured topology places the methanol molecules near the charged centers and the water molecules at the periphery [100]. The same technique was used to assess the stabilization energy of the proton-bound methylfluoride dimer, $CH_3F\cdots\overset{+}{H}\cdots FCH_3$. The result, 134 ± 8 kJ/mol [101] shows that the two methylfluoride molecules are held together by a very strong hydrogen bond.

### Neutralization Reionization Mass spectrometry

The technique of Neutralization Reionization Mass Spectrometry (NRMS) allows the generation of neutral species AB by neutralizing a beam of fast, mass selected ions $AB^+$ in the charge transfer (electron transfer) reaction : $AB^+ + G_1 \rightarrow AB^{\bullet} + G_1^+$, where G is a stationary target, a metal vapour or a permanent gas. Under the appropriate experimental conditions this is a relatively efficient process having a large cross section compared with momentum-transferring collisions. The neutrals thus generated have retained the high velocity of their parent ion and in about a microsecond they are characterized farther downstream by collision induced dissociative ionization (CIDI) with a stationary gas: $AB^{\bullet} + G_2 \rightarrow AB^+ + G_2 + e^-$, followed by $AB^+ \rightarrow A^+ + B$, etc. The first experiments were performed in the sixties on small species like $H_3^+$, $He_2^+$ and $HeH^+$, whose neutral ground states are repulsive or very weakly bound.

Although it was recognized in the earlier work that the IE of the charge transfer agent greatly influences the population of the bound (excited) electronic states in which the neutral is initially formed, it was not until the Porter group started their systematic studies using (alkali) metal vapours as target atoms in thermoneutral and exothermic charge transfer that the stability of a great many small neutral species with fascinating unorthodox binding properties were investigated. These include hypervalent species (Rydberg radicals) like $H_3^{\bullet}$, $NH_4^{\bullet}$, $CH_5^{\bullet}$, $H_3O^{\bullet}$, $H_2F^{\bullet}$, $H_2Cl^{\bullet}$ and their deuteriated analogues, related cluster type species like $NH_4(NH_3)_n^{\bullet}$ (n= 1-3), rare gas hydrides and the rare gas dimers $He_2$, $Ne_2$ and $Ar_2$. In recent work small organic radicals and molecules like $HCNH^{\bullet}$ [102], $DCO^{\bullet}$, $DOCO^{\bullet}$, $(CD_3)_2COD^{\bullet}$ [103] and the $C_2H_2/C_2H_3^{\bullet}$ system [104-106] have also been investigated.

The earlier results have been reviewed by Gellene and Porter [107] and the literature published on these species up to the middle of 1987 has also been reviewed [9,10,108]. Very recent results involve the observation of the unique metastability for NeH produced by the neutralization of $NeH^+$ with K vapour in a study of the rare gas hydrides HeH, NeH, ArH and XeH [109,110] and the observation of its discrete emission spectrum [111]. The first evidence for a metastable state of the deuteriated methyloxonium radical $CD_3OD_2^{\bullet}$ has been presented [112,103] as well as a remarkable result on the stability of $D_3^{16}O^{\bullet}$ but not $D_3^{18}O^{\bullet}$ in a study on the effect of ion excitation on the formation of metastable states in charge transfer reactions with Na [113]. The nature of the metastable state of $H_3^{\bullet}$ has been a matter of debate [114] and Selgren and Gellene have presented their first results on this system using the new technique of photoassisted fragmentation spectroscopy. These authors also report $CH_5^{\bullet}$ to be an observable species in neutralization experiments with Zn [115]. In the early experiments on these small species the beam of neutrals from charge transfer was not analyzed by collisional ionization but by examining the neutral beam profile. Later experiments involve both neutral beam profile measurements and with separate instrumentation collisional ionization with $NO_2$. The latter technique relates to the methodology currently in use for NRMS of organic ions which involves the higher mass and energy resolution required for studying larger systems. Using this instrumentation it has been claimed that $CH_5^{\bullet}$ and $H_3O^{\bullet}$ are generated as stable species in endothermic charge transfer reactions with Xenon [116,117] but very recent reports make this claim questionable and point to isotopic interferences [118-120].

### NRMS of organic ions.

From a chemist's point of view the possibility to reduce a given cation in the rarefied gas-phase of a mass spectrometer by charge transfer neutralization is very attractive. It allows the tailor made generation of unusually reactive but thermodynamically stable molecules which, because of intermolecular reactions, cannot be studied in solution or even in the solid state. The introduction of NRMS as a new asset to experimental physical organic chemistry has greatly benefited from the pioneering work of the McLafferty group at Cornell University. Its introduction in 1980 with the first results of neutralization with metal vapours on a home built multisector instrument, which has recently been improved and extended [121] was timely. The interpretation of NR spectra relies heavily upon comparisons with collision-induced dissociation (CID) mass spectra: by 1980 the CID technique, also pioneered by the McLafferty group, was already well developed.

Modification of the commercial VG ANALYTICAL ZAB 2f magnetic deflection type mass spectrometers of reversed geometry at Utrecht U. (Terlouw et al.) and Ottawa U. (Holmes et al.) with a home built neutralization cell for endothermic charge transfer with Xenon [122] quickly followed. Factory made cells, now a standard accessory, were installed on the Toronto U. (A.G. Harrison, R.E. March et al.) [123] and the TU Berlin (H. Schwarz et al.) ZAB type instruments. A further modification describing the use of a metal vapour cell in the VG ZAB instrument has just appeared [124].

The neutralization cell is situated in front of the standard collision gas cell for CID experiments in the field-free region between the magnet and the electric sector. Between the two cells a deflector electrode is present which, when charged, removes all ions which have not been neutralized from the beam so that only neutral species enter the standard collision cell.

In contrast with matrix isolation spectroscopy, the structure information about a neutral generated in a NR experiment is limited to assigning the connectivity of the atoms. This assignment is based on the chemistry observed in its collision-induced dissociative ionization (CIDI) mass spectrum. However, often the only (convenient) way to generate a highly reactive neutral is via its reduction (or when starting with negative ions its oxidation) in the very diluted gas-phase of the mass spectrometer and the NR procedure is quick and relatively simple.

From the massive body of information which has been acquired by the McLafferty group [9], Moran and co-workers [125-128] and the Holmes/Terlouw group [129] on establishing the optimum conditions for the charge transfer neutralization of organic ions, it follows that the nature of the target gas (i.e. its IE) is one of the most important parameters. For the neutralization reaction $AB^+ + G_1 + Q_N \rightarrow AB^{\bullet} + G_1^+$ a satisfactory yield of neutrals is generally obtained if the energy deficit $Q_N$ lies in the range of 0 - 4 eV.

$Q_N$ is defined as IE $(G_1)$ - RE $(AB^+)$, where IE = the ionization energy of the target and RE = the recombination energy for the electron capture reaction $AB^+ + e^- \rightarrow AB$. RE $(AB^+) \sim$ IE (AB), unless large geometry differences exist between the ground state of ion and neutral.

If $Q_N = 0$ then near resonant conditions obtain; if $Q_N < 0$, the process is exothermic which is generally not desirable in experiments on organic ions, whereas for $Q_N > 0$ the reaction is endothermic and the energy deficit (a few eV) is supplied by the translational energy ( 8 - 10 keV) of the fast beam of $AB^+$ ions.

The **neutralization** gases commonly employed for NR of organic ions are Hg and Xe (IE=10.4 and 12.1 eV resp.). Both gases yield appropriate $Q_N$ values in the neutralization of most organic ions (whose RE values are in the range 7 - 10 eV) and they produce an acceptably small degree of interference from neutrals generated by collision-induced dissociation of the ions $AB^+$. In this respect Hg vapours, extensively used by the McLafferty group, may have some advantages over Xenon which is however easier to manipulate. For **reionization** $O_2$ is preferred over He (commonly employed in CID studies) because it combines a high efficiency with "soft" ionization.The high efficiency may be aided by the electron transfer process : $AB + O_2 \rightarrow AB^+ + O_2^-$.

The effect of $O_2$ as a reionizing agent compared with He, a target which greatly increases the degree of fragmentation, is well exemplified by the Xe/He and the Xe/$O_2$ NR spectra of methyl formate ($HCOOCH_3$, M=60) [129]. The relative intensity of the "survivor" signal at m/z 60 is 20 % in the Xe/$O_2$ spectrum (c. 45 % when the energy resolution is enhanced to a few eV), whereas it is only c. 1% in the Xe/He spectrum. The results are closely similar to those obtained earlier by the McLafferty group [9,130] using Hg as the neutralization agent. This spectrum has been further analysed by these authors [130] and by Villeneuve and Burgers [131] to study the participation of the isomeric distonic ion $[HC(OH)OCH_2]^{+\bullet}$ in the rearrangement of methyl formate molecular ions.

Summarizing, it can be stated that Hg/$O_2$ and Xe/$O_2$ spectra of an organic ion may be expected to yield a sizeable fraction of reionized, non-dissociating neutrals and structure characteristic dissociation products, unless either the neutral lives in a shallow potential well and/or has a geometry which is considerably different from that of the ion.

The interpretation of NR spectra relies heavily upon knowledge from thermochemistry and/or ab initio molecular orbital theory calculations - see ref. 36 for an excellent theoretical study on the stability of simple ylid ions and their neutral counterparts - on the height of dissociation and isomerization barriers in both the original ion and its neutralized counterpart. In addition it is important to realize that NR efficiencies of "survivor" ions of neighbouring mass, or even among isomeric and isobaric ions, can easily differ by a factor of 100. It is therefore important to consider carefully the purity of the primary ion flux to be neutralized in terms of undesired species of the same m/z ratio. The NR mass spectra of $H_3O^+$ [117-120] and the ylid ions $CH_2OH_2^{+•}$ and $CH_2FH^{+•}$ [132] exemplify these problems, the neutral counterparts having mistakenly been identified as stable.

In the two recent review articles by Wesdemiotis and McLafferty [9] and Terlouw and Schwarz [10] a great many examples were given of (i) applications of NRMS for the search for hitherto elusive radicals and molecules, including ylides, betaines, enols, carbenes and also the the long-sought free carbonic acid molecule $H_2CO_3$, and (ii) the use of the technique as an aid in studying ion structures and isomerization mechanisms, like the classical versus bridged structure of the ethyl cation. In this section new results not discussed in these reviews will be briefly summarized.

In an NR study of isomeric ethyl halide radical cations [133] the neutral counterparts of the conventional isomer $CH_3CH_2Cl^{+•}$, the ylid ion $CH_3CHClH^{+•}$ and the ion-dipole complex $[CH_2CH_2/HCl]^{+•}$ were found to be stable. In contrast the isomeric ylid ion $CH_3ClCH_2^{+•}$ spontaneously dissociated into $CH_3^{•}$ + $CH_2Cl^{•}$. Similar results were obtained for the bromo analogues. Whereas ions $CH_3CH_2Cl^{+•}$, $CH_3CH_2Br^{+•}$ and $CH_3CH_2I^{+•}$ yielded the expected recovery signal in their NR spectra, $CH_3CH_2F^{+•}$ produced no such signal. This was attributed to the wholly vertical process of collision-induced ionization failing to regenerate stable $CH_3CH_2F^{+•}$ ions from stable $CH_3CH_2F$ molecules. A similar situation may obtain for $CCl_4$ which together with its ylid isomer could not be detected as stable species in NR experiments [134]. Neutralization of $ClHCl^-$ anions followed by reionization into positive ions produced no evidence for the stability of the $ClHCl^•$ radical, in agreement with the results of ab initio molecular orbital calculations [135]. Further experimental evidence has been presented by Wesdemiotis et al. for the stability of the methylenechloronium ylide $H_2CClH$, [136], a species for which theoretical calculations have indicated that at best it lies in a very shallow potential well [36]. The above experimental evidence is at odds with the detailed observations reported in ref. [132], but it is nevertheless possible that the ylide $H_2CClH$ survives for sufficient time to be weakly observable.

The NRMS technique has been successfully applied in the first "synthesis" of stable molecules of the type $FeCH_x$ (x=0-3), species which may well serve as intermediates in homogenous and heterogenous catalytic reactions [137]. The radicals $NH_2O^•$ and $NHOH^•$, which may play a role in a catalytic cycle destroying ozone in stratospheric chemistry, and the gaseous $HSO_3^•$ radical, the key intermediate in the oxidation of $SO_2$ in the atmosphere, have been characterized in the gas-phase [138],[139].

Hydroxycarbene, $H\ddot{C}OH$, was successfully identified by the McLafferty group [140] using a new variant of the NR technique, i.e. on the basis of its NR⁻ spectrum: a fast neutral produced in a conventional NR experiment may also be reionized by collisions to produce *anions*, which together with their *anionic* dissociation products, can be measured in the

conventional NR setup by simply reversing the polarity of the electrostatic analyser :
$AB^+ + G_1 \rightarrow AB + G_1^+$, followed by $AB + Xe \rightarrow AB^- \rightarrow A^- + B^{\bullet}, A^{\bullet} + B^-$.
If the structure of the neutral is in doubt, an instrument equipped with an additional collision cell and sector for mass analysis makes it possible to obtain its collisional activation (CA) spectrum for further structure analysis. This method has successfully been applied to the study of the remarkably stable sulfur analogue of the carbon monoxide dimer $(CO)_2$, ethylenedithione, S=C=C=S, which in a combined NR computational study [141] was shown to be cleanly generated by the dissociative ionization of the tetramethyldithio-amide of squaric acid. Such measurements, which may be sensitivity limited, were also performed in the studies of Wesdemiotis et al. [136] and Hop et al. [132]. Whereas the ketene isomer H-C≡C-OH$^{+\bullet}$ was shown in earlier work [142] to cleanly produce the stable neutral ynol H-C≡C-OH in the gas-phase, the existence of neutral oxirene as another stable ketene isomer remains doubtful [143]. From a NR study of $C_7H_7^+$ benzyl, tropyl and tolyl cations [144] it was concluded that the tolyl radicals produced by Xe neutralization of tolyl ions do not isomerize to benzyl radicals on the time scale of the NR experiment. The collisional reionization efficiencies into both positive (using $O_2$) and negative (using Xe) ions have been compared for a series of neutralized $C_3H_n$ ions (n = 0-6) [145].

Combining the information from the CID and the NR spectra of the product ion produced by loss of $C_2H_4$ from OD labelled ethanesulfonic acid molecular ions ,**1**, (see Scheme below) with the results of ab initio calculations clearly shows that ions **2** are generated and that their neutralization produces sulphurous acid, $(HO)_2S=O$, as a stable molecule in the gas-phase [146].

Recent applications of NR as an auxiliary technique in the study of ion structures and reaction mechanisms involve the characterization of the ion dipole complex $[CH_2NH_2/H_2O]^+$ generated as a stable species in the gas-phase from protonated β-alanine [147] and the extensive isomerization reactions of 1,2-propanediol molecular ions into hydrogen bridged radical cations prior to their metastable and collision-induced dissociations [148].

## 2: Structures and Reactions of Gas-phase Organic Ions 49

### CIDI (Collision-Induced Dissociative Ionization) spectra of neutral species N in the dissociation of metastable ions $m_1^+ \rightarrow m_2^+ + N$.

A CIDI experiment is performed with the instrumental setup for NRMS but without using the additional collision gas chamber for neutralization. A small fraction of the beam of fast moving ions $m_1^+$ under investigation will dissociate unimolecularly , $m_1^+ \rightarrow m_2^+ + N$, during the flight to the standard collision gas chamber in the second field-free region of a reversed geometry instrument. Upon arrival at the charged deflector electrode all ions, i.e. $m_1^+$ and the *ionic* dissociation products $m_2^+$, are deflected away so that the only species entering the collision chamber are the fast moving neutrals N. Within the cell the neutrals are ionized by collision with inert gas molecules. Part of the ionized neutrals dissociate and the spectrum of their ionic dissociation products, which can conveniently be registered with the mass spectrometer, is used to characterize the structure of N. The utility of the method is well exemplified by the unpredicted behaviour of ionized methylacetate.

It has been firmly established that the $C_2H_3O^+$ ion generated in the metastable dissociation of ionized methyl acetate, $CH_3COOCH_3^{+\bullet}$,4, is the acetyl cation , $CH_3C=O^+$. However, the question of the structure of the neutral generated in this reaction , which could be either the methoxy radical ($CH_3O^{\bullet}$) or the hydroxymethyl radical ($CH_2OH^{\bullet}$), was not settled until recently. Results of CIDI experiments from several laboratories [130,149-151] have given rise to some lively discussions about the relative proportions of $CH_2OH^{\bullet}$ vs $CH_3O^{\bullet}$ radicals generated in this unimolecular dissociation but the central point that $CH_2OH^{\bullet}$ radicals are produced is not in doubt. In addition , there now is a consensus that $CH_3O^{\bullet}$ and $CH_2OH^{\bullet}$ are generated in the above reaction in a ratio of ca. 4 : 1. Whereas $CH_3O^{\bullet}$ can readily be generated from $CH_3COOCH_3^{+\bullet}$ by simple α-cleavage, an intriguing mechanism must be associated with the formation of $CH_2OH^{\bullet}$, which is the more stable radical by ca. 40 kJ/mol. The $CH_2OH^{\bullet}$ radicals do not originate from isomerization of initially formed $CH_3O^{\bullet}$ radicals but are formed from methyl acetate molecular ions which have undergone an extensive skeletal rearrangement.

As elegantly shown in a recent ab initio molecular orbital theory study [152] the minimum energy path for the dissociation of $CH_3COOCH_3^{+\bullet}$, **4**, involves isomerization into the distonic ion, **5**, to yield the hydrogen bridged radical cation **6**. This ion then serves as the precursor to the formation of $CH_3C=O^+$ and $CH_2OH^{\bullet}$ by direct bond cleavage as shown in the diagram below. This multistep pathway is calculated to be energetically favoured over the direct dissociation of **4** into $CH_3C=O^+/CH_3O^{\bullet}$ by 17 - 21 kJ/mol. The latter reaction predominates in the decomposition of methyl acetate ions of higher internal energies. The isomerization behaviour of ionized methylacetate, **4**, has also been studied in detail [153] by the semi-empirical MINDO/3 method, which predicted that the loss of $CH_2OH^{\bullet}$ from **4** involved the reaction sequence **4** → $CH_3C(=O)O(H)CH_2^{+\bullet}$, **7**, → $CH_3\text{-}C=O^+ + CH_2OH^{\bullet}$. However, the above ab initio study clearly showed that the isomerization barrier **4**→ **7** is so high that this possibility is effectively ruled out.

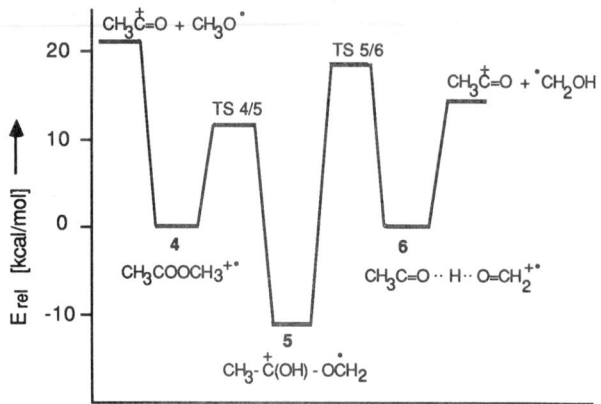

An interesting analogy with the behaviour of $CH_3COOCH_3^{+\bullet}$ is the observation [138] that metastable acetohydroxamic acid molecular ions, $CH_3CONHOH^{+\bullet}$, dissociate into $CH_3CO^+$ + $NH_2O^\bullet$ (not $NHOH^\bullet$).

The computational results of Heinrich et al. [152] offer some hope that the chemistry of ionized ethyl acetate, $CH_3COOCH_2CH_3^{+\bullet}$, may some day be understood. If some fragmentations can be termed disconcerting [154] then ethyl acetate's behaviour is downright bizarre. Despite extensive experimental work, including PEPICO experiments, the mechanism associated with loss of $H_2O$ remains shrouded in mystery. Obviously the reaction involves rearrangement and PEPICO experiments show that the most stable isomer involved has a $\Delta H_f$ of 460 kJ/mol [155] (vs $CH_3COOCH_2CH_3^{+\bullet}$ = 523 kJ/mol). All hydrogen atoms are involved as are both oxygen atoms.

Consider the hydrogen-bridged species 6, the key intermediate in the isomerization of ionized methyl acetate. Methyl substitution yields $CH_3-C=O\cdots H\cdots O=CH-CH_3^{+\bullet}$. A 1,6-H shift would yield the hydrogen bridged acetaldehyde-vinyloxy cation $CH_3-CH=O\cdots H\cdots O-CH=CH_2^{+\bullet}$, whose heat of formation, using the empirical Larson and McMahon formula [120] is estimated to be 456 kJ/mol ! A degenerate isomerization in this ion by way of a 1,7-H migration would account for the equivalence of the two oxygen atoms. Ab initio calculations are required to solve this problem, but considering the size of the system and the number of possible intermediates this will be a formidable task requiring all the skills of experienced theoreticians.

### Pyrolysis experiments

Another way to generate and characterize neutrals which because of intermolecular reactions can only be generated as stable species in the rarefied gas-phase is very low vapour pressure (VLVP) pyrolysis combined with tandem mass spectrometry. The scope of this technique is less wide than that of the NR technique but it allows the IE of the neutral to be determined and moreover it can be combined with other spectroscopic techniques to provide more structure information.

Using this procedure Wentrup et al. have prepared and identified $HC\equiv CNH_2$, $C_6H_5C\equiv CNH_2$, $H_2NCH=C=C=O$ and $HN=CHCH=C=O$ [156] as well as the $C_2H_2N_2$ isomers $CH_2=N-CN$ and $HN=CH-CN$ [157], all molecules of interest to cosmochemists. $C_6H_5C(OH)=CH_2$, the enol form of acetophenone, [158], (E)- and (Z)-1-hydroxybutadiene [159] and $CH_2=C(OH)C\equiv CH$, the enol form of $CH_3COC\equiv CH$, 3-butyne-1-one, [160] were prepared and identified by Tureček et al. who were further able to derive $\Delta H_f$ values for these neutral species from their measured IE values and $\Delta H_f$ values for the corresponding ions from dissociative ionization experiments. For the $C_4H_4O$ neutrals $CH_2=C(OH)C\equiv CH$ and $CH_3COC\equiv CH$ it was concluded [160] that the presence of the triple bond in the enol does not cause any significant stabilization of the enol against the keto form. Note that this conclusion is critically dependent on the $\Delta H_f$ values used for conjugated ethynyl carbonyl compounds (see section on ion and neutral thermochemistry). The energy difference between acetophenone and its enol, 41 kJ/mol, coincides with the experimental value for the $CH_3CHO/CH_2=CHOH$ system [11]. A larger energy difference, 57 kJ/mol, was computed for the latter system in a high level ab initio study [161] which further predicts that the enol of acetic acid, $CH_2=C(OH)_2$, is 133 kJ/mol higher in energy than the acid itself. This enol is a typical example of a molecule which can easily be generated by NRMS [162] but not by pyrolytic procedures.

## 3 ION STRUCTURES
### Distonic Ions

In gas-phase ion chemistry the word structure signifies connectivity i.e. which atoms are joined together and the probable formal locations of charge and radical sites [3]. It is becoming increasingly clear that so-called "distonic ions" are a well defined group of **stable** radical cations with characteristic properties [3]. The evocative term "distonic" was coined in 1984 by Radom and co-workers to describe those radical cations which formally have the charge and radical sites at separate heavy atoms, e.g. $\dot{C}H_2\text{-}CH_2\text{-}\overset{+}{O}H_2$ as opposed to $CH_3CH_2OH^{+\cdot}$. This terminology is rapidly gaining acceptance: for example no one gave, or apparently needed to give, a definition of the term at the 11th IMSC at Bordeaux (1988) although it was used repeatedly. The fear that the term could be jargon within the field of mass spectrometry [8a] is groundless since it has also been adopted in the field of ESR where such ions have long been studied. Distonic ions may also be referred to as [1,n] radical cations [3]. They may also be viewed as protonated radicals [3] but the term distonic ion is, in our opinion, an attractive and useful term.

Already there have appeared five reviews on the subject; the latest, a very thorough review by Hammerum [3] (with 418 references!) appeared in 1988 and covers the literature up to 1987. Hammerum has extended the nomenclature by denoting the relative positions of the charge and radical by Greek letters and we have adopted this nomenclature, e.g. $\dot{C}H_2\text{-}CH_2\text{-}\overset{+}{O}H_2$ is a β-distonic ion. However, ambiguity arises when the ion can be represented by resonance structures, i.e. $R_1\text{-}\overset{+}{C}(R_2)\text{-}X \leftrightarrow R_1\text{-}C(R_2)=X^+$. Thus for example the ion $\overset{\bullet}{C}H_2\text{-}CH_2\text{-}\overset{+}{C}H\text{-}OH$ was depicted [154] as as a carbenium ion radical, i.e. β-distonic whereas the analogue $\overset{\bullet}{C}H_2\text{-}CH_2\text{-}CH=\overset{+}{N}H_2$, visualized as an immonium ion radical [163], would be termed γ-distonic. There appears to be a reasonable consensus in the literature that oxygen containing ions are best viewed as carbenium ion radicals, rather than oxonium ion radicals. We further note that even when a distonic ion is depicted as an onium ion, the terminology adopted is that of a carbenium ion radical [3] and we propose to stick to this rule. This also implies that ring opened forms of epoxides like $^\bullet CH_2\text{-}O\text{-}CH_2^+ \leftrightarrow$ $^\bullet CH_2\text{-}\overset{+}{O}=CH_2$ are not α-distonic [3] but β-distonic ions.

Distonic ions are key intermediates in the fragmentation of many organic compounds and their reactions involve both the radical and the charged sites: often a charged moiety can easily migrate to the radical site thereby generating an isomeric distonic ion which in turn can shift its ionized part to the new radical site (see section reaction mechanisms). Such a sequence can rationalize many seemingly "puzzling" observations.

The Radom group [34] using ab initio computational results, has developed guidelines for assessing the stability of distonic radical cations $\overset{\bullet}{Y}(CH_2)_n\overset{+}{X}H$ relative to the conventional isomer $HY(CH_2)_nX^{+\bullet}$. It was inter alia concluded that the most favourable circumstances for a distonic radical cation to be preferred over its conventional isomer will occur for systems with a high proton affinity but which do not contain a group having a low ionization energy. Thus the distonic ion $CH_2NH_3^{+\bullet}$ is lower in energy than $CH_3NH_2^{+\bullet}$ because of the high proton affinity of $CH_3NH_2$; also the distonic ion $CH_2OH_2^{+\bullet}$ is more stable than the conventional form in this case largely due to the high ionization energy of $CH_3OH$. For the homologous series $^\bullet CH_2\text{-}(CH_2)_n\text{-}\overset{+}{N}H_3$ and $CH_3\text{-}(CH_2)_n\text{-}NH_2^{+\bullet}$ it was found that the distonic species was always thermodynamically preferred, but the energy difference appears to converge with increasing n towards a limit which is close to the energy differences between the system components, for n = 2 : $CH_3CH_2^\bullet + CH_3NH_3^+$ (representing the distonic isomer) and $CH_3CH_3 + CH_3NH_2^{+\bullet}$ (representing the conventional isomer).

Using this approach it was predicted that the distonic form of ionized ethylene glycol, i.e. $H_2\overset{+}{O}\text{-}CH_2\text{-}CH_2\text{-}O^\bullet$, is 90 kJ/mol more stable than its conventional counterpart $HOCH_2CH_2OH^{+\bullet}$. Subsequent ab initio calculations [164] definitely showed that $H_2\overset{+}{O}\text{-}CH_2\text{-}CH_2\text{-}O^\bullet$ is more stable than ionized ethylene glycol, by 69 kJ/mol, and that it is formed therefrom after geometric relaxation. Ionized glycol is metastable with respect to $CH_3OH_2^+ + HCO^\bullet$ and it was further suggested that a single 1,2-hydrogen shift with concomitant C-C bond cleavage could lead to the dissociation products. However, an alternative mechanism involving stable hydrogen-bridged structures has been proposed [165] and which is discussed in the next section.

### Hydrogen-bridged radical cations

Even-electron proton-bound molecule pairs $M_1\cdots H\cdots M_2^+$ are a well studied class of ions as they can be readily generated in ion/molecule reactions. By contrast, much less is known about the odd-electron counterparts, hydrogen-bridged radical cations $M\cdots H\cdots R^{+\bullet}$, formally proton-bound molecule-radical pairs. Undoubtedly such ions are harder to make than $M_1\cdots H\cdots M_2^+$ in ion/molecule reactions and attempts have been made to generate them by alternative routes, but there is an underlying problem. It is often not possible to assign a unique structure to an ion on the basis of reactivity, MI or CID experiments when its isomeric forms are reasonably expected to give closely similar characteristics. This is a well recognized problem in general, but it becomes particularly acute for ions of the type $M\cdots H\cdots R^{+\bullet}$ which could well, and frequently do dissociate very much like a distonic form. In such cases ab initio calculations are essential to obtain the height of the barriers separating the isomers as well as their relative energies. Thus although experiments unequivocally showed that the $C_2H_6O_2^{+\bullet}$ ions formed from ionized 1,4-butanediol, $HOCH_2CH_2CH_2CH_2OH^{+\bullet}$,**8**, comprised a vinyl alcohol, $CH_2=CHOH$, and water part, these experiments did not allow a differentiation between the hydrogen-bridged species $CH_2=C(H)-O\cdots H\cdots OH_2^{+\bullet}$ ,**9**, and the distonic forms $H_2\overset{+}{O}-CH_2-\overset{\bullet}{C}H-OH$ , **10**, and $\overset{\bullet}{C}H_2-CH(OH)-\overset{+}{O}H_2$ ,**11**. High level ab initio calculations revealed that the $-O\cdots H\cdots O-$ bridged ion **9** is not only thermodynamically by far the most stable isomer but that it is also very stable - by 100 kJ/mol - towards dissociation [166]. Ion **9** lies in a potential well of c. 60 kJ/mol [166]; at these and higher internal energies all [vinylalcohol/water]$^{+\bullet}$ isomers are extremely flexible and interconvert freely, a behaviour similar to $CH_2CH_2OH_2^{+\bullet}$ [167] and $CH_2=C=O^{+\bullet}/H_2O$ [168]. More than twenty years ago ion **10** was proposed as the key intermediate in the acid catalyzed dehydration of α,β-dihydroxyalkane radicals [3] : $HOCH_2\overset{\bullet}{C}HOH + H^+ \rightarrow H_2\overset{+}{O}CH_2\overset{\bullet}{C}HOH$, **10**, $\rightarrow CH_2=CHO + H_3O^+$, and the ab initio results strongly support this proposal.

Ab initio calculations show that at least three other $C_2H_6O_2^{+\bullet}$ hydrogen-bridged radical cations live in deep potential wells namely $CH_2-O(H)\cdots H\cdots O=CH_2^{+\bullet}$,**12**, $CH_3-O\cdots H\cdots O=CH_2^{+\bullet}$ ,**13**, and $CH_3-O(H)\cdots H\cdots O=CH^{+\bullet}$,**14**, and that their SE's are quite substantial. Ion **13** was generated by dissociative ionization, ion **12** by rearrangement of ionized 1,2-ethanediol, whereas **14** has not yet been generated. The ab initio calculations afforded a rationalization of the metastable behaviour of ionized 1,2-ethanediol in terms of the intermediacy of *stable* hydrogen-bridged species as shown in the following scheme [165] :

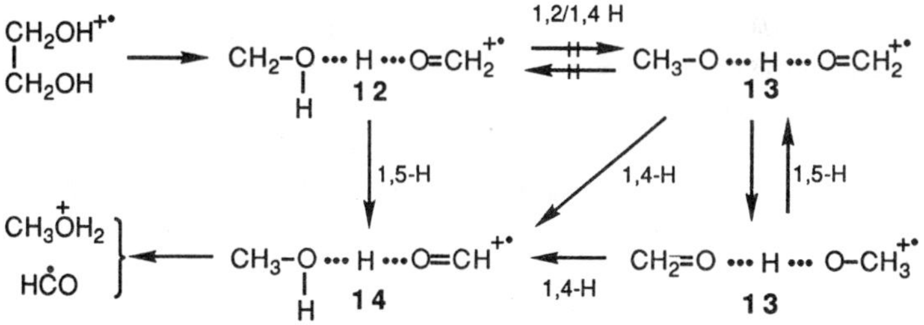

$$\begin{array}{c}
CH_2OH^{+\bullet} \\
| \\
CH_2OH
\end{array} \longrightarrow \begin{array}{c} CH_2-O\cdots H\cdots O=CH_2^{+\bullet} \\ | \\ H \quad \mathbf{12} \end{array} \underset{\longleftarrow}{\overset{1,2/1,4\,H}{\longrightarrow}} CH_3-O\cdots H\cdots O=CH_2^{+\bullet} \\ \mathbf{13}$$

$$\downarrow 1,5\text{-}H \qquad \swarrow 1,4\text{-}H \qquad \downarrow 1,5\text{-}H$$

$$\left.\begin{array}{c} CH_3\overset{+}{O}H_2 \\ H\overset{\bullet}{C}O \end{array}\right\} \longleftarrow \begin{array}{c} CH_3-O\cdots H\cdots O=CH^{+\bullet} \\ | \\ H \quad \mathbf{14} \end{array} \underset{1,4\text{-}H}{\longleftarrow} \begin{array}{c} CH_2=O\cdots H\cdots O-CH_3^{+\bullet} \\ \mathbf{13} \end{array}$$

Hydrogen-bridged species allow a rationalization of some mechanistically problematic reactions and they will undoubtedly provide stimulus for further investigations.

## "Destabilized" carbenium ions.

α-Acylcarbenium ions (R-C(O)-$\overset{+}{C}R_2$) are important intermediates in organic reactions [169-171]. While primary α-acylcarbenium ions, such as H-C(O)-$CH_2^+$ undergo extremely facile isomerization and can therefore not be generated even in the rarefied gas-phase [169], the secondary and tertiary carbenium ions, which contain electron-donating substituents in addition to the electron-withdrawing acyl group, might well be stable. Their potential generation by dissociative ionization is however often prevented by the preferred formation of a more stable rearrangement ion. Thus for example loss of $CH_3^\bullet$ from ionized methyl isobutyrate, $(CH_3)_2CHCOOCH_3^{+\bullet}$, does not produce the acyl carbenium ion, $CH_3\overset{+}{C}HCOOCH_3$, by simple cleavage but instead yields protonated methyl acrylate, $CH_2=CH\overset{+}{C}(OH)OCH_3$, via inter alia, a "hidden" hydrogen migration (see below). Indeed, Dommröse and Grützmacher [169] have shown that a prerequisite for the generation of acylcarbenium ions is a good leaving group, e.g. I$^\bullet$, which effectively suppresses the rearrangement process by virtue of the small C-I bond strength. This will undoubtedly be used to generate and study other destabilized carbenium ions.

## Multiply charged cations

Because of their remarkable structural features the chemistry of multiply charged cations has received considerable attention in recent years from both theoreticians and experimentalists [6]. Charge stripping (CS) mass spectrometry is the experimental method of choice for their study. As with NR mass spectrometry, the interpretation of CS spectra relies heavily upon ab initio molecular orbital theory calculations. Also, as with NR mass spectrometry, it is important to consider carefully the purity of the primary ion beam to be oxidized. Thus it had previously been concluded that CH$^+$ ions give rise to a signal at m/z 6.5 which was proposed to correspond to CH$^{++\bullet}$, but later high resolution experiments showed that this signal actually corresponded to $^{13}C^{++}$ [172]. From CS experiments the vertical ionization energy of a singly charged ion ($Q_{min}$) can be obtained by two different methods which have recently been compared [173]. Accurate measurement of $Q_{min}$ values can lead to surprising results. For example, Maquin and Stahl [174] deduced from $Q_{min}$ values that for polyphenyls the quantity { IE(cation) - IE(neutral) - E(HH) }, where E(HH) is the calculated Coulombic energy in the dication if the charges are at the *terminal H atoms*, is remarkably constant. This result, it was suggested, may relate to the semiconductor character of these polyphenyls.

In general multiply charged cations are thermodynamically unstable towards dissociation into two ions, but a significant kinetic stability associated with the weakening of a covalent bond may prevent dissociation [175]. Thus the exothermicity of the reaction $CHe_4^{4+} \rightarrow CHe_3^{3+} + He^+$ is enormous, 1605 kJ/mol, but the barrier, 72 kJ/mol, is sufficiently large to allow, at least in principle, experimental observation [176,177]. Such an extraordinary species is truly a tiger in a cage.

Multiply charged ions are stable because as the nuclei separate, initially the *decrease* in energy due to decreased Coulomb repulsion is smaller than the loss in covalent

## 2: Structures and Reactions of Gas-phase Organic Ions 55

binding energy. Ab initio calculation on the deprotonation of $AH_n^{++}$ ions has unearthed a perplexing phenomenon [175,177]. It was found that the point at which the Coulombic repulsion and covalent binding balance can lie far from the equilibrium point, with an extreme value of 12.45 Å for deprotonation of $MgH^{++}$. This result is puzzling because at 12.45 Å the covalent binding must be extremely small [176]. However, Gill and Radom [176] point out that the Coulombic repulsion energy in an asymmetric dication $AB^{++}$ is not $r^{-1}$ but $q_a q_b r^{-1}$, where $q_a$ and $q_b$ are the fractional charges on A and B and r the distance between them. Thus initially the Coulomb repulsion does not decrease as $r^{-1}$ and it might even, in principle, increase ! [175,176]. Hence the transition state may be a very late one. However, this electrostatic model [175] has limited applicability since in many systems the charges $q_a$ and $q_b$ approach 2.0 and 0.0 as the bond is stretched.

An improved model, by the same group [176,178], considered the potential energy curve along the reaction coordinate as arising from an avoided crossing between a repulsive state $[A^+ + B^+]$ and an attractive state $[A^{++} + B]$. A parameter $\Delta = IE(A^+) - IE(B)$ was introduced. Since $\Delta$ is generally positive, fragmentation to $A^+ + B^+$ and not to $A^{++} + B$ occurs. It was predicted that for large values of $\Delta$ the transition state occurs early and with a shorter bond length; for small values a late transition state obtains and the equation $r_{TS} = \Delta^{-1}$ is thus derived. Thus a semi-quantitative prediction of the distance between the separating partners in the transition structure can be made based solely on a knowledge of the ionization energy of the products $A^+$ and $B^+$, IE $A^+$ and IE $B^*$ [175]. In general for the reaction $A^{(n+1)+} \rightarrow A^{n+} + B^+$, the equation $r_{TS} = n \Delta_n^{-1}$, where $\Delta_n = IE(A^{n+}) - IE(B)$, is obtained. Also, for small values of $\Delta$, the kinetic energy released in dication fragmentations may be estimated from the simple equation $T \approx \Delta$ [178]. In a very recent paper entitled "How Does a Dication Lose a Proton ?" Gill and Radom [179] give an excellent discussion of the mechanism associated with deprotonation of dications. It appears that the reaction is best described as a two-step process : $AH^{++} \rightarrow A^{++}\cdots H^* \rightarrow A^+\cdots H^+ \rightarrow A^+ + H^+$. Implications for a proper theoretical description were delineated.

### Stability reversal by (multiple) ionization.

It has been noted repeatedly that a stabilization reversal may take effect upon ionization; for example ionized enols are thermodynamically more stable than their keto counterparts, in contrast with the neutral system. Similarly the methylene oxonium radical cation, $CH_2OH_2^{+\cdot}$, is found to be more stable than the the methanol ion, $CH_3OH^{+\cdot}$, whereas the ylid lies considerably higher in energy than methanol [36]. It is found that a continuation of this trend accompanies further ionization [ [180] + ref. 8 therein]. It can be seen that in all cases the thermodynamic stability of the ylid, relative to that of the conventional isomer, increases upon successive ionization. Also the well-depth of the ylid *increases* with successive ionization with $CH_2PH_3$ as a notable exception; that for the conventional isomers *decreases* upon successive ionization.

## 4 REACTION MECHANISMS

In a recent review on distonic ions [3] it was pointed out that such ions may undergo a number of characteristic reactions and that many "puzzling" observations can often be ascribed to the characteristics of distonic intermediate ions. Distonic ions are being increasingly proposed as key intermediates in the dissociation of aliphatic carbonyl compounds [3,4,154,181-194] to explain the observed skeletal rearrangement of the aliphatic chain. The work followed earlier labelling experiments by Schwarz and coworkers on aliphatic acids [see 191] and by Bouchoux and coworkers on ketones [4]. It appears that distonic ions formed from aliphatic carbonyl compounds frequently undergo two reactions, one of which is referred to as the [1,n]-hydroxycarbene-radical migration and the other as the [1,2]-enol-olefin shift [182].

### The [1,n]-hydroxycarbene-radical migration

The β-distonic ion $\overset{\bullet}{C}H_2$-$CH_2$-$\overset{+}{C}H$-OH (15) isomeric with $CH_3$-$CH_2$-$CH=O^{+\bullet}$, ionized propanal, has been generated and its $\Delta H_f$ assessed to be 757 kJ/mol. [195]. It was found experimentally [195] and theoretically [181] that the $\Delta H_f$ was close to that of ionized propanal (773 kJ/mol) but that a large barrier (137 kJ/mol) [181] separates the two ions. The reaction of lowest energy requirement appeared to be the degenerate isomerization reaction:

$$^{\bullet}CH_2\text{-}CH_2\text{-}\overset{+}{C}HOH \rightleftharpoons HO\overset{+}{C}H\text{-}CH_2\text{-}CH_2^{\bullet} \quad [a]$$

The associated transition structure, ionized cyclopropanol, lies ~30 kJ/mol above **15**. The reaction which can occur via the covalently bound cyclic transition state [182] may be viewed as a migration of ionized hydroxycarbene to the radical centre. At higher energies, ~117 kJ/mol above **15**, [181], isomerization takes place by way of a 1,2-hydrogen shift to produce the stable enol **16**:

$$^{\bullet}CH_2\text{-}CH_2\text{-}\overset{+}{C}HOH \text{ (15)} \rightarrow CH_3\text{-}CH=CHOH^{+\bullet} \text{ (16)} \quad [b]$$

Experiments are in full agreement with the theoretical prediction that reactions [a] and [b] take place below the dissociation threshold [195]. Reactions [a], a formal [1,2]-hydroxy carbene migration, and [b], a 1,2-hydrogen shift, are the prototype isomerization reactions that occur in many distonic carbonyl radical cations, and these reactions elegantly rationalize observed fragmentation patterns as discussed below. In γ- and δ-distonic ions the reaction may also take place via four- and five-membered cyclic transition structures corresponding to the [1,3]- and [1,4]-hydroxycarbene-radical migrations respectively [191].

(i) Schwarz and co-workers have extensively studied the dissociation mechanisms of aliphatic acids and esters [191]. Metastable carboxylic acid cation radicals are characterized by formation of n-alkyl radicals. For example, 2-methylbutyric acid cation radicals, $CH_3CH_2CH(CH_3)COOH^{+\bullet}$, **17**, dissociate almost exclusively to $CH_3^{\bullet}$ and protonated methacrylic acid, $CH_2=C(CH_3)\overset{+}{C}(OH)_2$. By integration of all available data Schwarz et al. arrived at the sequence in the following Scheme.

The most important features are : (a) the dissociation is **not** a simple cleavage process; (b) the sequence commences with a 1,4-hydrogen shift, which (see later) must be fast and irreversible, to produce the β-distonic ion **18**. The β-distonic ions do **not** lose $CH_3^{\bullet}$ although a more stable product ion, **21**, could be formed by this reaction; (c) combinations of reversible 1,2-hydrogen and [1,2]-hydroxycarbene-radical migrations take effect to lead to the stable enol cation radicals **19** and **19a**; (d) the enol ions undergo C-C cleavage to produce **20** and **20a** and a $CH_3^{\bullet}$ radical. Notably this sequence of events rationalizes the interesting observations that both original methyl groups are eliminated to the same extent and that the daughter ions are **20** & **20a** and not the more stable ions **21** & **21a** ! (From an ICR study of the gas phase basicity of carbonyl compounds [65] it follows that **21** is more stable than **20** by 9 kJ/mol.). Remarkable is the observation that it is the enol ion which dissociates and **not** the β-distonic isomer. It appears, so far without exception, that formation of alkyl radicals from carboxylic acids and esters occurs from enol ions and not from the distonic forms [191]. Weiske and Schwarz [191] point out that the solution to this puzzling result is surprisingly trivial when one analyses the reverse process : *addition of the nucleophilic alkyl radicals will prefer attack at $C_\beta$ and not at $C_\alpha$.* AE measurements of the metastable peak associated with the reaction **17→20** (and other alkyl losses from ionized acids and esters) would appear appropriate to ascertain whether reverse barriers are involved.

Similarly, based on an analysis of isotope effects in metastable 2-ethylbutanoic acid ions it was concluded [193] that they lose $C_2H_5^\bullet$ by means of a 1,4-hydrogen shift, a [1,2]-hydroxycarbene-radical migration and the rate-determining 1,2-hydrogen shift (acid ion→ β-distonic ion → β-distonic ion → enol ion) but from labelling and CID data it became clear that for high energy ions the hydroxycarbene-radical migration takes place to only a minor extent and the reaction sequence is : acid ion→ β-distonic ion → products [196].

(ii) Formation of $CH_3^\bullet$ from ionized methylisobutyrate, $(CH_3)_2CHCOOCH_3^{+\bullet}$, **22**, has always been considered the prototype of a reaction initiated by a hidden hydrogen migration. Earlier work showed that the reaction is not a simple cleavage process but is initiated by a 1,4-hydrogen shift; all experimental data available could be interpreted by the sequence :

**22** → 1,4-H → $\dot{C}H_2\text{-}CH(CH_3)\text{-}\overset{+}{C}(OH)\text{-}OCH_3$ **(23)** → $CH_2=C(H)\text{-}\overset{+}{C}(OH)\text{-}OCH_3 + CH_3^\bullet$ **(24)**.

Note that this sequence is contrary to the above generalization in that it is a β-distonic ion which loses $CH_3^\bullet$. However the alternative (correct) mechanism (distonic ion → enol ion → products) would not have been evident from the available data. This was recognized by the Schwarz group and from CID experiments on $^{13}C$ labelled **22**, i.e. by using $(CH_3)_2CHCOOCH_3^{+\bullet}$, it could be shown that the daughter ions **24** are $CH_2=\overset{*}{C}(H)\text{-}\overset{+}{C}(OH)\text{-}OCH_3$ and not $CH_2=C(H)\text{-}\overset{+}{C}(OH)\text{-}\overset{*}{O}CH_3$ and thus the above generalization does hold.

A (simplified) generalization for the losses of alkyl radicals $R_1^\bullet$ and $R_2^\bullet$ from ionized carboxylic acids and esters is :

$$\begin{array}{c}
\text{acid} \\ \text{ester}
\end{array} \Big\} \xrightarrow{\text{H shift}} \text{distonic ion} \xrightleftharpoons[]{\text{H shift}} \text{enol ion} \xrightarrow{-R_1^\bullet}$$

with $-R_1^\bullet$ loss upward from distonic ion, hydroxycarbene shift down to

$$\text{distonic ion} \xrightleftharpoons[]{\text{H shift}} \text{enol ion} \xrightarrow{-R_2^\bullet}$$

and $-R_2^\bullet$ loss downward from distonic ion.

(iii) Other incipient carbenes too can migrate to a radical centre. Bouchoux et al. [183] have shown that the loss of ethene from ionized 2-hexanone, $CH_3\text{-}CO\text{-}CH_2\text{-}CH_2\text{-}CH_2\text{-}CH_3^{+\bullet}$, **25**, involves rearrangement via a reversible 1,5-hydrogen shift into the γ-distonic ion ,**26**, $CH_3\overset{+}{C}(OH)\text{-}CH_2\text{-}CH_2\text{-}\overset{\bullet}{C}H\text{-}CH_3$. Next $^{\bullet}CH_2\text{-}CH_2\text{-}CH(CH_3)\text{-}\overset{+}{C}(OH)CH_3$ ,**27**, is formed via a [1,3]-$CH_3COH$ -migration, from which dissociation into $CH_2=CH_2 + CH_3CH=C(OH)CH_3^{+\bullet}$ occurs. The chemistry of **25** will be further discussed in the next section.

(iv) It is tempting to try to interpret other observations of longstanding interest in terms of the [1,n]-hydroxycarbene-radical migration. One of the important fragmentations of small ($<C_{10}$) straight-chain aliphatic aldehydes is loss of $C_2H_4$. Twenty years ago Liedtke and Djerassi [197] published labelling data on hexanal and heptanal. The observation that in the $C_2H_4$ loss from ionized hexanal, $CH_3CH_2CH_2CH_2CH_2CH=O^{+\bullet}$, **28**, the carbon atoms C(2) and C(3) are eliminated as a unit as are C(5) and C(6), is readily explained on the basis of a [1,3]-hydroxycarbene-radical migration coupled with a 1,5-hydrogen transfer: **28** → 1,5-H → $CH_3CH_2\overset{\bullet}{C}HCH_2CH_2\overset{+}{C}(H)\text{-}OH$, **29** → 1,3-HCOH → $C(6)H_3C(5)H_2C(H)(\overset{+}{C}HOH)C(3)H_2\overset{\bullet}{C}(2)H_2$ **30**. The branched γ-distonic ion **30** will lose ethylene as $C(2)H_2=C(3)H_2$ or following a second 1,5-hydrogen shift ( **30** → $^{\bullet}C(6)H_2C(5)H_2C(H)(\overset{+}{C}HOH)C(3)H_2C(2)H_3$ , **31**) as $C(5)H_2=C(6)H_2$.

The same mechanism predicts that for heptanal only C(2) and C(3) will be eliminated as ethylene, and the 1,5-hydrogen shift in the analogue of **30** will lead to loss of $C_3H_6$, in excellent agreement with the experimental data.

(v) Loss of $C_2H_4$ from metastable 4-octanone ions, $CH_3CH_2CH_2COCH_2CH_2CH_2CH_3^{+\bullet}$ , **32**, was recently reported to produce (a) 2-hydroxy-1-hexene ions via a McLafferty rearrangement in the n-propyl chain and (b) a "3-hexanone type" ion via loss of $C_2H_4$ from the n-butyl chain [198]. The latter $C_2H_4$ loss could be rationalized by a 1,5-H shift from the n-butyl chain to the carbonyl group (**32**→**33**) followed by a [1,3]-hydroxy carbene-radical shift (**33**→**34**):

(**32**) → $CH_3(CH_2)_2\overset{+}{C}(OH)CH_2CH_2\overset{\bullet}{C}HCH_3$ (**33**) → $CH_3(CH_2)_2\overset{+}{C}(OH)C(CH_3)HCH_2\overset{\bullet}{C}H_2$ (**34**) →
$CH_3CH=C(OH)CH_2CH_2CH_3^{+\bullet}$ (**35**) + $CH_2=CH_2$.

If this proposal is correct the product ion will be the "3-hexanone type" enol ion (**35**) and it is then further predicted that the loss of ethylene involves the carbon atoms 5 and 6, in addition to the carbon atoms 1 and 2.

### The [1,2]-enol-olefin shift.

Information on the prototype enol-olefin shift was not available in the literature up to July 1988. However at the 11th IMSC meeting in Bordeaux Bouchoux et al. [199] presented results of ab initio calculations on the prototype reaction and relevant parts are briefly discussed. The ion $\overset{\bullet}{C}H_2CH_2CH_2\overset{+}{C}HOH$, **36**, the next higher homologue of **15** is , in the ground state, a (covalently-bound) γ-distonic ion but the excited system comprises a positively charged vinyl alcohol moiety, $CH_2=CHOH^{+\bullet}$ with a non-polar ethylene molecule, an ion-neutral complex [5]. Below the threshold for the dissociation reaction of lowest energy requirement, loss of $C_2H_4$, a reversible [1,2]-enol-olefin shift takes place i.e. the degenerate isomerization: $^{\bullet}CH_2CH_2CH_2\overset{+}{C}HOH \rightleftharpoons HO\overset{+}{C}HCH_2CH_2CH_2^{\bullet}$.

Contrary to an earlier report [194] the dissociation into $CH_2=CHOH^{+\bullet} + C_2H_4$ was calculated to have no barrier for the reverse process. The occurrence of a (competing) [1,3]-hydroxycarbene migration was not considered in the above computational study. Earlier results had been interpreted implicitly in terms of such an enol-olefin shift, but only after recent detailed experiments [4,182,183,185,187,189,190] has it become clear that the reaction is ubiquitous:

(i) The unimolecular chemistry of 2-hexanone ions, **25**, is governed by the loss of a methyl radical. Extensive $^{13}C$ labelling experiments by Bouchoux et al. [183] elegantly revealed that the $CH_3^{\bullet}$ loss mainly comprises C(4) and C(6) *in equal proportions* and the following mechanism was proposed:

$CH_3COCH_2CH_2CH_2CH_3^{+\bullet}$, **25** $\to CH_3\overset{+}{C}(OH)\text{-}CH_2CH_2\overset{\bullet}{C}HCH_3$ ,**37**(distonic ion) $\to$
$CH_3\overset{+}{C}(OH)\text{-}CH_2\text{-}C(H)(CH_3)\overset{\bullet}{C}H_2$ ,**38** (distonic ion) $\to CH_3C(OH)=CH\text{-}C(H)(CH_3)_2^{+\bullet}$ ,**39** (enol ion)
$\to CH_3^{\bullet} + CH_3\overset{+}{C}(OH)CH=CHCH_3$ **(40)**.

The reaction commences with a (reversible) 1,5-hydrogen migration, **25→37**, the first step in the McLafferty rearrangement. A subsequent [1,2]-enol-olefin, **37→38**, and hydrogen shift, **38→39**, produce the enol species **39** which dissociates. An important observation is that it is the enol species which fragments and that it is formed, inter alia, via a 1,2-hydrogen migration in the alkyl radical moiety [183], in perfect agreement with the general mechanism shown above [191]. Loss of $CH_3^{\bullet}$ from ionized 2-hexanone had been discussed in the first volume of this series (p. 42-43); it was concluded that the reaction involved formation of a five-membered ring with concerted loss of $CH_3^{\bullet}$. The reaction now appears to be much more complex, but nevertheless chemically reasonable.

(ii) A different approach was taken by Grützmacher and co-workers [185] in a simple yet elegant experiment. The authors calculated similarity indices (SI) from the CID spectra of all methylketones which contain seven carbon atoms:

$CH_3COCH_2C(CH_3)_3$ ,**41**, $CH_3COCH_2CH_2CH(CH_3)_2$, **42**, $CH_3COCH_2CH_2CH_2CH_2CH_3$, **43**, $CH_3COCH_2CH(CH_3)C_2H_5$ ,**44**, $CH_3COCH(CH_3)CH_2CH_2CH_3$,**45**, $CH_3COCH(CH_3)CH(CH_3)_2$, **46**, $CH_3COC(CH_3)_2CH_2CH_3$, **47**, and $CH_3COCH(CH_2CH_3)_2$ , **48**.

Next a low SI was taken as an indication of identical structures. It then appeared that **41** and **42** were indistinguishable as were **43** and **44**. It was proposed that these species interconvert, pairwise, via ion-molecule complexes formed from the McLafferty products; mutual interconversion of stable ions is possible by a formal addition of the enol cation radical to the olefin molecule; the interconversion may be viewed as a [1,2]-enol-olefin shift. Note that for each of the ions **47** and **48** the respective [1,2]-enol-olefin shifts are degenerate and so high SI values obtain.

### Consecutive enol-olefin and hydroxycarbene-radical migrations.

Many puzzling, seemingly complex reactions can be rationalized by consecutive enol-olefin and hydroxycarbene-radical migrations.

(i) The ethylene molecule formed from 3-methylbutyric acid, **49**, contains C(2) and C(4) as a unit [200] and this result was rationalized by McAdoo et al. [189] on the basis of

consecutive enol-olefin (**50→51**) and [1,3]-hydroxycarbene-radical (**51→52**) shifts :
$(CH_3)(CH_3)CH-CH_2-COOH^{+\bullet}$ ,**49** → 1,5H → $(CH_3)(\dot{C}H_2)CH-CH_2-\overset{+}{C}(OH)_2$ ,**50** →
$CH_3-\dot{C}H-CH_2-CH_2-\overset{+}{C}(OH)_2$ ,**51** → $(HO)_2\overset{+}{C}-C(CH_3)(H)(CH_2-\dot{C}H_2)$,**52** → $CH_3-CH=C(OH)_2^{+\bullet} + CH_2=CH_2$.
Note that this mechanism is basically the same as that proposed in 1984 by Audier and Sozzi [200].

(ii) The molecular ions of the methyl ketones **41** and **42** are metastable with respect to the formation of $C_2H_4$, $C_2H_5^\bullet$ and $C_4H_7^\bullet$. The formation of $C_2H_4$ was interpreted by Bouchoux et al. [182] by a combination of enol-olefin and hydroxycarbene shifts as shown in the following Scheme:

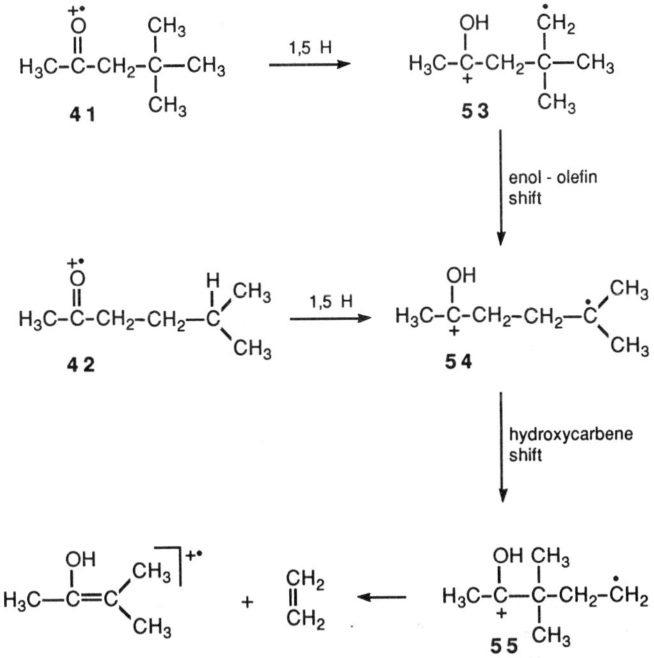

$\underset{\textbf{41}}{H_3C-\overset{+\bullet}{\overset{\displaystyle O}{\|}}-CH_2-\overset{CH_3}{\underset{CH_3}{\overset{|}{\underset{|}{C}}}}-CH_3} \xrightarrow{1,5\ H} \underset{\textbf{53}}{H_3C-\overset{OH}{\underset{+}{\overset{|}{C}}}-CH_2-\overset{\dot{C}H_2}{\underset{CH_3}{\overset{|}{\underset{|}{C}}}}-CH_3}$

enol - olefin shift ↓

$\underset{\textbf{42}}{H_3C-\overset{+\bullet}{\overset{\displaystyle O}{\|}}-CH_2-CH_2-\overset{H}{\underset{CH_3}{\overset{|}{\underset{\diagdown}{C}}}}{}^{\diagup CH_3}} \xrightarrow{1,5\ H} \underset{\textbf{54}}{H_3C-\overset{OH}{\underset{+}{\overset{|}{C}}}-CH_2-CH_2-\dot{C}{\overset{\diagup CH_3}{\diagdown CH_3}}}$

hydroxycarbene shift ↓

$H_3C-\overset{OH}{\overset{|}{C}}=C{\overset{\diagup CH_3}{\diagdown CH_3}}\Big]^{+\bullet} + \overset{CH_2}{\underset{CH_2}{\|}} \longleftarrow \underset{\textbf{55}}{H_3C-\overset{OH}{\underset{+}{\overset{|}{C}}}-\overset{CH_3}{\underset{CH_3}{\overset{|}{\underset{|}{C}}}}-CH_2-\dot{C}H_2}$

The authors also concluded that the energy requirement of the hydroxycarbene migration is larger than that for the enol-olefin shift. This may explain why the hydroxycarbene migration plays no role in the lower energy ions sampled in the CID experiments of Grützmacher et al. [185] discussed above.

(iii) It is amusing to note that loss of $C_2H_5^\bullet$ from the above ketones **41** and **42** may be rationalized along the lines developed by Schwarz. A 1,6-hydrogen shift in **42** leads to a δ-distonic ion which via a [1,4]-hydroxycarbene-radical migration and a 1,4-hydrogen shift may yield the species $CH_3-C(OH)=CH-CH(CH_3)-CH_2-CH_3^{+\bullet}$ ,**56**. This enol ion may lose $CH_3^\bullet$ and $C_2H_5^\bullet$ and according to the relative leaving group abilities obtained by Weiske and Schwarz [191] $CH_3^\bullet$ and $C_2H_5^\bullet$ should be lost in a ratio of 2 : 100 to be compared with the experimentally observed ratio of 4-5 : 100 [182].

(iv) A longstanding intriguing observation has recently been rationalized in terms of hydroxycarbene-radical and enol-olefin shifts [187]. This concerns the observed exchange of the β- and γ-hydrogens of butyric acid and its methyl ester, $CH_3CH_2CH_2COOR$ (R = H, $CH_3$), **57**, prior to the loss of $C_2H_4$ and the absence of any such equilibration reactions prior to the loss of $CH_3^{\bullet}$. In addition the methyl radical contains the γ- and α-hydrogens only. In the proposed mechanism the β- and γ-distonic ions $CH_3\dot{C}HCH_2\overset{+}{C}(OH)OR$, **58**, (R = H, $CH_3$) and $\dot{C}H_2CH_2CH_2\overset{+}{C}(OH)OR$, **59**, (R = H,$CH_3$) are key intermediates produced from **57** by 1,4- and 1,5-hydrogen shifts respectively. Since these distonic ions may **not** interconvert one has to assume that the 1,4-hydrogen shift (leading to **58**) is *fast* and *irreversible* whereas the 1,5-hydrogen shift is not, which is not easy to explain.

### The [1,2]-amine-radical shift.

Ab initio calculations have shown that the species $\dot{C}H_2\text{-}CH_2\text{-}\overset{+}{N}H_3$, **60**, the β-distonic form of ionized ethylamine, is the global minimum on the $C_2H_7N^{+\bullet}$ surface [35]. In contrast to the oxygen analogue [167] the barrier for the degenerate isomerization reaction :

$$\dot{C}H_2\text{-}CH_2\text{-}\overset{+}{N}H_3 \rightleftharpoons H_3\overset{+}{N}\text{-}CH_2\text{-}\dot{C}H_2$$

is quite large, 120 kJ/mol, but 130 kJ/mol below the dissociation pathway of lowest energy requirement. The above shift has been observed in other distonic amine ions too. Thus the CID mass spectra of $(CH_3)_2\dot{C}\text{-}CH_2\text{-}\overset{+}{N}H_3$, **61**, and $\dot{C}H_2\text{-}C(CH_3)_2\text{-}\overset{+}{N}H_3$, **62**, are indistinguishable and those of $CH_3\text{-}\dot{C}H\text{-}CH_2\text{-}\overset{+}{N}H_3$, **63**, and $\dot{C}H_2\text{-}C(H)(CH_3)\text{-}\overset{+}{N}H_3$, **64**, are very similar [201].

$$H_3\overset{+}{N}-\underset{CH_3}{\overset{CH_3}{\underset{|}{\overset{|}{C}}}}-\dot{C}H_2 \quad \underset{\text{amine shift}}{\rightleftarrows} \quad {}^{\bullet}\underset{CH_3}{\overset{CH_3}{\underset{|}{\overset{|}{C}}}}-CH_2\overset{+}{N}H_3$$

β-distonic **62**         β-distonic **61**

The migration of the $NH_3^+$ group (shift of protonated $NH_2$ [201]) can be viewed as a [1,2]-amine-radical migration, where the amine is ionized, in contrast to the oxygen analogue $\dot{C}H_2\text{-}CH_2\text{-}\dot{O}H_2$ where **neutral** $H_2O$ shifts along the charged ethene rod [167]. The [1,2]-amine-radical shift elegantly explains some interesting observations [201] e.g. the loss of $C_3H_7^{\bullet}$ from the distonic ion **65** shown in the next Scheme. Bjørnholm et al. [201] concluded that the amine distonic ions, at least those studied to date, cannot be considered as ion-dipole complexes; this is reflected in the large barrier associated with the degenerate isomerization $\dot{C}H_2\text{-}CH_2\text{-}\overset{+}{N}H_3 \rightleftharpoons H_3\overset{+}{N}\text{-}CH_2\text{-}\dot{C}H_2$. In addition they conclude that the radical site is directly involved in the reactions of distonic amine cation radicals and that the characteristic reactions are closely related to well-described reactions of neutral species [201,36].

## 2: Structures and Reactions of Gas-phase Organic Ions

$$\underset{\mathbf{65}}{\overset{H}{\underset{H}{\overset{|}{H_2\overset{\bullet}{C}}}}-CH_2-\overset{CH_3}{\underset{CH_3}{\overset{|}{\overset{+}{N}}}}-\overset{CH_3}{\underset{|}{\overset{|}{C}}}-CH_3} \quad \longrightarrow \quad \underset{\mathbf{66}}{\overset{H}{\underset{H}{\overset{|}{H_3C}}}-CH_2-\overset{\overset{\bullet}{CH_2}}{\underset{CH_3}{\overset{|}{\overset{+}{N}}}}-\overset{|}{\underset{|}{C}}-CH_3}$$

↓ [1,2]-amine shift

$$\underset{\text{loses }C_3H_7^{\bullet}}{\underset{\mathbf{68}}{\overset{H}{\underset{\downarrow}{\overset{|}{H_3C-CH_2-N-CH_2}}}}\overset{\overline{CH_3}}{\underset{CH_3}{\overset{|}{\underset{\backslash}{\overset{|}{C}}}-H}}\overset{+\bullet}{}} \quad \longleftarrow \quad \underset{\mathbf{67}}{\overset{H}{\underset{H}{\overset{|}{H_3C-CH_2-\overset{+}{N}-CH_2-C\overset{\bullet}{\overset{CH_3}{\underset{CH_3}{\diagdown}}}}}}}$$

**Intermediate ion-neutral complexes.**

Unimolecular reactions of solitary cation radicals and closed shell cations are being increasingly rationalized in terms of models that assume the intermediacy of ion- (induced) dipole complexes with a relatively strong electrostatic bond between the incipient fragments [5]. Two distinct but related phenomena [202-204] may be interpreted by means of ion-dipole interactions, namely (a) isomerization of the incipient carbenium ion $R_1^+$ into $R_2^+$, and (b) hydrogen transfer reactions between the formal components of the complex :

$$R_1X^{+\bullet} \rightarrow R_1^{+}\cdots X^{\bullet} \rightarrow R_2^{+}\cdots X^{\bullet}$$
$$\rightarrow [R_1\text{-H}]^{+}\cdots [X+H]$$

Their participation allows a simple rationalization of otherwise disconcerting observations [154]. It has been argued [202] that although many reactions may conveniently be written as stepwise this should not be taken to imply that the intermediate ion-molecule complexes represent local minima on the potential energy surface. Celebrated ion-neutral mediated reactions are the dissociations of onium ions [203] and of alkylphenyl ether radical cations [204] , e.g. n-butyl phenyl ether dissociating into $C_4H_8$ and ionized phenol, $C_6H_5OH^{+\bullet}$. However it was recently shown that sec-butyl phenyl ether displays a higher degree of regiochemical selectivity among specifically labelled analogues than the n-butyl ether [205], possibly indicating that the reaction is not ion-neutral mediated. Kondrat and Morton [206] observe for the sec-butyl ether reproducible differences in the dissociation of the labelled threo- and erythro-isomers, ruling out a complex mediated reaction.

McAdoo in his review on this subject [5] has suggested criteria to indicate when a reaction would be complex mediated but it was also noted that none of the listed properties needs be unique to complex mediated reactions. Hammerum [207] has also outlined some criteria for complex mediation but it does not yet seem possible to formulate clear predictive rules which can generally be applied. Very small kinetic energy release values (~1 meV) have been linked [46] to ion-dipole fragmentations but much larger values have been observed where the final step is exothermic [208].

Specific hydrogen exchange has provided experimental support for the intermediacy of ion-neutral complexes in the reactions of simple alcohol and ether type radical cations [208].

It has been expressed that claims for ion-neutral mediated reactions are not always justified. Thus Weiske et al. [209] argue that loss of $CH_4$ from ionized isopropyl methyl ether, $(CH_3)_2CHOCH_3^{+\bullet}$, can be explained by traditional kinetic arguments rather than by an ion-neutral model as originally proposed.

In their discussion of the very different dissociation behaviour of metastable isobutylamine and neopentylamine radical cations [$(CH_3)_2CHCH_2NH_2^{+\bullet}$ mainly yields $CH_2NH_2^+$ by direct bond cleavage whereas $(CH_3)_3CCH_2NH_2^{+\bullet}$ mainly dissociates into $CH_3NH_3^+$ via a double hydrogen migration] Hammerum and Derrick [202] argue from thermochemical considerations that in this case competition between direct bond cleavage and rearrangement reactions is not governed by the energetic accessibility of the ion-molecule complexes [$C_3H_6/CH_3NH_2^{+\bullet}$] and [$C_4H_8/CH_3NH_2^{+\bullet}$], but by the kinetics of competing reactions. When ionized isobutyl alcohol, $(CH_3)_2CHCH_2OH^{+\bullet}$, which dissociates via a double hydrogen migration into $CH_3OH_2^+$, and isobutylamine are compared it appears attractive to rationalize the different dissociation behaviour in terms of the inaccessibility, energetically, of the [$C_3H_6/CH_3NH_2^{+\bullet}$] complex vis à vis the accessibility of the [$C_3H_6^{+\bullet}/CH_3OH$] complex. However, since the [$C_3H_6/CH_3NH_2^{+\bullet}$] and [$C_4H_8/CH_3NH_2^{+\bullet}$] complexes have similar relative energies then from the above argument the latter complex ought not to be accessible from neopentylamine. Yet this compound does fragment by a double hydrogen migration. It was proposed [202] that crossing of the k vs E curves associated with direct bond cleavage and rearrangement occurs at lower energies for isobutylamine than for neopentylamine.

At very long lifetimes (> $10^{-4}$s) isobutylamine surprisingly fragments by loss of $C_2H_5^{\bullet}$ for which a mechanism, which includes an "ylid ion" shift akin to the hydroxycarbene-radical migration, was proposed. By combining the results on $C_4$ distonic amine ions [201,202] we derive the following Scheme :

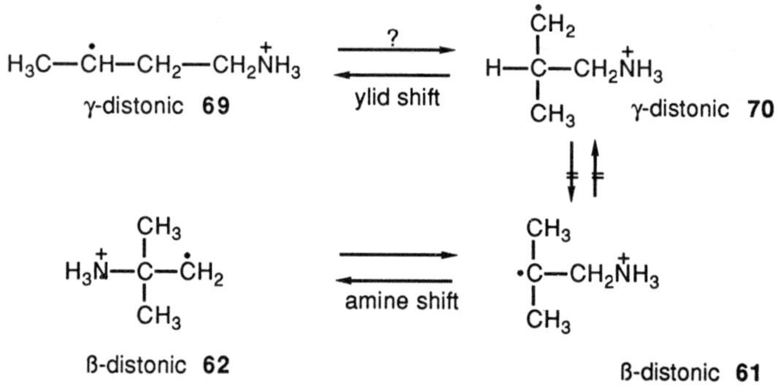

$$H_3C-\overset{\bullet}{C}H-CH_2-CH_2\overset{+}{N}H_3 \quad \underset{\text{ylid shift}}{\overset{?}{\rightleftarrows}} \quad H-\underset{\underset{CH_3}{|}}{\overset{\overset{\bullet}{C}H_2}{\underset{|}{C}}}-CH_2\overset{+}{N}H_3$$

γ-distonic **69**        γ-distonic **70**

$$H_3\overset{+}{N}-\underset{\underset{CH_3}{|}}{\overset{\overset{CH_3}{|}}{C}}-\overset{\bullet}{C}H_2 \quad \underset{\text{amine shift}}{\rightleftarrows} \quad \overset{\bullet}{\underset{\underset{CH_3}{|}}{\overset{\overset{CH_3}{|}}{C}}}-CH_2\overset{+}{N}H_3$$

ß-distonic **62**        ß-distonic **61**

It would appear that the β-distonic ions can interconvert as do the γ-distonic ions, but that there is no interconversion possible between the *pairs* of γ- and β-distonic ions. Hence it appears that in this system the charged group, $-NH_3^+$ or $-CH_2NH_3^+$, has a higher migratory aptitude towards a radical centre than a hydrogen atom.

In addition to the well known McLafferty rearrangement many an organic radical cation undergoes what has been termed "a McLafferty rearrangement with charge reversal" [189]. For example, ionized n-hexanal, $CH_3CH_2CH_2CH_2CH_2CH=O^{+\bullet}$, forms $[M-44]^{+\bullet}$ and m/z 44 ($CH_2=CHOH^{+\bullet}$) ions and in a mechanism involving ion-molecule interactions proposed by Morton [204], an explanation is given for the earlier observation that the neutral species of mass 44 u formed together with m/z 56 is $CH_3CHO$ and not $CH_2=CHOH$. Bouchoux et al. [154] propose an alternative mechanism in which the ionized enol/olefin complex $[CH_2=CHOH^{+\bullet}/CH_3CH=CH_2]$ is the key intermediate. Similarly, and at the same time, the formation of $C_4H_8^{+\bullet}$ ions and $C_2H_4O_2$ neutral species from ionized n-hexanoic acid, $CH_3CH_2CH_2CH_2CH_2COOH^{+\bullet}$, was rationalized on the basis of long-lived ion-neutral complexes which interconvert by way of hydrogen migrations between the formal components. This leads to the [ionized enol/neutral olefin] → [neutral keto/charged olefin] transformation. In this work the structure of the neutral $C_2H_4O_2$ species, acetic acid and not its enol, was unequivocally established from collision-induced dissociative ionization, CIDI, experiments [162].

McAdoo and co-workers [5] also rationalize the formation of the [McLafferty + H] ions from carboxylic acids on the basis of long-lived ion-neutral complexes coupled with hydrogen migrations between the formal components of the complexes. Ion-neutral complexes have also been invoked in the unimolecular chemistry of ionized n-butanol, to account for the observation that all hydrogen atoms scramble in ions decomposing after $10^{-5}$ s. [188].

Filges and Grützmacher [210,211] have thoroughly investigated the breakdown patterns of protonated benzaldehydes and acetophenones. The protonated species $R-\overset{+}{C}(OH)-C_6H_5-CH_2OCH_3$, (R = H, $CH_3$) undergoes an interesting unimolecular reaction, namely loss of $RCOOCH_3$. From labelling results and energy considerations it is shown that the reaction involves migration of the formyl (R = H) or acetyl (R = $CH_3$) group via intermediate ion-neutral complexes.

The extrusion of CO from $CH_3COCH_2COC(CH_3)_3^{+\bullet}$, ionized 2,2-dimethyl-3,5-hexanedione, specific loss of the CO moiety adjacent to the tert-butyl group, has also been interpreted in terms of the intermediacy of ion-neutral complexes [212]; enolic species are not involved. In general, as shown by GC/MS experiments, the keto and enol forms of 1,3-diketone ions do not interconvert [213].

Loosely bound hydrogen-bridged radical cation complexes are also being increasingly proposed as intermediates in the dissociation of radical cations. Ab initio calculations reveal that ionized acetic acid, $CH_3COOH^{+\bullet}$, [214,215], acetamide, $CH_3CONH_2^{+\bullet}$, [215,216] and acetone, $CH_3COCH_3^{+\bullet}$, [215], all fragment via such complexes. For the losses of OH$^{\bullet}$ and $H_2O$ from acetic acid ions,71, see the next Scheme, it was found that elongation of the C--OH bond does not immediately lead to dissociation but upon separation the OH group can migrate within the electrostatic field of the acetyl cation, $CH_3\overset{+}{C}=O$, leading to ion-dipole complexes **72** and **73**.

$$H_3C-C\overset{O^{+\bullet}}{\underset{OH}{\diagdown}} \longrightarrow H-\underset{\underset{O}{\overset{|}{H}}}{\overset{H}{\underset{|}{C}}}-C=\overset{+\bullet}{O} \longrightarrow \underset{H}{\overset{H}{\diagdown}}C=C=O^{+\bullet}$$

71　　　　　　　　72　　　　　　　　$\underset{H}{\overset{|}{O}}\diagdown H$　73

　　　　　　　　　$\downarrow -OH^\bullet$　　　　　　$\downarrow -H_2O$

The rate-determining step was found to be 72→73. Complex 73 was made independently by dissociative ionization of 1,3-dihydroxyacetone, $HOCH_2COCH_2OH$, [168]. The transition structure for the conversion 71→72 is only ~1 kJ/mol above 72. Species 73 is a good example to illustrate problems associated with the choice of the theoretical methods for a proper description of loosely bound complexes [168,214,215]. Within the UHF formalism and using the 3-21G basis set, optimization resulted in a covalently bound distonic ion, $H_2O-CH_2-C=O^{+\bullet}$. However, using a 4-31G [168] or 6-31G [214,215] basis set optimization resulted in complex 73, which was found to be, after single point calculations, 42 kJ/mol more stable than the distonic form. At internal energies of > 10 kJ/mol the water molecule can move from carbon to carbon. At internal energies close to the threshold for dissociation into $CH_2=C=O^{+\bullet}$ and $H_2O$ (64 kJ/mol by theory and 68 kJ/mol by experiment, [168]) the water molecule in 73 can move freely about the charged ketene "rod", showing a behaviour much like that of the $[C_2H_4/H_2O]^{+\bullet}$ complex associated with the distonic $C_2H_6O$ isomer $CH_2-CH_2-OH_2^{+\bullet}$, [167]. Drewello et al. [216] rationalized the loss of CO from ionized acetamide, $CH_3CONH_2^{+\bullet}$, in terms of complexes between $CH_2=C=O^{+\bullet}$ and $NH_3$ and the distonic ion $H_3N-CH_2-C=O^{+\bullet}$ which dissociates into the ylid ion $CH_2NH_3^{+\bullet}$ and CO. These authors also conclude that nucleophilic substitution of $NH_3$ to $CH_2=C=O^{+\bullet}$ is a very facile process, contrary to earlier suggestions. The stabilization energy of the hydrogen-bridged complexes of the type $X\cdots H\cdots CH_2-C=O^{+\bullet}$ (X = OH, $NH_2$, $CH_3$) was found to increase with increasing polarizability of $X^\bullet$ [214].

### The dissociation behaviour of ion-dipole complexes.

It has been observed that ion-dipole complexes may dissociate extremely slowly near onset, orders of magnitude slower than expected [44] and that the KER may be exceedingly small [166-168, 46, 217, 5,3, 208].

Ruttink has applied the adiabatic channel theory to the ion-dipole complex $C_2H_4^{+\bullet}/H_2O$ (KER < 0.4 meV) and his calculations indicate that the excess of energy for the metastable ions is not larger than 1.8 meV [46], which is a factor 100-1000 smaller than what is usually encountered. That dissociations, occurring at such exceptionally small energies, are at all observed is perhaps not surprising as metastable ions are selected by rate-constant, not by energy. Shao et al. [44] , by integration of results from various techniques including RRKM/QET and ab initio calculations, have convincingly argued that in the dissociation of ionized n-propanol, $CH_3CH_2CH_2OH^{+\bullet}$, an intermediate ion-dipole complex can slow down the dissociation rate by a factor of $10^4$, not by virtue of the greater thermodynamic stability of the complex but because of the extreme anharmonic potential associated with ion-dipole forces which can accommodate a large density of states: ionized n-propanol is metastable with respect to ionized cyclopropane (and not ionized propene) and water, obviously a rearrangement reaction producing one of the most intense metastable peaks known to ourselves. The initial formation of $[CH_2-CH_2-CH_2]^{+\bullet}$ which collapses to $[c-C_3H_6]^{+\bullet}$ can be ruled out on energetic grounds [218]. The AE as measured by PEPICO measurements is far too small to allow the ion to be metastable. There is a barrier for the reverse reaction (0.25 eV) which as PEPICO-PD experiments show is associated with the first step. To account for the observed slow rates the rearranged form should lie 1.38 eV below ionized propanol, but no such low energy isomer appears to exist. The ion-dipole complex $c-C_3H_6^{+\bullet}/H_2O$ does lie below ionized propanol but only by 0.23 eV (in fact the activation energy for the dissociation of the ion-dipole complex is equal to, or nearly so, to the activation energy for its formation!). It was concluded that the very anharmonic potential associated with ion-dipole forces may slow down the reaction sufficiently for it to be readily observed in the metastable time frame.

For other examples of ion-neutral mediated reactions the reader is referred to the review by McAdoo [5].

## Methylnitrite

The unimolecular chemistry of methylnitrite, $CH_3ONO^{+\bullet}$, ions has attracted considerable interest in recent years due to the reported non-statistical behaviour and a large isotope effect [219-222,40]. Close to the threshold $CH_3ONO^{+\bullet}$ fragments to $NO^+ + [C/H_3/O]^{\bullet}$ and to $[C/H_3/O]^+ + NO^{\bullet}$, but the rate, as measured with respect to the appearance of the products [222], is much lower ($10^6$ s$^{-1}$) for the formation of $[C/H_3/O]^+$ **ions** than for $[C/H_3/O]^{\bullet}$ **neutrals** ($>10^7$ s$^{-1}$) and it becomes even lower upon isotopic substitution of H by D. Also the measured critical energy for the slower process is only 0.5 eV [222] (0.31 eV [221]) and so the minimum rate is predicted to be $10^8$ s$^{-1}$, much larger than that observed. These findings were taken as indicating non-statistical behaviour. It appears that at $10^{-5}$ s, formation of $NO^+$ is collision induced [221]. However, it has been confirmed [222] that the ion $[C/H_3/O]^+$ is not $CH_3O^+$ but $CH_2OH^+$ and so the slower process must be formulated as a rearrangement reaction. Baer and Hass [222] propose that the low rate of formation of $CH_2OH^+$ results from the competition between direct dissociation ($NO^+ + CH_3O^{\bullet}$) and isomerization into a *low energy* isomer of $CH_3ONO^{+\bullet}$ followed by slow dissociation into $CH_2OH^+ + NO^{\bullet}$. By integration of results obtained from ab initio molecular orbital calculations and statistical theory calculations they concluded that the isomer $HCN(OH)_2^{+\bullet}$ has the correct structure and energy to account for the dissociation data. However, from $^{18}O$ labelling experiments by Egsgaard and Carlsen [219] it becomes clear that participation of any species with equivalent oxygen atoms can be ruled out. Ferguson [223] has argued that the hydrogen migration involved in the formation of $CH_2OH^+$ is due to tunneling through an adiabatic barrier and it was suggested that m/z 34 from $CD_3ONO^{+\bullet}$ would not be $CD_2OD^+$ but $D_2NO^+$. However, subsequent experiments, inspired by this suggestion, showed that the m/z 34 ions do have the $CD_2OD^+$ structure [219]. In an exhaustive theoretical study Leyh-Nihant and Lorquet [224] explain the peculiarities associated with the dissociation of ionized methylnitrite by a *statistical*, non-adiabatic model. Ab initio calculations as well as multipolar expansions reveal that the diabatic energy surfaces which correlate to each dissociation asymptote cross at a large value (c. 8 Å) of the reaction coordinate. Stretching of the ON bond leads to a situation where the $NO^+$ ion interacts with the dipole moment of the $CH_3O^\bullet$ radical. The internal energy of 0.5 eV [222] (0.34 eV [221]) allows some distortion of the incipient $CH_3O^\bullet$ radical; at a separation of 8 Å between the components of the complex electron transfer takes effect, leading to $CH_3O^+$ ions born with a deformed geometry, which immediately rearrange to the $CH_2OH^+$ structure as predicted by earlier ab initio calculations :

$$H_3C-O-N=\overset{+\bullet}{O} \longrightarrow H-\overset{H}{\underset{H}{C}}-\overset{\bullet}{O}----\overset{\bullet}{N}=O$$

$$\overset{+}{C}H_2OH + NO^\bullet \longleftarrow H-\overset{H}{\underset{H}{C}}-\overset{+}{\overset{\bullet}{O}}----\overset{\bullet}{N}=O$$

"Metastability" results from the weakness of the non-adiabatic interaction taking place at such large internuclear distances. The authors note that the peculiarities of the system can be accounted for by statistical theory.

Lifshitz et al. [225] have concluded that prior to decomposition, ionized nitromethane, $CH_3NO_2^{+\bullet}$, rearranges to both aci-nitromethane, $CH_2=N(O)OH^{+\bullet}$ and methylnitrite and that the associated barriers are of (near) equal height. It was tentatively proposed that an ion-dipole complex $CH_3^+\cdots NO_2$ is formed, through which both isomerizations may take place and that prior to dissociation into $CH_2NO_2^+ + H^\bullet$ intramolecular charge transfer takes effect. If so, the resemblance with the mechanism of Leyh-Nihant and Lorquet for methylnitrite is striking. Note that such intramolecular charge transfer must also take effect in the [ionized enol/neutral olefin] → [neutral keto/charged olefin] transformation [162] discussed above.

### Remote charge site fragmentations

Gross and co-workers [226,227] have continued their investigations on an unusual and exciting class of reactions of even-electron ions, called remote charge site fragmentations. Such reactions are of great analytical importance as they give rise to spectra which clearly show the location of structural features and the length of the hydrocarbon chain in functionalized alkanes. Strictly speaking, the word remote refers to through-bond and not necessarily to through-space distances. In long chain functionalized hydrocarbons charge remote fragmentations, if they occur, lead to losses of the elements $C_nH_{2n+2}$ (and for unsaturated chains also by loss of alkenes), whereas other, better known, processes lead to $C_nH_{2n+1}^+$ and $C_nH_{2n-1}^+$. For saturated compounds containing no functional group(s) both type of reactions yield smooth patterns in CID spectra, but when a structural feature is present - e.g. a double bond - this smooth distribution is disrupted *only* for the charge remote fragmentations allowing the location of the structural feature. A strongly localized charge is required for the initiation of remote fragmentations [227]. Protonated 9-octadecenol, $CH_3$-$(CH_2)_7$-CH=CH-$(CH_2)_7$-$CH_2OH_2^+$ dissociates uniformly to the $C_nH_{2n-1}^+$ series but with no indication as to the location of the double bond ( these ions presumably arise from [M + H - $H_2O$] [227]). However, $CH_3$-$(CH_2)_7$-CH=CH-$(CH_2)_7$-$CH_2OHLi^+$, the lithiated molecular ion, dissociates by charge remote fragmentations i.e. by losses of the elements of alkanes (high-mass ions) and by losses of alkenes (low-mass ions) with a clear gap at the cross-over point , i.e. at the position of the double bond. Cooks and co-workers [228] have concluded that the internal energy required to display remote site fragmentation is compound dependent, and that the energy required is not low. The mechanisms of such reactions are not well understood and considering the analytical importance of these processes further research is highly desired. MS/MS/MS experiments could give valuable information as to the mechanisms involved: thus we propose that structure analysis of the [M + Li - $C_nH_{2n+2}]^+$ ions from saturated functionalized alkanes will give the location of the double bond in the daughter ion.

**Acknowledgement.**
Contributions in the preparation of this manuscript from Dr W. Heerma, Prof. J.L. Holmes, Prof. J.C. Lorquet, Dr P.J.A. Ruttink and Prof. H. Schwarz are gratefully acknowledged.

**REFERENCES**

1. C. Lifshitz, *Int. Rev. Phys. Chem.*, 1987, **6**, 35.
2. J.H. Futrell (ed.), "Gaseous Ion Chemistry and Mass Spectrometry", Wiley, New York , 1986.
3. S. Hammerum, *Mass Spectrom. Rev.*, 1988, **7**, 123.
4. G. Bouchoux, *Mass Spectrom. Rev.*, 1988, **7** ,1; ibid., 203 (part II)
5. D.J. McAdoo, *Mass Spectrom. Rev.*, 1988, **7**, 363.
6. W. Koch and H. Schwarz, in "Structure/Reactivity and Thermochemistry of Ions", eds. P. Ausloos and S.G. Lias, D. Reidel, Holland, 1987, p. 413.
7. W.J.Hehre, L.Radom, P.v.R.Schleyer and J.A.Pople, Ab Initio Molecular Orbital Theory, Modern Quantum Chemistry. Wiley-Interscience, New York, 1986.
8. (a) A.L. Burlingame, T.A. Baillie and P.J. Derrick, *Anal. Chem.*, 1986, **58**, 165R; (b) A.L. Burlingame, D. Maltby, D.H. Russell and P.T. Holland, *Anal. Chem.*, 1988, **60**, 294R.
9. C. Wesdemiotis and F.W. McLafferty, *Chem. Rev.*,1987, **87**, 485.
10. J.K. Terlouw and H. Schwarz, *Angew. Chem.,Int. Ed. Engl.*, 1987, **26**, 805 ; ibid. *Angew. Chem.*, 1987, **99**, 799.
11. S.G. Lias, J.E. Bartmess, J.F. Liebman, J.L. Holmes, R.D. Levin and W.G. Mallard, *J. Phys. Chem. Ref. Data*, 1988, **17**, supplement 1.

12. J.B. Pedley, R.D. Naylor and S.P. Kirby, Thermochemical Data of Organic Compounds, 2nd. ed., Chapman and Hall, London, 1986.
13. F. Tureček, Collect. Czech. Commun., 1987, **52**, 1928.
14. I.Powis, Acc. Chem. Res., 1987, **20**, 179.
15. J.B. Westmore and M.M. Alauddin, Mass Spectrom. Rev., 1986, **5**, 345.
16. C.N. McEwen, Mass Spectrom. Rev., 1986, **5**, 521.
17. P. Jena, B.K. Rao and S.N. Khanna, "Physics and Chemistry of Small Clusters", Plenum Press, New York, 1987.
18. S. Sugano, Y. Nishina and S. Ohnishi, "Microclusters", Springer-Verlag, Berlin, 1987.
19. T.D. Märk, Int. J. Mass Spectrom. Ion Processes, 1987, **79**, 1.
20. J.E. Campana, Mass Spectrom. Rev., 1987, **6**, 395.
21. S.T. Graul and R.R. Squires, Mass Spectrom. Rev., 1988, **7**, 263.
22. W. Worthy, C & EN, 1987, December 7, 25.
23. H. Halim, N. Heinrich, W. Koch, J. Schmidt and G. Frenking, J. Comp. Chem., 1986, **7**, 93.
24. M.S. Dewar and K.M. Dieter, J. Am. Chem. Soc., 1986, **108**, 8075.
25. J.J. Dannenberg and L.K. Vinson, J. Phys. Chem., 1988, **92**, 5635.
26. D. Higgins, C. Thomson and W. Thiel, J. Comp. Chem., 1988, **9**, 702.
27. J.A. Pople and L.A. Curtiss, J. Phys. Chem., 1987, **91**, 155.
28. J.A. Pople and L.A. Curtiss, J. Phys. Chem., 1987, **91**, 3637.
29. D.J. de Frees and A.D. McLean, J. Comp. Chem., 1986, **7**, 321.
30. J.H.O.J. Wijenberg, J.H. van Lenthe, P.J.A. Ruttink, J.L. Holmes and P.C. Burgers, Int. J. Mass Spectrom. Ion Processes, 1987, **77**, 141.
31. M. Sana, G. Leroy, D. Peeters and C. Wilante, J. Mol. Structure (Theochem), 1988, **164**, 249.
32. Y. Apeloig and M. Karni, J. Chem. Soc., Perkin Trans. II, 1988, 625.
33. P. von R. Schleyer, Pure Appl. Chem., 1987, **59**, 1647.
34. B.F. Yates, W.J. Bouma and L. Radom, Tetrahedron, 1986, **22**, 6225.
35. B.F. Yates and L. Radom, Org. Mass Spectrom., 1987, **22**, 430.
36. B.F. Yates, W.J.Bouma and L. Radom, J. Am. Chem. Soc., 1987, **109**, 2250.
37. R.H. Nobes, W.J.Bouma, J.K.McLeod and L.Radom, Chem. Phys. Lett., 1987, **135**, 78.
38. P.M.W. Gill and L. Radom, Chem. Phys. Lett., 1986, **132**, 16.
39. R. Postma, P.J.A. Ruttink, F.B. van Duijneveldt, J.K. Terlouw and J.L. Holmes, Can. J. Chem., 1985, **63**, 2798.
40. M.L. McKee, J. Phys. Chem., 1986, **90**, 2335.
41. M.C. Blanchette, J.L. Holmes, C.E.C.A. Hop, F.P. Lossing, R. Postma, P.J.A. Ruttink and J.K. Terlouw, J. Am. Chem. Soc., 1986, **108**, 7589.
42. T.L. Bunn and T. Baer, J. Chem. Phys.,1986, **85**, 6361.
43. T.L. Bunn, A.M. Richards and T. Baer, J. Chem. Phys., 1986, **84**, 1424.
44. J-D. Shao,T. Baer,J.C. Morrow and M.L. Fraser-Monteiro, J. Chem. Phys., 1987, **87**, 5242.
45. T.L. Bunn and M.T. Bowers, J. Phys. Chem., 1988, **92**, 1813.
46. P.J.A. Ruttink, J. Phys. Chem., 1987, **91**, 703.
47. B.A. Rumpf and P.J. Derrick, Int. J. Mass Spectrom. Ion Processes, 1988, **82**, 239.
48. Z.V.I. Zaretskii, J.M. Curtis, A.G. Brenton and J.H. Beynon, Rapid Commun. Mass Spectrom., 1987, **1**, 45; id., ibid., 1987, **1**, 117.
49. Z.V.I. Zaretskii, J.M. Curtis, A.G. Brenton, J.H. Beynon and C. Djerassi, Org. Mass Spectrom.,1988, **23**, 453; id., ibid., 1988, **23**, 460.
50. G. Bouchoux, Y. Hoppilliard and P. Jaudon, Org. Mass Spectrom., 1987, **22**, 98.
51. G. Bouchoux, F. Djazi, Y. Hoppilliard, P. Jaudon and N. Nouts, Org. Mass Spectrom., 1988, **23**, 33.
52. J.M. Curtis, A.G. Brenton, J.H. Beynon and R.K. Boyd, Org. Mass Spectrom., 1987, **22**, 779.
53. P. Jonathan, A.G. Brenton and J.H. Beynon, Org. Mass Spectrom., 1988, **23**, 114.
54. Md. A. Mabud, T. Ast and R.G. Cooks, Org. Mass Spectrom., 1987, **22**, 418.
55. M.J. Hayward, Md. A. Mabud and R.G. Cooks, J. Am. Chem. Soc., 1988, **110**, 1343.
56. T. Ast, Md. A. Mabud and R.G. Cooks, Int. J. Mass Spectrom. Ion Processes, 1988, **82**, 131.

57. Md. A. Mabud, T. Ast, S. Verma, Y.-X. Jiang and R.G. Cooks, *J. Am. Chem. Soc.*, 1987, **109**, 7597.
58. M. Vincenti and R.G. Cooks, *Org. Mass Spectrom.*, 1988, **23**, 317.
59. M.T. Kinter and M.M. Bursey, *J. Am. Chem. Soc.*, 1986, **108**, 1797.
60. S. Okada, Y. Abe, S. Taniguchi and S. Yamabe, *J. Am. Chem. Soc.*,1987, **109**, 295
61. E.L. White, J.-C. Tabet and M.M. Bursey, *Org. Mass Spectrom.*, 1987, **22**, 132.
62. J.-P. Schmit, S. Beaudet and A. Brisson, *Org. Mass Spectrom.*, 1986, **21**, 493.
63. G. Bouchoux, F. Djazi, Y. Hoppilliard, R. Houriet and E. Rolli, *Org. Mass Spectrom.*, 1986, **21**, 209.
64. J.-P. Morizur, C. Monteiro, J. Tortajada and G. Bouchoux, *Org. Mass Spectrom.*, 1986, **21**, 774.
65. G. Bouchoux, F.Djazi, R. Houriet and E. Rolli, *J. Org. Chem.*, 1988, **53**, 3498.
66. K. Hiraoka, H. Takimoto, K. Morise, T. Shoda and S. Nakamura, *Bull. Chem. Soc. Jpn.*, 1986, **59**, 2247.
67. J.A. Hunter, C.A.F. Johnson, I.J.M. McGill, J.E. Parker and G.P. Smith, *J. Chem. Soc. Faraday Trans. 2*, 1987, **83**, 2025.
68. A. Kamar, A.B. Young and R.E. March, *Can. J. Chem.*, 1986, **64**, 1979.
69. J.C. Sheldon, G.J. Currie and J.H. Bowie, *Aust. J. Chem.*, 1986, **39**, 839.
70. J.E. Szulejko, J.J. Fisher, T.B. McMahon and J. Wronka, *Int. J. Mass Spectrom. Ion Processes*, 1988, **83**, 147.
71. A. Maquestiau, C. Jortay, D. Beugnies, R. Flammang, R. Houriet, E. Rolli and G. Bouchoux, *Int. J. Mass Spectrom. Ion Processes*, 1988, **82**, 33.
72. J.F. Garvey and R.B. Bernstein, *J. Am. Chem. Soc.*, 1986, **108**, 6096.
73. D. Farcasiu and R.G.Pancirov, *Int. J. Mass Spectrom. Ion Processes*, 1986, **74**, 207
74. R. Houriet, E. Rolli, R. Flammang, A. Maquestiau and G. Bouchoux, *Org. Mass Spectrom.*, 1987, **22**, 770.
75. J.N. Robinson and J.M. Tedder, *Org. Mass Spectrom.*, 1987, **22**, 154.
76. K. Stanney, J.M. Tedder and A.L. Mitchell, *J. Chem. Soc., Perkin Trans II*, 1986, 1383.
77. S. Suzuki and I. Koyano, *Int. J. Mass Spectrom. Ion. Processes*, 1987, **80**, 187.
78. R.S. Mason, M.T. Fernandez and K.R. Jennings, *J. Chem. Soc. Faraday Trans. II*, 1987, 89.
79. M.T. Fernandez, K.R. Jennings and R.S. Mason, *ibid.*, 1987, 159.
80. R.S. Mason, D. Milton and F.M. Harris, *J. Chem. Soc. Chem. Commun.*,1987, 1453.
81. M. Attina, F. Cacace and M. Yanez, *J. Am. Chem. Soc.*, 1987, **109**, 5092.
82. M. Bensimon and R. Houriet, *Int. J. Mass Spectrom. Ion Processes*, 1986, **72**, 93.
83. P. Pfaendler, G. Bodenhausen, J. Rapin, R. Houriet and T. Gäumann, *Chem. Phys. Lett.*, 1987, **138**, 195.
84. T. Drewello, C.B. Lebrilla,H. Schwarz, L.J. de Koning, R.H. Fokkens, N.M.M. Nibbering, E. Anklam and K.D. Asmus, *J. Chem. Soc. Chem. Commun.*, 1987, 1381.
85. C.A. Deakyne, M.Meot-Ner,T.J.Buckley and R.Metz, J. Chem. Phys.,1987, **86**, 2334.
86. M. Meot-Ner, Z. Karpas and C.A. Deakyne, *J. Am. Chem. Soc.*, 1986, **108**, 3913.
87. J.S. Knight, C.G. Freeman and M.J. McEwan, *J. Am. Chem. Soc.*, 1986, **108**, 1404.
88. M. Meot-Ner, J.F. Liebman and J.E. Del Bene, *J. Org. Chem.*, 1986, **51**, 1105.
89. M. Meot-Ner and Z. Karpas, *J. Phys. Chem.*, 1986, **90**, 2206.
90. S. Dheandhanoo, L. Forte, A. Fox and D.K. Bohme, *Can. J. Chem.*, 1986, **64**, 641.
91. S. Fornarini, R. Gabrielli and M. Speranza, *Int. J. Mass Spectrom. Ion Processes*, 1986, **72**, 137.
92. F. Ozturk, G. Baykut, M. Moini and J.R. Eyler, *J. Phys. Chem.*, 1987, **91**, 4360.
93. W.D. Reents, H.D. Roth, M.L. Schilling and C.J. Abelt, *Int. J. Mass Spectrom. Ion Processes*, 1986, **72**, 155.
94. W.J. van der Hart, L.J. de Koning, N.M.M. Nibbering and M.L. Gross, *Int. J. Mass Spectrom. Ion Processes*, 1986, **72**, 137.
95. R.W. Holman, M.D. Rozeboom, M.L. Gross and W.D. Warner, *Tetrahedron*, 1986, **42**, 6235.
96. C.G. de Koster, J.J. van Houte, J.B. Shadid and J. van Thuijl, *Rapid Commun. Mass Spectrom.*, 1988, **2**, 97.
97. J.C. Tabet, *Tetrahedron*, 1987, **43**, 3413.
98. S.F. Nelsen, D.T. Rumack and M. Meot-Ner, *J. Am. Chem. Soc.*, 1987, **109**, 1373.
99. J.M. van Doren, S.E. Barlow, C.H. DePuy, V.M. Bierbaum, I. Dotan and E.E. Ferguson, *J. Phys. Chem.*, 1986, **90**, 2772.

100. M. Meot-Ner, *J. Am. Chem. Soc.*, 1986, **108**, 6189.
101. T.B. McMahon and P. Kebarle, *J. Am. Chem. Soc.*, 1986, **108**, 6502.
102. A.B. Raksit, D.M. Hudgins, S. Buchau and R.F. Porter, *Rapid Commun. Mass Spectrom.*, 1987, **1**, 57.
103. A.B. Raksit and R.F. Porter, *Org. Mass Spectrom.*, 1987, **22**, 410.
104. D.M. Hudgins, A.B. Raksit and R.F. Porter, *Org. Mass Spectrom.*, 1988, **23**, 375.
105. R.E. March and A.B. Young, *Int. J. Mass Spectrom. Ion Processes*, 1987, **76**, 11.
106. R.E. March and J.G. McMillan, *ibid.*, 1988, **85**, 91.
107. G.I. Gellene and R.F. Porter, *Acc. Chem. Res.*, 1983,**16**, 200.
108. S.F. Selgren and G.I. Gellene, *Anal. Instrum. (N.Y.)*, 1988, **17**, 113.
109. S.F. Selgren and G.I. Gellene, *Chem. Phys. Lett.*, 1987, **141**, 508.
110. S.F. Selgren, D.E. Hipp and G.I. Gellene, *J. Chem. Phys.*, 1988, **88**, 3116.
111. W. Ketterle and H. Walther, *Chem. Phys. Lett.*, 1988, **146**, 180.
112. A.B. Raksit and R.F. Porter, *J. Chem. Soc. Chem. Commun.*, 1987, 500.
113. A.B. Raksit and R.F. Porter, *Int. J. Mass Spectrom. Ion Processes*, 1987, **76**, 299.
114. (a) G.I. Gellene and R.F. Porter, *J. Chem. Phys.*, 1988, **88**, 5894; (b) J.F. Garvey and A. Kuppermann, *ibid.*, 1988, **88**, 5985.
115. S.F. Selgren and G.I. Gellene, *J. Chem. Phys.*, 1987, **87**, 5804.
116. W.J. Griffiths, F.M. Harris. A.G. Brenton and J.H. Harris, *Int. J. Mass Spectrom. Ion Processes*, 1986, **74**, 317.
117. W.J. Griffiths, F.M. Harris and J.H. Beynon, *ibid.*, 1987, **77**, 233.
118. R.E. March and A.B. Young, *Int. J. Mass Spectrom. Ion Processes*, 1988, **85**, 237.
119. J. Bordas-Nagy, J.L. Holmes and C.E.C.A. Hop, *ibid.*, 1988, **85**, 243.
120. J.L. Holmes, *Adv. Mass Spectrom.*, 1989, **11**, 53.
121. R.Feng, C.Wesdemiotis, M.A.Baldwin and F.W. McLafferty, *Int. J. Mass Spectrom. Ion Processes*, 1988, **86**, 95.
122. J.L. Holmes, A.A. Mommers, J.K. Terlouw and C.E.C.A. Hop, *Int. J. Mass Spectrom. Ion Processes*, 1986, **68**, 249.
123. A.G. Harrison, R.S. Mercer, E.J. Reiner, A.B. Young, R.K. Boyd, R.E. March and C.J. Porter, *Int. J. Mass Spectrom. Ion Processes*, 1986, **74**, 13.
124. M.C. Blanchette, J. Bordas-Nagy, J.L. Holmes, C.E.C.A. Hop, A.A. Mommers and J.K. Terlouw, *Org. Mass Spectrom.*, 1988, **23**, 804.
125. B. Sedgewick, P.R. Nelson, P.A. Steiner IV and T.F. Moran, *Org. Mass Spectrom.*, 1988, **23**, 256.
126. G.C. Shields, P.A. Steiner IV, P.R. Nelson, M.C. Trauner and T.F. Moran, *Org. Mass Spectrom.*, 1987, **22**, 64.
127. J.B. Sedgewick, P.R. Nelson, C.A. Jordan, L.E. Abbey, Y. Xu and T.F. Moran, *Chem. Phys. Lett.*, 1988, **146**, 113.
128. J.B. Sedgewick, B.P. Paulson, C.G. Shields and T.F. Moran, *Int. J. Mass Spectrom. Ion Processes*, 1987, **79**, 127.
129. J.K. Terlouw, *Adv. Mass Spectrom.*, 1989, **11**, 984.
130. C. Wesdemiotis, R. Feng, E.R. Williams and F.W. McLafferty, *Org. Mass Spectrom.*,1986, **21**, 689.
131. S. Villeneuve and P.C. Burgers, *Org. Mass Spectrom.*, 1986, **21**, 733.
132. C.E.C.A. Hop, J. Bordas-Nagy, J.L. Holmes and J.K. Terlouw, *ibid.*., 1988, **23**, 155.
133. M.C. Blanchette, J.L. Holmes and F.P. Lossing,*ibid.*, 1987, **22**, 701.
134. C.E.C.A. Hop, J.L. Holmes, F.P. Lossing and J.K. Terlouw, *Int. J. Mass Spectrom. Ion Processes*, 1988, **83**, 285.
135. C.E.C.A. Hop and J.L. Holmes, *ibid.*, 1988, **84**, 331.
136. C. Wesdemiotis, R. Feng, M.A. Baldwin and F.W. McLafferty, *Org. Mass Spectrom.*, 1988, **23**, 166.
137. C.B. Lebrilla, T. Drewello and H. Schwarz, *Organometallics*, 1987, **6**, 2268.
138. Ch. Lifshitz, P.J.A. Ruttink, G. Schaftenaar and J.K. Terlouw, *Rapid Commun. Mass Spectrom.*, 1987, **1**, 61.
139. H. Egsgaard, L. Carlsen, H. Florencio, T. Drewello and H. Schwarz, *Chem. Phys. Lett.*, 1988, **148**, 537.
140. R. Feng, C. Wesdemiotis and F.W. Mclafferty, *J. Am. Chem. Soc.*, 1987, **109**, 6521.
141. D. Sülzle and H. Schwarz, *Angew. Chem.*, 1988, **100**, 1384.
142. B.L. M. van Baar, T. Weiske, J.K. Terlouw and H. Schwarz, *Angew. Chem.*, 1986, **98**, 275; ibid., *Angew. Chem. Int. Ed. Engl.*, 1986, **25**, 282.

143. B.L.M. van Baar, N. Heinrich, W. Koch, R. Postma, J.K. Terlouw and H. Schwarz, *Angew. Chem.*, 1987, **99**, 153; ibid. *Angew. Chem. Int. Ed. Engl.*, 1987, **26**, 40.
144. J. Buschek and J.L. Holmes, *Org. Mass Spectrom.*, 1988, **23**, 765.
145. C. Wesdemiotis and R. Feng, *Org. Mass Spectrom.*, 1988, **23**, 416.
146. D. Sülzle, M. Verhoeven, J.K. Terlouw and H. Schwarz, *Angew. Chem.*, 1988, **100**, 1591.
147. W. Heerma, W. Kulik, P.C. Burgers and J.K. Terlouw, *Int. J. Mass Spectrom. Ion Processes*, 1988, **84**, R1.
148. B.L.M. van Baar, P.C. Burgers, J.L. Holmes and J.K. Terlouw, *Org. Mass Spectrom.*, 1988, **23**, 355.
149. J.L. Holmes, C.E.C.A. Hop and J.K. Terlouw, *Org. Mass Spectrom.*, 1986, **21**, 777.
150. R.S. Mercer and A.G. Harrison, *Org. Mass Spectrom.*, 1987, **22**, 710.
151. P.C. Burgers, J.L. Holmes, C.E.C.A. Hop and J.K. Terlouw, *Org. Mass Spectrom.*, 1986, **21**, 549.
152. N. Heinrich, J. Schmidt, H. Schwarz and Y. Apeloig, *J. Am. Chem. Soc.*, 1987, **109**, 1317.
153. R. Caballol, J. Igual, J.M. Poblet and J.P. Sarasa, *Int. J. Mass Spectrom. Ion Processes*, 1986, **71**, 75.
154. G. Bouchoux, Y. Hoppilliard and P. Longevialle, *Rapid Commun. Mass Spectrom.*, 1987, **1**, 94.
155. L. Freiser-Monteiro, M.L. Freiser-Monteiro, J.L. Butler and T. Baer, *J. Phys. Chem.*, 1982, **86**, 752.
156. C. Wentrup, H. Briehl, P. Lorencak, U.J. Vogelbacher, H.W. Winter, A. Maquestiau and R. Flammang, *J. Am. Chem. Soc.*, 1988, **110**, 1337.
157. C. Wentrup, P. Lorenčak, A. Maquestiau and R. Flammang, *Chem. Phys. Lett.*, 1987, **137**, 241.
158. F. Tureček, *Tetrahedron Lett.*, 1986, **27**, 4219.
159. F. Tureček, Z. Havlas, F. Maquin, N. Hill and T. Gäumann, *J. Org. Chem.*, 1986, **51**, 4061.
160. F. Tureček, Z. Havlas, F. Maquin and T. Gäumann, *Helv. Chim. Acta*, 1986, **69**, 683.
161. M. Rodler, *Chem. Phys.*, 1986, **105**, 345.
162. B.L.M. van Baar, J.K. Terlouw, S. Akkök, W. Zummack and H. Schwarz, *Int. J. Mass Spectrom. Ion Processes*, 1987, **81**, 217.
163. X.-Z. Qin and F. Williams, *J. Am. Chem. Soc.*, 1987, **109**, 595.
164. B.F. Yates, W.J. Bouma, J.K. McLeod and L. Radom, *J. Chem. Soc. Chem. Commun.*, 1987, 204.
165. P.C. Burgers, J.L. Holmes, C.E.C.A. Hop, R. Postma, P.J.A. Ruttink and J.K. Terlouw, *J. Am. Chem. Soc.*, 1987, **109**, 7315.
166. R. Postma, S.P. van Helden, J.H. van Lenthe, P.J.A. Ruttink, J.K. Terlouw and J.L. Holmes, *Org. Mass Spectrom.*, 1988, **23**, 503.
167. R. Postma, P.J.A. Ruttink, B.L.M. van Baar, J.K. Terlouw, J.L. Holmes and P.C. Burgers, *Chem. Phys. Lett.*, 1986, **123**, 409.
168. R. Postma, P.J.A. Ruttink, J.K. Terlouw and J.L. Holmes, *J. Chem. Soc. Chem. Commun.*, 1986, 683.
169. A.-M. Dommröse and H.-Fr. Grützmacher, *Int. J. Mass Spectrom. Ion Processes*, 1987, **76**, 95.
170. A.-M. Dommröse and H.-Fr. Grützmacher, *Org. Mass Spectrom.*, 1987, **22**, 437.
171. R. Wolf, A.-M. Dommröse and H.-Fr. Grützmacher, *ibid.*, 1988, **23**, 26.
172. W. Koch, B. Liu, T. Weiske, C.B. Lebrilla, T. Drewello and H. Schwarz, *Chem. Phys. Lett.*, 1987, **142**, 147.
173. M.L. Langford, D. Mathur and F.M. Harris, *Rapid Commun. Mass Spectrom.*, 1988, **2**, 167.
174. F. Maquin and D. Stahl, *Chem. Phys. Lett.*, 1988, **145**, 447.
175. P.M.W. Gill and L. Radom, *Chem. Phys. Lett.*, 1987, **136**, 294.
176. L. Radom, P.M.W. Gill, M.W. Wong and R.H. Nobes, *Pure & Appl. Chem.*, 1988, **60**, 183.
177. M.W. Wong, R.H. Nobes and L. Radom, *J. Chem. Soc. Chem. Commun.*, 1987, 233.
178. L. Radom, P.M.W. Gill and M.W. Wong in "The Structure of Small Molecules and Ions", Eds. R. Naaman and Z. Vager, Plenum, New York, 1989, in press.
179. P.M.W. Gill and L. Radom, *J. Am. Chem. Soc.*, 1988, **110**, 5311.
180. B.F. Yates, W.J. Bouma and L. Radom, *J. Am. Chem. Soc.*, 1986, **108**, 6545.

181. G. Bouchoux and J. Tortajada, *Rapid Commun. Mass Spectrom.*, 1987, **1**, 86.
182. G. Bouchoux, F. Bidault, F. Djazi, B. Nicod and J. Tortajada, *Org. Mass Spectrom.*, 1987, **22**, 748.
183. G. Bouchoux, J. Tortajada, J. Dagaut and J. Fillaux, *ibid.*, 1987, **22**, 451.
184. G. Bouchoux, Y. Hoppilliard and J. Tortajada, *Int. J. Mass Spectrom. Ion Processes*, in press.
185. M. Masur, A. Sprafke and H.-Fr. Grützmacher, *Org. Mass Spectrom.*, 1987, **22**, 307.
186. D.J. McAdoo and C.E. Hudson, *Int. J. Mass Spectrom. Ion Processes*, 1986, **70**, 57.
187. C.E. Hudson, T. Lin and D.J. McAdoo, *Org. Mass Spectrom.*, 1987, **22**, 311.
188. D.J. McAdoo and C.E. Hudson, *Org. Mass Spectrom.*, 1987, **22**, 615.
189. D.J. McAdoo, C.E. Hudson, M. Skyiepal, E. Broido and L.L. Griffin, *J. Am. Chem. Soc.*, 1987, **109**, 7648.
190. D.J. McAdoo, *Org. Mass Spectrom.*, 1988, **23**, 350.
191. T. Weiske and H. Schwarz, *Tetrahedron*, 1986, **42**, 6245.
192. H. Audier and G. Sozzi, *Nouv. J. Chim.*, 1986, **10**, 579.
193. H.E. Audier and G. Sozzi, *Rapid Commun. Mass Spectrom.*, 1987, **1**, 25.
194. T. Ha, C. Radloff and M.T. Nguyen, *J. Phys. Chem.*, 1986, **90**, 2991.
195. D.J. McAdoo, J.C. Traeger and C.E. Hudson, *Org. Mass Spectrom.*, 1988, **23**, 760.
196. E. Kluft, N.M.M. Nibbering, M.B. Stringer, P.C.H. Eichinger and J.H. Bowie, *Int. J. Mass Spectrom. Ion Processes*, 1988, **85**, 215.
197. R.J. Liedtke and C. Djerassi, *J. Am. Chem. Soc.*, 1969, **91**, 6814.
198. J.H. Beynon, R. Flammang, E.E. Kingston and A. Maquestiau, *Bull. Soc. Chim. Belg.*, 1986, **95**, 1099.
199. G. Bouchoux, *Adv. Mass Spectrom.*, 1989, **11**, 812.
200. H.E. Audier and G. Sozzi, *Org. Mass Spectrom.*, 1984, **19**, 150.
201. T. Bjørnholm, S. Hammerum and D. Kuck, *J.Am. Chem. Soc.*, 1988, **110**, 3862.
202. S. Hammerum and P.J. Derrick, *J. Chem. Soc. Perkin Trans II*, 1986, 1577.
203. R.D. Bowen and D.H. Williams, *J. Chem. Soc. Chem. Commun.*, 1981, 836.
204. T.H. Morton, *Tetrahedron*, 1982, **38**, 3195.
205. G. Sozzi, H.E. Audier, P. Mourgues and A. Millet, *Org. Mass Spectrom.*, 1987, **22**, 746.
206. R.W. Kondrat and T.H. Morton, *Org. Mass Spectrom.*, 1988, **23**, 555.
207. S. Hammerum, *J. Chem. Soc. Chem. Commun.*, 1988, 858.
208. S. Hammerum and H.E. Audier, *J. Chem. Soc. Chem. Commun.*, 1988, 860.
209. T. Weiske, S. Akkök and H. Schwarz, *Int. J. Mass Spectrom. Ion Processes*, 1987, **76**, 117.
210. U. Filges and H.-Fr. Grützmacher, *Org. Mass Spectrom.*, 1986, **21**, 673.
211. U. Filges and H.-Fr. Grützmacher, *Org. Mass Spectrom.*, 1987, **22**, 444.
212. H.-Fr. Grützmacher and M. Masur, *Org. Mass Spectrom.*, 1988, **23**, 223.
213. M. Masur, H.-Fr. Grützmacher, H. Münster and H. Budzikiewicz, *Org. Mass Spectrom.*, 1987, **22**, 493.
214. N. Heinrich and H. Schwarz, *Int. J. Mass Spectrom. Ion Processes*, 1987, **79**, 295.
215. N. Heinrich in "Structure/Reactivity and Thermochemistry of Ions, eds. P. Ausloos and S.G. Lias, D. Reidel, Holland, p.271.
216. T. Drewello, N. Heinrich, W.P.M. Maas, N.M.M. Nibbering, T. Weiske and H. Schwarz, *J. Am. Chem. Soc.*, 1987, **109**, 4810.
217. P.J.A. Ruttink in "The Structure of Small Molecules and Ions", Eds. R. Naaman and Z. Vager, Plenum, New York, 1989.
218. P. Du, D.A. Hrovat and W. T. Borden, *J. Am. Chem. Soc.*, 1988, **110**, 3405.
219. H. Egsgaard and L. Carlsen, *Chem. Phys. Lett.*, 1988, **147**, 30.
220. M.P. Irion, A. Selinger, A.W. Castleman Jr, E.E. Ferguson and K.G. Weil, *Chem. Phys. Lett.*, 1988, **147**, 33.
221. H. Egsgaard, L. Carlsen and S. Ebel, *Ber. Bunsenges. Phys. Chem.*, 1986, **90**, 369
222. T. Baer and J.R. Hass, *J. Phys. Chem.*, 1986, **90**, 451.
223. E.E. Ferguson, *Chem. Phys. Lett.*, 1987, **138**, 450.
224. B. Leyh-Nihant and J.C. Lorquet, *J. Chem. Phys.*, 1988, **88**, 5606.
225. C. Lifshitz, M. Rejwan, I. Levin and T. Peres, *Int. J. Mass Spectrom. Ion Processes*, 1988, **84**, 271.
226. J. Adams, L.J. Deterding and M.J. Gross, *Spectros. Int. J.*, 1987, **5**, 199.
227. J. Adams and M.J. Gross, *J. Am. Chem. Soc.*, 1986, **108**, 6915.
228. V.H. Wysocki, M.E. Bier and R.G. Cooks, *Org. Mass Spectrom.*, 1988, **23**, 627.

# 3
# Developments and Trends in Instrumentation

BY F. A. MELLON

## 1 Introduction

The tremendous stimulus given to mass spectrometry by the introduction of fast-atom bombardment (FAB) in 1981 has had effects on instrumental development beyond the immediate ones of engendering improvements in the mass range and detection capabilities of magnetic sector mass spectrometers. These secondary consequences include increased interest in all forms of soft-ionization (e.g. lasers, plasma desorption) and in alternative designs of high mass analyser (e.g. Time-of-flight, Fourier-transform Ion Cyclotron Resonance, Wien filter). A further effect has been to boost the development of tandem mass spectrometers, which may be used to induce diagnostic fragmentation in the energy-deficient molecular ions produced by FAB and other soft-ionization techniques.

Whilst much recent progress in the development of mass spectrometry instrumentation may sometimes appear to be "technology led", the appearance of soft-ionization, high mass methodology has coincided with significant advances in biology and biotechnology (particularly in protein engineering). These biological advances are capable of exploiting fully the improved mass-spectrometric methods and are often stretching even the most sophisticated instruments up to, and even beyond, their limits. The interaction between the disciplines of physics, chemistry and biology has undoubtedly been synergistic and is expected to yield even more dividends in the future. Instrumental developments have not been confined solely to advances in high mass performance. Improvements in sensitivity and detection limits are very apparent in the performance figures given by most commercial manufacturers. This is another factor which has stimulated the development and general acceptance of tandem mass spectrometers. Increased sensitivity generates more intense mass spectra, not only from the analyte(s) but also from interferents, encouraging the use of MS/MS techniques to recover analytical selectivity. Selectivity has also been improved, in a growing number of examples, by elegant use of laser techniques such as resonance-enhanced multiphoton ionization.

"Bench top" mass spectrometers are dominating the commercial market in terms of volume sales. Five instrument manufacturers now produce (or are about to produce) at least one type of bench-top GC/MS system.[1] The performance of these instruments is often very impressive; many outstrip the low resolution detection limits and data handling capabilities of 5-10 year old magnetic sector mass spectrometers.

Interest in the combination of mass spectrometry with other spectroscopic techniques is on the increase. Instruments which combine chromatography, mass spectrometry and infrared spectroscopy are now commercially available and there have been attempts to incorporate other forms of spectroscopy into research instruments, with varying degrees of success.

## 2 Ionization methods

2.1 **Lasers** - Although the first observation of two-photon absorption took place in 1953[2] and multiphoton ionization(MPI) was first demonstrated in 1964[3], it was not until the late 1970s that MPI and mass spectrometry were first combined.[4,5] Since then, interest in the use of lasers in mass spectrometry has increased significantly. This interest arises from several different properties which make lasers attractive devices to the mass spectrometrist: their ability to desorb atoms and molecules from surfaces, their ionizing properties, their very high sensitivity and selectivity, and their potential as sophisticated probes into the electronic structure of ions and ionic fragments. The high ionization efficiency achievable with lasers has been of particular interest to mass spectrometrists. This has yet to be translated into very high sensitivities in routine mass spectrometric analysis, for reasons which will be discussed later.

A useful review of analytical methodology in laser mass spectrometry is provided by an overview of techniques, instruments and applications.[6] MPI mass spectrometry has been reviewed extensively[7,8] and a brief, but nevertheless very useful, survey of inorganic applications of Resonance

Ionization Mass Spectrometry (RIMS) has also appeared.[9]

An interesting recent development in MPI mass spectrometry has been the construction of "Double-barrelled" instruments in which desorption and ionization steps are separated.[10-13] Both systems described in these reports employ a $CO_2$ laser for desorbing solid samples from a target. The vaporized neutral molecules are cooled in a supersonic molecular beam, ionized by resonance-enhanced multiphoton ionization (REMPI) using a tunable dye laser, and mass analysed by either linear [10,11], or reflectron [12] time-of-flight (TOF). Mass resolutions of the order of 500 [10,11] and 10,000[12] at full width half maximum (FWHM) were obtained. Studies of small, thermally labile biomolecules, including tyrosine and its analogues[11,20] were reported using the lower resolution instrument. The reflectron mass spectrometer yielded impressive data on several biomolecules,[12-18] including, significantly, bovine insulin (molecular weight 5729 Da).[19] The potential of these commercially available designs[21,22] of laser mass spectrometer is still being explored, however a number of analytical advantages are already apparent: mass spectra can be obtained from crude biological extracts, molecular ions can be produced from labile molecules, and the degree of fragmentation can be controlled by varying the power of the ionizing laser. The major disadvantage of the technique appears to be the reliance on REMPI; molecules which do not possess a resonance state cannot be efficiently ionized. This may exclude some biomolecules from investigation, unless some form of "chromophoric derivatization" is employed. It also appears that full advantage of the high ionization efficiency capabilities of lasers has yet to be realized in these instruments. For maximum ionization efficiency, both temporal and spatial overlap of the sample plume with the ionizing pulse must be maximised. This does not appear to have been achieved in the double-barrelled instruments (partly because ionization pulses must be very short in order to maintain TOF resolution), nevertheless ionization yields appear to be significantly higher than those obtained by electron ionization. Two-step laser desorption REMPI/TOF mass spectrometry has also been used to investigate PTH-amino acids introduced without jet cooling.[23]

Double-barrelled laser mass spectrometric techniques have been applied to the analysis of thin films.[24] The method, dubbed post-ablation ionization (PAI), entails the addition of a second laser to a commercial reflection mode laser microprobe, and does not employ beam cooling (unnecessary for elemental analysis). PAI yielded enhanced detection limits for elements difficult to analyse by conventional laser microprobe mass spectrometry, and simplified the spectra by reducing cluster ion formation. The method generated useful analytical data for diverse materials, including thin film dielectrics, semiconductors and metals.

Separation of desorption and ionization steps has also been accomplished by argon ion beam sputtering of neutral molecules into the gas phase, where they are ionized by a 10.6 eV UV laser, and detected by reflectron TOF mass spectrometry.[25] At this wavelength, ionization takes place by <u>single</u> photon absorption. This circumvents two of the most significant problems associated with multi-photon absorption, its non-uniform detection sensitivity and the uncontrolled fragmentation which often results when MPI is applied under conditions which maximize sensitivity. The combined sputtering/laser ionization method has been christened SALI (Surface Analysis by Laser Ionization). An overview of SALI provides a useful summary of its applications and potential.[26] A variation on this theme combines sputter atomization (using $Ar^+$, $N_2^+$ or $O_2^+$ ion beams) with RIMS by interfacing a commercial microprobe mass analyzer with a tunable dye laser.[27] The useful yield and sensitivity of laser MPI mass spectrometry of sputtered neutrals has been explored.[28] Three systems were compared: SALI (mentioned above), sputter-initiated resonance ionization spectroscopy (SIRIS), and surface analysis by resonance ionization of sputtered atoms (SARISA). High useful yields (atoms detected/atoms sputtered) and high sensitivity for the detection of minority neutral species were claimed for the SARISA instrument. A review of the analytical capabilities and development of inorganic RIMS contains much useful information regarding the types of ion source and mass analyser currently in use.[29] A complementary review gives an excellent account of techniques and instrumentation used in REMPI mass spectrometry of <u>molecules,</u> from small aromatics to labile biomolecules.[30]

A potentially ultrasensitive technique for analysing up to half the elements in the periodic table, Photon Burst Mass Spectrometry (PBMS), has been described.[31] The principles of PBMS derives from a property of single ions which possess an optically isolated pair of energy levels. Such ions can absorb and emit photons many hundreds of times in the short (micro- or milli-second) period in which they are induced to traverse a laser beam. Collection and recording of the "photon-burst" which occurs under these conditions allows (in principle) the detection of <u>single</u> ions. Ions subjected to PBMS can be mass-selected using either conventional magnetic sector or accelerator mass spectrometers. The technique is capable of generating very high levels of isobaric and isotopic selectivity and sensitivity; rare isotopes of relative concentration of $10^{-15}$ should be measurable. Photon burst and double resonance ionization mass spectrometry have been compared in an overview which assesses their applicability to trace isotope analysis.[32]

Most laser ionization techniques utilize pulsed lasers, however, continuous wave lasers also have their merits, as demonstrated by their success in producing efficient, isotopically selective ionization of barium in a quadrupole mass spectrometer.[33, 34] An isotopic selectivity of 800 was obtained between $^{137}Ba$ and $^{138}Ba$. Although continuous wave lasers produce lower ionization efficiencies, compared with pulsed lasers, this is more than compensated for by the improved duty cycle. Ionization efficiency was estimated to be only $10^{-4}$, however, this may be improvable to a theoretical value close to unity.

Other applications of RIMS in isotopic analysis include isotope selective photoionization and laser-induced fluorescence of Lu and Mg[35], separation of rare earth isotopes[36], and the detection of very small samples of refractive elements by RIMS of a pulsed thermal atomic beam formed by laser desorption.[37] The difficulties caused by the small spatial and temporal overlap of atomic plumes and laser beams in RIMS have been tackled in a novel way: atoms are confined inside a cylindrical interaction chamber during laser illumination.[38] Because approximately 100 wall collisions occur before any neutral atom is lost, the probability of ionization is increased greatly.

A spectrograph coupled with an Nd:YAG laser produces improved resolution and generates fewer multiply-charged ions compared with spectra produced by conventional spark source ionization.[39] The main advantage of the laser ionization method lies in its ability to ionize insulators without the necessity for extensive sample pre-treatment. This type of instrument, if developed further, might provide an interesting alternative to inductively coupled plasma and glow discharge mass spectrometry.

An excellent review of laser microprobe mass spectrometry contains descriptions of the ionization methods currently in use, or undergoing development.[40] Selected analytical applications are also presented.

The first demonstration of high resolution laser desorption mass spectrometry of high mass organics has been claimed.[41] Masses up to 9,700 Da were detected in the high mass region of PEG-8000 by a Fourier-transform mass spectrometer equipped with a 7-Tesla magnet. Resolution was mass dependent, ranging from 160,000 at m/z 3,200 to 60,000 at m/z 5,922.

An interesting variant on laser desorption techniques is provided by a UV laser-induced surface ionization source with prism internal reflection.[42] The prism causes the laser beam to reflect on the underside of a silvered surface upon which the analyte is deposited. This unusual arrangement produces a very sharp ionization pulse and prevents any gas-phase ionization (which degrades mass resolution in TOF mass spectrometers). Resolutions of 3,900 and 11,000 were produced in linear and reflectron TOF mass spectrometers respectively. A theoretical resolution of over 1,000,000 is possible with this design.

Matrices have been used to enhance mass spectra produced by ultraviolet laser desorption of non-volatile compounds.[43] Solid matrices of tryptophan and nicotinic acid, and liquid matrices of 3-nitrobenzyl alcohol and 2-nitrophenyl octylether, significantly improved desorption of non-volatile compounds, resulting in high yields of protonated and cationized molecular ions with little or no fragmentation. Protonated molecular ions were observed for molecules of mass up to 2845 Da (mellitin in a nicotinic acid matrix). Although outside the report period of this review, it should be noted that even more impressive data have been obtained since this work was published, with detection of masses approaching 200,000 Da (glucoseisomerase tetramer).[44] High mass detection by laser desorption TOF mass spectrometry of molecules of up to 34,472 Da molecular weight (carboxypeptidase A) has been reported by other workers.[45]

An optical fibre interface has been used as a simple and inexpensive means for conducting light from an Nd:YAG laser into the ion source of a triple quadrupole mass spectrometer.[46] Isobutane gas enhanced the ion currents produced from neutral molecules (including tetrapeptides) deposited in the ion volume. The main disadvantage of the method arises from the incompatibility of a pulsed laser with a conventional scanning mass spectrometer.

The importance of the time-lag parameter, the time allowed to elapse between ion formation and the ion draw-out pulse in laser-TOF mass spectrometry, has been investigated comprehensively.[47] Pronounced effects on fragmentation patterns resulted from variations in time-lag, emphasizing the importance of reporting this parameter when recording fragment ion abundances.

**2.2 Secondary Ion sources** - Developments in secondary ionization sources (ion and atom guns) have largely consisted of steady improvements and consolidation of existing techniques rather than any revolutionary advances.

Although the potential of raster-scanned microfocused neutral beams for analysing and imaging small areas of a sample has been demonstrated using a FAB-type source[48], the lack of fine-focusing and the high gas load imposed on the mass spectrometer vacuum system imposes some limitations. A novel design of neutral molecular beam SIMS gun has overcome many of these limitations by utilizing a unique neutralization mechanism: autoneutralization.[49] The gun employs $SF_6$ as the bombarding species and takes advantage of the sequence:

$$SF_6 + e^- \rightarrow SF_6^{*-} \xrightarrow{k} SF_6 + e^-$$

where the autoneutralizing step proceeds with a mean lifetime (1/k) of 10-20 µs. By accelerating and focusing $SF_6^-$ ions down a flight tube of suitable length, up to 30-40% neutralization efficiency was attained. Residual ions were removed by electrostatic deflection. Focusing down to a spot size of

0.5 mm was possible, with neutral fluxes of about 100 pA equivalent. The practicability of the device was demonstrated by the analysis of thin films of insulating polymers, with quadrupole mass analysis of the secondary ions produced. The device has potential in Fourier-transform mass spectrometry where the vacuum requirements and strong magnetic fields make both conventional FAB and ion guns impracticable.

Details of a patent describing a primary ion beam raster gating technique for SIMS have been disclosed.[50] A primary ion gun raster scans an area of the surface under test producing a microscopic crater. A "beam blanking" circuit then selectively generates a scan of a smaller area at the bottom of the primary crater. This technique prevents artefacts which can be produced in conventional SIMS instruments by deposition and re-sputtering of material on the side walls of the crater. Loss of spatial resolution in three-dimensional SIMS because of sputter deposition/re-ionization is a general problem, and has been discussed in some detail.[51] Theoretical and experimental studies showed that a trade-off between local resolution, analytical sensitivity and geometric feature height exists. Some unconventional possibilities for avoiding redeposition problems were suggested.

Liquid-metal ion sources possess some attractive properties as primary beam guns for SIMS, particularly their capacity for producing extremely high brightness ion beams. In view of their potential as ion sources, investigation of their operating characteristics is highly desirable and such studies have been carried out on a capillary-type liquid-metal ion source.[52] The mass and energy spectra of the most prominent ions produced by the gun ($In^+$, $In_2^+$, $In_2^+$ and $In_3^+$) revealed significant differences in the energy spectra of molecular species. These results have important consequences for the interpretation of peak height ratios in SIMS, because relative abundance of atomic species and molecular clusters ejected from the analyte are strongly energy-dependent.

Interfacing a SIMS source to Nier-Johnson double-focusing mass spectrometer produces a relatively low-cost instrument, suitable for characterizing electrode surfaces and involatile organic compounds.[53]

Sputtered neutral mass spectrometry (SNMS) is becoming increasingly important as a technique for elemental and isotopic analysis. The fundamentals and some selected applications of SNMS have been described in recent reviews.[54,55] Neutral atomic species sputtered into the gas phase can be ionized by a variety of methods, including Penning, hot electron gas, electron beam, and non-resonant multiphoton ionization (the last mentioned has been discussed above [24-28]). SNMS has several advantages over SIMS, including its relatively uniform sensitivity for all elements, absence of matrix effects, and good detection limits (ppm to sub-ppb).SNMS is operable in several different modes and charge compensation regimes[56], and can also be combined with SIMS in a single instrument.[57]

Modification of electron ionization (EI) sources fitted to FAB-equipped commercially-available mass spectrometers enables acquisition of spectra which have both EI and FAB characteristics.[58] In the design presented, the FAB gun, probe and target are incorporated in a single unit. The advantages of the combined EI/FAB source arise from its ability to ionize sputtered neutrals and reference compounds (by EI) for accurate mass determination of analytes ionized by FAB.

**2.3 Plasma Desorption** - Plasma desorption mass spectrometry (PDMS) has enjoyed a relatively slow growth since its introduction in 1974[59], until recently. A turning point was provided by the introduction of the first commercial $^{252}$Cf PDMS instrument in 1984[60], and since that time applications have increased more rapidly, if not spectacularly. The title and contents of a review article[61] "Plasma Desorption Mass Spectrometry: Coming of Age", accurately reflect the current status of this interesting technique.

$^{252}$Cf-PDMS was originally developed using apparatus optimized for the study of very thin samples (a few thousand Å). Thick samples of either solids or liquids require a slightly different approach: higher pumping speeds, and orientation of the sample in the horizontal rather than the vertical plane, are necessary. An instrument incorporating these features has been designed and tested on a number of solid and liquid samples, ranging in molecular weight from glycerol (92 Da) to bovine insulin (5733 Da)[62]. Molecular ion yields for samples presented in either solid or liquid forms (dissolved in glycerol) were very similar. Combined $^{252}$Cf ionization and Fourier-transform mass spectrometry is increasing in importance, as exemplified by a report of the mass spectra of compounds of up to 2,000 Da molecular weight.[63] Improvements in ion-trapping efficiency and sample presentation are necessary before higher masses can be successfully detected (this is discussed in more detail in section 3.3).

An unusual application of PDMS is exemplified by the surface analysis of glasses and plastics.[64] The low intensity of the bombarding flux obviates charging effects and the low escape depth (10Å) of desorbed species helps to maintain sample integrity. The feasibility of microscopic analysis with a collimated 84-MeV Kr$^{7+}$ beam was also evaluated and produced 11 µm spatial resolution in this preliminary study.

An interesting sample bombardment technique which, although non-mass spectrometric in the published application, may ultimately have potential as a sample vaporization/ionization method for desorbing large molecules, has been described.[65] Cluster ions containing 50-200 water molecules and one proton were accelerated to 275kV and used to bombard viruses dispersed on thin carbon films. Subsequent microscopic examination provided morphological evidence for disassembly of single virus particles (cylindrical segments of virus were removed).

**2.4 Rare gas Plasma Sources** - Following the first demonstration of inductively coupled plasma mass spectrometry (ICP/MS) in 1980[66], the technique has established itself very rapidly as a sensitive, versatile method for elemental analysis. In order to dispel any impression that ICP/MS is inherently difficult to operate for multi-element analyses, an investigation of the effect of several ion source operating parameters on overall performance has been undertaken.[67] When the major variables (plasma power, gas flows, and aperture load-coil spacing) were kept constant, an adequately uniform response was obtained on a range of elements of differing atomic masses and chemistries. Effects caused by variation in gas flow rate were examined and setting this parameter so as to maximise elemental response also minimised the effects of interfering species, such as polyatomic, oxide, and doubly charged ions. A similar exercise has been carried out to determine the optimum operating

3: *Developments and Trends in Instrumentation*

conditions for precise measurement of isotope ratios of Li, Fe, Cu, and Zn (enriched isotopes of which are used in human metabolic studies).[68] Stable measurements were obtained over a relatively wide range of RF powers and the optimum RF power for maximising the ion abundance of each element was also established. Optimum flows of argon gas, and of analyte solution (Cu only) were also determined. The study concluded that accurate stable isotope ratios were measurable by ICP/MS for all the elements studied (<1%RSD; Li>1.5%RSD), sufficient for routine application in many human metabolic studies.

An improved ion source/ion extraction system for ICP/MS has been disclosed.[69] A combination of improved vacuum system (three-stage) and ion extraction electrodes results in a simplified and compact instrument, compared to conventional designs. Significantly improved analytical sensitivity is claimed for this instrument.

Argon gas consumption represents the major running cost in ICP/MS, so any successful ion source design which reduces gas flow is to be welcomed. The high flow rates (ca. 14 L/min.) are necessary because the argon gas serves a dual purpose in conventional plasma torches, acting both as plasma gas and coolant. A water-cooled torch has allowed reduction of argon consumption to levels of 2-3 L/min.[70] The torch performed well when tested under a variety of analytical regimes, however, the necessity for further investigations into its operating characteristics was acknowledged.

A completely different approach to the reduction of rare gas consumption is illustrated by the development of microwave-induced plasma (MIP) sources.[71,72] Analyte ions are distributed non-homogeneously throughout the MIP plasma, unfortunately necessitating sampling from a position where oxide and hydroxide formation is favoured.[71] Elemental detection limits were dependent on ionization potential and ranged from 7-70 ng/ml. The ion source could not be recommended in the form used ($Al_2O_3$ discharge tube) because the small volume of the analytically useful sampling zone, the poor stability of the plasma and the necessity for daily replacement of the discharge tubes are excessive drawbacks. An alternative MIP source design, incorporating a silica discharge tube, was investigated as a potential soft ionization source for **organic** mass spectrometry.[72] Detection limits of 10-40 ng/s were found for a variety of organic compounds (by monitoring the $M^+$· ions). Spectra were similar to EI spectra, apart from increased molecular abundance in the MIP source system. Fragmentation of perfluorotributylamine increased with increasing MIP power until the molecular ion disappeared completely at a power level of 45W.

**2.5 Glow Disharge** - Glow discharge mass spectrometry (GDMS), a very old technique, has undergone something of a renaissance in recent years and is fast displacing spark source mass spectrometry for the elemental analysis of bulk solids. A very detailed review of GDMS, covering the fundamentals of the method and experimental and analytical applications, has appeared.[73] The review also contains useful descriptions of commercial instrumentation available at the time of writing. An overview of GDMS, largely devoted to a description of the performance of a commercial instrument, provides another useful introduction to the technique.[74] A review of the state-of-the-art in glow discharge lamp spectrometry includes an account of advances in GDMS.[75]

GDMS has, in the main, been implemented on double-focusing magnetic sector mass spectrometers. This is because the discharge produces molecular and doubly-charged species which must be resolved from elemental species. Such instruments are, however, very expensive and an alternative approach has been attempted by coupling a glow discharge source to a quadrupole mass spectrometer.[76,77] Although a low resolution device such as this is more prone to interference effects, the presence of interferences may be inferred by plotting a set of relative sensitivity factors (RSF) of selected elements (Ti, V, Co and P in steel for example) against relative concentration. Any marked change in RSF with concentration is evidence for interference, and more detailed studies may allow correction for these effects. Interference effects may also be reduced by using neon, instead of argon, as this shifts the interferents to different masses, removing ions such as $ArN^+$ (which is isobaric with the $^{54}Fe$ isotope in iron). Full details of GDMS ion source designs used in conjunction with low resolution mass spectrometers are given in another publication.[78]

## 3: Developments and Trends in Instrumentation

**2.6 Atmospheric Pressure Ionization (API)** - API mass spectrometry has, until recently, mainly been limited to quadrupole instruments. This is because of the technical difficulty of connecting a device which operates at atmospheric pressure to a high voltage (magnetic-sector) mass analyser. An electrical discharge suppression system, mounted in the pumping line of an API source, has circumvented these problems and allowed operation at source potentials of 6-10kV.[79]

By combining API mass spectrometry with laser ionization and ion mobility measurements (as a means of separating ions produced at atmospheric pressure in a rapid and continuous manner), an analytical device which combines ultra high sensitivity with speed of analysis has been realized.[80] The results of UV laser and Ni β source ionization were compared and showed that the laser produces lower abundance background spectra and was less prone to memory effects than the radioactive source. Detection limits of down to ca. 0.3 ng were found in laser desorption API of selected (high proton affinity) non-volatile species.[81] The laser desorption ion mobility spectroscopy/API technique is particularly suited to the detection of explosives because such materials usually possess high electron affinities and thus produce abundant negative ions.[82] Detection limits of about 200 pg were obtained, a particularly impressive result for a technique applied directly to crude or impure samples.

The characteristics of an API mass spectrometer suitable for operation with a liquid chromatograph have been described.[83]

**2.7 Chromatography/Mass Spectrometry** - Some advances in ion source design for chromatography/mass spectrometry are noted here. Detailed discussion of chromatography/mass spectrometry can be found in Chapters 6 and 7.

A laser-based apparatus for thermal micro-region vaporization from a TLC plate has been developed.[84] Mass spectra were recorded in two modes, electron or direct thermal ionization.

Numerous GC/MS studies of alkylbenzenes have been carried out with the aid of electron ionization. A novel approach to this type of analysis is exemplified by _laser_ ionization mass spectrometry where changes in laser light intensity can be used to promote or reduce fragmentation.[85] In another unusual variation on routine GC/MS analysis, carboxylic acids eluting from a GC column were conducted to a target and ionized by FAB.[86] Analytically useful data were obtained with or without a matrix coating the target surface.

A new type of apparatus for the mass spectrometric analysis of fluids, suitable for on-line chromatography/mass spectrometry, is based on the electrically-induced ionization of nebulized droplets.[87] The ionization technique differs from both electrospray and ion evaporation techniques in that much higher electrode potentials are used. Potentials were induced at a needle tip by a 150 kV power supply. Aqueous solvents required voltages of 15-60 kV, and aprotic solvents more than 110 kV, to induce ionization. Liquid flow rates of up to 4 ml/min could be accommodated. On-line mass spectrometric methods for detection of components separated by capillary zone electrophoresis (CZE) have been introduced(see Chapter 7).[88,89]

**2.8 Miscellaneous** - A new desorption/ionization mass spectrometric technique, based on thermionic emission from a potassium "glass" ($SiO_2$: $Al_2O_3$:$K_2O$), coated onto a resistively heated rhenium wire, shows promise for the analysis of labile molecules.[90] The method, $K^+$ ionization of desorbed species ($K^+$IDS), generates $(M + K)^+$ ions and (for compounds with sufficiently low ionization energies) $M^{+\cdot}$ ions. Negative-ion spectra could also be generated, yielding $(M - H)^-$ ions. Fragment ions, consistent with thermal decomposition products, were also observed. The method, although insensitive, shows considerable promise and is inexpensive to implement. Examples of $K^+$IDS spectra showed the wide applicability of the method, which produced molecular weight and fragment ion data for saccharides, antibiotics, peptides, steroids, polyethylene and polypropylene glycol mixtures and xanthins. The $K^+$IDS technique may benefit from combination with chemical ionization, which may assist desorption of surface absorbed material, or induce diagnostic reactions with suitable reagent gases.

The $K^+$IDS ionization method operates, with certain types of material, as a surface ionization (SI) source and SI has been combined, beneficially, with TOF mass spectrometry.[91] Application of pulse voltages to the ion source plates

maintained the resolution and sensitivity of the TOF mass spectrometer. The instrument is intended for use in the analysis of organic compounds in microparticulate matter.
Efficient, labour-saving ion source cleaning methods, which do not degrade performance, are few and far between, so the development of an effortless method which produces a mirror finish on ion source components is especially welcome.[92]

## 3 Mass Analysers

A fascinating discourse on the nature of the "ideal" mass analyser concluded (very properly!) that no such device exists.[93] The article encompasses discussion of quadrupole, TOF, Fourier-transform ion cyclotron resonance, magnetic sector and "hybrid" mass spectrometers in a manner which is both instructive and entertaining.

**3.1 Magnetic sector instruments** - Magnetic sector mass spectrometers, powerful and useful though they are, have always been inferior to quadrupole instruments in two areas of performance: the speed and the reproducibility of scan functions. In an effort to reproduce the ease of calibration and high scan speed of the quadrupole mass spectrometer on a high resolution instrument, a novel design of commercial sector instrument has been devised.[94] The spectrometer incorporates an electromagnet with an air-core. Electrostatic lenses are disposed on either side of the magnet, so as to shorten the focal length (which would otherwise be inordinately long, due to the low field strength of the magnet). The elimination of hysteresis effects allows rapid and reproducible scanning, removing the need for frequent calibration of the mass range. Resolution is adjustable non-mechanically by varying the object and image distance (whilst maintaining resolution) by the use of electrostatic zoom lenses.
One of the main problems facing mass spectrometrists is that changes in research interests, company policy etc. might leave them with an instrument which is not wholly suitable for, or is even incapable of, pursuing the new goals. With the advent of a novel, modular design of mass spectrometer, upgradeable

according to requirements, it should no longer be necessary to resort to the expensive option of acquiring an entirely new instrument to meet new demands.[95] Two different types of operating console and magnetic (B) analyser (2,000 or 10,000 Da mass range at 8 kV accelerating potential), an electric sector (E) and a quadrupole (Q) mass analyser can be combined in a variety of configurations. These range from ordinary double-focusing magnetic sector, through hybrid tandem (magnetic/quadrupole combination), to four sector (EBEB) tandem. A total of eleven different instruments can be constructed from the individual modules. Designs are interconvertible with minimum redundancy. An advanced design of focusing collision cell, capable of varying ionic velocity in order to facilitate energy transfer, is incorporated in the EBEB instrument.

Computer-aided design is now an *essential* tool in the development of new spectrometers. Recent progress in this area has been exemplified by proposals for the improvement of mass spectrometers.[96] Several designs of high performance mass spectrometer were proposed, and two were described in detail: a single-focusing magnetic-sector mass spectrometer of novel field arrangement (QQBQC, where C is a cylindrical electric sector) and a single-focusing instrument with large incidence and exit angles. The highly developed TRIO computer program[97] was used to investigate the effects of second and third order aberrations on simulated peak shape. The smaller instrument, designed for $^3$He/$^4$He isotope ratio determination, was able to measure ratios of less than $10^{-7}$. The high mass instrument was claimed to attain both high mass and high resolution, despite its small size. Full performance figures were, unfortunately, not given.

A compact, double-focusing mass spectrometer, comparable in size to a quadrupole, has been developed as a test-bed for ion sources, and as a SIMS instrument in its own right.[98] It comprises magnetic and electric sector fields, with opposing deflections of 30°. Fields of this type are only weakly focusing, consequently an Einzel lens is incorporated for angle focusing. The instrument produces a mass range of 700 Da at 1000 V accelerating potential, can be housed in a 30 cm length vacuum chamber, and weighs only 20 kg.

3: *Developments and Trends in Instrumentation*  91

The objectives of attaining both high transmission and high performance in a magnetic mass spectrometer, within a relatively small size, are particularly difficult to achieve. An interesting approach to this problem has been proposed: a "multipassage" instrument in which electrostatic mirrors and lenses are used to reflect the ion beams through a magnetic field several times, thereby improving resolving power.[99] Ion entry and exit are electrostatically controlled. An electrostatic mirror (in a single passage system) has also been used in an improved design of achromatic mass spectrometer, proposed as the principal component of an ion microscope.[100]

A theoretical study of the properties of magnetic/electric sector field mass spectrometers whose mean planes of deflection are tilted with respect to one another, has indicated that such designs should yield good mass resolution, stigmatic and energy focusing.[101]

A review of recent developments in tandem mass spectrometry encompasses discussion of instrumentation, with emphasis on the role of "hybrid" instruments. These unite the advantages of magnetic sectors and quadrupoles in design configurations such as the BEQQ.[102] A related review concludes with the interesting suggestion of combining low energy (eV or keV) mass spectrometry with high energy (MeV) accelerator mass spectrometry for some types of molecular analysis.[103] A tandem mass spectrometer, based on two magnetic sectors and intermediate electric and magnetic quadrupole lenses, has been devised as a disease diagnosis aid.[104] The instrument is capable of identifying micro-organisms by pyrolysis mass spectrometry (normal and/or collision-induced), followed by chemometric analysis of the resulting data. This type of instrument has considerable potential as a rapid method for the identification of microbial and tissue samples.

Four-sector tandem mass spectrometers are very powerful analytical instruments which, if they are to be exploited to their full potential, require calibration of the second stage of mass analysis over extended mass ranges. The calibration technique reported in the previous volume in this series[105] suffers from a number of drawbacks.[106] Most prominent among these are the cumbersome nature of the method, and the necessity for re-calibration if the mass values of different precursor ions are not close together. An improved technique

has been devised by ionizing a suitable calibrant in a second
ion source which is interposed between the two double-focusing
segments of an EBEB tandem mass spectrometer.[107] A data set is
generated by scanning the second magnet (with the second
electrostatic analyser voltage held constant). This data set
can then be used to generate linked scans at constant B/E.
Assignments of better than 0.3 amu for masses up to 3,500 Da
were achievable.

Time-resolved ion momentum spectrometry (TRIMS) is an
elegant technique based on an integration of magnetic sector
and TOF methodologies.[108] Product ions, arising from
dissociations in the first field-free region which precedes the
magnet of a single focusing magnetic-sector mass spectrometer,
have the same detector arrival times, but are separated from
each other according to momentum, and from the precursor ion by
time. To improve the time-resolving stage of TRIMS a post-
sector beam deflector was devised to form ion "packets".[109]
This system proved to be superior to ion source pulsing for
maintaining good time and mass resolution.

3.2 Quadrupoles and quadrupole ion storage traps - A
theoretical investigation of the effects of variations in the q
value (the Mathieu parameter) on the performance of quadrupole
mass spectrometers is an important contribution to the
understanding of the limitations imposed by imperfect rod
alignment and very rapid scanning.[110] This work is
complemented by a related study on the effects of bent or bowed
rods on quadrupole performance.[111] Both these papers are of
fundamental importance to instrument design and manufacture.
The relationship of system design to overall performance in
quadrupole residual gas analysers has been reviewed, applying
the lessons of the aforementioned studies, as well as other
criteria.[112]

Quadrupoles which operate in RF-only mode possess several
advantages over "normal" quadrupoles where mass separation is
achieved by combined RF and DC voltages. RF-only quadrupoles
have the ability to produce better transmission of high masses,
consistent peak shape, and favourable cost/performance ratio.
An improved approach to the design of RF-only quadrupoles is
claimed in a patent describing a scanning system which relies
on low frequency modulation of a very small DC voltage, with

ion detection using an annular system.[113] The advantages of this approach appear to be improved resolution and elimination of background noise due to photons, excited neutrals, etc.

Assessment of the performance of mass spectrometers (especially with regard to quantitative measurements) is particularly desirable for quadrupoles, given the limitations imposed by their low resolution characteristics. For example, a quadrupole mass spectrometer has been evaluated for potential use in "on-site" verification of materials entering nuclear fuel re-processing plants.[114] The external precision of the major isotope ratio measurements was found to be better than 1%.

A detailed investigation of small quadrupole mass spectrometers used for partial pressure measurement has concluded that great care must be exercised, both in the operation of the instruments and the interpretation of results.[115] Performance characteristics of particular interest were the limits of the useful pressure range, measurement accuracy, and long-term stability. Differences in performance were found between instruments of similar design and similar (manufacturer's) specification. The authors make a plea for the adoption of an agreed practice for defining performance characteristics, a laudable aim of which manufacturers of all types of mass spectrometer should take note.

The performance of residual gas analysers[116], and specifications for reliable measurements with these instruments[117], have been discussed. ICP/MS presents particular problems with regard to the ion optics of quadrupole mass spectrometers. Because one of the main functions of ICP/MS is to detect a wide range of elements in a single analysis, it is important to maintain ion transmission efficiency over a wide range of masses. Unfortunately, this parameter can vary significantly with mass, depending on the lens bias potentials. A detailed investigation of these effects has revealed that such limitations on performance are not critical, due to the general need for qualitative full scan analyses only[118]. It was nevertheless possible (when sample size was sufficient) to improve performance by dividing the mass range into three separate regions, each with different optimum lens bias settings. Modifications to instrument design were suggested where sample size is likely to be limited.

A special type of quadrupole mass analyser, designed for the measurement of ionized cluster beams, may be of general use in the detection of large complex molecules of masses above 1,000 Da.[119] The instrument was used to measure charge-to-mass ratios of ions, clusters and microdroplets emitted by a liquid metal ion source.

Triple quadrupole (QQQ) mass spectrometers continue to receive attention as important tools for characterizing complex mixtures. One of the factors preventing realization of the full potential of QQQ instruments is the lack of reference libraries of "standard" collision-induced decomposition (CID) spectra. This situation has arisen because of the lack of standardized instrument operating conditions. As a consequence, very different CID spectra may be reported for the same molecule. In an effort to overcome some of these problems, a QQQ instrument which is "kinetically well-behaved" has been constructed.[120-122] A kinetically well-behaved mass spectrometer is defined as one in which the charge-transfer reaction $A^+ + B \rightarrow C + D^+$ produces data which exhibit identical dependence of reaction cross-section for reactant ion decay, and product ion formation, when plotted against collision energy. The instrument is configured to operate with either a molecular beam of gas or a collision chamber target.

Proposals for a novel design of organic SIMS instrument incorporate both QQQ and reflectron TOF mass analysers.[123] Ions desorbed by the primary (Ar or Xe) ion beam can be channeled into either mass analyser. This configuration has the ability both to produce structural information at low masses (using the QQQ) and to detect high mass ions (TOF). Facilities for post-ionizing desorbed neutrals by non-resonant MPI are also proposed.

A combination of quadrupole mass filter and TOF mass analyser in a **sequential** arrangement has been constructed.[124] The system is designed for the detection of nitroaromatic compounds and other electronegative species of environmental concern in ambient air. Precursor ions produced by API ($^{63}$Ni) are mass selected in a quadrupole filter, focused by an Einzel lens and activated in a collision cell, whence the product ions are mass analysed by TOF. This unusual arrangement ensures compact size, mobility, and relative cheapness of construction.

Another unusual hybrid mass spectrometer consists of a combined quadrupole/high resolution magnetic sector instrument in which the quadrupole provides the first stage of mass analysis.[125,126] Some of the capabilities offered by this design include the ability to separate product ions with high resolution, access to high and low energy CID regimes, low eV collision energies for ion/molecule studies, and a variety of sequential reaction experiments. This type of hybrid instrument is likely to be of more use in the study of ion chemistry than as an analytical tool. Analytical MS/MS is most often applied to the study of complex mixtures, where high resolution is more appropriate in the first stage of mass analysis.

MS/MS experiments generally rely on activation of selected ions by collision with neutral gas molecules. This has several disadvantages: impact energy is not totally controllable, the vacuum system is heavily loaded, and energetic neutral particles can produce detector noise. A different approach, surface-induced dissociation (SID[127-129]), has been proposed to circumvent some or all of these problems. The principle of SID, collision of a mass-selected ion beam with a metal target, was reported in Volume 9 of this series.[105] Since this initial report a triple quadrupole SID mass spectrometer has been constructed and its properties have been evaluated.[129] The metal target is interposed between the second and third quadrupoles, which are set at right angles to one another. Ion beams are induced either to strike the target or undergo deflection without collision (useful for optimizing some aspects of instrument performance). This instrument was particularly efficient (ca. 16%) at converting translational energy into average internal energy over a wide range of collision energies. The system also has the potential for studying ion/surface reactive collisions and sequential gas and surface induced decompositions.

The development of the ion trap detector (ITD) as a viable GC/MS instrument, and its introduction as a commercial mass spectrometer, has provided an interesting alternative to quadrupole bench-top mass spectrometers.[130-133] One of the essential performance requirements of such instruments (which are often aimed at relatively unskilled users) is that they should produce spectra and GC/MS chromatograms equivalent in appearance to those obtained using more conventional designs. Problems with earlier versions of the ITD (apparently caused by the production of protonated molecules in ion/molecule reactions and by poor dynamic range) now appear to have been overcome and the instrument has established itself alongside conventional bench-top mass spectrometers. The performance of the ion trap has been enhanced by the incorporation of chemical ionization facilities.[134] Tailored RF scans are used to store CI reagent ions, which are then reacted with sample molecules. Detection limits were comparable to, or even better than, those found in EI mode. Development of a "research" version of the ITD, the ion trap mass spectrometer (ITMS), has introduced another dimension into the performance of ion traps: the possibility of conducting MS/MS experiments.[135, 136] Ions generated by EI or CI are stored in the ion trap and undesired masses are then ejected by tailored RF voltage scans, leaving the precursor ion of interest. The precursor ions are then induced to oscillate by an AC voltage applied to the end cap electrodes of the ion trap. This deposits internal energy in the ions by colliding them with the helium buffer gas. The CID product ions are subsequently mass-analyzed. Very high CID efficiencies (approaching 100%) are attainable and the implementation of more elaborate scan modes enable recording of sequential (MS/MS/MS) decompositions. An alternative to CID, laser-induced photodissociation, has also been introduced successfully to the ion trap, with the aid of a fibre optic interface.[137] The preliminary results of a comparison of ITMS with a reversed-geometry magnetic sector mass spectrometer in an MS/MS study of prostaglandin derivatives (derived from crude biological samples) have been reported.[138] Both techniques have their strengths and weaknesses, however, the high CID efficiency of the ITMS instrument may ultimately produce the better detection limits.

A comprehensive review of ion-trapping techniques and applications includes discussion of quadrupole ion storage traps.[139] This article is recommended reading for the breadth of applications it covers; these include spectroscopic and sub-atomic investigations, measurement of fundamental constants, and more familiar mass spectrometric studies.

3: *Developments and Trends in Instrumentation* 97

A novel ion-trapping technique for the study of ion/molecule reactions has been developed on a triple-quadrupole mass spectrometer.[140] Ions are confined in the centre quadrupole by maintaining a positive potential (for positive ions) on an aperture lens interposed between the centre and third quadrupole. Reversal of the lens potential ejects trapped ions for subsequent mass analysis.

### 3.3 Fourier-transform Ion Cyclotron Resonance

The Fourier-transform ion cyclotron resonance (FT-ICR) mass spectrometer is also an ion trap, although one with significantly different properties to quadrupole ion storage instruments. FT-ICR has yet to enjoy widespread acceptance, not only because of the high cost of the instrumentation, but also because of technical difficulties such as poor dynamic range, and limitations on sample introduction. These problems are rapidly being overcome, but often by using measures which increase the complexity and cost of the instrument. Cost comparisons are rather subjective exercises, however! FT-ICR is sometimes compared unfavourably to standard double-focusing magnetic sector mass spectrometry (in terms of cost versus performance). The true magnetic sector "equivalent" of FT-ICR is the 4-sector tandem mass spectrometer; an instrument which is likely to be similarly priced.

FT-ICR does possess some advantages over magnetic sector mass spectrometers. These include compatibility with pulsed ionization techniques (e.g. lasers), simultaneous detection of all the ions in the spectrum (multichannel advantage), high resolution with no sensitivity loss, ion storage and manipulation facilities, and ease of carrying out sequential tandem ($MS^n$) experiments.

FT-ICR has been the subject of a symposium from which a book containing many excellent articles has been produced.[141] Of special note are chapters dealing with the principles and features of FT-ICR[142], analytical applications[143], and routes to instrument improvement.[144] The development of FT-ICR instrumentation and its application to a number of problems has been reviewed.[145,146]

The problems of resolution degradation in FT-ICR cells because of gas loading are being tackled in a number of ways.

Many of these approaches are based on the generation of ions in an external source, followed by ion injection via tandem quadrupole mass spectrometers[147], or by electrostatic means.[148] An elegant solution, which does not rely on external ionization, entails the use of a dual-cell technique.[149] Several new approaches are also showing promise as methods for separating the sample introduction/ionization and mass analysis steps in FT-ICR. One design of remote ion source relies on an arrangement of electrostatic lenses for transferring ions into the trapping cell via a conductance limit.[150] A novel feature of the system is the inclusion of a ferromagnetic ring which encircles the analyser cell. This perturbs the magnetic field of the superconducting magnet, creating a magnetic bottle which greatly improves ion trapping efficiency. Ion transfer modes eliminate unwanted masses by using a TOF effect. The use of TOF in ICR mass spectrometry was first demonstrated in another design of commercial instrument.[151] The resolution of TOF mass selection was low (m/Δm ≈5) due to the kinetic energy spread of the ions. However, this is sufficient for mass selective transfer of ions to the high-resolution ion cyclotron cell.

Commercial dual-cell FT-ICR designs are now sufficiently well established for a substantial portfolio of applications to be described.[152] Commercial external source designs are also proving to be successful.[143] The combination of capillary supercritical fluid chromatography (SFC) with FT-ICR presents a more severe technical challenge than GC/FT-ICR. Nevertheless, the successful coupling of SFC with dual-cell[153] and differentially pumped transmissive cell FT-ICR[154] has been reported (see Chapter 7).

Following the first report of $^{252}$Cf PDMS with FT-ICR detection[155], several improvements in methodology have been devised.[156,157] Alterations of source to sample-cell configuration have greatly improved ion trapping efficiencies [156], and application of a small voltage to the tip of a sample probe (which incorporates the $^{252}$Cf radioemitter) has improved the appearance of spectra, eliminating anomalous peaks.[157] There is still considerable scope for the improvement of $^{252}$Cf/FT-ICR techniques (trapping efficiencies are low and resolution is poor compared to laser desorption FT/ICR). The method is, nevertheless, very promising and, given its possible use in high resolution and MS/MS applications, does have some potential advantages over $^{252}$Cf/TOF. The most notable successes in high mass analysis of organic molecules by FT/ICR have, however, been achieved by laser desorption[41] and caesium-ion SIMS[158] (where molecular ions were observed for polypeptides of up to 12 kDa molecular weight).

The feasibility of inducing electron impact excitation of trapped ions has been demonstrated by gating the electron beam in a dual-cell instrument.[159] The technique can be executed more rapidly than CID (which requires several hundred milliseconds in FT-ICR), making it an attractive method for combining with GC/FT-ICR. Laser photodissociation of trapped ions in FT-ICR generates fragments similar to those seen in CID experiments on a 15-residue peptide.[160] Data were obtained on 10 pmol of sample and enabled the full sequence to be determined. This technique has potentially greater selectivity than electron-impact or CID dissociation, but is likely to be inefficient for molecules which do not contain strong chromophores.

A number of new excitation and detection techniques have been developed to enhance and extend the performance of FT-ICR. These have been the subject of a comprehensive review, to which the reader is referred for more detailed discussion.[161] Some recent advances of note are aimed at ways of avoiding spectrum degradation when waveforms are sampled at frequencies below the Nyquist limit.[162,163]

**3.4 Time of Flight** - The re-emergence of TOF mass analysers as "mainstream" mass spectrometers is well attested by a special issue of "Analytical Instrumentation" devoted to this topic.[164] TOF mass spectrometry has made considerable advances since it was first introduced: the prototype instrument was equipped with a flight tube 10 m long and had a resolving power of "about 2".[165, 166] The resurgence of interest has largely been fuelled by the appearance of ionization techniques, such as $^{252}$Cf and laser desorption/ionization, which are ideally suited to the time-scale and multichannel recording abilities of TOF, as discussed earlier in this report.

The performance and design of TOF mass spectrometers have been discussed in a comprehensive theoretical study of several different types of instrument configuration.[167]

An overview of recent advances and applications of $^{252}$Cf PDMS contains details of improved performance with an extended flight tube TOF mass analyser [168] (variable between 30 and 140 cm, somewhat shorter than the instrument mentioned above!). The analyser produced significantly increased resolution (1550, FWHM definition at m/z 393). Resolution in TOF mass spectrometry is directly proportional to mass, so this value is very encouraging. The influence of sample loading technique on the nature of the mass spectrum was emphasised in this report.

A new generation of $^{252}$Cf TOF mass spectrometer incorporates features aimed at increasing both the flexibility of the instrument and its performance in high mass applications.[169] Target position was tightly controlled and two lengths of flight tube were available (26 cm for studying kinetic energy distributions; 55 cm for high accelerating voltage, high mass studies). Electrostatic plates inserted into the flight tube were used to deflect ions onto an off-axis detector; undeflected neutrals were detected on-axis. Improvements in timing electronics and detector performance yielded more accurate mass measurements than earlier designs. An alternative $^{252}$Cf TOF instrument from the same laboratory employs low voltage acceleration and a reflection flight tube to produce velocity focusing, and hence higher resolution.[170] A theoretical mass resolution of 20,000 was estimated for this configuration of mass analyser.

Reflection flight tubes have also been used in conjuction with SIMS, yielding mass resolution of 13,000 (FWHM) and dynamic range covering five orders of magnitude.[171, 172] This design was suitable both for the analysis of inorganic materials and for organic trace analysis at high mass (10,000 Da). Energy-compensated TOF mass analysers may be used in SIMS and laser analysis of surfaces, yielding high sensitivity image resolution of better than 0.25 μm.[173]

Conversion of a conventional Wiley-McLaren TOF mass analyser for laser desorption studies entails modification of timing circuitry and use of post-acceleration detection for high mass acquisition.[174] Detection of 1 pmol of gramicidin S indicated that the modified instrument was capable of carrying out analyses at very high sensitivities. The conversion, with minor modifications, was also suitable for undertaking SIMS of samples dissolved in liquid matrices.[175]

A new mode of operation in TOF mass spectrometry, Fourier transformation, is designed to increase sensitivity and scan recording speeds without compromising resolution.[176,177] Ions are subjected to a sinusoidally modulated accelerating voltage, and detector current is modulated in phase with the ion source. The normal time domain signal is recovered by Fourier transformation. Although the predicted increase in detector current was not realized in practice, this was not attributed to deficiencies in the Fourier-transform methodology.

The ion optics of the mass spectrometer with multiple focusing (two to four electric fields) have been studied in the third-order approximation.[178] A four-sector toroidal field instrument produced mass resolution of 2,000 (FWHM) on the gramicidin S molecular ion (SIMS ionization). Given that better resolution has been obtained using much simpler TOF designs,[171,172] this particular multiple-focusing system does not appear to represent a significant advance.

A tandem TOF mass spectrometer for surface-induced dissociation studies has been constructed.[179] Product ion spectra were similar to those observed in a magnetic sector/quadruple SID instrument. It was noted that further improvements in sensitivity, resolution and precision of mass scale calibration would be necessary before the instrument was capable of rivalling previously developed SID mass spectrometers.

### 3.5 Accelerator Mass Spectrometry

Accelerator mass spectrometry (AMS) describes a collection of techniques, all of which entail the use of a high energy (MeV) second stage of acceleration, for the measurement of very small isotope ratios (down to less than $10^{-12}$). AMS is a well established method which could easily fill a review chapter in its own right, consequently only a few leading references will be discussed here.

The first stage of mass analysis in AMS is similar to normal mass spectrometry in that ions (often generated by caesium-ion bombardment of a solid target) are accelerated to keV translation energies. The ions are then focused and separated by electric and magnetic fields, prior to acceleration to MeV energies. Charge-stripping, by passage through a gas or thin-foil target, and a second stage of mass analysis and detection then follow. The very high selectivity of AMS is a result both of the efficient removal of molecular interferences (by the high energy, charge-stripping collisions) and of the separation of isobaric species (by energy loss measurement). In some cases, measurement of $^{14}C^-$ for example, high sensitivity and selectivity are also favoured because the interfering isobar ($^{14}N^-$ in this case) is unstable and rapidly reverts to a neutral species.

The fundamentals of AMS have been described [180], and its promise in trace element analysis has been discussed.[181] AMS has been applied to problems in geology, archaeology, cosmology, oceanography and several other disciplines.[182] Opportunities for important new applications in biology and toxicology are also emerging [183], and the present and future prospects of AMS in a host of scientific disciplines have recently been discussed.[184]

**3.6 Ion Imaging** - Ion imaging by SIMS has already been mentioned briefly in other sections of this report. Only a few leading references will be discussed. The technique is a rapidly evolving one, already well established in the micro-characterization of metallic and geological material, which has considerable potential in the biological sciences. The status and prospects of SIMS imaging has recently been reviewed from the standpoint of biological microanalysis.[185] Lateral resolution of 0.5 μm is attainable with current instrumentation. A new high spatial resolution SIMS instrument employs Mattauch-Herzog geometry and a focal plane detection system to achieve sub-micron resolution with high sensitivity.[186] This type of mass spectrometer is particularly effective in analysing small volume samples because rapid and distinctive intensity changes are observable. A new method for determining accurate masses in SIMS microscopy entails use of an external standard, sequential acquisition of high resolution spectra, and superposition of the standard and unknown spectra.[187] The technique makes high resolution SIMS (an indispensable technique in materials science) a routine operation and enables detection of transient peaks through real-time operation.

3.7 **Miscellaneous** - Tests of instrument accuracy, although important in assessing mass spectrometer performance, are often difficult to establish. Synthetic mixtures of highly enriched uranium isotopes have been used as probes to investigate the feasibility of producing a routine method for testing the accuracy of isotope ratio measurement.[188] The effectiveness of the isotope sets was demonstrated by an evaluation of the accuracy of a thermal ionization mass spectrometry instrument.

The increasing use of automated features and digital control, together with the growing technical complexity of mass spectrometric equipment, can make fault diagnosis very difficult. A possible solution to this problem is provided by an expert system for tracing faults in the vacuum system of a mass spectrometer.[189]

## 4 Detectors

Soft ionization methods (such as FAB) tend to produce ions which decrease in abundance as the molecular weight of the sample increases. The difficulty of detecting the feeble ion currents produced is compounded on scanning instruments because only a small fraction of the total ion current is sampled during any brief time interval. Improved data can be obtained by multichannel analysis of limited mass range scans in the molecular ion region.[190] A more elegant (if considerably costlier!) solution is to use a multichannel detection system in place of a conventional single-channel detector.[191, 192] One design is attached to a Nier-Johnson type mass spectrometer which is equipped with specially shaped entrance and exit pole faces on the magnetic sector.[191] This arrangement produces a flat focal plane perpendicular to the ion beam. A microchannel plate detector is positioned at this focus, yielding an increase in detection efficiency of several orders of magnitude at mass resolution of 3,000 (FWHM) and over a $\Delta m/m$ range of 11%. An alternative, now commercially available, employs an electro-optical "array" detector which provides a $\Delta m/m$ mass window of 4%.[192] Ions impinging on a pair of chevron mounted microchannel plates produce secondary electrons which are accelerated towards a phosphor screen coating the surface of an extended fibre-optic faceplate. Photons emitted by the phosphor are conducted to a photodiode array detector.

Abundant signals were observed for the FAB-produced molecular ions of insulin (down to levels of 1 pmol), proinsulin I and bovine trypsin.[192,193] An additional advantage of the array detector is its ability to detect transient signals produced by "dry" FAB.[194] The concomitant reduction in chemical noise produced improved detection limits. Further developments in the applications of focal plane detectors to analytical problems are expected in the near future. For example, computer controlled magnet switching, with suitable overlap of focal plane detector mass ranges, would enable complete mass spectra to be acquired with high sensitivity. Array detectors might even allow the recording of $^{252}$Cf spectra on magnetic sector mass spectrometers.

Preliminary studies of the nature of secondary particles produced in post-acceleration dectors (PADs) indicate that positive ions produce electrons and up to 50% negative ions.[195] These investigations should assist in the production of improved designs of PAD.

PDMS of high mass ions places stringent demands on detection efficiency. These demands are not always met by conventional secondary ion detectors. Simply increasing the velocity of ions in the conventional PADs fitted to TOF instruments unfortunately produces high background count rates. More sophisticated approaches have been based on increasing secondary electron production.[196, 197] Both detectors described in these reports induce post-acceleration of ions towards an efficient producer of secondary electrons. These are accelerated away from the target and are bent towards a set of channel plates by a magnetic field, producing the detection signal.

A major performance limitation of many $^{252}$Cf PDMS TOF mass spectrometers is the mis-match between mass spectrometer and data recording system resolution. A powerful computer system has been used to circumvent this problem by recording up to 800,000 $5/8$ ns data channels with very high dynamic range.[198] This allows examination of fine details in all regions of the mass spectrum and also yields reproducible mass assignment accuracies of the order of 100 ppm.

Difficulties in measuring fast pulses of ion current produced in resonance ionization mass spectrometry has hindered the application of RIMS to the measurement of isotope ratios with high precision. Improvements in ion counting methodology by allowing overlap of pulse counting and analogue regions of the detector have improved absolute calibration and dynamic range of a tandem mass spectrometer.[199] Electronic pulse reflections, which degrade the performances of fast pulse recording with micro-channel plate detectors, can be reduced by proper impedance matching.[200]

A Scintillation-type ion detector for ICP/MS produces improved precision in isotope ratio measurement, less degradation of gain in the long term and comparable performance in other respects when evaluated against a channeltron type detector.[201]

## 5 Sample Introduction

Use of flow injection as a sample introduction technique in ICP/MS can alleviate many of the problems which arise when analysing intractable samples of clinical, biological, or environmental origin.[202] Flow injection facilitates the determination of elements in solutions high in salt content, acidity and viscosity, without interface blockage and other deleterious effects. Sample throughput was also increased by a factor of four.

An improved design of micro-nebulizer for liquid sample introduction produces uniformly sized droplets, is resistant to clogging, and can be quickly disassembled for inspection and cleaning.[203]

Coolable probes have been used to improve performance in FAB mass spectrometry.[204] These probes enable volatile samples and matrices to be examined by FAB.

Enzyme reactions may be characterized by comparison of $^{252}Cf$ PDMS spectra of substrates, which have been non-covalently bound to nitrocellulose, before and after incubation.[205] Samples absorbed on the nitrocellulose matrix were introduced directly into the PDMS instrument.

## 6 Integrated Techniques

Mass spectrometers are frequently linked to chromatographs (GC/MS, LC/MS, etc.) or to other mass analysis devices (MS/MS), however, combination with other spectroscopic or microchemical techniques, referred to here as "integrated techniques", is considerably rarer, although of increasing interest and importance. This type of methodology was classically exemplified (in 1981) by the combination of Fourier-transform infrared spectroscopy and mass spectrometry.[206] The acquistion of both IR and MS data by a single injection onto a gas chromatograph allowed direct correlation of both types of information, greatly facilitating sample identification. The main drawback of the technique has been the disparity between MS and FT-IR sensitivities. MS detection limits are a factor of 10 to 100 better than those for FT-IR when light-pipe detector cells are used.[207, 208] The analysis of fuel spills and oxidative degradation products [207], and hazardous wastes [208], has benefited from the application of integrated FT-IR and MS techniques. An alternative approach, which produces FT-IR sensitivities much closer to those of mass spectrometry, entails "matrix isolation" of eluting GC peaks on a cooled rotating metal disk prior to FT-IR analysis.[209] The FT-IR spectra obtained from samples isolated in this manner contain narrow, intense peaks and yield sub-nanogram detection limits. This commercially available GC/FT-IR/MS instrument utilises a three way splitter which conducts 40% of the sample to the FT-IR isolation disk, 40% to a compact quadrupole mass spectrometer, and 20% to a flame ionization detector. Recent technological advances in GC/FT-IR/MS have been reviewed.[210]

A further refinement of GC/FT-IR/MS involves incorporation of a radioactivity detector into the instrument.[211] Although FT-IR detection limits necessitated sample loadings of 100-200 ng per component, the integrated system produced useful analytical data in metabolic studies of a number of herbicides. A serial GC/FT-IR/Ion-trap mass spectrometer system has also been constructed and provides a relatively inexpensive alternative to other published designs.[212]

Techniques other than FT-IR have been integrated with mass spectrometry; thermogravimetry for example.[213] This technology is particularly useful in forensic analysis. A more complex system is exemplified by the combination of differential scanning calorimetry (DSC), X-ray diffraction (XRD), and mass spectrometry in an instrument designed for characterizing materials heated in a controlled atmosphere.[214] Further development of integrated techniques can be expected in the future. Methodologies involving direct measurement of spectroscopic properties within the mass spectrometer are certainly feasible. For example laser techniques, already well established as ionization methods, may be adaptable to yield UV and fluorescence spectra. Sample introduction techniques which produce rotational and vibrational cooling of samples [10-22] also have potential for increasing the sensitivity and selectivity of molecular spectroscopic methods. If the combination of these technologies proves to be synergistic, there will be many benefits to both analytical and physicochemical research.

## REFERENCES

1. These are, in alphabetical order, Finnigan MAT, Hewlett-Packard, Nermag, Shimadzu and VG Masslab.

2. J.Brossel, B, Cagnac, and A. Kaslter, C.R.Acad. Sci., 1953, 237, 984.

3. P.D. Maker, R.W. Terhune, and C.M.Savage in 'Quantum Electronics', Columbia University Press, New York, 1964, p 1559.

4. M. Klewer, M.J.M. Oeerlage, J. Los, and M.J. van der Wiel, J. Phys. B, 1977, 10, 2809.

5. V.S. Antonov, I.N. Knyazov, V.S. Letokhov, V.M. Matjiuk, B.G. Moshev and V.K. Potapov, Opt. Lett, 1978, 3, 37.

6. R.J. Cotter, Anal. Chim. Acta., 1987, 195, 45.

7. D.M.Lubman, Prog. Anal. Spectrosc., 1987, 10, 529.

8. J. Grotemeyer and E.W. Schlag, Angew. Chem., 1988, 4, 461.

9. J.C. Travis, J.D. Fassett, and T.B. Lucatorto, Inst. Phys. Conf. Ser., 1986, 84, 91.

10. D.M. Lubman and R.M. Jordan, Rev. Sci. Instrum., 1985, 56, 573.

11. D.M. Lubman and R. Tembreull, Anal. Instrum., 1987, 16, 117.

12. U. Boesl, J. Grotemeyer, K. Walter, and E.W. Schlag, Inst. Phys. Conf. Ser., 1986, 84, 223.

13. U. Boesl, J. Grotemeyer, K. Walter, and E.W. Schlag, Anal. Instrum., 1987, 16, 151.

14. J. Grotemeyer, K. Walter, U. Boesl, and E.W. Schlag, Int. J. Mass Spectrom. Ion Processes, 1987, 78, 69.

15. J. Grotemeyer, U. Boesl, K. Walter, and E.W. Schlag, Org. Mass Spectrom., 1986, 21, 595.

16. J. Grotemeyer, U. Boesl, K. Walter, and E.W. Schlag, Org. Mass Spectrom., 1986, 21, 645.

17. J. Grotemeyer and E.W. Schlag, Org. Mass Spectrom., 1988, 23, 388.

18. J. Grotemeyer, U. Boesl, K. Walter and E.W. Schlag, J. Am. Chem.Soc., 1986, 108, 4239.

19. J. Grotemeyer and E.W. Schlag, Org. Mass Spectrom., 1987, 22, 758.

20. L. Li and D.M. Lubman, Appl. Spectrosc., 1988, 42, 418.

21. TOF-1; Bruker Analytische Messtechnik GmbH, Rheinstetten, FRG.

22. R.M. Jordan Co., Mt. View, CA., USA.

23. F. Engelke, J.H. Hahn, W. Henke, and R.N. Zare, Anal. Chem., 1987, 59, 909.

24. R.W. Odom and B. Schueler, Thin Solid Films, 1987, 154, 1.

25. U. Schühle, J.B. Pallix, and C.H. Becker, J. Am. Chem. Soc., 1988, 110, 2323.

26. J.B. Pallix, C.H. Becker, and N. Newman, Mat. Res. Bull., 1987, 12(6), 52.

27. D.L Donohue, W.H. Christie, D.E. Goeringer, D.H. Smith, and H.S McKown, in "Analytical Applications of Spectroscopy", ed. W.R. Laing, Lewis publ. Inc., 1986, p113.

28. C.E. Young, M.J. Pellin, W.F. Calaway, B. Jørgensen, E.L. Schweitzer, and D.M. Gruen, Nucl. Instrum. Methods Phys. Res., 1987, B27, 119.

29. J.P. Young, R.W. Shaw, D.E. Goeringer, D.H. Smith, and W.H. Christie, Anal. Instrum., 1988, 17, 41.

30. D.M. Lubman, Anal. Chem., 1987, 59, 31A.

31. W.M. Fairbank jr., Nucl. Instrum. Methods Phys. Res., 1987, B29, 407.

32. T.J. Whitaker, B.D. Cannon, and B.A. Bushaw, Laser Focus, 1988, 24, 88.

33. T.J. Whitaker, B.D. Cannon, and B.A. Bushaw, in "Analytical Chemistry Instrumentation", ed. W.R. Laing, Lewis Publ. Inc., 1986, p107.

34. B.A. Bushaw, B.D. Cannon, G.K.Gerke, and T.J. Whitaker, Inst. Phys. Conf. Ser., 1986, 84, 103.

35. R.Engelmann jr., R.A. Keller, C.M. Miller, D.C. Parent, W.M. Fairbank jr., R.D. LaBelle, S.-A. Lee, and E. Riis, Inst. Phys. Conf. Ser., 1986, 84,127.

36. D.P. Armstrong, W.H. McCulla, and G.K. Schweitzer, in "Analytical Chemistry Instrumentation", ed. W.R Laing, Lewis Publ. Inc., 1986, p119.

37. U. Krönert, St. Becker, Th. Hilberath, H.-J. Kluge, and C. Schulz, Appl. Phys., 1987, A44, 339.

38. F. Ames, A. Becker, H.-J. Kluge, H. Rimke, W. Ruster, and N. Trautmann, Fresenius Z. Anal. Chem., 1988, 331, 133.

39. R.W. Bonham and J.C. Quattlebaum, Spectroscopy (Oregon), 1988, 3(4), 27.

40. D.S. Simons, Appl. Surface Sci., 1988, 31, 103.

41. C.F. Ijames and C.L. Wilkins, J.Am. Chem. Soc., 1988, 110, 2687.

42. M. Yang and J.P. Reilly, Anal. Instrum., 1987, 16, 133.

43. M. Karas, D. Bachmann, U. Bahr, and F. Hillenkamp, Int. J. Mass Spectrom. Ion Processes, 1987, 78, 53.

44. M. Karas and F. Hillenkamp, Presented at the 11th International Mass Spectroscopy Conference, Bordeaux, Aug./Sept. 1988.

45. K. Tanaka, Y. Ido, S. Akita, Y. Yoshida, and T. Yoshida, Iyo Masu Kekyukai Koenshu, 1987, 12, 219.

46. W.B. Emary, K.V. Wood, and R.G. Cooks, Anal. Chem., 1987, 59, 1069.

47. W.B. Martin and R.M. O'Malley, Int J. Mass Spectrom. Ion Processes, 1987, 77, 203.

48. A.J. Eccles, J.A. van den Berg, A. Brown, and J.C. Vickermann, J. Vac. Sci. Technol., 1986, 4, 1888.

49. A.D. Applehans, J.E. Delmore, and D.A. Dahl, Anal. Chem., 1987, 59, 1685.

50. D.G. Welkie, US Patent 4,661,702, Apr. 28, 1987.

51. F.G. Rüdenauer and W. Steiger, Ultramicroscopy, 1988, 24, 115.

52. F.G. Rüdenauer, W. Steiger, H. Studnicka, and P. Pollinger, Int J. Mass Spectrom. Ion Processes, 1987, 77, 63.

53. S. Daolio, B. Facchin, and C. Pagura, Int. J. Mass Spectrom Ion Processes, 1987, 76, 277.

54. U. Kaiser and J.C. Huneke, Mat. Res. Bull., 1987, 12(6), 48.

55. H. Oechsner, Scanning Microsc., 1988, 2, 9.

56. J.F. Geiger, M. Kopnarski,H. Oechsner, and H. Paulus, Mikrochim. Acta, 1987, 1, 497.

57. R. Jede, K. Seifert, and G. Dunnebier, Fresenius Z. Anal.Chem., 1987, 329, 116.

58. C.J. Metral and R.A. Day, Anal. Lett., 1986, 19, 217.

59. D.F. Torgerson, R.P. Skowronski, and R.D. MacFarlane, Biochem. Biophys. Res. Commun., 1974, 60, 616.

60. BIO-ION Nordic, Uppsala, Sweden.

61. R.J. Cotter, Anal. Chem., 1988, 60, 781A.

62. M. Salehpour, P. Håkansson, B.U.R. Sundqvist, and A.G. Craig, Int. J. Mass Spectrom. Ion Processes, 1987, 77, 173.

63. J.A. Loo, E.R. Williams, J.J.P. Furlong, B.H. Wang, F.W. McLafferty, B.T. Chait, and F.H. Field, Int. J. Mass Spectrom. Ion Processes, 1987, 78, 305.

64. E.A. Schweikert, W.R. Summers, M.U.D. Beug-Deeb, P.E. Filpus-Luyckx, and L. Quinones, Anal. Chim. Acta, 1987, 195, 163.

65. M.C. Ledbetter, R.J. Beuhler, and L. Friedman, Proc. Natl. Acad. Sci. USA, 1987, 84, 85.

66. R.S. Houk, V.A. Fassel, G.D. Flesch, H.J. Svec, A.L. Gray, and C.E. Taylor, Anal. Chem., 1980, 52, 2283.

67. A.L. Gray and J.G Williams, J.Anal. At. Spectrom., 1987, 2, 599.

68. B.T.G. Ting and M. Janghorbani, J. Anal. At. Spectrom., 1988, 3, 325.

69. T. Osawa and T. Ito, U.S Patent 4,740,696, Apr. 26, 1988.

70. J.S. Gordon, P.S.C. van der Plas, and L. de Galan, Anal. Chem., 1988, 60, 372.

71. R.D. Satzger, F.L. Fricke, and J.A. Caruso, J. Anal. At. Spectrom., 1988, 3, 319.

72. E. Poussel, J.M. Mermet, D. Deruaz, and C. Beaugrand, Anal. Chem., 1988, 60, 923.

73. W.W. Harrison and B.L. Bentz, Prog. Anal. Spectrosc., 1988, 11, 53.

74. N.E. Sanderson, E. Hall, J. Clark, P. Charalambous, and D. Hall, Mikrochim. Acta, 1987, 1, 275.

75. J.A.C. Broekaert, J. Anal. At.Spectrom., 1987, 2, 537.

76. N. Jakubowski, D. Stuewer, and W. Vieth, Mikrochim. Acta, 1987, 1, 302.

77. N. Jakubowski, D. Stuewer, and W. Vieth, Fresenius Z. Anal. Chem., 1988, 331, 145.

78. N. Jakubowski, D. Stuewer, and G. Toelg, Int. J. Mass Spectrom. Ion Processes, 1986, 71, 183.

79. A.H. Grange, R.J. O'Brien, and D.F. Barkofsky, Rev. Sci. Instrum., 1988, 59, 656.

80. L. Kolaitis and D.M. Lubman, Anal. Chem., 1986, 58, 1993.

81. L. Kolaitis and D.M.Lubman, Anal. Chem., 1986,58, 2137.

82. S.D. Huang, L. Kolaitis, and D.M. Lubman, Appl.Spectrosc., 1987, 41, 1371.

83. M. Sakairi and H. Kambara, Anal. Chem., 1988, 60, 774.

84. L. Binghuan and L. Huazhang, Kexue Tongbao, 1987, 32, 231.

85. R.B. Opsal and J.P. Reilly, Anal. Chem., 1988, 60, 1060.

86. E.J. Gallegos, J. Chromatogr. Sci., 1987, 25, 296.

87. W.M. Lagna, U.S. Patent No. 4,667,100, May 19, 1987.

88. J.A. Olivares, N.T. Nguyen, C.R. Yonker, and R.D. Smith, Anal. Chem., 1987, 59, 1230.

89. R.D. Smith, J.A. Olivares, N.T. Nguyen, and H.R. Udseth, Anal. Chem., 1988, 60, 436.

90. D.D. Bombick and J. Allison, Anal. Chem., 1987, 59, 458.

91. D.A. Chatfield and M. Ajami, Int. J. Mass Spectrom. Ion Processes, 1987, 77, 241.

92. L.R. Alexander, V.L. Maggio, V.E. Green, J.B. Gill, E. R. Barnhart, D.G. Patterson jr., and L.C. Nicolaysen, Anal. Chem., 1987, 59, 2543.

93. C. Brunée, Int. J. Mass Spectrom. Ion Processes, 1987, 76, 125.

94. R.H. Bateman, U.S. Patent 4,723,076, Feb. 2, 1988.

95. J. Warburton, Int. Biotech. Lab., 1988, 6(2), 14.

96. H. Matsuda, Nucl. Instrum. Meth. Phys. Res., 1987,A258, 310.

97. T. Matsuo, H. Matsuda, Y. Fujita, and H. Wollnik, Mass Spectrosc. (Japan), 1976, 24, 19.

98. H. Liebl, Nucl. Instrum. Meth. Phys. Res., 1987, A258, 323.

99. C. Berger and M. Baril, Nucl. Instrum. Meth. Phys. Res., 1987, A258, 335.

100. M. Baril and M. Noël, Nucl. Instrum. Meth. Phys. Res., 1987, A258, 318.

101. H. Waldrich and H. Ewald, Nucl. Instrum. Meth. Phys. Res., 1988, A263, 414.

102. J.C. Schwartz and R.G Cooks, Spectrosc. Int. J. , 1987, 5, 49.

103. R.G. Cooks and O.W. Hand, Nucl. Instrum. Meth Phys. Res., 1987, B29, 427.

104. E.J. Sjoberg, U.K. Patent Appl. GB 2,187,035, 26 Aug. 1987.

105. T.R. Kemp, in "Mass Spectrometry, Volume 9", ed. M.E. Rose, RSC, London, 1986, p122.

106. R.K. Boyd, P.A. Bott, D.J. Harvan, and J.R. Hass, Int. J. Mass Spectrom. Ion Processes, 1986, 69, 251.

107. K. Sato, T. Asada, M. Ishihara, F. Kunihiro, Y. Kammei, E. Kubota, C.E. Costello, S.A. Martin, H.A. Scoble, and K. Biemann, Anal. Chem., 1987, 59, 1652.

108. J.T. Stults, C.G. Enke, and J.F. Holland, Anal. Chem., 1983, 55, 1323.

109. B.A. Eckenrode, J.T. Watson, C.G. Enke, and J.F. Holland, Int. J. Mass Spectrom. Ion Processes, 1988, 83, 177.

110. P.H. Dawson, Int. J. Mass Spectrom. Ion Processes, 1988, 83, 295.

111. P.H. Dawson, Int. J. Mass Spectrom. Ion Processes, 1988, 84, 185.

112. P.H. Dawson, J. Vac.Sci. Technol., 1986, A4, 1709.

113. P.H. Dawson, U.S. Patent No. 4,721,854, Jan. 26, 1988.

114. R. Fiedler, in "Analytical Chemistry Instrumentation", ed. W.R. Laing, Lewis Publ. Inc., 1986, p127.

115. W.E. Austin, F.M. Mao, J.M. Yang, and J.H. Leck, J. Vac. Sci Technol., 1987, A5, 2631.

116. J.H. Batey, Vacuum, 1987, 37, 659.

117. W.G. Bley, Vacuum, 1988, 38, 103.

118. J.P.Schmit and M.Chtaib, Canad. J. Spectrosc., 1987, 32, 56.

119. D. Nayak, K. Pourrezaei, M. Francois, and A. Bahasadri, Rev. Sci. Instrum., 1987, 58, 2249.

120. R.I. Martinez and S. Dheandhanoo, Int. J. Mass Spectrom. Ion Processes, 1986, 74, 241.

121. R.I. Martinez and S. Dheandhanoo, J. Res. Natl. Bur. Stand., 1987, 92, 229.

122. R.I. Martinez, Rev. Sci. Instrum., 1987, 58, 1702.

123. B.L. Bentz and R.E. Honig, Springer Proc. Phys., 1986, 9, 192.

124. G.L. Glish, S.A. McLuckey, and H.S. McKown, in "Analytical Chemistry Instrumentation", ed. W.R. Laing, Lewis Publ. Inc., 1986, p33.

125. G.L. Glish, S.A. McLuckey, and H.S. McKown, Anal. Instrum., 1987, 16, 191.

126. E.H. McBay, G.L. Glish, S.A. McLuckey, and L.K. Bertram, in "Analytical Chemistry Instrumentation", ed. W.R. Waring, Lewis Publ. Inc., 1986, p199.

127. Md. A. Mabud, M.J. DeKrey, and R.G. Cooks, Int. J. Mass Spectrom. Ion Processes, 1985, 67, 285.

128. M.J. DeKrey, Md. A. Mabud, J.E.P. Syka, and R.G. Cooks, Int. J. Mass Spectrom. Ion Processes, 1985, 67, 295.

129. M.E. Bier, J.W. Amy, R.G Cooks, J.E.P. Syka, P. Ceja, and G. Stafford, Int. J. Mass Spectrom. Ion Processes, 1987, 77, 31.

130. G.C. Stafford, P.C. Kelley, and D.C. Bradford, Am. Lab., 1983, 15, 51.

131. G.C. Stafford, P.E. Kelley, J.E.P. Syka, W.E. Reynolds, and J.F.J. Todd, Int. J. Mass Spectrom. Ion Phys., 1984, 60, 85.

132. G.C. Stafford, P.E. Kelley, and D.R. Stephens, U.S. Patent 4,540,884, Sept. 10th, 1985.

133. P. Bishop, Anal. Proc., 1987, 24, 368.

134. J.S. Brodbelt, J.N. Louris, and R.G. Cooks, Anal. Chem., 1987, 59, 1278.

135. P.E. Kelley, G.C. Stafford jr., and J.E.P. Syka, U.S. Patent 4,749,860, Jun. 7 1988.

136. J.N. Louris, R.G. Cooks, J.E.P. Syka, P.E. Kelley, G.C. Stafford jr., and J.F.J. Todd, Anal. Chem., 1987, 59, 1677.

137. J.N. Louris, J.S. Brodbelt, and R.G. Cooks, Int. J. Mass Spectrom. Ion Processes, 1987, 75, 345.

138. R.J. Strife, P.E. Kelley, and M. Weber-Grabau, Rapid Commun. Mass Spectrom., 1988, 2, 105.

139. J. Allison, and R.M. Stepnowski, Anal. Chem., 1987, 59, 1072A.

140. G.G. Dolnikowski, M.J. Kristo, C.G. Enke, and J.T.Watson, Int. J. Mass Spectrom. Ion Processes, 1988, 82, 1.

141. "Fourier Transform Mass Spectrometry - Evolution, Innovation, Applications", ed. M.V. Buchanan, ACS, Washington DC, 1987,

142. M.V. Buchanan and M.B. Comisarow, in "Fourier Transform Mass Spectrometry - Evolution, Innovation, Applications", ed. M.V. Buchanan, ACS, Washington DC, 1987, p1.

143. F.H. Laukien, M. Allemann, P. Bischofberger, P. Grossman, Hp. Kellerhalls, and P. Köfel, in "Fourier Transform Mass Spectrometry - Evolution, Innovation, Applications", ed. M.V. Buchanan, ACS, Washington DC, 1987, p81.

144. R.P. Grese, D.L. Rempel, and M.L. Gross, in "Fourier Transform Mass Spectrometry - Evolution, Innovation, Applications", ed. M.V. Buchanan, ACS, Washington DC, 1987, p34.

145. M. Johnston, Spectroscopy (Oregon), 1987, 2(2), 14.

146. M. Johnston, Spectroscopy (Oregon), 1987, 2(3), 14.

147. R.T. McIver, R.L. Hunter, and W.D. Bowers, Int. J. Mass Spectrom. Ion Processes, 1985, 64, 67.

148. P. Köfel, M. Allemann Hp. Kellerhals, and K.-P. Wanczek, Int. J. Mass Spectrom. Ion Processes, 1985, 65, 97.

149. R.B. Cody, J.A. Kinsinger, S. Ghaderi, I.J. Amster, and F.W. McLafferty, Analyt. Chim. Acta, 1985, 178,43.

150. S. Ghaderi, O. Vosburger, D.P. Littlejohn, and J.L. Shohet, U.S. Patent 4,739,165, Apr. 19, 1988.

151. P. Köfel, M. Allemann, Hp. Kellerhals, and K.-P. Wanczek, Int. J. Mass Spectrom. Ion Processes, 1986, 72, 53.

152. R.B. Cody jr. and J.A. Kinsinger, in "Fourier Transform Mass Spectrometry - Evolution, Innovation, Applications", ed. M.V. Buchanan, ACS, Washington DC, 1987, p59.

153. E.D. Lee, J.D. Henion, R.B. Cody, and J.A. Kinsinger, Anal. Chem., 1987, 59, 1309.

154. D.A. Laude jr., S.L. Pentoney jr., P.R. Griffiths, and C.L. Wilkins, Anal. Chem., 1987, 59, 2283.

155. J.C. Tabet, J. Rapin, M. Poreti, and T. Gaumann, Chimia, 1986, 40, 169.

156. S.K. Viswanadham, D.M. Hercules, R.R. Weller, and C.S. Giam, Biomed. Env. Mass Spectrom., 1987, 14, 43.

157. J.A. Loo, E.R. Williams, I.J. Amster, J.J.P. Furlong,, B.H. Wang, F.W. McLafferty, B.T. Chait, and F.H. Field, Anal. Chem., 1987, 59, 1880.

158. D.F. Hunt, J. Shabanowitz, J.R. Yates, N.-Z. Zhu, D.H. Russell, and M.E. Castro, Proc. Natl. Acad. Sci. USA, 1987, 84, 620.

159. R.B. Cody and B.S. Freiser, Anal. Chem., 1987, 59, 1054.

160. D.F. Hunt, J. Shabanowitz, and J.R. Yates, J. Chem. Soc. Chem. Commun., 1987, 548.

161. A.G. Marshall, T.-C.L. Wang, L. Chen, and T.L. Ricca, In "Fourier Transform Mass Spectrometry - Evolution, Innovation, Applications", ed. M.V. Buchanan, ACS, Washington DC, 1987, p21.

162. F.R. Verdun, T.L. Ricca, and A.G. Marshall, Appl. Spectrosc., 1988, 42, 199.

163. M. Wang and A.G. Marshall, Anal. Chem., 1988, 60, 341.

164. "Advances in Time-of-Flight Mass Spectrometry", ed. J.E. Campana, Anal. Instrum., 1987, 16(1).

165. A.E. Cameron and D.F. Eggers jr., Rev. Sci. Instrum., 1948, 19, 605.

166. J.E. Campana, Anal. Instrum., 1987, 16, 1.

167. H. Wollnik, Springer Proc. Phys., 1986, 9, 184.

168. I. Kamensky and A.G. Craig, Anal. Instrum., 1987, 16, 71.

169. R.D. Macfarlane, J.C. Hill, and D.L. Jacobs, Anal. Instrum., 1987, 16, 51.

170. P.W. Geno, and R.D. Macfarlane, Int. J. Mass Spectrom. Ion Processes, 1987, 77, 75.

171. E. Niehuis, T. Heller, H. Feld, and A. Benninghoven, J. Vac. Sci Technol., 1987, A5, 1243.

172. E. Niehuis, T. Heller, H. Feld and A. Benninghoven, Springer Proc. Phys. 1986, 9, 198.

173. A.R. Waugh, D.R. Kingham, C.H. Richardson, and M. Goff, J. de Phys., 1987, 48, suppl. C6, 577.

174. J.K.Olthoff, I. Lys, P. Demirev, and R.J. Cotter, Anal. Instrum., 1987, 16, 93.

175. J.K. Olthoff, J.P. Honovich, and R.J. Cotter, Anal. Chem., 1986, 59, 999.

176. F.J.Knorr, M. Ajami, and D.A. Chatfield, Anal. Chem., 1986, 58, 690.

177. F.J.Knorr, U.S. Patent 4,707,602, 17 Nov. 1987.

178. T. Matsuo, T. Sakurai, and H. Matsuda, Nucl. Instrum Methods Phys. Res., 1987, A258, 327.

179. K. Schey, R.G. Cooks, R. Grix, and H. Wöllnik, Int. J. Mass Spectrom. Ion Processes, 1987, 77, 49.

180. A.E. Litherland, Phil. Trans. R. Soc. Lond., 1987, A323, 5.

181. J C. Rucklidge, A.E. Litherland, and L.R. Kilius, J. Trace Microprobe Tech., 1987, 5, 23.

182. W. Wölfli, Nucl. Instrum. Meth. Phys. Res., 1987, B29, 1.

183. D. Elmore, Biol. Trace Element Res., 1987, 12, 231.

184. W. Kutschera, Nucl. Instrum. Meth. Phys. Res., 1988, A268, 552.

185. M.S. Burns, Ultramicrosc., 1988, 24,269.

186. Y. Nihei, H. Satoh, S. Tatsuzawa, M. Owari, M. Ataka, R. Aihara, K. Azuma, and Y. Kammei, J. Vac. Sci. Technol., 1987, A5, 1254.

187. N.A. Thorne and F. Degrève, Surf. and Interface Anal., 1988, 11, 189.

188. K.J.R. Rosman, W. Lycke, R. Damen, R. Werz, F. Hendrickx, L. Traas, and P. De Bièvre, Int. J. Mass Spectrom. Ion Processes, 1987, 79, 61.

189. W. Staringer, H. Groiss, Ch. Travniczek, and K. Varmuza, Mikrochim. Acta, 1986, 2, 187.

190. R.L. Johnson and L.C.E. Taylor, Org. Mass Spectrom., 1987, 22, 807.

191. C.E.D. Ouwerkerk, A.J.H. Boerboom, T. Matsuo, and T. Sakurai, Int. J. Mass Spectrom. Ion Processes., 1986, 70, 79.

192. J.S. Cottrell and S. Evans, Anal. Chem., 1987, 59, 1990.

193. A.E. Ashcroft, R.S. Brown, A.D. Coles, S. Evans, D.J. Milton, and B. Wright, Spectroscopy (Oregon), 1988, 3, 57.

194. L.C.E. Taylor, D.A. Brent, and J.S. Cottrell, Biochem. Biophys. Res. Commun., 1987, 145, 542.

195. G.H. Wang, W. Aberth, and A.M. Falick, Int. J. Mass Spectrom Ion Processes, 1986, 69, 233.

196. A. Hedin, P. Håkansson, and B.U.R. Sundqvist, Int. J. Mass Spectrom. Ion Processes, 1987, 75, 275.

197. S. Della-Negra, C. Deprun, and Y. Le Beyec, Rapid Commun. Mass Spectrom., 1987, 1, 10.

198. L.I. Grace, B.T. Chait, and F.H. Field, Biomed. Env. Mass Spectrom., 1987,14, 295.

199. M.J. Kristo and C.G. Enke, Rev. Sci. Instrum., 1988, 59, 438.

200. L.-Q. Huang, R.J. Conzemius, G.E. Holland, and R.S. Houk, Anal. Chem., 1988, 60, 1635.

201. L.-Q. Huang, S.-J. Jiang, and R.S. Houk, Anal. Chem., 1987, 59, 2316.

202. J.R. Dean, L. Ebdon, H.M. Crews, and R.C. Massey, J. Anal.At. Spectrom., 1988, 3, 349.

203. P.C. Goodley, and D.H.Loucks, Eur-Pat. Appl. No. 87112173.7 (Publ. No. 0265 617) 21.08.87.

204. K. Heckles, R.A.W. Johnstone, and A.H. Wilby, Tet. Lett., 1987, 28, 103.

205. B.T. Chait, T. Chaudhary, and F.H. Field, in "Methods of Protein Sequence Analysis", ed. K.A. Walsh, Humana Press, Clifton, NJ,1987, p483.

206. C.L. Wilkins, G.N. Giss, G.M. Brissey, and S. Steifer, Anal. Chem., 1981, 51, 113.

207. J.C. Demirigian, Trends Anal. Chem., 1987, 6, 58.

208. D.F. Gurka and R. Titus, in "Analytical Chemistry Instrumentation", ed. W.R. Laing, Lewis Publ. Inc., 1986, p17.

209. D. Robinson and R. Presser, Presented at the 35th ASMS Conference on Mass Spectrometry and Allied Topics, Denver, Co. 1987, p285.

210. C.L. Wilkins, Anal. Chem., 1987, 59, 571A.

211. H. Fujiwara, R.T. Solsten, and S.J. Wratten, Spectroscopy (Oregon), 1987, 2(10), 24.

212. E.S. Olson and J.W. Diehl, Anal. Chem., 1987, 59, 443.

213. W. Schwanebeck and H.W. Wenz, Fresenius Z. Anal. Chem., 1988, 331, 61.

214. T. Fawcett, Chemtech., 1987, 17, 564.

# 4
# Application of Computers and Microprocessors in Mass Spectrometry

BY J. R. CHAPMAN

## 1 Introduction

This review is again based on a computer search of appropriate keywords in the Mass Spectrometry Bulletin data base now accessed via the ESA Information Retrieval Service. This computer search was supplemented by a manual search of "CA Selects - Mass Spectrometry" which increased the number of references collected by almost two-thirds.

Areas which provide considerable current activity, measured in terms of papers published, include inorganic mass spectrometry and pyrolysis mass spectrometry as before together with a noticeable upsurge in the publication of relatively sophisticated applications of personal computers. Publications in the area of library search have diminished to be replaced by a consideration of more interpretative schemes, particularly those which make use of other types of spectroscopic data. Networking of computer systems has also become an important topic, to some extent in conjunction with the need to use data from other analytical techniques. The powerful multivariate analysis methods of principal components analysis and factor analysis have been applied extensively during this period.

The present review covers the scientific literature from July 1986 to June 1988. Within the main sections of the review, the references are arranged, as far as possible, under headings devoted to distinct analytical or instrumental techniques. In addition to these references, four books[1,2,3,4], which include chapters on data acquisition[1], data processing[1,2], library searching[3,4], and process control[2] in mass spectrometry, have been published during this period.

## 2 Instrumentation (Instrument Control and Data Acquisition)

**2.1 Computer Instrumentation.** - The period reviewed has seen the publication of a large number of papers describing the use of personal computers in mass spectrometry often to perform all of the required instrument control, data acquisition and data processing tasks. Generally these applications have been in more specialised areas which do not involve repetitive fast scanning and where there is no need for the manipulation of huge data matrices or large libraries of reference spectra as in routine GC/MS operation. Examples are the use of a quadrupole mass spectrometer and an Apple IIe computer to analyse volatiles in geological inclusions[5] (section 2.6), experiments based on electrostatic analyzer scanning under the control of an Apple IIe[6], the recording of transient phenomena in heterogeneous catalysis again using an Apple IIe and a quadrupole mass spectrometer[7] and a complete system for control, data acquisition and processing in ionic collision experiments which also includes a digitizer for input of graphical data from text[8]. Interfacing of a quadrupole mass spectrometer with a commercial ADALAB system for the Apple IIe and with a "home-built" ADAS system for the Commodore PET has also been described[9].

Another article describes data acquisition and instrument control with a negative-ion research mass spectrometer using a system based on an Apple II[10]. An important part of this system is a versatile interface card which can control a range of instrument scanning modes and simultaneously accumulate and display the data produced. Overall, the system was better than any commercially available multichannel analyzer. A paper from the University of Liege describes, in detail, the emulation of the functions of a commercial multichannel analyzer as part of a system based on hardware built round a Sinclair ZX Spectrum computer[11]. The use of a low cost computer for control and data acquisition, in this case for recording ionization efficiency curves, is justified by the existence of more sophisticated central computing facilities for data processing and storage.

Personal computers have also been applied in a number of other situations, for example linked thermogravimetry-mass spectrometry experiments (section 2.6) where two systems[12,13] based on UTI-100C quadrupoles have been used together with an IBM PC. In addition, the use of an IBM PC for data acquisition from a photoplate system in spark source mass spectrometry has been reported (section 2.4).

Another application of personal computers is to the measurement of nitrogen isotope abundance[14]. An interesting publication[15] describes the use of an 8-bit computer, based on a Z-80 chip and built from a kit, to control a commercial 1200 amu mass range quadrupole analyzer for organic analysis. Conversion to a 16-bit computer which provides more RAM for high mass (2000 amu) analysis is now in progress.

In a number of cases a personal computer is used for data analysis only. A particularly convenient configuration for this type of application is one in which the personal computer, whilst situated on the user's desk, is able to communicate with other processing stations and sources of data, including instruments, via a local area network (LAN)[16]. A local area network provides high speed data transfer between a mass spectrometer and the data processing station as well as making every computer in the network accessible to the user.

2.2 Expert Systems and Robotics.- In a recent review of artificial intelligence in chemistry, Gray[17] suggests that current expert system technology is appropriate to frequently occurring, limited-domain problems such as experiment planning and instrument optimization and control. Structure elucidation by combined spectral techniques, although an early application of artificial intelligence in chemistry, is not an ideal area because of the complexity of the task and the fact that it does not occur with sufficient frequency. One example of a suitable area is a continuation of the efforts of Wong and co-workers[18] in the application of expert systems to the operation of a triple quadrupole mass spectrometer (TQMS). This presentation discusses both suitable knowledge representation schemes and the design of an interface between the expert system and TQMS. Preliminary results on the optimization of the TQMS on chemical standards by this method are given.

Another area in which assistance to the initial instrument setting-up can be given is calibration for mass conversion. Jamieson and co-authors have described a simplex method for the optimization of time-to-mass calibration procedures using an Incos data system[19]. Another presentation[20] describes an automated computer based method for the determination of multiplier gain, an important parameter in routine operation. This article also includes a good discussion of the theory of signal amplification and the source and treatment of errors in this area. In the context of a critical assessment of methods for the construction of expert systems, Varmuza and co-workers[21] have implemented an expert system for computer-aided fault diagnosis in the vacuum system of a mass spectrometer.

The preparation and presentation of samples for analysis can benefit from the application of techniques for automation. Although strictly outside the timescale of this review, papers at the recent Bordeaux International meeting provided important examples of the application of robotics to sample handling. In one example[22,23], the authors found that a robotic system could be successfully used to load solid samples in solution, or in suspension, onto the filament of a direct chemical ionization (DCI) probe for automated analysis. A report has also appeared[24] which describes software used to control a smart autosampler for GC/MS analyses in an EPA laboratory where strict tuning and calibration criteria must be met before any samples can be analysed within a 12 hour period. An automatic sample loader for GC/MS on a LKB 9000 has been described in detail[25]. In this case a 8085 based microcomputer controls movement to vacuum, for example to position the sample in the vaporization zone, starts the mass spectrometer scan, dumps the sample when complete and requests the next sample. Other publications in the area of automation in GC/MS deal with interfacing a Finnigan OWA-30 mass spectrometer and a Carlo Erba Mega series gas chromatograph[26] and the automation of headspace concentration and sampling for GC/MS[27]. The automation of sample preparation for isotopic analysis based on a cryogenically operated gas trapping line has also been described[28].

2.3 Tandem Mass Spectrometry (MS/MS).— A new technique demonstrates the flexibility of a FORTH-based software control system for a triple quadrupole mass spectrometer (TQMS)[29]. Ion-molecule reactions can be studied with good sensitivity by trapping the reaction products in the central quadrupole of a conventional TQMS. Microcomputer controlled voltages applied to appropriate lenses are used to trap and then extract ions from this central quadrupole. A simple modification to the existing data acquisition algorithm permits this trap and pulse technique to be used in conjunction with conventional MS/MS methods such as daughter ion and parent ion scanning. Another article[30], again concerned with triple quadrupole instrumentation, describes the construction of a flexible data acquisition system, using the C programming language, that can fully exploit the present and future potential of experimentation with this type of instrument. This interesting article also discusses a number of practical aids, for example the use of colour, in processing "multi-dimensional" data from MS/MS experiments as well as the overall logic of processing this type of data.

A method whereby fragment ion spectra from a tandem magnetic sector MS/MS instrument may be acquired and processed by a standard mass spectrometry data system has been published[31]. The method is sufficiently general that the cell in which the precursor ions are collisionally activated may be floated at any desired potential. Other publications relevant to the use of MS/MS instruments based on high performance magnetic sector analyzers include a description of a computer-controlled instrument of hybrid geometry[32] and a program to calculate mass discrimination arising from energy release in the ion decomposition process[33]. Another paper[34] comments on the inability of data systems to recognise and deal with metastable peaks in normal spectra.

A feature of all MS/MS methods is the ability to screen out chemical noise through the additional analyzer stage. As a consequence, there is a requirement to be able to record a large dynamic range of signals at the final collector. A recent publication[35] describes a microcomputer based control system for a dual-output Channeltron detector that can collect analog ion current and ion-counting information simultaneously. Ion current and ion-counting data are scaled in software and provide consistent ion intensity values from both signals. This system, which provides a useful dynamic range over nine orders of magnitude, also features automatic over-current protection for the pulse-counting section of the multiplier.

2.4 Inorganic Mass Spectrometry.- A number of publications during the review period have discussed hardware and software associated with the recording of topological and depth profiling information using secondary ion mass spectrometry (SIMS) techniques. A video camera based digital imaging system supported by an extensive package of image acquisition and processing software has been developed for a Cameca ion microscope[36]. This system has been used to obtain the three-dimensional distribution of selected trace elements over single particles which are only a few micrometers in diameter. This system is based on software improvements that allow the optimization of both data acquisition time and microchannel plate multiplier gain in response to the changing intensity of the ion image[37]. In this way a considerably increased dynamic range of signal (8 orders of magnitude) may be recorded. Another paper[38] also discusses computer control software for the Cameca instrument but uses a resistive anode encoder system as detector.

Another paper using a Cameca ion microscope[39] describes the adoption of techniques routinely used in organic mass spectrometry to inorganic SIMS. In this case, accurate mass measurement is carried out by sequential analysis of the unknown and a reference sample followed by superposition of the two spectra. A relative precision in mass measurement of the order of 10 ppm is achieved by this means. These authors also describe the use of a sensitive spectrograph system which, although covering a restricted mass range, decreases acquisition times and allows the observation of transient phenomena.

Other papers describe data systems for depth profiling using ion microprobe instruments. In one example[40,41], a single centered gated area is used to sample only secondary ions from the centre of the crater. This gated area is divided in software into 8x8 sub-areas for use as a variable gate. By this means, post-acquisition optimization of dynamic range and detection limit becomes possible and sample alignment problems are reduced. A framestore data system has been used in SIMS analysis to store all relevant data during a depth profile experiment[42]. Various forms of depth profile may be reconstructed after data collection and this system also offers improvements to dynamic range in software. Two references[43,44] provide general descriptions of data acquisition and control facilities for quadrupole SIMS instrumentation. A customized atom probe instrument is described in two papers which cover the IBM PC-AT based computer system and the various instrumental operating modes[45] and the data analysis software which is written in C language[46].

Another form of topological analysis is available through the use of a laser microprobe. A LAMMA 1000 laser microprobe has recently been automated and programmed to produce ion maps with high lateral resolution from both organic and inorganic samples[47]. Hardware modifications are mainly concerned with the addition of stepper motors to the sample stage manipulators. Software is used to control stage movement and to acquire and manipulate data.

Currently the system has an upper limit of a 100x100 matrix for display of ion maps. Lastly, an old-established method of trace inorganic analysis, spark source mass spectrometry, has been updated. A system based on an IBM PC-AT controls a spectrum plate comparator for the automatic evaluation of data recorded on photographic plates[48]. Compared with the previous system, where data reduction was necessary because of limited memory, the present system can record complete transmittance information at 2 micron intervals. This ability to work with data from even the weakest lines results in improved detection of trace impurities.

2.5 Isotopic Analysis.- A new computer controlled detection system for high precision isotope ratio measurements from a double-collector instrument has been described[49]. In this method, resistive amplifiers have been replaced by low noise capacitative integrating detectors. A standard digital voltmeter is then used to alternately sample the two outputs at precise intervals over a period of time. When each output is plotted, the computer uses a linear regression method to calculate both slopes which are then ratioed to give the isotope ratio. A particular advantage of this method is that the standard error of the slope gives a measure of the system stability at the time of measurement. Other recent reports describe the automated measurement of carbon dioxide concentration and $^{13}C$ enrichment in carbon dioxide from breath and blood[50] and $^{13}C$ enrichment in carbon dioxide derived from leucine[51].

2.6 Thermal Desorption.- A number of "home-built" data systems for combined thermogravimetric-mass spectrometric analysis, all used with instrumentation based on quadrupole mass spectrometers, have been described. Two of these[12,13] are based on an IBM PC and offer data acquisition, instrument control and data display facilities. In one case[12], the C programming language is used. Another "home-built" quadrupole thermogravimetric system is based on a DEC PDP-11[52]. The description of this instrument also includes a detailed discussion of a fast digital filtering method for intensity data recorded during restricted mass range scanning. Another interesting computer-controlled thermogravimetric instrument presents material derived from the sample as a molecular beam[53]. By mechanical modulation of this sample beam and the use of appropriate data acquisition procedures a considerable enhancement of signal-to-noise characteristics can be achieved.

Two instruments provide computer based analysis of volatiles from vacuum degassing of various materials. In one case, the analysis is of volatiles in geological inclusions[5]. The computer (Apple IIe) recognises the bursting of a fluid inclusion from intensity changes at specified masses and then stores the relevant mass spectral scan data. Facilities for instrument control and data manipulation are also provided. In the other example[54], a computer assisted system is used to analyse gases released from bubbles in glass samples by vacuum degassing.

2.7 Time-of-flight Mass Spectrometry.- A special issue of the journal **Analytical Instrumentation**[55] has been devoted to advances in time-of-flight mass spectrometry. A number of the articles in this issue discuss instrumentation and, in particular, data systems for use with time-of-flight instruments. An article by Macfarlane and co-workers is one of three[56,57,58] which deal with high mass analysis following energetic particle bombardment. Macfarlane[57] describes how the use of more (e.g. five) calibration points for time-to-mass conversion leads to a more meaningful standard deviation for measured masses and how the continuous recording of these calibration masses during the entire experiment leads to increased accuracy. The measurement of the molecular ion of the peptide eglin at a mass of 8134.06±0.1 (literature value 8134.03) is given as an example of this new method. Improved methods of digital background correction are also described.

Publications from the groups of Standing[56] and Chait[58] offer more detail of data acquisition hardware. Both systems use time-to-digital converters for initial measurement of the flight time and then transfer this data to 32-bit systems, a Capro 68K[56] and a MicroVAX I respectively[58]. Both authors are able to demonstrate improvements in resolution over a wide mass range, in counting efficiency and in dynamic range, as a result of these hardware improvements. A very similar data acquisition system has been used by Glish and co-workers[59] to record data from a tandem quadrupole/time-of-flight instrument. The whole system is controlled by an IBM PC-XT and based around a time-to-digital converter and a CAMAC crate controller.

Allison et al[60] have reported on the further development of an integrating transient recorder for time array detection (TAD) in time-of-flight mass spectrometry in the same issue of **Analytical**

Instrumentation. This device is designed to cope with the very high
data rates anticipated when fast scanning time-of-flight techniques
are used in conjunction with high performance gas chromatography.
An eight-bit flash ADC is used to digitize the incoming signal at a
rate of 200 Msamples per second. Subsequently, at least ten
consecutive spectra are summed to reduce data rates and achieve a
maximum scan repetition rate of 1000 per second. Although much
processing is presently carried out post-run, transfer of this
activity to real time is in hand. Further development of this
system is awaited with interest.

2.8 Fourier Transform Mass Spectrometry (FTMS).- A number of
developments in this area have been recorded during the period of
the review. Following the derivation of expressions for the
expected precision in measurement of peak height, width and
position, a direct comparison reveals that experimental data is
significantly poorer than expected[61]. These results provide a
direct test of the nature of noise in Fourier transform spectra. An
automatic peak-unfolding routine, based on the use of a look-up
table, for low mass detection in FTMS has been reported[62]. This
method permits the use of a reduced digitizer rate to increase
resolution without sacrificing low mass information. A paper from
Wilkins and co-workers[63] describes the recording of accurate mass
data from GC/FTMS as an adjunct to the use of library search and
other spectral data (section 3.3).

A further report[64] of the maximum entropy method, which is
proposed as an alternative to the Fourier transform method of
frequency measurement, has appeared. This method may have some
potential as a means of achieving higher resolution than is normally
available. The Hadamard transform has been proposed as the basis of
a method for handling FTMS data from MS/MS experiments where several
primary ions have been simultaneously selected and reacted[65]. A
simple but versatile digital pulse generator has been developed
which can be used to control a FTMS instrument[66]. This generator
fills a single slot on an IBM PC and costs less than 200$ to
produce.

2.9 Process Control.- Two recent papers[67,68] have demonstrated
true feedback control of an aerobic digestion process by mass
spectrometry. For example, controlled addition of a carbon source,
regulated by the hydrogen signal from a mass spectrometer,

resulted in a high steady state rate of methanogenesis[67]. Another publication[69] describes a microprocessor controlled quadrupole mass spectrometer for fermentor gas analysis and control that is also easily interfaced to multiplexing schemes. Another microprocessor controlled quadrupole mass spectrometer for fermentor gas analysis has recently been described[70]. Additionally, Cottee and Blackwell[71] have discussed the potential of pyrolysis mass spectrometry for fermentation quality control.

Other examples of process control by mass spectrometry that have appeared during this period have been the control of multicomponent chemical streams related to the oil industry[72], process monitoring and quality assurance of polymeric materials by computer controlled pyrolysis-mass spectrometry[73] and the use of a multichannel quadrupole instrument for control of the deposition rate from an electron beam evaporation system[74]. Other related publications describe a commercial system intended for on-line process analysis[75], the real time analysis of gases used in general anaesthesia[76] and an ion mobility spectrometer used as the basis of an alarm system for chemical vapours[77](section 2.10).

2.10 Miscellaneous Instruments. - A few publications relating to data processing in ion mobility spectrometry have been recorded during this period. Analytical models of the ion drift process have been applied to the improvement of overall operation and data processing, including signal averaging[78], and a compact digital signal averager has been designed for these instruments[79]. Another publication[77] describes a complete system which comprises an ion mobility cell and a microprocessor and which uses pattern recognition methods to recognise the presence of specific chemical vapours. Miscellaneous instruments reported in the literature include a computer controlled mass spectrometer for energy and angular analysis of sputtered products[80] and a computerized two-colour picosecond laser mass spectrometer[81].

### 3 Data Analysis

#### 3.1 GC/MS

3.1.1 Target Compound Analysis.- A fully automated procedure for the targetted analysis of polychlorobiphenyls (PCB's), based on a reverse library search and described in the preceding review (SPR

Vol.9), has been extensively tested with PCB samples prepared using a number of work-up procedures[82,83] and extended to include the determination of chlorinated pesticides[83,84]. Whereas the targetted analysis procedure was developed to quantitatively determine specific compound types, a more recent publication reports on the use of a low resolution mass spectrometer as a selective detector for all chlorinated compounds[85]. Based on previous work by Anderegg[86], this program exhaustively searches recorded spectra for chlorine isotope patterns and then plots their presence as a specific "total chlorine chromatogram".

Varmuza and Lohninger[87] have used pattern recognition methods to define new features that can be used to recognise compounds belonging to a specific chemical class. In this case, these methods have been applied to GC/MS data to search for polycyclic aromatic hydrocarbons (PAH's). Again the graphical presentation used is in the form of a selective chromatogram, in this case for PAH's. A computerized GC/MS method for the identification of petroporphyrins which makes extensive use of GC retention index data has also been described[88].

3.1.2 GC/MS Profiling.- GC/MS profiling has been particularly applied in two areas - metabolic profiling and crude oil analysis. Two extensive reviews of metabolic profiling in the clinical field, which both contain important sections on automated GC/MS methods, have recently appeared[89,90]. The principal GC/MS method described in both reviews is MSSMET developed by Sweeley and co-workers[91]. Both MSSMET and a similar method[89] use a compound identification method based on library search and GC retention data and also provide statistical routines for profile comparison. Another publication[92] concentrates on the automated GC/MS profiling of organic acids in abnormal urines that have previously been screened by GC alone. The automated profiling described uses a dedicated organic acid library and will identify about 80% of all peaks.

In the field of crude oil profiling, an interesting paper describes the use of metastable ion monitoring to locate sterane and triterpane biomarkers[93]. Principal components analysis of data which summarises the abundance of these biomarkers allows good differentiation of oils from different fields and geological formations. The first three principal components could be correlated with geochemical processes. In another study[94] the distributions of tricyclic- and pentacyclic terpanes were

determined for 216 crude oils world-wide. A similar multivariate
analysis of this data again led to factors that could be correlated
with geochemical processes. Other authors[95] have suggested that
principal components analysis on original scanning GC/MS data is
more suitable than the analysis of data condensed in the form of
biomarker ratios for correlation with geochemical data.

### 3.1.3 Quantitative Determinations.-
Computer based methods have
been used in a number of cases to improve quantitative data acquired
by selected ion monitoring procedures. For example, following their
description of a linear regression method, used to obtain a response
ratio with improved precision from analyte and internal standard
selected ion monitoring data[96], Gaskell and co-workers have
investigated more commonly applied data smoothing methods and their
effect on precision and accuracy[97]. In general, these authors
found that data smoothing based on the moving average principle was
only able to achieve a minor improvement in precision and also
introduced a significant bias into the response ratio calculations.
Use of the linear regression method[96] achieved superior precision
without bias.

An article by Delaney[98] compares a multivariate approach with
conventional methods for the estimation of detection limits in
selected ion monitoring GC/MS. The multivariate method is based on
the use of principal components analysis to obtain a composite
response from selected ion monitoring data obtained at a number of
masses. This response is then regressed onto sample concentration
to evaluate instrumental and method detection limits. Principal
components analysis (in the form of the SIMCA program) has also been
used to analyse data from selected ion monitoring determinations of
polychlorinated dibenzodioxins and dibenzofurans[99]. The method
was used to define error boundaries for ion ratios which could be
used to accept or reject sample and calibration data. Principal
components analysis was also used in this investigation to classify
residue profiles according to sample origin. Schoeller[100] has
presented a general review of mass spectrometric calculations
covering isotope dilution, curve fitting and statistical techniques.

QSIMPS (a Quantitative Selected Ion Monitoring and Processing
System) is a collection of data acquisition and analysis methods
which use selected ion monitoring in drug pharmacokinetics[101].
The whole system has been operational for over 24 months and
demonstrates automation from sample injection to final data

reporting. Other recent reports cover automation of the
quantitative analysis of PCB's[102] and fluphenazine[103] by
selected ion monitoring methods.

Two articles[104,105] have covered quantitative analysis of
unseparated gas mixtures by scanning mass spectrometry. In a
theoretical study[104], a procedure for improving the accuracy of
quantitation by calibration with standard mixtures rather than pure
compounds and by correction of calibration coefficients is
described. The use of cross-correlation values at zero
displacement, rather than peak heights or areas, for the calculation
of ratios has been shown to lead to improved precision[105].

3.2 Library Search and Related Techniques.- A recent report[106]
provides an update on the status of the EPA-NIH data base of
evaluated electron impact spectra. A single spectrum for each
substance is selected from multiple spectra of the same substance by
an automated method with the aid of a quality index (QI). QI is the
product of a number of individual quality factors designed to
measure different properties of the mass spectrum. In this report,
some modification of the original quality factor definitions[107]
has been made and the redefined QI then used to select a new data
base which now contains 42,261 spectra. A special effort has been
made to provide a reference data base free of bias that can be used
with a variety of search systems.

Another collection of reference spectra[108] provides 2000
spectra, which represent 1817 compounds, as a tool for the analysis
of water pollutants by GC/MS. This collection includes a wide range
of organic and organometallic compounds. Although most modern
instruments provide spectra that are highly compatible with spectra
from standard data bases a word of caution has been offered[109]
with regard to spectra produced by the ion trap detector where a
tendency towards protonation under GC/MS conditions can be
misleading.

In some cases, spectra other than those produced by electron
impact ionization are more useful for library searching. For
example, positive ion chemical ionization spectra are better for the
identification of trichothecenes[110]. This publication describes a
library of such spectra as well as automatic routines for carrying
out searches on GC/MS data. Another publication[111] offers
software for the Atari 520 ST+ which can be used for processing mass
spectrometric data in forensic-toxicological studies.

A number of publications have stressed the use of GC retention
index information in conjunction with library search on GC/MS data.
For targetted analysis of specific compounds or compound types, the
most logical use of retention index data is as a pre-search filter.
A description of a library search system limited to environmental
chemicals[112] demonstrates exactly this procedure, i.e. a pre-
search based on GC retention index windows and a main search based
on a spectrum similarity index. Improved reliability and fast
retrieval is demonstrated. Another composite search which uses
Kovats indices as a pre-search filter[113] demonstrates greatly
reduced retrieval times. An application of composite searching to
hydrocarbon analysis describes a special algorithm which allows the
use of Kovats indices even when n-alkanes are absent[114]. Further
demonstrations of the additional use of retention index data have
been made in the analysis of volatiles in fragrances[115] and of
petroporphyrins[99].

A thorough investigation of various symmetric distance measures
for the comparison of unabbreviated mass spectra has been
published[116]. No one measure is optimal in all situations
especially in the presence of different types of noise.
Normalization of the spectrum to unit vector length or standard
measure can improve the results. Other publications in this area
include a molecular weight adjustment to the well-known McLafferty
probability based matching system[117] and an application of the
fitness index method to the differentiation and identification of
isomeric methylindoles[118].

### 3.3 Spectrum Interpretation; Use of Other Spectroscopic Data. -
Siegel and co-workers[119] have described a method of structural
analysis by mass spectrometry in which masses of fragment ions in
the spectrum are correlated with the masses of possible molecular
substructures. A complete set of substructures is generated most
efficiently from a set of so-called superatoms by a bond removal
method. This approach to structural analysis, which can be
supplemented by the use of fragmentation rules, has been applied to
a wide range of spectral types including the thermospray spectra of
oligomycin antibiotics. Two recent papers are relevant to
structural analysis using primary neutral losses. In one case[120]
the statistical properties of neutral losses from the molecular
mass of compounds with normal and "absent" molecular ions are
compared. Another paper[121] offers a program for calculating

possible compositions of fragments lost from the molecular mass. A
paper by Hippe and co-workers[122] describes a number of programs
for comparison of mass spectra and for structural analysis.

A review of the combination of FTIR and FTMS has appeared[123]
and a further paper in this area specifically describes the use of
accurate mass data to restrict library searches of FTIR and FTMS
data[63]. A time warping algorithm for matching chromatograms from
separate GC/FTIR and GC/MS analyses has also been described[124].
An interesting software package allows the origin of natural
products to be recognised through isotopic analysis using NMR and MS
data[125]. An extensive system of structural analysis by mass
spectrometry which, in turn, is part of an overall system which uses
other types of spectroscopic data has been described[126,127]. This
system uses a search based on characteristic fragment ions, neutral
losses or ion series to provide a "hit-list" of tentative matches.
Since the structures of these compounds are all fully coded, the
codes may then be searched to find any common substructures. The
use of mass spectral data in this way is only part of a multi-
dimensional spectroscopy system so that, for example, sub-structures
provided by the MS analysis may be subsequently input to a $^{13}C$ NMR
substructure generator to further limit the possibilities.

A number of other authors have discussed the computerized
interpretation of a combination of IR, NMR and MS data. For
example, Moldoveanu[128] has described a system whose data base of
substructures is limited to the information contained in common
correlation charts and can be held in a small computer. The system
tries to fit substructures suggested by NMR and IR within a
structure whose molecular weight is given by mass spectrometry. MS
fragment ion information can be used optionally to refine the
output. An expert system described by Hippe and co-workers[129]
covers a number of spectroscopic techniques and includes a structure
generator program. These same authors have presented a related
paper which describes a program for computer simulation of mass
spectra[130]. Other papers in this area cover the identification
of substructures from MS and $^{13}C$ NMR data[131], a clustering study
of IR, $^{13}C$ NMR and MS data[132] and a report on the use of an IBM
mainframe computer for the analysis of IR, NMR and MS data[133]. A
review of automated spectrum interpretation[134] covers conventional
library search as well as indirect database methods, such as
artificial intelligence, pattern recognition and spectrum
simulation, in IR, NMR and MS.

3.4 Interpretation of MS/MS Data.- In a recent review[135], Enke and co-workers have pointed out some limitations of the conventional library search approach, i.e. that the database can never be complete and that the technique loses some precision as the database expands and more similar spectra are entered. Whilst recognising the great value of library search in routine analysis, these authors go on to discuss the use of indirect database or interpretative methods for the analysis of complete unknowns and further stress the value of MS/MS data in this area. For example, following on from an earlier matching system for MS/MS spectra[136], these authors have developed[135] a computer method (MAPS) that searches for and identifies, as rules, the relationships between MS and MS/MS spectral features and chemical substructures. This MAPS system is part of a larger automated chemical structure elucidation system (ACES). In ACES, substructures from the MAPS system, together with possible empirical formulae determined by a new program from low resolution MS and MS/MS data, are fed into a structure generating system which outputs candidate structures. The ultimate goal of this work is to provide an expert system that includes a TQMS which carries out "intelligent" experiments via a feedback loop. Although the MS/MS matching scheme described[136] was able to mask some variations in spectra due to changes in instrumental conditions, MS/MS spectra are often very dependent on instrumental conditions and there is still no general database of MS/MS spectra or any standard conditions for collecting such spectra[137,138,139].

3.5 Pattern Recognition.- A report[140] on the classification and identification of toxic organic compounds in air by pattern recognition methods has appeared. The aim of this study was to develop pattern recognition procedures to supplement or complement the identification of target compounds by library search procedures. Reference spectra were converted to their autocorrelation spectra and these spectra classified by SIMCA procedures to locate natural data classes within the data. Unknowns could be assigned to one of these classes and then a k-nearest-neighbour technique used for identification within this class. The method worked well on target compounds but a major problem with untargetted compounds was misclassification of compounds that were not members of any modelled class[140,141]. An improvement to the k-nearest-neighbour algorithm has been tested on data sets of chemical ionization spectra recorded using transition metal ions as the ionizing agent[142]. The evaluation of the use of these metal ion reagents with various organic compounds showed that Sc+ gave a perfect classification of ketones, aldehydes and ethers[143]. In this same study, features selected by pattern recognition methods were examined to shed some light on reaction mechanisms.

Principal components analysis has been applied to the analysis of involatile organic compounds using a laser microprobe mass spectrometer[144]. These mathematical techniques are particularly useful in helping to overcome problems in chemical identification due to the relatively low mass resolution of the instrument. A theoretical paper by Malinowski[145] discusses the distribution of error eigenvalues resulting from principal components analysis of spectral data. A fuzzy classification approach is appropriate when the classes concerned do not have sharply defined boundaries. An initial investigation of the use of this method in mass spectrometry has been directed towards the differentiation of double bond isomers of dodecadienol and derivatives[146].

Pattern recognition methods have also been applied to MS/MS data. For example, principal components analysis has been recommended for the analysis of small but significant differences in mass spectra and used to differentiate isomeric $C_9H_{12}$ alkylbenzenes using MS/MS and conventional electron impact data[147]. Similar methods have been proposed for hydrocarbon analysis in jet fuels using MS and MS/MS data[148,149]. Several other applications of pattern recognition methods to GC/MS analysis[87,93,94,95,99] have already been discussed in section 3.1. Pattern recognition methods are now universally used in processing pyrolysis-mass spectrometry data (section 3.8) and a number of interesting applications in inorganic analysis[150,151,152,153] can be found in section 3.9.

3.6 Biopolymer Sequencing.- A number of authors have discussed the sequencing of peptides from FAB(MS) data. Ishikawa and Niwa[154] have described a program which, using a background corrected FAB spectrum, searches for a three amino acid starting sequence and then extends this sequence whilst checking against the spectrum. The correct sequence was always the best fit when amino acid analysis data was included, but otherwise the results were much more ambiguous. The use of CID spectra from FAB quasimolecular ions provides data that is more suitable for sequencing and Hamm et al[155] have given details of a program in which all possible amino acid sequences are generated from a known amino acid composition

and checked against the CID spectrum. Scoble et al[156] have
commented that the requirement for an amino acid analysis removes
one advantage of a CID experiment, i.e. the ability to directly
analyse impure or mixed peptides. Following a detailed examination
of CID spectra of peptides, these authors have developed a program
in which sequence details are elicited from the CID data alone. An
emphasis on fast graphics displays and on user interaction adds to
the flexibility and utility of this approach. Another algorithm
which can be applied to both FAB and CID-FAB data and which also
does not require the input of amino acid analysis data has been
described more recently[157].

Several other papers relating to computerized biopolymer
sequencing have been published during this period. Jankowski and co-
workers[158] have described a program for sequence searching in the
FAB spectra of oligonucleotides whilst two publications[159,160]
offer programs for deriving the amino acid sequence of cyclic
peptides from electron impact data. Erickson and Jardine[161] have
implemented the program PEPALG, originally reported by Biemann and
co-workers, in a form suitable for use on a personal computer. In
this case, peptide sequences are derived from a GC/MS analysis of
suitable derivatives following an enzyme digest.

3.7 Miscellaneous Organic Analysis.- A general method for the
deconvolution of overlapping chromatographic peaks based on factor
analysis has been applied to poorly resolved GC/MS as well as LC/UV
and GC/FTIR data[162]. Estimates of compound concentrations and
their full spectra are available without any prior assumptions about
chromatographic peak shape. Factor analysis has also been applied
to the analysis of unseparated mixtures using MS and $^{13}$C NMR
data[163]. Geladi and Wold[164] have proposed a new method for pre-
processing more complex mixture spectra prior to the application of
curve resolution methods such as principal components analysis.
Another method for the analysis of superimposed mass spectra, based
on the location of characteristic masses, has also been
reported[165].

Blom and co-workers have published a number of papers on the so-
called "average mass" approach to isotopic analysis from mass
spectra. A first paper[166] describes the theoretical basis of the
calculation of total isotopic content directly from the average mass
or centroid of an isotopic cluster and applies the method to the
calculation of $^{13}$C content. A subsequent paper[167] considers

the applications and limitations of the same method when significant
interfering ions, e.g. $M^+$ and $(M-H)^+$ ions, are present. A third
paper[168] describes the application of this method as a means of
producing a linear calibration plot in quantitative determinations
using stable isotope dilution. Practical applications of this
method are awaited with interest.

A further report by Blom[169] has extended his work into the
determination of elemental compositions from an analysis of isotopic
abundance data in low resolution spectra. Tenhosaari[170] has also
published details of a program for the calculation of elemental
compositions from low resolution data, although this pattern
matching approach is less rigorous and, unlike Blom, does not
attempt to analyse the sources of error in the method. Other papers
concerned with the determination of elemental compositions discuss a
new algorithm for the determination of compositions from accurate
mass data using a personal computer[171] and the determination of
accurate mass measurements from high resolution ion counting
data[172]. A new program which eliminates the distortion of
ionization efficiency data due to the energy spread of the electron
beam and which can locate breaks in the ionization efficiency curve
to within a few hundredths of an electron volt has been
reported[173].

3.8 Pyrolysis-Mass Spectrometry.- Data processing of Curie-point
pyrolysis low-voltage electron impact mass spectra, which was
discussed in some detail in an earlier review (SPR Vol.8), has again
been the subject of a large number of publications. Applications of
this established technique have included the characterisation of
such diverse materials as fossil fuels[174,175,176,177],
microorganisms[178,179,180,181,182], soils[183], orange juice[184],
and organics in chalk[185]. The same multivariate statistical
techniques have been employed in data processing even when low-
voltage electron impact ionization has been replaced by soft
ionization techniques, particularly field
ionization[186,187,188,189,190,191], as well as in other similar
experiments which have investigated atmospheric pressure
ionization[192] and laser desorption/ionization[193].

The standard data processing procedure uses normalization and
autoscaling followed by factor analysis which transforms the
original pyrolysis-MS data into a new set of independent variables.
Factor analysis is an efficient method of reducing data

dimensionality and provides a very suitable method for the
differentiation of sample classes. However, other methods, more
recently introduced, have then to be used to provide data that
offers a chemical interpretation of the differences underlying the
separate classes. Principal amongst these methods is the variance
diagram (VARDIA) method[175,194] which has recently been extended to
the analysis of time resolved pyrolysis-MS data to extract the time
profiles and spectra of separate components as they are
evolved[175,196]. Unsupervised data analysis methods have also been
used by Voorhees and co-workers to extract information on chemical
characteristics[197]. The VARDIA method has been applied by Windig
and co-workers to non-autoscaled spectra to extract pure component
spectra and quantitative data from relatively simple mixtures[198].
Canonical variates analysis has been used by some authors to
correlate pyrolysis-MS data with other sample related data[175,199].
Programs designed to analyse and correlate carbohydrate pyrolysis
products in a thermogravimetric-MS experiment have also been
reported[200].

3.9 Inorganic Analysis.- Factor analysis methods have been applied
to data from a number of surface analysis techniques[152,153,201].
For example, factor analysis of Auger data acquired during depth
profiling experiments offers an improvement in detection limits over
conventional methods since information from all data channels is
used[153]. Factor analysis does not offer the same improvement of
detection limits in secondary ion mass spectrometry (SIMS) but does
handle interfering species in this technique and in Auger
spectroscopy[153]. The application of factor analysis in SIMS,
Auger spectroscopy and XPS has been reviewed recently[152]. Pattern
recognition methods have been applied to the complex SIMS spectra of
aluminium and manganese borides to select features that are compound
specific[151]. A software package for the quantitative analysis of
SIMS images has been presented[202]. The package uses digital image
processing and pattern recognition techniques. Another
publication[203] describes digital image processing software that is
now available for the direct correlation of ion and electron
micrographs. Data obtained on different instruments and with
different detectors may be compared in this case.

Data processing techniques have also been applied to a number of
mass spectrometric techniques used for bulk rather than for surface
analysis of inorganic materials. Two papers[204,205] have been
published which describe improved methods for the calculation of

sensitivity coefficients from standard materials in the evaluation of spark source mass spectrometry data. Principal components and cluster analysis of spark source data has been used to categorise metallic samples on the basis of trace element distributions[150]. The calculation of quantitative data in analysis by glow discharge mass spectrometry has also been discussed recently[206]. Vaughan and Horlick[207] have presented details of a computerized reference manual for spectral data and interferences in inductively coupled plasma-MS (ICP/MS) written for the Apple Macintosh. A statistical treatment of lead isotope composition data leads to an improved correlation with age in geochronological studies[208].

## 4 Other software

**4.1 Instrument Design.**— A computer program has been used to calculate ion flight times, velocities, and energies for ions with different initial starting positions, kinetic energies, and directions in a time resolved momentum spectrometer (TRIMS)[209]. This MS/MS instrument, based on a magnetic analyzer, uses the magnet to provide momentum data and time-of-flight measurements to provide velocity data. Other published applications to magnetic sector instruments include the use of the CHEOPS program to calculate ion trajectories in the extraction and acceleration regions of a desorption ionization source[210] and the use of the TRIO program to investigate the ion optics of a two magnet analyzer system which offers focal plane detection with a linear mass scale and a wide mass range[211].
A number of papers by Szabo and Haegg have presented computer simulations of ion trajectories in multipole analyzers operating in conventional[212,213] and RF-only[214] modes. Other published examples of computers used in instrument design include a Monte-Carlo simulation of charged particle analyzer line shapes[215], design of a lens system for a quadrupole SIMS instrument[216], considerations of the effect of ion formation conditions on transmission in a laser microprobe mass analyzer (LAMMA)[217], and the calculation and simulation of a resonance magnetic mass spectrometer[218].

**4.2 Miscellaneous.**— A number of papers which present computer simulations of the sputtering process have appeared. A Monte-Carlo simulation of sputtering (TRIM) has been reviewed and various applications of the program, including element mapping with the scanning ion microprobe, discussed[219]. A computer code (ITMC) has been developed for a detailed study of the transport of charged particles in solid materials and of surface related phenomena such as sputtering[220]. Different models for computer simulation of the sputtering of metal targets by keV ions have also been compared[221]. Two papers which deal with collision processes, i.e. ion-atom collisions[222] and electron scattering by molecules[223], have been published. In teaching, a simulator for training in the use of a mass spectrometer[224] and a scheme for training in interpretative mass spectrometry[225] have been described.

## References

1. J.T.Watson, "Introduction to Mass Spectrometry", Raven Press, New York, 1985.
2. F.A.White and G.M.Wood, "Mass Spectrometry. Applications in Science and Engineering", John Wiley and Sons, New York, 1986.
3. S.R.Heller, "Computer-Supported Spectroscopic Databases", Ellis Horwood, Chichester, 1986.
4. G.Vernin, M.Petitjean, J.Metzger, D.Fraisse, K.N.Suon, and C.Scharff, "Capillary Gas Chromatography in Essential Oil Analysis", Dr. Alfred Heuthig Verlag, Heidelberg, 1987.
5. C.Barker and M.P.Smith, Anal. Chem., 1986, 58, 1330.
6. J.C.Traeger and A.A.Mommers, Org. Mass Spectrom., 1987, 22, 592.
7. D.Bianchi and J.P.Joly, Bull. Soc. Chim. Fr., 1985, 668.
8. C.Badrinathan and D.Mathur, Indian J. Pure Appl. Phys., 1987, 25, 308.
9. A.Viste, B.Prouse, and M.Renner, Proc. S. D. Acad. Sci., 1985, 64, 62.
10. R.Dressler, M.Gremaud, P.H.Chassot, and M.Allan, Chimia, 1985, 39, 327.
11. C.Servais, R.Locht, and J.Momigny, Int. J. Mass Spec. Ion Proc., 1986, 74, 179.
12. A.C.Liu and C.M.Friend, Rev. Sci. Instrum., 1986, 57, 1519.
13. T.O'Connor and J.A.Schreifels, Rev. Sci. Instrum., 1986, 57, 1213.
14. K.Samukawa and Y.Komai, Radioisotopes, 1987, 36, 440.
15. A.Slomp, G.Chiasera, C.Mezzena, and F.Pietra II, Rev. Sci. Instrum., 1986, 57, 2786.
16. J.R.Chapman and P.A.Ryan, TrAC, Trends Anal. Chem., 1988, 7, 244.
17. N.A.B.Gray, Anal. Chim. Acta, 1988, 210, 9.
18. H.R.Brand and C.M.Wong, Proc. 5th Conf. Artif. Intell. (AAAI-86), Philadelphia, 11-15 Aug. 1986.
19. W.D.Jamieson and R.Guevremont in "Advances in Mass Spectrometry 1985, Part B", John Wiley, Chichester, 1986, p.1223.
20. W.J.Fies Jr., Int. J. Mass Spec. Ion Proc., 1988, 82, 111.
21. W.Staringer, H.Groiss, C.Travniczek, and K.Varmuza, Mikrochim. Acta, 1986, 2, 187.
22. D.J.Martin, 11th International Mass Spectrometry Conference; Bordeaux, 1988, Abstract TUE 28.
23. D.C.Smith, V.Gould, M.J.Hammond, and D.E.Todd, 11th International Mass Spectrometry Conference; Bordeaux, 1988, Abstract TUE 41.
24. C.A.Koch and J.DeWald, LC-GC, 1988, 6, 150.
25. G.S.Kath, W.J.McKeel, J.L.Smith, and J.M.Liesch, Rev. Sci. Instrum., 1986, 57, 3114.

26. R.J.Pell and H.L.Gearhart, HRC CC, J. High Resolut. Chromatogr. Chromatogr. Commun., 1987, 10 365.
27. J.M.Zechman, S.Aldinger, J.N.Labows Jr., J. Chromatogr., 1986, 377, 49.
28. W.A.Brand, M.Weber-Grabau, and K.Habfast, "Advances in Mass Spectrometry 1985, Part B", John Wiley, Chichester, 1986, p.1065.
29. G.G.Dolnikowski, M.J.Kristo, C.G.Enke, and J.T.Watson, Int. J. Mass Spec. Ion Proc., 1988, 82, 1.
30. D.Jaquen, N.Morin, and C.Rolando, Spectra 2000, 1987, 15, 37.
31. R.K.Boyd, P.A.Bott, B.R.Beer, D.J.Harvan, and J.R.Hass, Anal. Chem., 1987, 59, 189.
32. D.Williams and C.Porter, Am. Lab. (Fairfield, Conn.), 1988, 20(2), 98.
33. B.A.Rumpf, C.E.Allison, and P.J.Derrick, Org. Mass Spectrom., 1986, 21, 295.
34. J.Meili, Org. Mass Spectrom., 1986, 21, 299.
35. M.J.Kristo and C.G.Enke, Rev. Sci. Instrum., 1988, 59, 438.
36. X.B.Cox, S.R.Bryan, R.W.Linton, and D.P.Griffis, Anal. Chem., 1987, 59, 2018.
37. S.R.Bryan, R.W.Linton, and D.P.Griffis, J. Vac. Sci. Technol., 1986, 4A, 2317.
38. R.L.Crouch and C.J.Hitzman, Microbeam Anal., 1986, 21, 101.
39. N.A.Thorne and F.Degreve, SIA, Surf. Interface Anal., 1988, 11, 189.
40. S.M.Daiser, H.Frenzel, and J.L.Maul, Mikrochim. Acta, 1987, 1, 371.
41. C.Scholze, H.Frenzel, and J.L.Maul, J. Vac. Sci. Technol., 1987, 5A, 1247.
42. M.Preuss, J.Wolstenholme, and V.Dammann, Fresenius' Z. Anal. Chem., 1987, 329, 211.
43. W.Hilgers and J.Herion, Surf. Interface Anal., 1986, 9, 71.
44. M.G.Dowsett, J.W.Heal, H.Fox, and E.H.C.Parker, Springer Ser. Chem. Phys., 1986, 44, 176.
45. M.K.Miller, J. Phys. Colloq. N.C2, 1986, 493.
46. M.K.Miller, J. Phys. Colloq. N.C2, 1986, 499.
47. Z.A.Wilk and D.M.Hercules, Anal. Chem., 1987, 59, 1819.
48. L.Radermacher, Mikrochim Acta., 1986, 2, 325.
49. B.R.McCord and J.W.Taylor, Anal. Chem., 1986, 58, 2589.
50. C.M.Scrimgeour and M.J.Rennie, Biomed. Environ. Mass Spectrom., 1988, 15, 365.
51. C.M.Scrimgeour, K.Smith, and M.J.Rennie Biomed. Environ. Mass Spectrom., 1988, 15, 369.
52. G.Varhegyi, F.Till, and T.Szekely, Thermochim. Acta, 1986, 102, 115.
53. J.Behrens Jr., Rev. Sci. Instrum., 1987, 58, 451.
54. N.Stoll, E.Hartung, B.Keinert, and K.Heide, ZfI-Mitt., 1986, 115, 195.
55. Analytical Instrumentation, 1987, 16(1).
56. K.G.Standing, R.Beavis, G.Bolbach, W.Ens, F.Lafortune, D.Main, B.Schueler, X.Tang, and J.B.Westmore, Anal. Instrum. (N.Y.), 1987, 16, 173.
57. R.D.Macfarlane, J.C.Hill, and D.L.Jacobs, Anal. Instrum. (N.Y.), 1987, 16, 51.
58. L.I.Grace, B.T.Chait, and F.H.Field, Biomed. Environ. Mass Spectrom., 1987, 14, 295.
59. G.L.Glish, S.A.McLuckey, and H.S.McKown, Anal. Instrum. (N.Y.), 1987, 16, 191.
60. J.Allison, J.F.Holland, C.G.Enke, and J.T.Watson, Anal. Instrum. (N.Y.), 1987, 16, 207.
61. L.Chen, C.E.Cottrell, and A.G.Marshall, Chemom. Intell. Lab. Syst., 1986, 1, 51.
62. R.B.Cody, J.A.Kinsinger, and S.D.Goodman, Anal. Chem., 1987, 59, 2567.
63. D.A.Laude Jr., C.L.Johlmann, R.S.Brown, and C.L.Wilkins, Fresenius' Z. Anal. Chem., 1986, 324, 839.
64. A.Rahbee, Int. J. Mass Spec. Ion Proc., 1986, 72, 3.
65. F.W.McLafferty, D.B.Stauffer, S.Y.Loh, and E.R.Williams, Anal. Chem., 1987, 59, 2212.
66. E.Y.Sidky, E.A.Wachter, and T.C.Farrar, Rev. Sci. Instrum., 1988, 59, 806.
67. T.N.Whitmore, D.Lloyd, G.Jones, and T.N.Williams, Appl. Microbiol. Biotechnol., 1987, 26, 383.

| | |
|---|---|
| 68 | D.Lloyd and T.N.Whitmore, Lett. Appl. Microbiol., 1988, 6, 5. |
| 69 | S.J.Coppella and D.Prasad, Biotechnol. Bioeng., 1987, 29, 679. |
| 70 | I.Berecz, S.Bohatka, Z.Dios, I.Gal, L.Kiss, R.M.Kovacs, G.Langer, J.Molnar, A.Paal, K.Sepsy, I.Szabo, and G.Szekely, Vacuum, 1987, 37, 85. |
| 71 | F.H.Cottee and I.G.Blackwell, J. Anal. Appl. Pyrolysis, 1987, 11, 549. |
| 72 | J.Shen, Arabian J. Sci. Eng., 1986, 11, 45. |
| 73 | H.L.C.Meuzelaar, W.Windig, S.M.Huff, and J.M.Richards, Anal. Chim. Acta, 1986, 190, 119. |
| 74 | W.Sevenhans, J-P.Locquet, and Y.Bruynseraede, Rev. Sci. Instrum., 1986, 57, 937. |
| 75 | V.H.Adams, Am. Lab. (Fairfield, Conn.), 1986, 18(12), 72 |
| 77 | J.E.Roehl, Opt. Eng., 1985, 24, 985. |
| 78 | E.A.Aronson, Rept. SAND-87-0072, 1987. |
| 79 | A.Prini, A.H.Lawrence, and S.Laframboise, J. Phys. E, 1987, 20, 1422. |
| 80 | P.Viaris de Lesegno, L.Naze, and J-F.Hennequin, "Advances in Mass Spectrometry, Part B", John Wiley, Chichester, 1986, p.925. |
| 81 | J.D.Simon, D.M.Szaflarski, and M.A.El-Sayed, Proc. Int. Conf. Lasers, 1984, 176. |
| 82 | A.L.Alford-Stevens, T.A.Bellar, J.W.Eichelberger, and W.L.Budde, Anal. Chem., 1986, 58, 2014. |
| 83 | A.L.Alford-Stevens, J.W.Eichelberger, and W.L.Budde, Environ. Sci. Technol., 1988, 22, 304. |
| 84 | A.L.Alford-Stevens, T.A.Bellar, J.W.Eichelberger, and W.L.Budde, Anal. Chem., 1986, 58, 2022. |
| 85 | S.Johnsen and K.Kolset, J. Chromatogr., 1988, 438, 233. |
| 86 | R.J.Anderegg, J. Chromatogr., 1983, 275, 154. |
| 87 | H.Lohninger and K.Varmuza, Anal. Chem., 1987, 59, 236. |
| 88 | J.P.Gill, R.P.Evershed, and G.Eglinton, J. Chromatogr., 1986, 369, 281. |
| 89 | T.Niwa, J. Chromatogr., 1986, 379, 313. |
| 90 | J.F.Holland, J.J.Leary, and C.C.Sweeley, J. Chromatogr., 1986, 379, 3. |
| 91 | C.C.Sweeley, J.Vrbanac, D.Pinkston, and D.Issachar, Biomed. Mass Spectrom., 1981, 8, 436. |
| 92 | M.F.Lefevere, B.J.Verhaege, D.M.Declerck, and A.P.De Leenheer, Biomed. Environ. Mass Spectrom., 1988, 15, 311. |
| 93 | N.Telnaes and B.Dahl, Org. Geochem., 1986, 10, 425. |
| 94 | J.E.Zumberge, Geochim. Cosmochim. Acta, 1987, 51, 1625. |
| 95 | E.Oberrauch, T.Salvatori, L.Novelli, and S.Clementi, Chemom. Intell. Lab. Syst., 1987, 2, 137. |
| 96 | G.C.Thorne, S.J.Gaskell, and P.A.Payne, Biomed. Environ. Mass Spectrom., 1984, 11, 415. |
| 97 | G.C.Thorne and S.J.Gaskell, Biomed. Environ. Mass Spectrom., 1986, 13, 605. |
| 98 | M.F.Delaney, Chemom. Intell. Lab. Syst., 1988, 3, 45. |
| 99 | D.L.Stalling, P.H.Peterman, L.M.Smith, R.J.Norstrom, and M.Simon, Chemosphere, 1986, 15, 1435. |
| 100 | D.A.Schoeller, J. Clin. Pharmacol., 1986, 26, 396. |
| 101 | W.A.Garland, J.Hess, and M.P.Barbalas, TrAC, Trends Anal. Chem., 1986, 5, 132. |
| 102 | J.Schulz, LaborPraxis, 1987, 11, 648. |
| 103 | M.Jemal, E.Ivashkiv, D.Both, R.Koski, and A.I.Cohen, Biomed. Environ. Mass Spectrom., 1987, 14, 699. |
| 104 | F.V.Bablievski, Anal. Chim. Acta, 1987, 201, 241. |
| 105 | A.Celikkaya and S.Suzer, Anal. Chem., 1986, 58, 3256. |
| 106 | D.T.Terwilliger, A.L.Behbehani, J.C.Ireland, and W.L.Budde, Biomed. Environ. Mass Spectrom., 1987, 14, 263. |
| 107 | G.W.A.Milne, W.L.Budde, S.R.Heller, D.P.Martinsen, and R.G.Oldham, Org. Mass Spectrom., 1982, 17, 547. |
| 108 | R.Massot and J.C.Leclerc, Comm. Eur. Communities, (Rep.) EUR, 1986, EUR 10388, 107. |
| 109 | W.M.N.Ratnayake, A.Timmins, T.Ohshima, and R.G.Ackman, Lipids, 1986, 21, 518. |
| 110 | D.W.Hewetson and C.J.Mirocha, J.-Assoc. Off. Anal. Chem., 1987, 70, 647. |

111 P.Roesner, R.Kuehnle, T.Junge, and P.Mueller, Mitteilungsbl.-Ges. Dtsch. Chem. Fachgruppe Chem.-Inf., 1987, 12, 10.
112 K.Masumoto and T.Yamamoto, Annu. Rep. Osaka City Inst. Public Health Environ. Sci., 1984, 47, 1.
113 C.Boniface, G.Vernin, and J.Metzger, Analusis, 1987, 15, 564.
114 E.M.Steward and E.W.Pitzer, J. Chromatogr. Sci., 1988, 26, 218.
115 H.Yamada, K.Harada, S.Nakamura, and S.Mihara, J. SCCJ, 1987, 21, 30.
116 F.Drablos, Anal. Chim. Acta, 1987, 201, 225.
117 D.B.Stauffer and F.W.McLafferty, Org. Mass Spectrom., 1986, 21, 313.
118 Y.Kuwahara, Y.Yonekawa, and T.Suzuki, Agric. Biol. Chem., 1987, 51, 573.
119 M.M.Siegel, N.Bauman, and G.T.Carter, Anal. Chim. Acta, 1986, 186, 163.
120 B.G.Derendyev and S.A.Nekhoroshev, Izv. Sib. Otd. Akad. Nauk SSSR, Ser. Khim. Nauk, 1987, 90.
121 G.J.Kleywegt and H.A.Van't Klooster, TrAC, Trends Anal. Chem., 1987, 6, 55.
122 J.Koziol and Z.Hippe, Chem. Anal. (Warsaw), 1986, 31, 597.
123 C.L.Wilkins, Anal. Chem., 1987, 59, 571A.
124 C.P.Wang and T.L.Isenhour, Anal. Chem., 1987, 59, 649.
125 G.G.Martin, F.J.C.Pelissolo, and G.J.Martin, Comput. Enhanced Spectrosc., 1986, 3, 147.
126 R.Neudert, W.Bremser, and H.Wagner, Org. Mass Spectrom., 1987, 22, 321.
127 W.Bremser and R.Neudert, Eur. Spectrosc. News, 1987, N.75, 10.
128 S.Moldoveanu and C.A.Rapson, Anal. Chem., 1987, 59, 1207.
129 Z.Hippe, J.Duliban, J.Koziol, and M.Mazur, Pr. Nauk. Inst. Chem. Nieorg. Metal Pierwiastkow Rzadkich Politech. Wroclaw, 1986, 55, 148.
130 E.Sorkau, B.Adler, G.Fic, and Z.Hippe, Chem. Anal. (Warsaw), 1986, 31, 377.
131 I.I.Strokov, I.V.Gritsenko, and K.S.Lebedev, Izv. Sib. Otd. Akad. Nauk SSSR, Ser. Khim. Nauk, 1987, 78.
132 J.Zupan and M.Novic, Vestn. Slov. Chem. Drus., 1986, 33, 163.
133 B.Hohne, T.Cozzolino, and W.Stapelkamp, Am. Lab. (Fairfield, Conn.), 1988, 20(2), 148.
134 G.Small, Anal. Chem., 1987, 59, 535A.
135 C.G.Enke, A.P.Wade, P.T.Palmer, and K.J.Hart, Anal. Chem., 1987, 59, 1363A.
136 K.P.Cross and C.G.Enke, Comput. Chem., 1986, 10, 175.
137 R.I.Martinez and S.Dheandhanoo, J. Res. Natl. Bur. Stand. (U.S.), 1987, 92, 229.
138 R.I.Martinez, Rapid Comm. in Mass Spectrom., 1988, 2, 8.
139 R.I.Martinez and R.G.Cooks, 35th Annual Conference on Mass Spectrometry and Allied Topics; Denver, June 1987, p.1175.
140 W.J.Dunn III, M.G.Koehler, S.L.Emery, and D.R.Scott, Chemom. Intell. Lab. Syst., 1987, 1, 321.
141 D.R.Scott, W.J.Dunn, and S.L.Emery, Environ. Sci. Technol., 1987, 21, 891.
142 R.A.Forbes, E.C.Tews, B.S.Freiser, M.B.Wise, and S.P.Perone, J. Chem. Inf. Comput. Sci., 1986, 26, 93.
143 R.A.Forbes, E.C.Tews, Y.Huang, B.S.Freiser, and S.P.Perone, Anal. Chem., 1987, 59, 1937.
144 R.A.Fletcher and L.A.Currie, Microbeam Anal., 1987, 22nd, 369.
145 E.R.Malinowski, J. Chemometrics, 1987, 1, 33.
146 Y.Gu, Org. Mass Spectrom., 1988, 23, 487.
147 J.J.Weber, J.VanThuijl, and H.J.DeJong, Anal. Chim. Acta, 1986, 188, 195.
148 H.L.Meuzelaar and W.H.McClennen, Gov. Rep. Announce. Index (U.S.), 1986, 86, Abstr.No. 651,882.
149 H.L.Meuzelaar and W.H.McClennen, Gov. Rep. Announce. Index (U.S.), 1986, 86, Abstr.No. 651,881.
150 X.D.Liu, F.Michiels, P.VanEspen, and F.Adams, Mikrochim. Acta, 1987, 3, 49.
151 P.Wilhartitz and M.Grasserbauer, Mikrochim. Acta, 1986, 2, 313.
152 J.S.Solomon, Thin Solid Films, 1987, 154, 11.
153 S.W.Gaarenstroom, Appl. Surf. Sci., 1986, 26, 561.
154 K.Ishikawa and Y.Niwa, Biomed. Environ. Mass Spectrom., 1986, 13, 373.
155 C.W.Hamm, W.E.Wilson, and D.J.Harvan, Comput. Appl. Biosci., 1986, 2, 115.
156 H.A.Scoble, J.E.Biller, and K.Biemann, Fresenius' Z. Anal. Chem., 1987, 327, 239.

157 M.M.Siegel and N.Bauman, Biomed. Environ. Mass Spectrom., 1988, 15, 333.
158 F.Soler, K.Jankowski, H.Virelizier, D.Gaudin, J.Ulrich, and R.E.A.Teoule, J. Bioelectr., 1986, 4, 43.
159 A.A.Tuinman, Bioact. Mol., 1986, 1, 215.
160 K.Ishikawa, Y.Niwa, K.Hatakeda, and T.Gotoh, Org. Mass Spectrom., 1988, 23, 290.
161 B.J.Erickson and I.Jardine, Biomed. Environ. Mass Spectrom., 1986, 13, 343.
162 R.F.Lacey, Anal. Chem., 1986, 58, 1404.
163 R.A.Hearmon, J.H.Scrivens, K.R.Jennings, and M.J.Farncombe, Chemom. Intell. Lab. Syst., 1987, 1, 167.
164 P.Geladi and S.Wold, Chemom. Intell. Lab. Syst., 1987, 2, 273.
165 L.Alder, R.Donau, and C.Krueger, ZfI-Mitt., 1986, 135.
166 K.Blom, C.Dybowski, B.Munson, B.Gates, and L.Hasselbring, Anal. Chem., 1987, 59, 1372.
167 K.F.Blom, Anal. Chem., 1988, 60, 966.
168 K.F.Blom, Org. Mass Spectrom., 1987, 22, 530.
169 K.F.Blom, Org. Mass Spectrom., 1988, 23, 194.
170 A.Tenhosaari, Org. Mass Spectrom., 1988, 23, 236.
171 B.V.Ioffe and I.G.Zenkevich, Zh. Org. Khim., 1986, 22, 2245.
172 V.V.Raznikov and M.O.Raznikova, Int. J. Mass Spec. Ion Proc., 1988, 85, 1.
173 V.V.Raznikov, A.F.Dodonov, and V.V.Zelenov, Int. J. Mass Spec. Ion Proc., 1986, 71, 1.
174 H.L.C.Meuzelaar, W.Windig, J.H.Futrell, A.M.Harper, and S.R.Larter, ASTM Spec. Tech. Publ., 1986, 902, 81.
175 G.S.Metcalf, W.Windig, G.R.Hill, and H.L.C.Meuzelaar, Int. J. Coal Geol., 1987, 7, 245.
176 M.Nip, W.Genuit, J.J.Boon, J.W.DeLeeuw, P.A.Schenck, M.Blaszo, and T.Szekely, J. Anal. Appl. Pyrolysis, 1987, 11, 125.
177 T.Chakravarty, W.Windig, K.Taghizadeh, and H.L.C.Meuzelaar, Energy Fuels, 1988, 2, 191.
178 A.M.Donnison, C.Gutteridge, J.R.Norris, H.W.Morgan, and R.M.Daniel, J. Anal. Appl. Pyrolysis, 1986, 9, 281.
179 J.B.M.Droege, W.J.Rinsma, H.A.Van't Klooster, A.C.Tas, and J.Van der Greef, J. Chemometrics, 1987, 1, 231.
180 L.A.Shute, C.S.Gutteridge, J.R.Norris, and R.C.W.Berkeley, J. Appl. Bacteriol., 1988, 64, 79.
181 I.Brondz, J. Chromatogr., 1986, 379, 367.
182 C.S.Gutteridge, L.Vallis, and H.J.H.Macfie, Spec. Publ. Soc. Gen. Microbiol., 1985, 15, 369.
183 J.J.Boon, L.M.Dupont, and J.W.DeLeeuw, Peat Water, 1986, 215.
184 R.E.Aries, C.S.Gutteridge, and R.Evans, J. Food Sci., 1986, 51, 1183.
185 J.M.Bracewell, N.Pacey, and G.W.Robertson, J. Anal. Appl. Pyrolysis, 1987, 10, 199.
186 H-R.Schulten, J. Anal. Appl. Pyrolysis, 1987, 12, 149.
187 H-R.Schulten, N.Simmleit, and R.Mueller, Anal. Chem., 1987, 59, 2903.
188 N.Simmleit and H-R.Schulten, Fresenius' Z. Anal. Chem., 1986, 324, 9.
189 H-R.Schulten, R.Hempfling, and W.Zech, Geoderma, 1988, 41, 211.
190 H-R.Schulten and N.Simmleit, Comm. Eur. Communities (Rep.) EUR, 1988, EUR 11244, 602.
191 H-R.Schulten, N.Simmleit, and A.Marzec, Fuel, 1988, 67, 619.
192 A.P.Snyder, J.H.Kremer, H.L.C.Meuzelaar, W.Windig, and K.Taghizadeh, Anal. Chem., 1987, 59, 1945.
193 B.Lindner and U.Seydel, "Advances in Mass Spectrometry,Part B", John Wiley, Chichester, 1986, p.951.
194 W.Windig, W.H.McClennen, H.Stolk, and H.L.C.Meuzelaar, Opt. Eng., 1986, 25, 117.
195 W.Windig, E.Jakab, J.M.Richards, and H.L.C.Meuzelaar, Anal. Chem., 1987, 59, 317.
196 W.Windig, T.Chakravarty, J.M.Richards, and H.L.C.Meuzelaar, Anal. Chim. Acta, 1986, 191, 205.
197 S.J.Deluca, K.J.Voorhees, and E.W.Sarver, Anal. Chem., 1986, 58, 2439.

198 W.Windig, W.H.McClennen, and H.L.C.Meuzelaar, Chemom. Intell. Lab. Syst., 1987, 1, 151.
199 T.Chakravarty, H.L.C.Meuzelaar, P.R.Jones, and M.R.Khan, Prepr. Pap.-Am. Chem. Soc. Div. Fuel Chem., 1988, 33, 235.
200 A.E.Pavlath and K.S.Gregorski, J. Anal. Appl. Pyrolysis, 1987, 11, 341.
201 M.E.Kargacin and B.R.Kowalski, Anal. Chem., 1986, 58, 2300.
202 Y-C.Ling, M.T.Bernius, and G.H.Morrison, J. Chem. Inf. Comput. Sci., 1987, 27, 86.
203 L.K.Turner, Y-C.Ling, M.T.Bernius, and G.H.Morrison, Anal. Chem., 1987, 59, 2463.
204 W.Vieth, Spectrochim. Acta, 1987, 42B, 1085.
205 X.D.Liu, P.Van Espen, and F.Adams, Anal. Chim. Acta, 1987, 198, 71.
206 N.Jakubowski, D.Struewer, and W.Vieth, Anal. Chem., 1987, 59, 1825.
207 M.A.Vaughan and G.Horlick, Appl. Spectrosc., 1987, 41, 523.
208 O.G.Koshevoi, Izv. Akad. Nauk. Kaz. SSR, Ser. Geol., 1986, 75.
209 J.T.Stults, J.F.Holland, J.T.Watson, and C.G.Enke, Int. J. Mass Spec. Ion Proc., 1986, 71, 169.
210 C.E.D.Ouwerkerk and A.J.H.Boerboom, Int. J. Mass Spec. Ion Proc., 1986, 71, 59.
211 Z-H.Hu, H-N.Chen, A.J.H.Boerboom, and H.Matsuda, Int. J. Mass Spec. Ion Proc., 1986, 71, 29.
212 C.Haegg and I.Szabo, Int. J. Mass Spec. Ion Proc., 1986, 73, 277.
213 C.Haegg and I.Szabo, Int. J. Mass Spec. Ion Proc., 1986, 73, 237.
214 C.Haegg and I.Szabo, Int. J. Mass Spec. Ion Proc., 1986, 73, 295.
215 R.E.Negri and J.W.Taylor, Rev. Sci. Instrum., 1986, 57, 2780.
216 M.G.Dowsett, R.M.King, H.Fox, and E.H.C.Parker, Springer Ser. Chem. Phys., 1986, 44, 179.
217 M.De Wolf, T.Mauney, E.Michiels, and R.Gijbels, Scanning Electron Microsc., 1986, 799.
218 N.N.Aruev, E.L.Baidakov, B.A.Mamyrin, and A.V.Yakovlev, Sov. Phys.-Tech. Phys., 1987, 32, 303.
219 J.P.Biersack, Nucl. Instrum. Methods Phys. Res., 1987, B27, 21.
220 A.M.Hassanein and D.L.Smith, Nucl. Instrum. Methods Phys. Res., 1986, B13, 225
221 M.M.Jakas and D.E.Harrison, Nucl. Instrum. Methods Phys. Res., 1986, B14, 535
222 C.Bottcher and M.R.Strayer, Bull. Am. Phys. Soc., 1986, 31, 985.
223 R.K.Nesbet, Bull. Am. Phys. Soc., 1986, 31, 935.
224 M.Statheropoulos, Fresenius' Z. Anal. Chem., 1987, 328, 595.
225 J.R.Dias, Match, 1986, 17, 175.

# 5
# Organic Negative Ions: Structure, Reactivity, and Mechanism

BY R. A. J. O'HAIR AND J. H. BOWIE

## 1 Introduction

The material contained in this review has been selected from published work listed in the Mass Spectrometry Bulletin[1] for the period May 1986 - June 1988. Major journals have also been searched for the period January - June 1988. The policy adopted in previous reviews[2] has been retained - *selective* references are listed in each section; only those of *particular* interest in the context of this Chapter are treated in any detail.

In past reviews,[2] analytical methods using negative ions, and the basic negative ion fragmentations of major classes of compounds (e.g. peptides, carbohydrates, nucleotides, nucleosides), have been considered in detail. We have not done this in the present review: in certain sections lists of compounds studied by a particular analytical procedure are all that appear. Analytical Chemists can find references to their areas of interest by scanning the lists at the beginning of sections 2-5.

The emphases of this review are i) structures and fragmentations of organic and (selected) organometallic negative ions, and ii) the negative ion molecule chemistry of both organic and (selected) organometallic systems. Readers interested in these areas should skip the compound lists at the beginning of each of sections 2-5 and concentrate on the latter parts of sections 2-4 and all of section 6.

A variety of books[3-22] and comprehensive reviews[23-26] has appeared during the reviewing period. Reviews on subjects described in this Chapter are referred to at the appropriate place in the text. Other reviews of interest not specifically dealt with in the text, are listed here.[27-43]

## 2 Negative Ions formed by Electron Capture (or Dissociative Electron Capture): Experimental and Theoretical

The following species have been investigated:- small negative ions including $CH_3^-$,[44,46] $SiH_3^-$,[45-47] and $CH_5^-$;[48] ethane,[49] ethene,[50] ions in hydrocarbon flames,[51,52] carbon clusters,[53-55] fulvenes,[56] haloalkanes,[57-59] halobenzenes,[60,61] catecholamines,[62] nitroaromatics,[63] $(CO_2)_2^-$ ions,[64,65] the chloroacetate anion,[66] pentafluorobenzyl esters of unsaturated fatty acids,[67] succinylacetone,[68] methylamine,[69] dimethoxyamine,[70] alkylnitriles and isonitriles,[71,72] cyanogen,[73] amides of 2-diazomalonic acid,[74] 1,6-diazobicyclo[3.1.0]hexanes,[75] N-substituted α-pyrrolidones,[76] N-[(-)-jasmonoyl]-S-trytophan from *Vicia faba*,[77] 4-substituted 1-phenyl-3-methylpyrazol-5-ones,[78] 1,3,4,5-tetraarylimidazolidine-2-thiones,[79] 1,2-dihydro-3H-1,4-benzodiazepin-2-ones,[80] 8-hydroxyquinolines,[81] brominated 1-benzyl-1,2,3,4,-tetrahydroisoquinolines,[82] propafenone,[83] arachidonic acid,[84] griseochelin,[85] pipecolic acid,[86] methylbenzoquinones,[87] flavonoids,[88] 11-deoxyprostaglandins,[89] scapanin diterpenoids,[90] an iridoid glucoside from *Randia dumetorum*,[91] gibberelins,[92] brassinosteriods,[93] cyclopentanyl acetic acid glycopyranose esters,[94] dialkyl disulphides,[95,96] sulphones,[97,98] dinitrobenzoate diglyceride derivatives of phospholipids,[99] metal derivatives of dialkyl dithiophosphates,[100] cyclic borates,[101] and hetero-organic derivatives of group IV elements.[102]

Photoelectron spectroscopy of $SiH_3^-$ indicates that the electron affinity $(SiH_3)$ is $1.41 \pm 0.01$ eV, the bond dissociation energy $(H_3Si-H)$ is $378 \pm 10$ kJ mol$^{-1}$, and that $SiH_3^-$, like $CH_3^-$ is pyramidal.[45] *Ab initio* calculations (MP2/6-31++G$^{**}$) indicate that the most stable structure for the hypothetical species $CH_5^-$ is $[H^-(CH_4)]$(binding energy $H^-...C$ is 25 kJ mol$^{-1}$) and that this structure should not convert to $[CH_3^-(H_2)]$.[48] Studies using intersecting electron and molecular beams indicate ion pair processes **1** and **2** for ethane,[49] whereas ethene produces the ions shown in eqns **3-5**.[50] Premixed acetylene and benzene oxygen flames yield a number of interesting negative ions including $C_{60}^-$, $C_{74}^-$, $C_{78}^-$ and $C_{84}^-$.[51] It is suggested that the C60 species corresponds to the truncated icosahedron named buckminsterfullerene;[53] photoelectron spectra of $C_n^-$ species indicate that linear chains are formed for n=2-9, while monocyclic ring structures are formed when n=10-29.[54,55]

*Ab initio* calculations (MP3/6-31+G$^*$) indicate that the $D_{2D}$ structure **6** is the most stable $(CO_2)_2^-$ species,[64] and (3-21+d+sp) that the chloroacetate anion should decompose to $Cl^-$ plus an α-lactone (eqn **7**).[66] MINDO 3 calculations suggest that there

# 5: Organic Negative Ions: Structure, Reactivity, and Mechanism

$$C_2H_6 + e^- \longrightarrow (C_2H_6^{-\cdot})^* \begin{cases} C_2H_5^+ + H^- + e^- & (1) \\ CH_3^+ + CH_3^- + e^- & (2) \end{cases}$$

$$C_2H_4 + e^- \longrightarrow (C_2H_4^{-\cdot})^* \begin{cases} C_2H_3^+ + H^- + e^- & (3) \\ CH_4^{+\cdot} + C^{-\cdot} + e^- & (4) \\ CH_3^+ + CH^- + e^- & (5) \end{cases}$$

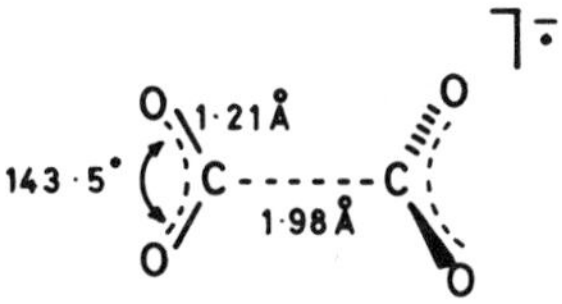

(6)

$$ClCH_2CO_2^- \longrightarrow \underset{O}{\triangle}{=}O + Cl^- \quad (7)$$

$$R^1CH_2CO\bar{C}HR^2 \longrightarrow {}^\cdot CH_2CO\bar{C}HR^2 + R^{1\cdot} \quad (8)$$

$$RCH_2CO\bar{N}H \longrightarrow {}^\cdot CH_2CO\bar{N}H + R^\cdot \quad (9)$$

$$Ar\bar{C}HOR \longrightarrow Ar\,CHO\rceil^{\bar{\cdot}} + R^\cdot \quad (10)$$

$$RCH_2CH_2O^- \longrightarrow [H^-(RCH_2CHO)] \longrightarrow R\bar{C}HCHO + H_2 \quad (11)$$

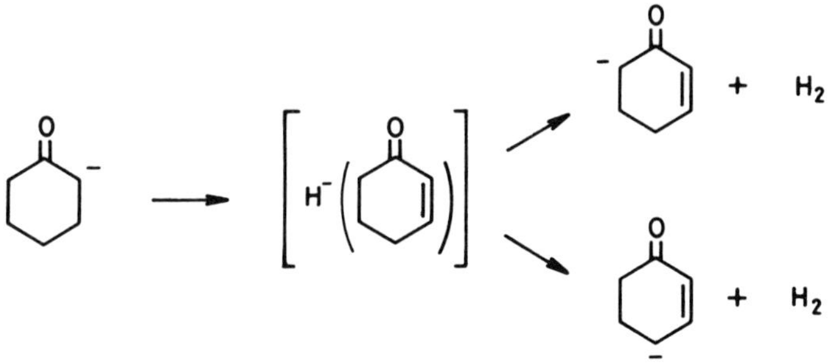

(12)

(13)

are two electronic states of the methylamine radical anion which undergo fragmentation.[69] Photoelectron spectroscopy of the thioformaldehyde radical anion indicates the electron affinity ($CH_2S$) to be $0.465 \pm 0.02$ eV.[96]

## 3 Negative Ion Chemical Ionization Mass Spectrometry

Reviews on various aspects of NICI have been published during the review period.[104-112] The different techniques of NICI have been detailed previously.[2,103] The following paragraph lists, in compound type, those systems whose negative ion spectra have been recorded by *any* NICI technique:- homosubstituted bicyclooctenes,[113] conjugated dienes,[114] toluene derivatives,[115] alkylbenzenes,[116] various aromatic hydrocarbons,[117-123] haloalkanes,[124-126] chloroepoxides,[127-128] toxaphene,[129] chlorophenols and chloroanilines,[130] aryl halo compounds,[131-136] halobenzofurans and benzodioxins;[137-141] alkoxide ions from methanol,[142] ethanol,[143] t-butanol,[144,145] aryl substituted alcohols;[146] polyhydroxy compounds,[147,148] phenols,[149] epoxides,[150] ethers,[151-153] 3-ethylpentan-2-one,[154] 3,3-dimethylheptan-4-one,[155] cyclohexanones,[156] acetylacetones,[157] conjugated ketones,[158-160] anthraquinone,[151] alkyl carboxylic acids,[162] various alkyl and aryl carboxylic acids,[163-168] alkyl esters,[169] dimethyl succinates,[170] alkyl malonates,[171] alkyl acetoacetates,[172] allyl phenyl acetates,[173] various alkyl and aryl esters,[174-178] leukotrienes,[179-183] prostaglandins,[184-198] thromboxanes,[199] arachidonic acid,[200] tricothelenes,[201-205] triterpene ethers,[206] friedelane derivatives,[207] cannabinoids,[208] 5-hydroxyisoavrainvilleol,[209] cholesteryl esters,[210] steriod sulphates,[211] annulated furanocoumarins,[212] flavanoid glycosides,[213,214] *myo*-inositols,[215] glycosides,[216] amines,[217-219] a pyrrole lactone from *Pisum Sativum*,[220] pyrrolizidine alkaloids,[221,222] ergotamine,[223] benzimidazoles,[224] tetrahydroisoquinolines,[225] quinolinic acid,[226] promazine,[227] triazines,[228] iormetazepan,[229] benzodiazepines,[230] purine,[231] dinucleotides,[232] amino acids,[233] formamide,[234] alkyl and aryl amides,[235] ochratoxin A,[236] 14-azadispiro(5,1,5,2,)pentadec-9-ene-7,15-diones,[237] peptides,[238-240] arylhydroxy oximes,[241] nitrosoamines,[242] aliphatic nitro compounds,[243,244] aryl nitro compounds,[245-255] nitrates,[256-258] nitriles,[259-262] isocyanates,[263] sulphur anion chemistry in flames,[264,265] phenyl 1,3-dithianes,[266] thiobenzoyl compounds,[267] sulphones,[268]

5: *Organic Negative Ions: Structure, Reactivity, and Mechanism*     149

sulphonated azo dyes,[269-271] various sulphonyl derivatives,[272,273] organophosphorus pesticides,[274-276] the trimethyl silyl anion,[277] anions containing multiple bonds to silicon,[278] and miscellaneous organometallics.[279-283]

The CA mass spectra of many $(M-H)^+$ ions derived by deprotonation ($HO^-$ or $NH_2^-$) of organic molecules have been studied, and general fragmentation rules for *even electron ions* have been reported.[104,146,235] The general rules are outlined below for negative ions including carbanions, enolate ions, and those ions where the negative charge is localised on nitrogen, oxygen or sulphur. Several examples of each type of fragmentation will be indicated as each fragmentation type is enunciated.

The four *basic* fragmentation types are:-

i)  simple cleavage reactions where loss of a radical forms a stable radical anion* e.g. eqns **8** (enolates),[154,155] **9** (amides)[235] and **10** (ethers)[151,152]

ii) reactions which occur by initial formation of an anion complex**,† which may then undergo a variety of reactions involving the incipient anion including deprotonation, $S_N2$ reactions, elimination reactions and direct displacement of the anion, e.g. eqns **11** (alkoxides),[143-147,178] **12** and **13** (cyclohexanones),[156,285,286] and **14-17** (esters).[170, cf 169,171,172,287]

iii) reactions which are not directed by the first-formed deprotonated species, but where proton transfer forms a new anion which may fragment through an ion complex [as in ii) above], e.g. **18** (ketones),[154,155,284-286] **19** (carboxylic-

---

\* There have been reports of fragmentations remote from the charged centre which are apparently not influenced by the charge.[284,308,309] A particular example is shown in section **4**, formula **29**.[309] While such fragmentations are certainly possible, we have not observed them (to date) in our simple systems.

\*\* The intermediacy of radical/radical anion complexes cannot be excluded in certain cases.[144,145,151,235]

† The possibility of concerted reactions must also be considered. The only substantiated concerted reaction that we have studied is $Me_3Si^- \longrightarrow [Me^-(Me_2Si)]^{\ddagger} \longrightarrow MeSiCH_2^- + CH_4$.[277]

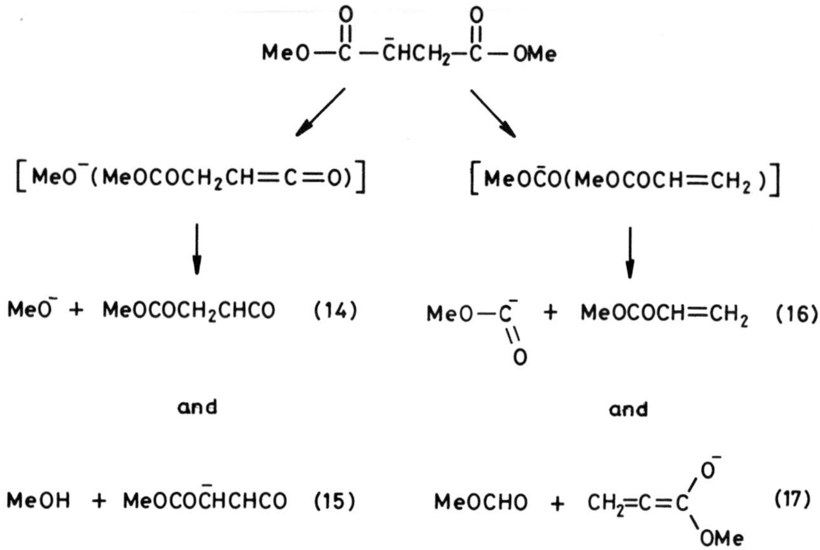

$$\text{MeO}-\overset{\overset{\text{O}}{\|}}{\text{C}}-\bar{\text{C}}\text{HCH}_2-\overset{\overset{\text{O}}{\|}}{\text{C}}-\text{OMe}$$

$[\text{MeO}^-(\text{MeOCOCH}_2\text{CH}=\text{C}=\text{O})]$

$\downarrow$

$\text{MeO}^- + \text{MeOCOCH}_2\text{CHCO}$ (14)

and

$\text{MeOH} + \text{MeOCO}\bar{\text{C}}\text{HCHCO}$ (15)

$[\text{MeO}\bar{\text{C}}\text{O}(\text{MeOCOCH}=\text{CH}_2)]$

$\downarrow$

$\text{MeO}-\underset{\text{O}}{\overset{-}{\text{C}}}_{\|} + \text{MeOCOCH}=\text{CH}_2$ (16)

and

$\text{MeOCHO} + \text{CH}_2=\text{C}=\text{C}\overset{\text{O}^-}{\underset{\text{OMe}}{\diagdown}}$ (17)

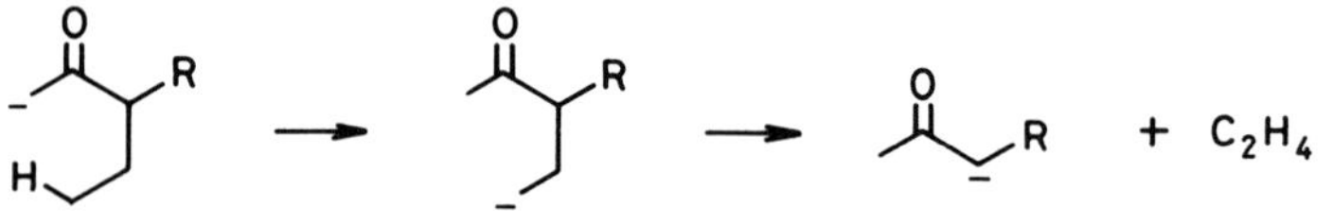

$\underset{\text{H}}{\overset{\text{O}}{\|}}\diagdown\text{R} \longrightarrow \overset{\text{O}}{\|}\diagdown\text{R} \longrightarrow \overset{\text{O}}{\|}\diagdown\text{R} + \text{C}_2\text{H}_4$ (18)

$^-\text{O}_2\text{CCHEt} \longrightarrow \text{HOCO}\bar{\text{C}}\text{Et}_2 \longrightarrow \text{HO}^- + \text{Et}_2\text{CCO}$ (19)

$\text{EtCO}\bar{\text{N}}\text{H}$

$\downarrow$

$\text{EtC(OH)}=\text{N}^- \longrightarrow [\text{HO}^-(\text{EtCN})] \longrightarrow \text{Me}\bar{\text{C}}\text{HCN} + \text{H}_2\text{O}$ (20)

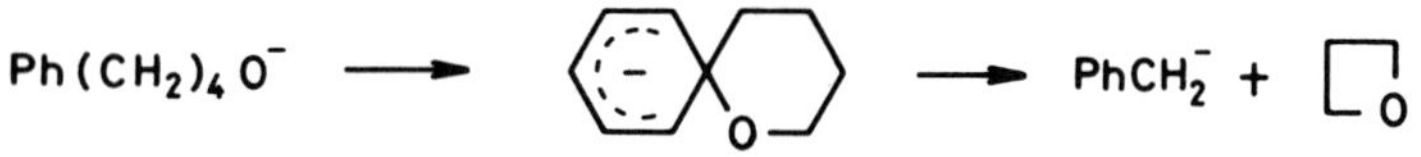

$\text{Ph(CH}_2)_4\text{O}^- \longrightarrow$ [spiro intermediate] $\longrightarrow \text{PhCH}_2^- +$ [cyclopentanone] (21)

5: *Organic Negative Ions: Structure, Reactivity, and Mechanism*  151

acids),[162] and **20** (amides).[235]
iv) rearrangement reactions, including internal nucleophilic substitution/displacement, e.g. eqn **21**,[146] and skeletal rearrangement reactions, e.g. eqn **22** (Wittig rearrangement),[151,152] **23** (Wittig -oxy Cope rearrangement),[153,288] and **24** (Claisen ester enolate rearrangement).[173]

Most of the other studied reactions fall into one or other of the categories outlined above. Several other reactions merit specific mention. The elimination of $H_2$ from $MeO^-$ occurs by the stepwise process $MeO^- \rightleftharpoons [H^-(CH_2O)] \longrightarrow HCO^- + H_2$, with both steps being rate determining.[142] The first specific double proton transfer in a negative ion reaction has been reported (eqn **25**).[146] Aryl ions $PhX^-(X=CHPh,$[115] $CPh_2$[115] or $Ph_2SiO$[278]) eliminate $C_4H_4$ from a benzene ring without either H or C scrambling. A proposed mechanism is shown in eqn **26**. Deprotonated alkyl aryl sulphones undergo an interesting proton transfer as evidenced by the processes shown in eqns **27** and **28**.[268]

4 Negative Ion Fast Atom Bombardment Mass Spectrometry

Several reviews[289-291] have appeared (other reviews will be listed at the appropriate place in the text), several papers have reported models for ionization in FAB,[292,293] and a CI/FAB source has been described.[294] FAB spectra have been reported for the following systems: glycols,[295,296] polyether dendrimes,[297] peat lubricants,[298] fatty alcohols,[299] hexanal metabolites,[300] anthraquinone,[301] carboxylic acids,[302-309] lipids,[310-320] leukotrienes,[321] oligomers,[322] diterpenes,[323-325] triterpene glycosides,[326-344] cardiac glycosides,[345] steroid analysis(review),[346] steroids and steroid sulphates,[347-350] steroid glycosides,[351-356] alimycins,[357] phthalide glycosides,[358] flavones and flavone glycosides,[359-368] anthocyanins,[369] resin glycosides,[370] tannins,[371,372] carbohydrates,[373-380] tetra-alkyl ammonium salts,[381,382] amino acids,[383-387] peptide sequencing (reviews),[388,389] various peptides;[390-402] nucleoside/nucleotide sequencing (review),[403] (reports);[404-412] nitroaromatics,[413] pyrrolizidine alkaloids,[414] nitrogen dyestuffs,[415] porphrins,[416] thiols,[417] sulphates and sulphonates,[418-421] organophosphorus pesticides,[422,423] and organometallics (review),[424] (reports).[425-429]

FAB is the usual m.s. method for peptides, carbohydrates,

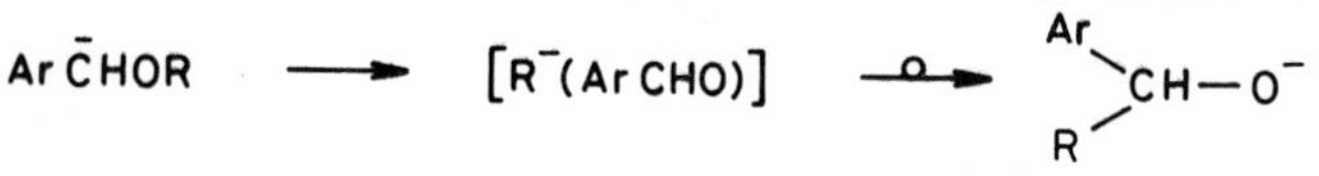

$$Ar\bar{C}HOR \longrightarrow [R^-(ArCHO)] \xrightarrow{\quad} \begin{array}{c} Ar \\ R \end{array}\!\!\!CH-O^- \qquad (22)$$

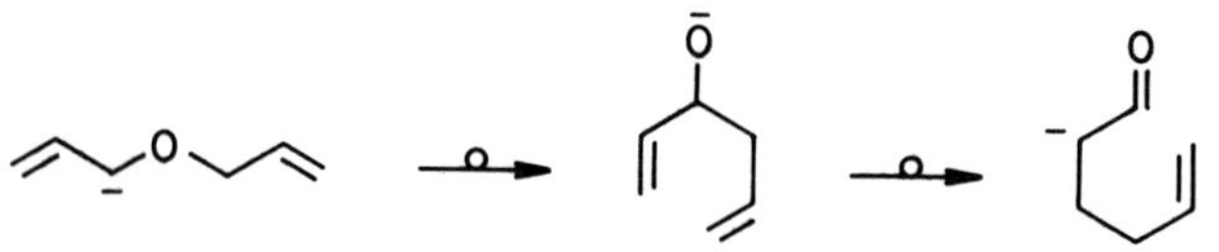

(23)

(24)

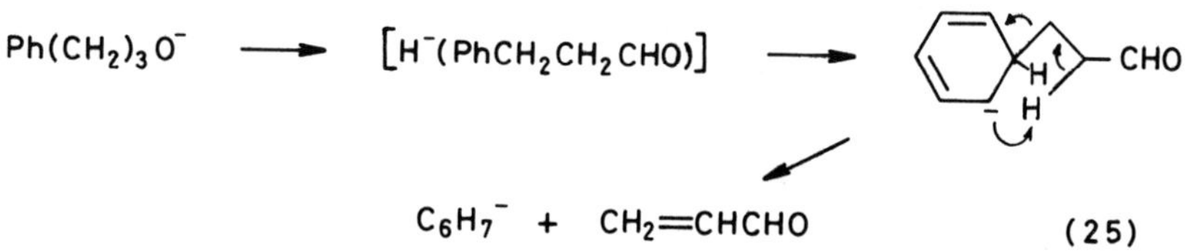

$$Ph(CH_2)_3O^- \longrightarrow [H^-(PhCH_2CH_2CHO)] \longrightarrow$$

$$C_6H_7^- + CH_2=CHCHO \qquad (25)$$

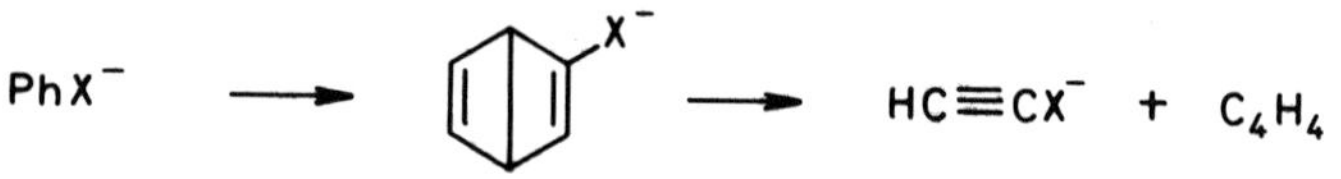

$$PhX^- \longrightarrow \qquad \longrightarrow HC\equiv CX^- + C_4H_4 \qquad (26)$$

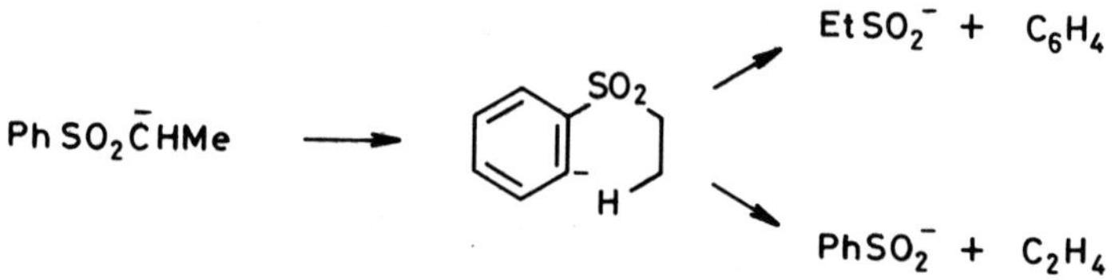

$$PhSO_2\bar{C}HMe \longrightarrow \qquad \begin{array}{c} \nearrow EtSO_2^- + C_6H_4 \quad (27) \\ \searrow PhSO_2^- + C_2H_4 \quad (28) \end{array}$$

(29)

glycosides, nucleotides and nucleosides, and their negative ion fragmentations have been reviewed in previous reports.[2] Many classes of compound yield (M-H)⁻ ions which undergo little fragmentation. In such cases, collisional activation is often applied in order to provide structural information. Particular examples chosen from the above literature which have used the collisional activation technique include fatty acids and lipids,[305,307-309] amino acids,[386] peptides,[395] nucleotides (sequencing),[412] and alkaloids.[413] Some interesting fragmentations include i) the collision induced cleavages of the long chain carboxylate species shown in **29**.[309] It is interesting to speculate whether these are 'remote' fragmentations (i.e. fragmentations neither initiated nor influenced by the centre of charge), or fragmentations initiated by specific proton transfers to the carboxylate centre (cf eqn **19**, sect **3**), ii) the collision induced retro Claisen reaction (eqn **30**) of carnitine derivatives.[386]

Some ions, particularly those derived from compounds containing C-O or P-O bonds undergo facile decomposition without the requirement of collisional activation. An example is avilamycin A whose negative ion cleavages are shown in formula **31**.[357] In other cases, a combination of positive ion and negative ion cleavages has aided structure determination, e.g. the β-lactam antibiotic **32** gave the positive (∿∿∿) and negative (────) fragmentations shown.[401]

### 5 Other Ionization Techniques

Several reviews on SIMS have been published during the review period.[430,431] Articles on negative ion SIMS include - fluorocarbon clusters,[432] carboxylic acids,[433,434] polymers,[435,436] lipids,[437] amines,[438] amino acids,[439,440] peptides[441,442] and nucleotides.[443,444]

A review on laser spectroscopy has appeared.[445] Laser desorption mass spectrometry has been used to determine the negative ion mass spectra of the following compounds:- carboxylic acids,[446,447] polyphenylenes and various aromatics,[448-451] amino acids,[452-454] glycosides,[455] porphrins,[456] organophosphorus pesticides,[457] and carbon cluster ions from boron carbide.[458]

$^{252}$Cf Plasma Spectroscopy has been used to form negative ions from:- aromatics,[459] lipids,[460,461] saccharides,[462] surfactants,[463] hydrazium salts,[464] nucleosides and nucleotides,[465-467]

(30)

(31) R = COCHMe$_2$

(32)

and porphrin sulphonates.[468] A review on field desorption mass spectrometry has been published,[469] together with an article on matrix effects in FD ms.[470] The nomenclature 'desorption/ionization' mass spectrometry has been used to describe the technique whereby ions are formed from a thermally labile species placed on a potassium thermionic emitter.[471] A review on ion photofragment spectroscopy is available.[472] Photoelectron spectra of radical anions may be determined.[473] The neutralisation-reionization mass spectra of alkoxide negative ions have been determined,[474] and further examples of the use of charge inversion reactions of negative ions have been reported.[475-477]

## 6 Ion Molecule Reactions and Related Topics

The work described in this section has been performed using ion cyclotron resonance, flowing afterglow or high pressure mass spectrometric techniques. There is some overlap with the chemical ionization work described earlier. The ion molecule work is divided arbitrarily into the following areas:- physical parameters, carbon chemistry, solvated (cluster) ions, silicon chemistry and organometallic chemistry.

Models for the interactions of anions with peptides have been reported,[478,479] pulsed electron high pressure mass spectrometry has been used to determine free energies for electron transfer equilibria,[480] the $\Delta H^o_{acid}$ values of 3 and 5 hydrogens of cyclohex-2-en-1-one indicate that deprotonation is more favoured at the 3 position by 27 kJ mol$^{-1}$,[481] and there have been several theoretical reports of acid base equilibria.[482-484] The reactions of many anions with dimethylfulvene have been reported; the general reaction sequence is shown in eqns **33** and **34**.[485] The most important property controlling anion reactivity in these reactions was found to be the extent of charge delocalisation of the anion. For example, localised anions like HO$^-$ and MeO$^-$ react exclusively by proton abstraction (eqn **33**), while delocalised ions like $CH_2=CHCH_2^-$ and $PhCH_2^-$ yield mainly adduct (eqn **34**).[485] Azole acidities have been determined.[486]

Laser photon spectroscopy of the formyl anion **35** (geometries at 6-311++G level[142]) indicates an electron affinity (EA) for the formyl radical of 0.313±0.005 eV (DCO$^\cdot$=0.301±0.005).[487] Ions resulting from deprotonation of MeCN and MeNC have been studied by a number of groups.[488-493] *Ab initio* calculations

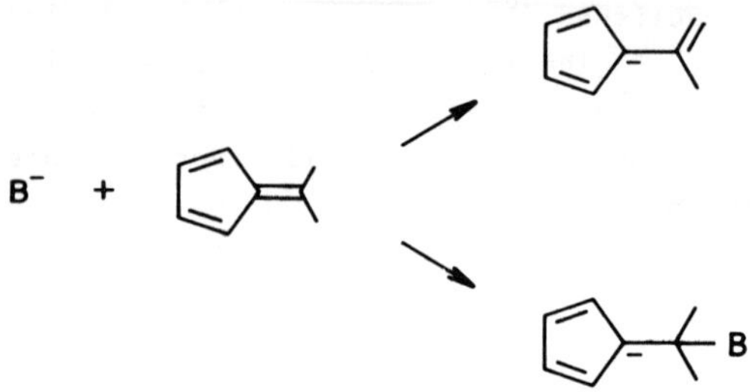

$$B^- + \text{[fulvene-methylene]} \longrightarrow \text{[cyclopentadienyl]}=\text{CH}_2 + BH \quad (33)$$

$$\searrow \text{[cyclopentadienyl]}-\text{C(Me)}_2-B \quad (34)$$

$$H \overset{1.18}{\underset{109.6°}{-}} \overset{-}{C} \underset{1.25\,\text{Å}}{=} O \quad (35)$$

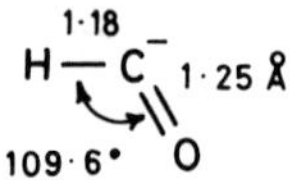

$$HO^- + MeNC \xrightarrow{0.67} {}^-CH_2NC + H_2O \quad (36)$$
$$\xrightarrow{0.33} CN^- + MeOH \quad (37)$$

$${}^-CH_2CN + PhCHO \longrightarrow [{}^-CH_2CN(PhCHO)] \quad (38)$$

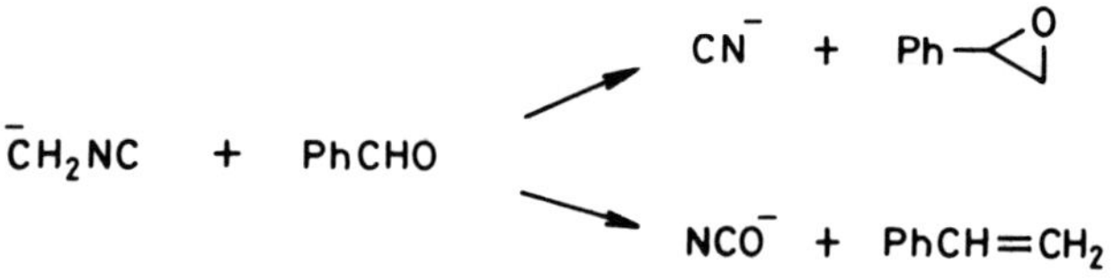

$${}^-CH_2NC + PhCHO \nearrow CN^- + Ph\text{-}\underset{O}{\triangle} \quad (39)$$
$$\searrow NCO^- + PhCH=CH_2 \quad (40)$$

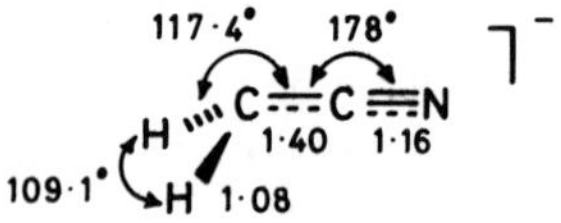

(41) $[H\text{-}C=C=N$, angles 117.4°, 178°, 109.1°, bonds 1.40, 1.16, 1.08]$^-$

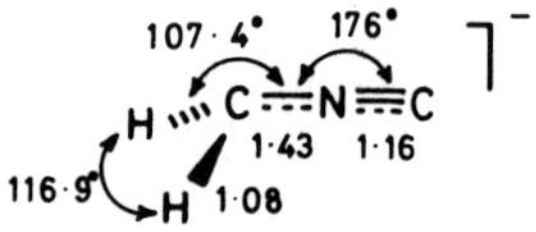

(42) $[H\text{-}C=N=C$, angles 107.4°, 176°, 116.9°, bonds 1.43, 1.16, 1.08]$^-$

(MP2/6-31+G$^*$//6-31+G$^*$) suggest that $^-CH_2CN$ is more stable than $^-CH_2NC$ by 38 kJ mol$^{-1}$.$^{(488)}$ Flowing afterglow studies indicate that i) the $\Delta H^o_{acid}$ values of MeCN and MeNC are 1557$\pm$12 and 1565$\pm$12 kJ mol$^{-1}$ respectively, ii) while MeCN is deprotonated exclusively by HO$^-$, MeNC can also undergo S$_N$2 reactions (eqns **36** and **37**), and iii) the two ions have different reactivities (e.g. eqns **38-40**).$^{489}$ Photoelectron spectroscopy has been used to determine EA $\cdot CH_2CN$ and $\cdot CH_2NC$ as 1.543$\pm$0.014$^{490}$ and 1.059$\pm$0.024 eV.$^{491}$ *Ab initio* calculations$^{488,490,491}$ and photoelectron spectroscopy$^{490-493}$ both indicate the slightly pyramidal geometries shown in formulae **41** and **42** (*ab initio* geometries$^{490,491}$ recorded). The following electron affinities have been reported:- ethyldiazoacetate;$^{494}$ o-benzyne ($C_6H_4$=0.56 eV),$^{495}$ cyanoethylenes and benzonitriles,$^{496}$ perfluorobenzene (0.52 eV),$^{497}$ and various quinones.$^{498}$

Several reviews on gas phase ion chemistry have been published.$^{499-501}$ The methylene radical anion$^{502,503}$ is a strong base ($\Delta H^o_{acid}$ $CH_3 \cdot$=1707$\pm$4 kJ mol$^{-1}$)$^{503}$ and a very reactive species; for example its reactions with CO and $N_2O$ are shown in eqns **43-46**.$^{503}$ The analogous ion $Ph_2C^{-\cdot}$ is also very reactive as evidenced by its reactions with $CS_2$ and $HCO_2Me$ (eqns **47-49**).$^{504}$ Decarboxylation of ions $RCO_2^-$ can be used to produce $R^-$ species when EA of $R\cdot$ is positive.$^{505}$ Some reactions of fluorocarbanions$^{506}$ and the allyl anion$^{507}$ have been discussed, homoconjugative stabilization in carbanions has been described,$^{508}$ and there have been a number of reports of the theoretical aspects of nucleophilic reactivity.$^{509-511}$

The reaction of HO$^-$ with diethyl ether has been shown to proceed exclusively by an E2 mechanism as shown by the evidence summarised in eqns **50** and **51**.$^{512}$ Elimination reactions of thioethers have been studied;$^{513,514}$ deprotonation of dialkyl sulphides followed by elimination occurs as shown in eqn **52** (cf eqn **18**).$^{513}$ The reaction between acetone and its enolate anion proceeds as shown in eqn **53**.$^{515}$ Further reactions of $O^{-\cdot}$ have been described.$^{516,517}$ The hydroperoxide ion $HO_2^-$ may be formed in the flowing afterglow by the reaction sequence shown in eqns **54** and **55**.$^{518}$ The ion is a powerful oxidizing agent, oxidising, for example $CO_2$ and $SO_2$ (eqns **56** and **57**) in fast reactions. It can also be used to form the peroxyformate ion (eqn **58**), which may undergo gas-phase Baeyer-Villiger reac-

$$CH_2^{-\bullet} + CO \longrightarrow CH_2=C=O + e^- \quad (43)$$

$$\begin{array}{c} CH_2N^- + NO^\bullet \quad (44) \\ \nearrow \\ CH_2^{-\bullet} + N_2O \longrightarrow [^\bullet CH_2-N=N-O^-] \longrightarrow CN^- + H_2 + NO^\bullet \quad (45) \\ \searrow \\ CN_2^{-\bullet} + H_2O \quad (46) \end{array}$$

$$\begin{array}{c} [Ph_2C^{-\bullet}(CS_2)] \quad (47) \\ \nearrow \\ Ph_2C^{-\bullet} + CS_2 \\ \searrow \\ Ph_2CS^{-\bullet} + CS \quad (48) \end{array}$$

$$Ph_2C^{-\bullet} + HCO_2Me \longrightarrow Ph_2C=CHO^- + MeO^\bullet \quad (49)$$

$$\begin{array}{c} EtO^- + H_2^{18}O + C_2H_4 \quad (50) \\ \uparrow \\ ^{18}OH^- + Et_2O \xrightarrow{E2} [H_2^{18}O \cdot C_2H_4 \cdot C_2H_5O^-]^\dagger \\ \downarrow \\ [EtO^-(H_2^{18}O)] + C_2H_4 \quad (51) \end{array}$$

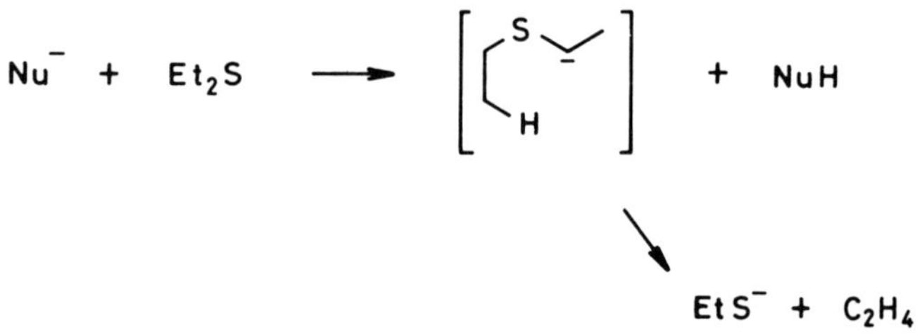

$$Nu^- + Et_2S \longrightarrow \left[ \begin{array}{c} S \diagdown \\ \diagdown \\ H \end{array} \right] + NuH$$

$$\downarrow$$

$$EtS^- + C_2H_4 \quad (52)$$

$$\underset{}{\overset{O^-}{\diagup\!\!\!\diagdown}} + MeCOMe \longrightarrow \underset{}{\overset{O\;\;\;\;O^-}{\diagup\!\!\!\diagdown\!\!\!\diagup\!\!\!\diagdown}} + CH_4 \quad (53)$$

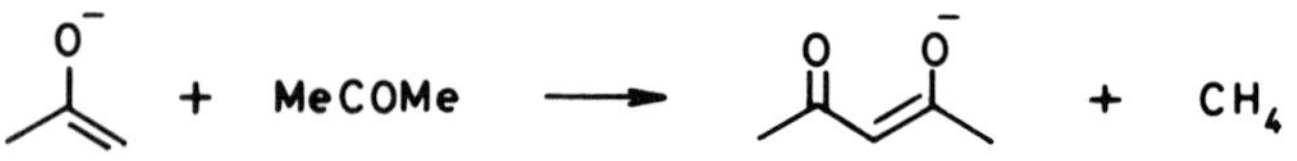

$$NH_2^- + \text{>}\!\!-\!\!\text{<} \longrightarrow \text{-}\!\!>\!\!-\!\!\text{<} + NH_3 \quad (54)$$

$$\text{-}\!\!>\!\!-\!\!\text{<} + O_2 \longrightarrow HO_2^- + \text{>}\!\!=\!\!\text{<} \quad (55)$$

$$HO_2^- + CO_2 \longrightarrow [HOOCO_2^-] \longrightarrow CO_3^{\bar{}} + HO^\bullet \quad (56)$$

$$HO_2^- + SO_2 \longrightarrow [HOOSO_2^-] \longrightarrow SO_3^{\bar{}} + HO^\bullet \quad (57)$$

$$HO_2^- + HCONMe_2 \longrightarrow H\overset{\overset{O}{\|}}{C}-OO^- + HNMe_2 \quad (58)$$

$$H\overset{\overset{O}{\|}}{C}OO^- + MeCOMe \longrightarrow \left[ H\overset{\overset{O}{\|}}{C}OO\overset{\overset{O^-}{|}}{C}Me_2 \right]$$

$$\downarrow$$

$$H-\overset{\overset{O}{\|}}{C}-O^- + MeOCOMe \quad (59)$$

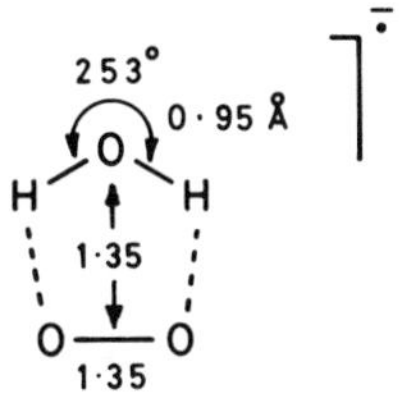

(60)

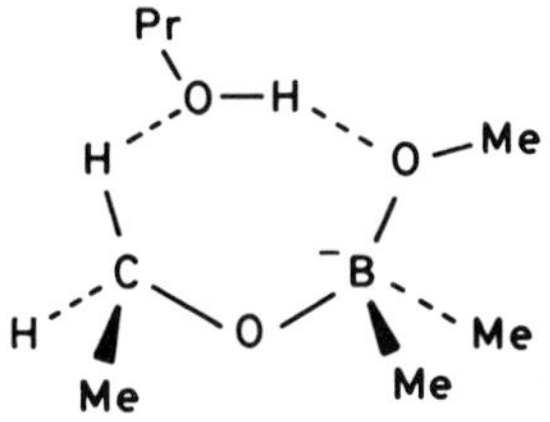

(63)

$$[Cl^-(HOMe)] + CF_3COCl \longrightarrow [Cl^-(CF_3COCl)] + MeOH \quad (61)$$

tions (eqn **59**).[518] Several reports of ion molecule reactions of halo systems have been published.[519,520]

Much effort has been focussed on solvated (cluster) anions during the review period. It is generally accepted that $H_3O^-$ corresponds to $[H^-(HOH)]$,[521,522] but it has also been suggested that there may be two forms of $H_3O^-$.[(523)] The ground state structure of $[O_2^{-\cdot}(HOH)]$ is computed (MP3/6-31G$^*$) to correspond to **60** (dissociation energy 86 kJ mol$^{-1}$).[524] Various physical parameters have been determined for $HO^-$ and $MeO^-$ solvated species;[525-531] for example i) the dissociation energies of $[HO^-(HOH)]$, $[MeO^-(HOH)]$ and $[MeO^-(HOMe)]$ have been determined experimentally to be 112, 99 and 121 kJ mol$^{-1}$ respectively,[529] and ii) the efficiencies of the deprotonation reactions between $[MeO^-(HOMe)_n]$ and acetone are 0.93 (n=0), 0.08 (n=1) and 0.001 (n=2) respectively.[531] Solvated halogen anions have also been studied;[519,532-540] the reaction between $[Cl^-(HOMe)]$ and $CF_3COCl$ is of particular interest[519] since it produces an ion $CF_3COCl_2^-$ which could either be a tetrahedral species or an ion complex. An i.c.r. pulse experiment showed the two chlorines in the product ion to be non-equivalent, thus the reaction is that shown in eqn **61**.[519] Bond energies of a variety of other solvated species have been determined.[541,542]

When unsymmetrical monosolvated alkoxide ions react with boron, carbon or silicon ethers the major nucleophilic substitution reaction is that where the smaller alkoxide reacts at the central atom; see e.g. eqn **62** for a boron example.[543] A combination of experiment and theory has shown that reaction **62** is collision controlled, exothermic with no internal barriers, and that the specificity is due to long range interaction between the approaching reactants. Reaction proceeds through **63** where the basic part of the incoming nucleophile H- bonds to a methyl hydrogen and thus delivers MeO to the boron centre. It is likely that the specificity of many fast ion molecule reactions is caused by similar long range interaction between reactants.

A review of the gas-phase anion chemistry of organosilanes is available.[545] Ions $H_3Si^-$ and $Me_3Si^-$ readily form SiO and SiS bonds in ion molecule reactions; e.g. the reactions of $H_3Si^-$ with $CO_2$ and $SO_2$ are summarised in eqns **64** and **65**.[546] Reactions of the following organosilanes have been reported:- trigonal bipyramidal silicon species,[547,548] methyl silanes (gas phase

$$[\text{PrO}^-(\text{HOMe})] + \text{Me}_2\text{BOEt} \longrightarrow [\text{PrO}^-(\text{HOEt})]$$
$$+$$
$$\text{Me}_2\text{BOMe} \quad (62)$$

$$\text{H}_3\text{Si}^- + \text{CO}_2 \longrightarrow [\text{H}_3\text{SiCO}_2^-] \xrightarrow{\text{o}} \text{H}_3\text{SiO}^- + \text{CO} \quad (64)$$

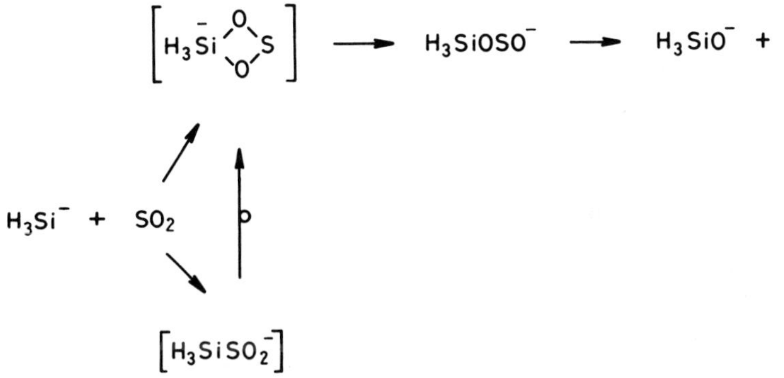

$$\left[ \text{H}_3\bar{\text{Si}} \underset{\text{O}}{\overset{\text{O}}{\diagup}} \text{S} \right] \longrightarrow \text{H}_3\text{SiOSO}^- \longrightarrow \text{H}_3\text{SiO}^- + \text{SO}$$

$$\text{H}_3\text{Si}^- + \text{SO}_2 \qquad (65)$$

$$[\text{H}_3\text{SiSO}_2^-]$$

$$[\text{HCSi}]^- + {}^{13}\text{CO}_2$$
$$\downarrow$$
$$\left[ \underset{\text{Si}}{\overset{\text{H}}{\diagdown}} \text{C} = {}^{13}\text{CO}_2^- \right] \xrightarrow{\text{o}} \left[ \begin{array}{c} \text{H}\bar{\text{C}} - \text{Si} \\ |_{13} \quad | \\ \text{C} - \text{O} \\ \diagup\diagup \\ \text{O} \end{array} \right]$$

$$\downarrow$$

$$\text{HC}\equiv{}^{13}\text{C}-\text{O}^- + \text{SiO} \quad (66)$$

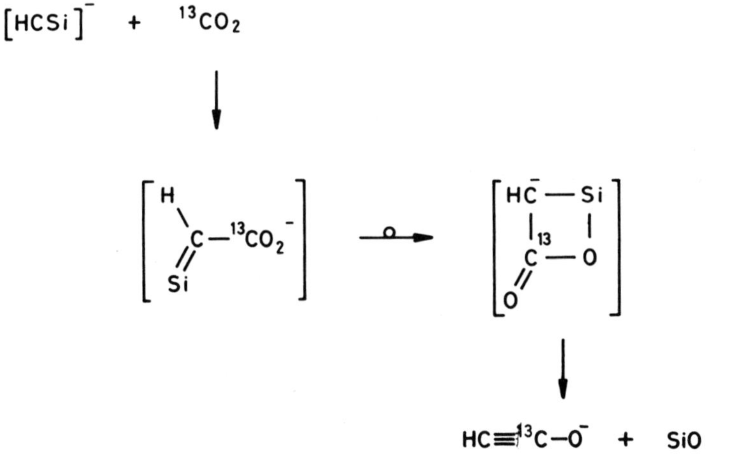

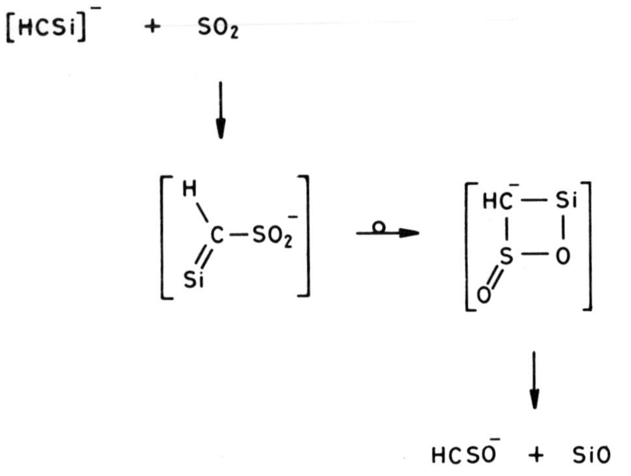

$$[HCSi]^- + SO_2$$

$$\downarrow$$

$$\begin{bmatrix} H \\ \phantom{}_{Si}^{}C-SO_2^- \end{bmatrix} \xrightarrow{\phantom{o}} \begin{bmatrix} HC^- - Si \\ |\phantom{xx}| \\ O^{\phantom{x}}S-O \\ \phantom{xx}\|\phantom{xx} \\ \phantom{xxx}O \end{bmatrix}$$

$$\downarrow$$

$$HCSO^- + SiO \qquad (67)$$

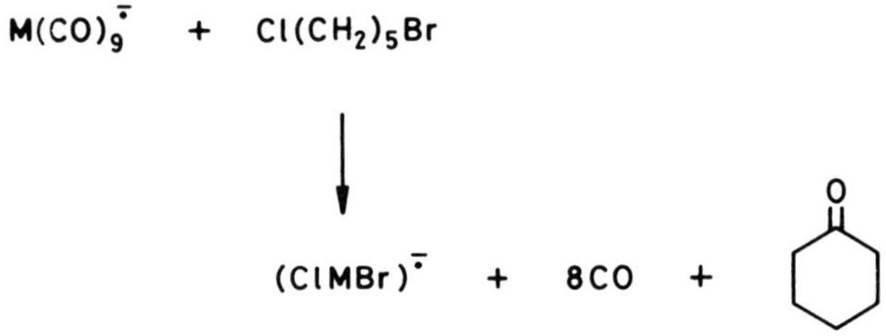

$$M(CO)_9^- + Cl(CH_2)_5Br$$

$$\downarrow$$

$$(ClMBr)^- + 8CO + \text{cyclohexanone} \qquad (68)$$

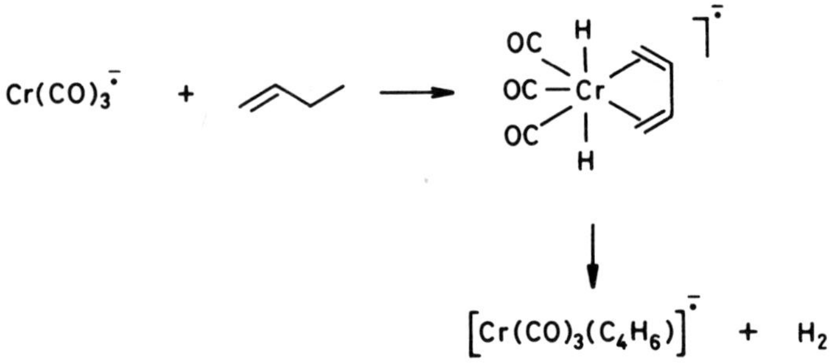

$$Cr(CO)_3^- + \text{CH}_2\text{=CHCH}_2\text{CH}_3 \longrightarrow \begin{bmatrix} OC \phantom{xx} H \\ OC-Cr \\ OC \phantom{xx} H \end{bmatrix}^-$$

$$\downarrow$$

$$[Cr(CO)_3(C_4H_6)]^- + H_2 \qquad (69)$$

$$Mn(CO)_3^- + MeOH \longrightarrow [(CO)_3Mn(H)_2(CH_2O)]^-$$

$$\downarrow$$

$$[(CO)_3Mn(CH_2O)]^- + H_2 \qquad (70)$$

acidities),[549] cyclic silanes (and germanes),[550] fluorosilanes,[551] and the formation of multiply bonded silicon species.[278,552,553] Of particular interest are the reactions of the silacetylide anion $HC\equiv Si^{-} \longleftrightarrow H\bar{C}=Si$; for example the reaction with $^{13}CO_2$ and $SO_2$ are summarised in eqns **66** and **67**.[553]

Finally, there have been a number of reports of ion-molecule reactions of metal containing anions.[554] Of most relevance here are those reactions of metal carbonyl ions which effect some modification of organic neutrals. For example, the possible conversion of alkyl dihalides into ketones (eqn **68**, M=Fe, Cr, Co),[554] alkenes to hydrogen (eqn **69**),[557] and methanol to hydrogen (eqn **70**).[561] While it must be stressed that the neutral products in such reactions are yet to be identified, this is nevertheless a fruitful area of research since it may lead to the discovery of viable catalytic reactions in the condensed phase.

## 7 Conclusion

Even a cursory comparison of this and the previous review[2] will indicate the increase in the number of papers in analytical negative ion mass spectrometry within the past few years. No diminution in output is likely and Vol. **11** of this series may need a separate review of this area. Notwithstanding the increase in the volume of analytical work, the fundamental work on structure, mechanism and reactivity of negative ions is still being carried out by the familiar few research groups, although there have been several notable converts in the review period. I (J.H.B.) have been reviewing this area for more than 20 years: negative ion chemistry is still a relatively unexplored topic in a number of major areas - much more work needs to be done.

### Acknowledgement

Much of the literature survey for this review was carried out at the University of Colorado, Boulder, by R.A.J.O. as an exchange graduate student. We are indebted to C.H. DePuy and the University of Colorado for the provision of facilities.

### References

1 **Mass Spectrometry Bulletin**, Mass Spectrometry Data Centre, The Royal Society of Chemistry, The University of Nottingham.
2 J.H. Bowie, in 'Mass Spectrometry', (ed.) M.E. Rose (Specialist Periodical Reports), The Royal Society of Chemistry, London, 1987, Vol. 9, p. 172: also appropriate sections in Vols. 1-8.

3   A. Temkin, 'Autoionization. Recent Developments and Application', Plenum Press, New York and London, 1985.
4   S. Facchetti,(ed.), 'Mass Spectrometry of Large Molecules', Elsevier Publ. Co., Amsterdam, 1985.
5   F.W. Karasek, O. Hutzinger and S. Safe (eds.), 'Mass Spectrometry in Environmental Sciences', Plenum Press, New York, 1985.
6   P.A. Lyon (ed.), 'Desorption Mass Spectrometry: Are SIMS and FAB the Same?'(ACS Symposium Series No. 291), American Chemical Society, Washington D.C., 1985.
7   M.J. Coggiola, D.L. Huestis and R.P. Saxon, 'Electronic and Atomic Collisions', (14th Int. Conf. Phys. Electron At. Collisions, Palo Alto, California, July, 1985), North-Holland Publ. Co., Amsterdam, 1985.
8   A.L. Burlingame and N. Castagroli (eds.), 'Mass Spectrometry in the Health and Life Sciences', Elsevier Science Publ. Co., Amsterdam, 1985.
9   J.R. Chapman, 'Practical Organic Mass Spectrometry', J. Wiley and Sons, New York, 1985.
10  H.E. Duckworth,R.C. Barber, and V.S. Venkatasubramanian, 'Mass Spectroscopy', 2nd Edit., Cambridge University Press, Cambridge, U.K., 1986.
11  V.N. Reinhold, 'Mass Spectrometry in Biomedical Research', J. Wiley and Sons, New York, 1986.
12  F.A. White and G.M. Wood, 'Mass Spectrometry. Application in Science and Engineering', J. Wiley and Sons, New York, 1986.
13  'Secondary Ion Mass Spectrometry. SIMS V', Springer Ser. Chem. Phys. Vol. 44, Springer-Verlag, Berlin, 1986.
14  C.J. McNeal 'Mass Spectrometry in the Analysis of Large Molecules', J. Wiley and Sons, New York, 1986.
15  'Advances in Mass Spectrometry, 1985', (10th Int. Mass Spectrom. Conf. Swansea, U.K., Sept., 1985), J. Wiley and Sons, Chichester, 1986.
16  'Mod. Methods Plant. Anal., New Ser. Vol. 3, Springer-Verlag, Berlin, 1986.
17  'Structure, Reactivity and Thermochemistry of Ions', NATO ASI Ser. Vol. 193, Reidel Publ. Co., Dordrecht, 1987.
18  J.D. Rosen, 'Applications of New Mass Spectrometric Techniques in Pesticide Chemistry', J. Wiley and Sons, New York, 1987.
19  A. Kuksis, 'Chromatography of Lipids in Biomedical Research and Clinical Diagnosis', J. Chromatogr. Libr. Vol. 37, Elsevier Publ. Co., Amsterdam, 1987.
20  A. Benninghoven, F.G. Ruderauer and H.W. Werner, 'SIMS: Basic Concepts, Instrumental Aspects and Trends', J. Wiley and Sons, New York, 1987.
21  J. Gilbert, (ed.), 'Applications of Mass Spectrometry in Food Science', Elsevier Applied Science, Barking, U.K., 1987.
22  H. Jaeger, 'Capillary Gas Chromatography - Mass Spectrometry in Medicine and Pharmacology', A. Heuthig, Heidelberg, 1987.
23  M.E. Rose (ed.), 'Mass Spectrometry' (Specialist Periodical Reports), The Royal Society of Chemistry, London, Vol. 9, 1987.
24  A.L. Burlingame, T.A. Baillie and P.J. Derrick, Anal. Chem., 1986, 58, 165 R.
25  A.L. Burlingame, D. Mattly, D.H. Russell, and P.T. Holland, Anal.Chem., 1988, 60, 294 R.
26  Various authors, Chem.Rev., 1987, 87, 483-647.
27  D.W. Kopperaal, 'Atomic Mass Spectrometry', Anal.Chem., 1988, 60, 113 R.
28  K.D. Cook, 'Electrohydrodynamic Mass Spectrometry', Mass Spectrom.Rev., 1986, 5, 467.
29  C.N. McEwen, 'Radicals in Analytical Mass Spectrometry', Mass Spectrom. Rev., 1986, 5, 521.

30  R.P. Evershed, 'Analysis of Mixtures by Mass Spectrometry - Developments and New Applications of GC/MS', M.E. Rose (ed.), 'Mass Spectrometry' (Specialist Periodical Reports), The Royal Society of Chemistry, London, 1987, 9, 196.
31  M.E. Rose, 'Analysis of Mixtures by Mass Spectrometry - LC/MS and SFC/MS', in 'Mass Spectrometry' (Specialist Periodical Reports), The Royal Society of Chemistry, London, 1987, 9, 264.
32  W.M.A. Niessen, 'Direct Liquid Interfacing for LC/MS', Chromatographia, 1986, 21, 342.
33  S.E. Unger and B.M. Warrack, 'Thermospray LC/MS and MS/MS', in Spectroscopy (Springfield Oregon), 1986, 1, 33 (Chem.Abs., 1986, 105, 01 003032j).
34  J.A. Settlage and H. Jaeger, 'Recent Developments in the Economy of Quantitative GC/MS', in 'Chromatographic Methods - Advances in Capillary Chromatography', A. Heuthig Verlag, Heidelberg, 1986.
35  J. De Graeve, F. Berthou, M. Prost, P. Azpino and J.-C. Prome, 'Application of GC/MS to the Identification of Volatile Organic Compounds', in 'Chromatographic - Mass Spectrometric Methods. Application to Environmental, Pharmacological and Biochemical Fields', Masson, Paris, 1986, pp 239-329.
36  N.J. Jensen and M.L. Gross, 'Mass Spectrometric Methods for Standard Determination and Analysis of Fatty Acids', Mass Spectrom.Rev., 1987, 6, 497.
37  C.G. Smith, R.A. Nyquist, N.H. Mahle, P.B. Smith, S.J. Martin and A.J. Pasztor, 'Analysis of Synthetic Polymers', Anal.Chem., 1987, 59, 119 R.
38  J. Sherma, 'Pesticides', Anal.Chem., 1987, 59, 18 R.
39  K. Levsen, 'Mass Spectrometry in Environmental Organic Analysis', Org. Mass Spectrom., 1988, 23, 406.
40  N.J. Jensen and M.L. Gross, 'Phospholipid Analysis', Mass Spectrom.Rev., 1988, 7, 41.
41  D.J. Harvey, 'Cannabinoids', Mass Spectrom.Rev., 1987, 6, 135.
42  A.P. Bruins, 'GC/MS of Essential Oils', in 'Capillary GC in Essential Oil Analysis', A. Heuthig Verlag, Heidelberg, 1987, pp 329-357.
43  B.L. Milman, 'Mass Spectrometry of Organic Salts', J.Anal.Chem.U.S.S.R. (Engl.Trans.), 1986, 41, 1346.
44  J. Kalcher and R. Janoschek, Chem.Phys., 1986, 104, 251.
45  M.R. Nimros and G.B. Ellison, J.Am.Chem.Soc., 1986, 108, 6522.
46  J.V. Ortiz, J.Am.Chem.Soc., 1987, 109, 5072.
47  J. Kalcher, Chem.Phys., 1987, 118, 273.
48  D. Cremer and E. Kraka, J.Phys.Chem., 1986, 90, 33.
49  P. Plessis and P. Marmet, Can.J.Chem., 1987, 65, 1424.
50  P. Plessis and P. Marmet, Can.J.Phys., 1987, 65, 165.
51  P. Gerhardt, S. Loiffler and K.H. Homann, Chem.Phys.Lett., 1987, 137, 306.
52  A.N. Hayhurst and H.R.N. Jones, J.Chem.Soc. Faraday Trans II, 1987, 1.
53  Y. Liu, S.C. O'Brien, Q. Zhang, J.R. Heath, F.K. Tittel, R.F. Curl, H.W. Kroto and R.E. Smalley, Chem.Phys.Lett., 1986, 126, 215.
54  S.H. Yang, C.L. Pettiette, J. Conceicao, O. Cheshrovsky and R.E. Smalley, Chem.Phys.Lett., 1987, 139, 233.
55  S.H. Yang, K.J. Taylor, M.J. Crayeraft, J. Conceicao, C.L. Pettiette, O. Cheshrovsky and R.E. Smalley, Chem.Phys.Lett., 1988, 144, 431.
56  I.I. Furlei, E.A. Burmistrov, F.Z. Galin, V.N. Iskandarova, V.K. Mavrodiev and G.A. Tolstikov, Bull.Acad.Sci.U.S.S.R. Div.Chem.Sci.(Engl.Trans), 1986, 35, 1618.
57  S.H. Alajajian and A. Chutjian, J.Phys.B., 1987, 20, 5567.
58  R.B. Metz, T. Kitsopoulos, A. Weaver and D.M. Neumark, J.Chem.Phys., 1988, 88, 1463.
59  A.A. Christodoulides, L.G. Christophorou and D.L. McCorkle, Chem.Phys. Lett., 1987, 139, 350.

60  K. Su and H. Wei, Fenxi Huaxue, 1986, 14, 246; (Chem.Abst., 1987, 105, 12 107688µ).
61  D.S. Waddell, P.G. Sim and R.K. Boyd, Rapid Commun. Mass Spectrom., 1987, 1, 106.
62  A.P.J.M. De Jong, R.M. Kok, C.A. Cramers and S.K. Wadman, J.Chromatogr. Biomed.Appl., 1986, 382, 19.
63  S.A. McLuckey and G.L. Glish, Int.J. Mass Spectrom. Ion Proc., 1987, 76, 41.
64  S.H. Fleishman and K.D. Jordan, J.Phys.Chem., 1987, 91, 1300.
65  M. Tsukada, N. Shima, S. Tsuneyuki, H. Kageshima and T. Kondow, J.Chem. Phys., 1987, 87, 3927.
66  D. Antolovic, V.J. Shiner and E.R. Davidson, J.Am.Chem.Soc., 1988, 110, 1375.
67  J.-C. Prome, H. Aurelle, F. Couderc and A. Savagnac, Rapid Commun. Mass Spectrom., 1987, 1, 50.
68  C. Jakobs, L. Dorland, B. Wikkerink, R.M. Kok, A.P.J.M. De Jong and S.K. Wadman, Clin.Chim.Acta, 1988, 171, 223.
69  C.W. Sweeney, Int.J. Mass Spectrom. Ion Proc., 1988, 82, 207.
70  V.I. Khvostenko, O.G. Khvostenko, N.L. Asfandiarov and G.A. Tolstikov, Dokl.Phys.Chem.(Engl.Trans.), 1986, 291, 1147.
71  M. Heni and E. Illenberger, Int.J. Mass Spectrom. Ion Proc., 1986, 73, 127.
72  M. Heni, E. Illenberger and D. Lentz, Int.J. Mass Spectrom. Ion Proc., 1986, 71, 199.
73  A. Kuehn, H.-P. Fenzlaff and E. Illenberger, Chem.Phys.Lett., 1987, 135, 335.
74  A.T. Lebedev, A.G. Kazaryan, V.A. Bakulev, Yu.T. Shafran, V.S. Fal'ko, V.G. Lukin and V.S. Petrosyan, Khim.Geterotsikl.Soedin., 1987, 941.
75  O.G. Khvostenko, B.G. Zykov, N.L. Asfandiarov, V.I. Khostenko, S.N. Denisenko, G.V. Shuskov and R.G. Kostyanovskii, Khim.Fiz., 1985, 4, 1366; (Chem.Abst., 1986, 104, 16 138703d).
76  A.I. Ermakov, A.P. Pleshkova, A.A. Sorokin, S.Ya. Skachilova, M.G. Pleshakov and A.P. Zuev, J.Org.Chem.U.S.S.R.(Engl.Trans.), 1985, 21, 1819.
77  C. Brueckner, R. Kramell, G. Schneider, J. Schmidt, A. Preiss, G. Sembdner and K. Schruber, Phytochemistry, 1988, 27, 275.
78  A.I. Ermakov, A.A. Sorokin and V.G. Voronin, Khim.Geterotsikl.Soedin, 1985, 12, 1663.
79  K.P. Madhusudaran, R. Pratap, D.S. Bhakuni, G. Prasad, G. Singh, A.K. Upadhyaya and K.N. Mehrotra, Indian J.Chem.(Sect. B), 1987, 26, 794.
80  V.I. Khvostenko, O.G. Khvostenko, G.S. Lomakii, B.G. Zykov, N.L. Asfandiarov, V.A. Mazurov, S.A. Androrati, A.S. Yavorskii, L.N. Yakubovskaya and T.N. Voronira, Bull.Acad.Sci.U.S.S.R. Div.Chem.Sci. (Engl.Trans.), 1987, 36, 1175.
81  A.I. Ermakov, O.G. Khvostenko, V.G. Voronin, A.A. Sorokin, N.L. Asfandiarov and V.I. Khvostenko, J.Org.Chem.U.S.S.R.(Engl.Trans.), 1986, 22, 2140.
82  K.P. Madhusudanan, P. Kumar and D.S. Bhakuni, Ind.J.Chem., 1985, 24, 1188.
83  G.L-Y. Chan, J.E. Axelson, F.S. Abbot, C.R. Kerr and K.M. McErlane, J.Chromatogr.Biomed.Appl., 1987, 417, 295.
84  J.S. Hadley, A. Fradin and R.C. Murphy, Biomed.Envir. Mass Spectrom., 1988, 15, 175.
85  W. Schade, V. Graefe and J. Schmidt, Biomed.Envir. Mass Spectrom., 1988, 15, 359.
86  R.M. Kok, L. Kaster, A.P.J.M. De Jong, B. Poll-The, J.-M. Saudubray and C. Jakobs, Clin.Chem.Acta., 1987, 168, 143.
87  P.B. Comita and J.I. Brauman, J.Am.Chem.Soc., 1987, 109, 7591.
88  M.H.A. Elgamal, D. Voigt and G. Adam, J.Prakt.Chem., 1986, 328, 893.

89  I.I. Furlei, M.S. Miftakov, V.K. Mavrodiev, A.S. Vorobev, N.A. Danilova and G.A. Tolstikov, Chem.Nat.Compd.U.S.S.R.(Engl.Trans.), 1985, 21, 716.
90  J. Schmidt, S. Huneck, P. Franke and J.D. Connolly, Org. Mass Spectrom., 1987, 22, 359.
91  O.P. Sati, D.C. Chaukiyal, M. Nishi, K. Miyahara and T. Kawasaki, Phytochemistry, 1986, 25, 2658.
92  G. Adam, A. Preiss, P.D. Hung and L. Kutschabsky, Tetrahedron, 1987, 43, 5815.
93  J. Schmidt, H.-M. Vorbrodt and G. Adam, Biomed.Envir. Mass Spectrom., 1986, 13, 663.
94  O. Miersch, B. Wrobel and S. Sembdner, Z.Chem., 1987, 27, 261.
95  V.S. Shmakov, I.I. Furlei, N.K. Lyapina, V.I. Khvostenko, A.A. Polyakova and G.A. Tolstikov, Dokl.Chem.(Engl.Trans.), 1986, 287, 38.
96  S. Moran and G.B. Ellison, Int.J. Mass Spectrom. Ion Proc., 1987, 80, 83.
97  I.I. Furlei, V.K. Mavrodiev, E.E. Shul'ts, R.V. Kunakova, U.M. Dzhemilev and G.A. Tolstikov, Bull.Acad.Sci.U.S.S.R. Div.Chem.Sci.(Engl.Trans.), 1985, 34, 2291.
98  N.K. Lyapina, V.S. Shmakov, I.I. Furlei, A.D. Ulendeeva, A.S. Vorob'ev and G.A. Tolstikov, Bull.Acad.Sci.U.S.S.R. Div.Chem.Sci.(Engl.Trans.), 1987, 36, 1570.
99  P.E. Haroldson and R.C. Murphy, Biomed.Envir. Mass Spectrom., 1987, 14, 573.
100 C. Kajdas, R. Tummler, H. von Ardenne, W. Schwarz, T.H. Radom and V.R. Polen, ZfI-Mitteilungen Leipzig, 1986, 115, 107.
101 G.P. Boldrini, L. Lodi, E. Taghavini, C. Trombini and A. Umani-Ronchi, J.Organomet.Chem., 1987, 336, 23.
102 I.I. Furlei, V.K. Mavrodiev, I.M. Salimgareeva and N.G. Bogatova, Bull. Acad.Sci.U.S.S.R. Div.Chem.Sci.(Engl.Trans.), 1986, 35, 514.
103 J.H. Bowie, in 'Mass Spectrometry', (ed.), M.E. Rose (Specialist Periodical Report), The Royal Society of Chemistry, London, 1985, Vol. 8, p. 165.
104 J.H. Bowie, M.B. Stringer, R.N. Hayes, M.J. Raftery, G.J. Currie and P.C.H. Eichinger, 'Collision Induced Dissociations of Enolate Negative Ions in the Gas Phase', Spectroscopy Int.J., 1985, 4, 277.
105 J.H. Bowie, 'Collisional Processes of Negative Ions. Analytical and Mechanistic Aspects', Adv. Mass Spectrom., 1986, 10A, 553.
106 J.B. Westmore and M.M. Alauddin, 'Ammonia Chemical Ionization Mass Spectrometry', Mass Spectrom.Rev., 1986, 5, 381.
107 H. Budzikiewicz, 'Negative Chemical Ionization of Organic Compounds', Mass Spectrom.Rev., 1986, 5, 345.
108 M. Ryska and I. Koruna, 'Negative Ion Chemical Ionization', ZfI-Mitteilungen Leipzig, 1986, 115, 145.
109 N.P.E. Vermeulen, W. Onkentout, M. Van Der Graaf, B.J. Xu and A.G.L. Burm, 'Glass Capillary GLC/MS - Applications in the Analysis of Drugs', Chromatogr. Methods, 1987, 107, see also M.S. Lee and R.A. Yost, 'Rapid Identification of Drug Metabolites by MS/MS', Biomed.Envir. Mass Spectrom., 1988, 15, 193.
110 A.F. Gross, P.S. Given and A.K. Athnasios, Anal.Chem., 1987, 59, 212 R.
111 F. Turecek, 'Stereochemistry of Organic Ions in the Gas Phase', Coll. Czech.Chem.Commun., 1987, 52, 1928.
112 H. Jaeger, W. Gielsdorf, S.W. Sanders, N. Haeming, N. Michaelis and J. Pasper, 'Performing BV-BA/BE Studies with Highly Sophisticated Analytical Methods - Practical Examples', Chromatogr. Methods, 1987, 302.
113 M. Grassi, G. Audisio and P. Traldi, Org. Mass Spectrom., 1987, 22, 85.
114 C. Lange, Org. Mass Spectrom., 1987, 22, 55.
115 G.J. Currie, J.H. Bowie, R.A. Massy-Westropp and G.W. Adams, J.Chem.Soc. Perkin Trans II, 1988, 403.
116 A.G. Harrison and H.Y. Tong, Org. Mass Spectrom., 1988, 23, 135.

117 E.A. Stemmler, R.A. Hites, B. Arbogast, W.L. Budde, M.L. Deinzer, R.C. Dougherty, J.W. Eichelbeiger, R.L. Foltz, C. Grimm, E.P. Grimsrud, C. Sakashita and L.J. Sears, Anal.Chem., 1988, 60, 781.
118 S.A. Brotherton and W.M. Gulick, Anal.Chim.Acta., 1986, 186, 101.
119 P.P. Wickramayake, K.W.M. Sui and S.S. Berman, Org. Mass Spectrom., 1986, 21, 279.
120 G.W. Dillow and I.K. Gregor, Org. Mass Spectrom., 1986, 21, 386.
121 M.V. Buchanan, I.B. Rubin, M.B. Wise and G.L. Glish, Biomed.Envir. Mass Spectrom., 1987, 14, 395.
122 L.R. Hilpert, Biomed.Envir. Mass Spectrom., 1987, 14, 383.
123 R.H. Bieri and J. Greaves, Biomed.Envir. Mass Spectrom., 1987, 14, 555.
124 D.H. Evans, R.G. Keesee and A.W. Castleman, J.Chem.Phys., 1987, 86, 2927.
125 D.B. Kassel, K.A. Kayganich, J.T. Watson and J. Allison, Anal.Chem., 1988, 60, 911.
126 J.D. MacNeil, J.R. Patterson, A.C. Fasser, C.D. Salisbury, S.V. Tessaro and C. Gates, Int.J.Envir.Anal.Chem., 1987, 30, 145.
127 M. Oehme, D. Stocke and H. Knoppel, Anal.Chem., 1986, 58, 554.
128 W.A. Korfmacher, L.G. Rushing, P.H. Sutonen, C.J. Branscomb and C.L. Holder, J. High Resolut.Chromatogr.Chromatogr.Commun., 1987, 10, 332.
129 D.L. Swackhamer, M.J. Charles and R.A. Hites, Anal.Chem., 1987, 59, 913.
130 T.M. Trainor and P. Vouros, Anal.Chem., 1987, 59, 601.
131 T. Ramdahl, G.E. Carlberg and P. Kolsaker, Sci. Total Environ., 1986, 48, 147.
132 N. Haering, Z. Salara, I. Todesko and H. Jaeger, Azneim-Forsch., 1987, 37, 1402.
133 P.K. Freeman, R. Srinivasa, J.-A. Campbell and M.L. Deinzer, J.Am.Chem. Soc., 1986, 108, 5531.
134 H.-R. Buser, Anal.Chem., 1986, 58, 2913.
135 E.A. Stemmer and R.A. Hites, Anal.Chem., 1988, 60, 787.
136 T.M. Trainor and P. Vouros, Anal.Chem., 1987, 59, 601.
137 J.A. Laramee, B.C. Arbogast and M.L. Deinzer, Anal.Chem., 1986, 58, 2907.
138 C. Rappe, M. Nygren, G. Lindstrom, H.R. Buser, O. Blaser and C. Wuethrich, Envir.Sci.Techol., 1987, 21, 964.
139 D.S. Waddel, H.S. McKinnon, B.G. Chittim, S. Safe and R.K. Boyd, Biomed. Envir. Mass Spectrom., 1987, 14, 457.
140 J.R. Donnelly, W.D. Munslow, T.L. Vonnature, N.J. Nunn, C.M. Hedin, G.W. Sovocool and R.K. Mitchum, Biomed.Envir. Mass Spectrom., 1987, 14, 465.
141 B.K. Afghan, J. Carron, P.D. Goulden, J. Lawrence, D. Leger, F. Onuska, J. Sherry and R. Wilkinson, Can.J.Chem., 1987, 65, 1086.
142 J.C. Sheldon, J.H. Bowie and D.E. Lewis, Nouveau J.Chim., 1988, 12, 269.
143 R.N. Hayes, J.C. Sheldon, J.H. Bowie and D.E. Lewis, J.Chem.Soc.Chem. Commun., 1984, 1431; Aust.J.Chem., 1985, 38, 1197.
144 W. Tumas, R.F. Foster, M.J. Pellerite and J.I. Brauman, J.Am.Chem.Soc., 1987, 109, 961.
145 W. Tumas, R.F. Foster and J.I. Brauman, J.Am.Chem.Soc., 1988, 110, 2714.
146 M.J. Raftery, J.H. Bowie and J.C. Sheldon, J.Chem.Soc. Perkin Trans II, 1988, 563.
147 W.C. Brumley, D. Andrzejewski and J.A. Sphon, Org. Mass Spectrom., 1988, 23, 204.
148 B.P.-Y. Lau and D. Weber, J.Agric. Food Chem., 1987, 35, 412.
149 M.G. Stiachan, G.B. Anderson, Q.N. Porter and R.B. Johns, Org. Mass Spectrom., 1987, 22 670.
150 G. Bouchoux, Y. Hoppiliard, P. Jaudon and J.-M. Pechine, Rapid Commun. Mass Spectrom., 1987, 1, 20.
151 P.C.H. Eichinger, J.H. Bowie and T. Blumenthal, J.Org.Chem., 1986, 51, 5078.
152 P.C.H. Eichinger and J.H. Bowie, J.Chem.Soc. Perkin Trans II, 1988, 497.
153 P.C.H. Eichinger and J.H. Bowie, J.Chem.Soc. Perkin Trans II, 1987, 1499.

154 M.B. Stringer, J.H. Bowie and G.J. Currie, J.Chem.Soc. Perkin Trans II, 1986, 1821.
155 G.J. Currie, M.B. Stringer, J.H. Bowie and J.L. Holmes, Aust.J.Chem., 1987, 40, 1365.
156 M.J. Raftery and J.H. Bowie, Int.J. Mass Spectrom. Ion Proc., 1987, 79, 267.
157 R.N. Hayes, J.H. Bowie and J.C. Sheldon, Int.J. Mass Spectrom. Ion Proc., 1986, 71, 233.
158 K.P. Madhusudanan, U.S. Murthy and D. Fraisse, Org. Mass Spectrom., 1987, 22, 665.
159 K.P. Madhusudan, R. Jain, S. Mittal, S. Durani and R.S. Kapil, Org. Mass Spectrom., 1986, 21, 781.
160 K.P. Madhusudan, Neelima and A.P. Bhaduri, Ind.J.Chem.(Sect. B), 1986, 25, 1224.
161 C.L. Johlman, L. Spencer, D.T. Sawyer and C.L. Wilkins, J.Org.Chem., 1987, 52, 3027.
162 M.B. Stringer, J.H. Bowie, P.C.H. Eichinger and G.J. Currie, J.Chem.Soc., Perkin Trans II, 1987, 385.
163 A. Tunlid, H. Er, G. Westerdahl and G. Odham, J.Microbiol. Methods, 1987, 7, 77.
164 M. Guichordant, M. Lagarde, M. Lesieur and F. De Maack, J.Chromatogr. Biomed.Appl., 1988, 425, 25.
165 S.S. Cutie, G.J. Kallos and P.B. Smith, J.Chromatogr., 1987, 408, 349.
166 M. Dawson, C.M. McGee, P.M. Brooks, J.H. Vine, E. Lacey and T.R. Watson, J.Chromatogr.Biomed.Appl., 1987, 420, 129.
167 P. Reymard, M. Saugy and P.E. Pilet, Plant Physiol., 1987, 85, 8.
168 R.A. Creelman, D.A. Gage, J.T. Stults and J.A.D. Zeevaart, Plant Physiol., 1987, 85, 726.
169 R.N. Hayes and J.H. Bowie, J.Chem.Soc. Perkin Trans II, 1986, 1827.
170 M.J. Raftery and J.H. Bowie, Aust.J.Chem., 1987, 40, 711.
171 R.N. Hayes and J.H. Bowie, Org. Mass Spectrom., 1986, 21, 425.
172 P.C.H. Eichinger and J.H. Bowie, Org. Mass Spectrom., 1987, 22, 103.
173 P.C.H. Eichinger, J.H. Bowie and R.N. Hayes, J.Org.Chem., 1987, 52, 5224.
174 A.G. Netting, B.V. Millowar, G.T. Vaughan and R.O. Lidgard, Biomed.Envir. Mass Spectrom., 1988, 15, 375.
175 H. Brevard, F. Cozzolino, R. Fellas and G. George, Parfumes, Cosmet. Aromes, 1987, 75, 83. Plant Physiol., 1987, 85, 726.
176 G. Takeuchi, M. Weiss and A.G. Harrison, Anal.Chem., 1987, 59, 918.
177 A.M. Bambagiotti, S.A. Coran, F.F. Vincieri, T. Petrucciani and P. Traldi, Org. Mass Spectrom., 1986, 21, 485.
178 G. Podda, L. Corda, A.M. Maccioni and P. Traldi, Org. Mass Spectrom., 1986, 21, 395.
179 M. Balazy and R.C. Murphy, Anal.Chem., 1986, 58, 1098.
180 J.Y. Westcott, K.R. Stenmark and R.C. Murphy, Prostaglandins, 1986, 31, 227.
181 V. Kaever, M. Martin, J. Fauler, K.-H. Marx and K. Resch, Biochem.Biophys. Acta., 1987, 922, 337.
182 W.R. Mathews, G.L. Bundy, M.A. Wynalda, D.M. Guido, W.P. Schneider and F.A. Fitzpatrick, Anal.Chem., 1988, 60, 349.
183 C.R. Pace-Asciak, Prostaglandins, Leukotrienes Med., 1986, 22, 1.
184 J.Y. Westcott, S. Chang, M. Balazy, D.O. Stene, P. Pradilles, J. Maclouf, N.F. Voelkel and R.C. Murphy, Prostaglandins, 1986, 32, 857.
185 A.B. Schilling, I.M. Zulah, M.L. Putteman, E.R. Hall and D.L. Venton, Biomed.Envir. Mass Spectrom., 1986, 13, 545.
186 J.A. Yergey, H.-Y. Kim and N. Salem, Anal.Chem., 1986, 58, 1344.
187 A. Martineau and P. Falardeu, J.Chromatogr.Biomed.Appl., 1987, 417, 1.
188 S. Fischer, A. Vischer, V. Preac-Mursic and P.C. Weber, Prostaglandins, 1987, 34, 367.
189 D.A. Herold, B.J. Smith, R.M. Ross, F. Marquis, C.R. Ayers, M.R. Wills and J. Savoy, Prostaglandins, 1987, 33, 599.

190 R.D. Voyksner, E.D. Bush and D. Brent, Biomed.Envir. Mass Spectrom., 1987, 14, 523.
191 H.J. Leis, E. Hohenester, H. Gleispach, E. Malle and B. Mayer, Biomed. Envir. Mass Spectrom., 1987, 14, 617.
192 E. Idaka, H. Kaiya and H. Horitsu, J.Chromatogr.Biomed.Appl., 1987, 420, 373.
193 B. Mayer, H. Gleispach and W.R. Kukovetz, Biochem.Biophys.Acta, 1987, 918, 209.
194 J.J. Vrbarac, T.D. Eller and D.R. Knapp, J.Chromatogr.Biomed.Appl., 1988, 425, 1.
195 J.C. Bordet, M. Guichardant and M. Lagarde, Biochem.Biophys.Acta, 1988, 958, 460.
196 D.F. Werdellorn, K. Seibert and L.J. Roberts, Proc.Natl.Acad.Sci.U.S.A., 1988, 85, 304.
197 N. Shindo, T. Saito, K. Murayama, Biomed.Envir. Mass Spectrom., 1988, 15, 25.
198 H. Schweer, H.W. Seyberth, C.O. Meese and O. Fuerst, Biomed.Envir. Mass Spectrom., 1988, 15, 143.
199 H. Schweer, C.O. Meese, O. Fuerst, P.G. Kuche and H.W. Seyberth, Anal. Biochem., 1987, 164, 156.
200 M.L. Schwartzman, M. Balazy, J. Masfurer, N.G. Abraham, J.C. McGift and R.C. Murphy, Proc.Natl.Acad.Sci.U.S.A., 1987, 84, 8125.
201 T. Krishnamurthy and E.W. Sarver, J.Chromatogr., 1986, 355, 253; Biomed. Envir. Mass Spectrom., 1988, 15, 11, 185.
202 P. Begley, B.E. Foulger, P.D. Jeffery, R.M. Black and R.W. Read, J.Chromatogr., 1986, 367, 87.
203 T. Krishnamurthy and E.W. Sarver, Anal.Chem., 1987, 59, 1272.
204 T. Krishnamurthy, E.W. Sarver, S.L. Greene and B.B. Jarvis, J.Assoc.Off. Anal.Chem., 1987, 70, 132; T. Krishnamurthy, M.B. Wasserman and E.W. Sarver, Biomed.Envir. Mass Spectrom., 1986, 13, 503.
205 R. Kostiainen and A. Rizzo, Analytica Chim.Acta., 1988, 204, 233.
206 K.P. Madhusudanan, R. Mehrotra, C. Singh and S.P. Popli, Ind J.Chem.(Sect.B), 1985, 24, 1047.
207 K.P. Madhusudanan, H.S. Garg and D.S. Bhakuni, J.Ind.Chem.Soc., 1985, 62, 613.
208 J.M. Rosenfeld, R.A. McLeod and R.L. Foltz, Anal.Chem., 1986, 58, 716.
209 M. Colon, P. Guevara, W.H. Gerwick and D. Ballantine, J.Nat.Prod., 1987, 50, 368.
210 R.P. Everskere and L.J. Goad, Biomed.Envir. Mass Spectrom., 1987, 14, 131.
211 L.O.G. Weidolf, E.D. Lee and J.D. Henion, Biomed.Envir. Mass Spectrom., 1988, 15, 283.
212 B. Pelli, P. Traldi, P. Rodighiero and A. Guiotto, Biomed.Envir. Mass Spectrom., 1986, 13, 417.
213 A. Sakushima and S. Nishibe, Phytochem., 1988, 27, 915.
214 S. Calcamese, V. Amico and M. Hardy, Bull.Soc.Chem. France, 1988, 91.
215 J. Turk, B.A. Wolf and M.L. McDaniel, Biomed.Envir. Mass Spectrom., 1986, 13, 237.
216 N.J. Carrington, G. Vaughan and B.N. Milborrow, Phytochem., 1988, 27, 673.
217 W.G. Stillwell, M.S. Bryant and J.S. Wishnok, Biomed.Envir. Mass Spectrom., 1987, 14, 221.
218 J. Berthelot, C. Guette and C. Lange, Org. Mass Spectrom., 1988, 23, 52.
219 C. Lange, C. Guette, J. Berthelot and J.J. Basselier, Org. Mass Spectrom., 1987, 22, 123.
220 D.G. Lynn, K. Jaffe, M. Cornwall and W. Trarrontaro, J.Am.Chem.Soc., 1987, 109, 5858.

221 H.J. Huizing, F. De Boer, H. Hendriks, W. Balraadjsing and A.P. Bruins, Biomed.Envir. Mass Spectrom., 1986, 13, 293.
222 H. Hendriks, W. Balraadjsing, H.J. Huizing and A.P. Bruins, Planta.Med., 1987, 53, 456.
223 N. Haering, J.A. Settlage and R. Shubert, Chromatogr. Methods, 1987, 165.
224 I. Koruna, E. Ryska and E. Kleinerova, Z.f.I.-Mitteilungen, Leipzig, 1986, 115, 149.
225 P. Pant, P. Madhusudanan and D.S. Bhakuni, Ind.J.Chem.(B), 1986, 25, 630.
226 M.P. Heyes and S.P. Markey, Biomed.Envir. Mass Spectrom., 1988, 15, 291.
227 T.R. Covey, E.D. Lee and J.D. Henion, Anal.Chem., 1986, 58, 2453.
228 J.S.M. De Wit, C.E. Parker, K.B. Tomer and J.W. Jorgenson, Anal.Chem., 1987, 59, 2400.
229 S. Takahashi, Biomed.Envir. Mass Spectrom., 1987, 14, 257.
230 F. Rubio, S. Chen, T. Crews, F. De Grazia, W.A. Garland and M. Barbalas, J.Chromatogr.Biomed.Appl., 1987, 421, 281.
231 K.J. Volk, M.S. Lee, R.A. Yost and A. Brajter-Toth, Anal.Chem., 1988, 60, 720.
232 I. Isern-Flecha, X.-Y. Jiang, R.G. Cooks, W. Pfleiderer, W.-G. Chae and C. Chang, Biomed.Envir. Mass Spectrom., 1987, 14, 17.
233 A.P.J.M. De Jong, R.M. Kok, C.A. Cramers, S.K. Wadman and E. Haan, Clin. Chim.Acta., 1988, 171, 49.
234 J.P. Kiplinger, A.T. Maynard and M.M. Bursey, Org. Mass Spectrom., 1987, 22, 534.
235 M.J. Raftery and J.H. Bowie, Int.J. Mass Spectrom. Ion Proc., 1988, 85, 167.
236 D. Abramson, J.Chromatogr., 1987, 391, 315.
237 K.P. Madhusudanan, Neelima and A.P. Bhudiori, Ind.J.Chem.(Sect. B), 1986, 25, 1224.
238 K. Stachowiak, C. Wilder, M.L. Vestal and D.F. Dyckes, J.Am.Chem.Soc., 1988, 110, 1759.
239 R.D. Plattner and R.G. Powell, J.Nat.Prod., 1986, 49, 475.
240 M. Tornovist, J. Mowrer, S. Jensen and L. Ehrenberg, Anal.Biochem., 1986, 154, 255.
241 G. Fu, Y. Chen, Y. Wu, C. Zhon, X. Xu and H. Shen, Shitsuryo Bunseki, 1987, 35, 300.
242 Y.Y. Wigfield, N.P. Gurprasad, M. Lanouette and S. Ripley, J.Assoc.Off. Anal.Chem., 1987, 70, 792.
243 J.A. Ballantine, D.J. Barton, J.F. Carter, J.P. Davies, K. Smith, G. Stedman and E.E. Kingston, Org. Mass Spectrom., 1988, 23, 1.
244 K. Michaelis, J.A. Settlage and H. Jaeger, Chromatogr. Methods, 1987, 273.
245 Y. Tokuma, T. Kujiwara and H. Noguchi, J.Pharm.Sci., 1987, 76, 310.
246 R.D. Voyksner and J. Yinon, J.Chromatogr., 1986, 354, 393.
247 S.A. McLuckey and G.L. Glish, Org. Mass Spectrom., 1987, 22, 224.
248 E.A. Stemmler and R.A. Hites, Biomed.Envir. Mass Spectrom., 1987, 14, 417.
249 B.N. Pramanik and P.R. Das, Org. Mass Spectrom., 1987, 22, 742.
250 L.A. Krieger and E.P. Grimsrud, Int.J. Mass Spectrom. Ion Proc., 1988, 83, 189.
251 T.D. Williams, L. Vachon and R.J. Anderegg, Anal.Biochem., 1986, 153, 372.
252 W.A. Korfmacher and L.G. Rushing, J. High Resolut.Chromatogr.Chromatogr. Commun., 1986, 5, 293.
253 D. Mitchell, R.D. Bowen, K.R. Jennings, R.S. Varma and G.W. Kabalka, J.Chem.Soc. Perkin Trans II, 1987, 1495.
254 W.A. Korfmacher, L.G. Rushing, R.J. Engelbach, J.P. Freeman, Z. Djeric, E.K. Fifer and F.A. Beland, J. High Resolut.Chromatogr.Chromatogr. Commun., 1987, 10, 43.
255 W.A. Korfmacher, L.G. Rushing, J. Arey, B. Zielinska and J.N. Pitts, J. High Resolut.Chromatogr.Chromatogr.Commun., 1987, 10, 641.
256 E. Atlas, Nature, 1988, 331, 426.
257 A.H. Lawrence and P. Neudorfl, Anal.Chem., 1988, 60, 104.
258 T. Terada, C. Sakata, K. Ishiboshi, T. Nakamura, R. Ishimura and T. Tsuchiya, Xenobiotica, 1988, 18, 291.

259  G. Knop and F. Arnold, J.Geophys.Res.Lett., 1987, 14, 1262.
260  W.E. Wentworth, C.F. Batten, D'Sa.E. Desai and E.C.M. Chen, J.Chromatogr., 1987, 390, 249.
261  G.-R. Her, G.G. Dolnikowski and J.T. Watson, Org. Mass Spectrom., 1986, 21, 329.
262  H. Budzikiewicz and A. Poppe, Org. Mass Spectrom., 1988, 23, 338.
263  O. Suzuki and H. Brandenberger, Fresenius' Z.Anal.Chem., 1987, 326, 228.
264  N.S. Karellas and J.M. Goodings, Can.J.Chem., 1986, 64, 2412.
265  J.M. Goodings, D.K. Bohme, K. Elguindi and A. Fox, Can.J.Chem., 1986, 64, 689.
266  J.H. Bowie, P.Y. White and T. Blumenthal, Org. Mass Spectrom., 1987, 22, 541.
267  M. Jemaz, S. Black and A.I. Cohen, Rapid Commun. Mass Spectrom., 1987, 1, 129.
268  M. Morin, C. Rolando and J.N. Verpeau, Adv. Mass Spectrom., 1985, 10A, 779.
269  D.A. Flory, M.M. McLean, M.L. Vestal and L.D. Betowski, Rapid Commun. Mass Spectrom., 1987, 1, 48.
270  A.P. Bruins, L.O.G. Weindolt, J.D. Henion and W.L. Buddle, Anal.Chem., 1987, 59, 2647.
271  T.R. Covey, A.P. Bruins and J.D. Henion, Org. Mass Spectrom., 1988, 23, 178.
272  R.J. Terjeson, J. Mohtasham, R.M. Sheets and G.L. Gard, J.Fluorine Chem., 1988, 38, 3.
273  T.A. Wehner, J.S. Wood, R. Walker, G.V. Downing, W.J.A. Vanderheuvel, J.Chromatogr., 1987, 399, 251.
274  S.V. Hummel and R.A. Yost, Org. Mass Spectrom., 1986, 21, 785.
275  D. Barcelo, F.A. Maris, R.B. Geerdink, R.W. Frei, G.J. De Jong and V.A.T. Brinkman, J.Chromatogr., 1987, 394, 65.
276  S. Tarabe, N. Karran, T. Wakinoto and R. Tatsukawa, Int.J.Environ.Anal. Chem., 1987, 29, 277.
277  J.C. Sheldon, J.H. Bowie and P.C.H. Eichinger, J.Chem.Soc. Perkin Trans II, 1988, 1263.
278  R.A.J. O'Hair, J.H. Bowie and G.J. Currie, Aust.J.Chem., 1988, 41, 57.
279  G.W. Dillow, I.K. Gregor and M. Gilhaus, Org. Mass Spectrom., 1986, 21, 151.
280  G. Fu, Y. Qian, Y. Xu and S. Chen, J.Organomet.Chem., 1986, 314, 113.
281  L.D. Detter and R.A. Walton, Polyhedron, 1986, 5, 1321.
282  K. Knoll and G. Huttner, J.Organomet.Chem., 1987, 329, 369.
283  G.W. Dillow and I.K. Gregor, Inorg.Chem., 1988, 27, 2102.
284  J. Adams, L.J. Deterding and M.L. Gross, Spectroscopy Int.J., 1987, 5, 199.
285  D.F. Hunt, J. Shabanowitz and A.B. Giordani, Anal.Chem., 1980, 52, 386; Envir.Hlth.Perspect., 1980, 36, 33.
286  D.F. Hunt, A.B. Giordani, J. Shabanowitz and G. Rhodes, J.Org.Chem., 1982, 47, 738.
287  S.W. Froelicher, R.E. Lee, R.R. Squires and B.S. Freiser, Org. Mass Spectrom., 1985, 20, 4.
288  M.D. Rozeboom, J.P. Kiplinger and J.E. Bartmess, J.Am.Chem.Soc., 1984, 106, 1025.
289  V.K. Kapoor and A.S. Chawla, 'FAB Mass Spectrometry', Ind.J.Chem.(Sect. B), 1986, 25, 573.
290  J. Belanger and J.R.J. Pare, 'FAB MS in the Pharmaceutical Analysis of Drugs', J.Pharm.Biomed.Anal., 1986, 4, 415.
291  M. Hattori and T. Namba, 'Analysis of Crude Drug Components by LC/MS', Jeol. News, 1987, 23A, 2.
292  J. Sunner, A. Morales and P. Kebarle, Anal.Chem., 1988, 60, 98.
293  D.H. Williams, A.F. Findeis, S. Naylor and B.W. Gibson, J.Am.Chem.Soc., 1987, 109, 1980.
294  J.E. Campana, M.M. Ross and J.H. Callahan, Int.J. Mass Spectrom. Ion Proc., 1987, 78, 195.

295 J.P. Kiplinger and M.M. Bursey, Org. Mass Spectrom., 1988, 23, 342.
296 J. Rivera, D. Fraisse, F. Ventura, J. Caixach and A. Figueras, Fresenius' Z.Anal.Chem., 1987, 328, 577.
297 A.B. Padias, H.K. Hall, P.A. Tomalia and J.R. McConnell, J.Org.Chem., 1987, 52, 5305.
298 R.B. Freas and J.E. Campala, Anal.Chem., 1986, 58, 2434.
299 J. Adams and M.L. Gross, J.Am.Chem.Soc., 1986, 108, 6915.
300 C.K. Winter, H.J. Segall and A.D. Jones, Drug Metab.Dispos., 1987, 15, 608.
301 J.R. Lloyd and M.L. Cotter, Biomed.Envir. Mass Spectrom., 1986, 13, 447.
302 E.J. Gallegos, J.Chromatogr.Sci., 1987, 25, 296.
303 D.M. Peterson, R.A. Martinez, N. Satsangi, S.T. Weintraub, P.L. Stotter and S.J. Friedberg, J. Lipid Res., 1988, 29, 94.
304 H. Aurelle, M. Tzeilhon, D. Prome, A. Savagnac and J.C. Prome, Rapid Commun. Mass Spectrom., 1987, 1, 65.
305 N.J. Jensen and M.L. Gross, Lipids, 1986, 21, 362.
306 R.C. Seid, W.M. Bone and L.R. Phillips, Anal.Biochem., 1986, 155, 168.
307 J. Adams and M.L. Gross, Anal.Chem., 1987, 59, 1576.
308 R.L. Cerny, K.B. Tomer and M.L. Gross, Org. Mass Spectrom., 1986, 21, 655.
309 K.B. Tomer, N.J. Jensen and M.L. Gross, Anal.Chem., 1986, 58, 2429.
310 M.M. Ross, R.A. Neihof and J.E. Campana, Anal.Chim.Acta., 1986, 181, 149.
311 D.N. Heller, C. Fenselau, R.J. Cotter, P. Demirev, J.K. Olthoff, J. Honovich, M. Uy, T. Tanaka and Y. Kishimoto, Biochem.Biophys.Res. Commun., 1987, 142, 194.
312 J. Muething, H. Egge, B. Kriep and P.F. Muehlradt, Eur.J.Biochem., 1987, 163, 407.
313 Y. Ishikawa, S. Gasa, R. Minami and A. Maketa, J.Biochem.(Tokyo), 1987, 101, 1369.
314 G. Ferrante, I. Ekiel and G.D. Sprott, Biochem.Biophys.Acta., 1987, 921, 281.
315 D.K.H. Chou, G.A. Schwarting, J.E. Evans and F.B. Jungalwala, J.Neurochem., 1987, 49, 865.
316 H. Morii, M. Nishihara, M. Ohga and Y. Koga, J. Lipid Res., 1986, 27, 724.
317 H. Muenster, J. Stein and H. Budzikiewicz, Biomed.Envir. Mass Spectrom., 1986, 13, 423.
318 N.J. Jensen, K.B. Tomer and M.L. Gross, Lipids, 1987, 22, 480, 580.
319 H. Muenster and H. Budzikiewicz, Rapid Commun. Mass Spectrom., 1987, 1, 126.
320 D.N. Heller, R.J. Cotter, C. Fenselau and O.M. Uy, Anal.Chem., 1987, 59, 2806.
321 L. Orning, Eur.J.Biochem., 1987, 170, 77.
322 A. Ballistreri, D. Garozzo, M. Guiffrida, G. Mondando, A. Filippi, C. Guaita, P. Manaresi and F. Pilate, Macromolecules, 1987, 20, 1029.
323 J. Jurenitsch, J. Maurer, U. Rain and W. Robien, Phytochemistry, 1988, 27, 626.
324 R. Lorenzi and N. Ceciarelli, Phytochemistry, 1986, 25, 817; M.J. Ackland, J. Gordon, J.R. Hanson, B.L. Yeoh and A.H. Ratcliffe, Phytochemistry, 1988, 27, 1031.
325 S.B. Mahato and B.C. Pal, Phytochemistry, 1986, 25, 909.
326 J. Gunzinger, J.D. Msouthi and K. Hostettmann, Phytochemistry, 1986, 25, 2501.
327 C. Borel, M.P. Gupta and K. Hostettmann, Phytochemistry, 1987, 26, 2685.
328 D.A.C. Dorsaz and H. Hostettmann, Helv.Chim.Acta., 1986, 69, 2038.
329 S.B. Mahato and B.C. Pal, J.Chem.Soc. Perkin Trans I, 1987, 629.
330 J.C. Burrows, K.R. Price and G.R. Fenwick, Phytochemistry, 1987, 26, 1214.
331 K.W. Glombitza and H. Kurth, Planta.Med., 1987, 53, 548.
332 S.B. Mahato, N.P. Sahu, P. Luger and E. Mueller, J.Chem.Soc. Perkin Trans II, 1987, 10, 1509.

333  C. Pizza, Z. Zhang-Liang and N. De Tommasi, J.Nat.Prod., 1987, 50, 927.
334  R. Chenili, A. Babadjainian, R. Faure, K. Boukef, G. Balansara and
     E. Vidal, Phytochemistry, 1987, 26, 1785.
335  F. Gafner, J.-C. Chapuis, J.D. Msonthi and K. Hostettmann, Phytochemistry, 1987, 26, 2501.
336  H.-U. Marschall, H. Matern, B. Egestad, S. Matern and S. Sjovall, Biochem.Biophys.Acta., 1987, 921, 392.
337  A.A. Ahmed and N.A.M. Saleh, J.Nat.Prod., 1987, 50, 256.
338  R. Maffei Facino, M. Carini, P. Traldi, B. Pelli, B. Gioia and
     E. Arlandini, Biomed.Envir. Mass Spectrom., 1987, 14, 187.
339  V.U. Ahmad, V. Suttara, S. Arif and Q. Najmus-Saquib, Phytochemistry, 1988, 27, 304.
340  V.U. Ahmad, N. Baro, I. Fatima and S. Baro, Tetrahedron, 1988, 44, 247.
341  R. Higuchi, Y. Tokimitsu and T. Komori, Phytochemistry, 1988, 27, 1165.
342  M.A. Dubois, R. Bauer, M.R. Cagiotti and H. Wagner, Phytochemistry, 1988, 27, 881.
343  C.L. Curl, K.R. Price and G.R. Fenwick, J.Sci. Food Agric., 1988, 43, 229.
344  K.R. Price, J. Eagles and G.R. Fenwick, J.Sci. Food Agric., 1988, 42, 183.
345  R. Isobe, T. Komori, F. Abe and T. Yamauchi, Biomed.Envir. Mass Spectrom., 1986, 13, 585.
346  C. Shackleton, S. Gaskell and D.J. Liberato, Chromatogr. Methods, 1987, 185.
347  K.B. Tomer and M.L. Gross, Biomed.Envir. Mass Spectrom., 1988, 15, 89.
348  A.M. Lawson, M.J. Madigan, D. Shortland and P.T. Clayton, Clin.Chim.Acta., 1986, 161, 221.
349  M.V. D'Auria, R. Riccio, L. Mirale, S. La Barre and J. Pusset, J.Org. Chem., 1987, 52, 3947.
350  S.J. Gaskell, Biomed.Envir. Mass Spectrom., 1988, 15, 99.
351  M. Adinolfi, G. Barone, M.M. Corsaro, R. Lanzetta, L. Mangoni and
     M. Parrilli, Can.J.Chem., 1987, 65, 2317.
352  Y. Fujimoto, T. Yamada, N. Ikekawa, I. Nishiyama, T. Matsui and P. Hoschi, Chem.Pharm.Bull.(Tokyo), 1987, 35, 1829.
353  Y. Itakura and T. Komori, Liebigs Ann.Chem., 1986, 359, 499.
354  R. Riccio, M. Iorizzi and L. Minale, Bull.Soc.Chim.Belg., 1986, 95, 869.
355  F. Abe, T. Nagao, Y. Mori, T. Yamauchi and Y. Saiki, Chem.Pharm.Bull. (Tokyo), 1987, 35, 4087.
356  Y. Fujimoto, T. Yamada, N. Ikekawa, I. Nishiyama, T. Matsui and M. Hoshi, Chem.Pharm.Bull.(Tokyo), 1987, 35, 1829.
357  J.L. Mertz, J.S. Peloso, B.J. Barber, G.E. Babbitt, J.L. Occolowitz, V.L. Simson and R.M. Kline, J. Antiobiotics, 1986, 39, 877.
358  A.J. Chulia, J. Garcia and A.M. Mariotte, J.Nat.Prod., 1986, 49, 514.
359  F.A. Tomas-Barberan, J.B. Harborne and R. Self, Phytochemistry, 1987, 26, 2281.
360  F.W. Crow, K.B. Tomer, J.H. Looker and M.L. Gross, Anal.Biochem., 1986, 155, 286.
361  A.G.R. Nair, T.R. Seetharaman, B. Voirin and J. Favre-Bonvin, Phytochemistry, 1986, 25, 768.
362  B. Wald, R. Galensa, K. Herrmann, L. Grotjahn and V. Wray, Phytochemistry, 1986, 25, 2904.
363  A.A. Ahmed and N.A.M. Saleh, J.Nat.Prod., 1987, 50, 256.
364  D. Schaufelberger, M.P. Gupta and K. Hostettmann, Phytochemistry, 1987, 26, 2377.
365  D. Barron and R.K. Ibrahim, Phytochemistry, 1987, 26, 2085.
366  D. Strack, J. Heilemann, E.-S. Klinpott and V. Wray, Z.Naturforsch.C., 1988, 43, 37.
367  N.A.M. Saleh, S.A. Maksoud, W.M.M. Amer, K.R. Markham and D. Barrow, Phytochemistry, 1988, 27, 309.

368  C.A. Williams, J.B. Harborne, F.A. Tomas-Barberan, Phytochemistry, 1987, 26, 2553.
369  F.A. Tomas-Barberan, J.B. Harborne and R. Self, Phytochemistry, 1987, 26, 2759.
370  N. Noda, M. Ono, K. Miyahara, T. Kawasaki and M. Okabe, Tetrahedron, 1987, 43, 3889.
371  R. Self, J. Eagles, G.C. Galletti, I. Mueller-Harvey, R.D. Hartley, A.G.H. Lea, D. Magnolato, U. Richli, R. Gujer and E. Haslam, Biomed. Envir. Mass Spectrom., 1986,·13, 449.
372  H. Feng, G.I. Nonaka and I. Nishioka, Phytochemistry, 1988, 27, 1185.
373  J.-C. Prome, H. Aurelle, D. Prome and A. Savagnac, Org. Mass Spectrom., 1987, 22, 6.
374  A. Proliac and J. Raynaud, Pharmazie, 1987, 42, 557.
375  F.M. Kelleher and V.P. Bhavanandan, Carbohydr.Res., 1986, 155, 89.
376  K. Koizumi, T. Utamura, M. Sato and Y. Yagi, Carbohydr.Res., 1986, 153, 55.
377  P. Wetzel, K. Kunish, F. Kruggel, H. Stein, J. Scherkenbeck, A. Hiltmann, H. Duddeck, D. Mueller, J.E. Maggio, H.-W. Fehlhaber, G. Seibert, Y. Van Heijenoort and J. Van Heijenoort, Tetrahedron, 1987, 43, 585.
378  V.N. Reinhold, S.A. Carr, B.N. Green, M. Petiton, J. Choay and P. Sinay, Carbohydr.Res., 1987, 161, 305.
379  N.-Y. Chen, N. Chen, H. Li, Y.-Z. Chen, F.Z. Zhao, C.-X. Chen, C.-R. Yang, Hua. Hsueh Hsuech Pao, 1987, 45, 682.
380  P. Tsai, A. Dell and C.E. Ballou, Proc.Natl.Acad.Sci., 1986, 83, 4119.
381  E. Tolun and J.F.J. Todd, Int.J. Mass Spectrom. Ion Proc., 1987, 79, 237.
382  E. Tolun and J.F.J. Todd, Org. Mass Spectrom., 1988, 23, 98, 105.
383  T. Jain, C. Kaiser, D.M. Ackerman, D.L. Ladd, K.-L. Fong, G.D. Roberts, D.B. Staiger, I.L. Davis, R.L. Webb and L.M. Jackman, Can.J.Chem., 1986, 64, 2418.
384  S. Naylor, R.P. Mason, J.K.M. Sanders, D.H. Williams and G. Moreti, Biochem.J., 1988, 249, 573.
385  L.D. Arnold, R.G. May and J.C. Vederas, J.Am.Chem.Soc., 1988, 110, 2237.
386  A. Lignori, G. Sindona and N. Uccella, J.Am.Chem.Soc., 1986, 108, 7488.
387  G. Vernin, C. Boniface, J. Metzger, T. Obreterar, J. Kantasubrata, A.M. Siouffi, J.L. Larice and D. Fraisse, Bull.Soc.Chim.Fr.,1987, 681.
388  D.H. Williams, A.C.S.Sym.Ser.N 291, Am.Chem.Soc., Washington D.C., 1985, 217.
389  K. Biemann and S.A. Martin, Mass Spectrom.Rev., 1987, 6, 1.
390  K. Yokoi, K. Nagaoka and T. Nakashima, Chem.Pharm.Bull.(Tokyo), 1986, 34, 4554.
391  A.E. Ashcroft, J.R. Chapman and J.S. Cottrell, J.Chromatogr., 1987, 394, 15.
392  C. Dass and D.M. Desiderio, Anal.Biochem., 1987, 163, 52.
393  L. Poneter and D.H. Williams, J.Nat.Prod., 1986, 49, 26.
394  R. Isobe, I. Fujii and K. Kanematsu, Trends Anal.Chem., 1987, 6, 78.
395  M.J. Bertrand and P. Thebault, Biomed.Envir. Mass Spectrom., 1986, 13, 347.
396  L.A. Savoy, R.M.L. Jones, S. Pochan, J.G. Davies, A.U. Muir, R.E. Offord and K. Rose, Biochem.J., 1988, 249, 215.
397  E. Arlandini, B. Gioia, G. Perseo, B. Danieli and F.M. Rubiro, Biomed. Envir. Mass Spectrom., 1987, 14, 487.
398  E. Tolun, C. Dass and D.M. Desiderio, Rapid Commun. Mass Spectrom., 1987, 1, 77.
399  T. Matsuo, Nippon Kagaku Kaishi, 1986, 11, 1671.
400  Y. Kawano, R. Higuchi, R. Isobe and T. Komori, Liebigs Ann.Chem., 1988, 19.
401  R. Cooper and S. Unger, J.Org.Chem., 1986, 51, 3942.
402  L.A. Actis, W. Fish, J.H. Crosa, K. Kelleman, S.R. Ellenberger, F.M. Hauser and J. Sanders-Loehr, J.Bacteriol., 1986, 167, 57.
403  K.H. Sehram, Trends Anal.Chem., 1988, 7, 28.

404  M. Poc, J.I. Germershausen and M. MacCoss, Anal.Biochem., 1987, 164, 450.
405  A. Sandstrom and J. Chattopadhyaya, J.Chem.Soc.Chem.Commun., 1987, 862.
406  D.L. Slowikowski and K.H. Schram, Nucleosides Nucleotides, 1985, 4, 347, (Chem.Abs., 1985, 103, 25 215698q).
407  R.L. Cerny, M.L. Gross and L. Grotjahn, Anal.Biochem., 1986, 156, 424.
408  A. Wolter, C. Moehringer, H. Koester and W.A. Koenig, Biomed.Envir. Mass Spectrom., 1987, 14, 111.
409  F. Söler, K. Jankowski, H. Vireliziev, D. Gaudin, J. Ulrich and R.E.A. Teoule, J.Bioelectr., 1985, 4, 43, (Chem.Abs., 1986, 104, 05 031138w).
410  F. Söler and K. Jankowski, Spectroscopy Int.J., 1985, 4, 35.
411  J.J. Dino, C.R. Guerat, K.B. Tomer and D.G. Kaufman, Rapid Comm. Mass Spectrom., 1987, 1, 69.
412  R.L. Cerny, K.B. Tomer, M.L. Gross and L. Grotjahn, Anal.Biochem., 1987, 165, 175.
413  K. Balasanmugam and J.M. Miller, Org. Mass Spectrom., 1988, 23, 267.
414  K.B. Tomer, M.L. Gross and M.L. Deinzer, Anal.Chem., 1986, 58, 2527.
415  R. Haessner, R. Borsdorf, G. Dube, A. Lehmann, H. Ruotsalainen and G. Bach, Org. Mass Spectrom., 1986, 21, 473.
416  E. Forest, J. Ulrich, J.C. Marehon and H. Virelizier, Org. Mass Spectrom., 1987, 22, 45.
417  E.J. Breaux, J.E. Patanella, E.F. Sanders and H. Fujiwara, Biomed.Envir. Mass Spectrom., 1988, 15, 123.
418  N. Patarasakulchai, F. Leem, H. Tjandra, D. Nelson and P.T. Southwell-Keely, Org. Mass Spectrom., 1986, 21, 375.
419  D. Dugat, G. Just and S. Sahoo, Can.J.Chem., 1987, 65, 88.
420  K. Balasanmugam, S.K. Viswanadham, D.M. Hercules, R.J. Cotter, D. Heller, A. Benninghoven, W. Sichtermann, V. Anders, T. Keough, R.D. Macfarlane and C.J. McNeal, Appl.Spectrosc., 1987, 41, 821.
421  S.L. Hunt, F.E. Behr, C.D. Winter, P.A. Lyon, R.L. Cerny, K.B. Tomer and M.L. Gross, Anal.Chem., 1987, 59, 2653.
422  Y. Tordeur, G.W. Sovocool, R.K. Mitchum, W.J. Neiderhoot and J.P. Donnelly, Biomed.Envir. Mass Spectrom., 1987, 14, 733.
423  G.M. Allmaier and E.R. Schmidt, Rapid Commun. Mass Spectrom., 1987, 1, 42.
424  M.I. Bruce and M.J. Liddell, Appl.Organomet.Chem., 1987, 1, 191.
425  S.E. Unger, T.J. McCormick, E.N. Treher and A.D. Nunn, Anal.Chem., 1987, 59, 1145.
426  M.I. Bruce, D.N. Duffy, M.J. Liddell, M.R. Snow and E.R.T. Tiekirk, J.Organomet.Chem., 1987, 335, 365.
427  K.R. Jennings, T.J. Kemp and B. Sieklucka, Inorg.Chim.Acta., 1988, 141, 163.
428  C.-M.T. Hayward and J.R. Shapley, Organometallics, 1988, 7, 448.
429  G. Cerveau, C. Chuit, R.J.P. Corriu, L. Gerbier, C. Reye, J.L. Aubagnac and B.E. Amrani, Int.J. Mass Spectrom. Ion Phys., 1988, 82, 259.
430  W.J.A. Van der Heuvel, R.W. Walker and J.R. Carlin, Chromatogr. Methods, 1987, 139.
431  J.C. Vickerman, Chem. in Brit., 1987, 23, 969.
432  M. Leleyler, J.Phys., 1987, 48, 1963.
433  J.M. Rynkowski, S. Affrossman, J.M.R. MacAllister and R.A. Pethrick, Surf.Sci., 1987, 182, 1.
434  G. Bolbock, R. Beavis, S. Della Negra, C. Deprun, W. Ers, Y. Lebeyec, D.E. Main, B. Schueler and K.B. Standing, Nucl.Instrum. Methods Phys.Res. (Sect. B), 1988, 30, 74.
435  M.J. Hearn and D. Briggs, Surf, Interface Anal., 1988, 11, 198.
436  K. Wittmaack, Surf. Interface Anal., 1987, 10, 311.
437  E. Matsui, K. Ogura and S. Handa, J.Biochem.(Tokyo), 1987, 101, 423.
438  W.V. Ligon and S.B. Dorn, Anal.Chem., 1986, 58, 1889.
439  O.W. Hard, B.H. Hsu and R.G. Cooks, Org. Mass Spectrom., 1988, 23, 16.
440  K.V. Malakhov, M.Ya. Tuckina and S.L. Dobychin, J.Anal.Chem.U.S.S.R. (Engl.Trans), 1986, 41, 406.

441  W.V. Ligon, Anal.Chem., 1986, 58, 485.
442  B.W. Gibson, A.M. Falick, A.L. Burlingame, L. Nadasdi, A.C. Nguyen and G.L. Kenyon, J.Am.Chem.Soc., 1987, 109, 5343.
443  W.V. Ligon and S.B. Dorn, Fresenius' Z.Anal.Chem., 1986, 325, 626.
444  F. Lafortune, K.G. Standing, J.B. Westmore, M.J. Damha and K. Ogilvie, Org. Mass Spectrom., 1988, 23, 228.
445  'Laser Spectroscopy VII', T.W. Hansch and Y.R. Shen (eds.), Springer Ser. Opt. Sci. Vol 49, Springer Verlag, Berlin, (Proc. 7th Int. Comf. Laser Spectros., Hawaii, 1985).
446  G.D. Byrd, A.J. Fatiadi, D.S. Simons and E. White, Org. Mass Spectrom., 1986, 21, 63.
447  P. Coad, R.A. Coad, C.L.C. Yang and C.L. Wilkins, Org. Mass Spectrom., 1987, 22, 75.
448  C.E. Brown, P. Kovacic, C.A. Wilkie, J.A. Kinsinger, R.E. Hein, S.I. Yaniger and R.B. Cody, J.Polym.Sci.Polym.Chem.Ed., 1986, 24, 255, (Chem.Abs., 1986, 104, 20 168968 ).
449  K. Balasanmugam, S.K. Viswanadham and D.M. Hercules, Anal.Chem., 1986, 58, 1102.
450  L.L. Miller, A.D. Thomas, C.L. Wilkins and D.A. Weil, J.Chem.Soc.Chem. Commun., 1986, 661.
451  A.D. Thomas and L.L. Muller, J.Org.Chem., 1986, 51, 4160.
452  C.H. Watson, G. Baykut and J.R. Eyler, Anal.Chem., 1987, 59, 1133.
453  Y.H. Paik and P. Dowd, J.Org.Chem., 1986, 51, 2910.
454  C.D. Parker and D.M. Hercules, Anal.Chem., 1986, 58, 25.
455  M.L. Coates and C.L. Wilkins, Biomed.Envir. Mass Spectrom., 1986, 13, 199.
456  R.S. Brown and C.L. Wilkins, Anal.Chem., 1986, 58, 3196.
457  J.J. Morelli, S.K. Viswanadham, A.G. Sharkey and D.M. Hercules, Int.J. Envir.Anal.Chem., 1987, 31, 295.
458  S. Becker and H.J. Dietze, Int.J. Mass Spectrom. Ion Phys., 1988, 52, 287.
459  J.H. Zoeller, R.A. Zingaro and R.D. Macfarlane, Int.J. Mass Spectrom. Ion Proc., 1987, 77, 21.
460  Y.M. Yang, E.A. Sokoski, H.M. Fales and L.K. Pannell, Biomed.Envir. Mass Spectrom., 1986, 13, 489.
461  P.A. Demirev, Biomed.Envir. Mass Spectrom., 1987, 14, 241.
462  I. Jardine, S.W. Hunter, P.J. Brennan, C.J. McNeal and R.D. Macfarlane, Biomed.Envir. Mass Spectrom., 1986, 13, 273.
463  C.J. Neal and R.D. Macfarlane, J.Am.Chem.Soc., 1986, 108, 2132.
464  D.L. Fished and J.E. Hunt, Org. Mass Spectrom., 1987, 22, 799.
465  K.A. Jacobsen, L.K. Pannell, K.L. Kirle, H.M. Fales and E.A. Sokoloski, J.Chem.Soc. Perkin Trans II, 1986, 2143.
466  A. Viari, J.-P. Ballini, P. Vigny, D. Shire and P. Dousset, Biomed.Envir. Mass Spectrom., 1987, 14, 83.
467  A. Viari, J.P. Ballini, P. Vigny, C. Blonski, P. Dousset and D. Shire, Tetrahedron Lett., 1987, 28, 3349.
468  Y. Sun, A.E. Martell, D. Chen, R.D. Macfarlane and C.J. McNeal, J. Heterocyclic Chem., 1986, 23, 1565.
469  H.M. Schiebel and H.-R. Schulten, Mass Spectrom.Rev., 1986, 5, 249.
470  F.W. Roellgen, P. Daehling, E. Bramer-Weger, F. Okuyama and M. Subhan, Org. Mass Spectrom., 1986, 21, 623.
471  D.D. Bombick and J. Allison, Anal.Chem., 1987, 59, 458.
472  J.T. Mosely, Adv.Chem.Phys., 1985, 60, 245.
473  J.M. Oakes and G.B. Ellison, Tetrahedron, 1986, 42, 6263.
474  R.S. Mercer and A.G. Harrison, Org. Mass Spectrom., 1987, 22, 710.
475  W.J. Griffiths and F.M. Harris, Chem.Phys.Lett., 1987, 142, 7.
476  W.J. Griffiths and F.M. Harris, Int.J.Mass Spectrom. Ion Proc., 1987, 77, 127.
477  W.J. Griffiths and F.M. Harris, Org. Mass Spectrom., 1987, 22, 559.
478  M. Meot-Ner, J.Am.Chem.Soc., 1988, 110, 3071.
479  M. Meot-Ner, J.Am.Chem.Soc., 1988, 110, 3075.

480 S. Chowdhury, T. Heinis, E.P. Grimsrud and P. Kebarle, J.Phys.Chem., 1986, 90, 2747.
481 J.E. Bartmess and J.P. Kiplinger, J.Org.Chem., 1986, 51, 2173.
482 C.S. Ewig and J.R. Van Wazer, J.Phys.Chem., 1986, 90, 4360.
483 C.C. Han and J.I. Brauman, J.Am.Chem.Soc., 1988, 110, 4048.
484 D. Smith and N.G. Adams, J.Phys.B., 1987, 20, 4903.
485 M.D. Brickhouse and R.R. Squires, J.Am.Chem.Soc., 1988, 110, 2706.
486 J. Catalan, R.M. Claramunt, J. Elgerero, J. Layrez, M. Meredez, F. Anvia, J.H. Quian, M. Taagepera and R.W. Taft, J.Am.Chem.Soc., 1988, 110, 4105.
487 K.K. Murray, T.M. Miller, D.G. Leopold and W.C. Lineberger, J.Chem.Phys., 1986, 84, 2520.
488 J. Kaneti, P.v.R. Schleyer, T. Clark, A.J. Kos, G.W. Spitznagel, J.G. Andrade and J.B. Moffat, J.Am.Chem.Soc., 1986, 108, 1481.
489 J. Filley, C.H. DePuy and V.M. Bierbaum, J.Am.Chem.Soc., 1987, 109, 5992.
490 S. Moran, H.B. Ellis, D.J. DeFrees, A.D. McLean and G.B. Ellison, J.Am. Chem.Soc., 1987, 109, 5996.
491 S. Moran, H.B. Ellis, D.J. DeFrees, A.D. McLean, S.E. Paulson and G.B. Ellison, J.Am.Chem.Soc., 1987, 109, 6004.
492 K.R. Lykke, D.M. Neumark, T. Andersen, V.J. Trapa and W.C. Lineberger, J.Chem.Phys., 1987, 87, 6842.
493 J. Marks, D.M. Wetzel, P.B. Comita and J.I. Brauman, J.Chem.Phys., 1986, 84, 5284.
494 R.N. McDonald and A.K. Chowdhury, Tetrahedron, 1986, 42, 6253.
495 D.G. Leopold, A.E.S. Miller and W.C. Lineberger, J.Am.Chem.Soc., 1986, 108, 1379.
496 S. Chrowdhury and P. Kebarle, J.Am.Chem.Soc., 1986, 108, 5453.
497 S. Chowdhury, G. Nicol and P. Kebarle, Chem.Phys.Lett., 1986, 127, 130.
498 T. Heinis, S. Chowdhury, S.L. Scott and P. Kebarle, J.Am.Chem.Soc., 1988, 110, 400.
499 N.M.M. Nibbering, Recl.Trav.Chim.Pays-Bas., 1986, 105, 245.
500 S.T. Graul and R.R. Squires, Mass Spectrom.Rev., 1988, 7, 263.
501 N.M.M. Nibbering, Adv.Phys.Org.Chem., 1988, 24, in press.
502 M. Okumura, L.I. Yeh, D. Normand, J.J.H. Van Der Biesen, S.W. Bustamente, Y.T. Lee, T.J. Lee, N.C. Handy and H.F. Schaefer, J.Chem.Phys., 1987, 86, 3807.
503 C.H. DePuy, S.E. Barlow, J.M. Van Doren, C.R. Roberts and V.M. Bierbaum, J.Am.Chem.Soc., 1987, 109, 4414.
504 R.N. McDonald and W.Y. Gurg, J.Am.Chem.Soc., 1987, 109, 7328.
505 S.T. Graul and R.R. Squires, J.Am.Chem.Soc., 1988, 110, 607.
506 R.N. McDonald, W.D. McGhee and A.K. Chowdhury, J.Am.Chem.Soc., 1987, 109, 7334.
507 J.M. Van Doren, S.E. Barlow, C.H. DePuy and V.M. Bierbaum, Int.J. Mass Spectrom. Ion Proc., 1987, 81, 85.
508 R.E. Lee and R.R. Squires, J.Am.Chem.Soc., 1986, 108, 5078.
509 E. Buncel, S.S. Shaik, I.-H. Um and S. Wolfe, J.Am.Chem.Soc., 1988, 110, 1275.
510 R. Osman, K. Namboodiri, H. Weinstein and J.R. Rabinowitz, J.Am.Chem.Soc., 1988, 110, 1701.
511 C.-C. Han, J.A. Dodd and J.I. Brauman, J.Phys.Chem., 1986, 90, 471.
512 L.J. DeKoning and N.M.M. Nibbering, J.Am.Chem.Soc., 1987, 109, 1715.
513 W.W. Van Berkel, L.J. DeKoning and N.M.M. Nibbering, J.Am.Chem.Soc., 1987, 109, 7602.
514 L.J. DeKoning and N.M.M. Nibbering, J.Am.Chem.Soc., 1988, 110, 2066.
515 G. Bouchoux and Y. Hoppilliard, Tetrahedron Lett., 1987, 28, 4537.
516 J.J. Grabowski and S.J. Melby, Int.J. Mass Spectrom. Ion Proc., 1987, 81, 147.
517 J.M. Van Doren, S.E. Barlow, C.H. DePuy and V.M. Bierbaum, J.Am.Chem.Soc., 1987, 109, 4412.
518 J.H. Bowie, C.H. DePuy, V.M. Bierbaum and S.A. Sullivan, Can.J.Chem., 1986, 64, 1046.

519  C.-C. Han and J.I. Brauman, J.Am.Chem.Soc., 1987, 109, 589.
520  T. Su, A.C.L. Su, A.A. Viggiano and J.F. Paulson, J.Phys.Chem., 1987, 91, 3683.
521  G. Chalasinski, R.A. Kendall and J. Simons, J.Chem.Phys., 1987, 87, 2965.
522  W. DeLarge and N.M.M. Nibbering, Int.J. Mass Spectrom. Ion Proc., 1987, 80, 201.
523  W.J. Griffiths and F.M. Harris, Org. Mass Spectrom., 1987, 22, 812.
524  L.A. Curtiss, C.A. Melendres, A.E. Read and F. Weinhold, J.Comput.Chem., 1986, 7, 294.
525  E. Carbonell, J.L. Andrés, A. Liedós, M. Duran and J. Bertrán, J.Am.Chem.Soc., 1988, 110, 996.
526  K. Ohta and K. Morokuma, J.Phys.Chem., 1985, 89, 5845,
527  P.M. Hierl, A.F. Ahrens, M. Henchman, A.A. Viggiano and J.F. Paulson, Int.J. Mass Spectrom. Ion Proc., 1987, 81, 101.
528  M. Meot-Ner, J.Am.Chem.Soc., 1986, 108, 6189.
529  M. Meot-Ner and L.W. Sieck, J.Phys.Chem., 1986, 90, 6687.
530  M. Meot-Ner and L.W. Sieck, J.Am.Chem.Soc., 1986, 108, 7525.
531  J.H. Bowie, R.N. Hayes, J.C. Sheldon and C.H. DePuy, Aust.J.Chem., 1986, 39, 1951.
532  M. Meot-Ner, J.Am.Chem.Soc., 1988, 110, 3854.
533  K. Hiraoka, Bull.Chem.Soc.Jap., 1987, 60, 2555.
534  K. Mitsuke, H. Tada, F. Misaizu, T. Kordau and K. Kuchitsu, Chem.Phys.Lett., 1988, 143, 6.
535  K. Hiraoka and S. Mizise, Chem.Phys., 1987, 118, 457.
536  J.W. Larson and T.B. McMahon, J.Am.Chem.Soc., 1988, 110, 1087.
537  C.R. Moylan, J.A. Dodd, C.-C. Hau and J.I. Brauman, J.Chem.Phys., 1987, 86, 5350.
538  S. Chowdhury and P. Kebarle, J.Chem.Phys., 1986, 85, 4989.
539  L.W. Sieck, J.Phys.Chem., 1986, 90, 6684.
540  K. Hiraoka, S. Mizise and S. Yamabe, J.Chem.Phys., 1987, 86, 4102.
541  M. Moet-Ner, J.Am.Chem.Soc., 1988, 110, 3858.
542  J.W. Larson and T.B. McMahon, J.Am.Chem.Soc., 1987, 109, 6230.
543  R.N. Hayes, J.H. Bowie and G. Klass, J.Chem.Soc., Perkin Trans.II, 1984, 1167; R.N. Hayes and J.H. Bowie, J.Chem.Soc., Perkin Trans.II, 1985, 567.
544  H. Van Der Wel, N.M.M. Nibbering, J.C. Sheldon, R.N. Hayes and J.H. Bowie, J.Am.Chem.Soc., 1987, 109, 5823.
545  C.H. DePuy, R. Damrauer, J.H. Bowie and J.C. Sheldon, Accts.Chem.Res., 1987, 20, 127.
546  J.C. Sheldon, J.H. Bowie, C.H. DePuy and R. Damrauer, J.Am.Chem.Soc., 1986, 108, 6794.
547  J.C. Sheldon, R.N. Hayes, J.H. Bowie and C.H. DePuy, J.Chem.Soc., Perkin Trans II, 1987, 275.
548  J.R. Damewood and C.M. Hadad, J.Phys.Chem., 1988, 92, 33.
549  R. Damrauer, S.R. Kass and C.H. DePuy, Organometallics, 1988, 7, 637.
550  R. Damrauer, W.P. Weber and G. Manuel, Chem.Lett., 1987, 235.
551  L.M. Babcock, W.S. Taylor and C.R. Herd, Int.J. Mass Spectrom. Ion Proc., 1987, 81, 259.
552  W. Tumas, K.E. Salomon and J.I. Brauman, J.Am.Chem.Soc., 1986, 108, 2541.
553  R. Damrauer, C.H. DePuy, S.E. Barlow and S. Gronert, J.Am.Chem.Soc., 1988, 110, 2005.
554  S.W. McElvany and J. Allison, Organometallics, 1986, 5, 416.
555  K.R. Lane, L. Sallans and R.R. Squires, J.Am.Chem.Soc., 1986, 108, 4368.
556  K.R. Lane and R.R. Squires, J.Am.Chem.Soc., 1986, 108, 7187.
557  I.K. Gregor, J.Organomet.Chem., 1987, 329, 201.
558  D. Wang and R.R. Squires, Organometallics, 1987, 6, 905.
559  I.K. Gregor, Org. Mass Spectrom., 1987, 22, 644.
560  D.L. Bricker and D.H. Russell, J.Am.Chem.Soc., 1987, 109, 3910.
561  R.N. McDonald and M.T. Jones, Organometallics, 1987, 6, 1991.

562  D. Wang and R.R. Squires, J.Am.Chem.Soc., 1987, 109, 7557.
563  R.A. Forbes, L.H. Laukien and J. Wronka, Int.J. Mass Spectrom. Ion Proc., 1988, 83, 23.
564  A.K. Chowdhury and C.L. Wilkins, J.Am.Chem.Soc., 1987, 109, 5336.
565  G.E. Streit and L.M. Babcock, Int.J. Mass Spectrom. Ion Proc., 1987, 75, 221.

# 6
# Analysis of Mixtures by Mass Spectrometry Part I: Developments in Gas Chromatography/Mass Spectrometry

BY R. P. EVERSHED

## 1 General Considerations

**1.1 Introduction.-** A substantial increase in publication rate in the field of GC/MS has been seen during the current review period (June 1986 - June 1988; Figure 1). These statistics have been compiled from information supplied by the Royal Society of Chemistry, Mass Spectrometry Data Centre. They serve, not only to further emphasize the prominence of GC/MS in routine mixture analysis, but also to confirm that substantial scope still exists for the development and application of new methodologies. Pinpointing a single factor responsible for this increase is no simple matter; although the wider use of bench-top GC/MS systems, incorporating quadrupole or ion trap mass analysers, has probably been influential.

The enormous volume of published material which has appeared requires a more stringent reviewing policy than that previously adopted. Hence, whilst the aim remains to cover the full range of subject areas, citations will be confined largely to those which advance methodology and represent novel applications. A listing of published books, reviews, abstracts and conference proceedings together with general subject areas is given in Table 1. Readers should note the appearance of a new journal 'Rapid Communications in Mass Spectrometry'.[69] First published in 1987, and as its title suggests, it provides a forum for rapid publication of preliminary accounts on all aspects of the science of gas-phase ions.

For comparative purposes Figure 1 also shows the rate of publications in the LC/MS field. The low rate compared to GC/MS is a clear reflection of the considerable technical difficulties in LC/MS interfacing. GC/MS on the other hand can be regarded as the 'natural combination', as both techniques deal with volatile and semivolatile compounds, and show optimum performance with sample sizes in the nanogram range or lower. The interfacing of SFC with MS, although not trivial, is easier than HPLC and shows considerable promise, offering an alternative particularly in the case of relatively non-polar or intermediate polarity thermally labile substances. More detailed discussions of LC/MS and SFC/MS are to be found in Chapter 7.

**1.2 Instrumentation.-** Developments and trends in mass spectrometry instrumentation are reviewed in detail in Chapter 3. Developments especially pertinent to GC/MS are considered below.

Progress has been made in the development of a GC/MS system comprising a time-of-flight (TOF) mass analyser.[70] The incorporation an electrostatic sector, serving as a kinetic energy filter, between the ion source and TOF analyser overcame some of the problems previously associated with GC/TOFMS. Data system developments were also described. A major aim in this work was to exploit the potential of high scan rates offered by TOF analysers in meeting the demands of very high resolution GC analyses.

A GC/MS system comprising a laser ionization source and a reflectron TOFMS was employed in an investigation of analytical selectivity, for the analysis of nitro- and nitroso-containing compounds,[71] and ionization processes in alkylbenzenes.[72] In the latter study accurate comparisons were only made possible as a result of the quantitative reproducibility of GC.[72]

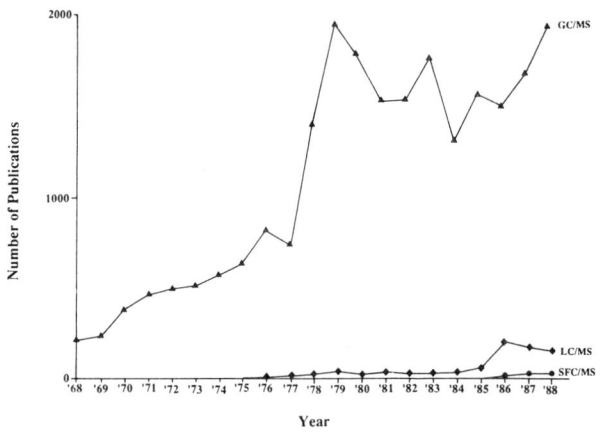

Figure. Graph showing numbers of papers per year in the GC/MS field, from 1968 to the present day. These statistics were supplied by the Royal Society of Chemistry, Mass Spectrometry Data Centre. Publication rates in the LC/MS and SFC/MS fields are presented for comparative purposes.

The connection of a GC to an inductively coupled-plasma source *via* a short (0.5m) stainless steel tube provided an element/isotope specific detection system holding much promise for studies of environmental/clinical metal organic species.[73] The potential of this combination was demonstrated in the separation and detection of nanogram quantities of alkyl tin compounds. Specificity in the analysis of carboxylic acids in oil-well production water has been achieved through the unlikely combination GC/FABMS.[74] The FAB target and column exit were arranged so as to allow the GC eluting components entering the selvedge region to condense in the $Ca(OH)_2$/triethanolamine matrix. Positive-ion (PI) spectra showed complex adduct ions, while negative-ion (NI) spectra were more useful, displaying prominent [M-H]$^-$ and characteristic fragment ions.

A GC/MS system comprising an EI reactor to permit the isolation of reaction products induced by accelerated electron beams has been described.[75] A specialized ion source for GC/atmospheric pressure ionisation-MS has been developed so as to reproduce exactly conditions exisiting in the thermionic ionization GC detector operating in the $N_2$ carrier gas mode.[76] Unexpected results from this system provided a basis for speculation on the mechanism of thermionic ionization response.

Table 1 Books, reviews, conference proceedings and abstracts

| Subject Areas | References |
| --- | --- |
| General | 2,4,5,6,9,10,12,14,58 |
| Chemical ionization GC/MS | 3 |
| Abstacts, conference proceedings | 4,7,8,19,22,64,66 |
| Biomedical, clinical, metabolic | 11,15,46-55 |
| Pharmacology | 11,55,56 |
| Tandem MS | 16-18 |
| GC/FTIR/MS | 19-21 |
| SIM | 16-18 |
| GC aspects | 25-31,102 |
| Derivatization | 32 |
| Long-chain compounds | 33-35,37 |
| Steroids | 36-40 |
| Amines | 41,42 |
| Amino acids, peptides | 43,44 |
| Carbohydrates | 45 |
| Food | 57 |
| Environmental, pesticides | 19,59-62 |
| Forensic | 63 |
| Organic geochemistry | 64-66 |
| Fuel | 67 |
| Pyrolysis-GC/MS | 68 |

Chemical ionization (CI) has been successfully demonstrated in an ion trap mass spectrometer for the first time, hence greatly extending the versatility of this analyser.[77] The spectra obtained using a variety of reagent gases (methane, isobutane, ammonia and argon) were comparable to those obtained using conventional mass spectrometers. The problem of EI interferences was removed by using the unique selective mass storage facility.

The interfacing of GC/Fourier transform infrared (GC/FTIR) with benchtop mass spectrometers, such as the ion trap[78] and mass selective detector[79] serves to emphasise the continuing interest, and special complementarity of this combination for complex mixture analysis.[20,21] The addition of a radioactivity detector to the GC/FTIR/MS combination increased flexibility still further in metabolic studies.[80] The analytical capabilities of a GC/FTIR/FTMS system have been illustrated with accurate mass measured compositions used as a post-search filter to eliminate incorrect IR and MS search results.[81] The advent of GC/FTMS incorporating differentially pumped ion sources has opened up the possibility of using all conventional interfacing and ionization methods, while maintaining the high resolution capability of FTICR instruments.[82] Fast array data processors and Fourier transformations allow acquisition and storage of broad-band spectra, and avoid data overloading.

Tandem mass spectrometry is finding increasingly wider application as researchers acquire the necessary instrumentation. Applications are described in detail in the appropriate sections below. Recent advances in tandem mass spectrometry instrumentation have been reviewed.[16]

With a view to correlating analytical and olfactory data, two reports have appeared which describe apparatus designed to perform parallel MS and olfactory 'sniffing' analyses of GC column eluent.[83,84] Both systems employed packed GC columns and split the GC efluent between the MS and an olfactory 'sniffing' port. The systems have been employed to analyse pollutants[84] and flavour compounds.[83]

Since the introduction of flexible-fused silica capillary GC columns, which made possible the direct mode of GC/MS coupling, investigations in the field of interfacing have slowed dramatically. Although the direct inlet is generally preferred, the open-split interface is used in situations where columns are changed frequently, to avoid venting the MS. An open-split GC/MS interface constructed of flexible-fused silica boasts maximum sensitivity, reproducibility and minimal adsorption losses compared alternative arrangements.[85]

**1.3 The Role of Data Systems.-** Applications of computers and microprocessors in mass spectrometry are reviewed in detail in Chapter 4. Hence, only developments especially pertinent to GC/MS are reviewed here.

Automated computerized identification alogorithms are especially valuable for the determination of unknown mass spectra from GC/MS analyses of complex mixtures. An improvement has been made to the most widely used of these, the Probability Based Matching (PBM) algorithm, by modifying predicted reliability values on the basis of the principle that the match of a compound of low molecular weight has higher probability of being correct.[86]

A microcomputer program has been developed which combines low resolution 70eV EI information from isotope patterns, and that from recognition of common fragment ions, to increase reliability in calculating the elemental compositions of unknowns.[87] Low resolution mass spectra have also been processed by a chemometric method to define compound class.[88] The potential of retention indices as filters to reduce the time required to search library spectra has been emphasised once more.[89]

An effective computer-based algorithm has been developed which overcomes the problem of matching FTIR interferograms to the correct mass chromatograms from GC/FTIR/MS data.[90] A time-warping algorithm was used to eliminate time axis stretch and squeeze in the chromatogram so that peaks could be reliably matched. The method adopted achieved a better match than when a linear regression procedure was employed.

**1.4 Quantification.-** The inherently high sensitivity and selectivity of SIM combined with high degree of accuracy and precision attainable has led to mass spectometry being regarded as a reference technique for other analtyical methods employed in a variety of fields.[23,24]

Especially pertinent to modern quantitative GC/MS is an investigation of the expected errors in SIM assays using capillary columns.[91] Problems in this area arise from insufficient data collection owing to non-continuous sampling of the GC eluent. This problem is exacerbated where very high resolution GC columns produce very narrow chromatographic peaks. Theoretical treatment showed that to ensure an error in GC/MS peak height of <3% at least five samples need to be observed over the top half of the peak, and for 1.5% at least seven. In another study several smoothing routines were rigorously evaluated for the optimization of SIM data in dual ion monitoring.[92] It was found that smoothing achieved only minor improvements in precision and introduced bias into peak high ratio calculations. On the other hand, linear regression of response ratios of analyte and internal standard yielded superior precision without introducing bias. Caution was expressed in the use of smoothing of multiple ion monitoring data.

A short review has appeared introducing the principles of isotope dilution MS.[93] Significant to improving the accuracy and precision of isotope dilution MS has been the development of a new method of relating the composition of isotopic mixtures to the average mass of the mass spectral pattern.[94] This approach overcomes problems associated with uncertainties in the isotopic purity of labelled internal standards, natural isotope patterns and the presence of interfering fragment ions. GC/MS has also been used to derive the specific activities of $^{14}$C-labelled compounds from the relative peak sizes of $M^{+\cdot}$ ions or other ions containing all of the $^{14}$C.[95] The method boasts practical accuracy, precision and convenience compared to conventional methods of specific activity measurements (eg. radioactivity-weight method).

Relevant to quantitative GC/MS are theoretical treatments of factors affecting the limit of detection in chromatography.[96,97] Attention has been drawn to the implications of the solvent effect in the quantification of minor constituents by GC and GC/MS in the analysis of deuterium labelled and unlabelled benzofluoroanthanes.[98]

**1.5 Sampling.-** Injection systems for GC have been extensively reviewed,[25-27] and factors affecting quantitative performance of cold on-column and splitless injection systems assessed.[99] On-column injection is generally to be preferred for quantitative work as this circumvents the problems involved with extra-column vaporization (ie. discrimination, degradation, etc.). In the ideal mode of operation of on-column injection, flash volatilization of the solvent due to too high a temperature must be avoided. The necessarily low GC oven temperature required inevitably increases overall analysis time; hence considerable interest exists in the development of on-column injectors for sample introduction at high temperatures.[100-102]

The attraction of capillary column heartcutting GC/GC/MS (so-called multidimensional GC/MS) lies in its ability to maximize the separation of selected components and reduce overall analysis time.[30,103,104] A recent advance in this field is the direct coupling of microbore HPLC to high-resolution GC/MS.[105] A novel feature of this system was the direct LC/GC coupling *via* a newly developed low dispersion isotachic (constant linear velocity) stream splitter,[106] which obviated the need for a GC retention-gap. The potential of the system was demonstrated in the analysis of a solvent refined coal.[105]

A preconcentration step is prerequisite to the GC/MS analysis of trace components. For volatile organic compounds, preconcentration can be achieved by cyrogenic trapping[107,108] or dynamic headspace analysis.[104,109] Where higher analyte concentrations are involved, such as in the case of vapours contained in chemical process streams, sampling to the GC/MS can be achieved *via* a valve injection system, while maintaining analytes in the vapour phase.[110,111] The volatile constituents of Chinese medicinal herbs have been determined by heating 3-15mg herb samples at 250ºC for 80s in a specially constructed vaporizer.[112] Advantages over traditional steam distillation methods include minimal sample preparation, small sample size, and avoidance of potential interferences from solvents.

A promising development has been the demonstration of on-line supercritical fluid extraction (SFE) GC and GC/MS analysis of natural products from a range of matrices. Coupling of the SFE cell to the GC was achieved by inserting the SFE outlet restrictor capillary (150µm o.d.) into the GC column through the on-column injection port. Cooling of the GC oven ensured thermal focussing of the analytes at the restrictor outlet. For 0.5mg samples extractions were performed with 300 atm. $CO_2$ at 45ºC for 10 mins. After removal of the restrictor and flushing the $CO_2$ from the GC column, analyses were performed in the normal way.

Purge and trap is widely used for the isolation of volatile organics prior to GC/MS analysis.[114,115] An effective approach to the concentration of trace organic components from aqueous solutions or solvent extracts proceeds with adsorption on small (5.4cm x 0.4mm i.d.) Tenax cartridges. Transfer of analytes to the GC column was achieved using a custom built thermal desorption unit. High recoveries were achieved for the wide range of compounds tested.[116,117]

**1.6 Chromatographic Aspects.-** The selection of the most suitable gas chromatograph for GC/MS work should not be undertaken lightly. A reference source for commercial instrumentation is available,[29] as is a general review of gas chromatography,[31] and an overview of modern high resolution GC column technology.[28]

Developments in stationary phase technology largely centre on modifications of polymethylsilicones, through the addition of polar moieties. Immobilization of stationary phases with a view to increasing the versility of high-temperature GC continues to attract substantial interest. However, relatively little high-temperature GC/MS work has been attempted.[118-124]

Flexible fused-silica metal-clad capillary columns offer substantially superior mechanical and thermal robustness compared to their polyimide-coated counterparts.[125,126] However, applications using such columns for GC/MS have yet to appear. Attention of potential users is drawn to the hazards of high voltage breakdown when attempting direct connection into the sources of magnetic sector instruments.

Considerations with respect to improving speed and efficiency in GC and GC/MS have been discussed.[127,128,129] Chromatographic theory predicts enhanced efficiency, sensitivity and speed through reduced internal diameter of an open tubular capillary column. WCOT capillary columns with 100μm i.d. yield high plate numbers per unit length, thus permitting complex separations to be achieved more rapidly and with improved sensitivity.[129,130] Limitations which might preclude the general use of such columns in GC/MS work include: their low sample capacity, and problems in maintaining chromatographic integrity of narrow GC peaks in slower scanning mass spectrometers. However, successful applications have been demonstrated in the analysis of toxicological and environmental samples employing a fast scanning quadrupole instrument. Problems of column overload and injection irreproducibility were overcome by automated injection of submicrolitre sample volumes.[130]

### 1.7 Derivatization.-

Method development guidelines for chemical derivatization in gas chromatography have been presented.[32] In GC/MS, as in GC, a common aim of derivatization is the blocking of protic sites (eg. $-CO_2H$, $-OH$, $-SH$, $-NH_2$) to allow polar compounds to elute from the GC column. However, in GC/MS selective derivatization can serve also to increase the structural information content of mass spectra by producing characteristic fragmentations. Additionally, derivatization can serve to increase sensitivity in trace analyses by say, increasing the proportion of ion current carried by a single fragment ion, or by introducing strongly electrophoric group(s) to enhance NICI response.

Silylation remains the most common means of blocking protic sites and conferring desirable mass spectrometric properties. The novel diethylhydrogensilyl-cyclic diethylsilylene derivative, introduced during the last review period, has been shown to be useful for the analysis of a range of biologically important prostanoids[131-134] and steroids.[135-137] Low picogram detection limits appear readily attainable in either low or high resolution SIM modes. The routine use of these derivatives awaits their commercial availability.

The novel allyldimethylsilyl ether (ADMS) derivative of steroids, leukotrienes and prostaglandins have also been prepared by reaction with $N,O$-bis(allyldimethylsilyl)-trifluoroacetamide at room temperature.[138-140] The mass spectra were characterized by abundant high mass ions arising from elimination of an allyl radical ($[M - 41]^+$). SIM yielded low picogram detection limits for prostaglandin $E_2$, $F_{2\alpha}$ and 5β-pregnane-3α,20α-diol. The ADMS derivatives of leukotriene methyl esters appear to be less desirable for trace analyses.[138,139] Catalytic hydrogenation prior to derivatization to trimethylsilyl (TMS), t-butyldimethylsilyl (TBDMS) or ADMS ethers was found to increase GC sensitivity and MS high mass ion abundance.[139,140] TBDMS ester derivatives were deemed not to offer major advantages over methyl esters in the GC/MS analysis of monohydroxyeicosatetraenoic acids and leukotrienes for reasons of relatively poor hydrolytic stability and diminished high mass ion intensity,[141] However, methoxime/TBDMS ether/TBDMS ester derivatives were prefered for the stable isotope dilution GC/MS determination of urinary prostanoids.[142] Limits of detection of 50pg ml$^{-1}$, free from interferences, were achieved by high resolution (m/$\Delta$m 10,000; 10% valley) SIM of the abundant [M - C$_4$H$_9$]$^+$ ion.

Treatment of 15-ketoprostaglandin F$_{2\alpha}$ (15-keto-PGF$_{2\alpha}$) with diazomethane at -78°C yielded only the desired methyl ester, preventing pyrazoline adduct formation.[143] Subsequent dimethylisopropylsilylation afforded a derivative achieving a GC/MS SIM limit of detection (s/n = 10) of 12pg. The applicability of the method was demonstrated following the *in vitro* enzymic conversion of PGF$_{2\alpha}$ to 15-keto-PGF$_{2\alpha}$.

TBDMS esters were the derivative of choice for the GC/MS analysis of volatile and non-volatile organic acids,[144] and mono-, di- and tricarboxylates.[145] These derivatives are characterized by the familiar [M - 57]$^+$ fragment ion. The relative usefulness of the CI of TMS ethers versus EI of TBDMS ethers for the GC/MS determination of 2-ketoacids as their quinoxalinol derivatives have been discussed.[146,147] The major advantage of the TBDMS over TMS derivatization is that the former did not require CI for the trace analysis of 2-ketoacids in biological samples. An alternative derivative, namely the *N*-methyl-quinoxalone, has also been proposed.[148] This is formed readily using *N,N*-dimethylformamide dimethylacetal, and readily determined by EI using SIM. By avoiding silylation, low enrichments of stable isotope tracers were more readily quantified through this approach.

A new silylating reagent, (2-cyanoethyl)dimethyl(diethylamino)silane, has been introduced[149] which is especially suited to blocking acidic groups.[150] While being developed principally for use with nitrogen-phosphorus detection the derivative also displays mass spectrometric properties desirable for use in trace analyses. The GC/MS determination of chlorophenoxy acids and related herbicides as their (cyanoethyl)dimethyl derivatives has been performed.[151] Their spectra display distinctive [M - 54]$^+$, [M - 82]$^+$ and [M - 98]$^+$ ions. High purity mass spectra of TMS enol ethers of various carbonyl compounds have been recorded by GC/MS following on-column silylation with *N,O*-bis(TMS)trifluoroacetamide (BSTFA).[152]

Cyclic ferroceneboronate derivatives have been readily obtained from a range of diols and related compounds.[153] This reagent extends the range of boronic acid reagents and shows particular promise for diols with masses below *ca.* 300 daltons. In addition to their favourable GC properties, the mass spectra (EI) are dominated by M$^{+\cdot}$ ions displaying the characteristic isotope clusters of iron and boron.

Considerable interest exists in the use of mass spectrometry for the structure analysis of fatty acids. Methods available up to 1987 have been reviewed.[33] Dimethyldisulphide derivatives are finding increasing application.[154-156] As shown previously in the case of long-chain alkenes,[157] monounsaturated double-bond position and geometry are readily determined.[155,156] Alkylthiolation of *Z*- and *E*-isomers leads to *threo* and *erythro* adducts which are well separated by GC. Characteristic EI (70eV) fragmentations pinpoint double-bond position unequivocally. The determination of polyunsaturated compounds by this technique is more complicated. Depending upon the number of -$CH_2$- groups separating the double-bonds in di-unsaturated compounds, derivatization leads to linear or cyclic polyethers.[154] Only in the case of conjugated dienes were mass spectral interpretations complicated. The location of a conjugated diene unit in a fatty acid is very straightforward following conversion to the methyl tetra(trimethylsiloxy)alkanoate.[158] A potentially more desirable approach, for reasons of much reduced wet chemical manipulations, involves formation of the 4-phenyl-1,2,4-triazoline-3,5-dione Diels-Alder adduct (**1**).[159] The reaction is rapid and readily applied to sub-microgram sample sizes. The mass spectra display fragmentations characteristic of the position of the conjugated diene unit.

**1**

Investigations of the mass spectra of picolinyl ester derivatives of isomeric unconjugated octadecadienoic acids indicated that in some compounds unambigous assigments might be precluded where the full range of authentic compounds was not available.[160] Problems seemingly exist in interpretation of mass spectra where the double-bond is close to a carboxyl group.[161] For the analysis of natural samples picolinyl esters were preferred to pyrrolide ester derivatives for reasons of producing more acceptable mass spectral data, especially in the cases of polyunsaturated components.[162] A major problem in the analysis of complex fatty acids mixtures as picolinyl esters, namely the overlap of components with different degrees of unsaturation, has been overcome by silver ion HPLC pre-fractionation prior to GC/MS.[163,164] A feasibility study on the use of deuteriated pyrrolidides for double-bond location in fatty acids concluded that severe limitations exist in pinpointing double bonds close to the pyrrolidyl groups. An additional disadvantage of this approach, for microchemical work, lies in the need to purify individual components.[165] In the continuing search for the 'ideal' heterocyclic derivative for charge stabilization, to aid in double-bond location, a new technique has been proposed.[168] Fatty acids are condensed with 2-amino-2-methylpropanol to yield 2-alkenyl-4,4-dimethyloxazolines (**2**) which possess very good GC characeristics (comparable to methyl esters), and EI spectra displaying diagnostic ions for double-bond position in the chain of mono- and polyenoic acids (containing up to six double bonds). The technique is also useful for localizing cyclopropane rings in long-chain fatty acids.[167] Picolinyl esters have been used to determine cyclopropane and unsaturated fatty acids in *Campylobater* species by GC/MS.[168]

By analogy with picolinyl ester derivatives for double-bond location in fatty acids, novel picolinyldimethylsilyl ethers have been developed for the structure determination of unsaturated fatty alcohols by GC/MS.[169] The mass spectra display abundant $M^{+\cdot}$ ions (base peak) and radical-induced fragment ions which provide information on double-bond position in mono- and polyunsaturated alcohols.

$$\text{structure with NH}_2, \text{OH} + \text{HO-CO-C}_n\text{H}_{2n-x} \longrightarrow \text{oxazoline with C}_n\text{H}_{2n-x}$$

x = 1, 3, 5, 7, ....

2

### 1.8 Stereo- and Positional Isomeric Assignments.

- Papers describing derivatization methods for making stereo- and positional isomeric assignments have been reviewed above ('Derivatization').

A novel new approach to the analysis of alkene isomers employs capillary GC/FTMS using $Fe^+$ CI.[170] Multiphoton dissociation and ionization (MPD/MPI) of iron pentacarbonyl was used to produce atomic iron ions for the GC/FTMS experiment. For $C_7$ - $C_{13}$ alkenes 0.6-11ng (20-280ng injected onto the GC column) respectively, produced acceptable mass chromatograms and FT mass spectra. These detection limits compare favourably to those achieved when using 'wet-chemical' derivatization prior to GC/MS analysis.[170]

The use of CI methods for the structure investigation of long-chain aliphatic compounds has been the subject of much recent research acitivity. $(CH_3)_3CCl$, $(CH_3)_2CHCH_2Cl$ and $C_2H_5C(CH_3)_2Cl$ have been used as alternatives to iso-butane to locate conjugated double-bonds in long-chain alkenes, aldehydes and alcohols.[171] The reactions of plasmas of $CH_4$, i-$C_4H_{10}$ and $NH_3$ with double- and triple-bonds in aliphatic compounds have been reviewed.[172] i-$C_4H_{10}$ was concluded to be the most useful of these three protonating reagent gases, yielding abundant $[M + H]^+$ ions for functionalized compounds and ions diagnostic of location of unsaturation in monounsaturated and conjugated dienes.

Negative-ion chemical ionization (NICI) using $CH_4/N_2O$ (90:10) as reagent gas allowed the location of the epoxide ring in aliphatic epoxides to be readily determined, with the added possibility of stereoisomeric assignments.[173] GC/MS with NICI ($OH^-$ or $NH_2^-$) has been used to generate $[RCO_2]^-$ anions from fatty acid methyl esters.[174] Subsequent high energy collisionally induced dissociation (CID) with B/E, or mass-analysed ion kinetic energy scanning (MIKES), of daughter ions allowed double-bond location in monounsaturated fatty acid methyl esters. Experiments on a hybrid (EBQQ) instrument confirmed that low energy collisions do not generate the characteristic daughter ions.

Provided a full range of authentic isomers is available location of double-bonds in unsaturated long-chain compounds can be determined directly from EI spectra even without prior chemical modification. Determinations are generally based on ratios of selected characteristic ions.

This approach has been successfully demonstrated in the cases of tetradecenyl acetates,[175] methyl undecanoates[176] and methylene interrupted polyenoic acids.[177]

Determination of specific isomers in the case of substances of environmental interest is of special importance owing to their widely differing toxicities. The GC retention characteristics and mass spectral data (EI, CI (positive- and negative-ion)) have been recorded for thirty seven of the thirty eight possible isomers of tetrachlorodibenzofurans (TCDF) produced by unambiguous synthesis.[178] The CI spectra yielded little structural information not obtainable by EI, and problems of lack of reproducibility and sensitivity limit the usefulness of CI for trace analysis of TCDFs. In another study twenty one of the twenty eight theoretically possible pentachlorodibenzofurans were synthesised and their GC and MS (EI) characteristics recorded.[179] While their retention indices correlated well with those derived from a predictive mathematical model, their mass spectra lacked structural information. High resolution narrow bore (100µm i.d.) open tubular capillary columns offer potential for isomer-specific separation and quantification of TCDDs.[180] The twenty two TCDD isomers tested were resolvable on a new cyanopropyltolylallylsiloxane coated column. A smectic liquid-crystalline stationary phase offered unique selectivity for the majority of the biologically active, 2,3,7,8-class congeners of polychlorodibenzodioxins (PCDDs) and dibenzofurans (PCDFs) in capillary GC/MS analyses.[181] Similarly, a polymeric liquid crystalline phase coated capillary effected complete separation of the most toxic 2,3,7,8-TCDD from TCDD isomers and PCB isomers.[182] OV-240-OH, an OH-terminated polysiloxane stationary phase containing 33% 3-cyanopropyl groups has shown promise for isomer specific separations of PCDDs and PCDFs. The high degree of immobilization achievable with this stationary phase opens up the possibility of analysis of high boiling compounds, e.g. mixed brominated/chlorinated dibenzo-p-dioxins and dibenzofurans.[183]

A study of the CID spectra of the [M - Cl + O]⁻ ion of 40 PCB congeners on a triple quadrupole mass spectrometer indicated that reliable identification is possible even when GC separation is incomplete.[184] The 'ortho effect', observed for chlorinated biphenyls (PCBs) and brominated biphenyls (PBBs) having 2,2', 2,2',6 or 2,2',6,6' halogens, can be combined with GC retention indices for isomer specific identifications by GC/MS.[185] The proposed technique avoids multiple GC determinations employing different stationary phases in enviornmental monitoring.

Charge exchange CIMS (CECIMS) can be used to predict the correct isomers in multimethylated PAHs, high molecular weight PAHs and fluorinated benzenes in the absence of standard compounds.[186,187] GC/CECIMS with a thermally stable stationary phase revealed the possible occurrence of several isomeric PAHs in carbon black.[186] In a GC/MS investigation of PAHs employing PI and NICI, hydrogen in the PI mode offered greatest potential for isomer differentiation.[188] Capillary GC/MS using NI atmospheric pressure ionization (API) has been employed to distinguish nitronaphthalene and aminonitropyrene isomers.[189,190] Detection limits in the SIM mode for these compounds were in the low picogram range. Isomers were differentiated on the basis of differing relative retention times. GC/NICIMS has been shown to be well suited to separation and differentiation of chlorinated styrene isomers in environmental samples.[191] Synthetic n-pentadecylpyridine isomers, analogues of long-chain alkyl pyridines, components of crude shale oil, are readily distinguishable on the basis of their differing GC retention times and EI spectra.[192]

## 2. Applications

**2.1 Long-chain Compounds.**- Recent methods for the structure investigation of long-chain alkyl compounds are reviewed above ('Derivatization' and 'Stereo- and Positional Isomeric Assignments'). Extensive reviews of analytical methods, including GC/MS, employed in lipid anlaysis in clinical research and biomedical diagnosis have appeared.[34,35]

Long-chain α-mycolic acids, isolated from various *Mycobacterium* spp. are probably the longest-chain natural lipids yet determined by GC/MS.[119] These substances, bacterial cell-wall components, were separated on a short (0.3-0.4m x 3.0mm i.d.) packed (1% OV-101 on Gaschrom Q) column operated isothermally at 320-340°C. The TMS ether derivatives of methyl α-mycolates (**3**), containing up to 86 carbon atoms and two double-bonds, eluted after over 2 hours. Interest continues in the analysis of intact long-chain fatty acyl lipids by GC/MS.[120,123,124] Highly immobilized GC stationary phases are generally preferred for high-temperature GC/MS analyses of this nature.

$$\underset{C_{20}-C_{60}}{\longleftarrow \beta\text{-unit} \longrightarrow} \underset{C_8-C_{26}}{\longleftarrow \alpha\text{-unit} \longrightarrow}$$

$$CH_3C_mH_{2m}\text{---}CH\text{---}CHCOOCH_3$$
$$\underset{OTMS}{|} \quad \underset{\underset{CH_3}{|}}{\overset{|}{C_nH_{2n}}}$$

(m=18-58; n=5-23)

**3**

Rather than high-temperature GC/MS analysis of the intact lipids, a conventional degradative (transmethylation) approach was applied to the profiling of fatty acids derived from sebaceous triglycerides by GC/MS.[193] Resolution of double-bond positional and geometric isomers was achieved on a 50m fused-silica column coated with a highly polar cyanopropylsiloxane stationary phase. In all, forty seven individual components were characterized. Fatty acids have been determined at the picogram level by GC/NICIMS following conversion to the electron capturing pentafluorobenzyl esters.[194,195] The high sensitivity achieved permitted detection of *ca* 600 bacteria the size of *Escherichia coli*.[194] The characeristic 3-hydroxy fatty acid of the lipopolysaccharide lipid A of *E. coli* infecting human urine was readily determined. The phenylcarbamate derivative was preferred for the GC/MS characterization of the 3-hydroxyalkanoic acids in the lipopolysaccharides of *Rhizobium trifolii*.[196]

**2.2 Prostaglandins and Related Eicosanoids.**- A major advance in improving selectivity in the analysis of prostanoids has been the application of tandem mass spectrometry using a triple quadrupole instrument.[197-201] Monitoring daughter ions produced by CID, of EI fragment ions from methyl ester/methyloxime/TMS ether derivatives of prostaglandin $E_2$ and 6-oxo-prostaglandin $F_{1\alpha}$ and the methyl ester/TMS ether derivative of prostaglandin $F_{2\alpha}$, increased sensitivity by an order of magnitude compared to conventional SIM.[197] A notable advantage of MS/MS is the reduced sample preparation, as exemplified in the determination 11-dehydrothromboxane $B_2$ in urine and plasma.[198] Detection limits in the low picogram range were readily achieved. Both pentafluorobenzyl ester/methoxime/TMS[199] and methyl ester/methoxime/TMS[200] derivatives show considerable promise for routine use in the sensitive and selective detection of prostanoids by GC/MS/MS from biological matrices. The CID daughter ion spectra of methyl ester/methyloxime/TMS ether derivatives of thromboxanes were somewhat less useful, in that $TXB_2$ and 2,3-dinor-$TXB_2$ display non-specific ions. 11-Dehydro-$TXB_2$ yielded a complex daughter ion spectrum.[201]

In the absence of tandem MS, refinements in sample preparation can improve selectivity and sensitivity in trace analyses. Immunoaffinity chromatography offers substantially greater selectivity compared to conventional adsorption techniques. Antibodies have been raised against thromoboxane $B_2$[202] ($TXB_2$) and 6-keto-prostaglandin $F_{1\alpha}$[203] and immobilized on $N$-hydroxysuccimidyl silica gel. Urinary extracts were readily purified in one step on immunoaffinity columns and determined, interference-free, by capillary GC/NICIMS.[202,203] The cross-reactivity of an antibody raised against $TXB_2$ with 2,3-dinor-$TXB_2$ was used to advantage in the simultaneous extraction of both compounds from urine, by immunoaffinity chromatography, for subsequent determination by GC/NICIMS.[204] Rapid sample clean-up prior to GC/NICIMS of free arachidonic acid, monohydroxyeicosatetraenoic acid and prostaglandins has been achieved on mini-columns prepared from a special silica gel (Silicar CC-4).[205]

GC/NICIMS of pentafluorobenzyl ester-o-methyloxime TMS ether and pentafluorobenzyl ester TMS ether derivatives is finding routine application in the investigation of wide range of prostanoids, and now constitutes one of the most sensitive methods available for quantitative analysis.[199,205-211] The high sensitivities attainable (low picogrȧm) make it a suitable choice for use in the GC/MS validation of immunoassay procedures.[208,212] A general recommendation to emerge from these studies is that GC/MS should be used routinely to determine specificities in radioimmunoassay (RIA) and enzymeimmunoassay (EIA) procedures. In this, as in other fields, the use of stable isotopically labelled internal standards has become a minimum acceptable requirement to ensure reliable quantification.[132,142,143,197-204,206-215] Deuterium labelled methoximes should be largely avoided as internal standards in prostaglandin determinations owing to exchange of isotopic label on derivatization to pentafluorobenzyl ester TMS ethers.[216]

New derivatives for the analysis of prostanoids have been reviewed above (see 'Derivatization').

**2.3 Steroids.-** The role of GC/MS in biomedical analysis of steroids has been reviewed.[36,37] A symposium volume has appeared containing papers covering a range of applications of GC/MS in the steroid field.[40] As in the case of prostanoids (see above), isotope dilution GC/MS has achieved the status of a reference methodology for the assay of steroid hormones,[36,37,217-221] cholesterol[218,222] and bile acids.[223] The prerequisites of a reference method have been succinctly delineated.[218]

Corticosteroids have been determined in a variety of biological fluids by isotope dilution GC/MS.[224-226] The ethical acceptability of stable isotopes was taken advantage of in the development of a cortisol assay suitable for monitoring its production rate in babies and children.[226] A notable feature of the method was the determination of ($^{13}C_4$) isotope at very low (>0.1%) enrichments. Thermospray LC/SIMMS was found to provide adequate sensitivity for the determination of cortisol in serum extracts.[227] A significant, though perhaps not unexpected finding was the much poorer sensitivity and precision compared to that achieved by GC/SIMMS of the bis(methyl)oxime, tris(TMS) ether derivative. LC/MS had the notable advantages of decreased sample handling, and elimination of the two-stage derivatization.[227] Some applications of GC/MS to corticosteroid and anabolic steroid analysis in the horse have been presented.[39,228] Measurements of equilin and estrone in human plasma have been carried out by GC/NICIMS following their conversion to volatile pentafluorophenyldimethylsilyl ethers which yield abundant molecular anions.[229] The potential of the 3,5-bis-trifluoromethylbenzoyl derivative for the quantitative analysis of mono- and dihydroxysteroids has been assessed.[230] With the exception of corticosterone, a single product was formed for a range of androgens, estrogens, pregnanes, corticosteroids and sterols. NICI yielded abundant M⁻˙ anions with detection limits in SIM mode in the low picogram range.

The pentafluorobenzyl ester-dimethylethylsilyl ethers (eg. **4**) offer considerable potential in the analysis of bile acids. Lithocholic acid, deoxycholic acid, chenodeoxycholic acid, unsodeoxycholic acid and cholic acid were well separated on an apolar methylsilicone coated capillary column. Detection limits with SIM of the [M - 181]⁻ ion in NICI mode (isobutane) were in the low femtogram range.[231] These exceptionally low detection limits offer considerable promise for use in the determination of bile acids in tissues. Other derivatives that have been used in determinations in this field include methyl ester-TMS ethers,[232-235] ethyl ester-dimethylethylsilyl ethers[236] and isobutyl ester-TMS ethers.[237]

**4**

Stable isotopes ($^2H$ and $^{13}C$) are finding increasing application in the study of bile acid biosynthesis,[238] metabolism,[233,235,237] pool size determinations[232,234,235] and turnover rates.[235]

6: *Analysis of Mixtures by Mass Spectrometry; Part I*

(24-$^{13}$C) Labelled bile acids have been used as internal standards for the first time.[234,235] Although corrections must be made for incomplete labelling, the use of $^{13}$C eliminates the problems of *in vivo* exchange, commonly encountered in the case of deuterium and tritium.[234]

GC/MS has been used as a complement to RIA in the determination of ecdysteroids (arthropod moulting hormones) at the picogram level.[38,121] The high molecular weight of the TMS ether derivatives of these polyhydroxylated sterols (M+ *m/z* 912 for hexa-TMS-20-hydroxyecdysone) necessitated the use of capillary columns coated with a thin film (0.1µm) of apolar cross-linked polydimethylsiloxane stationary phase. Short (12m) capillary columns coated with a similarly thin film of cross-linked apolar stationary phase were employed in the GC/MS analysis of steryl fatty acyl esters.[38,120,239] A notable aspect of this work was the use of NICI with ammonia reagent gas at a source temperature of 300ºC, required to generate structural information and maintain the integrity of the gas chromatographic profile. The method is of particular use in the case of sector instruments where the heating effects of the high source potentials preclude of low-temperature PICI.[120] High-temperature GC/MS has also been used in the characterization of intact novel steryl ferulates from seeds.[122]

Despite the considerable potential of tandem mass spectrometry for the determination of steroid hormones and their metabolites, applications during the current review period have been limited. GC/MS/MS experiments employing a triple stage quadrupole showed that steroidal 4-en-3-ones exhibit [M + H]+ under PICI, which undergo CIDs to give daughter ions highly characteristic of ring A stereochemistry.[240]

For quantitative analyses and tracer studies low resolution SIM continues to be the method of choice with deuteriated internal standards generally preferred.[219,241-243] Although, the *in vivo* conversion of testosterone into its metabolites, dihydrotestosterone and 5α-androstane-3α,17β-diol, has been studied by GC/SIM (MS) following *in vivo* infusion of 20mg [$^{13}$C]-testosterone.[244]

**2.4 Carbohydrates.**- GC/MS still retains a significant role in carbohydrate analysis. Even in the structure elucidation of complex polysaccharides GC/MS is usually employed as a complement to FAB in the analysis of enzyme digests or chemical degradation products. The characterization of mucin oligosaccharide from rat intestine, following trypsin digestion, has offered up a notable application of high-temperature GC/MS.[118] Permethylated neutral oligosaccharides with up to seven sugars (molecular weights upto 1553.8; **5**) were separated and determined by GC/MS on a fused-silica capillary column (10m x 0.25mm i.d.) coated with 0.05µm of cross-linked SE-54. Elution of the heptasaccharide required temperature programming of the GC oven to 370ºC. The analysis revealed that the linkage GalNAc was substituted at both positions 3 and 6. GC/MS of permethylated alditol acetates and FABMS of permethylated oligosaccharide alditols lead to characterization of oligosaccharides released by endo- β -galactosidase of human blood group O erythrocyte glycopeptides.[245] The core region *Citrobacter* lipopolysaccharide from strain PCM 1487 has been established by methylation GC/MS, chemical degradation and one- and two-dimensional $^1$H-NMR.[246]

The chain length and sugar content of oligosaccharides attached to phenolic lignin or flavolan polymers has been determined by GC/MS following enzymic hydrolysis.[247] The hemiacetal ends of the oligosaccharide fragments, still attached to the phenolic polymer, were reduced with NaBH$_4$. Following acid hydrolysis and reduction with NaBD$_4$ the mixed isotopic products were determined as their alditol acetate derivatives by capillary GC/MS with SIM. A similar approach has been employed to assay enzymes that cleave glycosidic bonds.[248] Enzymically treated substrates were reacted with NaBH$_4$, then following acid hydrolysis of glycosidic bond, the new glycoses were reduced with NaBD$_4$. GC/MS of the alditol acetate derivatives was used to assess the amounts of $^2$H and $^1$H, and hence deduce the proportion of glycosidic bonds in each enzymically hydrolysed sugar.

$$\begin{array}{l}
\phantom{Fuc-O-Hex-O-HexNAc-O-}CH_2OCH_3 \\
\phantom{Fuc-O-Hex-O-HexNAc-O-}|\phantom{HC-}CH_3 \\
\phantom{Fuc-O-Hex-O-HexNAc-O-}HC-N-CO-CH_3 \\
Fuc-O-Hex-O-HexNAc-O-CH \\
\phantom{Fuc-O-Hex-O-HexNAc-O-}HC-O-CH_3 \\
\phantom{Fuc-O-Hex-O-HexNAc-O-}HC-O-CH_3 \quad M=1553.8 \\
Fuc-O-Hex-O-HexNAc-O-CH_2
\end{array}$$

**5**

The strengths and weaknesses of GC/EIMS and CIMS (with various reagent gases) have been assessed in relation to their use in determining the position and extent of $^{18}$O- and $^{13}$C-labeling in synthetic permethylated acyclic sugars. PI ammonia CI was deemed the method of choice for determining the extent-of-labelling while EI was preferred for determining the position.[249] An isotope dilution GC/MS method in which [$^{13}$C]glucose was used as the internal standard meets the requirements of a definitive method for the determination of sera.[250]

Reduction of disaccharide methoximes with methanolic sodium cyanoborohydride and subsequent permethylation with methylsulphinyl-methyl iodide gives deoxy(methylmethoxy-amino)alditol glycosides. GC/MS yields EI spectra which enable, through SIM, detection of aldose or ketose containing disaccharides, with different types of O-glycosidic linkages. The method is claimed to be especially suitable for the assay of complex biological mixtures.[251]

A systematic analysis of inositol mono- and polyphosphates by GC/MS and FAB has been conducted.[252] EI spectra of all the positional isomers of myo-inositol monophosphate and myo-inositol-1,2-cyclic phosphate were determined by GC/MS as their TMS ethers.

**2.5 Amines.-** Methodologies, including GC/MS, for the determination of polyamines have been reviewed.[42] The most prominent form in which spermidine is excreted in urine, $N$-(3-acetamidopropyl)pyrrolidin-2-one, has been determined by isotope dilution GC/MS using a trideuteriated analogue as internal standard.[253] Ammonia CI afforded abundant [M + H]$^+$ ions for SIM determination. Putrescine has been determined in several regions of rat brain tissue at the picomole level by GC/MS.[254] Acid extraction and ion exchange chromatography were followed by derivatization with pentafluoropropionic anhydride. A moving-needle injection system was used to enhance sensitivity. A sensitive method of determining $^{15}NH_3$ by GC/MS has been developed in order to measure putrescine oxidation in plasma.[255] Ammonia was trapped from sample solution by microdiffusion and converted to pentafluorobenzamide for determination by SIM. Detection limits were at the picomole level for a number of derivatives that were tested for the determination of acid metabolites of biogenic amines in plasma and urine. A combination of the methyl or trifluoroethyl esters of the pentafluoropropionyl derivatives was deemed most useful.[256] Under certain derivatizing conditions back-exchange of deuterium in internal standards was observed. Amines in small (0.5ml) urine samples have been readily extracted using a cartridge containing 100mg of octadecyl bonded porous silica.[257] Amines were eluted with methanol and converted to pentafluorobenzyl derivatives prior to GC/SIMMS. Amine-ethanethiol-O-phthalaldehyde-TMS ether derivatives of biogenic amines show promise for EI GC/MS determination at the picogram level.[258] Capillary GC/NICIMS was preferred for the determination of tryptophan, tryptamine and serotonin using deuterium labelled internal standards.[259] The acidic metabolites (conjugated and unconjugated) of p-tyramine have been determined by using high resolution (m/$\Delta$m 5000) SIM following ingestion of p-tyramine-$\beta,\beta$-$^2H_2$ hydrochloride.[260]

GC/MS plays a major role in the quantitative analysis and metabolic profiling of catecholamines and related compounds.[41] The NICI spectra per-$O$-acetyl carboxypentafluorobenzyl ester derivative of acidic catecholamine metabolites contain abundant molecular and diagnostic fragment anions.[261] In addition the presence of the strongly electrophoric carboxyl-PFB group affords high ionization efficiency, thus allowing GC/MS profiling of metabolites in lumber cerebrospinal fluid and plasma at the sub-nanomole level with good precision ($\sigma_{rel}$<5%).

Melatonin, the hormone playing a key role in circadian and biological rhythms in vertebrates, has been determined in plasma at low concentration (<10-100pg ml$^{-1}$).[262] Derivatization with pentafluoropropionic anhydride was believed to give the tris-pentafluoropropionyl product initially which was then transformed, pyrolytically, following injection into the GC to the spirocyclic pentafluoropropionyl (PFP) derivative. GC resolution was optimized by employing a capillary column switching device. Quantifications were performed by medium resolution (m/$\Delta$m 3000) SIM of the M$^{+\cdot}$ ion (EI, 70eV). A two-step gradient elution afforded separation of 6-hydroxymelatonin sulphate and 6-hydroxymelatonin glucuronide (excreted in urine) on silica-packed Sep-Pak cartridges. 6-Hydroxymelatonin was then released by enzymic hydrolysis, and determined as the PFP derivative by isotope dilution GC/MS.[263]

An isotope dilution GC/MS technique, employing $^{13}C_2$-creatinine as internal standard has been proposed as a definitive method for the detection and quantification of serum creatinine.[264] A 'bracketing' technique involving measuring standards with isotope ratios slightly higher and slightly lower than that of the analyte ensured higher precision.

Pyrrolizidine alkaloids are phytotoxins produced by several plant species world wide which have been responsible for livestock and human poisonings. GC/MS employing PICI and NICI has provided an aid to the rapid structure elucidation of pyrrolizidine alkaloids as their TMS derivatives.[265] PICI (NH$_4^+$) yielded information concerning molecular mass and nature of the necine base. NICI, on the other hand, afforded structurally diagnostic anions formed by nucleophilic displacement reaction with [OH]⁻. Pyrrolizidine alkaloid metabolites from mouse liver microsomes have been determined by the complementary use of GC/MS and FAB/MS/MS.[266] The identification of trace amounts of the metabolites dihydropyrrolizine, senecic acid and monocrotalic acids in hepatic microsomal extracts relied on determination by capillary GC/MS of their TMS derivatives.

The phytohormones, indole 3-acetic acid and abscisic acid, have been quantitatively determined in small samples of plant tissue by GC/MS using deuteriated internal standards. Detection limits of 1ng g⁻¹ were achieved by methane PICI. The technique was applied to determinations in various tissues in citrus, prunes and apples.[267]

**2.6 Amino Acids and Peptides.-** While tandem MS will eventually obviate the need for conventional MS procedures for the determination of blocking groups, GC/MS still has a role to play until instrumentation becomes widely available. Methods of protein microcharacterization, including GC/MS, have been complied into a single volume.[42] Examples of the complementary use of GC/MS, Edman degradation and FAB-MS include the structure elucidation of A21978, a complex of new acidic peptide antibiotics,[268] and recombinant hepatitis B surface antigen protein.[269] Both studies required MS for reasons of N-terminal acylation. N-terminal amino acid sequences in two analogues of bovine growth hormone, produced in *E. coli* by recombinant DNA techniques, have been determined by GC/MS.[270] Metabolites of cyclosporine (a cyclic undecapeptide with potent immunosuppressive properties useful for preventing rejection of transplanted organs), have been investigated by MS with view to identifying factors responsible for side-effects.[271] Following molecular weight determination by FAB or thermospray LC/MS, metabolites were hydrolysed to component amino acids, which were esterified, acetylated, and identified by GC/MS. To distinguish metabolism at four identical methylleucines partial hydrolysis to smaller peptide fragments was performed, followed by trimethylsilylation and GC/MS determination. A peptide sequencing program is available in FORTRAN and PASCAL for off-line analysis of GC/MS data.[272]

N-Heptafluorobutyryl methyl esters derivatives have been employed to determine dipeptides in human urine following extraction by cation-exchange chromatography.[273] Under the chromatographic conditions employed, tryptophan and arginine containing dipeptides were too involatile to be determined. TBDMS derivatives offer a potential alternative for dipeptide determinations. This latter derivative is formed readily in 90min at 60°C. The TBDMS derivatives show more favourable chromatographic behaviour and stability during storage compared to TMS derivatives. Unlike TMS derivatives no cyclization of N-glycyl dipeptides to diketopiperizine was detectable by GC/MS. The TBDMS derivatives display highly favourable EI and PICI behaviour.[274]

TBDMS derivatization of amino acids, introduced during the previous review period, is finding widespread application.[275-281] Derivatization is generally achieved in one step under mild conditions with the notable advantage, in the case of acid labile amino-acids (eg. glutamine and asparagine), that degradations are avoided.[275,276,278,279] Furthermore, the histidine derivative formed a stable compound showing good chromatographic properties, in contrast to the perfluoroacyl ester derivatives.[276] It seems likely that the latter group of derivatives will be superseded by TBDMS derivatives for routine amino acid analysis by GC/MS and GC. Notable feature of the TBDMS derivatives is the abundant high mass ions, $[M - 57]^+$ in their EI mass spectra, thus removing the need for CI.

Stable isotopes have found wide application in this area for the investigation of various aspects of the metabolism and biosynthesis of amino acids and proteins. Deuterium,[277,282-284] $^{15}N$[278,285-287] and $^{13}C$[281,284,288] labels have all been used. Specific examples of applications of stable isotopes in this area include: the automated quantification and isotope enrichment determination of glutamine and glutamate in plasma,[278] stable isotope ratio analysis of leucine and ketoisocaproic acid in plasma,[281] analysis of amino acids in rat plasma and brain samples,[287] determination of $^{13}C$-label enrichment in mice brain tissue of 4-aminobutyric acid, aspartate, glutamate and glutamine,[284] determination of phenylalanine and tyrosine in human plasma,[282] quantification of homocysteine, cysteine and methionine in serum,[277] determination of histidine in human plasma,[286] measurement of apolipoprotein B synthesis,[285] determinations of L-DOPA and 3-o-methylDOPA in plasma and cerebrospinal fluid,[283] and the development of a definitive method for the measurement of thyroxine in serum.[288]

**2.7 Clinical and Metabolic Studies.-** Drug metabolism, Pharmocokinetics and Toxicity are reviewed in Chapter 9, so will not be dealt with here.

Reviews pertinent to this area include: those relating to advances in instrumentation and strategies for metabolic profiling,[49] general approaches to metabolic profiling,[48] the differential chemical diagnosis of lactic acidosis,[45] profiling ketone bodies and organic acids in diabetes,[47] microbial taxonomy,[52-54] isotope dilution MS in clinical chemistry,[50] isotopic ratio measurements in biomedical and nutrition research,[51] and the analysis of lipids in biomedical research and clincial diagnosis.[34,35] A symposium volume has appeared presenting papers from a variety of areas of medicine and pharmacology.[55]

The GC/MS analysis of urinary organic acids for the purpose of diagnosing inborn errors in metabolism is becoming increasingly routine in clinical laboratories.[289-291] The widening availability of reliable and inexpensive bench-top GC/MS systems requiring relatively unskilled operators has contributed to this trend.[290] Rapid sample throughout is a key feature of effective screening programs, therefore automation of aspects of sample preparation and/or analysis become a necessity.[289-290]

In a study of urinary organic acids automated profiling using a dedicated library resulted in about 80% of all peaks being identified.[289] Mini-columns prepared from equivolumes of Porapak Q and Porapak T showed considerably more reliability, in terms of the range of compounds isolated from urine, compared to solvent extraction.[289] GC/MS was preferred to GC/FID as important diagnostic compounds which co-eluted were readily resolved by mass chromatography in quantitative analyses. This approach has been used in one laboratory to process >4000 samples over a five year period to successfully detect many metabolic disorders.

In this as in other areas GC/MS can be used as a reference method for other less expensive or time-consuming assays. For example an isotope dilution GC/MS method has been developed as a reference method for determining urinary oxalate, and used to evaluate existing enzyme, HPLC and GC assays.[292]

An elegant solution to a longstanding problem of co-elution in the analysis of $\alpha$-hydroxy- and $\alpha$-ketoacids in serum has been found.[293] Sodium borodeuteride reduction of the ketoacids *in situ* yielded more stable deuterium labelled hydoxyacids. Deuteriated hydroxyacids and endogenous non-deuteriated hydroxyacids were then readily resolved and quantified by GC/MS with SIM, a more extensively $^2$H labelled analogue being added as internal standard. Sodium borodeuteride treatment has been further employed in investigations of urines for pyruvic and acetoacetic acids (which are unstable under acidic extraction conditions).[294] The reduction produced deuteriated lactic and 3-hydroxybutyric acids which were readily differentiated by GC/MS SIM from their endogenous counterparts. Hence, reliable simultaneous determination of lactic, pyruvic, 3-hydroxybutyric and acetoacetic acids were readily perofrmed. GC/MS (SIM) of [M + NH4]$^+$ was used to confirm, following infusion of ($^{13}$C)threonine, that L-threonine is the precursor of urinary 2-hydroxybutyric, 4-deoxyerythronic and 4-deoxythreonic acids in diabetes mellitus sufferers.[295]

Diagnosis of adrenoleukodystophy (ALD) has been achieved by analysis of fatty acids from a dried blood spot on filter paper.[296] The fatty acids from the blood spot were simultaneously extracted and transmethylated, then analysed by GC/MS. Fatty acid distributions were found not to change within a week at room temperature; this would be a notable advantage in screening programmes as samples could be conveniently sent by mail. The abundances of long-chain ($C_{24:0}$ and $C_{26:0}$) fatty acids were significantly greater in ALD patients than controls.

The GC/MS analysis of the volatile constituents of saliva has provided a new approach to metabolic profiling.[297] Solvent extraction and trimethylsilylation followed by GC/MS revealed over 150 components for each saliva sample. The approach showed promise in diagnosing certain pathological conditions, e.g. diabetes mellitus. The detection of significant levels of cholesterol suggested a close serum-saliva cholesterol correlation. Hence, this approach may provide a non-invasive means of monitoring serum cholesterol.[297]

A reference method for the determination of total glycerol, as the heptafluorobutyric ester, in serum is based on isotope dilution GC/MS employing [1,3-$^{13}$C$_2$]glycerol as internal standard.[298] [1,2,3-$^2$H$_5$]Glycerol was used in the GC/MS determination of steady-state and nonsteady-state glycerol kinetics. The validity of using [1,2,3-$^2$H$_5$]glycerol was tested against [2-$^{13}$C]glycerol.[299]

A stable isotope tracer GC/MS method was utilized to quantify the secretion of diet-derived fatty acids into human milk.[300] [$^2$H$_6$]Tripalmitin, [$^2$H$_{18}$]triolein and [$^2$H$_{12}$]trilinolein were administered to healthy lactating women. No differences were seen between the isotopic enrichments of the individual fats. Sex pheromone biosynthesis in the red-banded leafroller moth

(*Argyrotaenia velutinana*) has been studied using stable isotope labelled tracer with GC/MS analysis using CI and SIM.[301] Incubations were conducted with sex pheromone glands with multiply labelled precursors, $^2$H or $^{13}$C labelled fatty acids or triacylglycerols. While synthetic triacylglycerols were shown not be precursors for pheromone components, hexadecanoic and tetradecanoic acids were found to be intermediates in pheromone biosynthesis. Advantages of using multiple stable isotopes in conjunction with GC, as markers for biosynthetic studies, over radiolabelled compounds, include: avoidance of misleading results owing to recycling of acetyl CoA from mitochondrial degradation, and allowing alternative biosynthetic pathways to be distinguished in the same experiment. An isotope dilution MS assay for oleic acid in plasma used the [1-$^{13}$C] analogue as internal standard.[302] GC/MS and FAB/MS have been compared for the quantitative analysis of lysoplatelet activating factor from human neutrophils.[303] GC/MS was performed following cleavage of the phosphocholine moiety, followed by condensation of the monoglyceride with acetone to yield the 1-*O*-alkyl-2,3-isopropylidene. A deuterium labelled internal standard was used in both GC/MS and FAB/MS determinations; detection limits were 200pg and 5ng respectively.

β-Hydroxymyristic acid has been suggested as a marker substance for endotoxins (lipopolysaccharide constituents of Gram-negative bacteria cell walls). Of a number of derivatives that were tested the 3-*O*-pentafluorobenzoyl-methyl ester was preferred for GC/MS with NICI in SIM mode for detection of the M⁻· ion.[302] A detection limit of 0.3pg was achieved in the analysis of water samples.

**2.8 Food and Agricultural Chemistry.-** Applications of mass spectometry in food sicence have been compiled into a single volume.[57] Additionally, GC/MS methods for the analysis of plant components have been comprehensively presented in a multi-author work.[58]

The health hazard of trichothecene mycotoxins produced by imperfect fungi associated with foodstuffs, continues to stimulate interest in the development of sensitive analytical techniques for their detection and quantification.[305-308] Of the methods investigated to date GC/MS with NICI of heptafluorobutyryl ester derivatives has been used for trace analysis.[305,307] Source temperature was found to produce dramatic effects on NICI spectra and only low temperatures (313K optimum) produced abundant M⁻· ions for use in MS/MS analyses employing CID.[307] GC/MS with NICI offers the most sensitive technique currently available, with detection limits for SIM[305] or selected reaction monitoring (SRM)[307] in the low picogram range. A library of PICI mass spectra of trichothecenes has been developed.[308] Macrocyclic trichothecenes, the most toxic of the known trichothecenes, present problems to GC/MS owing to involatility compared to simple trichothecenes. The most successful approach developed to date relies on analysis of hydrolysates as HFB esters by GC/MS with NICI.[306] Methodologies for the detection of trichothecenes in human fluids are reviewed below (see below 'Toxicology and Forensic Science').

Concern over the occurrence of trace levels of ethyl carbamate (urethane) in various alcoholic beverages has prompted investigations aimed at improving quantification.[309-311] Comparison of high resolution SIM and MS/MS SRM showed both approaches to produce acceptable reproducibility and specificity at the ppb level.[311] One of the CID MS/MS studies drew attention to a possible flaw in reliable quantification *via* isotope dilution at low concentrations owing to inadequate sampling of the GC peak.[310] However, Brumley *et al.*[311] observed no such problem in sampling peaks which were 3s wide. In this latter study good agreement was seen between independent determinations by GC/MS/MS, GC/matrix isolation FTIR and GC/nitrogen/thermal energy analysis.[311] A fast and sensitive method for the determination of diethylene glycol in wines has been developed. The method relies on decolorization by charcoal followed by dilution with $CH_3CN$, then filtration to achieve a minimum detection limit of 0.1mg/L (ppm).[312] Procedures for the analysis of plasticizers, acetyltributyl citrate[313] and di-(2-ethylhexyl)adipate,[314] by stable isotope dilution GC/MS have been reported. Procedures allow reliable determination of levels of migration of these compounds from packagings into various foods.

The detection of potentially toxic drug residues in meat products is readily performed by GC/MS. For example, a range of macrolide antibiotics was determined in beef and pork following acid hydrolysis, to remove sugars, and in some cases hydrogenation, followed by acetylation.[315] Also, diethylstilbestrol (a synthetic estrogen) has been determined in meat samples by isotope dilution GC/MS, to serve as a reference method for positive RIA results.[316]

**2.9 Environmental Science and Toxicology.-** This continues to be the most prolific area of application of GC/MS in its various forms.

**2.9.1 Papers of General Interest.-** An assessment of the ion trap detector in the context of environmental monitoring has been conducted.[317] The investigation covered precision of SIM, full scan relative abundance measurements, sensitivity and detection limits in SIM and full scan data acquisition, and the linear range of concentration calibration. An analytically useful model for the estimation of EI response factors on a quadrupole MS has been developed.[318] Good agreement was achieved between observed and estimated response factors for a wide range of priority pollutants. Directly linked GC/FTIR/MS has been shown to be superior to either stand-alone technique for the non-target screening of environmental extracts, providing rapid on-line identification and compound class confirmations.[19,319] The development of a method for screening for 49 priority pollutants, including aromatics, phenols, phthalate esters, pesticides, haloethers and chlorinated hydrocarbons, has been presented.[320] Both $CH_2Cl_2$ extractable acidic and base/neutral fractions were anlaysed by GC and GC/MS without prior sample clean-up. Kováts' and Lee retention indices have been recorded by GC/MS on a DB-5 (similar to SE-54) column for more than 200 compounds of environmental interest.[321] These retention indices are useful in GC/MS work where compounds are indistinguishable on the basis of mass spectra alone.

High selectivity and sensitivity has been achieved in the detection of polycyclic aromatic compounds (PACs) by a GC/MS system incorporating a tunable UV laser beam.[322] Photoionization products (cations, electrons and photons) were simultaneously monitored in a TOFMS. For 21 compounds tested detection limits were in the low picogram range. Polyaromatic hydrocarbons

(PAHs) from a coal tar standard reference material (SRM 1597) have been investigated by GC, GC/MS and HPLC.[323] More than 60 of the 104 components detected were identified by GC/MS, while quantitative determinations were made for the 30 most abundant constituents. The coal tar standard is intended as reference material for making inter-laboratory comparisons. GC/MS analyses of extracts of wooden railway sleepers have revealed the presence of creosote residues.[324] The presence of high levels of carcinogenic and cocarcinogenic PAHs were taken to be a health risk in instances where old railway sleepers are installed as playground equipment. The characterization of PAHs isolated from carbon black has been achieved by GC/MS. Isomeric PAHs were differentiated on the basis of $MH^+$ to $M^{+\cdot}$ ion ratios, using a semiemperical predictive method for high molecular weight PAHs, for which experimental first ionization energies were not available. The high pressure electron capture mass spectrometer offers a highly specific and sensitive GC detector owing to the fortuitous characteristics of the reactions of thermalized gas-phase electrons with organic molecues.[326] A kinetic model has been presented for production of unconventional negative ions, and the system's operation demonstrated in the analysis of environmentally important compounds.

The GC/MS analysis of organics leached from the human finger tip by methylene chloride, provides a useful reference source for potential experimental artifacts.[327] Comparison was also made between impurities in three brands of glass distilled methylene chloride.

### 2.9.2 Air and Airborne Particulate Pollution.-

Alkyl nitrates have been detected in remote atmospheres for the first time by GC/MS with NICI.[328] This observation raises concern as these long-lived alkyl nitrates, transported in the troposphere from polluted areas, can be converted to NOx, a noted catalyst in ozone formation and destruction. Significant to gaining a greater understanding of the atmospheric chemistry of sulphur has been the development of isotope dilution GC/MS techniques for the determiantion of trace levels of sulphur dioxide[329] and carbonyl sulphide.[330] Concern over the occurrence of potentially hazardous substances in the indoor environemnt has prompted the development of a GC/MS method for the determination of tetra- and pentachlorinated dibenzofurans and dibenzo-p-dioxins.[331] The method employs high resolution ($m/\Delta m$ 10,000) SIM to assess both gaseous and particulate-bound components at pg $m^{-3}$ concentrations. Although thermal desorption GC/MS analysis was able to distinguish between smokers' and nonsmokers' offices, on the basis of airborne nicotine levels, it was not possible to positively attribute organic contaminants to tobacco smoke.[332] Isotope dilution GC/MS utilizing a headspace assay has used to determine background levels of benzene in the breath of urban smokers and nonsmokers.[333] Malodorous short-chain fatty acids (containing 4-6 carbon atoms) have been effectively determined by GC/MS (SIM) at the picogram level as their benzyl esters.[334]

Airborne particulate matter has been tested for the presence of a variety of aromatic compound types. Phenolic compounds adsorbed on airborne particles require complex separation from interferences prior to GC/MS analysis.[335] Selectivity and sensitivity in the analysis of mononitropyrenes and mononitrofluoranthenes in air particulate matter has been enhanced by using capillary GC with NI atmospheric pressure ionization (API) MS. The presence of 4-nitropyrene in an air particulate extract was confirmed for the first time.[336] This technique is also very sensitive for nitronaphthalenes, the detection limit for 1-nitronaphthalene being 0.3pg.[337] The selectivity of NICIMS for specific PAHs allowed determinations of isomeric PAHs and alkylated-PAHs in the absence of authentic standards in two paticulate Standard Reference Materials (SRM), namely: SRM 1650 diesel particulate matter and SRM 1649 urban particulate matter.[338] Quantitative determinations were performed for methylfluoranthene, methylpyrenes, methylbenzofluoranthenes and methylbenzo(a)pyrenes. High resolution GC/MS formed the basis of analyses of organic components of aerosols for environmental monitoring purposes.[339,340]

Statistical analyses of GC/MS data by pattern recognition techniques have been employed both in the classification and identification of hazardous organic compounds,[341] and in the classification of samples from different locations.[342]

**2. 9.3 Water Pollution and Effluents.-** An interlaboratory comparison of precision and accuracy in multicomponent analysis of organics, present at the low nanogram level, in drinking water showed PAHs, organochlorines and chlorinated aromatics to be reliably determined by GC/MS.[343] The large standard deviations associated with analyses of organophosphates were suggested to derive from irreproducibilities associated with the GC analyses. A systematic analytical approach involving the complementary use of GC/MS, HPLC and FABMS, including FAB-CID-MIKES, provided a thorough analysis of organic pollutants of Barcelona's water supply over a two-year period.[344] Cronic pollutants of the Llobregat river included surfactants, plasticizers, ethyleneglycol derivatives, phosphates, hydrocarbons, etc. Products of weathering of benzthiazoles (**6** and **7**), commonly used as anti-oxidants in the manufacture of rubber tyres, have been identified in coastal sediments using GC/MS.[345] These compounds have been proposed as specific indicators of the contribution of street runoff to the contamination of sediments in coastal areas.

**6**          **7**

Growing concern over the fate of persistent organotin compounds in the environment necessitates increasingly sensitive detection techniques. A wide range of organotin compounds has been determined from environmental samples by GC and GC/MS, using ethylation to enhance the volatility of ionic components.[346] Alkyl derivatives may be preferred to stannanes for reasons of enhanced stability to handling. However, a highly sensitive method for the determination of methyltins in sediment uses a purge and trap-GC/MS approach.[347] Stannanes are generated *in situ*

6: *Analysis of Mixtures by Mass Spectrometry; Part I* 205

by treatment with NaBH$_4$ following borate buffering of the sample, eliminating the need for excessive sample handling. Heated purge and trap GC/MS has been compared to charcoal tube adsorption GC/MS for the analysis of 1,4-dioxane in water.[348] Both methods yielded similar results for the 15 samples analysed. An adapted purge and trap GC/MS technique was found to be effective for the analysis of 51 volatile organics, including the purgeable priority pollutants listed by the U.S. E.P.A.[349] The importance of employing control samples in surveys was emphasized by the occurrence of anomalous results.

Pyrolysis-GC/MS has been used for the determination of arsenocholine and acetylarsenocholine in aquatic organisms.[350] The compounds were determined in fish from arsenic-polluted and unpolluted water. The existence of a general metabolic pathway for these compounds in aquatic ecosystems was suggested. On-line flash evaporation/pyrolysis GC/MS appears to offer a potentially rapid method of screening for anthropogenic pollutants in sediments and soils.[351] While GC/MS offers the most effective means of monitoring, connection to other GC detectors was also suggested.

A report has appeared describing the results of the certification analysis of a suite of marine sediment reference materials, intended to serve as an aid to the determination of PAHs in such matrices.[352] Reliable values were established for concentrations of the 16 priority pollutant PAHs by complementary use of GC/MS and HPLC/MS.

Surfactants and their degradation products occur widely in wastewaters. Non-halogenated octylphenol polyethoxylate residues have been determined by GC/MS with CI and EI.[353] Determinations of nonylphenoxy carboxylic acids in sewage and polluted river water employed GC/MS and HPLC.[354] The environmental occurrence and fate of long-chain alkylbenzenes has been investigated by GC/MS.[355] These latter compounds have obvious potential as detergent waste-specific molecular tracers.

The aerobic laboratory degradation of an Arabian light oil residue by a pure culture of *Pseudomonas* was followed by GC, GC/MS and emission spectroscopy.[356] Unusual methylthio metabolites of polychlorinated biphenyls have been identified in sediment samples by GC/MS with NICI.[357] Derivatives of dicyclopentadiene have been identified from ground water by means of GC/high resolution MS, deutrium exchange of active hydrogen by CI and GC/MS following on-column base catalyzed deuteration of enolizable hydrogens.[358] Pyridine bases have been determined in river water and bottom sediments by GC/MS, following their distillation from environmental samples.[359] In all, twenty pyridine bases were determined at the picogram level by GC/MS SIM on a 15m x 0.53mm i.d. DBWAX coated capillary column (equivalent to PEG 20M).

**2.10.4 Halogenated Residues and Pesticides.-** A symposium volume devoted to chlorinated dioxins and related compounds is available.[59] Investigations concerning isomer determinations of TCDDs, TCDFs, PCBs and related compounds have been reviewed above (see above 'Stereo- and Positional Isomeric Assignments').

Of general interest to researchers working with chlorinated analytes is a report of enhancement of electron capture NI responses following pretreatment of the ion source with $CCl_4$.[360] Abundant $[M + Cl]^-$ ions are observed, serving as indicators of molecular weight in unknowns. Also of general interest in this field is an investigation of electron capture negative ion CI mass spectra for electrophoric derivatives of chlorophenols and chloroanilines.[361] The formation of anlayte-specific anions was found to be strongly influenced by the nature of the electrophoric group and summed electron-donating or withdrawing properties of aromatic ring substituents. Electron capture NIMS has been found to be a sensitive and reproducible technique for the determination of nitro compounds. The ECNI mass spectra of a range of 2,6-dinitroaniline and 2,4-dinitrophenol herbicides and related derivatives have been investigated using isotope labelling and accurate mass measurements.[362]

Tandem mass spectrometry offers immense potential for the detection of trace organic contaminants in environmental samples. A rigorous study compared hybrid high resolution GC/MS/MS (HRGC/MS/MS) to HRGC/high resolution MS (HRGC/HRMS) for the characterization of TCDDs.[362] Instrument performance was evaluated in terms of sensitivity, linearity, reproducibility and selectivity. Selectivity was found to be superior with HRGC/MS/MS, although comparable performance was observed in all other respects. Another study using a triple quadrupole mass spectrometer demonstrated the importance of optimizing collision cell parameters using the analytes of interest.[364] Using this approach detection of 2,3,7,8-TCDD averaged 0.45pg over a period of four days without retuning. The enhanced sensitivity of GC/MS/MS SRM can greatly simplify sample clean-up protocols.[363]

In the absence of GC/MS/MS, GC/HRMS provides a valuable alternative, as was demonstrated in the analysis of 2,3,7,8-TCDD in serum at a limit of detection of 1.25 part-per-quadrillion (ppq).[365] A straightforward procedure for the preparation of chlorinated derivatives of dibenzo-p-doxin and related compounds for use as reference compounds has been described.[365]

Indications are that polybrominated dibenzodioxins (PBDDs) and dibenzofurans (PBDFs) constitute a significant environmental hazard. Consequently interest focuses on the development of methods for routine trace analysis. Methodologies employed are largely analogous to those used for their chlorinated counterparts.[367-369] Recommended criteria to ensure maximum sensitivity and selectivity from GC/MS SIM analyses include: (i) monitoring two ions of the molecular ion cluster; (ii) measuring their ratio; (iii) monitoring a reasonably abundant (usually >14% rel. abund.) fragment ion.[369] Selective detection of brominated and brominated/chlorinated aromatic compounds by GC/MS with NICI is achieved by monitoring characteristic bromide ions ($m/z$ 79 and 81). Improved selectivity in the isomer specific GC/MS analyses of PCDDs and PCDFs has been achieved by coupling SP 2331 and DP 17 coated capillary columns in series.[370] A detailed systematic approach to the analysis of ultratrace levels of dioxins and related compounds has been developed with a view to standardizing methodology for a wide range of sample types.[371] The varying nature of the sample matrices requires carefully tailored clean-up procedures. Detection of PCDDs and PCDFs rested on GC/MS employing EI or CI modes.

6: *Analysis of Mixtures by Mass Spectrometry; Part I* 207

Used in the original discovery of polychlorinated biphenyls (PCBs), GC/MS remains the primary tool for their investigation. Carbon-13 labelled polychlorobiphenyls have been synthesised for use as standards in isotope dilution GC/MS analyses.[372] Unusual toxic non-ortho chlorine substituted PCBs have been detected in higher mammals for the first time.[373] Metabolites of PCBs also attract attention as these can be more toxic than the parent compounds. PFP derivatives were the preferred of six reagents tested for use in GC/MS and GC/ECD analyses of chlorinated hydroxybiphenyls and hydroxybiphenyls.[374] The total content and isomer distribution of methylsulphone metabolites of PCBs in human tissues have been determined by GC/MS and GC/ECD.[375] The high sensitivity of an ion trap detector (comprising automatic gain control software) in the full scanning mode has been used to advantage in the determination of tetrachlorobenzyltoluenes (Uglilec, a PCB substitute) in fish.[376] An effort has been made to define a simple measure of the degree of metabolism of PCB isomers in animals based on GC/MS analysis.[377]

Gel-permeation chromatography has been effective in the separation of organochlorines from lipids in tissues, prior to GC/MS[371,378,379,380] while lipophilic gel lipidex 5000 has been used for the extraction of organochlorine pesticides, PCBs, DBDs and DBFs from human milk.[381]

Problems associated with the reproducibility and interpretability of ECNI mass spectra have prompted a systematic investigation of parameters controlling the nature and magnitude of signals in the spectra of aromatic pollutants.[382] A method, based on the principle that MS responses of congeners of the same degree of chlorination vary only slightly, has been used in the compositional analysis of polychlorinated mixtures.[377,383]

GC/MS remains the method of choice for the rigorous confirmation and quantification of many pesticide residues.[60] A multi-author work has appeared devoted to the analysis of pesticides by modern mass spectrometric techniques.[61]

As in other fields of trace analysis tandem MS offers considerable potential in the determination of pesticide residues. Two possible approaches include: screening for specific compounds by selected reaction monitoring, or screening for compound classes through neutral loss or parent scans. Tandem MS data, recorded on a tandem quadrupole, for 26 organophosphate and 16 carbamate pesticides have been reported.[384] Procedural details relating to the optimization of ionization and CID conditions were also discussed. A cautionary note on the use of alcohols as solvents in the GC/MS analysis of phenylurea pesticides has appeared. Esters of *N*-(3,4-dichlorophenyl)carbamic acid are formed during the GC analysis, leading to erroneous conclusions about the original pesticide.[385] The highly desirable NICI behaviour of many pesticides and herbicides, owing to the presence of electrophoric moieties, makes this the ionization mode of choice for many applications. Toxaphene has been quantified in environmental samples by NICI with SIM.[386] The detection limit was found to be 75pg with the NICI response linear over 4 orders of magnitude. Heptachlor epoxide and octachlor have been identified in milk at the 0.1ppm level using GC/MS with NICI.[387] Methane chemical ionziation of pyrethrin pesticides results in fragmentations at the ester oxygen single bond to yield characteristic anions and cations which are conveniently detected for the purposes of identification or quantification using a pulsed positive ion negative ion CI (PPINICI) module.[388] Triclopyr (3,5,6-trichloro-2-pyridinyloxyacetic acid, a systemic herbicide) yields an abundant characteristic carboxylate anion under ECNICI conditions affording a detection limit of 70pg in SIM mode.[389] Organophosphorous pesticides have been detected in plasma and urine using capillary GC/MS in EI and CI modes.[390] Recoveries of pesticides from plasma were reduced owing to protein binding. A stable isotope dilution GC/SIMMS assay has been developed for the determination of Dicamba (3,6-dichloro-2-methoxybenzoic acid) and 2,4-D (2,4-dichloro-2-phenoxyacetic acid) in water and soil.[391] The technique was successfully applied to the analysis of 300 water and 300 soil samples. Detection limits were 0.1-1.0ppb for waters and 1-10 ppb for soils. Mass spectrometry and thermal energy analyzer (TEA) detection have been compared for the determination of N-nitrosodiethanolamine in Dinoseb formulations.[392] The detection limit of 0.02ng achieved in Townsend discharge mode exceeded that of TEA by more than two orders of magnitude.

An interlaboratory comparison of an automated method for the determination (USA EPA Method 680) of PCBs and chlorinated pesticides in water, soil and sediment identified several factors affecting data quality.[393] Extraction, clean-up procedures and GC performance were all pinpointed as sources of error; MS performance escaped criticism.

**2.9.5 Toxicology and Forensic Science**.- GC/MS, often with SIM, is used extensively for the determination of drugs, metabolites, pollutants and pesticides in biological tissues and fluids. Occupational exposure to solvents and manufacturing material gives cause for particular concern. Acrylamide is a potent cummulative neurotoxin in man and animals, and is widely used for the production of polyacrylamide products. Covalent adduction to hemaglobin *via* cysteine residues occurs readily as evidenced by the release of S-(2-carboxyethyl)cysteine on hydrolysis, which is readily detected and quantified (stable isotope dilution) by GC/MS.[394] Amino acid extracts have been determined as N-HFB methyl esters using iso-butane CI (positive ion). The same research group used a similar approach to monitor human exposure to ethylene oxide, determining $N^T$-(2-hydroxyethyl)histidine and $N$-(2-hydroxyethyl)valine by GC/MS after derivatization.[395] In the case of solvent exposure (eg. benzene[396] and methanol[397]) headspace GC/MS assays with deuteriated internal standards have been used. In support of toxicological studies a wide-bore (0.53mm i.d.) thick film (3.0µm) capillary column has been used to overcome problems of column overloading in the GC/MS analysis of low concentration impurities in technical grade solvents.[398] The analysis of plasma or urine extracts by GC/MS has been used as a means of monitoring exposure to a variety of amines including: piperazine,[399] ethylenediamine,[400] 4,4'-diaminodiphenylmethane and analogues,[401] and amine metabolites of degradation products from rigid polyurethane.[402]

Methods for the detection of trichothecene mycotoxins in foodstuffs have been reviewed above (see 'Food and Agricultural Chemistry'). Considerable interest also exists in their detection in human body fluids, especially since their alleged use as chemical warfare agents. Trichothecenes of widely varying polarity have been considered.[402-405] GC/NICIMS of trichothecenes extracted from blood as their PFP esters achieved detection limits in the 0.1-5ppb range.[403] The method was used in a collaborative study involving the analysis of 42 samples. In another study underivatized tricothecenes were readily determined human blood by GC/MS with PICI using ammonia reagent gas.[405] Trichothecenes have been detected in urine as their HFB esters using EI ionization, with no advantage being gained using PICI or NICI.[404] Contaminants of the chemical warfare agent O-ethyl S-[2-(diisopropylamino)ethyl] methylphosphonothiolate (VX) have been identified by GC/MS under both EI and CI conditions.[406] Thiodiglycol (2,2'-thiodiethanol) a metabolite of the vesicant mustard gas has been determined in urine by headspace isotope dilution GC/MS analysis following conversion to mustard gas by HCl treatment.[407] Comparison of the thiodiglycol concentrations urine of victims of an alleged mustard gas attack, with those of a control group, provided evidence consistent with the allegation.

Some applications of GC/MS have been presented in a review of techniques for the trace analysis of drugs in forensic chemistry.[63] RIA followed by GC/MS confirmation provides a very rigorous test procedure for the detection of THC metabolites in urine.[408] The ion trap detector (ITD) has been evaluated for forensic use in the detection of 11-nor-$\Delta^9$-THC carboxylic acid and was found to be suitable for corroborating the results of immunoassays.[409] $^2$H$_6$-11-nor-$\Delta^8$-THC-9-COOH has been proposed as a new internal standard for the quantitative analysis of THC in biological fluids in order to avoid problems of isobaric interference in isotope dilution assays when using $^2$H$_3$-9-THC-COOH.[410] Solid phase extraction using Mini Bed Amberlite XAD-2 columns has been effective in isolating cocaine and one of its major metabolites, benzoylecgonine from urine. Subsequent isotope dilution GC/MS assay was based on deuterated internal standards.[411] Results obtained from the RIA analysis of LSD in urine were constantly higher than those obtained by GC/MS or HPLC/fluorescence assays, emphasizing the need for reliable reference methodologies.[412] The reliability of using isotope labelling and subsequent MS isotope ratio measurement as the basis of an illicit compound tracing mechanism has been discussed with particular reference to methamphetamine.[413]

**2.10 Organic Geochemistry and Fuel.-** GC/MS in its various forms continues to be the workhorse technique of this field. A two-volume symposium proceedings presents numerous papers containing examples of applications of GC/MS.[64] Two books have appeared compiling mass spectra and applications of GC/MS to the analysis of biological marker compounds.[65,66] New developments in the analysis of PACs in Fuels have been reviewed.[67] PAHs in C$_8$ isomer aromatic feed have been determined by complementary use of GC, GC/MS and GC/FTIR.[414] Polyaromatic carbonyl compounds have been identified from a neutral coal oil fraction.[415] GC/CIMS using tetramethylsilane as reagent gas has facilitated the selective detection of oxygenated components of gasoline, through the reaction of reagent TMS ions with alcohols and ethers, but not with hydrocarbons.[416]

Many novel sedimentary components have been characterized by GC/MS, examples of these include novel acyclic sesterterpenoids,[417] mono-, di- and trimethyl-2-(4,8,12-trimethyltridecyl)chromans[418] and organic sulphur compounds.[419,420] In the latter study,[420] and in an investigation of sulphur compounds in a pyrolysate of sulphur-rich kerogen,[421] classifications of homologous series were based on Kováts' plots derived from GC/MS data together with comparisons with standard compounds. Further applications of pyrolysis-GC/MS in the geochemical field are reviewed below (see 'Pyrolysis-GC/MS'). GC/MS derived Kováts' plots and co-injections were also used in the characterization of petroporphyrins from Boscan crude oil and La Luna Mara Shale, an oil-source rock pair.[422] While this latter study determined demetallated petroporphyrins as their bis-(TBDMS)Si(IV) derivatives, the possibility of high-temperature GC/MS analysis of naturally occurring free-base and metalloporphyrins has also been demonstrated.[423]

Statistical analysis of complex GC/MS data is being used increasingly to establish geochemical relationships. In one example oil-oil correlation has been established using multivariate analysis of metastable ion monitoring data on steranes and triterpanes.[424] Amongst other statistical treatments are a prediction of source rock characteristics based on terpane biomarkers[425] and the characterization of sea water samples based on biomarker fatty acids.[426]

**2.11 Pyrolysis-GC/MS.-** The current review period has seen an increase in interest in this area. A symposium volume has appeared containing numerous examples of applications of pyrolysis-GC/MS and pyrolysis-MS.[66] Areas of applications of pyrolysis-GC/MS include: man-made polymers, biopolymers, organic geochemistry, environmental science, microorganisms, foodstuffs, etc.

On-line pyrolysis is most commonly achieved using Curie-point or heated coil apparatus. A simple modification to a Curie-point pyrolyser allows rapid pressure equilibration, immediately after pyrolysis, between the chamber and the capillary column.[427]

In the geochemical field pyrolysis-GC/MS is currently being employed in the study of geopolymers, i.e. kerogen, asphaltenes, coal components, etc. and other fossil materials. A review has appeared of biomarker compounds in kerogens, as revealed by pyrolysis-GC/MS and pyrolysis-GC.[428] A comparison has been made between biomarker (steranes and triterpanes) distributions of the free saturated hydrocarbons, and flash pyrolysates of asphaltenes, kerogen concentrates and extracted sediments.[429] Curie-point pyrolysis GC/photoionization-MS of Hungarian brown coals revealed homologous series of various classes of compound.[430] Significant differences in pyrolysis products allowed clear distinctions to be drawn between organic inputs to the various coals. Rapid pyrolysis has the advantage that the radicals formed immediately leave the heated zone with the inert carrier gas, thus reducing secondary reactions. Rapid pyrolysis-GC/MS of an asphaltene revealed a high concentration of xylene products, consistent with the thoery of bond dissociation energy in carbon structure that supports the cleavage of C-C bond to the aromatic ring.[431] Analytical pyrolysis has also been used to study soil[432] and aqueous[433] humic materials. In the former example Curie-point pyrolysis-GC/MS of soil humic acid yielded pyrolysis products of polysaccharides, proteins and lignins, in addition to fatty acids and aliphatic hydrocarbons.[432] A large proportion of functionalized aromatic products was observed in the case of aqueous humic material.[433]

Woody and cuticular plant materials of both buried[434-436] and recent origin[435-438] have also been investigated by pyrolysis-GC/MS. The data obtained from such materials is generally highly complex, for example, more than 300 components were observed to arise from an oligotrophic peat.[434] In the case of fossil plant cuticles[435] evidence derived from pyrolysis-GC/MS suggested the existence of a new, highly aliphatic and resistant biopolymer, which may explain the occurrence of straight-chain aliphatic moieties in organic-matter-rich sediments and coals.

Pyrolysis-GC/MS studies on man-made polymers include analysis of thermal decomposition products of liquid crystal polymers,[439] unsaturated polyester resins containing various chlorinated norbornene dicarboxylic acid units in the backbone,[440] pseudo-ladder polymers,[441] phenol-formaldehyde condensates,[442] and the determination of reactive comonomers and/or amino resins in acrylic copolymers.[443]

Sequence information on pairs of aliphatic amino acid moieties (valine, leucine, isoleucine, alanine) can be derived from flash pyrolysis-GC/MS of polyamino acids, proteins and complex proteinaceous material. The method is rapid and sensitive.[444] The very complex pyrolyzates obtained from human hair could only be deconvoluted by recourse to pyrolysis-GC/MS/MS. EI and CI with CID provided data supporting the formation of 2,4-imidazolidinediones and pyrrolidino[1,2a]-3,6-piperazinediones. None of the data showed potential usefulness for the forensic examination of hair.[445]

Pyrolysis-GC/MS of polysaccharides produced unique compounds (anhydrosugars) which are characteristic of the nature and abundance of neutral saccharide units making up the complex biopolymer.[448] Group A and Group B streptococci have been differentiated on the basis of pyrolysis-GC/MS owing to the presence of a group B specific polysaccharide which gave rise to a single characteristic chemical marker in the pyrograms of group B organisms.[447] Direct in-source pyrolysis-MS and pyrolysis-GC/MS have been used sequentially for fast profiling, and then characterization of specific products of thermal degradation of a wide range of biological materials, e.g. chitin, cellulose, hemicellose, lignin, wood, peat, coal and foodstuffs.[448]

Various flame retardant components of plastics and textiles have been investigated by pyrolysis-GC/MS in order to screen for potentially toxic thermal degradation products. In the case of triglycidyl isocyanurate, carbon dioxide, ethene oxide, acrolein, isocyanates, pyridines and quinones were detected following pyrolysis at 600°C.[449] Partially phosphated triglycidyl isocyanurate showed rather different behaviour, producing much higher levels of acrolein (70-90%). In the presence of polypropylene the level of acrolein was much reduced (7-5%).[450] Pyrolysis-GC/MS of brominated diphenyl ethers at various temperatures (600-900°C) produced bromobenzenes, bromophenols and brominated dibenzofurans; dioxins were not observed.[451]

# References

1. R.P. Evershed, in 'Mass Spectrometry', e.d. M.E. Rose (Specialist Periodical Reports), The Royal Society of Chemistry, London, 1987, Vol. 9, p.196.
2. M. Warner *Anal. Chem.*, 1987, **59**, 855A.
3. A.P. Bruins, in 'Advances in Mass Spectrometry 1985', e.d. J.F.J. Todd, John Wiley and Sons, N.Y., 1986, 119.
4. *Anal. Chem.*, 1987, **59**, 701A.
5. F.A. Settle, Jr., and M.A. Pleva 'Gas Chromatography-Mass Spectrometry: A Knowledge Base (electronic module)', Elsevier, Amsterdam, 1988.
6. F.W. Karasek and R.E. Clement 'Basic Gas Chromatography-Mass Spectrometry: Principles and Techniques', Elsevier, Amsterdam, 1988.
7. Proc. 15th Meeting British Mass Spectrom. Soc., Brighton, U.K., 14-17th Sept. 1986.
8. Proc. 16th Meeting British Mass Spectrom. Soc., York, U.K., 6-9th Sept. 1987.
9. A.L. Burlingame, D. Maltby, D.H. Russel and P.T. Holland *Anal. Chem.*, 1988, **60**, 294R.
10. K.R. Jennings *Spectros. Int. J.*, 1987, **5**, 21.
11. E.C. Horning in 'Advances in Mass Spectrometry 1985', ed. J.F.J. Todd, John Wiley and Sons, Chichester.
12. M.A. Grayson *J. Chromatogr. Sci.*, 1986, **24**, 529.
13. J.F.J. Todd, ed. 'Advances in Mass Spectrometry 1985', John Wiley and Sons, Chichester, 1986.
14. F. White and G. Wood 'Mass Spectrometry: Applications in Science and Engineering', John Wiley and Sons, Chichester, 1986.
15. D.L. Smith in 'Gaseous Ion Chemistry and Mass Spectrometry',ed. J.H. Futrell, John Wiley and Sons, Chichester, 1986, 305.
16. J.C. Schwartz and R.G. Cooks *Spectros. Int. J.*, 1987, **5**, 49.
17. F.W. McLafferty in 'Advances in Mass Spectrometry 1985',ed. J.F.J. Todd, John Wiley and Sons, Chichester, 1986, 493.
18. J.F.J. Todd in 'Advances in Mass Spectrometry 1985', ed. J.F.J. Todd, John Wiley and Sons, Chichester, 1986, 35.
19. D.F. Gurka and R. Titus 'Analytical Chemistry Instrumentation', ed. W.R. Laing, Lewis Publishers Inc., Michigan, 1986, 17.
20. C.L. Wilkins *Anal. Chem.*, 1987, **59**, 571.
21. J.C. Demirgian *Trends Anal. Chem.*, 1987, **6**, 58.
22. Proc. 34th Annu. Conf. Mass Spectrom. Allied Topics, Cincinati, Ohio, USA. 8-13th June 1986.
23. W.J.A. Vanden Heuvel and J.M. Liesch *Chem. Anal. (N.Y.)*, 1986, **85**, 91.
24. E. Gelpi in 'Advances in Mass Spectrometry 1985', ed. J.F.J. Todd, John Wiley and Sons, Chichester, 1986, 397.
25. P. Sandra ed. 'Sample Introduction in Capillary Gas Chromatography', Dr. Alfred Huethig Verlag, Heidelberg, 1985.
26. W. Jennings and M.F. Mehran, *J. Chromatogr. Sci.*, 1986, **24**, 34.
27. J.V. Hinshaw, *J. Chromatogr. Sci.*, 1987, **25**, 49.
28. M.A. Kaiser and M.S. Klee, *J. Chromatogr. Sci.*, 1986, **24**, 369.
29. F.L. Bayer *J. Chromatogr. Sci.*, 1986, **24**, 549.
30. W.V. Ligon and R.J. May *J. Chromatogr. Sci.*, 1986, **24**,2
31. R.E. Clement, F.I. Onuska, G.A. Eiceman and H.H. Hill *Anal. Chem.*, 1988, **60**, 279R.
32. D.H. McMahon *J. Chromatogr. Sci.*, 1985, **23**, 426.
33. N.J. Jensen and M.L. Gross *Mass Spectrom. Rev.*, 1987, **6**, 497.
34. A. Kuksis and J.J. Myher *J. Chromatogr. Biomed. Appln.*, 1986, **379**, 57.
35. A. Kuksis ed. 'Chromatography of Lipids in Biomedical Research and Clinical Diagnosis', Elsevier, Amsterdam, 1987.
36. C.H.L. Shackleton *J. Chromatogr. Biomed. Appln.*, 1986, **379**, 91.
37. A.P. DeLeenheer, M.F. Lefevere and L.M.R. Thienpont *J. Pharmaceut. Biomed. Anal.*, 1986, **4**, 735.
38. R.P. Evershed, M.C. Prescott, L.J. Goad and H.H. Rees *Biochem. Soc. Trans.*, 1987, **15**, 175.
39. E. Houghton, M.C. Dumasia and P. Teale *Analyst*, 1988, **113**, 1179.
40. S. Gorog ed. 'Advances in Steroid Analysis '84', Elsevier, Amsterdam, 1985.
41. A.M. Krstulovic ed. 'Quantitative analysis of Catacholamines and Related Compounds', John Wiley and Sons, N.Y., 1986.
42. N. Seiler *J. Chromatogr. Biomed. Appln.*, 1986, **379**, 157.
43. J.E. Shively ed. 'Methods of Protein Microcharacterisation', Humana Press, Clifton, New Jersey, 1986.
44. R.M. Caprioli *Mass Spectrom. Rev.*, 1987, **6**, 237.

45. K. Kakehi and S. Honda *J. Chromatogr. Biomed. Appln.*, 1986, **379**, 27.
46. I. Matsumoto and T. Kuhara *Mass Spectrom. Rev.*, 1987, **6**, 77.
47. H.M. Liebich *J. Chromatogr. Biomed. Appln.*, 1986, **379**, 347.
48. T. Niwa *J. Chromatogr. Biomed. Appln.*, 1986, **379**, 313.
49. J.F. Holland, J.J. Leary and C.C. Sweeley *J. Chromatogr. Biomed. Appln.*, 1986, **379**, 3.
50. A.P. De Leenheer, M.F. Lefevere, W.E. Lambert and E.S. Coline *Adv. Clin. Chem.*, 1985, **24**, 111.
51. D.L. Hachey, W.W. Wong, T.W. Boutton and P.D. Klein *Mass Spectrom. Rev.*, 1987, **6**, 289.
52. I. Brondz and I. Olsen *J. Chromatogr. Biomed. Appln.*, 1986, **379**, 367.
53. I. Brondz and I. Olsen *J. Chromatogr. Biomed. Appln.*, 1986, **380**, 1.
54. M.W. D. *Anal. Chem.*, 1986, **58**, 1310A.
55. H. Jaeger ed. 'Capillary Gas Chromatography - Mass Spectrometry in Medicine and Pharmacology' Dr. Alfred Huthig Verlag, Heidelberg, 1987.
56. W.J.A. Vandenheuvel *Xenobiotica*, 1987, **17**, 397.
57. J. Gilbert 'Application of Mass Spectrometry in Food Science Elsevier, Barking, 1987.
58. H.F.Linskers and J.F.Jackson 'Modern Methods of Plant Analysis New Ser. Vol. 3' Springer-Verlag, Berlin, 1986.
59. O. Hutzinger, W. Crummett, F.W. Karasek, E. Merian, G. Reggiani, M. Reissinger and S. Safe *Chemosphere*, 1986, **15**, 1079-2132.
60. J. Gilbert, J.R. Startin and C. Crews *Pestic. Sci.*, 1987, **18**, 273
61. J.D. Rosen ed. 'Appln. of New Mass Spectrometry Technique in Pesticide Chemistry' John Wiley and Sons, N.Y., 1987.
62. G. Blomkvist *Var Foda Suppl.*, 1986, **2**, 125.
63. I.S. Lurie, J.M. Moore and D.A. Cooper *Chem. Anal. (N.Y.)*, 1986, **85**, 319.
64. D. Leythaeuser and J. Rullkotter eds. 'Advances in Organic Geochemistry 1985' Pergamon Press, 1986.
65. R.P. Philip 'Fossil Fuel Biomarkers: Applications and Spectra', Elsevier, Amsterdam, 1985.
66. R.B. Johns 'Biological Markers in the Sedimentary Record' Elsevier, Amsterdam, 1986.
67. C.E. Miller and D.E. Honigs *Spectroscopy*, 1987, **2**, 30.
68. H.-R. Shulten ed. *J. Anal. Appl. Pyrol.*, 1987, **11**.
69. 'Rapid Communications in Mass Spectrometry' Heyden and Son Ltd.
70. J. Allison, J.F. Holland, C.G. Erke and J.F. Watson *Anal. Inst.*, 1987, **16**, 207.
71. R.B. Opsal and J.P. Reilly *Anal. Chem.*, 1986, **58**, 2919.
72. R.B. Opsal and J.P. Reilly *Anal. Chem.*, 1988, **60**, 1060.
73. J.C. Van Loon, L.R. Alcock, W.H. Pinchin and J.B. French Spectros. Letts., 1986, 19, 1125.
74. E.J. Gallegos *J. Chromatogr. Sci.*, 1987, **25**, 296.
75. K. Yamaguchi, A. Ohsawa and H. Igeta *Chem. Pharm. Bull.*, 1986, **34**, 426.
76. C.S. Jones and E.P. Grimsrud *J. Chromatogr.*, 1987, **409**, 139.
77. J.S. Brodbelt, J.N. Louris and R.G. Cooks *Anal. Chem.*, 1987, **59**, 1278.
78. E.S. Olson and J.W. Diehl *Anal. Chem.*, 1987, **59**, 443.
79. J.R. Looper, I.C. Bowater and C.L. Wilkins *Anal. Chem.*, 1986, **58**, 2791.
80. H. Fujiwara, R. Thomas and S.J. Wratten *Spectroscopy*, 1987, **2**, 24.
81. D.A. Laude, C.L. Johlman, R.S. Brown and C.L. Wilkins *Fresenius Z. Anal. Chem.*, 1986, **324**, 839.
82. P. Grossmann, P. Caravatti, St. Dutsch, M. Allemann and H.P. Kellerhals *Lab. Prac.*, 198, **36**, 63.
83. S. Nitz, F. Drawert and U. Gellert *Chromatographia*, 1986, **22**, 51.
84. H. Eustache, J. Allois, C. Bonnefond, A. Jacquot, A. Lorant and V. Mevel *Analusis*, 1988, **16**, 37.
85. J.F. Pankow and L.M. Isabelle *J. High Res. Chromatogr. Chromatogr. Commun.*, 1987, **10**, 617.
86. D.B. Stauffer and F.W. McLafferty *Organic Mass Spectrom.*, 1986, **21**, 313.
87. A. Tenhosaari *Organic Mass Spectrom.*, 1987, **23**, 236.
88. H. Lohninger and K. Varmuza *Anal. Chem.*, 1987, **59**, 236.
89. C. Boniface, G. Vernin and J. Metzger *Analusis*, 1987, **15**, 564.
90. C.P. Wang and T.L. Isenhour *Anal. Chem.*, 1987, **59**, 649.
91. B.R. Petit *Biomed. Environ. Mass Spectrom.*, 1986, **13**, 473.
92. G.C. Thorne and S.J. Gaskell *Biomed. Environ. Mass Spectrom.*, 1986, **13**, 605.
93. K.G. Heumann *Fresenius Z. Anal. Chem.*, 1986, **325**, 661.
94. K.F. Blom *Organic Mass Spectrom.*, 1987, **22**, 530.
95. H. Kanamaru, R. Takai, M. Horiba, I. Nakatsuka and A. Yoshitake *Radioisotopes*, 1985, **34**, 67.
96. J.E. Knoll *J. Chromatogr. Sci.*, 1985, **23**, 422.
97. T. Noy and C. Cramers *J. High Res. Chromatogr. Chromatogr. Commun.*, 1988, **11**, 264.

98. W.J. Cretney, F.A. McLaughlin and B.R. Fowler *J. High Res. Chromatogr. Chromatogr. Commun.*, 1987, **10**, 428.
99. R.P. Snell, J.W. Danielson and G.S. Oxborrow *J. Chromatogr. Sci.*, 1987, **2**, 225
100. K. Grob, Jr. and T. Laubli *J. Chromatogr.*, 1986, **357**, 34
101. K. Grob, Jr. and T. Laubli *J. Chromatogr.*, 1986, **357**, 357.
102. E. Geeraert in 'Sample Introduction in capillary gas chromatography Vol. 1', ed. P. Sandra Dr. Alfred Huethig Verlag, Heidlberg, 1985, p.133.
103. J.F. Elder, Jr., B.M. Gordon and M.S. Uhrig *J. Chromatogr. Sci.*, 1986, **24**, 26
104. A. Hagman and S. Jacobsson *J. Chromatogr.*, 1987, **395**, 241.
105. T.V. Raglione, J.A. Troskosky and R.A. Hartwick *J. Chromatogr.*, 1987, **409**, 213.
106. T.V. Raglione, J.A. Troskosky and R.A. Hartwick *J. Chromatogr.*, 1987, **409**, 205.
107. N. Schmidbauer and M. Oehme *J. High. Res. Chromatogr. Chromatogr. Commun.*, 1987, **10**, 398.
108. C.W. Bayer and M.S. Black *J. Chromatogr. Sci.*, 1987, **25**,60
109. D.M. Wyatt *J. Chromatogr. Sci.*, 1987, **25**, 257
110. V.M. Allenger, D.D. McLean and M. Ternan *J. Chromatogr. Sci.*, 1986, **24**,95
111. P.J. Baugh, A. Casson, M.W. Jones and A.C. Jones *J. Chromatogr.*, 1987, **411**, 445.
112. C. Yaozu, L. Zhaolin, X. Dunyuan and Q. Limin *Anal. Chem.*, 1987, **59**, 744.
113. S.B. Hawthorne, M.S. Krieger and D.J. Miller *Anal. Chem.*, 1988, **80**, 472.
114. Y. Lopez-Avila, R. Wood, M. Flanagan and R. Scott *J. Chromatogr. Sci.*, 1987, **25**,286
115. R. Otson and C. Chain *Int. J. Environ. Anal.Chem.*, 1987, **30**, 275.
116. K.M. Hart and J.F. Pankow *J. High Res. Chromatogr. Chromatogr. Commun.*, 1987, **10**, 484.
117. J.F. Pankow, M.P. Ligocki, M.E. Rosen, L.M. Isabelle and K.M. Hart *Anal. Chem.*, 1988, **60**, 40.
118. H. Karlson, I. Carlstedt and G.C. Hansson *FEBS Letts.,*1987, **226**, 23
119. K. Kaneda, S. Naito, S. Imaizumi, I. Yano, S. Mizuno, I. Tomiyasu, T. Baba, E. Kusunose and M. Kusunose *J. Clin. Microbiol.*, 1986, **24**, 1060.
120. R. P. Evershed and L. J. Goad *Biomed. Environ. Mass Spectrom.*, 1987, **14**, 131
121. R. P. Evershed, J.G. Mercer and H. H. Rees *J. Chromatogr.*, 1987, **390**, 357
122. R. P. Evershed, N. Spooner, M. C. Prescott and L. J. Goad *J. Chromatogr.*, 1988, **440**, 23
123. G. Audisio, A. Rossini, G. Bianchi and P. Avato *J. High Res. Chromatogr. Chromatogr. Commun.*, 1987, **10**, 594.
124. T. Rezanka and M. Podojil *J. Chromatogr.*, 1986, **362**, 399.
125. S. R. Lipsky and M. L. Duffy *J. High Res. Chromatogr. Chromatogr. Commun.*, 1986, **9**, 376
126. S. R. Lipsky and M. L. Duffy *J. High Res. Chromatogr. Chromatogr. Commun.*, 1986, **9**, 725
127. C.A. Cramer *J. High Res. Chromatogr. Chromatogr. Commun.*, 1986, **9**, 676.
128. M. Proot and P. Sandra *J. High Res. Chromatogr. Chromatogr. Commun.*, 1986, **9**, 618.
129. K.J. Hyver and R.J. Phillips *J. Chromatogr.*, 1987, **399**, 33.
130. K.J. Hyver *J. High Res. Chromatogr. Chromatogr. Commun.*, 1988, **11**, 69.
131. M. Ishibashi, K. Watanabe, H. Miyazaki and S. Krolik *Chem. Pharm. Bull.*, 1986, **34**, 3510.
132. M. Ishibashi, K. Watanabe, H. Miyazaki and S. Krolik *Yakugaku Zasshi*, 1986, **106**, 1118.
133. M. Ishibashi, K. Watanabe, K. Yamashita, H. Miyazaki and S. Krolik *J. Chromatogr.*, 1987, **391**, 183.
134. K. Yamashita, K. Watanabe, M. Ishibashi, H. Miyazaki, K. Yokota, K. Horie and S. Yamamoto *J. Chromatogr.*, 1987, **399**, 223.
135. M. Ishibashi, M. Itoh, K. Yamashita, H. Miyazaki and H. Nakata *Chem. Pharm. Bull.*, 1986, **399**, 223.
136. H. Nakata, M. Ishibashi, M. Itoh and H. Miyazaki *Organic Mass Spectrom.*, 1987, **22**, 23.
137. M. Ishibashi, T. Irie and H. Miyazaki *J. Chromatogr.*, 1987, **399**, 197.
138. S. Steffenrud, P. Borgeat, M.J. Evans and M.J. Bertrand *Biomed. Environ. Mass Spectrom.*, 1986, **13**, 657.
139. S. Steffenrud, P. Borgeat, M.J. Evans and M.J. Bertrand *Biomed. Environ. Mass Spectrom.*, 1987, **14**, 313.
140. S. Steffenrud, P. Borgeat, H. Salari, J.J. Evans and M.J. Bertrand *J. Chromatogr. Biomed. Appln.*, 1987, **416**, 219.
141. S. Steffenrud, P. Borgeat, M.J. Evans and M.J. Bertrand *J. Chromatogr. Biomed. Appln.*, 1987, **423**, 1.
142. D.A. Herold, J. Savory, M. Kinter, R. Ross and M.R. Wills *Analytica Chimica Acta*, 1987, **197**, 149.
143. K. Yamashita, K. Watanabe, M. Ishibashi, M. Katori and H. Miyazaki *J. Chromatogr. Biomed. Appln.*, 1988, **424**, 1.
144. D.L. Schooley, F.M. Kubiak and J.V. Evans *J. Chromatogr. Sci.*, 1985, **23**, 385
145. T.P. Mawhinney, R.S.R. Robinett, A. Atalay and M.A. Madson *J. Chromatogr.*, 1986, **361**, 117.
146. F. Rocchiccioli *Biomed. Environ. Mass Spectrom.*, 1986, **13**, 387.

147. U. Langenbeck and H. Luthe *Biomed. Environ. Mass Spectrom.*, 1986, **13**, 387.
148. A.A. Fernandes, S.C. Kalhan, F.G. Njoroge and G.S. Matousek *Biomed. Environ. Mass Spectrom.*, 1986, **13**, 569.
149. M.J. Bertrand, S. Stefanidis and B. Sarrasin *J. Chromatogr.*, 1986, **351**, 47.
150. M.J. Bertrand, S. Stefanidis, A. Donais and B. Sarrasin *J. Chromatogr.*, 1986, **354**, 331.
151. M.J. Bertrand, A.W. Ahmed, B. Sarrasin and N.N. Mallet *Anal. Chem.*, 1987, **59**, 1302.
152. A.E. Yatsenko, A.I. Mikaya, L.S. Glebov and V.G. Zaikin *Bull. Acad. Sci. USSR Div. Chem. Sci. (Engl. Transl.)*, 1986, **35**, 666.
153. C.J.W. Brooks and W.J. Cole *J. Chromatogr.*, 1987, **399**, 207.
154. P. Scribe, J. Guezzenec, J. Dagaut, C. Pepe and A. Saliot *Anal. Chem.*, 1988, **60**, 828.
155. M. Vincenti, G. Guglielmetti, G. Cassani and C. Tonini *Anal. Chem.*, 1987, **59**, 694.
156. P.D. Nichols, J.B. Guckert and D.C. White *J. Microbiol. Methods*, 1986, **5**, 49.
157. J.P.J. Billen, R.P. Evershed, A.B. Attygalle, E.D. Morgan and D.G. Ollett *J. Chem. Ecol.*, 1986, **12**, 669.
158. G. Janssen, A. Verhulst and G. Parmentier *Biomed. Environ. Mass Spectrom.*, 1988, **15**, 1.
159. D.C. Young, P. Vouros, B. Decosta and M. F. Holick *Anal. Chem.*, 1987, **59**, 1954.
160. W.W. Christie, E.Y. Brechany, F.D. Gunstone, M.S.F. Lie Ken Jie and R.T. Holman *Lipids*, 1987, **22**, 664.
161. W.W. Christie, E.Y. Brechany and R.T. Holman *Lipids*, 1987, **22**, 224.
162. W.W. Christie, E.Y. Brechany, S.B. Johnson and R.T. Holman *Lipids*, 1986, **21**, 657.
163. W.W. Christie and K. Stefanov *J. Chromatogr.*, 1987, **392**, 259.
164. W.W. Christie, E.Y. Brechany and K. Stefanov Chem. Phys. *Lipids*, 1988, **46**, 127.
165. R.A. Klein and B. Schmitz *Biomed. Environ. Mass Spectrom.*, 1986, **13**, 429.
166. J.Y. Zhang, Q.T. Yu, B.N. Liu and Z.H. Huang *Biomed. Environ. Mass Spectrom.*, 1988, **15**, 33.
167. J.Y. Zhang, Q.T. Yu and Z.H. Huang *Mass Spectroscopy*, 1987, **35**, 23.
168. R. Wait and M.-J. Hudson *Lett. Appl. Microbiol.*, 1985, **1**, 95.
169. D.J. Harvey *Biomed. Environ. Mass Spectrom.*, 1987, **14**, 103.
170. D. A. Peake, S. -K. Huang and M. L. Gross *Anal. Chem.*, 1987, **59**, 1557
171. J. Einhorn, H. Virelizier and J.C. Tabet *Spectrosc. Int. J.*, 1987, **5**, 171.
172. H. Budzikiewicz *Spectrosc. Int. J.*, 1987, **5**, 183.
173. G. Bouchoux, Y. Hoppilliard, P. Jandon and J.-M. Pechine *Spectrosc. Int. J.*, 1987, **5**, 247.
174. M. Bambagiotti, S.A. Coran, F.F. Vincieri, T. Petrucciani and P. Traldi *Organic Mass Spectrom.*, 1986, **21**, 485.
175. M. Horiike and C. Hirano *Biomed. Environ. Mass Spectrom.*, 1987, **14**, 183.
176. Y. Kuwahara, Y. Yonekawa, T. Kamikihara and T. Suzuki *Agric. Biol. Chem.*, 1986, **50**, 2017.
177. A.J. Fellenberg, D.W. Johnson, A. Poulos and P. Sharp *Biomed. Environ. Mass Spectrom.*, 1987, **14**, 127.
178. D.S. Waddell, H.S. McKinnon, B.G. Chittim, S. Safe and R.K. Boyd *Biomed. Environ. Mass Spectrom.*, 1987, **14**, 457.
179. B.G. Chittin, J.A. Madge and S.H. Safe *Chemosphere*, 1986, **15**, 1931.
180. F.L. Onuska, R.J. Wilkinson and K. Terry *J. High Res. Chromatogr. Chromatogr. Commun.*, 1988, **11**, 9.
181. M. Swerev and K. Ballschmiter *J. High Res. Chromatogr. Chromatogr. Commun.*, 1987, **10**, 544.
182. K.P. Naikwadi and F.W. Karasek *J. Chromatogr.*, 1986, **369**, 203.
183. P. Schmid and M.D. Muller *J. High Res. Chromatogr. Chromatogr. Commun.*, 1987, **10**, 548.
184. R. Guevremont, R.A. Yost and W.D. Jamieson *Biomed. Environ. Mass Spectrom.*, 1987, **14**, 435.
185. G.W. Sovocool, R.K. Mitchum and J.R. Donnelly *Biomed. Environ. Mass Spectrom.*, 1987, **14**, 579.
186. W.J. Simonsick and R.A. Hites *Anal. Chem.*, 1986, **58**, 2114.
187. W.J. Simonsick and R.A. Hites *Anal. Chem.*, 1986, **58**, 2121.
188. S.A. Brotherton and W.M. Gulick *Analytica Chimica Acta*, 1986, **186**, 101.
189. W. A. Korfmaccher, L. G. Rushing, R. J. Engelbach, J. P. Freeman, Z. Djuric, E. K. Fifer and F. A. Beland *J. High Res. Chromatogr. Chromatogr. Commun.*, 1987, **10**, 43.
190. W.A. Korfmacher and L.G. Rushing *J. High Res. Chromatogr. Chromatogr. Commun.*, 1986, **9**, 293.
191. T. Ramdahl, G.E. Carlberg and P. Kolsaker *Sci. Total Environ.*, 1986, **48**, 147.
192. D.R. Hardy, G.W. Mushbrush, W.M. Stalick, E.J. Beal and R.N. Hazlett *Rapid Commun. Mass Spectrom.*, 1988, **2**, 16.
193. Z.M.H. Marzouki, A.M. Taha and K.S. Gomaa *J. Chromatogr. Biomed. Appln.*, 1988, **425**, 11.
194. G. Odham, A. Tunlid, G. Westerdahl, L. Larsson, J.B. Guckert and D.C. White *J. Microbiol. Methods*, 1985, **3**, 331.
195. F.J.M. vanKuijk, D.W. Thomas, R.J. Stephans and E.A. Dratz *J. Free Radicals Biol. Med.*, 1985, **1**, 387.
196. R.I. Hollingsworth and F.B. Dazzo *J. Microbiol. Methods*, 1988, **7**, 295.
197. H. Schweer, H.W. Seyberth and R. Schubert *Biomed. Environ. Mass Spectrom.*, 1986, **13**, 611.

198. H. Schweer, C.O. Meese, O. Furst, P.G. Kuhl and H.W. Seyberth *Anal. Biochem.*, 1987, **164**, 156.
199. H. Schweer, H.W. Seyberth, C.O. Meese and O. Furst *Biomed. Environ. Mass Spectrom.*, 1988, **15**, 143.
200. H. Schweer, H.W. Seyberth and C.O. Messe *Biomed. Environ. Mass Spectrom.*, 1988, **15**, 129.
201. H. Schweer, H.W. Seyberth, C.O. Meese and O. Furst *Biomed. Environ. Mass Spectrom.*, 1988, **15**, 139.
202. H.L. Hubbard, T.D. Eller, D.E. Mais, P.V. Halushka, R.M. Baker, I.A. Blair, J.J. Vrbanac and D.R. Knapp *Prostaglandins*, 1987, **33**, 149.
203. J.J. Vrbanac, T.D., Eller and D.R. Knapp *J. Chromatogr. Biomed. Appln.*, 1988, **425**, 1.
204. C. Chiabrando, A. Benigni, A. Piccinelli, C. Carminati, E. Cozzi, G. Remuzzi and R. Fanelli *Anal. Biochem.*, 1987, **163**, 255.
205. B. Mayer, R. Moser, H.-J. Leis and H. Gleispach *J. Chromatogr. Biomed. Appln.*, 1986, **378**, 430.
206. J.A. Lawson, C. Patrono, G. Ciabattoni and G.A. Fitzgerald *Anal. Biochem.*, 1986, **155**, 198.
207. A. Martineau and P. Falardeau *J. Chromatogr. Biomed. Appln.*, 1987, 417, 1.
208. W.R. Mathews, G.L. Bundy, M.-A. Wynalda, D.M. Guido, W.P. Schneider and F. A. Fitzpatrick *Anal. Chem.*, 1988, **60**, 349.
209. N. Shindo, T. Saito and K. Murayama *Biomed. Environ. Mass Spectrom.*, 1988, **15**, 25.
210. H.J. Leis, E. Malle, R. Moser, J. Nimpf, G.M. Kostner, H. Esterbauer and H. Gleispach *Biomed. Environ. Mass Spectrom.*, 1986, **13**, 483.
211. H.J. Leis, E. Malle, B. Mayer, G.M. Kostner, H. Esterbauer and H. Gleispach *Anal. Biochem.*, 1987, **162**, 337.
212. J.Y. Westcott, S. Chang, M. Balazy, D.O. Stene, P. Pradelles, J. Maclouf, N.F. Voelkel and R.C. Murphy *Prostaglandins*, 1986, **32**, 857.
213. H.J. Leis, E. Hohenester, H. Gleispach, E. Malle and B. Mayer *Biomed. Environ. Mass Spectrom.*, 1987, **14**, 617.
214. A. Ferretti, V.P. Flanagan and V.B. Reeves *Anal. Biochem.*, 1987, **167**, 174.
215. J.S. Hadley, A. Fradin and R.C. Murphy *Biomed. Environ. Mass Spectrom.*, 1988, **15**, 175.
216. D.A. Herold, B.J. Smith, R.M. Ross, F. Marquis, C.R. Ayers, M.R. Wills and J. Savory *Prostaglandins*, 1987, **33**, 599.
217. H. Gleispach in 'Advances in Steroid Analysis'84', ed. S. Gorog, Elsevier, Amsterdam, 1985, 413.
218. L. Siekmann and R. Rohle *Fres. Z. Anal. Chem.*, 1986, **324**, 208.
219. L. Muller and Phillipou *Clin. Chem.*, 1987, **33**, 256.
220. D.J. Porubek and S.D. Nelson *Biomed. Environ. Mass Spectrom.*, 1988, **15**, 157.
221. M.C. Patricot, B. Mathian, S. Serpentie and A. Revol. *Clin. Chim. Acta*, 1986, **158**, 139.
222. O. Pelletier, L.A. Wright and w.C. Breckenridge *Clin. Chem.*, 1987, **33**, 1403.
223. F. Stellard and G. Paumgartner *Clin. Chim. Acta*, 1987, **162**, 45.
224. N. Hirota, T. Furuta and Y. Kasuya *J. Chromatogr. Biomed. Appln.*, 1988, **425**, 237.
225. L. Dehennin, K. Nahoul and R. Scholler *J. Steroid Biochem.*, 1987, **26**, 337.
226. T.E. Chapman, G.P.B. Kraan, N.M. Drayer, G.T. Nagel and B.G. Wolthers *Biomed. Environ. Mass Spectrom.*, 1987, **14**, 73.
227. S.J. Gaskell, K. Rollins, R.W. Smith and C.E. Parker *Biomed. Environ. Mass Spectrom.*, 1987, **14**, 717.
228. E. Houghton, M.C. Dumasia, P. Teale, M. Moss and S. Sinkins *J. Chromatogr. Biomed. Appln.*, 1986, **383**, 1.
229. P.R. Robinson, M.D. Jones and J. Maddock *J. High Res. Chromatogr. Chromatogr. Commun.*, 1987, **10**, 6.
230. S. Murray and D. Watson *J. Steroid Biochem.*, 1986, **25**, 255.
231. J. Goto, K. Watanabe, H. Miura, T. Nambara and T. Iida *J. Chromatogr.*, 1987, **388**, 379.
232. R. Mahara, H. Takeshita, T. Kurosawa, S. Ikegawa and M. Tohma *Anal. Sci.*, 1987, **3**, 449.
233. M. Tohma, H. Takeshita, R. Mahara and T. Kurosawa *J. Chromatogr. Biomed. Appln.*, 1987, 421, 9.
234. F. Stellard and G. Paumgartner *Clin. Chim. Acta*, 1987, **162**, 45.
235. F. Stellaard, M. Sackmann, F. Berr and G. Paumgartner *Biomed. Environ. Mass Spectrom.*, 1987, **14**, 609.
236. H.-U. Marschall, B. Egestad, H. Makern, S. Makern and J. Sjovall *FEBS Letts*, 1987, **213**, 411.
237. M. Aso, K. Miyazaki, J. Yanagisawa and f. Nakayama *J. Biochem.*, 1987, **101**, 1429.
238. C. Tsaconas, P. Padieu, G. Maume, M. Chessebeur, N. Hussein and N. Pitoizet *Anal. Biochem.*, 1986, **157**, 300.
239. R.P. Evershed, V.L. Male and L.J. Goad *J. Chromatogr.*, 1987, **400**, 187.
240. T. Cairns and E. G. Siegmund *Rapid Commun. Mass Spectrom.*, 1987, **1**, 108.
241. V. Papadopoulos, S. Carreau, E. Szerman-Joly, M.A. Drosdowsky, L. Dehennin and R. Scholler *J. Steroid Biochem.*, 1986, **24**, 1211.

242. G. Moneti, A. Costantantini, A. Guarna, R. Salerno, M. Pazzagli, A. Natali, A. Goti and M. Serio *J. Steroid Biochem.*, 1986, **25**, 765.
243. F. Jacolot, D. Picart, f. Berthou and H.H. Flock *Biomed. Environ. Mass Spectrom.*, 1986, **13**, 389.
244. H. Vierhapper, P. Nowotny and W. Waldhausl *J. Steroid Biochem.*, 1988, **29**, 105.
245. P. Scudder, A.M. Lawson, E.F. Hounsell, R.A. Carruthers, R.A. Childs and T. Feizi *Eur. J. Biochem.*, 1987, 168, 585.
246. E. Romanoska, A. Gamian and J. Dabrowski *Eur. J. Biochem.*, 1986, **161**, 557.
247. J.L. Minor and R.C. Pettersen *J. Agric. Food Chem.*, 1987, **35**, 993.
248. R.A. O'Neill, A.R. White, W.S. York, A.G. Darvill and P. Albersheim *Phytochem.*, 1988, **27**, 329.
249. D.B. Kassel and J. Allison *Anal. Chem.*, 1988, **58**, 1670
250. O. Pelletier and C. Arratoon *Clin. Chem.*, 1987, **33**, 1397.
251. H.J. Chaves Das Neves and A.M.V. Riscado *J. Chromatogr.*, 1986, **367**, 135.
252. W.R. Sherman, K.E. Ackerman, R.A. Berger, B.G. Glish and M. Zunbo *Biomed. Environ. Mass Spectrom.*, 1986, **13**, 333.
253. G.A. Van Den Berg, A.W. Kingma, H. Elzinga and F.A.J. Mustiet *J. Chromatogr. Biomed. Appln.*, 1986, **383**, 251.
254. T. Noto, T. Hasegawd, H. Kamimura, J. Nakas, H. Hashimoto and T. Nakajima *Anal. Biochem.*, 1987, **160**, 371.
255. S. Fujihara, T. Nakashima and Y. Kurogochi *J. Chromatogr. Biomed. Appln.*, 1986, **383**, 271.
256. B.A. Davis and D.A. Durden *Biomed. Environ. Mass Spectrom.*, 1987, **14**, 197.
257. M.J. Avery and G.A. Junk *J. Chromatogr. Biomed. Appln.*, 1987, **420**, 379.
258. P.A. Tippett, B.E. Clayton and A.I. Mallet *Biomed. Environ. Mass Spectrom.*, 1987, **14**, 737.
259. T. Hayashi, M. Shimamara, F. Matsuda, Y. Minatogawa, H. Naruse and T. Lida *J. Chromatogr. Biomed. Appln.*, 1986, **383**, 259.
260. A.A. Boulton and B.A. Davis *Biomed. Environ. Mass Spectrom.*, 1987, **14**, 207.
261. A.P.J.M. De Jong, R.M. Kok, C.A. Cramers and S.K. Wadman *J. Chromatogr. Biomed. Appln.*, 1986, **382**, 19.
262. C.R. Lee and H. Esnaud *Biomed. Environ. Mass Spectrom.*, 1988, **15**, 249.
263. P.L. Francis, A.M. Leone, I.M. Young, P. Stovell and R.E. Silman *Clin. Chem.*, 1987, **33**, 453.
264. M. J. Welch, A. Cohen, H. S. Hertz, K.J. Ng, R. Schaffer, P. Van Der Lijn and E. White *Anal Chem*, 1986, **58**, 1681.
265. H. Hendriks, W. Balraadjsing, H.J. Huizing and A.P. Bruins *Planta Medica*, 1987, 456.
266. C.K. Winter, H.J. Segall and A.D. Jones *Biomed. Environ. Mass Spectrom.*, 1988, **15**, 265.
267. J.H. Vine, D. Noiton, J.A. Plummer, C. Baleriola-Lucas and M.G. Mullins *Plant Physiol.*, 1987, **85**, 419.
268. M. Debono, M. Barnhart, C.B. Carrell, J.A. Hoffmann, J.L. Occolowitz, B.J. Abbott, D.S. Fukirda, R.L. Hamill, K. Biemann and W.C. Herlihy *J. Antibiotics*, 1987, **40**, 761.
269. M.E. Hemling, S.A. Carr, C. Capiau and J. Petre *Biochem.*, 1988, **27**, 699.
270. P.T. Wingfield, P. Graber, G. Buell, K. Rose, M.G. Simona and D. Burleigh *Biochem. J.*, 1987, **243**, 829.
271. N.R. Hartman and I. Jandine *Biomed. Environ. Mass Spectrom.*, 1986, **13**, 361.
272. B.J. Erickson and I. Jardine *Biomed. Environ. Mass Spectrom.*, 1986, **13**, 343.
273. J. Jandke and G. Spiteller *J. Chromatogr. Biomed. Appln.*, 1986, **382**, 39.
274. M. Corbett, C.M. Scrimgeour and P.W. Watt *J. Chromatogr. Biomed. Appln.*, 1987, **419**, 263.
275. G. Fortier, D. Tenaschuk and S.L. MacKenzie *J. Chromatogr.*, 1986, **361**, 253.
276. S.L. MacKenzie, D. Tenaschuk and G. Fortier *J. Chromatogr.*, 1987, **387**, 241.
277. S.P. Stabler, P.D. Marcell, E.R. Podell and R.H. Allen *Anal. Biochem.*, 1987, **162**, 186.
278. L.W. Anderson, D.W. Zaharevitz and J.M. Strong *Anal. Biochem.*, 1987, **163**, 358.
279. H.J. Chaves das Neves and A.M.P. Vasconcelos *J. Chromatogr.*, 1987, **392**, 249.
280. H.J. Chaves das Neves, A.M.P. Vasconcelos, J. Rueff and P. Noguecra Ramos *J. High Res. Chromatogr. Chromatogr. Commun.*, 1988, **11**, 12.
281. A.G. Calder and A. Smith *Rapid Commun. Mass Spectrom.*, 1988, **2**, 14.
282. T. Hayashi, Y. Minatogawa, K. Kamada, M. Shimamura and H. Naruse *J. Chromatogr. Biomed. Appln.*, 1986, **380**, 239.
283. A.P.J.M. de Jong, R.M. Kok, C.A. Cramers, S.K. Wadman and E. Haan *Clinica Chimica Acta*, 1988, **171**, 49.
284. I.M. Kapetanovic, W.D. Yonekawa and H.J. Kupferberg *J. Chromatogr. Biomed. Appln.*, 1987, **414**, 265.
285. D.R. Cryer, T. Matsushima, J.B. Marsh, M. Yudkoff, P.M. Coates and J.A. Cortner *J. Lipid Res.*, 1986, **27**, 508.
286. T. Furuta, Y. Kasuya, H. Shibasaki and S. Baba *J. Chromatogr. Biomed. Appln.*, 1987, **413**, 1.

287. A.K. Singh and H. Ashraf *J. Chromatogr. Biomed. Appln.*, 1988, **425**, 245.
288. L. Siekmann *Biomed. Environ. Mass Spectrom.*, 1987, **14**, 683.
289. M.F. Lefevere, B.J. Verhaeghe, D.M. Declerck and A.P. DeLeenheer *Biomed. Environ. Mass Spectrom.*, 1988, **15**, 311.
290. J. Greter and C.-E. Jacobson *Clin. Chem.*, 1987, **33**, 473.
291. S.F. Yeh, K.-J. Hsiao, S.-H. Hung and K.-T. Chang *J. Chin. Chem. Soc.*, 1986, **33**, 251.
292. W. Koolstra, B.G. Wolthers, M. Hayer and H. Elzinga *Clinica. Chimica. Acta*, 1987, **170**, 227.
293. O.A. Mamer, N.S. Laschic and C.R. Scriver *Biomed. Environ. Mass Spectrom.*, 1986, **13**, 553.
294. O.A. Mamer *Biomed. Environ. Mass Spectrom.*, 1988, **15**, 57.
295. D.B. Kassel, M. Martin, W. Schall and C.C. Sweeley *Biomed. Environ. Mass Spectrom.*, 1986, **13**, 535.
296. H. Nishio, S. Kodama, S. Yokogama, T. Matsuo, T. Mio and K. Sumino *Clin. Chim. Acta*, 1986, **159**, 77.
297. A. Lochner, S. Weisner, A. Zlatkis and B.S. Middleditch *J. Chromatogr. Biomed. Appln.*, 1986, **378**, 267.
298. L. Siekmann, A. Schonfelder and A. Siekmann *Fres. Z. Anal. Chem.*, 1986, **324**, 280.
299. M. Beylot, C. Martin, B. Beaufrere, J.P. Riou and R. Mornex *J. Lipid Res.*, 1987, **28**, 414.
300. D.L. Hachey, M.R. Thomas, E.A. Emken, C. Garza, L. Brown-Booth, R.O. Adlof and P.D. Klein *J. Lipid Res.*, 1987, **28**, 1185.
301. L.B. Bjostad and W.L. Roelofs *J. Chem. Ecol.*, 1986, **12**, 431.
302. F. Swanto, S. Humfeld and H. Reinauer *Chromatographia*, 1986, **21**, 693.
303. P.E. Haroldsen, K.L. Clay and R.C. Murphy *J. Lipid Res.*, 1987, **28**, 42.
304. A. Sonesson, L. Larsson, G. Wasterdahl and G. Odham *J. Chromatogr. Biomed. Appln.*, 1987, **417**, 11.
305. T. Krishnamurphy, M.B. Wasserman and E.W. Sarver *Biomed. Environ. Mass Spectrom.*, 1986, **13**, 503.
306. T. Krishanmurphy, E.W. Sarver, S.L. Greene and B.B. Jarvis *J. Assoc. Off. Anal. Chem.*, 1987, **70**, 132.
307. R. Kostianienen and A. Rizzo *Analytica Chim. Acta*, 1988, **204**, 233.
308. D.W. Hewetson and C.J. Mirocha *J. Assoc. Off. Anal. Chem.*, 1987, **70**, 647.
309. B.P.-Y. Lau, D. Weber and B.D. Page *J. Chromatogr.*, 1987, **402**, 233.
310. T. Cairns, E.G. Siegmund, M.A. Luke and G.M. Doose *Anal. Chem.*, 1987, **59**, 2055.
311. W.C. Brumley, B.J. Canas, G.A. Perfetti, M.M. Mossoba, J.A. Sphon and P.E. Corneliussen *Anal. Chem.*, 1988, **60**, 975.
312. B.P.-Y. Lau and D. Weber *J. Agric. Food Chem.*, 1987, **35**, 412.
313. L. Castle, J. Gilbert, S.M. Jickells and J.W. Grimshaw *J. Chromatogr.*, 1988, **437**, 281.
314. J.R. Startin, I. Parker, M. Sharman and J. Gilbert *J. Chromatogr.*, 1987, **387**, 509.
315. K. Takatsuki, I. Ushizawa and T. Shoji *J. Chromatogr.*, 1987, **392**, 249.
316. C.H. Van Peteghem, M.F. Lefevere, G.M. van Haver and A.P. De Leanheer *J. Agric. Food Chem.*, 1987, **35**, 228.
317. J.W. Eichelberger and W.L. Budde *Biomed. Environ. Mass Spectrom.*, 1987, **14**, 357.
318. A.D. Sauter, J.J. Downs, J.D. Buchner, N.T. Ringo, D.L. Shaw and J.G. Dulak *Anal. Chem.*, 1986, **58**, 1665.
319. D.F. Gurka and R. Titus *Anal. Chem.*, 1988, **58**, 2189.
320. P.H. Kiang and R.L. Grob *J. Environ. Sci. Health*, 1986, **A21/1**, 15.
321. C.E. Rostad and W.E. Pereira *J. High Res. Chromatogr. Chromatogr. Commun.*, 1986, **9**, 329.
322. R.L. Dobson, A.P. D'Silva, S.J. Weeks and V.A. Fassel *Anal. Chem.*, 1986, **58**, 2129.
323. S.A. Wise, B.A. Benner, G.D. Byrd, S.N. Chester, R.E. Rebbert and M.M. Schantz *Anal. Chem.*, 1988, **60**, 887.
324. W. Rotard and W. Mailahn *Anal. Chem.*, 1987, **59**, 65.
325. W.J. Simonsick and R.A. Hites *Anal. Chem.*, 1986, **58**, 2114.
326. L.J. Sears, J.A. Campbell and E.P. Grimsrud *Biomed. Environ. Mass Spectrom.*, 1987, **14**, 401.
327. H.G. Nowicki *J. High Res. Chromatogr. Chromatogr. Commun.*, 1986, **9**, 472.
328. E. Atlas *Nature*, 1988, **331**, 426.
329. A.R. Driedger, D.C. Thornton, M. Lalevic and A.R. Bandy *Anal. Chem.*, 1987, **59**, 1196.
330. E.E. Lewin, R.L. Taggart, M. Lalevic and A.R. Bandy *Anal. Chem.*, 1987, **59**, 1296.
331. R.M. Smith, P.W. O'Keefe, D.R. Hilker and K.M. Aldous *Anal. Chem.*, 1986, **58**, 2414.
332. C.W. Bayer and M.S. Black *Biomed. Environ. Mass Spectrom.*, 1987, **14**, 363.
333. L.D. Gruenke, J.C. Craig, R.C. Wester and H.I. Maibach *J. Anal. Toxicol.*, 1986, **10**, 225.
334. A. Yasuhara *Agric. Biol. Chem.*, 1987, **51**, 2259.
335. T. Tomingas, W. Monch and U. Matthiesen *Chromatographia*, 1986, **22**, 191.
336. W.A. Korfmacher, L.G. Rushing, J. Arey, B. Zielinska and J.N. Pitts *J. High Res. Chromatogr. Chromatogr. Commun.*, 1987, **10**, 641.
337. W.A. Korfmacher and I.G. Rushing *J. High Res. Chromatogr. Chromatogr. Commun.*, 1986, **9**, 293.

338. L.R. Hilpert *Biomed. Environ. Mass Spectrom.*, 1987, **14**, 383.
339. M.A. Mazurek, B.R.T. Simoneit, G.R. Cass and H.A. Gray *Int. J. Environ. Anal. Chem.*, 1987, **29**, 119.
340. M.-A. Sicre, J.-C. Marty, A. Saliot, X. Aparicio, J. Grimalt and J. Albaiges *Int. J. Environ. Anal. Chem.*, 1987, **29**, 73.
341. D.R. Scott, W.J. Dunn and S.L. Emery *Environ. Sci. Technol.*, 1987, **21**, 891.
342. N.B. Vogt, F. Brakstad, K. Thrane, S. Nordenson, J. Krane, E. Aamot, K. Kolset, K. Esbensen and E. Steinnes *Environ. Sci. Technol.*, 1987, **21**, 35.
343. F.M. Benoit and G.L. LeBel *Bull. Environ. Contam. Toxicol.*, 1986, **37**, 686.
344. J. Rivera, F. Ventura, J. Caixach, M. De. Torres and A. Figueras *Int. J. Environ. Anal. Chem.*, 1987, **29**, 15.
345. R.B. Spies, B.D. Andnesen and D.W. Rice *Nature*, 1987, **327**, 697.
346. M.D. Muller *Anal. Chem.*, 1987, **59**, 617.
347. C.C. Gilmour, J.H. Tuttle and J.C. Means *Anal. Chem.*, 1986, **58**, 1848.
348. P.S. Epstein, T. Mauer, M. Wagner, S. Chase and B. Giles *Anal. Chem.*, 1987, **59**, 1987.
349. R. Otson and C. Chan *Int. J. Environ. Anal. Chem.*, 1987, 30, 275.
350. H. Norin, A. Christakopoulos, L. Rondahl, A. Hagman and S. Jacobsson *Biomed. Environ. Mass Spectrom.*, 1987, **14**, 117; A. Christakopoulos, B. Hamasur, H. Norin and I. Nordgren *Biomed. Environ. Mass Spectrom.*, 1988, **15**, 67.
351. J.W. de Leeuw, E.W.B. de Leer, J.S. Sinninghe Damste and P.J.W. Schuyl *Anal. Chem.*, 1986, **58**, 1852.
352. P.G. Sim, R.K. Boyd, R.M. Gershey, R. Guevemont, W.D. Jamieson, M.A. Quilliam and R.J. Gergely *Biomed. Environ. Mass Spectrom.*, 1987, **14**, 375.
353. E. Stephanou, M. Reinhard and H.A. Ball *Biomed. Environ. Mass Spectrom.*, 1988, **15**, 275.
354. M. Ahel, T. Conrad and W. Giger *Environ. Sci. Technol.*, 1987, **21**, 697.
355. R.P. Eganhouse *Int. J. Environ. Anal. Chem.*, 1986, **26**, 241.
356. J.M. Bayona, J. Albaiges, A.M. Solanas, P. Pares, P. Garrigues and M. Ewald *Int. J. Environ. Anal. Chem.*, 1986, **23**, 289.
357. H.-R. Buser and M.D. Muller *Environ. Sci. Technol.*, 1986, **20**, 730.
358. R.B. van Breemen, C.C. Fenselau, R.J. Cotter, A.J. Curtis and G. Connolly *Biomed. Environ. Mass Spectrom.*, 1987, **14**, 97.
359. T. Tsukioka and T. Murakani *J. Chromatogr.*, 1987, **396**, 319.
360. D.B. Kassel, K.A. Kayanich, J.T. Watson and J. Allison *Anal. Chem.*, 1988, **60**, 911.
361. T.M. Trainor and P. Vuros *Anal. Chem.*, 1987, **59**, 601.
362. E.A. Stemmler and R.A. Hites *Biomed. Environ. Mass Spectrom.*, 1987, **14**, 417.
363. Y. Tondeur, W.N. Niederhut, J.E. Campana and S.R. Missler *Biomed. Environ. Mass Spectrom.*, 1987, **14**, 449.
364. D.H. Schellenberg, B.A. Bobbie, E.J. Reiner and V.Y. Taguchi *Rapid Commun. Mass Spectrom.*, 1987, **1**, 111.
365. D.G. Patterson, L. Hampton, C.R. Lapeza, W.T. Belser, V. Green, L. Alexander and L.L. Needham *Anal. Chem.*, 1987, **59**, 2000.
366. E.R. Barnhart, D.G. Patterson, D.L. Ashley, V. Maggio, C.C. Alley, L. Alexander and J.A.H. MacBride *Anal. Chem.*, 1987, **59**, 2248.
367. H.-R. Buser *Anal. Chem.*, 1986, **58**, 2913.
368. H.-R. Buser *Chemosphere*, 1987, **16**, 713.
369. J.R. Donnelly, W.D. Munslow, T.L. Vonnahme, N.J. Nunn, C.M. Hedin, G.W. Sovocool and R.K. Mitchum *Biomed. Environ. Mass Spectrom.*, 1987, **14**, 465.
370. M. Swere and K. Ballschmitter *Fres. Z. Anal. Chem.*, 1987, **327**, 51.
371. B.K. Afghan, J. Carron, P.D. Gouldan, J. Lawrence, D. Leger, F. Onuska, J. Sherry and R. Wilkinson *Can. J. Chem.*, 1987, **65**, 1086.
372. J.D. McCurry and C.D. Daves *Anal. Chem.*, 1986, **58**, 2785.
373. S. Tanabe, N. Kannan, T. Wakimoto and R. Tatsukawa *Int. J. Environ. Anal. Chem.*, 1987, **29**, 199.
374. A. Dekok, E. Rijmerse, M. Amoureus, R.B. Geerdink and U.A. Th. Brinkiman *Int. J. Environ. Anal. Chem.*, 1987, **29**, 227.
375. K. Haraguchi, H. Kuroki and Y. Masuda *J. Chromatogr.*, 1986, **361**, 239.
376. P. Furst, C. Kruger, H.-A. Meemken and W. Groebel *J. Chromatogr.*, 1987, **405**, 311.
377. W.A. Heidmann *Chromatographia*, 1986, **22**, 363.
378. R.J. Norstrom, M. Simon and M.J. Milvihill *Int. J. Environ. Anal. Chem.*, 1986, **23**, 267.
379. G.L. Lebel and D.T. Williams *J. Assoc. Off. Anal. Chem.*, 1986, **69**, 451.
380. M.P. Seymour, T.M. Jefferies, A.J. Floyd and L.J. Noturianni *Analyst*, 1987, **112**, 427.
381. K. Noren and J. Sjovall *J. Chromatography Biomed. Appln.*, 1987, **422**, 103.

382. D.S. Waddell, P.G. Sim and R.K. Boyd *Rapid Commun. Mass Spectrom.*, 1987, **1**, 106.
383. W.A. Heidmann *Chromatographia*, 1986, **22**, 363.
384. S.V. Hummel and R.A. Yost *Organic Mass Spectrom.*, 1986, **21**, 785.
385. T. Tamiri and S. Zitrin *Biomed. Environ. Mass Spectrom.*, 1987, **14**, 39.
386. D.L. Swackhamer, M.J. Charles and R.A. Hites *Anal. Chem.*, 1987, **59**, 913.
387. W.A. Karfmacher, L.G. Rushing, P.H. Siitonen, C.J. Branscomb and C.L. Holder *J. High Res. Chromatogr. Chromatogr. Commun.*, 1987, **10**, 332.
388. R.O. Lidgard, A.M. Duffield and R.J. Wells *Biomed. Environ. Mass Spectrom.*, 1986, **13**, 677.
389. P. Begley and B.E. Foulger *J. Chromatogr.*, 1988, **438**, 45.
390. A.K. Singh, D.W. Hewetson, K.C. Jordon and M. Ashraf *J. Chromatogr.*, 1986, **369**, 83.
391. V. Lopez-Avila, P. Hirata, S. Kraska and J.H. Taylor *J. Agric. Food Chem.*, 1986, **34**, 530.
392. Y.Y. Wingfield, N.P. Gurprasad, M. Lanouette and S. Ripley *J. Assoc. Off. Anal. Chem.*, 1987, **70**, 792.
393. A.L. Ashford-Stevens, J.W. Eichelberger and W.L. Budde *Environ. Sci. Technol.*, 1988, **22**, 304.
394. E. Bailey, P.B. Farmer, I. Bird, J.H. Lamb and J.A. Peal *Anal. Biochem.*, 1986, **157**, 241.
395. P.B. Farmer, E. Bailey, S.M. Gorf, M. Tornqvist, S. Osterman-Golkar, A. Kautiainer and D.P. Lewis-Enright *Carcinogenesis*, 1986, **7**, 637.
396. L.D. Gruenke, J.C. Craig, R.C. Wester and H.I. Maibach *J. Anal. Toxicol.*, 1986, **10**, 225.
397. E. Davoli, L. Cappellini, L. Airoldi and R. Fanelli *J. Chromatogr. Sci.*, 1986, **24**, 3.
398. A.T. Chatham, R.D. Brown and D.A. Mills *Bull. Environ. Contam. Toxicol.*, 1987, **38**, 789.
399. G. Skarping, T. Bellander and L. Mathiasson *J. Chromatogr.*, 1986, **370**, 245.
400. H. Heusler, E. Richter, J. Eping and M. Schmidt *J. High Res. Chromatogr. Chromatogr. Commun.*, 1986, **9**, 548.
401. J. Cocker, L.C. Brown, H.K. Wilson and K. Rollins *J. Anal. Toxicol.*, 1988, **12**, 9.
402. C. Rosenberg and H. Savolainen *J. Chromatogr.*, 1986, **358**, 385.
403. P. Begley, B.E. Foulger, P.D. Jeffery, R.M. Black and R.W. Read *J. Chromatogr.*, 1986, **367**, 87.
404. R.M. Black, R.J. Clarke and R.W. Read *J. Chromatogr.*, 1986, **367**, 103.
405. P.A. D'Agostino, L.R. Provost and D.R. Drover *J. Chromatogr.*, 1986, **367**, 77.
406. P.A. D'Agostino, L.R. Provost and J. Visentini *J. Chromatogr.*, 1987, **402**, 221.
407. E.R.J. Wils, A.G. Hulst and J. van Laar *J. Anal. Toxicol.*, 1988, **12**, 15.
408. M.L. Abercrombie and J.S. Jewell *J. Anal. Toxicol.*, 1986, **10**, 178.
409. H.H. McCurdy, L.J. Lewellen, L.S. Callahan and P.S. Childs *J. Anal. Toxicol.*, 1986, **10**, 175.
410. M.A. Elsohly, D.I. Stanford and T.L. Little *J. Anal. Toxicol.*, 1988, **12**, 54.
411. R.W. Taylor, N.C. Jain, M.P. George *J. Anal. Toxicol.*, 1987, **11**, 233.
412. P. Francom, D. Andrenyak, H.-K. Lim, R.R. Bridges, R.L. Foltz and R.T. Jones *J. Anal. Toxicol.*, 1988, **12**, 1.
413. I.A. Low, R.H. Liu, M.G. Legendre, E.G. Piotrowski *Biomed. Environ. Mass Spectrom.*, 1986, **13**, 531.
414. V.N. Garg, B.D. Bhatt, V.K. Kaushik and K.R. Murthy *J. Chromatogr. Sci.*, 1987, **25**, 237.
415. C.Y. Ma, E.H. McBay, C.-H. Ho and W.H. Griest *Int. J. Environ. Anal. Chem.*, 1987, **30**, 37.
416. R. Orlando and B. Munson *Anal. Chem.*, 1986, **58**, 2788.
417. J.N. Robson and S.J. Rowland *Nature*, 1986, **324**, 561.
418. J.S.S. Damste, A.C.K.-van Dalen, J.W. de Leeuw, P.A. Schenck, S. Guoying and S.C. Brassell *Geochim. Cosmochim. Acta.*, 1987, **51**, 2393.
419. J.C. Schmid, J. Connan and P. Albrecht *Nature*, 1987, **329**, 54.
420. J.S. Sinninghe, J.W. de Leeuw, A.C.K.-van Dalen, M.A. de Zeeuw, F. de Lange, W.I.C. Rijpstra and P.A. Schenck *Geochim. Cosmochim. Acta.*, 1987, **51**, 2369.
421. J.S. Sinninghe, A.C.K.-van Dalen, J.W. de Leeuw and P.A. Schenck *J. Chromatogr.*, 1988, **435**, 435.
422. J.P. Gill, R.P. Evershed and G. Eglinton *J. Chromatogr.*, 1986, **369**, 281.
423. W. Blum, W.J. Richter and G. Eglinton *J. High Res. Chromatogr. Chromatogr. Commun.*, 1988, **11**, 148.
424. N. Telnaes and B. Dahl *Organic Geochem.*, 1986, **10**, 425.
425. J.E. Zumberge *Geochim. Cosmochim. Acta.*, 1987, **51**, 1625.
426. M.-A. Sicne, J.-L. Paillasseur, J.-C. Marty and A. Saliot *Org. Geochem.*, 1988, **12**, 281.
427. J.M. Halket and H.-R. Schutten *J. High Res. Chromatogr. Chromatogr. Commun.*, 1986, **9**, 596.
428. R.P. Philip and T.D. Gilbert *J. Anal. Appl. Pyrol.*, 1987, **11**, 93.
429. G. van Graas *Org. Geochem.*, 1986, **10**, 1127.
430. M. Nip, W. Genuit, J.J. Boon, J.W. De Leeuw, P.A. Schenck, M. Blazso and T. Szekely *J. Anal. Appl. Pyrol.*, 1987, **11**, 125.
431. A. Mascherpa and A. Casallni *J. High Res. Chromatogr. Chromatogr. Commun.*, 1988, **11**, 296.
432. H.-R. Schulten, G. Abbt-Braun and F.H. Frimmel *Environ. Sci. Technol.*, 1987, **21**, 349.

433. C. Saiz-Jimenez and J.W. de Leeuw  *J. Anal. Appl. Pyrol.*, 1987, **11**, 367.
434. D.G. van Smeerdijk and J.J. Boon  *J. Anal. Appl. Pyrol.*, 1987, **11**, 377.
435. M. Nip, E.W. Tegelaar, H. Brinkhuis, J.W. de Leeuw, P.A. Schenck and P.J. Holloway  *Org. Geochem.*, 1986, **10**, 769.
436. C. Saiz-Jimenez, J.J. Boon, J.I. Hedges, J.K.C. Hessels and J.W. de Leeuw  *J. Anal. Appl. Pyrol.*, 1987, **11**, 437
437. O. Faix, D. Meier and I. Grobe  *J. Anal. Appl. Pyrol.*, 1987, **11**, 403.
438. A.D. Pouwels, A. Tom, G.B. Eijkel and J.J. Boon  *J. Anal. Appl. Pyrol.*, 1987, **11**, 417.
439. M. Blazso, B. Zelei, B.R. Gandhe and D. Sek  *J. Anal. Appl. Pyrol.*, 1987, **11**, 233.
440. G.H. Irzel, C.T. Vijayakumar, J.K. Fink and K. Lederer  *J. Anal. Appl. Pyrol.*, 1987, **11**, 277.
441. M. Blazso and E. Jakab  *J. Anal. Appl. Pyrol.*, 1987, **11**, 245.
442. L. Prokai  *J. Anal. Appl. Pyrol.*, 1987, **12**, 265.
443. A.M. Casanovas and X. Rovira  *J. Anal. Appl. Pyrol.*, 1987, **11**, 227.
444. J.J. Boon and J.W. de Leeuw  *J. Anal. Appl. Pyrol.*, 1987, **11**, 313.
445. T.O. Munson and D.D. Fetterolf  *J. Anal. Appl. Pyrol.*, 1987, **11**, 15.
446. R.J. Helleur  *J. Anal. Appl. Pyrol.*, 1987, **11**, 297.
447. C.S. Smith, S.L. Morgan, C.D. Parks, A. Fox and D.G. Pritchard  *Anal. Chem.*, 1987, **59**, 1410.
448. H.-R. Schulten and J.M. Halket  *Organic Mass Spectrom.*, 1986, **21**, 613.
449. G. Andisio, F. Severini, A.M. Maccioni and P. Traldi  *Biomed. Environ. Mass Spectrom.*, 1986, **13**, 519.
450. G. Andisio, A. Rossini, F. Severini and R. Gallo  *J. Anal. Appl. Pyrol.*, 1987, **11**, 263.
451. H. Thoma and O. Hutzinger  *Chemosphere*, 1987, **16**, 1353.

# 7
# Analysis of Mixtures by Mass Spectrometry Part II: Techniques Other than Gas Chromatography/Mass Spectrometry

BY D. A. CATLOW AND M. E. ROSE

## 1 Introduction

This review is divided into sections covering combinations of mass spectrometry with capillary zone electrophoresis (CZE/MS) and thin layer chromatography (TLC/MS), supercritical fluid chromatography (SFC/MS), high-performance liquid chromatography (LC/MS), and mass spectrometry itself (MS/MS). We do not claim that the review is comprehensive. Indeed, some of the above methods have become so routine that comprehensive coverage would include some mundane uses. In other words, we continue the approach of trying to highlight innovations in technique and application at the expense of routine applications.

## 2 Combinations With Capillary Zone Electrophoresis And Thin Layer Chromatography

This review period (mid-1986 to mid-1988) has seen the introduction of a new combined method, and the associated inevitable abbreviation: CZE/MS.[2-4] This particular newcomer will be very welcome if it fulfils its originators' anticipations[4] of providing the ability to "separate with high efficiency, detect and hopefully identify biologically important compounds with attomole range detection limits". The enormous potential of CZE/MS is a result of the separate yet compatible strengths of CZE,[5] the electrospray interface[2,3,6] and mass spectrometry. The CZE method is based on the migration of charged species in capillary columns (20-200 μm i.d.) under the influence of an applied electric field (10-40 kV). Supporting media like gels are not necessary and the narrowness of the tubes aids heat dissipation, both factors which reduce zone broadening.[5] Selectivity in electrophoretic mobility is gained by controlling pH or buffer concentration. Impressive results can be obtained with instrumentation that is simple and cheap. As one set of authors[5] put it, "Even a university can afford the investments".

CZE is an elution method that resembles chromatography and generates large numbers of theoretical plates ($10^5 - 10^7$).[3] Coupling with mass spectrometry was inevitable to provide CZE with a sufficiently selective, universal and sensitive detector (typical injection volumes are 1-100 nl). The two techniques are compatible because the ionic strength of the buffer and the flow rate needed for CZE (ca. 1 μl/min) match those needed for electrospray ionization. For CZE an electrical contact with the eluting liquid is needed, and this contact can form the basis for an electrically induced nebulization process that produces ions at atmospheric pressure. The electrospray of highly charged droplets is evaporated against a flow of dry nitrogen. The droplets break up until small enough for field-assisted ion desorption (i.e. ion evaporation) to occur from the droplet surface. Whilst the precise mechanism is still unclear, the 3-6 kV potential gradient produces positive or negative ions by a mild method that does not require heating. A three-stage vacuum system permits coupling to a quadrupole mass spectrometer.[2,3,6] The spectra obtained tend to be dominated by molecular or quasi-molecular ions, with cluster ions and multiply charged species being significant under certain conditions. The spectra are dependent on the solvent system and electrospray voltage. Sensitivity also seems to be critically dependent on the same variables.[6]

Presently reported work is restricted to analytes of quaternary ammonium salts and oligopeptides (mainly), and positive-ion mass spectra. The ionic and partially ionized analytes in aqueous solution ($H_2O/CH_3OH/salt$)[2,6] generally afforded detection limits in the fmol range and up to 620,000 theoretical plates. At present, spectral sensitivity and reproducibility seem best when desolvation is maximized (high nitrogen flow), but for the future a greater understanding of such effects, i.e. the electrospray mechanism and manipulation of solvent/buffer system, is required. Clearly, for polar analytes CZE/MS is a technique to watch in the next few years!

A patent[7] describes an interface probe which allows large and polar molecules separated electrophoretically to be analysed in real-time by a conventional mass spectrometer. Components separated by gel electrophoresis may be transferred by a blotting procedure to a nitrocellulose support and then examined by secondary-ion mass spectrometry (SIMS).[8] Applications included

organic dyes, steroid sulphates and small peptides. The same
group[8-11] has described direct analysis by SIMS of compounds
separated by TLC, e.g. phenothiazine drugs,[10] small peptides,
phosphonium salts and transition metal acetylacetonates. The
ultimate spatial resolution for molecular mapping was reported[8] to
be 1 μm, appropriate for imaging of overlapping components.
Microgramme sensitivity is routine and ng sensitivity appears to be
attainable. The spectra obtained were dependent on the phase-
transfer matrix (used to provide a steady supply of sample to the
surface from which ions are sputtered[8,9]) but independent of the
underlying support.[10] In a different device, mixtures of amino
acids[12] and dipeptides[13] as their dansylated methyl esters have been
examined at the pmol level by direct TLC/MS.

Thermaly labile, polar and non-volatile organic compounds
(phospholipids, antibiotics and oligopeptides) have been subjected
to TLC/SIMS by moving developed TLC plates automatically and
stepwise in the ion source under bombardment by a focussed $Xe^+$
beam.[14] Sample sizes of about 1 μg applied to the plate were
appropriate for the method. Conjugates of 4-nitrophenol and 4-
hydroxyantipyrene have also been determined by TLC/SIMS, this time
by cutting out the spots and attaching them to a SIMS probe.[15]
Quantitative analysis of the zwitterionic acetyl- and
propionylcarnitines was possible using a stable-isotope dilution
approach and TLC/SIMS.[16] Finally in this section, steroid hormones
have been separated and determined by TLC/SIMS using a matrix of
glycerol,[17] and phospholipids have been examined by both TLC/SIMS
(triethanolamine matrix)[17] and by TLC/FAB.[18]

### 3 Supercritical Fluid Chromatography/Mass Spectrometry

3.1    Overview. - Interest in SFC/MS has remained high. The
technique is carving itself a niche between GC/MS and LC/MS but
appears to be replacing neither.[4,19] This is because the three
methods are complementary, although there are areas where their
applications overlap.[19] It is also noticeable that in the last two
years SFC/MS, in its complementary packed-column and capillary-
column guises, has become a tool for solving real problems. In a
recent SFC workshop (January, 1988) 47% of attendees were from
industrial or governmental laboratories whilst a minority (31%) were
from academia, the remainder being instrument manufacturers.[20]

The two years have also been blessed with a large number of
articles and reviews on SFC,[21-28] and the publication of a timely

book.[29] There have been several reviews of SFC/MS,[4,23, 30-36] the substantial one by Smith's group being particularly recommended reading.[33] The analysis of complex mixtures of hydrocarbons[37] and of pharmaceutical products[38] by SFC and SFC/MS has also been reviewed. Articles of interest on SFC alone are very numerous indeed but include a review of gradient elution,[39] a discussion of packed versus capillary columns,[40, 41] the elution of polar and ionic compounds using reversed micelles in supercritical fluid mobile phases,[42] and aerosol formation at the point of decompression.[33,43] The routine coupling of SFC with supercritical fluid extraction of samples (SFE/SFC), where the need for solvent elimination prior to chromatography is obviated, will be of considerable advantage.[20,44]

3.2    Interfaces. - The high flow rate SFC/MS interface developed by Smith's group[45] for capillary column work was described (prematurely!) in Volume 9 of this series.[1] Work on this interface has continued[33,46] and several other approaches and devices have been explored. At the affordable end of mass spectrometry, SFC/MS has been effected with an ion trap detector (ITD).[47] The ITD's "big brother" has also been utilized, making feasible on-line SFC/FTMS.[48,49] In another interesting combination, ion mobility spectrometry has been used as an SFC detector.[50] Interfaces have been varied, including deposition onto a moving belt for packed-column SFC/MS,[51] the use of the thermospray approach for SFC/MS,[51-53] simply modified GC/MS-type interfaces for SFC/EIMS[54] and SFC/CIMS,[55] frit restrictors as decompression devices,[51,56] direct coupling via various restrictors to CI ion sources,[55-60] and a pulsed nozzle in conjunction with laser multiphoton ionization and a TOF mass spectrometer.[61-63] No one method has yet emerged as always superior; for example, the optimum design is dependent on the preferred type of column (long capillary columns for high-resolution chromatography of complex mixtures, or packed columns for speed, higher loading and versatility). Convenient switching between SFC/MS and GC/MS[54,55] or even simultaneous coupling of capillary column SFC and GC to a mass spectrometer[64] will be an important consideration for some analysts.

For a thorough discussion of Smith's direct fluid injection interfaces and the difficulties of transporting compounds through restrictors, the reader is referred elsewhere.[33,45,46] The conclusion was drawn[33] that the ideal restrictor, allowing pulse-free and complete transport of even involatile solutes without pyrolysis and without plugging, has not yet been developed.[33] Research activity into the expansion process and restrictors, though, remains high. Interface designs that allow routine analysis of volatile and medium polarity compounds by capillary-column SFC/MS are said[46] to be available. The direct fluid injection interface has been used with a magnetic sector mass spectrometer, allowing high resolution and high masses to be accessed,[46] and with a triple quadrupole instrument for on-line SFE/MS/MS.[65]

The coupling of SFC and FTMS has the potential to be a versatile and powerful technique, but the pressure in the analyser needs to be kept low to achieve the best FTMS performance. In both reported SFC/FTMS systems,[48,49] the gas load to the analyser was minimized by employing capillary columns and the differentially pumped dual-cell FTMS design. Using $CO_2$ as mobile phase and a transfer line (1 m x 50 µm i.d.) which is maintained at 100°C and terminates in a restrictor (tapered to 1 µm) at the entrance of the source cell, pressures of about $10^{-5}$ and $10^{-7}$ torr were achieved in the source and analyser cells, respectively.[48] These values are high for FTMS but similar to those tolerated in GC/FTMS. Under such conditions, caffeine (25 ng on-column) was successfully examined in both EI and CI modes and accurate mass data recorded (8300 fwhm). Precise temperature control of the transfer line and heating of the restrictor to at least 250°C were said to be required to aid the transport of involatile molecules.[48] In the other interface[49] supercritical $CO_2$ was introduced via a 2 m x 156 µm i.d. fused-silica capillary (45°C) with a 10 cm x 5 µm restrictor (over 250°C) into the source. Source and analyser pressures of about $10^{-4}$ and $10^{-6}$ torr were reported.[49] Under these conditions EI-like spectra were obtained by a 13-eV electron beam and continuous ejection of $CO_2^{+\cdot}$ ions. Detection limits at unit resolution were thought to fall in the mid-pg range. Higher resolution (pressure-limited to about 10,000) was achieved at the expense of sensitivity.[49] Barbiturates and pesticides were amenable to SFC/FTMS at the ng level. To realize the full potential of this new method, the problem of high gas loads must be addressed. A reduction of the analyser pressure by one to two orders of magnitude would make a very significant improvement.[49] Also, the transfer lines are rather long.

A capillary column for SFC has been interfaced to a UV detector and then, via a capillary restrictor, to an ion mobility spectrometer.[50] Carbon dioxide was used both as mobile phase and as drift gas. Even with volatile solutes, long residence times in the spectrometer were a problem. With improvements, though, the method may form a useful tunable, selective detector for SFC.

An improvement in the transport of low volatility components in capillary column SFC/MS has been noted when a 5 μm restrictor was replaced with a 50 μm restrictor with integral frit that can be heated up to 600°C.[32,51] A frit restrictor is also favoured by others for coupling capillary columns to a magnetic sector instrument.[56] A hot direct insertion probe placed opposite to the SFC frit restrictor heats the column effluent.[56] Such heating appears to be a key factor not only for preventing precipitation of involatile analytes but also for decomposing $CO_2$ solvent clusters, irrespective of restrictor design.[51,54,56] Whilst solvent-moderated CI mass spectra can be obtained readily, true EI spectra are difficult but not impossible[48;54] to achieve in directly coupled capillary column SFC/MS. In many cases EI-like spectra are probably a result of charge exchange with carbon dioxide ions.[49,51,56] Sensitivity of SFC/EIMS remains poorer than that observed in SFC/CIMS.[54] A previously described[1] SFC/CIMS system, based on a tapered restrictor, has been improved in scope by coupling to a quadrupole mass filter with a mass range up to 3000 daltons and by using splitless injection with a 50 cm x 50 μm i.d. retention gap.[60]

As an example of the use of a thermospray source as a basis for a SFC/MS interface,[51-53] half of the effluent from a packed-column SFC was introduced into a slightly modified, filament-on thermospray device.[52] The use of $CO_2$ as mobile phase allowed EI-like spectra to be recorded, but better sensitivity was achieved by SFC/CIMS using $CO_2/CH_3OH$ as the mobile phase. Initial work indicates that ng quantities of analytes are sufficient.[52]

In the molecular beam approach [61-63] the supercritical fluid is injected into a TOF analyser by a pulsed orifice (150 μm), producing vibrationally cool molecules that are ionized selectively by the resonant two-photon laser method. Thus, identifications are based on mass analysis in conjunction with laser spectroscopy.

A study of the partition of supercritical pentane into SE-30 and SE-54 stationary phases includes a design for modifying a GC/MS system for SFC operation.[66] A report on interactions between supercritical solvents (including modifiers) and solutes, among other studies, has implications for SFC and SFC/MS.[67] The ion chemistry of polycyclic aromatic hydrocarbons (PAHs) under NICI conditions with carbon dioxide is also pertinent to SFC/MS.[68]

3.3 Applications. - The solvating power of the supercritical mobile phase and the ability to transport non-volatile compounds through an interface determine the limits of application of SFC/MS. There is still doubt regarding the suitability of SFC/MS for handling really polar, hydrophilic and non-volatile compounds.[24,33] Of course, in some instances derivatization of the analyte to a less polar species can be effective, as with aromatic acids,[63] but this is not a universal solution. Application of reverse micellar mobile phases may extend the scope by allowing large, ionic, hydrophilic species to be dissolved and subjected to SFC,[42] but compatibility of such phases with SFC/MS has yet to be reported.[33] Time will tell whether CZE/MS is a better solution for examining hydrophilic materials.

Analytical applications of capillary column SFC/MS have been described by others.[69,70] Areas of application and classes of compound studied have been: characterization of polymeric materials;[60,71,72] investigations of ion/molecule reactions;[73] organophosphorus insecticides[58] and other pesticides;[49,51,52,55] large alkanes[51] and waxes;[55,74] PAHs;[49,55-57] alkaloids;[52] carbohydrates;[53,60] lipids;[53,57] antibiotics;[59] drugs;[38,49] mycotoxins;[65] vitamins;[75] non-ionic surfactants;[76] and ecdysteroids.[77]

When subjected to capillary-column SFC/CIMS, various waxes (esters, glycerides, acids, alcohols and alkanes with $M_r$ = 300-1000) showed good peak shapes without derivatization. Split injection was used.[74] Smith's high flow rate interface[45] and pressure programming have been utilized with a packed microbore HPLC column for the characterization of organophosphorus insecticides.[58] The greater loading permitted by a packed column ensured low detection limits (p.p.b. range). Low concentrations of a polar modifier (propan-2-ol) served to cover active sites on the column and as the reactant gas for acquiring CI spectra. Alternatively ammonia can be introduced independently into the source for ammonia CI.[58] The same SFC/MS interface[45] has been used for the determination of high $M_r$, biologically active compounds like ionic polyether antibiotics and cyclosporin A.[59] The latter was successfully handled with microbore or capillary columns. The polar ionophores tailed significantly on a capillary column. The self-spouting interface described before[1] has been used in conjunction with pressure-programmed packed columns for analysing water- and fat-soluble vitamins.[75] Both types could be separated under the same SFC conditions. Finally here, trichothecene mycotoxins have been determined by direct supercritical fluid extraction of wheat and tandem mass spectrometry with a triple quadrupole (SFE/CIMS/MS).[65]

## 4 High-performance Liquid Chromatography/Mass Spectrometry

4.1 Overview. - The period of this review has seen LC/MS emerge as a routine analytical tool in many laboratories, both academic and industrial, due in no small part to the continued development and ready availability of the thermospray (TSP) type of interface. Prior to the introduction of the thermospray technique LC/MS was very much the province of the specialist, who used either home-made devices or one of the few commercial interfaces then available. Although much valuable work was done at that time, the technique was not widespread. The popularity of TSP and the contribution it has made to the increased usage of LC/MS can be seen by examining the numbers of papers produced compared to, for example, GC/MS, where the number of papers has remained reasonably constant. During the period of the previous report[78] LC/MS papers represented about 3% of the number of GC/MS papers and of these approximately 25% were TSP based. For the period of this report the number of LC/MS papers has increased to over 10% of that for GC/MS and half of those LC/MS publications were based on TSP. (These figures are based on a computer search of the Mass Spectrometry Bulletin and Chemical Abstracts data bases). Most of the remaining papers are accounted for by direct liquid introduction (DLI) techniques and moving belt transport devices, with smaller numbers from newer techniques which are beginning to make a serious impact, such as continuous flow (or dynamic) FAB and the monodisperse aerosol generation (MAGIC) technique, and from the atmospheric pressure ionization (API) interface. All these approaches to LC/MS interfacing are considered in the relevant sections below.

Several reviews and review-type articles have been published during this period[79-86] along with specialised reviews on TSP,[83,87] continuous flow FAB,[81,88,89] DLI[90-93] and the moving belt interface.[83,94] The latest Analytical Chemistry review of mass spectrometry[95] also contains many useful references.

4.2 Thermospray LC/MS. - Recent advances in this technique have concentrated on improving the performance of the hardware and hence the range of applications, and increasing our understanding of the processes involved. The incorporation of various electrodes in a TSP ion source in order to analyse a wider range of compounds or to fragment selectively the ions produced has formed an important area of research during this two-year period. All these developments have been extensively documented.[96-111]

One publication[97] considered the various ways in which ions could be caused to fragment (vaporizer/source temperature variation, CID with a repeller electrode, use of a filament or discharge ionization, MS/MS techniques) and gave examples of each. In a further paper[98] the advantages of true MS/MS techniques on a triple quadrupole instrument over discharge-induced fragmentation with a single quadrupole were discussed. Using CID-MS-CID-MS with discharge ionization on a triple quadrupole instrument was shown to provide structural information not otherwise obtainable. Experiments have also been performed[99] which, although using only solvent ions, have increased our understanding of the processes occurring when retarding electrodes are used. The optimization of vaporizer temperature and repeller potential for five test compounds showed[100] that optimum conditions were about the same for each of the test compounds and that in these cases the solvent-buffer ions could be used to optimize vaporizer temperature but not repeller potential.

The factors affecting high-mass performance and ion current stability in a TSP ion source have recently been considered in some detail.[96] The use of repeller electrodes and various design changes, including improving the gas-tightness of the source, were shown to give a significant improvement in performance, particularly at high mass. It was also considered by these authors that the ability to optimize the vaporizer orifice for different analyses was of major importance, a factor which has since (Bordeaux Conference, Sept. 1988) become incorporated in a commercial device.

Improvements in sensitivity were claimed on a magnetic sector instrument by the incorporation of a drift region after the ion sampling orifice which reduced the unintentional collisional activation of ions and the corresponding loss of abundance of the protonated molecule for the peptide Gramicidin-S.[101] However, this approach does not seem to have been universally adopted.

7: *Analysis of Mixtures by Mass Spectrometry; Part II* 231

Solvent selection for thermospray has been considered[102] in some detail. The ability of aqueous ammonium acetate solutions to produce direct TSP ionization from a variety of added organic solvent modifiers was investigated and an apparently strong relationship between solvent ionic activity coefficient and total solvent ion current was demonstrated.

Calibration of TSP data systems has concerned several authors, with polyethylene glycols (PEGs) being popular calibrants.[103] The improved performance of the latest generation of interfaces and sources has been shown to give much improved high-mass performance from PEG1000,[96] enabling a higher mass range to be used on suitable instruments. The same paper also recommended the use of sodium acetate for calibration, with cluster ions from this compound being detected to over 1600 daltons. It was claimed that this calibrant avoided the troublesome memory effects often found with PEGs.

An intriguing dual-beam thermospray interface has been constructed[104,105] which allows a greater independence of HPLC conditions. In this interface two capillaries are employed, one carrying sample solution, the other the required electrolyte. Slightly higher sensitivities and lower levels of fragmentation compared with single-beam operation were claimed for water-soluble compounds, together with a less critical temperature dependence of the vaporizer. Microbore columns could also be used successfully with this interface.

Of the new probes described in the literature[106,107] one features an easily exchanged capillary insert which was said to compare well with commercially available probes.[107]

One of the early problems with TSP LC/MS was the difficulty in coping with gradient elution systems: as the organic content of the eluant increases, the vaporizer temperature needs to be reduced in order to maintain optimum conditions. An automatic system to accomplish this has been described[108] and versions are now commercially available.

Several papers have set out to explore the underlying nature of the thermospray process: a better understanding of the mechanism(s) involved will generate new improved sources and interfaces and increase the areas of application. It was shown[109] that gas-phase ion/molecule reactions have a strong influence on the relative ion abundances and that proton affinity considerations played a major part in determining the appearance of a TSP spectrum. The conclusions were that: (i) the observed positive ions in thermospray spectra of ammonium acetate are determined by the gas-phase ion/molecule equilibria between $NH_4^+$ and $H_2O$, $NH_3$ and $CH_3COOH$ for $10^{-2}$ to 1M ammonium acetate concentrations; (ii) the primary TSP product ions are clustered $NH_4^+$ ions, which protonate gaseous organic bases by hydrogen transfer. For a mixture of bases the one with the highest proton affinity will dominate the spectrum; (iii) gas-phase ion/molecule reaction kinetics can be applied semi-quantitatively to the gas-phase equilibria. Average reaction times are of the order of a few hundred microseconds.

Other authors[110] similarly considered buffered TSP systems when they examined the effect on the observed spectra of changing the charge on the analyte in solution: for nine ionic compounds examined little change was observed. Like the previous paper, these authors too found that gas-phase ionization was predominating and supported their conclusions by comparing the TSP and FAB mass spectra of compounds which gave a poor TSP response. The dependence of the FAB spectra of these compounds on the solution pH and the lack of any such dependence in TSP led them to the conclusion that different mechanisms were operating and that even where the solutes were ionized in solution gas-phase ionization was still the predominant process, even though ion evaporation might have been expected to be important. In another paper[111] the same group demonstrated that for negative-ion formation gas-phase acidities could be used to predict the spectra from several important compound classes. It seems quite clear as a result of these studies that the major mechanism in TSP ionization from buffered solutions is very similar to that operating in classical chemical ionization. It is similarly clear that this is not the whole story, particularly in the absence of buffer. It was shown[112] that TSP ion evaporation in the absence of buffer or any auxiliary ionization technique appeared to be dependent on the type and relative amount of ionic species in solution and whilst $[M+H]^+$ species were formed by gas-phase reactions in accordance with the above findings, $[M+Na]^+$ and similar adduct ions were derived from desolvated ionic complexes. Nucleosides in particular were found to show a strong dependence on the solution-phase chemistry.

From the work done to date on the TSP mechanism it would appear that ions can be formed by a combination of four possible processes: (i) gas-phase ion/molecule reactions; (ii) field-induced ion evaporation from charged droplets; (iii) decomposition of solid salt particles formed at high electrolyte concentration; and (iv) simple solvent evaporation from charged droplets in the absence of a high electrical field.

As mentioned earler, the range of applications of TSP has grown tremendously, with hardly any area of organic mass spectrometry left untouched. Accordingly, the examples quoted below are of necessity selective, but illustrate the depth of interest in the technique.

A comparison between GC/MS and TSP LC/MS for the quantitative determination of serum cortisol has been reported.[113] A relative standard deviation of 7% at a concentration of 190 ng/ml was obtained which although satisfactory was inferior to that from GC/MS. However, the latter involved a much lengthier sample work-up procedure. Another comparison, this time between TSP and DLI LC/MS, was made as part of a study of the possibilities of LC/MS in the pharmaceutical industry,[114] although results were only of a preliminary nature.

The use of an automated sample preparation system was reported[115] and successfully demonstrated for the quantitative determination of labetalol in human plasma. The process involved the use of a solid-phase extraction module and was shown to be robust and reliable, but the use of a stable isotope-labelled internal standard was necessary to compensate for the varying ionization parameters of the TSP ion source.

Further quantitative work was performed as part of an evaluation of the potential of TSP LC/MS in neurochemistry.[116] This work illustrated a common problem described by several authors, namely a wide variation in response from compound to compound: tricyclic anti-depressants could be detected by SIM at the 10 pg level, whereas catecholamines required several nanograms. For full scanning data, as in the determination of unknown components of mixtures, this variation can be problematical, and generally "tens of nanograms" are needed. For prostaglandins $E_1$ and $E_2$ in human semen, for example, detection levels varied from 5-20 ng on-column,[117] and for corticosteroids 1 µg was used, although good quality data were claimed from $\geqslant$30 ng.[118] Similarly, for acetaminophen in rat bile, the full scan detection limit was 1 ng. For its glucuronide conjugate 2 µg and for its glutathione conjugate 4 µg was required.[119] Full scan sensitivity remains a problem, particularly for those involved in the identification of trace impurities.

Numerous applications involving metabolic studies have been published including a review on the use of TSP LC/MS for quantitative analysis using the isotope dilution technique.[120] An application involving the use of MS/MS techniques to look for specific sub-structural units of the administered drug enabled metabolites to be identified rapidly in complex sample matrixes.[121] Other examples illustrating the power of LC/MS/MS have been reported for steroids,[122] dyes in waste water,[123] rapid processing of drug abuse samples,[124] the sequencing of proteolytic digests from recombinant Eglin $C^{125}$ and the identification of two urinary glucuronide metabolites of doxylamine.[126] Applications in this area are certain to increase in the near future as MS/MS becomes more widely available.

Thermospray LC/MS has been readily adopted in the area of environmental research as evidenced by papers on the detection of dyes in waste water from a municipal water treatment plant,[123] a consideration of the optimum conditions for sulphonated azo-dyes in such waste-water systems,[127] the identification of tri-alkyl lead species in wine,[128] the analysis of thermally labile herbicides,[129] its use in food and agricultural research,[130] the detection of mycotoxins in biological samples[131] and the detection of pesticides and herbicides with post-column addition of ammonium acetate to enhance sensitivity.[132]

Much of the emphasis in mass spectrometry development over the last couple of years has been on the solution of biological problems and TSP LC/MS has been no exception. The potential of the latest versions of TSP for the ionization of peptides in the mass range 1000-4000 daltons has been investigated;[133] it was shown that 3 nmol of glucagon ($[M + H]^+$ = 3481.6 daltons) would give peak profile data in agreement with calculated profiles by accumulating 20 scans over an 18 mu window. Similar data were obtained from the peptide mellitin. Thermospray LC/MS has also been used to identify protein digests following the use of immobilized enzymes; the method was demonstrated for basic pancreatic trypsin inhibitor and illustrated the potential of this approach.[134] It is to be expected that further work on the sequencing of biologically-derived peptides by TSP LC/MS will be reported in the near future.

4.3 Transport Devices. - The major transport device of recent years has been the moving belt interface which enables one to obtain not only CI spectra with a reagent gas of choice but also EI spectra, a major advantage of this technique (and one claimed by MAGIC as well, see below). Transport devices have, however, been very much overshadowed by TSP especially the latest versions using various electrodes to effect ionization and fragmentation and also in combination with MS/MS.

An interesting paper was presented[135] at the 16th meeting of the British Mass Spectrometry Society which asked the question "Who killed the moving belt interface?" and contained an appraisal of the relative merits of the belt interface compared with TSP, claiming the former to be a much under-rated device. Although the claimed merits of the belt interface are real there can be little doubt that the TSP interface is considerably easier to use and mechanically much simpler, factors which have certainly contributed to its rapid and widespread acceptance.

The identification of a budesonide metabolite using a gradient elution system and methane CI has been described.[136] The authors reported that phthalate ester background, partly derived from the polyimide belt, complicated the identification of minor metabolites.

A method enabling the rapid profiling of diketopiperazines in cocoa powder which utilized a moving belt interface has been reported[137] and a nebulizer deposition device was used to assist in the characterization of 2-nitrofluorene, its metabolites and related compounds.[138] Spectra of acceptable quality were obtained from $\geqslant 65$ ng of material.

Belt interfaces can readily accept normal-phase systems as was illustrated by the analysis of glycosphingolipids at the 1-5 µg level. Spectra obtained gave ions characteristic of the ceramide, fatty acid, long-chain base and carbohydrate components.[139] The same system was used for the determination of vitamin E in seed oils.[140]

The difficulties associated with obtaining a smooth deposition of eluant onto the belt, especially in reversed-phase chromatography, have been tackled in several ways including spray devices,[136] nebulizers[138] and an electrically heated capillary spray device[141] which showed considerable advantages over other designs.

The moving belt has been suggested as a viable means of obtaining FAB ionization of LC eluates and results have been reported in this way[142] but the technical difficulties associated with this approach and the advent of continuous flow FAB systems (see later) have led to the gradual demise of this approach.

The MAGIC (mono-disperse aerosol generation) technique has been known for some time now, and involves the transport (hence its inclusion in this section) of a molecular beam via a momentum separator. The latest design improvements have recently been described[143] and two commercial systems based on these designs and differing mainly in the initial aerosol production are now available.[144] The technique requires the sample molecule to be vaporized by impact with the hot walls of the ion source followed by conventional EI or CI. The EI spectra produced are claimed to be very similar to conventional EI spectra and hence can be used in computer-based library searches.[143] The vaporization stage may well prove to be a limitation of this technique possibly preventing its exploitation for those polar, biological molecules for which TSP has been so successful, but initial results suggest the technique will still handle a wide variety of compounds. Sensitivity at the moment leaves a little to be desired, due at least in part to sample losses in the interface with approximately 25% of the original sample reaching the mass spectrometer.[145] This will undoubtedly be improved in the near future as more experience is gained with the commercial systems, with the technique probably making a significant contribution to the next report in this series.

4.4 Direct Liquid Introduction Methods. - This section covers DLI, nebulizing devices (other than MAGIC) and the interfacing of capillary column LC with MS.

The DLI technique has suffered a marked decline in usage with the growth of TSP and only a few papers have reported the use of such devices. The use of a DLI interface to analyse urine for drugs[146] has been described; the method used MS/MS with selected reaction monitoring and was tailored for a fast analysis/high throughput situation. Negative-ion CI with dissociative electron capture was used[147] in an investigation of the metabolites of benzo[a]pyrene, which were separated using a microbore $C_8$ column and acetonitrile-water as the eluant which subsequently became the CI reactant gas. Both positive-ion and negative-ion CI were used in a comparison of ionization methods for the LC/MS analysis of 10 organophosphorus pesticides.[148] The addition of chloroacetonitrile to the eluant and its effect on chloride ion attachment at different source temperatures was studied : it was found that the relative abundance of the [M + Cl]⁻ ion formed from several of the compounds studied increased strongly as the source temperature was decreased.

The quantitative determination of ochratoxin A in barley extracts has been performed by DLI LC/MS.[149] Detection levels were of the order of 3 μg/kg sample and the calibration graph was shown to be linear from 15 to 4600 ng of the toxin.

Several of the above applications utilized small bore LC columns in order to avoid the large split ratios necessary when using conventional 5 mm i.d. columns. The use of true capillary columns interfaced directly to the MS has also been investigated, the very low flow rates from these columns (5-10 μm i.d.) facilitating EI ionization.[150,151] However, only relatively volatile compounds appear to be amenable to this approach. Packed fused-silica columns (0.32 mm i.d.) have also been used in a similar way,[152] with the electrostatic field generated between the column tip and the ion source being claimed as the major factor in nebulization. Interest in these columns has been renewed with the introduction of continuous flow FAB systems. Devices which use a vacuum to generate an aerosol from the liquid flow emanating from a microbore or capillary column have been examined by several groups, being particularly popular with Japanese workers,[153-155] but the applications of such devices, like MAGIC (at present), seem to be limited to relatively volatile compounds.

This review period has seen the introduction of two new approaches to LC/MS interfacing both of which appear to have considerable potential. One of these, continuous flow or dynamic FAB, has already attracted great interest and is rapidly becoming established particularly in the analysis of high-mass biomolecules. The other technique, ion spray, uses an atmospheric pressure ionization (API) source, which has limited its application to some extent, but it has been shown to be useful for several "difficult" compounds, both quantitatively and qualitatively.

One of the early papers[156] on continuous flow FAB in combination with microbore HPLC clearly demonstrated its potential for peptide analysis in the molecular weight range 900-6000 daltons and also showed the identification of 16 of the possible 19 peptides derived from a tryptic digest of whale myoglobin. Several further examples in the same vein have also appeared.[157-159] A comparison of continuous flow FAB and conventional FAB for the analysis of permethylated oligo-saccharides has been reported[160] which showed the superior performance of the continuous flow probe due to the improved signal-to-noise ratio associated with the reduced matrix level. The analysis of non-ionic detergents by combined LC/FABMS has also been described.[161] Improvements in both interfaces and in packed capillary column technology are taking place all the time and as continuous flow probes become more widespread - all the major manufacturers can now supply such devices - the list of applications will doubtless increase rapidly.

The ion spray technique combines the principles of ion evaporation and electrospray. The total effluent from a microbore LC column is pneumatically nebulized at atmospheric pressure and under high voltage and dispersed into charged droplets using dry nitrogen. Ions emitted from the charged droplets are sampled into a triple quadrupole mass spectrometer for mass analysis.[162] A comparison of the technique with TSP, both using an API source, indicated that the ion spray technique produced ions by the ion evaporation mechanism, whereas TSP produced ions by both ion evaporation and gas-phase ion/molecule reactions.[163] The authors pointed out, however, that as the TSP was operating at atmospheric pressure results may be different from those obtained in a vacuum. This approach has since been taken up by other groups and a recent paper described the determination of drugs in human serum by LC/API.[164] The progress of research into LC/API will undoubtedly be followed with great interest by many mass spectrometrists.

4.5 Summary. - The last two years have seen a consolidation of the position of TSP as the premier interfacing technique for LC/MS, challenged only by continuous flow FAB for high mass work. Further progress with this latter technique can be expected in the near future. The success of TSP has led to the demise of both DLI and moving belt interfaces, with the introduction of the MAGIC technique on a commercial basis also likely to contribute to their continuing fall from grace in some areas. Overall, the period June 1986-June 1988 has seen LC/MS emerge as a powerful, routine analytical technique capable of solving real problems in very many areas of analytical chemistry and there can be little doubt that improvements will be made (indeed, some have already been introduced at the Bordeaux conference in Sept. 1988) to further the

applications of LC/MS. The next review in this series will
certainly not be short of material on which to report.

5   Tandem Mass Spectrometry

5.1   Overview. - This section of the review concerns itself
only with the analytical applications of tandem mass spectrometry
(MS/MS). The instrumental developments of the last two years, e.g.
the emergence of 4-sector instruments with wide mass ranges, the
development of array detectors, the continuing developments in ion
trap and Fourier-transform technology and the combination with
LC/MS are all covered elsewhere in this review. For those readers
interested in these instrumental aspects several informative
reviews are available.[165-178] Where instrumental developments have
direct analytical relevance they have been included here. Also
omitted from this review are references to MS/MS performed on
reverse-geometry single mass spectrometers (MIKES) and to linked
scans on a conventional geometry instrument, except where these
have been reported as part of comparisons with "true" MS/MS
techniques.

5.2   Applications. - The general trend in mass spectrometry
towards the life sciences, noted in the preceeding section, is
evident also in this area, with many of the published applications
over the review period being concerned with biomolecules. The
problems associated with peptide sequencing and their solution
using MS/MS techniques have been reported by several groups. Many
of these reports[125,179-183] follow a standard analytical procedure:
degradation of the original peptide by proteolytic enzymes or
chemical hydrolysis, followed by HPLC separation of the resultant
smaller peptides and MS/MS analysis of isolated fractions. In some
cases[182,184-186] the separation step has been omitted and the
digest examined directly by FABMS. The use of the first,
hydrolytic, step enables the subsequent MS/MS experiments to be
performed on a low collision energy instrument, e.g. a triple
quadrupole system. The alternative approach, direct MS/MS analysis
of the $[M + H]^+$ species produced from the intact original peptide
would require the use of a high mass, high collision energy (and
high cost) 4-sector instrument. The question of collision energy
generally is discussed below, as are 4-sector instruments.
   The use of a triple quadrupole instrument to examine the

products of a triptic digest of human apolipoprotein B was reported[179] and a similar instrument was used in work on the structural characterization of recombinant eglin C.[125] In another series of experiments the digestion of thioredoxin, a peptide of molecular weight 11748 daltons, gave 14 peptides ranging in length from 2-18 amino acid residues. These fragments were sequenced by MS/MS and the primary structure of the original peptide determined.[180] An even larger thioredoxin-type molecule containing 108 amino acid residues was isolated from the bacterium Chlorobium thiosulphatophilum and was sequenced following the use of four enzyme treatments and Edman degradation prior to MS/MS.[181]

The use of tandem mass spectrometry for protein sequencing has been described in some detail in the published proceedings of an international conference held in 1986[182] and a review showing the advantages of MS/MS over conventional methods of protein sequencing has also been published.[183] It was shown to be particularly advantageous where, for example, the N-terminus was blocked (making Edman-type degradations impossible) or where the peptide had been subjected to phosphorylation or sulphation. The reduction in the need for extensive purification, due to the separative capability of the first analyzer in a tandem system, was also seen as a distinct advantage.[183] This separation of the ions of interest prior to activation (usually by gas collision) also serves to reduce the extensive background in a FAB spectrum such that structurally significant daughter ions are readily detected; the corresponding ions in the normal spectrum may be of such low abundance that they are obscured by the background.[184] Peptides containing N-alkylated amino acids are also troublesome with conventional amino acid analysis and MS/MS has been used successfully in such cases[185] as well as in the differentiation of isomeric leucine and isoleucine.[185,186]

The results of a comparison between metastable ion scanning on a reverse-geometry instrument and MS/MS with collisionally induced decomposition on a tandem instrument of EBE geometry, for endorphin and ACTH peptides up to 2000 daltons, have been published.[187] These authors showed that there were no major differences in the daughter ions obtained from either instrument but they did observe a decrease in the signal-to-background ratio in the CID spectra. Consequently, they recommended a metastable ion spectrum in preference to CID for peptides of the type studied.

In order to utilize fully the analytical potential of tandem mass spectrometers it is essential that the selected primary ions can not only be generated in high yield but also be caused to fragment reproducibly. That is, the added excess of internal energy should be under control. Various means of imparting this additional internal energy have been considered [183,188,172] including electron collision-induced dissociation, laser dissociation and surface induced-dissociation as well as the more commonly used collision with a neutral species, usually helium. For many applications collision-induced dissociation using a neutral gas seems likely to remain the main method for some time to come. The major factors affecting the internal energy of ions were also considered in one of these papers.[188] These factors become more acute as relative molecular mass increases and those affecting the dissociation of large ions, including a consideration of target gas excitation which has often been overlooked, have been discussed.[189] One major problem affecting the sequencing of high mass biopolymers generally is that the ionization yield for $[M + H]^+$ ions, normally generated by FAB, decreases with increasing mass, with a corresponding decrease in the daughter ion yield following collisional activation.[190] This has been attributed to an increase in the number of degrees of freedom of the molecule, i.e. the average energy/vibrational mode decreases with increasing ionic mass leading to a reduction in the number of fragment ions produced.[187]

It has become increasingly apparent over the last 1-2 years that for large (>1200 daltons) molecules high (keV) collision energies are necessary to provide sufficient energy for fragmentation to occur; such collision energies can be obtained with 4-sector instruments. An instrument of BEEB geometry was used in a study of two dynorphins, one a heptapeptide, the other a tridecapeptide. Considerable sequence information was obtained, enabling several product/precursor relationships to be derived which were useful for peptide identification and possible quantification.[191] Further examples of the uses of 4-sector instruments are to be found in references 181, 192 and 193. One of the problems with early versions of 4-sector instruments was the necessity to operate MSI at a relatively low resolution (500-1000) to obtain usable CID transmission, as the losses in the collision cell were considerable. The development of multi-channel array detectors has improved the detection of such low ion currents by a factor of about 100 compared with the use of a post acceleration detector and single electron multiplier.[194] Procedures for calibrating any instrument are obviously of vital importance for their analytical use and a paper describing the calibration of a 4-sector instrument up to a mass of 3500 daltons in both positive-ion and negative-ion modes has been published.[167]

Tandem mass spectrometry techniques have been widely applied in the study of metabolites of various drugs and biological compounds.[195-209] From these varied papers the three discussed here demonstrate the power of MS/MS, not only in the metabolite area but generally. The rapid identification of drug metabolites using a triple quadrupole system was greatly facilitated by making use of the fact that metabolites usually retain various substructures of the original compound leading to common daughter ions. By looking for parents of selected daughter ions, molecular ions of possible metabolites can be revealed, which can in turn be made to give daughter ion spectra and hence structural information. Rapid profiling of complex samples such as urine can be accomplished in a minimal time with little prior sample preparation.[199,201] The amount of the epoxide metabolite of carbamazepine necessary for its identification in this way was found to be 0.4 ng/µl.[199] Thermo-spray LC/MS and FABMS were both used in conjunction with MS/MS in the third paper[195] to detect glucuronide and sulphate conjugates of a catecholamine drug in wine samples. In this case constant neutral loss scans specific for O-glucuronides (176 daltons) and for aryl sulphate esters (80 daltons) were used to detect possible metabolites from the complex sample matrix. Parent ion scans of common daughter ions were also used as in the first two papers. Similar methods have been used for the rapid monitoring of biologically-active substances in medicinal plants, looking specifically for lignans[210] and for the screening, confirmation and quantification of sulphonamide residues in crude extracts of pig kidney.[211] Such methods should also be applicable to the analysis of other complex systems, as illustrated by the detection of satratoxins in fermentation broths[212] and may also have applications in the pharmaceutical industry for screening bulk drugs for related impurities.

With the current level of environmental concern it is hardly surprising that the latest and most advanced analytical techniques are very soon applied to environmental problems and MS/MS has not been an exception. It has been applied to the examination of water for a variety of contaminants;[213-216] to the detection of low levels of toxins in foods;[217-221] in the examination of and detection of pesticides and herbicides;[222-227] and in the detection of the ubiquitous dioxins.[216,228-234] One paper of particular interest from this latter group[234] used MS/MS to identify PCB congeners where capillary column GC resolution was insufficient for an unambiguous separation of all the various congeners. The authors used collision-induced dissociation of the $[M - Cl + O]^-$ ion resulting from oxygen-enhanced NICI. A Townsend discharge tube was used to effect ionization. The daughter ion spectra of six PCBs which had very similar retention times were sufficiently different for their identification to be effected (by comparison with reference spectra).

Two other papers of special interest to many mass spectrometrists dealt with the quantification of ethyl carbamate in various alcoholic beverages.[235,236] In the first of these, the ethyl carbamate was extracted and separated from the other components present using capillary GC. Chemical ionization ($^i$butane) produced the $[M + H]^+$ ion for collisional activation in the second stage of a triple quadrupole analyser. Quantification down to 1 p.p.b. was achieved using a labelled internal standard. Over 80 wines and spirits were examined using similar extraction procedures in the second paper.[236] In this case methane was utilized as the CI reagent gas in a triple quadrupole system. Detection levels were higher than in the previous paper and the authors also noted that quantitative accuracy was not as good as that found using CI and monitoring two ions; this they attributed, at least in part, to the time scale involved in monitoring parent and daughter ions from both sample and labelled internal standard for a GC peak only 10s wide. It would apear that the custom of taking some of these spirits with suitable aliquots of an aqueous diluent also needs to be treated with caution: samples of chlorinated water from a natural lake were shown[214] by MS/MS to contain significant amounts of both choral and dichloro-acetonitrile.

The continued development of Fourier-transform mass spectrometry (FTMS) has brought this type of instrumentation to the point where it is now being used for exploratory MS/MS experiments by several groups. McLafferty and Amster have reviewed[168] FTMS generally with special emphasis on the advantages perceived for MS/MS operation, including the possibility of $MS^n$, the high

resolution capability, the simultaneous detection of nearly all the
ions produced and the efficient trapping of daughter ions in a
dual-cell instrument. Most of the reports of FTMS to date have
dealt with relatively small molecules (<200 daltons) and have
concentrated on the extremely high daughter resolution possible
using a dual-cell instrument in which the daughter ions are pulsed
across the conductance limit between the high and low (analyser)
pressure cells for detection. Resolutions of >200,000 have been
demonstrated for isobaric m/z 105 ions[237] and for daughters of mass
118 from a 1:1 mixture of cyclopropylbenzene and di-$^n$propyl-
sulphide;[238] in this latter case a resolution of 35000 was needed
to separate the EI-produced parent ions. Similar experimental
procedures were used to mass measure to better than 5 p.p.m.
accuracy a variety of daughter ions from a synthetic mixture from a
commercial baby lotion and from peppermint oil.[239] Perhaps the
most spectacular demonstration[240] of the potential of FTMS/MS was
the sequencing of 10 pmol of a 15-residue tryptic peptide of mass
1772.9 daltons using photodissociation. The same authors have used
a tandem quadrupole instrument as the front end of a FTMS[241-243] in
other peptide studies. Clearly, FTMS/MS has a promising future if
some of the instrumental problems with FTMS generally (discussed
elsewhere) can be overcome.

Many other examples of the use of MS/MS to solve analytical
problems are to be found in the literature of the last two years,
including lipid analysis,[244-252] structural studies of nucleotides
and similar compounds,[253-257] and drugs.[258-261] Finally, mention
should be made of the wide range of applications contained in the
references to review articles at the beginning of this section and
which will provide the reader with a wider historical survey than
this report allows.

## References

1. M.E. Rose, in "Specialist Periodical Reports: Mass Spectrometry", ed. M.E. Rose, Royal Society of Chemistry, London, Volume 8, 1985, p. 210; Volume 9,1987, p.264.
2. J.A. Olivares, N.T. Nguyen, C.R. Yonker, and R.D. Smith, **Anal. Chem.**, 1987, 59, 1230.
3. R.D. Smith and H.R. Udseth, **Nature**, 1988, 331, 639.
4. R.D. Smith and H.R. Udseth, **Chem. Br.**, 1988, 24, 350.
5. W.Th. Kok and G.J.M. Bruin, **Eur. Chromatogr. News**, 1988, 2 (5), 22.
6. R.D. Smith, J.A. Olivares, N.T. Nguyen, and H.R. Udseth, **Anal. Chem.**, 1988, 60, 436.
7. B.D. Andresen and E.R. Fought, U.S. US 4,705,616, 1987 (**Chem. Abs.**, 1988, 108, 68033s).
8. M.S. Stanley, K.L. Duffin, S.J. Doherty, and K.L.Busch, **Anal. Chim. Acta**, 1987, 200, 447; K.L. Busch, **Trends Anal. Chem.**, 1987, 6, 95.
9. J.W. Fiola, G.C. DiDonato, and K.L. Busch, **Rev. Sci. Instrum.**, 1986, 57, 2294; G.C. Didonato and K.L. Busch, **Anal. Chem.**, 1986, 58, 3231.
10. M.S. Stanley and K.L. Busch, **Anal. Chim. Acta**, 1987, 194, 199.
11. M.S. Stanley, K.L. Busch, and A. Vincze, **J. Planar Chromatogr. - Mod. TLC**, 1988, 1, 76.
12. R. Kraft, A. Otto, H.J. Zoepfl, and G. Etzold, **Biomed. Environ. Mass Spectrom.**, 1987, 14, 1.
13. R. Kraft, D. Buettner, P. Franke, and G. Etzold, **Biomed. Environ. Mass Spectrom.**, 1987, 14, 5.
14. K. Shizukuishi, Y. Numaziri, and Y. Kato, **Iyo Masu Kenkyukai Koenshu**, 1986, 11, 85.
15. H. Iwabuchi, A. Nakagawa, and K. Nakamura, **J. Chromatogr.**, 1987, 414, 139.
16. S. Yamamoto, H. Kakinuma, T. Nishimuta, and K. Mori, **Iyo Masu Kenkyukai Koenshu**, 1986, 11, 151.
17. H. Kajiura, **Seirigaku Gijutsu Kenkyukai Hakoku**, 1986, 8, 14 (**Chem. Abs.**, 1987, 106, 152497n).
18. A. Hayashi, T. Matsubara, Y. Nishizawa, T. Hattori, and M. Morita, **Iyo Masu Kenkyukai Koenshu**, 1986, 11, 147.
19. M. Warner, **Anal. Chem.**, 1987, 59, 855A.
20. J.D. Pinkston, **Trends Anal. Chem.**, 1988, 7, 154.
21. M.L. Lee and K.E. Markides, **J. High Resolut. Chromatogr. Chromatogr. Commun.**, 1986, 9, 652.
22. M.L. Lee and K.E. Markides, **Science**, 1987, 235, 1342.
23. D.W. Later, D.J. Bornhop, E.D. Lee, J.D. Henion, and R.C. Wieboldt, **LC-GC**, 1987, 5, 804.
24. P.J. Schoenmakers and F.C.C.J.G. Verhoeven, **Trends Anal. Chem.** 1987, 6, 10.
25. P.J. Schoenmakers and L.G.M. Uunk, **Eur. Chromatogr. News**, 1987, 1 (3), 14.
26. T. Greibrokk, B.E. Berg, A.L. Blilie, J. Doehl, A. Farbrot, and E. Lundanes, **J. Chromatogr.**, 1987, 314, 429.
27. R. Wall, **Int. Analyst**, 1987 (8), 28.
28. P.R. Griffiths, **Anal. Chem.**, 1988, 60, 593A.
29. R.M. Smith (ed.), "Supercritical Fluid Chromatography", Royal Society of Chemistry, London, 1988.
30. D.E. Games, A.J. Barry, I.C. Mylchreest, J.R. Perkins, and S Pleasance, **Lab. Prac.**, 1987, 36 (2), 45.
31. D.E. Games, A.J. Berry, I.C. Mylchreest, J.R. Perkins, and S. Pleasance, **Eur. Chromatogr. News**, 1987, 1 (1), 10.
32. D.E. Games, A.J. Berry, I.C. Mylchreest, J.R. Perkins, and S. Pleasance, **Anal. Proc.**, 1987, 24, 371.
33. R.D. Smith, H.T. Kalinoski, and H.R. Udseth, **Mass Spectrom. Rev.**, 1987, 6, 445.
34. R.D. Smith, B.W. Wright, and H.R. Udseth, in "Chromatographic Methods. Advances in Capillary Chromatography", Huethig Verlag, Heidelberg, 1986, p. 56.

35. P.J. Arpino, J. Cousin, and J. Higgins, **Trends Anal. Chem.** 1987, $\underline{6}$, 69.
36. D.E. Games, A.J. Berry, I.C. Mylchreest, J.R. Perkins, and S. Pleasance, in ref. 29, p. 159.
37. B.W. Wright, H.R. Udseth, E.K. Chess, and R.D. Smith, **J. Chromatogr. Sci.** 1988, $\underline{26}$, 228.
38. S.J. Lane, in ref 29, p. 175.
39. F.P. Schmitz and E. Klesper, **J. Chromatogr.**, 1987, $\underline{388}$, 3.
40. H.E. Schwartz, **LC-GC**, 1987, $\underline{5}$, 14.
41. P.J. Schoenmakers, in ref. 29, p. 102.
42. R.W. Gale, J.L. Fulton, and R.D. Smith, **Anal Chem.**, 1987, $\underline{59}$, 1977.
43. S.R. Goates, N.A. Zabriskie, J.K. Simons, and B. Khoobehi, **Anal. Chem.**, 1987, $\underline{59}$, 2927.
44. M. Saito, T. Hondo, and Y. Yamauchi, in ref. 29, p. 203.
45. R.D. Smith and H.R. Udseth, **Anal. Chem.**, 1987, $\underline{59}$, 13.
46. H.T. Kalinoski, H.R. Udseth, E.K. Chess, and R.D. Smith, **J. Chromatogr.**, 1987, $\underline{394}$, 3.
47. J.F.J. Todd, I.C. Mylchreest, A.J. Berry, D.E. Games, and R.D. Smith, **Rapid Commun. Mass Spectrom.**, 1988, $\underline{2}$, 55.
48. E.D. Lee, J.D. Henion, R.B. Cody, and J.A. Kinsinger, **Anal. Chem.**, 1987, $\underline{59}$, 1309.
49. D.A. Laude, jun., S.L. Pentoney, jun., P.R. Griffiths, and C.L. Wilkins, **Anal. Chem.**, 1987, $\underline{59}$, 2283.
50. S. Rokushika, H. Hatano, and H.H. Hill jun., **Anal. Chem.**, 1987, $\underline{59}$, 8.
51. A.J. Berry, D.E. Games, I.C. Mylchreest, J.R. Perkins, and S. Pleasance, **J. High Resolut. Chromatogr. Chromatogr. Commun.**, 1988, $\underline{11}$, 61.
52. A.J. Berry, D.E. Games, I.C. Mylchreest, J.R. Perkins, and S Pleasance, **Biomed. Environ. Mass Spectrom.**, 1988, $\underline{15}$, 105.
53. J.R. Chapman, **Rapid Commun. Mass Spectrom.**, 1988, $\underline{2}$, 6.
54. S.D. Zaugg, S.J. Deluca, G.U. Holzer, and K.J. Voorhees, **J. High Resolut. Chromatogr. Chromatogr. Commun.**, 1987, $\underline{10}$, 100.
55. S.B. Hawthorne and D.J. Miller, **Fresenius' Z. Anal. Chem.**, 1988, $\underline{330}$, 235.
56. E.C. Huang, B.J. Jackson, K.E. Markides, and M.L. Lee, **Chromatographia**, 1988, $\underline{25}$, 51.
57. J. Cousin and P.J. Arpino, **J. Chromatogr.**, 1987, $\underline{398}$, 125.
58. H.T. Kalinoski and R.D. Smith, **Anal. Chem.**, 1988, $\underline{60}$, 529.
59. H.T. Kalinoski, B.W. Wright, and R.D. Smith, **Biomed. Environ. Mass Spectrom.**, 1988, $\underline{15}$, 239.
60. J.D. Pinkston, G.D. Owens, L.J. Burkes, T.E. Delaney, and D.S. Millington, **Anal. Chem.**, 1988, $\underline{60}$, 962.
61. C.H. Sin, H.M. Pang, D.M. Lubman, and J. Zorn, **Anal. Chem.**, 1986, $\underline{58}$, 487; H.M. Pang, C.H. Sin, D.M. Lubman, and J Zorn, **ibid.**, 1581.
62. C.H. Sin, H.M. Pang, and D.M. Lubman, **Anal. Instrum.**, 1988, $\underline{17}$, 87.
63. H.M. Pang, C.H. Sin, and D.M. Lubman, **Spectrochim. Acta**, 1988, $\underline{43B}$, 671.
64. W. Blum, K. Grolimund, P.E. Jordi, and R. Ramstein, **J. High Resolut. Chromatogr. Chromatogr. Commun.**, 1988, $\underline{11}$, 441.
65. H.T. Kalinoski, H.R. Udseth, B.W. Wright, and R.D. Smith, **Anal. Chem.**, 1986, $\underline{58}$, 2421.
66. M.I. Selim and J.R. Strubinger, **Fresenius' Z. Anal. Chem.**, 1988, $\underline{330}$, 246.
67. C.R. Yonker and R.D. Smith, **Gov. Rep. Announce. Index** (U.S.), 1987, $\underline{87}$, Abstr. No. 711,484.
68. P.G. Sim and C.M. Elson, **Rapid Commun. Mass Spectrom.**, 1988, $\underline{2}$, 137.
69. H.T. Kalinoski, H.R. Udseth, B.W, Wright, and R.D. Smith, **J. Chromatogr.**, 1987, $\underline{400}$, 307.
70. G.D. Owns, L.J. Burkes, J.D. Pinkston, T. Keough, J.R. Simms, and M. P. Lacey, in "Supercritical Fluid Extraction and Chromatography", B.A. Charpentier and M.R. Sevenants (eds.), American Chemical Society Symposium Series, Washington, 1988, $\underline{366}$, p. 191.
71. F.P. Schmitz and H. Hilgers, **Makromol. Chem. Rapid Commun.**, 1986, $\underline{7}$, 59.
72. T.L. Chester and D.P. Innis, **J High Resolut. Chromatogr.Chromatogr. Commun.**, 1986, $\underline{9}$, 209.

73. P.J. Arpino and J. Cousin, **Rapid Commun. Mass Spectrom.**, 1987, **1**, 29.
74. S.B. Hawthrone and D.J. Miller, **J. Chromatogr.**, 1987, **388**, 397.
75. K. Matsumoto, S. Tsuge, and Y. Hirata, **Chromatographia**, 1986, **21**, 617.
76. K. Matsumoto, S. Tsuge, and Y. Hirata, **Shitsuryo Bunseki**, 1987, **35**, 15.
77. E.D. Morgan, S.J. Murphy, D.E. Games, and I.C. Mylchreest, **J Chromatogr.**, 1988, **441**, 165.
78. See Ref 1.
79. J.F. Holland, J.J. Leary, C.C. Sweeley, **J Chromatogr.**, 1986, **379**, 3.
80. J.A. De Haseth, **Spectroscopy**, 1987, 2(10), 14.
81. P.J. Arpino, **Comm. Eur. Communities. [Rep]Eur, Eur 10388 Org. Micropollut. Aquat. Environ.**, 1986, p.26.
82. D.E. Games, in **Appl. Mass Spectrom. Food Sci.**, ed. J. Gilbert, Elsevier, London, 1987, p.193-237.
83. R.M. Caprioli, **Biochem. Soc. Trans.**, 1987, **15(1)**, 162.
84. L.R. Snyder and S. Ahuja, **Chem. Anal.** (N.Y.), 1986, **85**, (Ultratrace Anal. Pharm. Other Compd. Interest), 109.
85. T.R. Covey, D.E. Lee, A.P. Bruins and J.D. Henion, **Anal. Chem.**, 1986, **58(14)**, 1451A.
86. A.L. Burlingame, T.A. Baillie and P.J. Derrick, **Anal. Chem.** 1986, **58(5)**, 165R.
87. T. Cairns and E G Siegmund **Anal. Methods Pestic. Plant Growth Regul.**, 1986, **14**, 193.
88. J. Stroh, **LC-GC**, 1987, **5(7)**, 562.
89. R.L. Cochran, **Appl. Spectrosc. Rev.**, 1986, 22(2-3), 137.
90. L.E. Martin, M.S. Lant and J. Oxford, **Method. Surv. Biochem. Anal.**, 1986, **16**, 399.
91. W.M.A. Niessen, **Chromatographia**, 1986, 21(5), 277.
92. W.M.A. Niessen, **ibid.**, 1986, 21(6), 342.
93. D. Barcelo and J. Albaiges, **Rev. Agroquim. Technol. Aliment.**, 1987, 27(4), 471.
94. M.E. Rose, **Mass Spectrom.**, 1987, **9**, 264.
95. A.L. Burlingame, D. Maltby, D.H. Russell and P.T. Holland, **Anal. Chem.**, 1988, **60**, 294R.
96. R.H. Robins and F.W. Crow, **Rapid Commun. Mass Spectrom.**, 1988, 2(2), 30.
97. W.H. McFadden and S.A. Lammert, **J Chromatogr.**, 1987, **385**, 201.
98. W.H. McFadden, D.A. Garteiz and E.G. Siegmund, **J. Chromatogr.**, 1987, **394(1)**, 101.
99. F.A. Bencsath and F.W. Field, **Anal. Chem.**, 1988, 60(13), 1323.
100. C. Lindberg and J. Paulson, **J. Chromatogr.** 1987, **394(1)**, 117.
101. G. Kilpatrick, I.A.S. Lewis and J.F. Smith, **Biomed. Environ. Mass Spectrom.**, 1987, 14(4), 155.
102. D.J. Liberato and A.L. Yergey, **Anal. Chem.**, 1986, **58**, 6.
103. M.A. Baldwin and G.J. Langley, **Org. Mass Spectrom.**, 1987, 22(8), 561.
104. L. Buetfering, G. Schmelzeisen-Redeker and F.W. Roellgen, **J. Chromatogr.**, 1987, **394(1)**, 109.
105. L. Buetfering, G. Schmelzeisen-Redeker and F.W. Roellgen, **J. Chem. Soc., Chem. Commun.**, 1986, **7**, 579.
106. D.S. Jones, R.H. Bateman, M. Lancaster, T. Liska, S.T. Krolik and V.C. Parr, **Proc. 15th Meeting Br. Mass Spectrom. Soc.**, Brighton UK, Sept. 1986, 1987, p.129.
107. S.E. Unger, T.J. McCormick, M.S. Bolgar and J.B. Hunt, **Anal. Chem.**, 1987, 59(8), 1242.
108. C.H. Vestal and G.J. Fergusson, **Int. J. Mass Spectrom. Ion Processes**, 1986, 70(2), 185.
109. A.J. Alexander and P. Kebarle, **Anal. Chem.**, 1986, **58**, 471.
110. R.W. Smith, C.E. Parker, D.M. Johnson and M.M. Bursey, **J. Chromatogr.**, 1987, **394(1)**, 261.
111. C.E. Parker, R.W. Smith, S.J. Gaskell, and M.M. Bursey, **Anal. Chem.**, 1986, 58(8), 1661.

112. R.D. Voyksner, **Org. Mass Spectrom.**, 1987, 22, 513.
113. S.J. Gaskell, K. Rollins, R.W. Smith and C.E. Parker, **Biomed.Environ. Mass Spectrom.**, 1987, 14(12), 717.
114. K.H. Schellenberg, M. Linder, A. Groepellin and F. Erni, **J. Chromatogr.**, 1987, 394(1), 239.
115. M.S. Lant, J. Oxford and L.E. Martin, **J. Chromatogr.**, 1987, 394(1), 223.
116. F. Artigas and E. Gelpi, **ibid.**, p.123.
117. J. Abian and E. Gelpi, **ibid.**, p.147.
118. D. Watson, G.W. Taylor, S. Laird and G.P. Vinson, **Biochem. J.**, 1987, 242(1), 109.
119. L.D. Betowski, W.A. Korfmacher, J.O. Lay, D.W. Potter and J.A. Hinson, **Biomed. Environ. Mass Spectrom.**, 1987, 14(12), 705.
120. A.L. Yergey, N.V. Esteban and D.J. Liberato, **Biomed. Environ. Mass Spectrom.**, 1987, 14(11), 623.
121. P. Rudewicz and K.M. Straub, **Anal. Chem.**, 1986, 58(14), 2928.
122. S.E. Unger and B.M. Warrack, **Spectroscopy** (Springfield, Oregon), 1986, 1(3), 33.
123. L.D. Betowski, S.M. Pyle, M.J. Ballard and G.M. Shaul, **Biomed. Environ. Mass Spectrom.**, 1987, 14(7), 343.
124. T.R. Covey, E.D. Lee and J.D. Henion, **Anal. Chem.**, 1986, 58, 2453.
125. F. Raschdorf, R. Dahinden, B. Domon, D. Muller and W.J. Richter, in "Mass Spectrometry of Large Molecules", ed. C.J. McNeal, Wiley (N.Y.), 1986, 49-65.
126. W.A. Korfmacher, C.L. Holder, L.D. Betowski and R.K. Mitchum, **J. Anal. Toxicol.**, 1987, 11(4), 182.
127. D.A. Flory, M.M. McLean, M.L. Vestal and L.D. Betowski, **Rapid Commun. Mass Spectrom.**, 1987, 1(3), 48.
128. M. Blaszkewicz, G. Baumhoer, B. Neidhart, R. Ohlendorf and M. Linscheid, **J. Chromatogr.**, 1988, 439(1), 109.
129. M. Shalaby, **Chem. Anal. (N.Y.)**, 1987, 91, 161.
130. F.A. Mellon, J.R. Chapman and J.E. Pratt, **J. Chromatogr.**, 1987, 394(1), 209.
131. E. Rajakyla, K. Laasasenaho and P.J.D. Sakkers, **J. Chromatog.**, 1987, 384, 391.
132. R.D. Voyksner, **Chem. Anal. (N.Y.)**, 1987, 91, 146.
133. P.J. Rudewicz, **Biomed. Environ. Mass Spectrom.**, 1988, 15, 461.
134. K. Stachowiak, C. Wilder, M.L. Vestal and D.F. Dyckes, **J. Am. Chem. Soc.**, 1988, 110(6), 1758.
135. S.J. Lane, R.J. Dennis and W.P. Blackstock, **Proc. 16th Meeting Brit. Mass Spectrom. Soc.**, York U.K., 1987, p.98.
136. C. Lindberg, J. Paulson and S. Edsbaecker, **Biomed. Environ. Mass Spectrom.**, 1987, 14(10), 535.
137. J. Van der Greef, A.C. Tas, L.M. Nijssen, J.Jetten and M. Hoehn, **J. Chromatogr.**, 1987, 394(1), 77.
138. L. Moeller and J.A. Gustafsson, **Biomed. Environ. Mass Spectrom.**, 1986, 13(12), 681.
139. J.E. Evans and R.H. McCluer, **Biomed. Environ. Mass Spectrom.**, 1987, 14(4) 149.
140. J. Van der Greef, A.J. Speck, A.C. Tas, J. Schrijver, M. Hoehn and U. Rapp, **LC-GC**, 1986, 4(7), 636.
141. G.M. Kresbach, T.R. Baker, R.J. Nelson, J. Wronka, B.L. Karger and P. Vouros, **J. Chromatogr.**, 1987, 394(1), 89.
142. S. Santikarn, G.R. Her and V.N. Reinhold, **J. Carbohydr. Chem.**, 1987, 6(1), 141.
143. P.C. Winkler, D.D. Perkins, W.K. Williams and R.F. Browner, **Anal. Chem.**, 1988, 60(5), 489.
144. Systems available from Hewlett-Packard and from Extrel Corporation (Thermabeam).

145. Comment by R.F. Browner, 11th Int. Mass Spectrom. Conference, Bordeaux, Sept. 1988.
146. T.R. Covey, E.D. Lee and J.D. Henion, **Anal. Chem.**, 1986, 58(12), 2453.
147. R.H. Bieri and J. Greaves, **Biomed. Environ. Mass Spectrom.**, 1987, 14(10), 255.
148. D. Barcelo, F.A. Maris, R.B. Geerdink, R.W. Frei, G.J. De Jong and U.A.T. Brinkman, **J. Chromatogr.**, 1987, 394(1), 65.
149. D. Abramson, **J. Chromatogr.**, 1987, 391(1), 315.
150. N.M.A. Niessen and H. Poppe, **J. Chromatogr.**, 1987, 385, 1.
151. N.M.A. Niessen and H. Poppe, **J. Chromatogr.**, 1987, 394(1), 21.
152. H. Alborn and G. Stenhagen, **J. Chromatogr.**, 1987, 394(1), 35.
153. K. Matsumoto and S. Tsuge, **Shitsuryo Bunseki**, 1986, 34(1), 33.
154. K. Matsumoto and S. Tsuge, **Shitsuryo Bunseki**, 1986, 34(4), 243.
155. K. Matsumoto, S. Tsuge, and H. Hirose, **Shitsuryo Bunseki**, 1987, 35(5), 240.
156. R.M. Caprioli, B. DaGue, T. Fan and W.T. Moore, **Biochem. Biophys. Res. Commun.**, 1987, 146(1), 291.
157. D.W. Hutchinson, A.R. Woolfit and A.E. Ashcroft, **Org. Mass Spectrom.**, 1987, 22(5), 304.
158. D.E. Games, S. Pleasance, E.D. Ramsey and M.A. McDowall, **Biomed. Environ. Mass Spectrom.**, 1988, 15(3), 179.
159. D. Cho, G. Dielman, J. Lloyd, P. Jahnke, R. Pesch and E. Schroeder, **Proc. 39th Pittsburgh Conf.**, New Orlean, Feb. 1988, abstr. no. N1046.
160. P. Boulenguer, Y. Leroy, J. M. Alonso, and J. Montreuil et al., **Anal. Biochem.**, 1988, 168(1-4), 164.
161. T. Takeuchi, S. Watanabe, N. Kondo, M. Goto and D. Ishii, **Chromatographia**, 1988, 25(6), 523.
162. A.P. Bruins, T.R. Covey and J.D. Henion, **Anal. Chem.**, 1987, 59(22), 2642,
163. T.R. Covey, A.P. Bruins and J.D. Henion, **Org. Mass Spectrom.**, 1988, 23(3), 178.
164. M. Sakairi and H. Kambara, **Anal. Sci.**, 1988, 4(2), 199.
165. L.G. Wright, J.C. Schwartz and R.G. Cooks, **Trends Anal. Chem.** (Pers. ed.), 1986, 5(9), 236.
166. K.L. Busch and G.C. DiDonato, **Am. Lab.**, 1986, 18(8), 17.
167. K. Sato, T. Asada, M. Ishihara, F. Kunihiro et al., **Anal. Chem.** 1987, 59(13), 1652.
168. F.W. McLafferty and J.I. Amster, **Int. J. Mass Spectrom. Ion Processes**, 1986, 72(1-2), 85.
169. R.P. Grese, D.L. Rempel and M.L. Gross, **ACS Symp. Ser.**, 1987, 359 (Fourier Transform Mass Spectrom.), 34.
170. D. Williams and C. Porter, **Am. Lab.**, 1988, 20(2), 98.
171. R.G. Cooks and O.W. Hand, **Nucl. Instrum. Methods Phys. Res.**, Sect. B, 1987, B29(1-2), 427.
172. J.C. Schwartz and R.G. Cooks, **Spectroscopy** (Ottawa), 1987, 5(1-6), 49.
173. D.H. Russell, **Life Sci Res. Rep.**, 1986, 33. 631.
174. F.W. McLafferty, in "Mass Spectrom. Anal. Large Mol.", ed. D.C. McNeal, Wiley (UK), 1986, p.107.
175. S.P. Markey, **J. Clin. Pharmacol.**, 1986, 26(6), 406.
176. D.L. Smith, **Nat. Prod. Chem.**. Proc Int. Symp. Pak. - U.S. Binatl. Workshop, ed. A.U. Rahman, 1984, p,441.
177. W.N. Delgass and R.G. Cooks, **Science**, 1987, 235(4788), 545.
178. M.L. Gross, K.B. Tomer, R.L. Cerny and D.E. Giblin, in "Mass Spectrom. Anal. Large Mol.", ed. C.J. McNeal, Wiley (UK), 1986, 171.
179. D.F. Hunt, J.R. Yates III, J. Shabanowitz, S. Winston and C.R. Hauer, **Proc. Natl. Acad. Sci. U.S.A.**, 1986, 83(17), 6233.
180. R.S. Johnson and K. Biemann, **Biochemistry**, 1987, 26(5), 1209.

181. R.W. Mathews, R.S. Johnson, K.L. Cornwell, T.C. Johnson, B.B. Buchanan and K. Biemann, **J. Biol. Chem**, 1987, 262(16), 7537.
182. D.F. Hunt, J.R. Yates III and J. Shabanowitz, in "Methods Protein Sequence Anal.", ed. K. Walsh, Humana Press, 1987, p.149.
183. K. Biemann and H.A. Scoble, **Science**, 1987, 237(4818), 992.
184. U. Rapp, M. Hoehn and G. Dielmann, **Chim. Oggi**, 1987, 3, 76.
185. K. Eckart, H. Schwarz, M. Chorev and C. Gilon, **Eur. J. Biochem.**, 1986, 157(1), 209.
186. R.S. Johnson, S.A. Martin, K. Biemann, J.T. Stults and J.T. Watson, **Anal. Chem.**, 1987, 59(21), 2621.
187. K.B. Tomer, M.L. Gross, H. Zappey, R.H. Fokkens and N.M.M. Nibbering, **Biomed. Environ. Mass Spectrom.**, 1988, 15(12), 649.
188. V.H. Wysocki, H.I. Kenttamaa and R.G. Cooks, **Int. J. Mass Spectrom. Ion Processes**, 1987, 75, 181.
189. D.L. Bricker and D.H. Russell, **J. Am. Chem. Soc.**, 1986, 108, 6174.
190. J.P. Kiplinger and M.M. Bursey, **Org. Mass Spectrom.**, 1988, 23, 342.
191. D.M. Desiderio, **Int. J. Mass Spectrom. Ion Processes**, 1986, 74(2-3), 217.
192. K. Biemann, in "Methods in Protein Sequence Analysis", ed. K.A. Walsh, Humana Press, 1987, p.123.
193. K. Biemann, **Anal. Chem.**, 1986, 58, 1288A.
194. M.M. Shiel and P.J. Derrick, **Org. Mass Spectrom.**, 1988, 25, 429.
195. P. Rudewicz and K.M. Straub, **Anal. Chem.**, 1986, 58(14), 2928.
196. R.B. Cole, C.R. Guenat and S.J. Gaskell, **Anal. Chem.**, 1987, 59(8). 1139.
197. D.S. Millington, D.A. Maltby and C.R. Cole, **Clin. Chim. Acta**, 1986, 155(2), 173.
198. M.S. Lee, R.A. Yost and R.J. Perchalski, **Annu. Rep. Med. Chem.**, 1986, 21, 313.
199. M.S. Lee and R.A. Yost, **Biomed. Environ. Mass Spectrom.**, 1988, 15(4), 193.
200. K.B. Tomer and M.L Gross, **Biomed. Environ. Mass Spectrom.**, 1988, 15(2), 89.
201. J.E. Coutant, R.J. Barbuch, D.K. Satonin and R.J. Cregge, **Biomed. Environ. Mass Spectrom.**, 1987, 14(7), 325.
202. H. Schweer, C.O. Meese, O. Fuerst, K.P. Gonne and H.W. Seyberth, **Anal. Biochem.**, 1987, 164(1), 156.
203. R.J. Perchalski, M.S. Lee and R.A. Yost, **J. Clin. Pharmacol.**, 1986, 26(6). 435.
204. K.B. Tomer, N.J. Jensen, M.L. Gross and J. Whitney, **Biomed. Environ. Mass Spectrom.**, 1986, 13(6), 265.
205. H. Schweer, H.W. Seyberth and R. Schubert, **Biomed. Environ. Mass Spectrom.**, 1986, 13(11), 611.
206. K.M. Straub, P. Rudewicz and C. Garvie, **Xenobiotica**, 1987, 17(3), 413.
207. J.O. Lay, Jr., D.W. Potter and J.A. Hinson, **Biomed. Environ. Mass Spectrom.**, 1987, 14(9), 517.
208. S.J. Gaskell, **Biomed. Environ. Mass Spectrom.**, 1988, 15(2), 99.
209. P.E. Haroldsen, M.H. Reilly, H. Hughes, S.J. Gaskell and C.J. Porter, **Biomed. Environ. Mass Spectrom.**, 1988, 15(11), 615.
210. S.A. Coran, M. Bambagiotti-Alberti, F.F. Vincieri and G. Moneti, **J. Pharm. Biomed. Anal.**, 1987, 5(5), 509.
211. E.M.H. Finlay, D.E. Games, J.R. Startin and J. Gilbert, **Biomed. Environ. Mass Spectrom.**, 1986, 13(11), 633.
212. T. Krishnamurthy and E.W. Sarver, **Biomed. Environ. Mass Spectrom.**, 1988, 15(4), 185.
213. H.F. Schroeder, **Gewaesserschutz, Wasser, Abwasser**, 1987, 95, 347.
214. M.L. Trehy, R.A. Yost and C.J. Miles, **Environ. Sci. Technol.**, 1986, 20(11), 1117.

215. E. Schneider and K. Levsen, **Comm. Eur. Communities**, EUR 10388, Org. Micropollut. Aquat. Environ., 1986, p.14.
216. G.L. LeBel, D.T. Williams, J.J. Ryan and B.P.Y. Lau, in "Chlorinated Dioxins Dibenzofurans Perspect", ed. C Rappe, G. Choudhary and L.H. Keith, Lewis: Chelsea, Mich., 1986, p.329.
217. B.P.Y Lau, P. Michalik, C.J. Porter and S. Krolik, **Biomed. Environ. Mass Spectrom.**, 1987, $14(12)$, 723.
218. J.K. Porter, C.W. Bacon, R.D. Plattner and R.F. Arrendale, **J. Agric. Food Chem.**, 1987, 35(3), 359.
219. R.D. Plattner, **Bioact. Mol.**, 1986, $1$, 195.
220. H.T. Kalinowski, H.R. Udseth, B.W. Wright and R.D. Smith, **Anal. Chem.**, 1986, $58(12)$, 2421.
221. T. Krishnamurthy and E.W. Sarver, **Anal. Chem.**, 1987, $59(9)$, 1272.
222. S.V. Hummel and R.A. Yost, **Org. Mass Spectrom.**, 1986, $21(12)$, 785.
223. Y. Tondeur, G.W. Sovocool, R.K. Mitchum, W.J. Niederhut and J.R. Donnelly, **Biomed. Environ. Mass Spectrom.**, 1987, $14(12)$, 733.
224. J.A.G. Roach and D. Andrzejewski, **Chem. Anal. (NY)**, 1987, $91$. 187.
225. J.D. Rosen in "Pestic. Sci. Biotechnol., Proc. Int. Congr. Pestic. Chem. 6th Meeting 1986", ed. R. Greenhalgh and T.R. Roberts, Blackwell: Oxford U.K., 1987, p.301.
226. R.D. Voyksner, W.H. McFadden and S.A. Lammert, **Chem. Anal.(NY)**, 1987, 91,247.
227. J.A. Page, **Pestic. Sci.**, 1987, $18(4)$, 291.
228. R. Kleopfer, M. Gerken, A. Carasea and D. Morey, **ACS Symp. Ser.**, 1987, $338$, (Solving Hazard. Waste Probl.), 259.
229. Y. Tondeur, W.N. Niederhut, J.E. Campana and S.R. Muissler, **Biomed. Environ. Mass Spectrom.**, 1987, $14(8)$, 449.
230. R.E. Clement, B. Bobbie and V. Taguchi, **Chemosphere**, 1986, $15(9-12)$, 1147.
231. J.S. Smith, D.B. Hur, M.J. Urban et al., in "Chlorinated Dioxins Dibenzofurans Perspect", ed. G.Choudhary and L.H. Keith, Lewis: Chelsea, Mich., 1986, p.367.
232. R.D. Kleopfer, **ASTM Spec. Tech. Publ.**, 1986, $925$, 124.
233. G.W. Sovocool, R.K. Mitchum, Y. Tondeur, W.D. Munslow, T.L. Vonnahme and J.R. Donnelly, **Biomed. Environ. Mass Spectrom.**, 1988, $15(12)$, 669.
234. R. Guevremont, R.A. Yost and W.D. Jamieson, **Biomed. Environmental Mass Spectrom.**, 1987, $14$, 435.
235. W.C. Brumley, B.J. Canas, G.A. Perfetti, M.M. Mossoba, J.A. Sphon and P.E. Corneliussen, **Anal. Chem.**, 1988, $60(10)$, 975.
236. T. Cairns, E.G. Siegmund, M.A. Luke and G.M. Doose, **Anal. Chem.**, 1987, $59(17)$, 2055.
237. M.B. Wise, **Anal. Chem.**, 1987, $59(18)$, 2289.
238. L.J. DeKoning, R.H. Fokkens, F.A. Pinkse and N.M.M. Nibbering, **Int. J. Mass Spectrom. Ion Processes**, 1987, $77(1)$, 95.
239. R.B. Cody, **Anal. Chem.**, 1988, $60(9)$, 917.
240. D.F. Hunt, J. Shabanowitz and J.R. Yates III, **J. Chem. Soc., Chem. Commun.**, 1987, 548.
241. K. Eckart, H. Schwarz, G. Becker and H. Kessler, **J. Org. Chem.**, 1986, $51$, 483.
242. G. Perseo, R. Forino, M. Galantino, B. Giola, V. Malatesta and R. De Castiglione, **Int. J. Peptide Protein Res.**, 1986, $27$, 51.
243. B.W. Gibson, A.M. Falick, A.L. Burlingame, G.L. Kenyon, L. Poulter, D.H. Williams and P. Cohen, in "Methods in Protein Sequence Analysis 1986", ed. K.A. Walsh, Humana Press, Clifton N.J., 1987, p.463.
244. N.J. Jensen, K.B. Tomer and M.L. Gross, **Lipids**, 1987, $22(7)$, 480.
245. B. Domon and C.E. Costello, **Biochemistry**, 1988, $27(5)$, 1534.
246. K.B. Tomer, N.J. Jensen and M.L. Gross, **Anal. Chem.**, 1986, $58(12)$, 2429.
247. J.J. Fournie, M. Riviere and G. Puzo, **J. Biol. Chem.**, 1987, $262(7)$, 3174.

248. N.J. Jensen, K.B. Tomer and M.L. Gross, **Lipids**, 1986, 21(9), 580.
249. R.L. Cerny, K.B. Tomer and M.L. Gross, **Org. Mass Spectrom.**, 1986, 21(10), 655.
250. J. Adams, L.J. Deterding and M.L. Gross, **Spectroscopy** (Ottawa), 1987, 5(1-6), 199.
251. H. Schweer, H.W. Seyberth, C.O. Meere and O. Fuerst, **Biomed. Environ. Mass Spectrom.**, 1988, 15(3), 143.
252. N.J. Jensen and M.L. Gross, **Mass Spectrom. Rev.**, 1988, 7(1), 41.
253. J.J. Dino, C.R. Guenat, K.B. Tomer and D.G. Kaufman, **Rapid Commun. Mass Spectrom.**, 1987, 1(4), 69.
254. L.M. Mallis, F.M. Raushel and D.H. Russell, **Anal. Chem.**, 1987, 59(7), 980.
255. I. Isern-Flecha, X.Y. Jiang, R.G. Cooks, W. Pfleiderer, W.G. Chae and C.J. Chang, **Biomed. Environ. Mass Spectrom.**, 1987, 14(1), 17.
256. K.B. Tomer, M.L. Gross and M.L. Deinzer, **Anal. Chem.**, 1986, 58(12), 2527.
257. R.L. Cerny, M.L Gross and L. Grotjahn, **Anal. Biochem.**, 1986, 156(2), 424.
258. K. Hirayama, S. Akashi, T. Ando, I. Horino, Y. Etoh, H. Morioka, H. Shibai and A. Murai, **Biomed. Environ. Mass Spectrom.**, 1987, 14(7), 305.
259. R. Cooper and S Unger, **J. Org. Chem.**, 1986, 51(21), 3942.
260. R.M. Facino and M. Carini, **Ann. 1st. Super. Sanita**, 1986, 22(1), 19.
261. M.M. Siegel, W.J. McGahren, K.B. Tomer and T.T. Chang, **Biomed. Environ. Mass Spectrom.**, 1987, 14(1), 29.

# 8
# Mass Spectrometry Applied to Natural Products: Nucleosides, Nucleotides, and Nucleic Acids

BY P. VIGNY AND A. VIARI

1 Introduction

Mass spectrometry is playing an increasing role in the identification of naturally occurring and synthetic nucleic acid components and of their biologically mediated covalent interactions with small xenobiotic molecules such as drugs or mutagens. In the area of naturally modified nucleosides, the discovery of the so-called 'hypermodified' nucleosides having complex structures and often isolated in low microgram-level quantities constitutes an exalting challenge in which mass spectrometry has already played and continues to play a prominent role. The exponential quantitative and qualitative development in the chemical synthesis of various families of oligonucleotides, which are extensively used either as linkers and adaptators in recombinant DNA techniques or to perform site-directed mutagenesis or gene regulation, has also revealed the urgent need for rapid and reliable physical characterization techniques. The impressive shifts of the mass spectrometry frontiers, both in the detection of involatile compounds and in the analysis of large molecules, has allowed mass spectrometry to enter efficiently this domain. Concerning the elucidation of DNA lesions induced by physical agents -such as UV light or ionizing radiations- or by chemical agents, one can consider that most of the chemical structures of the major modifications of nucleosides are now known. Although minor ones are still being discovered, the present challenge is related to the ability to identify them, to detect them in a quantitative manner at very low doses either in cells or tissues under conditions where the biological responses are available. Although immunological methods have received an increasing attention in the detection of such DNA lesions, due to their exceptional sensitivity and to the ability to visualize the lesions within individual cells and tissues, the efficiency of gas chromatography / mass spectrometry (GC/MS) and the potentiality of thermospray mass spectrometry coupled with high performance liquid chromatography (thermospray LC/MS) begin to be recognized. These techniques now appear as potential candidates for the *in vivo* detection of such lesions. There is no doubt that their use will be extended in the near future.

The present review covers more than seventy papers dealing exclusively or only partly with mass spectrometry of nucleosides, nucleotides and oligonucleotides. Mass spectrometry of nucleic bases themselves is not covered. For additional information the reader should also consult review articles published during the same period and dealing totally[1,2,3] or partly[4] with the same subject.

2 Nucleosides

Mass spectrometry has been shown in recent years to be a suitable technique for the structural determination of 'hypermodified' naturally occurring nucleosides and the recent review by McCloskey has discussed the respective advantages of the GC/MS of silyl derivatives and of the thermospray LC/MS approaches[1]. Highly modified nucleosides are mainly found in the tRNAs. Their formation requires multistep enzymatic processing and the resulting base modifications are relatively complex. The five principal groups of such nucleosides are those related to queuosine (Q), wye nucleosides (Y), 2-thio-5-substituted uridines ($s^2x^5U$) and $N^6$-substituted adenosines with isopentenyl or threonyl side chains. In the past two years covered by the present review, the structural characterization of a highly fluorescent member of the Y nucleoside family, namely 3-(β-D-ribofuranosyl)-4,9-dihydro-4,6,7-trimethyl-9-oxoimidazo[1,2-a]purine or mimG (1a), as well as of four methylated nucleosides has been carried out from extremely thermophilic archaebacterium tRNAs[5,6]. The method used combines a direct LC/MS study of the tRNA digest followed by a fast atom bombardment (FAB) mass spectrometric examination of the isolated nucleoside, electron impact (EI) analysis of its silylated derivative and a comparison with the corresponding synthetic compound. A new nucleoside belonging to the Q family has been identified in tRNA$^{Tyr}$ from E.Coli strain MRE 600[7]. It has been characterized as an epoxy derivative of queuosine, namely 7-{5-[(2,3-epoxy-4,5-dihydroxycyclopent-1-yl)amino]methyl}-7-deazaguanosine (1b). This finding constitutes the first report of an epoxide formation during

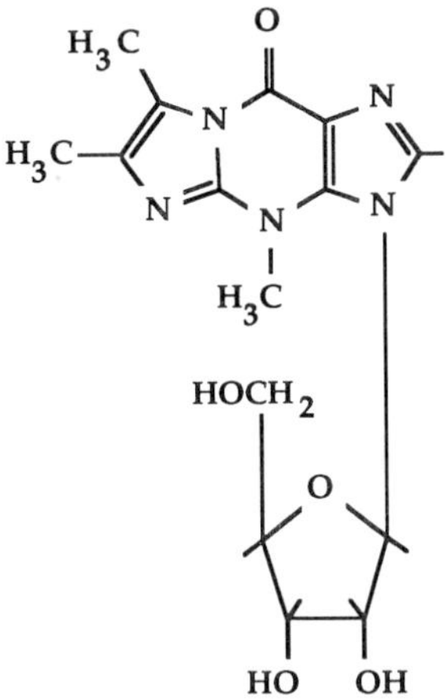

# 8: Mass Spectrometry Applied to Natural Products

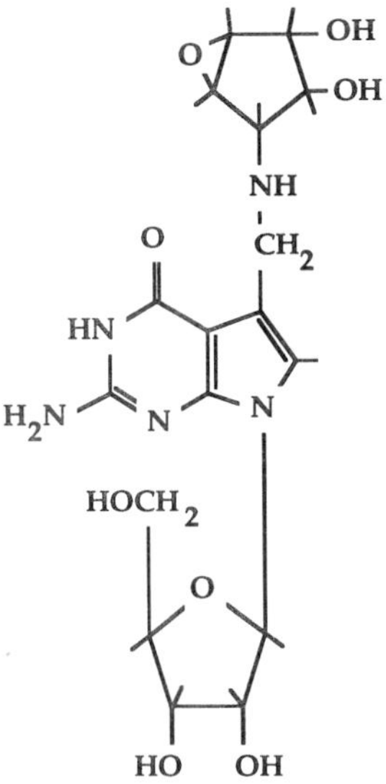

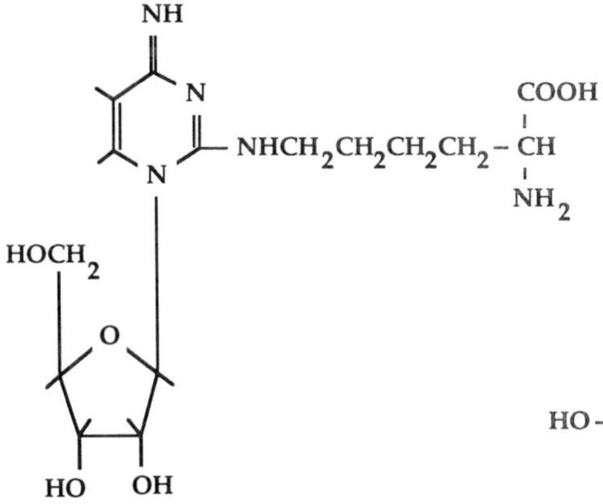

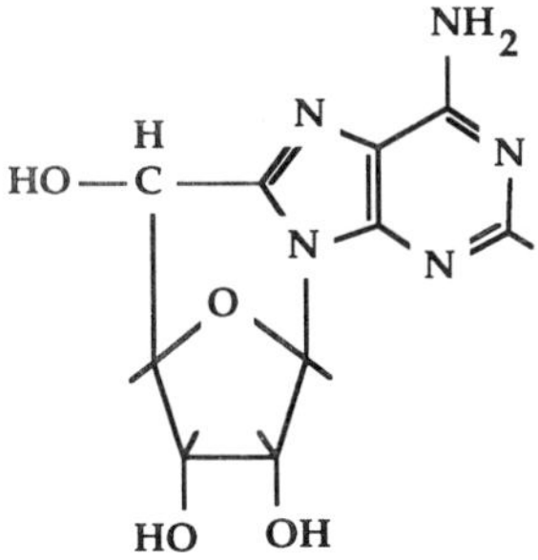

(1a)

(1b)

(1c)

(2)

the post transcriptional processing of RNA. A novel type of modified cytidine with a lysine moiety, namely 4-amino-2-($N^6$-lysino)-1-($\beta$-D-ribofuranosyl) pyrimidinium (1c), has also been characterized from a minor species of tRNA$^{Ile}$ from E. Coli A 19[8]. Recently the structure of the nucleoside antibiotic Liposidomycin B has been elucidated[9]. Another interesting aspect of nucleic acid research using mass spectrometry is its application to enzymatic activity monitoring. In this area, the enzymatic activity of a tRNA methyl transferase from E. Coli K-12 which is involved in the biosynthesis of a particular 2-thio-5-substituted uridine recalled above − namely the 5-methylaminomethyl-2-thiouridine (mnm$^5$s$^2$U) − has been approached by various biochemical and physico-chemical techniques including MS[10]. Mass spectrometry has in particular contributed to demonstrate that the two different intermediates in the biosynthesis of mnm$^5$s$^2$U are present in the tRNA of two mutants. The corresponding modified nucleosides have been found to be either a simpler (5-aminomethyl-2-thiouridine or nm$^5$s$^2$U) or a more complex (5-carbethoxymethylaminomethyl-2-thiouridine or cmnm$^5$s$^2$U) product than the final product mnm$^5$s$^2$U itself. The mechanism by which DNA (Ade-6-)methyltransferase catalyses the formation of m$^6$Ade was subject of discussion. The investigation carried out on a dodecamer substrate of the enzyme with a [6-$^{15}$N]adenine in the methylation site followed by isotopic analyses indicated that the methylase proceeds via a direct $N^6$-methylation and not by $N^1$-methylation followed by a rearrangement as expected from chemical considerations[11].

An increasing number of studies devoted to nucleosides are now dealing with the chemical characterization of nucleosides that have been modified by potentially mutagenic or/and carcinogenic chemicals or by physical agents (radiations). The studies are carried out on model compounds obtained *in vitro*. Most of them contain an exclusive mass spectrometric approach whereas a few combine the use of several physico-chemical techniques. In this latter case, mass spectrometry is mostly used as an accompanying technique to nuclear magnetic resonance (NMR). Amongst the former class of studies are nucleoside adducts formed with metabolites of pyrrolizidine alkaloids found in a variety of plants and recognized as hazardous to man. Several series of such adducts have been investigated by means of FAB/MS in combination with tandem mass spectrometry (MS/MS) which overcomes the poor fragmentation pattern generally observed with FAB[12]. The same techniques, including collision-induced decomposition (CID) mass analyzed ion kinetic energy (MIKE) measurements, were used to discriminate between various positional isomers of benzylated guanosines considered as model carcinogen-nucleic acid adducts[13], whereas metastable ion studies with a secondary ion time-of-flight (TOF) mass spectrometer allowed the discrimination of 2-O- and 4-O-alkylthymidines[14]. In the area of damage introduced to DNA by reactive oxygen species such as hydroxyl radicals when exposed to the $\gamma$-rays of $^{60}$Co, the use of GC/MS combined with microderivatization of DNA hydrolysates allowed their individual characterization[15,16,17,18]. The use of selected ion monitoring (SIM) facilitates the detection of such lesions at low quantities corresponding to radiation doses as low as 0.1 to 10 Gray which are relevant to studies on biological systems. GC/MS/SIM has therefore been suggested as an ideal analytical tool for the identification of such lesions in DNA, for their detection in biological fluids and for the study of their repair and biological consequences. In parallel to this approach, the high performance liquid chromatography/MS coupling (HPLC/MS) has been recognized as an alternative powerful route in the chemical identification of unknown modified nucleosides after digestion of nucleic acids. A detailed HPLC/MS/MS analysis of the various radiolysis products resulting from hydroxyl radical attacks on polyadenylic acid has been carried out[19]. It has lead to the identification of (R)- and (S)-8,5'-cycloadenosine shown in (2), 8-hydroxyadenosine and $\alpha$-adenosine, most of the information coming from the CID spectrum of the base-containing fragment $[B + 2H]^+$. Self-chemical ionization Fourier transform ion cyclotron resonance (FT-ICR) mass spectra of alkylated nucleosides and exocyclic adducts have also been reported[20], most of them showing a single peak corresponding to the protonated molecular ion. The high resolution capabilities of this technique combined with tandem mass spectrometric ($MS^n$) possibilities allows promising developments of FT-ICR in the identification of unknown adducts and in the differentiation of isobaric species. Amongst the studies in which mass spectrometry is used in conjunction with other physico-chemical techniques, one should mention a FAB with CID approach of the chemical carcinogen benzo[a]pyrene-guanine adduct formed by a one-electron oxidation pathway yielding a radical cation intermediate[21], a field desorption (FD) characterization of a guanosine-trans-4-N-acetoxy-N-acetylaminostilbene adduct[22], a FAB and EI study of a 5-bromouracil covalently bound to peptides considered as a model for bromouracil-DNA protein photocross-linking[23], and FAB characterizations of sensitized photo-oxidation products of thymidine[24] and of a covalent cross-link bis-adduct between mitomycin C and a B-DNA decamer.[25] Mass spectrometry has also been used as an aid in the synthesis of various modified nucleosides of potential interest prepared for pharmacological studies and as a tool to follow their fate in biological media. $^{252}$Cf-plasma desorption mass spectrometry (PDMS), because of its ease of use and of the straightforward interpretation of the spectra, appears to be a suitable technique to monitor the chemical synthesis process. An example is given with the synthesis of adenosine agonists ($N^6$-phenyladenosines) and antagonists (8-phenyl-1,3-dipropyl xanthine derivatives)[26], whereas HPLC/MS was used to follow and determine the stability of the potential anti-viral agent 2',3'-dideoxyadenosine in biological fluids at therapeutic doses[27], the deamination of the drug being controlled in plasma by thermospray HPLC/MS. The high degree of sensitivity and selectivity of GC/MS has been exploited in a new method for the quantification of 5-bromodeoxyuridine incorporated into DNA and used as a radiation sensitizer[28]. The method allows detection in 1µg DNA sample replacements as low as 1% and will permit the analysis of the presence of the radiosensitizer in cellular DNA from biopsy samples of normal or tumor tissues. Other studies deal with the electron impact mass spectrometric characterization of halo-sugar derivatives as potential tumor inhibitors and the ability to differentiate between 3'- and 5'-derivatives with respect to the 2'-derivatives[29], the synthesis and characterization of labelled [6-$^{15}$N] and [1-$^{15}$N]deoxyadenosines[30]. FAB/MS has been used to follow the chemo- and regioselective 5'-acylation of deoxyribonucleosides in view of the further preparation of nucleotide building blocks[31].

Nucleosides are also currently used as standard biological compounds in order to check new mass spectrometric arrangements and techniques. They have been used in particular as standard non-volatile species to check the detection limits of laser desorption (LD) atmospheric pressure mass spectrometry. In such a method, where the desorption of neutrals and the atmospheric pressure ionization process are performed in two separate steps, detection limits down to ~ 0.3 ng can be reached[32]. The fact that adenosine is not ionized in solution at neutral pH justified its use as a standard compound for a comparison of thermospray and ion spray mass spectrometry in an atmospheric pressure ion source[33]. Adenosine has also been used to illustrate the advantages of a dual-beam thermospray interface constructed with the aim of providing a mode of ionization almost independent of the operating conditions for high performance liquid chromatography[34], and to check the effect of a retarding electrode in the source of a thermospray mass spectrometer[35]. Further experimental conditions for LC/atmospheric pressure ionization/MS have been reported[36]. Another interesting approach is a statistical comparison between radioimmunoassay (RIA) and gas chromatography selected ion monitoring (GC/MS/SIM) methods in the quantitative analysis of adenosine[37]. This comparison is of particular interest in biological and pharmaceutical applications since it is obvious that physico-chemical techniques - including mass spectrometry - will compete with biochemical and immunological techniques in the next few years for the quantitative detection of various relevant molecules in biological fluids. Despite the advantages of simplicity, speed and a broad range of quantification of the immunological method and the disadvantages of time consumption and restricted range of quantification of mass spectrometry, the RIA measurements have been shown to incorporate a greater uncertainty than the GC-MS-SIM measurements. Finally, guanosine was used for preliminary studies on a reflection time-of-flight mass spectrometer modified to allow the detection of neutral species formed by the decomposition of metastable ions in the flight tube[38] and cytidine to check the advantages of amic acid derivative formation as charged surface-active derivatives in the mass spectrometric detection of amines[39].

### 3 Nucleotides

Because of their high polarity due to the phosphate moiety, nucleotides are generally not directly amenable to classical MS techniques and thus constitute a class of choice for 'soft ionization' techniques, particularly FAB[3]. Although their derivatization is not needed to observe molecular ions, this operation has been shown to restore structurally informative fragmentation absent in the positive-ion FAB spectra of adenosine and AMP[40]. The spectra of trimethylsilyl derivatives are then very similar to those obtained using EI.

The addition of different organic bases in the glycerol matrix is regularly proposed to enhance the negative molecular ion yield of molecules containing dissociable protons such as nucleotides. As an example, a significant improvement of the abundance of ATP quasi-molecular ions was obtained by the use of 1,8-diazabicyclo[5.4.0]-undec-7-ene (DBU)[41]. For the same reasons, acidic additives are suggested in positive mode[42]. Both the acidic and surfactant properties of p-toluenesulfonic acid contribute to improve the protonated molecular ion yield of a series of dinucleotides of the structure $P^1,P^n$-di(adenosine-5')-n-phosphate (n=di- through penta-). This chemical aspect of FAB/MS has been included in a review[43] and recently illustrated by the evidence of a FAB regioselective phosphate deprotection of fully protected phosphorothioate nucleotides[44]. Similarly, the use of a liquid matrix in negative-ion FDMS has been discussed[45]. The role of liquid phase chemistry is more evident in techniques such as thermospray MS which involve buffers in the ionization process itself. The influence of solution-phase chemistry on the ion evaporation process has been explored on various nucleosides and nucleotides[46]. It has been shown that the presence of buffer (ammonium acetate) reduces the solution-phase dependence (e.g. pH effects) and favours the protonated over cationized species. In the absence of buffer, the abundance of these two species directly follows the concentration of the acidic and salted forms in the solution, enabling the determination of pKa values from thermospray mass spectral data.

Fundamental mechanistic studies were undertaken by using metastable and collision-induced decomposition analysis[47]. All 3'- and 5'-monophosphate nucleosides of ribo- and deoxyribo-series were investigated in order to understand the role of the phosphate moiety in the loss of BH (B being the base) in negative-ion FAB/MS. As a general feature, this loss is favoured for the 3'-phosphate isomers. Other MS/MS results concerning di- and oligonucleotides will be detailed in the next sections.

The negative-ion mode is generally the better mode for observing the nucleotides which bear a naturally negatively charged phosphate moiety. Another reason for the preferred use of negative ions lies in the relatively greater complexity of the positive-ion spectra due to the addition of alkali-metal cations. The exclusive use of the negative-ion mode is however relaxed for synthetic uncharged species, zwitterionic or positively charged complexes and for MS/MS studies. As an example, positive-ion FAB/MS was used on a 1:1 mixture of cis—[Pt(NH$_3$)$_2$(OH$_2$)$_2$]$^{2+}$ and 5'-GMP to characterize a cyclic complex involving N$^7$ of the guanine moiety[48]. Another study with positive-ion FAB/MS and CID/MIKE analysis has given particularly straightforward evidence for the natural occurrence of cytidilate cyclase activity, a particularly controversial and confused biological problem[49]. Seven different products including cyclic CMP, formed from CTP in rat tissue extracts, were identified by this means. Four of them are novel cytidine products. The same technique (CID/MIKE) was used to study the dissociation reaction of [M + Na]$^+$ quasi-molecular ions of isotopically labelled UTP in order to determine the position of $^{18}$O before and after a positional isotope exchange (PIX) reaction[50]. In this particular case, advantage is taken of alkali-metal ion attachment to specific sites of the nucleotide to enhance structural information.

In the chemical area, fully and partially protected N$^6$-methyl adenosine 3'-phosphate used as key intermediates in the synthesis of modified oligonucleotides has been reported using negative-ion FAB/MS[51]. The technique has also proven to be helpful for the determination of sample purity of nucleosides 3'-phosphoramidites used as monomeric units in the synthesis of oligonucleotides via the phosphite triester approach[52]. The method is particularly useful for the characterization of unreacted nucleosides or phosphate-free impurities which would not have been detected by $^{31}$P NMR. Other examples of nucleotide characterization by FAB/MS in the nucleic acid synthesis domain have been reported including $^{13}$C labelled GDP[53] and nucleotide mono- and triphosphates containing various cytosine analogs[54], namely 2(1H)-pyrimidinone, 2(1H)-pyridinone and 4-amino-2(1H)-pyridinone. The behaviour of these unnatural bases incorporated into dinucleotides will be further discussed.

### 4 Dinucleotides

There is currently a great diversity in the nomenclature used to describe the fragments of di- and oligonucleotides. This often leads to misunderstandings or difficulties in the comparison of the results coming from different sources. For this reason, we have chosen in the present report the nomenclature depicted in (3). The standard notation has been adopted for the representation of oligonucleotides : e.g. ApApA stands for adenylyl (3'→5') adenylyl

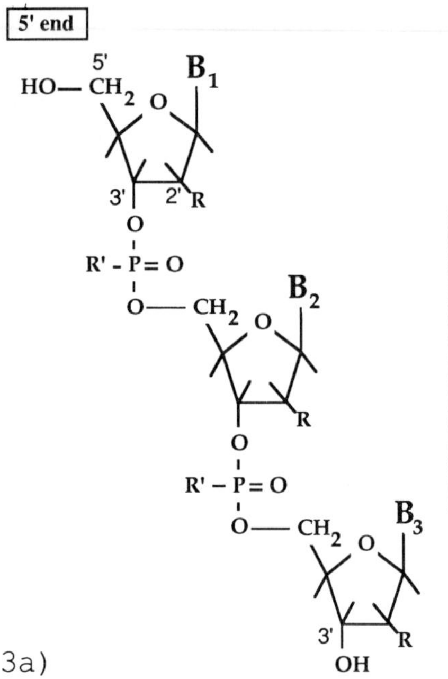

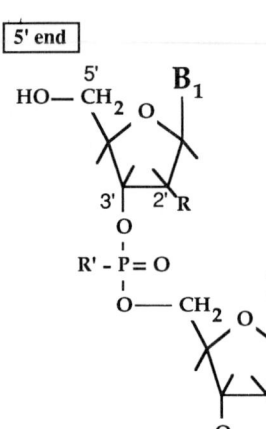

| R | sugar |
|---|---|
| H | deoxyribose |
| OH | ribose |

| R' | phosphate linkage |
|---|---|
| OH | phosphodiester (natural) |
| $CH_3$ | methylphosphonate |
| $OCH_3$ | methylphosphotriester |
| $OCH_2CH_3$ | ethylphosphotriester |
| SH | phosphorothioate |

(3a)

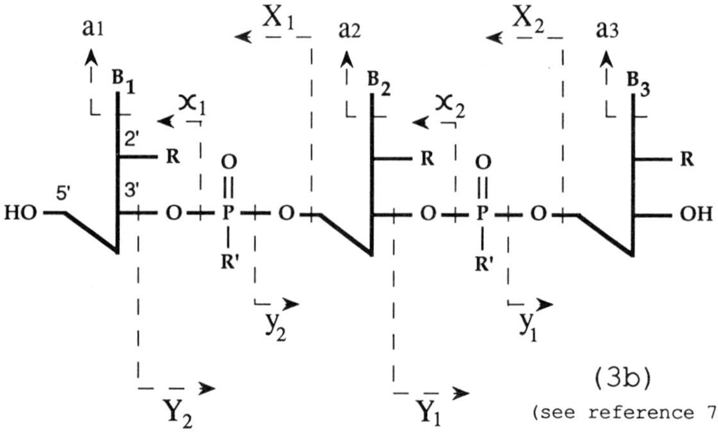

(3b) (see reference 73)

(3'→5') adenine. From this, it follows that the left end of the formula is the 5'-end (see fig.3a) and the right end is the 3'-end. The nomenclature for the fragments is given in the shorthand formula (3b). X denotes the fragments starting from the 5'-end and Y those starting from the 3'-end. The subscript indicates the number of base units retained in the fragment.

There are two closely related aspects in the characterization of oligonucleotides: i) the sequence determination and ii) the absolute orientation. Following the early work of Grotjahn and coworkers[55], the answer to question i) is given by specific 'sequence ions' (X and Y) arising from the cleavage of the phosphodiester bond (3b) and the answer to question ii) comes from the differences in the relative abundance of these ions. In the case of naturally occurring (negatively charged) oligomers, Y species were initially reported to be always more abundant than X species - for the same number of nucleoside units retained in the fragment. At the same time, the validity of this second rule was questioned[56] and this doubt still remains open. The two most sensitive points, in the comparison of the different results, seem to be the effects of base content and base sequence and of sample purity with a special emphasis on the influence of alkali-metal salts.

FAB MS/MS capabilities are of primary interest when studying the underlying mechanisms and fragmentation pathways of di- and oligonucleotides. Such a study was performed on all possible combinations of (3'→5') ribo- and deoxyribonucleoside monophosphates[47]. Metastable and collision-induced decompositions of $[M - H]^-$ are compared. The most abundant ion formed either by metastable or CID decomposition corresponds to the loss of BH from the 5'-end. This is true both for ribo- and deoxyribonucleotides. The X and Y sequence ions are also observed, the latter being always more abundant. A mechanism involving a six-membered ring intermediate is proposed (4) to account for this difference. For the Y fragment ion, two hydrogens (4' and 2') are thus available to build the ring (4a,b) whereas only the 2' hydrogen is available for the X fragment ion (4c). Although this type of study is of fundamental interest, direct negative-ion FAB spectra are often claimed to be sufficient to provide information on normal and modified dinucleotides. As an example eight combinations of unprotected dideoxyribonucleotides with isopropylphosphotriester linkage have been studied by this means[57]. As expected for such

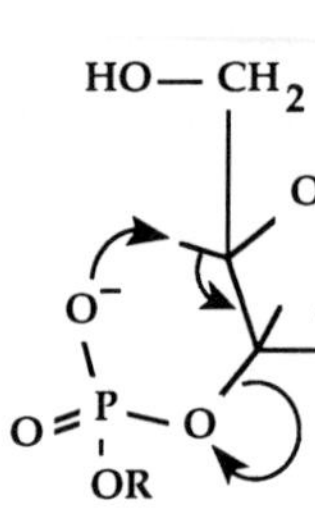

(4a) 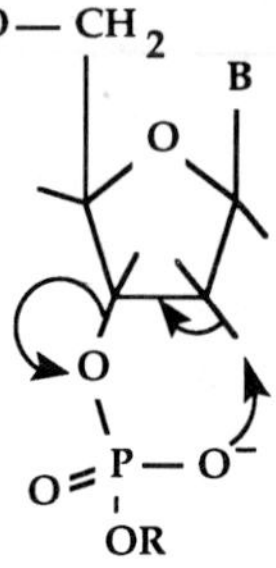   (4b)

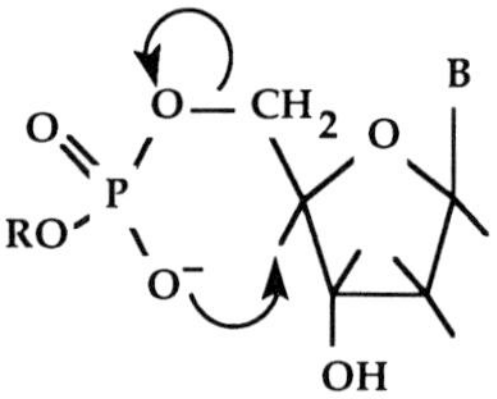

(4c) 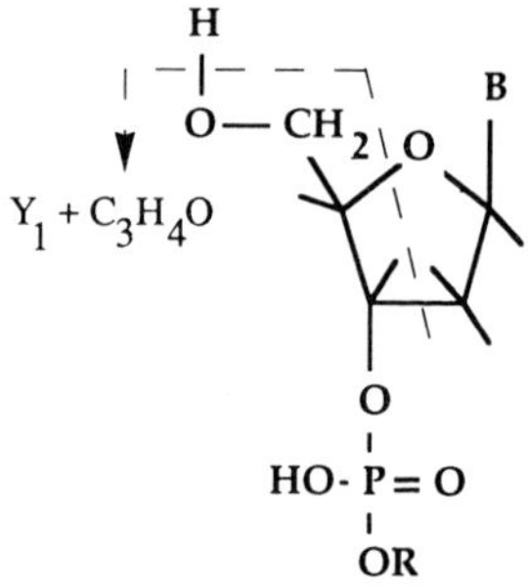   (5)  $Y_1 + C_3H_4O$

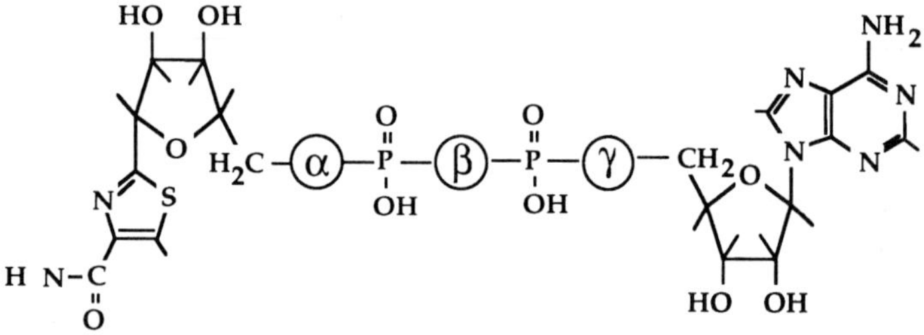

(6)

| α | β | γ |
|---|---|---|
| CH$_2$ | O | O |
| O | CH$_2$ | O |
| O | O | CH$_2$ |

uncharged species, quasi-molecular ions are observed in both polarities in addition to ions characteristic of phosphates and alkylated moiety. Sequence ions (X and Y species) are quoted to be of moderate abundance in negative mode. The Y species seem to be more abundant than the X ones. As mentioned in CID studies, the characteristic fragmentation of the N-glycosidic bond seems to be more likely from the 5'-end (a1 type) than from the 3'-end (a2 type).

Illustrations of the use of MS as an aid in the chemical synthesis of oligonucleotides have been given. Negative-ion FAB/MS is proposed as a very simple and rapid method for the characterization of dimeric phosphoamidite synthons which are used as uncharged key intermediates in oligonucleotide synthesis via the phosphite triester approach[58] and of a dinucleoside monophosphate containing the unnatural base 5,6-dihydro-5-azacytosine[59]. Again, in both cases, the X and Y fragment ions provide the information about the absolute orientation of the dimers. In the case of protected species[58], additional peaks are observed, resulting from the loss of the various protecting groups. The usefulness of the phosphate fragmentation has been clearly illustrated in a study of the NAD analog thiazole-4-carboxamide adenine dinucleotide (TAD) and of three phosphonate analogs[60] (6). When either the $\alpha$, $\beta$, or $\gamma$ oxygen is replaced by a methylene moiety, the corresponding fragment ions are absent from the negative FAB mass spectra. This property, together with the mass shift of these fragments, unambiguously indicates the position of the phosphonate linkage. Similar results were reported, but in the positive-ion mode, on two NAD derivatives bearing a sulfonatopropyl group on the adenine[61]. FAB/MS has also allowed the direct characterization of radiation-induced damage to dinucleosides monophosphates in the ultraviolet[62] and X-ray[63] domains. In the far ultraviolet domain the structure of a new d(ApA) photoproduct has been described.

Although the Y type fragments are generally observed to be more abundant than X type fragments, the generality of this rule remains questionable. As an example, a negative ion FAB mass spectrum of the ribodinucleoside monophosphate ApC has been presented where the X1 fragment ion is more abundant than the Y1 counterpart[54]. Another peak, corresponding to the 3'-end nucleotide plus a $C_3H_4O$ fragment (5) has thus been proposed as a more reliable indicator for the direction of the sequence. The same behaviour is observed for dinucleotides containing unnatural bases, mentioned in the nucleotide section, such as fully deprotected and partially protected d(M'pC) where M' stands for 2(1H)-pyrimidinone. Alternatives to FAB/MS for the analysis of dinucleotides have also been proposed. Positive-ion spectra of the dinucleoside monophosphate d(ApC) by using thermospray HPLC/MS with various experimental conditions were compared[46]. The $[M + H]^+$ quasi-molecular ion is much more abundant when the ion evaporation proceeds without buffer than with ammonium acetate buffer and this ion is totally absent when operating in CI mode (filament on). The advantages of desorption chemical ionization (DCI) over chemical ionization (CI) for the analysis of an unprotected dinucleoside with methylphosphotriester linkage (d(Tp(Me)T)) have been pointed out[64]. Quasi-molecular ions have been obtained in both polarities with DCI whereas no ion containing the methylated phosphate group could be observed by using CI with direct insertion probe. The best sensitivity is achieved in the negative-ion mode (≈ 1 ng limit). MS/MS data are also provided in order to ascertain peak assignments and to define a selected ion monitoring protocol suitable for mixture analysis and quantitative analysis at the subpicomole level. Finally, a so-called ERIAD technique has been used for the localization of the binding point of the chicken H4 histone to the depurinated dinucleotide d(TpG)[65]. A cross-linked nucleotide-peptide has thus been isolated and identified on the basis of the mono- and doubly charged positive molecular ions.

Special attention should be paid to dinucleotide model compounds in which the two bases are connected by a trimethylene chain instead of the sugar-phosphate link. This makes them directly amenable by classical MS techniques. Such analogs, containing 6-azathymine connected to 5-alkyluracils[66] and a cytosine-5-methoxyuracil cyclobutanic photoproduct[67], have been studied by EIMS.

## 5 Oligonucleotides

The pioneering work concerning the application of FAB/MS to the characterization of DNA fragments has been reviewed[68,69]. The negative-ion mode is the better mode for negatively charged oligomers since even traces of alkali metal salts ($Na^+$, $K^+$) lead to an abundant cationization, to more complex patterns and even to the loss of informative peaks. Fully protected oligomers cannot be directly studied by FAB and need to be partially deprotected at a terminal phosphate prior to the negative-ion FAB/MS analysis. This can be done either by chemical degradation of the phosphate protecting group or by direct reaction with the matrix. For such protected oligomers, bearing a free phosphate at the 3'-end, X fragment ions (with a monomethoxytrityl protecting group on the 5'-end) are the most abundant. The same behaviour is observed when normal, negatively charged, oligomers are only protected by a monomethoxytrityl group at the 5'-end whereas the Y fragments are dominating for completely deprotected species. Spectra of protected oligomers are quoted to be more complex because of additional losses of the protecting groups. No essential differences are found in the behaviour of ribo- and deoxyribonucleotides.

As mentioned before, the preferred cleavage of the phosphodiester bond from the secondary C(3') atom (Y sequence ions) rather than from the primary C(5') atom (X sequence ions) for unprotected species has been disputed[54]. In a study with tri-, tetra- and penta-nucleotides in the ribo- and deoxyribo-series, no evident correlation was found between the peak intensities and the position of the terminal phosphate group. In several examples, and particularly for longer oligomers, it was impossible to determine the absolute orientation. At the present time, no rational explanation has been proposed to account for this discrepancy. A way to avoid such difficulties is to use metastable and CID capabilities of FAB/MS. A study, similar to that reported in the dinucleotide section, has been undertaken by the same group[70] with oligoribo- and oligodeoxyribonucleotides of length up to 6-mers. The same approach, i.e. metastable and collision-induced decomposition of the $[M - H]^-$ ions produced by FAB desorption, was used. The preferential loss of the base moiety, from the 5'-end, which is, together with the loss of $H_2O$ and CONH, the most favourable process for di- and trinucleotides, tends to disappear in higher oligomers to the advantage of the X and Y fragmentation (the latter being the most abundant). However, for higher oligomers (penta and hexanucleotides) several other and more complex fragmentation processes compete with the sequence ion formation, leading to more complicated CID spectra. It must also be pointed out that, unlike their trideoxyribo- counterparts, triribonucleotides undergo an $y_2$ type, instead of $Y_2$, fragmentation (3b). This phenomenon is limited to the $y_2$ (and to a lesser extent $x_2$) type fragmentation and no $y_1$ (nor $x_1$) ions species are observed. Very recently, similar fragmentations for oligodeoxynucleotide carbamate analogs have been described allowing the determination of their sequences up to 6-mers[71].

FAB/MS/MS was also used to investigate the two tetradeoxyribonucleotides d(pCpTpCpT) and d(ApGpCpT) modified *in vitro* by BPDE, the ultimate metabolite of benzo[a]pyrene[72]. The CID spectra of the $[M + 3Na - 4H]^-$ ion from d(pCTCT) and of $[M - H]^-$ ion from d(ApGpCpT) exhibit the same features previously reported for oligonucleotides[70]. Although the spectra are much more complicated and rather noisy, particularly those obtained with the cationized species, the differences in the observed sequence ions, between modified and unmodified oligomers allow different possible sites of fixation of BPDE to be assigned.

Parallel to FAB/MS, Plasma Desorption mass spectrometry (PDMS) has been shown to provide diagnostic negative ion spectra of oligonucleotides. In order to understand in terms of sequence information, the main features of experimental PD spectra, ten tri deoxyribonucleoside monophosphates were chosen to be analyzed by this technique[73]. The main results are similar to those described for FAB/MS. The more informative ions are the $[M - H]^-$ quasimolecular ion, losses of the base moieties and X and Y sequence ions. With PDMS, the Y type ion species have always been observed to be more abundant than the X type. Another fragmentation has been suggested which leads to the central nucleotide 3'- and 5'-phosphate unit. This process is generally followed by water elimination. Although this fragmentation pathway is less favoured with trinucleotides, it has been shown to be more important as the length of the oligomers increases. As an example, it is predominant over $X_1$ and $Y_1$ sequence ions in the negative-ion PD spectra of the tetramer d(TpApTpA)[74] and the hexamer d(TpApTpApTpA)[75]. As heavy ion induced desorption occurs in the solid state, unlike liquid-phase FAB, the importance of the surface on which the samples are deposited (Mylar foil) has been pointed out and has been illustrated in the case of the diribonucleotide ApC and of the triribonucleotide ApApC[76]. A cationic surfactant, namely tridodecylmethylammonium chloride (TDMAC) was used to coat the surface in order to reduce both intermolecular binding forces and the desorption-quenching effect of matrix impurities. A significant enhancement in the yield of both molecular and fragment ions was obtained with the dimer and to a lesser extent with the trimer. In a similar way, a nitrocellulose coated surface - which is now routinely used in PDMS of proteins - was used to assist quasi-molecular ion desorption of the nona deoxyribonucleoside methylphosphonate d($T_9$)[75]. The positive and negative-ion PD spectra of these neutral oligonucleotides are complementary. The positive-ion mode generally gives rise to abundant cationized quasi-molecular ions whereas negative-ion spectra provide more diagnostic information about the sequence (Y type sequence ions) but no molecular ion[77]. The essential difference between normal negatively charged and uncharged oligomers lies in the absence of the X type sequence ions.

Secondary ion mass spectrometry (SIMS), in conjunction with TOF analysis, has also been applied to unprotected, partially and fully protected oligo-deoxyribonucleotides, with mixed results[78]. Fully protected oligomers (from di- to tetramers) display only positive cationized [M + Ag]$^+$ quasi-molecular ions (Ag arises from the substrate) but no negative molecular ions. Sequence ions are observed as X species. Partially protected oligomers (with the 3'-end phosphate protecting group removed) also display positive cationized quasi-molecular ions in addition to cationized X and Y sequence ions. In the negative-ion mode [M − H]$^−$ is now observed. Finally, fully deprotected species (tetra- to hexamers) display abundant [M + H]$^+$ and [M − H]$^−$ ions, but no sequence ions at all in either polarities. The same SIMS-TOF technique was used to differentiate between protected trinucleotides with vicinal 2'-5' and 3'-5' phosphodiester linkages[79]. A mechanism has been proposed to account for the difference in the fragmentation patterns observed in positive and negative ion spectra of two protected diastereoisomers A(2'p5'U)3'p5'U and arabino-A(2'p5'U)3'p5'U. The cleavage of the phosphodiester linkage is favoured when 2'- and 3'-phosphate oxygens are in *trans* rather than *cis* orientation with respect to the sugar plane although it is not possible to decide between a 2'-5' and 3'-5' cleavage because of the palindromic nature of the trinucleosides which were chosen.

Finally, a quadrupole mass spectrometer equipped with an ion spray interface was recently used to produce a negative-ion mass spectrum of an oligodeoxyribonucleotide with 14 bases[80]. This ionization technique produces multiply-charged molecular ions of the form [M − nH]$^{n-}$ with n ranging from 6 to 11 in this case. However no other information (sequence ions) is found in the spectra.

## References

1. J.A. McCloskey, in 'Mass Spectrometry in Biomedical Research', ed. S.J. Gaskell, John Wiley & Sons, Chichester, 1986, p.75.
2. K. Jankowski, J.R.Jocelyn Paré and R.H. Wightman, Adv. Heterocyl. Chem., 1986, 39, 79.
3. K.H. Schram, Trends Anal. Chem., 1988, 7, 28.
4. A.L. Burlingame, D. Maltby, D.H. Russell and P.T. Holland, Anal. Chem., 1988, 60, 294R.
5. J.A. McCloskey, P.F. Crain, C.G. Edmonds, R. Gupta, T. Hashizume, D.W. Phillipson and K.O. Stetter, Nucleic Acids Res., 1987, 15, 683.
6. C.G. Edmonds, P.F. Crain, T. Hashizume, R. Gupta, K.O. Stetter and J.A. McCloskey, J. Chem. Soc., Chem. Commun., 1987, 909.
7. D.W. Phillipson, C.G. Edmonds, P.F. Crain, D.L. Smith, D.R. Davis and J.A. McCloskey, J. Biol. Chem., 1987, 262, 3462.
8. T. Muramatsu, S. Yokoyama, N. Horie, A. Matsuda, T. Ueda, Z. Yamaizumi, Y. Kuchino, S. Nishimura and T. Miyazawa, J. Biol. Chem., 1988, 263, 9261.
9. M. Ubukata, K. Isono, K. Kimura, C.C. Nelson and J.A. McCloskey, J. Am. Chem. Soc., 1988, 110, 4416.
10. T.G. Hagervall, C.G. Edmonds, J.A. McCloskey and G.R. Björk, J. Biol. Chem., 1987, 262, 8488.
11. A.L. Pogolotti, A. Ono, R. Subramaniam and D.V. Santi, J. Biol. Chem., 1988, 263, 7461.
12. K.B. Tomer, M.L. Gross and M.L. Deinzer, Anal. Chem., 1986, 58, 2527.
13. Y. Tondeur, R.C. Moschel, A. Dipple and S.R. Koepke, Anal. Chem., 1986, 58, 1316.
14. F. Lafortune, W. Ens, F.E. Hruska, K.L. Sadana, K.G. Standing and J.B. Westmore, Int. J. Mass Spectrom. Ion Processes, 1987, 78, 179.
15. M. Dizdaroglu, J. Chromatogr., 1986, 367, 357.
16. M. Dizdaroglu, Biochem. J., 1986, 238, 247.
17. M. Dizdaroglu and D.S. Bergtold, Anal. Biochem., 1986, 156, 182.
18. M. Dizdaroglu, M.L. Dirksen, H. Jiang and J.H. Robbins, Biochem. J., 1987, 241, 929.
19. A.J. Alexander, P. Kebarle, A.F. Fuciarelli and J.A. Raleigh, Anal. Chem., 1987, 59, 2484.
20. R.R. Weller, J.A. Mayernik and C.S. Giam, Biomed. Environ. Mass Spectrom., 1988, 15, 529.
21. E.G. Rogan, E.L. Cavalieri, S.R. Tibbels, P. Cremonesi, C.D. Warner, D.L. Nagel, K.B. Tomer, R.L. Cerny and M.L. Gross, J. Am. Chem. Soc., 1988, 110, 4023.
22. R. Franz, H.R. Schulten and H.G. Neumann, Chem. Biol. Interactions, 1986, 59, 281.
23. T.M. Dietz, R.J. Von Trebra, B.J. Swanson and T.H. Koch, J. Am. Chem. Soc., 1987, 109, 1793.
24. C. Decarroz, J. R. Wagner, J. E. Van Lier, C. Murali Krishna, P. Riesz and J. Cadet, Int. J. Radiat. Biol., 1986, 50, 491.
25. M. Tomasz, R. Lipman, D. Chowdary, J. Pawlak, G.L. Verdine and K. Nakaniski, Science, 1987, 235, 1204.

26  K.A. Jacobson, L.K. Pannell, K.L. Kirk, H.M. Fales and E.A. Sokoloski, J. Chem. Soc., Perkin Trans.I, 1986, 2143.
27  P.A. Blau, J.W. Hines and R.D. Voyksner, J. Chromatogr., Biomed. Appl., 1987, 420, 1.
28  J. Maybaum, M.G. Kott, N.J. Johnson, W.D. Ensminger and P.L. Stetson, Anal. Biochem., 1987, 161, 164.
29  L. Alder, R. Donau, R. Stoesser and D. Cech, Biomed. Environ. Mass Spectrom., 1986, 13, 217.
30  X. Gao and R.A. Jones, J. Am. Chem. Soc., 1987, 109, 1275.
31  A. Liguori, E. Perri, G. Sindona and N. Uccella, Tetrahedron, 1988, 44, 229.
32  L. Kolaitis and D.M. Lubman, Anal. Chem, 1986, 58, 2137.
33  T.R. Covey, A.P. Bruins and J.D. Henion, Org. Mass Spectrom., 1988, 23, 178.
34  L. Bütfering, G. Schmelzeisen-Redeker and F.W. Röllgen, J. Chromatogr., 1987, 394, 109.
35  F.A. Bencsath and F.H. Field, Anal. Chem., 1988, 60, 1323.
36  M. Sakairi and H. Kambara, Anal. Chem., 1988, 60, 774.
37  K.D. Ballard, T.D. Eller, J.G. Webb, W.H. Newman, D.R. Knapp and R.G. Knapp, Biomed. Environ. Mass Spectrom., 1986, 13, 667.
38  D.M. Hercules, F.P. Novak, S.K. Viswanadham and Z.A. Wilk, Anal. Chim. Acta, 1987, 195, 61.
39  W.V. Ligon and S.B. Dorn, Anal. Chem., 1986, 58, 1889.
40  K.H. Schram and D.L. Slowikowski, Biomed. Environ. Mass Spectrom., 1986, 13, 263.
41  A. Sandström and J. Chattopadyaya, J. Chem. Soc., Chem. Commun., 1987, 862.
42  H. Moser and G.W. Wood, Biomed. Environ. Mass Spectrom., 1988, 15, 547.
43  C. Fenselau and R.J. Cotter, Chem. Rev., 1987, 87, 501.
44  A. Liguori, G. Sindona and N. Uccella, J. Chem. Soc. Perkin Trans.II,1988, 1661.
45  F.W. Röllgen, P. Dähling, E. Bramer-Weger, F. Okuyama and M. Subhan, Org. Mass Spectrom., 1986, 21, 623.
46  R.D. Voyksner, Org. Mass Spectrom., 1987, 22, 513.
47  R.L. Cerny, M.L. Gross and L. Grotjahn, Anal. Biochem., 1986, 156, 424.
48  M. Green and J.M. Miller, J. Chem. Soc., Chem. Commun., 1987, 1864.
49  R.P. Newton, N.A. Hakeem, B.J. Salvage, G. Wassenaar and E.E. Kingston, Rapid Commun. Mass Spectrom., 1988, 2, 118.
50  L.M. Mallis, F.M. Raushel and D. H. Russell, Anal. Chem., 1987, 59, 980.
51  A. Guy, D. Molko, L. Wagrez and R. Téoule, Helv. Chim. Acta, 1986, 69, 1034.
52  P.C. Toren, D.F. Betsch, H.L. Weith and J.M. Coull, Anal. Biochem., 1986, 152, 291.
53  M. Poe, J.I. Germershausen and M. MacCoss, Anal. Biochem., 1987, 164, 450.
54  A.M. Hogg, J.G. Kelland, J.C. Vederas and C. Tamm, Helv. Chim. Acta, 1986, 69, 908.
55  L. Grotjahn, R. Frank and H. Blöker, Nucleic Acids Res. 1982, 10, 4671.
56  M. Panico, G. Sindona and N. Uccella, J. Am. Chem. Soc., 1983, 105, 5607.
57  L.R. Phillips, K.A. Gallo, G. Zon, W.J. Stec and B. Uznanski, in 'Springer Ser. Chem. Phys. No SIMS V', ed. A. Benninghoven, R.J. Colton, D.S. Simons and H.W. Werner, Springer Verlag, Berlin, 1986, vol 44, p 518.

58  A. Wolter, C. Möhringer, H. Köster and W.A. König, Biomed. Environ. Mass Spectrom., 1987, 14, 111.
59  A.J. Goddard and V.E. Marquez, Tetrahedron Lett., 1988, 29, 1767.
60  V.E. Marquez, C.K.H. Tseng, G. Gebeyehu, D.A. Cooney, G.S. Ahluwalia, J.A. Kelley, M. Dalal, R.W. Fuller, Y.A. Wilson and D.G. Johns, J. Med. Chem., 1986, 29, 1726.
61  G. Carrea, G. Ottolina, S. Riva, B. Danieli, G. Lesma and G. Palmisano, Helv. Chim. Acta, 1988, 71, 762.
62  S. Kumar, N.D. Sharma, R.J.H. Davies, D.W. Phillipson and J.A. McCloskey, Nucleic Acids Res., 1987, 15, 1199.
63  C.R. Paul, C.A. Belfi, A.V. Arakali and H.C. Box, Int. J. Radiat. Biol., 1987, 51, 103.
64  I. Isern-Flecha, X.Y. Jiang, R.G. Cooks, W. Pfleiderer, W.G. Chae and C. Chang, Biomed. Environ. Mass Spectrom., 1987, 14, 17.
65  K.K Ebralidse, S.A. Grachev and A.D. Mirzabekov, Nature, 1988, 331, 365.
66  J. Jankowska, L. Celewicz and K. Golankiewicz, Org. Mass Spectrom., 1987, 22, 52.
67  B. Skalski, G. Wenska, S. Paszyc and Z. Stefaniak, Can. J. Chem., 1988, 66, 1027.
68  L. Grotjahn and H. Steinert, Biochem. Soc. Trans., 1987, 15, 164.
69  L. Grotjahn, in 'Mass Spectrometry in Biomedical Research' ed. S.J. Gaskell, Wiley, Chichester, 1986, p 215.
70  R.L. Cerny, K.B. Tomer, M.L. Gross and L. Grotjahn, Anal. Biochem., 1987, 165, 175.
71  D. Griffin, J. Laramée, M. Deinzer, E. Stirchak and D. Weller, Biomed. Environ. Mass Spectrom., 1988, 17, 105.
72  J.J. Dino, C.R. Guenat, K.B. Tomer and D.G. Kaufman, Rapid Commun. Mass Spectrom., 1987, 1, 69.
73  A. Viari, J.P. Ballini, P. Vigny, D. Shire and P. Dousset, Biomed. Environ. Mass Spectrom., 1987, 14, 83.
74  A. Viari, J.P. Ballini, P. Vigny, D. Shire and P. Dousset, in 'Mass Spectrometry in the Analysis of Large molecules' ed. C.J. McNeal, Wiley, Chichester, 1986, p 199.
75  A. Viari, J.P. Ballini, P. Meleard, P. Vigny, P. Dousset, C. Blonski and D. Shire, Biomed. Environ. Mass Spectrom., 1988, 16, 225.
76  C.J. McNeal and R.D. Macfarlane, J. Am. Chem. Soc., 1986, 108, 2132.
77  A. Viari, J.P. Ballini, P. Vigny, C. Blonski, P. Dousset and D. Shire, Tetrahedron Lett., 1987, 28, 3349.
78  A. Benninghoven, E. Niehuis, D. Greifendorf, D. van Leyen and W. Lange, in 'Springer Ser. Chem. Phys. No SIMS V', ed. A. Benninghoven, R.J. Colton, D.S. Simons and H.W. Werner, Springer Verlag, Berlin, 1986, vol 44, p 497.
79  F. Lafortune, K.G. Standing, J.B. Westmore, M.J. Damha and K.K. Ogilvie, Org. Mass Spectrom., 1988, 23, 228.
80  T.R. Covey, R.F. Bonner, B.I. Shushan and J. Henion, Rapid Comm. Mass Spectrom., 1988, 2, 249.

# 9
# The Use of Mass Spectrometry in Studies of Drug Metabolism and Pharmacokinetics

BY D. J. HARVEY

## 1 Introduction

1.1 <u>General</u>.- This review covers the period from July 1986 until June 1988 and is concerned with the use of mass spectrometry in studies of drug metabolism, pharmacokinetics and toxicity where the technique has been applied to the drug molecule itself; studies in which mass spectrometry has been used to investigate biochemical changes in response to drug treatment, to characterise synthetic intermediates, or to identify natural products, are largely beyond the scope of this review. In order to maintain as comprehensive a coverage of the literature as possible in the limited available space, much of the work on quantitative techniques and metabolism has been tabulated (Tables 1 and 2 respectively) with drugs listed in alphabetical order. References are given in the following textual subdivisions to compounds listed in the Tables where relevant.

1.2 <u>Books and Reviews</u>.- Several books containing substantial relevant material have been published[1,2], including the second edition of Clarke's "Isolation and Identification of Drugs"[3]. General reviews[4] include those on sample preparation for biological studies[5], applications of capillary GC/MS[6], and the biennial <u>Analytical Chemistry</u> reviews on mass spectrometry[7] and forensic science[8]. Several general symposia[9] including the Fourth Symposium on LC/MS and MS/MS (Montreux, October 1986)[10] and the Sixth International Symposium on Mass Spectrometry in the Life Sciences (Ghent, August/September 1986)[11] also contain papers of interest.

## 2 General Comments

2.1 <u>Current Trends</u>.- The most significant advances in mass spectrometric techniques applied to drug research during the review period have been in the area of liquid sample introduction, FAB ionization and the use of MS/MS. Several chapters and short reviews have appeared on applications of LC/MS[12-16] and it has been pointed out that the interface used in studies of drug metabolites must be capable of handling gradients in order to cope with the wide range of compounds of different polarity usually present[17]. Thermospray appears to have emerged as the method of choice for LC/MS coupling[18]; spectra are generally similar to those from ammonia CI[18]. Barbiturates cause problems in the positive ion mode but give [M - H]⁻ ions in negative mode[18]. Optimum conditions for thermospray have been found to be similar for many compounds[19,20]; the relative abundance of [M + H]⁺ ions maximises at lower vaporization temperatures, whereas higher temperatures tend to induce fragmentation. For studies on drug metabolites, model compounds can be used to obtain reasonable conditions but for quantitative studies it is best to optimize on the compound to be measured in order to ensure high precision. Fragmentation can also be induced by inclusion of a discharge ionization facility[20]. The ability of compounds to form negative ions under thermospray conditions, either by ion attachment or proton abstraction, has been predicted from gas-phase acidities[21]. A mixed-mode column simulating ion pair separations has been described for compounds not separated by conventional reverse-phase columns and has been used for thermospray analysis of a common-cold remedy containing 5 drugs[22]. For compounds that are difficult to handle by GC/MS, LC/MS with an atmospheric pressure ion source is an alternative[23]; detection of 5 pg of xanthines has been reported with the technique. An ion spray interface on an atmospheric pressure instrument has given detection limits of 10 pg for some steroids[24]. Supercritical fluid chromatography/MS (SCF/MS) is showing promise, particularly for polar compounds such as xanthines and sulphonamides[25-27]; barbiturates have been reported to give spectra similar to those recorded under 70 eV EI conditions[28]. CI spectra have been recorded by using iso-pentane as the reagent gas[27]. Laser desorption/FT gives comparable molecular ion abundance to FD or FAB with erythromycin and daunorubicin, and greater abundance with amoxicillin and digoxin[29]. Potassium adduct ions of these drugs have been obtained by adding potassium bromide to the sample[29]. The use of matrices showing strong resonance absorption at the laser wavelength extends the applicability of the technique[30]. LAMMA offers promise for analysis of some drugs but good spectra are only obtained from targets in the upper surface layer[31].

The use of FAB in drug research has been reviewed[32] and many recent examples, particularly for the analysis of conjugates[33], are listed in Table 2. Negative ion FAB usually gives cleaner spectra than FAB in the positive ion mode for these conjugates[34]. Neutral loss scans of, for example, 176u or 80u give rapid identification of glucuronide and sulphate conjugates respectively[35,36]. This has been achieved directly from untreated urine under thermospray conditions. Pyrolysis mass spectrometry also shows promise for conjugate identification[37], and the number of exchangeable hydrogen atoms in conjugates has been determined under FAB conditions by the use of $^2$H-glycerol as the matrix[38]. p-Nitrophenol and 4-hydroxyphenazone conjugates have been examined by FAB after separation by TLC and transference of the spot to the FAB probe[39]. Ionization efficiencies of many drugs under FAB conditions can be improved by factors of four to ten following quaternization[40], in direct contrast to conventional EI techniques where dequaternization is necessary. Continuous-flow FAB has been demonstrated to give abundant [M + H]$^+$ ions with drugs such as cyclosporin A[41]. Short reviews on the use of tandem mass spectrometry to characterise drug metabolites have been published[42,43], and it has been emphasised that success of the method relies on the metabolite retaining some structural features of the original drug[44]. The method is sufficiently selective for crude urine and tissue samples to be examined directly[45], and has been used for screening plant material for biologically active substances[46]. SIMS has been used to study lubricant methyl stearate films on pharmaceutical powders[47,48] and for phospholipid films[49] where 30KV gallium ions were used.

## 2.2 Artifacts and Contamination Problems.

Reports continue to appear on detection of artifacts encountered in drug research; the presence of these may hinder or confuse an analysis. A short review on artifacts in chromatography has been written by Middleditch and Zlatkis[50] and a book is to follow. Several contaminants encountered in drug-related products have been reported; these include pentamidine isothionate in bulk antimicrobial drugs, identified by FAB[51], 2-mercaptobenzothiazole, dibutyldithiocarbamate and several amines leached from rubber stoppers[52], and the plasticiser, acetyltributyl citrate leached from plastic film[53]. Concentrations in the 0.5 ppm range of the plasticiser di-2-ethylhexylphthalate, leached from blood collection apparatus, have been measured by methane CI in the blood from blood donors[54]. Polyols such as sorbitol have been shown to transesterify parabens[55], and w and w-1 hydroxylauric acids have been reported to undergo acetylation when extracted with ethyl acetate, an observation with implications for drug metabolite extractions[56]. Replacement of a fluorine atom from a pentafluorobenzoyl derivative by a methoxy group has been observed when the compound was reacted with methyl iodide in the presence of dimethylformamide, but not when acetonitrile was used as solvent[57];

formation of methoxy groups is a common metabolic route. Butylated hydroxytoluene metabolism has been studied[58].

2.3 Stable Isotopes.- A considerable volume of work has been published on applications of stable isotopes[59], much of which is discussed below for individual drugs. Several conferences have been devoted to the subject[60-62] and the use of isotopes in drug metabolism studies[63-68], pharmacokinetics[69,70], and clinical medicine[71,72] have been reviewed briefly. Reviews have also appeared on methods of labelling[73], effects caused by the presence of a label[74], quantitative aspects[75,76], and applications to anticancer drug research[77]. The most common uses of labelled compounds, other than as tracers and standards for quantification, are for the identification of metabolites in mixtures following administration of 1:1 molar mixtures of labelled and unlabelled compound (isotope cluster technique), for following drug kinetics during chronic drug therapy by substitution of one dose with a labelled drug sample (pulse labelling), for studying bioavailability by administration of labelled and unlabelled drug by different routes, and for investigating metabolism of diastereoisomers following administration of "pseudoracemates" consisting of a racemic mixture of differently labelled diastereoisomers.

### 3 Quantitative studies

Two reviews on quantitative aspects of drugs and other biologically interesting compounds have appeared[78,79]. It is apparent from Table 1, which lists major quantitative methods, that there has been a development in methods other than GC/MS for drug quantification in biological fluids during the review period, although GC/MS using small quadrupoles remains the method of choice. Nevertheless, for drug quantification in general, HPLC assays are usually more convenient as sample volatility is not limiting. No significant advances in methods for achieving high sensitivity appear to have been reported, 0.0005-0.001 ng/ml seems to have been the lower limit for assays of this type for several years. Of the methods available, such as high resolution SIM and multiple reaction monitoring, the advantages of the latter have been stressed[80]. Carbon tetrachloride in anhydrous ammonia has been used to achieve high sensitivity for measurement of adrenergic $alpha_2$ antagonists[81] under CI conditions. The sensitivity of detection of nicardipine has been reported to be about 2 orders higher with filament-on thermospray than with DLI but, in both modes, positive ions gave higher sensitivity than negative ones[82]. However, the sensitivity

Table 1. Methods for the quantification of drugs and their metabolites

| Drug | In[a] | Deriv.[b] | Standard | MS[c] | Method[d] | Det Limit[e] | Ref |
|---|---|---|---|---|---|---|---|
| Alphaprodine | P | - | Analogue | Q | GC/MS (SE-30/OV-17), EI | <2.0 ng/ml | 84 |
| Amphetamine | U | TFA | Analogue | Q | GC/MS (F-Si, MeSi), EI | 25 ng/ml | 85 |
| Anagrelide | P | Me | Mono-Cl Analogue | Q | GC/MS (F-Si, DB-1) EI (35 eV) | 0.5 ng/ml | 86 |
| Benzbromarone | P,U | Me | Warfarin | Q | GC/MS, EI | 5.0 ng/ml | 87 |
| Benzofenac | B | Me | - | - | CI, EI | 10 ng/ml | 88 |
| Bz-ecgonine | U | PFP | Scopolamine | Q | GC/MS (F-Si, MeSi), EI | 50 ng/ml | 85 |
| " | U | Pr | n-Bu-ester | Q | GC/MS (F-Si, DB-5), EI | 34 ng/ml | 89 |
| " | U | - | $^2H_3$ | Q | GC/MS (F-Si, MeSi), EI | 50 ng/ml | 90 |
| Betamethasone | J | MO,TMS | $^2H_4$ | Q | GC/MS(1%OV-1),CI(-),$CH_4$ | 0.1 ng | 91 |
| Biphenylacetic acid | P | PFB | $^2H_8$ | QQQ | GC/MS (F-Si, BP-1) CI (-ve, He) | 0.01 ng/ml | 92 |
| Bromocriptine | P | - | - | Q | GC/MS (F-Si, CP-Sil-8) | 0.02 ng/ml | 93 |
| Budralazine | P,U | - | $^2H_4$ | B | GC/MS (3% OV225), EI(30) | 10 ng/ml | 94 |
| Caffeine | S | - | Theophylline | Q | LC/MS (TSP, +ve) | 0.2 ng | 95 |
| $^2H_2$-Cannabinol | G P | TBDMS | $^2H_7$ | B | GC/MS (SP-2100), EI | 100 ng/ml | 96 |
| Captopril | B U | N-Et-mal | S-Bz-Captopril | B | GC/MS (2%,OV-1) EI (70 eV) | 0.1 ng | 97 |
| Carbamazepine | S | Me | Hexobarbitone | Q | GC/MS (3% OV-17) EI (70eV) | <1 µg/ml | 98 |
| CGS 14824A De-Et metab | P,U | Me | $^2H_5$ | Q | GC/MS (3% OV-101), EI | 30 pmole(P) 80 pmole(U) | 99 |
| 6-Cl-3-Me-$H_4$-Benzazepine | P | - | $^2H$ | Q | GC/MS (F-Si, DB-5) CI ($NH_3$, $CCl_4$) | 0.1 ng/ml | 81 |
| - des-Me | P | - | $^2H_5$ | Q | GC/MS (F-Si, DB-5) CI ($NH_3$, $CCl_4$) | 0.1 ng/ml | 81 |
| Clomipramine | B | PFP | $^2H_4$ | Q | GC/MS (F-Si, MeSi), EI | 2 nM/L | 100 |
| Clonidine | P | PFB | $^2H_4$ | QQQ | GC/MS (F-Si, OV-1) CI (-ve, $NH_3$) | 0.01 ng/ml | 101 |
| Clozapine N-De-Me | P | - | Me-homologue | Q | GC/MS (F-Si, OV-1701) EI | 1 ng/ml 0.5 ng/ml | 102 |
| Cocaine | U | TMS | $^2H_3$ | Q | GC/MS (F-Si, MeSi), EI | 50 ng/ml | 90 |
| Codeine | U | PFP | Nalorphine | Q | GC/MS (F-Si, MeSi), EI | 50 ng/ml | 85 |
| Cotinine | P | - | Methyprylone | Q | GC/MS (F-Si, SE-30), EI | 100 nM/1 | 103 |
| CP-55,940 (cannabinoid) | P | TFA | Lower-homologue | Q | GC/MS (F-Si, DB-5) EI (70 eV) | 0.5 ng/ml | 104 |
| CS-570 (Carbacyclin) | P | PFB TMS | $^2H_6$ | Q | GC/MS (F-Si, DB-5), CI (-ve, $NH_3$) | 0.1 ng/ml | 105 |
| Deoxyspergualin | P | Pyrim | $C_8$-analogue | B | GC/MS (Cap), EI(20), HR | 0.5 ng/ml | 106 |
| Detomidine | B P,U | - | Propranolol | Q | GC/MS (F-Si, DB-5) EI (70 eV), CI ($CH_4$) | 0.2 pM 0.02 pM | 107 |
| Dexamethasone | J | MO,TMS | $^2H_4$ | Q | GC/MS(1%OV-1),CI(-,$CH_4$) | 0.1 ng | 91 |
| Diclophensine | P | - | $^2H$ | - | GC/MS (SE-54) | 300 ng/ml | 108 |
| Di-Et-stilbestrol | T | TMS | $^2H_8$ | Q | GC/MS (F-Si, MeSi) EI (70 eV) | 200 ppt | 109 |
| Dihydroegotoxine | P | TMS | $^2H_3$ | BE | GC/MS (F-Si, MeSi) EI (70 eV), HR (5000) | 0.002 ng | 110 |
| Dihydroqinghaosu | B | - | Cedrol | Q | GC/MS (OV-1), CI ($CH_4$) | 10 ng/ml | 111 |
| Diltiazem | P | TBDMS | $^2H_4$ | B | GC/MS (OV-17), CI(i-Bu) | - | 112 |
| Dimethoxycoumarine | X | - | $^2H_3$ | EB | GC/MS (5% Thermon 3000) EI (70 eV) | 40 ng/ml | 113 |
| Dimetridazole | F | - | - | Q | GC/MS (F-Si, OV-101), EI | 0.1 ppm | 114 |
| Diphenylhydantoin | S | Me | Hexobarbitone | Q | GC/MS (3% OV-17) EI (70 eV) | <1 µg/ml | 98 |
| Emepronium Br | S,P | De-Me | $^2H_5$ | B | GC/MS (OV-1), EI | <5 ng/ml | 115 |

Table 1. (cont.)

| Drug | In[a] | Deriv.[b] | Standard | MS[c] | Method[d] | Det Limit[e] | Ref |
|---|---|---|---|---|---|---|---|
| Emepronium Br | S,P | De-Me | $^2H_5$ | B | GC/MS (F-Si, OV-1), EI | 3 ng/ml | 115 |
| Erythromycin | T | Ac | - | EB | GC/MS (OV-101), EI | 0.001 ppm | 116 |
| " | P | | $^2H_3$ | EB | FAB (Xe) | 50 ng/ml | 117 |
| -Et-succinate | P | - | $^2H_5$ | EB | FAB (Xe) | 100 ng/ml | 117 |
| Ethytoin | S | Me | Hexo-barbitone | Q | GC/MS (3% OV-17) EI (70 eV) | <1 µg/ml | 98 |
| Felodipine | P | - | $^2H_6$ | Q | GC/MS (F-Si, SE-54), EI | 0.2 nM/l | 118 |
| Fenbendazole + Metabolites | T | - | - | QQQ | Probe, EI CI | 200 ng/g 3.0 µg/g | 119 |
| Flecainide | S | - | Isomer | Q | GC/MS (F-Si, MeSi), EI | 25 ng/ml | 120 |
| Fludalanine | S,U | TMS MO | $^2H_3$ | Q | GC/MS (7%SE-30), CI($CH_4$) | 0.3 nmole /ml | 121 |
| Fluphenazine | P | TBDMS | $^2H_4$ | Q | GC/MS (F-Si, CPSi15), EI | 0.05 ng/ml | 122 |
| 5-Fluorouracil | P | Me | $^{15}N_2$ | EB | GC/MS (SF-96), EI9(70eV) | 2 ng/g | 123 |
| Forphenicinol + Metabolites | S,U | TMS, MO | Analogue | EB | GC/MS (3%OV-17), EI(70) | 100 ng/ml | 124 |
| Haloperidol | P | TFA | $^2H_4$ | QQQ | GC/MS (F-Si, DB-5) CI (-ve, $NH_3$) | 0.1 ng/ml | 125 |
| Ibuprofen | S | - | $^2H_3$ | Q | GC/MS (F-Si,MeSi),EI(70) | <2.4 nmole | 126 |
| " | P,U | S(-)Ph | Analogue | - | GC/MS (DB-5), EI | 10 mg/ml | 127 |
| Imipramine | P,U | TFA | $^2H_8$ | Q | GC/MS (F-Si), CI | 0.1 mg/ml | 128 |
| Isofloxythepin | B,U | TMS | $^2H_7$ | B | GC/MS (OV-101), EI(70eV) | 5.0 ng/ml | 129 |
| Isoniazid + Metabolites | S | - | $^2H_6$ | Q | GC/MS (OV-1701 cap) CI ($NH_3$) | 0.1 nM/ml | 130 |
| Isradipine + Metabolites | P,U | Me | $^{13}C_4$ Homologue | Q | GC/MS (F-Si, MeSi) | 0.04 ng/ml 0.15 ng/ml | 131 |
| Ketocyclazocine | P | PFP | Et Homologue | Q | GC/MS (3% OV-11) CI (+/-ve) | 0.1 ng/ml | 132 |
| Kitasamycin | T | - | - | EB | GC/MS (OV-101), EI | 0.5 ppm | 150 |
| Labetalol | P | - | $^2H_7$ | Q | LC/MS (ODS-2,TSP) | 5 ng/ml | 133 |
| Lidocaine | P | - | Mepivacaine | Q | GC/MS (SE-54), CI($NH_4$) | 0.2 nM/ml | 134 |
| Lormetazepam | P | TMS | $^2H_4$ | Q | GC/MS (F-Si, OV-1),CI(-) | - | 135 |
| LSD[f] | U | TMS | $^2H_3$ | Q | GC/MS (F-Si, DB-5), EI | 125 pg | 136 |
| " | U | TMS | $^2H_3$ | Q | GC/MS (F-Si, MeSi), EI | 0.5 ng/ml | 137 |
| Metaproterenol | U | TMS | t-Bu-analogue | Q | GC/MS (F-Si, DB-5) CI ($NH_3$) | 0.5 ng/ml | 138 |
| Methamphetamine | B | - | $^2H$ | - | GC/MS, EI | 0.05 ng/ml | 139 |
| " | U | TFA | Ph-cyclo-hexylamine | Q | GC/MS (F-Si, MeSi), EI | 25 ng/ml | 85 |
| Morphine | U | COEt | Nalorphine | Q BEB | GC/MS (F-Si, DB-5), EI Probe, EI M* | 15 ng/ml 50 ng | 140 |
| " | S | TMS | $^2H_3$ | Q | GC/MS (F-Si, BP-1) | | 141 |
| " | B | TFA | $^2H_3$ | Q | GC/MS | - | 142 |
| " | U | PFP | Nalorphine | Q | GC/MS (F-Si, MeSi), EI | 50 ng/ml | 85 |
| Moxonidine | P | PFB | $^2H_3$ | QQQ | GC/MS (F-Si, OV-1) CI (-ve, $NH_3$) | 0.1 ng/ml | 143 |
| MPPP[f] | P R | i-Bu-carb | $^2H_5$ | B | GC/MS (1% OV-1), EI | 5 ng/ml | 144 |
| MPTP[f] | R | - | $^2H_3$ | Q | GC/MS (PhMeSi), EI | 0.5 µg/g | 145 |
| Nadolol | P | TMS | $^2H_9$ | Q | GC/MS (F-Si, DB-17), EI | 0.2 ng/ml | 146 |
| Nicotine -N-oxide | U | Anis-ole | Me analogue | Q | GC/MS (F-Si, Me-Si), EI | - | 147 |
| Nilvadipine (+ and -) | P | - | $^2H_3$ | Q | GC/MS (F-Si, MePhSi), EC (-ve) | 0.025 ng/ml | 148 |
| Nitrendipine | P | - | $^2H_8$ | Q | GC/MS(F-Si,MePhSi),CI(-) | 0.1 ng/ml | 149 |
| Oleandomycin | T | - | - | EB | GC/MS (OV-101), EI | 0.1 ppm | 150 |

Table 1. (cont.)

| Drug | In[a] | Deriv.[b] | Standard | MS[c] | Method[d] | Det Limit[e] | Ref |
|---|---|---|---|---|---|---|---|
| Oxcarbazine | P | Enol-TMS | Carbamazepine-diol | B | GC/MS (F-Si, Me-Si) EI (20 eV) | 0.1 ng/ml | 151 |
| -10-OH | " | " | " | " | " | " | " |
| -Diol | " | " | " | " | " | 1.0 ng/ml | " |
| Oxyprocaine(M) | U | TMS | Methaqualone- | | GC/MS (2% OV-101) EI(20) | – | 152 |
| Pancuronium | S | – | $^2H_6$ | EB | Probe, CI (i-Bu) | 1.0 ng/ml | 153 |
| Nor-phenazone ($^{13}C^{15}N_2$) | T | TMS | Unlabelled | B | GC/MS (2% OV-225), EI | 60 ng/ml | 154 |
| 3-OH-Phenazone ($^{13}C^{15}N_2$) | T | TMS | Unlabelled | B | GC/MS (2% OV-225), EI | 25 ng/ml | 154 |
| 4-OH-Phenazone ($^{13}C^{15}N_2$) | T | TMS | Unlabelled | B | GC/MS (2% OV-225), EI | 69 ng/ml | 154 |
| Phenobarbitone | S | Me | Hexobarbitone | Q | GC/MS (3% OV-17) EI (70 eV) | <1 µg/ml | 98 |
| " | P | Me | $^{15}N_2, ^{13}C$ | Q | GC/MS (F-Si, OV-1), EI | 40 ng/ml | 155 |
| " | U | Me | $^{15}N_2, ^{13}C$ | Q | GC/MS (F-Si, OV-1), EI | 50 ng/ml | 155 |
| -p-OH | U | Me | $^{15}N_2, ^{13}C$ | Q | GC/MS (F-Si, OV-1), EI | 30 ng/ml | 155 |
| Phencyclidine | U | – | Ketamine | Q | GC/MS (F-Si, MeSi), EI | 10 ng/ml | 84 |
| Prednisolone | J | MO,TMS | $^2H_4$ | Q | GC/MS(1%OV-1)CI,(-ve,$CH_4$) | 0.1 ng | 91 |
| Primidone | S | Me | Hexobarbitone | Q | GC/MS (3% OV-17) EI (70 eV) | <1 µg/ml | 98 |
| Propiverine | P,U | TFA TMS | n-Bu-O-Pr-benzilate | EB | GC/MS (F-Si, MeSi) EI (70 eV) | 1-2 ng/ml | 156 |
| Quinupramin | P,U | – | S-analogue | Q | GC/MS (F-Si, OV-1), EI | 0.5 ng/ml | 157 |
| Rimantidine | P U | PFB | $^2H_4$ | EB | GC/MS CI (-ve) | 4.2 ng/ml 21 ng/ml | 158 |
| " | – | TBDMS | $^2H_3$ $^2H$ | – | GC/MS (cap), Res = 3000 | – | 159 |
| Rutoside metabolites | S | PFP/ HFIP | $^2H$ | EB | GC/MS (2% OV-105) EI (25 eV) | 1-2 ng/ml | 160 |
| S-3341[f] | P | $F_3C$-benz | $^2H_4$ | B | GC/MS (F-Si, OV-1) EC (-ve) | 2.0 ng/ml | 161 |
| " | P,U | TFA | $^2H_4$ | Q | GC/MS (F-Si, OV-1701) EC (-ve, $NH_3$) | 0.2 ng/ml | 162 |
| Sertralin | P | – | Br-analogue | B | GC/MS (S-10C), EI (20eV) | 1.0 ng/ml | 163 |
| Spiramycin | T | – | – | EB | GC/MS (OV-101), EI | 0.5 ppm | 150 |
| SQ 28,668[f] | P,U | TMS | $^2H_3$ | Q | GC/MS (F-Si),CI(-ve,$CH_4$) | 0.01 pg | 164 |
| Sulphamethazine | T | – | $^2H_4$ | EBQ | LC/MS (Belt), EI, M* | 100 ng/g | 165 |
| Tamoxifen | P | TMS,Me HFB | Androstanediol | Q | GC/MS (F-Si, OV-101) EI (70 eV) | 0.05 ng/ml | 166 |
| " -4 OH | T | HFB TMS | Pentene homologue | Q | GC/MS (F-Si, BP-1), EI | 0.25 ng | 167 |
| Terbutaline | P | – | $^2H_6$ | Q | LC/MS (cyclobond, TSP) | 4 pmole | 168 |
| Tertalolol | P,U | TMS | $^2H_9$ | Q | GC/MS (3% SE30), CI($NH_3$) | 1.0 ng/ml | 169 |
| -4-OH | P,U | TMS | $^2H_9$ | Q | GC/MS (3% SE30), CI($NH_3$) | 4.0 ng/ml | 169 |
| D-9-THC-7-COOH | U | TMS | $^2H_3$ | Q | GC/MS (F-Si, MeSi), EI | 10 ng/ml | 85 |
| " | U | PFP | – | Q | GC/MS (F-Si, DB-5), EI | 1.8 ng/ml | 170 |
| " | U | Me | $5'-^2H_3$ | Q | GC/MS (F-Si, DB-5), EI | 2 ng/ml | 171 |
| Theophylline | P | PFB | 3-i-Bu-1-Me-xanthine | EB | GC/MS (F-Si, SE-42), EI (70 eV) | – | 172 |
| Tilidine | P | – | Pethidine | Q | GC/MS (F-Si, Sil 19) EI | 10 mg/ml | 173 |
| Trenbolone | T | TMS | 19-Nor-testosterone | Q | GC/MS (F-Si,MeSi), EI | 0.5 ppb | 174 |
| Triazolam | B | – | $^2H_6$ | Q | GC/MS(F-Si,DB-1),CI($CH_4$) | 0.5 ng/ml | 175 |
| Trimetrexate | P | TMS | Analogue | Q | GC/MS (F-Si, MeSi), EI | 0.005 µM | 176 |
| Tulobuterol | P | HFB | Des-Cl | Q | GC/MS (F-Si, PhMeSi), EI | 0.17 ng/ml | 177 |
| Tylosin | T | – | – | EB | GC/MS (OV-101), EI | 0.5 ppm | 150 |

Table 1. (cont.)

| Drug | In[a] | Deriv.[b] | Standard | MS[c] | Method[d] | Det Limit[e] | Ref |
|---|---|---|---|---|---|---|---|
| Valproic acid -metabolites | S,U | TMS | Undecylenic acid | EB | GC/MS (3% OV-1) EI (70 eV) | 5-10 ng/ml | 178 |
| -4-ene | P,U | TBDMS | Di-n-Bu-acetic acid | Q | GC/MS (F-Si), EI (70 eV) | 100 ng/ml | 179 |
| Vecuronium | S | - | $^2H_6$ | EB | Probe, CI (i-Bu) | 1.0 ng/ml | 153 |
| Verapamil | S | - | D-517 | Q | GC/MS (F-Si, MeSi), EI | 50 ng/ml | 180 |
| Vinpocetine | P | - | Me ester | Q | GC/MS (F-Si, OV-1), EI | 0.1 ng/ml | 181 |
| Zeralanone | T | TMS | $^2H_8$ | Q | GC/MS (F-Si, MeSi) | 200 ppt | 108 |
| Zeralenol | " | " | " | " | EI (70 eV) | " |  |
| Zeralenone | " | " | " | " | " | " |  |
| Zofenopril + Metabs. | P | N-Et-mal | F-analogue | Q | GC/MS (F-Si, DB-17), EI | 1 ng/ml | 182 |

a) A = Bile, B = Blood, C = Cells, D = Lung, E = Leukocytes, F = Faeces, G = Saliva, H = Hepatocytes, I = Synovial fluid, J = Aqueous humour, L = Liver, M = Milk, P = Plasma, R = Brain, S = Serum, T = Tissue, U = Urine, W = Perfusate, X = Supernatant, Y = Cytochrome P-450, Z = Microsomes.

b) (Includes derivatives from Table 2) Ac = Acetyl, AlcC = Alkoxycarbonyl, Anisole = Derivative with anisole, Carb = Carbamate, CB = Carbobenzoxy, CN = Cyanide addition product, $d_9$-TMS = [$^2H_9$]TMS, EDMS = Ethyldimethylsilyl, Et = Ethyl derivative, $F_3C$-benzoyl = Trifluoromethylbenzoyl, $F_6C_5$-di-Co = Derivative with hexafluoroacetyl acetone, De-Me = Demethylation, i-Bu-carb = iso-Butyl-carbamate, HFIP = Heptafluoro-iso-propyl, HFP = Heptafluoropropyl, Me = Methyl derivative, N-Et-Mal = N-Ethylmaleimide, MO = Methyloxome, PFA = Perfluoroacyl, PFB = Pentafluorobenzoyl, PFP = Pentafluoropropyl, Pr = Propyl derivative, Pyrim = Pyrimidine, Redn = Reduction, S(-)-Ph = S(-)-Phenylethylamine, S(+)-A = S(+)-Amphetamine, TBDMS = tert-Butyldimethylsilyl, TFA = Trifluoroacetyl, TFE = Trifluoroethyl ester, $TiCl_3$ = Derivative with titanium chloride, TMS = Trimethylacetyl, TMS = Trimethylsilyl.

c) E = Electric sector, M = Magnetic sector, Q = Quadrupole.

d) Cap = Capillary, F-Si = Fused silica, Columns not indicated as cap or F-Si are packed. Electron energy normally 70 eV unless stated.

e) ng are used where appropriate. Otherwise, units are those in the publication.

f) LSD = Lysergic acid diethylamide, MPPP = 1-Methyl-4-phenyl-4-propionoxy-piperidine, MPTP = 1-Methyl-4-phenyl-1,2,3,6-tetrahydropyridine. S-3341 = 2-Dicyclopropylmethylamino-2-oxazoline, SQ 28,668 = {1S-[1-alpha,2-beta(5Z),3-beta(1E,3R,4S),4-alpha]}-7-{3-(3-hydroxy-4-phenyl-1-pentenyl)-7-oxabicyclo[2,2-1]-hept-2-yl}-5-heptanoic acid.

of thermospray has been reported to be limited because of noise. DLI, although more difficult to operate, appears less affected by noise[82]. Discussion on the merits of the various types of internal standard have continued. There has apparently only been one report that standards labelled with stable isotopes increase accuracy but substantial evidence that precision of measurement is improved[83]. Although some, but debatable, evidence exists that stable isotope-labelled compounds can act as carriers, large concentrations of these compounds can be detrimental as they lead to memory effects and unacceptable confidence limits. To overcome non-linear calibration problems inherent in assays with stable isotopes, the use of a plot of average mass against mole fraction of the labelled compound has been reported to give a straight line[183].

## 4. Metabolic studies

4.1 Model compounds.- General drug biotransformation pathways are frequently studied in controlled model systems. The use of FAB has contributed significantly to studies on glutathione conjugation as exemplified by investigations with the model substrate alpha-bromo-isovalerylurea[184]. Bromobenzene, which readily forms such conjugates following oxidation to an epoxide, is another well studied classical example[185-188]. Phenols, also formed by epoxidation, have been shown to undergo similar metabolism[189] as well as being converted into quinones[189,190], and both naphthalene[191] and phenanthrene[192] form glutathione conjugates via epoxides, the former reaction showing species stereoselectivity[191]. Taurine conjugation has been investigated using benzoic acid[193] as substrate, and chain elongation of carboxylic acids has been modeled with pyrimidine and thiazole derivatives[194,195].

4.2 Drug metabolism Studies.- These are summarised in Table 2. Drugs are listed in alphabetical order and tabulated drugs are also named in the sections below where relevant.

## 5. Anticancer Drugs

The study of anticancer drugs has been a fruitful area for the application of new mass spectrometric techniques on account of the generally involatile, unstable, and complex nature of many of the compounds. Included in Table 2 are metabolic studies on amino-glutethimide[204], amonafide[212], benfluron[218], busulfan[233-236], clomiphene[257], cyclophosphamide[198], elliptinium acetate[289,290], etoposide[299-302], 5-fluorouracil[312,313], ifosfamide[328], melphalan[347,348], mitonafide[369], mitoxantrone[370], pinafide[369], pyridoglutethimide[429,430], and tamoxifen[166,457-459],. Conjugation of a bile acid with 5-fluorouracil has identified a new role for these acids[313,314], and the identification of intact glucuronidase resistant glucuronides of etoposide by FAB MS has indicated that earlier studies have underestimated the role of this metabolic route[302]. Quantitative studies (Table 1) have been reported for deoxyspergualin[105], 5-fluorouracil[123], tamoxifen[167], and

Table 2. Mass spectrometric studies on metabolism

| Drug | In[a] | Metabolite[b] | MS[c] | Technique[d] | Deriv.[e] | Ref |
|---|---|---|---|---|---|---|
| 450191-S[f] | R | A 2,31 | Q | GC/MS(F-Si,SE54),EI,FAB | TMS, Ac | 196 |
| " | D | P 2,22,41 | - | Probe, EI, FD | | 197 |
| Alcophosphamide | R | U 25 | Q | GC/MS(OV-1), CI($NH_3$), S | TMS | 198 |
| Alfaprostol | O,P,R | U 12,16,52,74 | QQQ | GC/MS(F-Si,DB-5),CI($CH_4$ -ve), Isotope ($^{14}C$) | PFB TMS | 199 |
| Alfentanil | D,R | A,F 20,51,57 U | Q | GC/MS (F-Si, CP-Sil-19) Probe, EI, CI | TMS TFA | 200 |
| " | D,R | H 20,56 | Q | Probe, EI, CI | - | 201 |
| Alpidem | H,R | ABU 1,2,3,56 | Q | GC/MS(SE-30 Cap),EI(70) | TMS | 202 |
| 4-Amino-5-Et-3-T[f] | D,R | U 58,74 | B | Probe, EI (70 eV) | - | 203 |
| Aminoglutethimide | A,H I,R | U,F 3,7,8,26,27 39 | Q | Probe, EI (70 eV) | - | 204 |
| Amitriptyline | R | A 1,4,18,19,22 84 | EB | Probe, EI (12, 70 eV) | - | 205 |
| " | H | U 1,4,18,22,76 85 | EB | Probe, EI (70 eV) | - | 206 |
| " | R | A 1,19,22,30,71 | EB | FAB (Xe, 8 KV) | - | 207 |
| Amlodipine | H | U,F 2,14,55,57 74,75,79 | Q | GC/MS (F-Si, Me-Si), EI (70 eV), CI ($CH_4$) | Me,TMA TBDMS | 208 |
| " | D,R | U,F 2,4,14,52, 57,74,79 | Q | GC/MS (F-Si, Me-Si) EI (70 eV), CI ($CH_4$) | Me,TBDMS TMA,TFA | 209 |
| Amobarbitone | H | U 38 | - | GC/MS | Me | 210 |
| Amodiaquine | H | E 48,59 | - | FAB | - | 211 |
| Amonafide | H | U 9,22,29,30,39 | Q | LC/MS, TSP | - | 212 |
| Arbaprostil | D | U 2,12,30,84,89 | Q EB | GC/MS(3%OV-101)EI(40eV) GC/MS(F-Si,DV-1)CI($NH_3$) | Me,TMS MO | 213 |
| Aristolochic acid | A,D I,R | U,F 19,53,69,100 101 | B | Probe, EI (70 eV) | - | 214 |
| Aspirin | H | P,U 1,73 | EB | GC/MS (BP10Cap), EI(20) | Me | 215 |
| Atropine | H | U 9,22,74 | BE | Probe, EI | - | 216 |
| Bamipine | R | U 1,22,30 | BE | Probe, EI, FAB | - | 217 |
| Benfluron | A,H R,Q | U 1,9,14,17,18 22 | - | GC/MS, EI, CI, S(D) | TMS | 218 |
| Benoxinate | H | U 21,30,37,39 70,74 | - | GC/MS (2% OV-1201) EI (20 eV) | TMS | 219 |
| Benzarone | D,R | U,F 1,3, A | B | Probe, EI (70 eV) GC/MS (3% OV-17) | TMS TFA | 220 |
| Benzazepine | D | U 9,7,8,22,31 84,86 | EB QQQ | LC/MS (TSP), FAB (+/-) GC/MS (F-Si, DB-5) | - | 221 |
| Bz-chinozolin | - | U 29 | | EI, CI, DCI | - | 222 |
| Benzphetamine | H | U 1,19,22,56 | B | GC/MS (OV-17), CI, SIM | TFA | 223 |
| Benzylpenicillin | R | A 25 | EB | FAB (Xe, +ve, glycerol) | - | 224 |
| Bepridil | D,H J,R | P 1,4,7,23,30 | - | Probe, EI, CI ($NH_3$), FD FAB | - | 225 |
| " | A,J,H M,R | U,F 1,2,18,39,56 57,58,71,80 | EB | Probe, EI, DCI ($CH_4$) | | 226 |
| Bromazepam | R | F,L 25 | Q | GC/MS, (OV-17), EI | - | 227 |
| Bromerguride | D,R | LW 59 | Q | Probe, EI, CI ($CH_4$) | - | 228 |
| Budesonide | R | L 4,11,17,71, | - | LC/MS, CI($CH_4$), S($D^{18}O$) | - | 229 |
| " | H,M,R | LX 4,16,17,18,90 | - | LC/MS (C-18), CI | - | 230 |
| " | H,M,R | E 90 | Q | LC/MS (C-18), CI($CH_4$) GC/MS (3% OV-1) | TMS | 231 |
| " | M,R | LZ 4,90 | Q | LC/MS, CI($CH_4$),S($D^{18}O$) | - | 232 |
| Busulfan | R | U 4,10,28,40 | B | GC/MS (F-Si,CP-Sil18), EI | - | 233 |
| " | R | L 40 | TOF | $^{252}Cf$ | - | 234 |
| " | R | A 40 | TOF | $^{252}Cf$ | - | 235 |
| " | " | U 7,10,28 | -- | GC/MS | - | " |
| " | R | A 40 | - | FAB (+/-) | - | 236 |

Table 2. (cont.)

| Drug | In[a] | | Metabolite[b] | MS[c] | Technique[d] | Deriv.[e] | Ref |
|---|---|---|---|---|---|---|---|
| Bu-Delta-8-THC | M | L | 3,4,13,55 | B | GC/MS(SE 30), EI(25), S | Me/TMS | 237 |
| Bu-Delta-9-THC | M | L | 3,4,13,55 | B | GC/MS(SE-30), EI(25), S | Me/TMS | 237 |
| Calactin | R | L | 24 | Q | Probe, CI ($CH_4$) | - | 238 |
| Ca hopantenate | D | U | 38 | B | GC/MS (3% OV-1), EI(20) | TMS,Me | 239 |
| Camazepam | R | P | 2,22,74 | - | Probe, EI | - | 240 |
| CB-3717[f] | M | F | 41 | EB | Probe, EI (70 eV) | Me | 241 |
| Ceffepime | R | U | 9,56 | - | Probe, EI | - | 242 |
| 6-Cl-2-Pyridyl-methyl nitrate | D,R | U | 30,37,40,55 81 | Q | Probe, CI ($CH_4$, i-Bu) | Me | 243 |
| Chloroquin | H | U | 9,14,24,39,56 59,60 | Q | GC/MS, EI, CI | Me, Ac | 244 |
| Chlormethiazole | H | U | 9,10,103 | - | GC/MS (Cap), EI | $TiCl_3$ | 245 |
| Chlormezanone | H | P | 24,25 | Q | GC/MS(3%OV17),EI,CI($CH_4$) | Me | 246 |
| Chlorpheniramine | R | LZ | 1,9,14,17,22 60 | Q EB | GC/MS (F-Si,OV-1,DB-5) EI(70), H.Res,CI($CH_4$) | TMS,HFB TBDMS | 247 |
| Chlorphenoxamine | H | U | 1,9,14,17,24 22,57 | Q | GC/MS (F-Si, SE-54), EI CI ($CH_4$) | - | 248 |
| " | H | U | 1,9,19,22,30 57 | - | Probe, EI (70 eV) | - | 249 |
| Chlorpromazine | H | U | 31 | EB | FAB (Xe, 9kV, glycerol) | - | 250 |
| Cinobufagin | R | S | 73,91 | B | Probe, EI | - | 251 |
| Ciprofloxicin | H | U | 7,18,25,34,66 | BE | Probe, EI (70 eV), FAB | - | 252 |
| Cisapride | R | A | 34 | Q | LC/MS (RP-8, TSP) | - | 253 |
| " | D,H | U,F | 1,7,14,18,20 31,34,56,57 86,107 | Q | Probe, EI (70 eV) DCI ($CH_4$) | - | 254 |
| " | R | U,F | 1,7,14,18,20 31,34,56,57 86,107 | Q BE | Probe, EI (70 eV) DCI ($CH_4$), FAB (Xe) | - | 255 |
| Clavulanic acid | H | U | 6,18,25,60 | B | Probe, EI | - | 256 |
| Clomiphene | R | F | 1,19 | Q | GC/MS (F-Si), EI, FAB | TMS | 257 |
| CM-DXR, CM-DNR | H | T | 42 | Q | DCI ($NH_3$) | - | 258 |
| Cocaine | R | U | 1,9,19,22, 74,109 | B | GC/MS (3%OV-1), EI (70) | TMS | 259 |
| Cyclosporine | A | A | 4,55 | EB Q | GC/MS (Cap), EI FAB, LC/MS (TSP, -ve) | TMS | 260 |
| " | A | A | 22,29 | EB | FAB | - | 261 |
| Cyproheptadine | R | U | 76 | Q | GC/MS(CPSi15Cap),EI,DCI | - | 262 |
| Dapsone | H | T | 8,27 | Q | GC/MS (OV-1), CI(-ve) | - | 263 |
| Debrisoquin | H | U | 7 | Q | GC/MS (SE-30, Cap) CI (-ve, $CH_4$), S ($^2$H) | $F_6,C_5$-di-CO | 264 |
| (+)-Delta-9-THC | M | L | 2,3,4,13,55 | B | GC/MS(SE-30), EI(25), S | Me/TMS | 265 |
| (+)-Delta-8-THC | M | L,Z | 3,4,12,55,70 | B | GC/MS(SE-30), EI(25), S | Me/TMS | 266 |
| N-(De-Me-Py)-Bz-sulphonamide[f] | R | U | 1,2,6,30,34 | B BE | Probe, EI, FAB, CI DCI ($NH_3$) | - | 267 |
| 7-Deoxy-daunomycinone | E | - | 43,44,45 | EB | FAB | - | 268 |
| Detomidine | R | U | 1,2,30,40,55 | EB | Probe, GC/MS (F-Si) EI (70 eV), FAB | TMS | 269 |
| Dexamethasone | H | U | 71 | Q | GC/MS (SP-2100), EI(70) | TMS | 270 |
| " | H | U | 18,16 | Q | GC/MS(Cap), EI(70 eV) LC/MS,nebulizer,CI(+) | - | 271 |
| Dextromorphan | H | U | 1,18,20,22,39 70,71 | Q | GC/MS (F-Si, SE-54), EI CI ($CH_4$) | Ac | 272 |
| 2,4-Di-amino-6-T[f] | DIJR | A,U | 1,9,28,103 | - | Probe, EI (70 eV) | - | 273 |
| 3,4-Di-Cl-BzOAc[f] | R | U | 37,57,99 | Q | Probe, EI, LC/MS (TSP) | - | 274 |

Table 2. (cont.)

| Drug | In[a] | Metabolite[b] | MS[c] | Technique[d] | Deriv.[e] | Ref |
|---|---|---|---|---|---|---|
| Diethylpropion | H | U | 14,17,37,59 | B | GC/MS (Carbowax-20M, Apiezon-L), EI | - | 275 |
| Diethyl-stilbestrol | K | H | 1,2,15,30,34 | Q | GC/MS (F-Si,OV-1),EI(70)TMS | 276 |
| Diltiazem | R | LX | 73 | Q | GC/MS (SE-54), EI(70eV) TMS | 277 |
| Diltrizem | R,D,H | A,U | 1,14,19,20,22 | B | GC/MS (F-Si, DB-1), EI TMS | 278 |
|  |  | P | 74 |  | CI (-ve) TFA | 279 |
| Dimethyl-amphetamine | H,R | U | 1,9,22,30 | B | GC/MS (2% Apiezon-L) TMS CI (i-Bu) TFA | 280 |
| Di-Pr-NH$_2$-Et-I[f] | D,R | U | 1,30,56 | QQQ | FAB (Xe), LC/MS (TSP) | - | 281 |
| Doxylamine | R | U | 1,22,30 | Q EBE | FAB (+ve), DCI | - | 282 |
| " | J.R | L,U | 1,2,9,18,22 39,57 | B | GC/MS(OV-1,OV-17,OV-25) TFA EI(70eV), CI(NH$_4$),FAB | 283 |
| " | H | U | 22,39 | B | GC/MS (3% OV-1), EI TFA | 284 |
| " | J | U | 1,9 | QQQ | LC/MS | - | 285 |
| " | R | U,F H | 9,20,22 | Q | GC/MS (OV-17/OV-210) EI, CI(CH$_4$, 10%NH$_3$/N$_2$) | - | 286 |
| " | R | U | 30 | QQQ | TSP | - | 287 |
| Ebsilin | R | L | 1,25,30,77 32,104 | BE | Probe, EI (70 eV) FAB (Xe) | - | 288 |
| Elliptinium | H,R | U | 30,43,46 | Q | Probe, CI (NH$_3$) | - | 289 |
| " | R | U,A | 39,40 | EB | LC/MS (C$_{18}$), FAB | - | 290 |
| Enalapril(at) | R | U | 53 | Q | GC/MS (F-Si,QC2/BP1),EI Me | 291 |
| " | R | U | 41 | Q | GC/MS (F-Si,BP-1),EI,CI Me | 292 |
| Enciprazine | R | U | 23 | B | GC/MS (3% OV-1), EI | - | 293 |
| Enisoprost | H | U | 2,12,52,74,84 | Q | GC/MS (F-Si,DB1),EI(50) TMS,PFB CI(NH$_3$,-ve) MO,Me | 294 |
| Ethimizol | H | U,S | 58 | EB | Probe, EI (70 eV) | - | 295 |
| Ethinyloestradiol | H | LZ | 1,71 | - | Probe, EI | - | 296 |
| Etodolac | H | U,P | 15,94 | EB | GC/MS (F-Si, Me-Si), EI S-(+)-A | 297 |
| " | DHRM | U | 95 | EB | Probe, EI, High Res. Me | 298 |
| Etoposide | M | LZ | 20,47,48 | BE | Probe, EI (70 eV) | - | 299 |
| " | R | LZ | 20 | BE | Probe, EI, FD | - | 300 |
| " | M | Z | 20,49 | - | Probe, EI | - | 301 |
| " | A,R | APU | 30 | EB | FAB (Xe, +/-) | - | 302 |
| Felodipine | D,H,R | LZ | 2,52,74,75,79 | Q | GC/MS(F-Si,CP-Sil18),EI TMS | 303 |
| Fenetylline | H | U | 1,5,14,17,24 56 | Q | GC/MS (F-Si, DB5) TMS EI (70 eV) | 304 |
| Fenofibrate | H,R | U,F | 17,30,74 | - | Probe | - | 305 |
| " | D,I,R | U,F | 17,74 | EB | Probe, EI (70 eV) | - | 306 |
| Fenprofen | R | H | 96 | EB | Probe, EI | - | 307 |
| Flecainide | H | U | 7,18,41,57 | Q BEB | GC/MS (F-Si, DB-5) Ac EI, CI (-ve) | 308 |
| Flumecinol | H | U,F | 1,4, | B,EBG | C/MS (2% SE-30), EI | - | 309 |
| Flunixin | N | U,P | 6 | Q | GC/MS(SE-30Cap), EI(70) Me | 310 |
| Fluoromethol one | H | U | 15,71 | Q | GC/MS(F-Si,MeSi),EI(70) TMS,MO | 311 |
| 5-Fluorouracil | H | A | 35,50 | BE | GC/MS (F-Si), EI, FAB Me/TMS | 312 |
| " | R | L | 35,50 | BE | FAB, GC/MS | - | 313 |
| Furazolidone | P | LZ | 67 | EB | Probe, FD | - | 314 |
| " | R | U | 6 | EB | Probe, EI | - | 315 |
| Furfenorex | H | U | 1,22,52,56 | B | GC/MS (OV-17), CI, SIM TFA | 223 |
| Galanthamine | A,R | LX | 18,91 | B | GC/MS (SE30,OV17),EI,HR | - | 316 |
| Geranylgeranyl-lactone | R | U | 2,12 | EB | GC/MS (3% OV-1), EI(25) Me,TMS | 317 |
| Gomisin A | R | L | 19,70,71,108 | EB | Probe, EI | - | 318 |
| Gomphogenin | R | A,L | 18,71 | QQQ | Probe, CI (NH$_3$) | Me/Ac | 319 |
| Gomphoside | R | A,1 | 2,18,30,68,91 | QQQ | Probe, CI (NH$_3$) | Me/Ac | 319 |

Table 2. (cont.)

| Drug | In[a] | Metabolite[b] | MS[c] | Technique[d] | Deriv.[e] | Ref |
|---|---|---|---|---|---|---|
| Hexamethylene bisacetamide | H | U | 14,24,51,52 | - | GC/MS (D-1701), EI, CI | TFA,TFE | 320 |
| " | H | C | 14,24,51,53 | Q | GC/MS (F-Si), EI | TFE | 321 |
| HI-6 | R | U | 6,18,11 | EB | Probe, EI, CI (i-Bu), FD | - | 322 |
| Hydralazine | R | LZ | 2,18,39,60,82 83 | Q | Probe, EI (20 eV) | - | 323 |
| 7-OH-Delta-8-THC | I | LZ | 18 | EB | GC/MS(SE-30), EI, S($^2$H) | TMS | 324 |
| OH-Br-Bz-amino-cyclohexanol[f] | D,H,N O,R | U,P T | 17,89 | Q | GC/MS(F-Si,CP-Si15), EI | TFA | 325 |
| OH-Et-Bz-Sulph. | R | P | 18 | B | GC/MS (OV-1), EI(30) | TMS | 326 |
| Hydroxypropyl-aminotetraline | R | R,L | 1,19 | EB BE | GC/MS (F-Si,SE54),EI(70) Probe (HR), EI (60 eV) | - - | 327 |
| Ifosfamide | H | U | 54 | BE | FAB (Xe) | - | 328 |
| Imipramine | R | L | 86 | - | FD | - | 329 |
| Indeloxazine | R | P,U | 1,7,18,25,30 38,39,85 | EB | GC/MS (1% OC-22) EI (70 eV) | TMS | 330 |
| " | R | U | 18,38,85 | EB | FAB (Xe) | - | 331 |
| Isoniazid | H | U | 39 | Q | GC/MS, EI, S ($^{15}$N) | - | 332 |
| Ivermectin | O,R,S | T | 2,68 | B | Probe, EI (70 eV) | TMS | 333 |
| Ketamine | A,H,R | LZ | 22,29 | EB | GC/MS (F-Si, DB-5), EI | PFP | 334 |
| " | N | U | 15,22 | Q | GC/MS(F-Si,DB-1),EI(70) | - | 335 |
| Ketanserin | H | F,U | 1,14,18,30 | Q | Probe, EI | - | 336 |
| Ketoprolac tromethamine | A,J,H M,R | B,F U | 1 | - | GC/MS, EI | Me | 337 |
| Ketotifen | A,H,R | H | 9,17,22,31,86 | Q | GC/MS (SE-52), EI(70eV) | - | 338 |
| Laudanosine | A,D,H | A,U | 20,94 | Q | GC/MS (F-Si, DB-5), EI | TMS,TFA TBDMS | 339 |
| Lidocaine | A | L | 1,8,9,41,59 78 | Q | GC/MS (F-Si, OV-1) EI (70 eV) | TFA | 340 |
| " | R | U | 1,2,19,59,41 | EB | Probe, EI, CI (i-Bu) | - | 341 |
| Lorazepam | J,H,R | E | 33 | EB | FAB (Xe) | - | 342 |
| LU-253[f] | R | A,F B,P | 18,19,41,75 76 | BE Q | GC/MS (F-Si, OV-1), EI | TMS | 343 |
| Maprotiline | H | B | 39 | Q | GC/MS (F-Si,PhMeSi), EI | TFA | 344 |
| Meclomen | H | U | 1,2,55,75 | Q | GC/MS (2% OV-1), EI(70) | HFIP,TFA | 345 |
| Melperone | H | U | 4,7,14,18,15 | Q | GC/MS(F-Si,SE-54),EI,CI | Me,Ac | 346 |
| Melphalan | A,H | E | 40 | EB | FAB (Xe) | TMS | 347 |
| " | J | E | 40 | EB | FAB (Xe) | - | 348 |
| (S)-Mephenytoin | H | U | 1 | - | GC/MS, EI, S($^2$H) | - | 349 |
| Mespirenone | R | U,F | 10,71,92 | Q,EB | Probe, EI, CI (NH$_3$) | - | 350 |
| Methadone | R | A,U | 1 | B | GC/MS (1%OV-1), EI (70) | - | 351 |
| Methamphetamine | I,R | LX | 8,9,22,72 | EB | Probe, EI (70 eV) | - | 352 |
| Methapyrilene | F | - | 9,2 | Q | Probe, DCI, CI | - | 353 |
| " | R | U | 1,9,14,15,18 22,56,60 | EB BE,Q | GC/MS (F-Si,DB-5,packed) LC/MS, TSP, CI, EI | - | 354 |
| " | A,I,R | LZ | 8,6,9,18,22 55,56 | Q | GC/MS (F-Si), EI(70 eV) | TMS TBDMS | 355 |
| " | R | L | 6,9,55,56,65 | Q | GC/MS(F-Si,OV-1),EI(70) | TMS | 356 |
| 8-Methoxypsoralin | R | U | 1,25,20,34 | Q BE | LC/MS (C-18,TSP), Probe MS/MS, EI (70), FAB (Xe) | - | 357 |
| Me-delta-8-THC | M | L | 2,3,4,13,55 | B | GC/MS(SE-30), EI(25), S | Me/TMS | 358 |
| Me-delta-9-THC | M | L | 2,3,4,13,55 | B | GC/MS(SE-30), EI(25), S | Me/TMS | 358 |
| abn-Me-D-8-THC | M | L | 2,3,4,13,55 | B | GC/MS(SE-30), EI(25), S | Me/TMS | 358 |
| N-Me-formamide | M,R | U | 43 | EB | Probe, CI (i-Bu) | - | 359 |
| " | M | A | 43 | EB,BSIMS | (Cs),S,Probe,EI,CI | - | 360 |
| Me-Prednisolone | H | U | 15,17,18 | Q | GC/MS(F-Si,MeSi),EI,SIM | TMS,MO | 361 |
| 2-S-Pr-glycine[f] | H | U | 41 | B | GC/MS(3% OV-17), EI(70) | Me | 362 |
| Methtryptoline | R | U,H | 1 | B | GC/MS(F-Si,Chiral), EI | PFP | 363 |

Table 2. (cont.)

| Drug | In[a] | Metabolite[b] | | MS[c] | Technique[d] | Deriv.[e] | Ref |
|---|---|---|---|---|---|---|---|
| Metioprim | G,P | - | 10,22 | - | Probe, EI, FAB $^{252}$Cf | - | 364 |
| Metoprolol | H | U | 4,55 | Q | GC/MS (F-Si), EI (70eV) | TMS | 365 |
| " | R | LZ | 4 | EB | GC/MS(F-Si,DB-5),EI(70) | Carb | 366 |
| Mexilitine | A | L | 57 | Q | GC/MS (SE-54), EI(70eV) | - | 367 |
| Midaglizole | D | U | 6,18,25,84 | EB | Probe, EI | - | 368 |
| Mitonafide | R | U | 39,69 | Q | GC/MS (MeSi, Cap), EI | - | 369 |
| Mitoxantrone | H | U | 52 | Q | Probe, EI (44 eV) | - | 370 |
| MDL 257 | H | U | 11,12,25,83 | QQQ | Probe, EI(70),CID(20eV) | - | 371 |
| Molsidomine | A,D,R | U | 25 | EB | GC/MS, Probe, EI (70eV) | Me,TMS | 372 |
| Mopidralazine | R | U | 1,25,39,60 | BE | Probe, EI(70eV), S($^{13}$C) | - | 373 |
| Morantel | D,O,R | L,B | 6,7,17,18,24 | Q | LC/MS, TSP, ./- | - | 374 |
| | | U | 25,40 | EB | | | |
| MOTP[f] | D,R | U | 12,105 | - | GC/MS (1% OV-17),EI(70) | TMS | 375 |
| Nicotine | I | LZ | 9,106 | B | GC/MS (OV-1), EI (20eV) | - | 376 |
| " | R | U | 22,106 | EB | Probe, EI | - | 377 |
| " | A | DZ | 106 | Q | GC/MS (OV-17), EI, HR | - | 378 |
| R-(+)-Nicotine | I | U | 9 | Q | TSP | - | 379 |
| (S)-Nicotine | A | E | 6 | EB | GC/MS (F-Si, DB-5), EI | CN | 380 |
| Nitrofurantoin | R | U | 6 | B | Probe, EI | - | 315 |
| Nitrofurazone | R | U | 6 | B | Probe, EI | - | 315 |
| Nilvadipine | D,R | A,U | 2,69,74,79 | EB | Probe, EI(70eV),FAB(Xe) | - | 381 |
| Norcodeine | R | H | 20 | Q | GC/MS (F-Si, Me-Si), EI | PFP | 382 |
| Norethindrone | H | U,P | 16,17,30,34 | Q | GC/MS(F-Si,SE30),EI(40) | TMS,MO | 383 |
| " | H | M | 16,17,34 | Q | GC/MS(F-Si,SE30),EI(40) | Ac | 284 |
| 19-Nor-testosterone | H | U | 17,30,34 | Q | GC/MS (F-Si, Me-Si), EI | TMS,MO | 385 |
| Nortriptyline | R | A | 1,4,18,19,22 | EB | Probe, EI (12, 70 eV) | - | 205 |
| " | R | A | 1,19,22,30,71 84 | EB | FAB (Xe, 8 KV) | - | 207 |
| ONO-802[f] | H | U | 2,4,12,17,30 34,52,55 | B | GC/MS (1% OV-1) EI (25 eV) | MeO,Me TBDMS,TMS | 386 |
| Oxaminozoline | R | H | 15,25,106 | Q | GC/MS (F-Si,CP-Si15),EI | - | 387 |
| Oxaprotiline | D,R | U | 1,14,19,22,30 | BE,B | Probe, EI(70), FAB(Xe) | TMS | 388 |
| Oxazepam | J,H,R | E | 33 | EB | FAB (Xe) | - | 342 |
| " | M | U,F | 1,26,30,49 | B | GC/MS (3% SE-30), EI | Me | 389 |
| Oxcarbazepine | H | U | 1,7,17,30,87 88 | BE | Probe, EI (70 eV), FAB | TMS | 390 |
| Paracetamol | M | U | 40 | - | Probe,EI,CI, FD, FAB, S | AlcC | 391 |
| " | - | - | 40 | EB | FAB (Xe), MS/MS | - | 392 |
| Penberol | R | U | 1,57 | - | EI | | 393 |
| Penbutolol | H | U | 1,15,30,34 | EB | FAB, EI | TMS | 394 |
| Penicillamine | H | U,F | 39,61,62,63 | EB | FAB (Ar) | - | 395 |
| " | R | U | 39,46,61,62, | EB | FAB | - | 396 |
| Penticainide | B,D,H R | U | 4,14,18,52,53 56 | Q | GC/MS (3% OV-17), DCI EI (70eV),CI(CH$_4$,NH$_3$) | - | 397 |
| Perhexiline | R | LZ | 71 | Q | GC/MS (SE-30, Cap),EI | TFA | 398 |
| " | R | A | 71 | EB | GC/MS, EI (70 eV), SIM | HFB | 399 |
| Perindopril(at) | R | U | 53 | Q | GC/MS (F-Si,QC2/BP1),EI | Me | 291 |
| Phenacetin | R | A | 30,40 | QQQ EB | LC/MS (RP-18), TSP FAB (Xe) | - | 400 |
| " | R | Bi | 40 | BEQ | FAB, CID, MS/MS | - | 401 |
| Phenazone | R | E | 1,2,22 | B Q | Probe, EI, S($^{13}$C, $^{15}$N) GC/MS (2% OV-225), SIM | - | 154 |
| Phencyclidine | R | LZ | 1 | EB | GC/MS (OV-17), EI, S(D) | Me | 402 |
| " | R,M | U,L | 7,71 | Q | GC/MS (F-Si, MeSi), EI | TMS | 403 |
| Phendimetrazine | H | U | 9,22 | B | GC/MS (C/wax 20M+KOH),EI | TMS | 404 |
| Phenelzine | T | - | 39,82 | Q | GC/MS(F-Si,DB-5),EI(70) | Et | 405 |
| " | R | LZ | 14,82 | EB | GC/MS(DB-1), EI, S($^{18}$O) | - | 406 |

Table 2. (cont.)

| Drug | In[a] | Metabolite[b] | MS[c] | Technique[d] | Deriv.[e] | Ref |
|---|---|---|---|---|---|---|
| Pheneturide | H,R | U 1,3,4,19,41 53,84 | EB | Probe, EI (70 eV) | – | 407 |
| Pheniprazine | T | – 39,82 | Q | GC/MS(F-Si,DB-5),EI(70) | Et | 405 |
| Phenprocoumon | H | P,U 1 | Q | GC/MS (F-Si, OV 1), EI | Me | 408 |
| Phenylbutazone | R | E 6,97,98 | EB | Probe, EI (30 eV) | – | 409 |
| Picolyl-di-Me-Bz[f] | R | LZ 1,2,9,23 | EB | Probe, EI | – | 410 |
| Pinacidil | H | U 2,9,29,30, | – | FAB, MS/MS | – | 411 |
| Pivampicillin | H | U 64 | – | FAB, GC/MS | – | 412 |
| Prethcamide | H | U 22 | Q | GC/MS(F-Si,PhMeSi)EI,CI | – | 413 |
| Procainamide | H,R | LZ 8 | EB | Probe, CI (i-Bu) | – | 414 |
| Promazine | N | U 9,10,29 | QQQ | LC/MS (C-18), API | – | 415 |
| " | A | LX 4,9,14,22,56 | B EB | GC/MS, (3% OV-17, OV 1) Probe, EI | TMS,TFA | 416 |
| Propiverin | R | – 1,3,9,52,57,74 | EB | Probe, EI (70 eV) | – | 417 |
| Propofol | H | P,U 1,30,34 | EB | GC/MS (3% OV-17), EI | – | 418 |
| Propranolol | R | LZ 2 | EB | GC/MS(F-Si,DB5),EI,S(D) | TFA | 419 |
| " | H,R | L 1,56 | EB | GC/MS(F-Si,DB1),EI,S(D) | TFA | 420 |
| " | I,R | L 40 | Q | LC/MS (C-18), TSP, S(D) | – | 421 |
| " | R | LX 1 | EB | GC/MS (F-Si, DB-5) EI (20eV), S($^2$H, $^{18}$O) | TMS | 422 |
| " | H,R | U 1 | EB | GC/MS (F-Si, DB-5),SIM EI (20 eV), S ($^2$H) | TMS | 423 |
| Propranolol | H,R | U 1,14,17,19 | EB | GC/MS (F-Si, DB-5) EI (70 eV), S ($^2$H) | TFA | 424 |
| Pr-delta-8-THC | M | L 2,3,4,13,55 | B | GC/MS(SE-30), EI(25), S | Me/TMS | 425 |
| Pr-delta-9-THC | M | L 2,3,4,13,55 | B | GC/MS(SE-30), EI(25), S | Me/TMS | 425 |
| Pyrazinamide | H,R | U 6 | Q | Probe, EI (70 eV) | Me | 426 |
| " | H | U 6 | – | Probe, EI (20 eV) | – | 427 |
| " | R | U 6 | Q | Probe, EI | – | 428 |
| Pyrido-glutethimide | R | LZ 4,18 | EB | Probe, EI(70), CI, S(D) | – | 429 |
| " | H,R | U 9 | EB | Probe, EI (70 eV) | – | 430 |
| Pyrilamine | F | – 9,20,22 | Q | DCI, CI (+/-) | – | 431 |
| " | R | T 9,20,30 | Q | DCI, FAB | – | 432 |
| " | R | U 1,19,20,22,23 56 | Q | GC/MS(3% OV-11),CI(CH$_4$) | TMS | 433 |
| " | R | U 9,20,22 | QQQ | LC/MS (TSQ) | – | 285 |
| " | R | U 9,22 | Q | DCI,CI (CH$_4$,NH$_3$), EI(70) | – | 434 |
| Quinpirole | D,J M,R | U 4,30,31,53,55 56 | BE | Probe, EI(70eV), FAB, FI | – | 435 |
| R-7000[f] | R | A,U 1,20,25,69 | Q | – | Me,CB | 436 |
| Ramipril(at) | R | U 53 | Q | GC/MS (F-Si,QC2/BP1),EI | Me | 291 |
| Ranitidine | H | U 14,24 | Q | LC/MS, DLI, CI(-) S($^2$H) | Me | 437 |
| Rilmazafone | J | U 1,7,14,22,41 60 | – | Probe, EI | – | 438 |
| Rosaprostol | H,R | U 2,3,4.12,18 | EB | GC/MS, EI | MeO TMS,PFB | 439 |
| Roxatidine | D,R | LZ 7,18 | B | GC/MS(2% OV-1),EI,S($^2$H) | TMS | 440 |
| " | R | U 1,7,14,18,20 39,41,57,83 | B | GC/MS(2% OV-1),EI(20eV) S($^2$H$_{10}$) | TMS,Me TFA,HFP | 441 |
| Rulziracetam | D,J,R | P,U 25 | Q | GC/MS (3% OV-225), EI | – | 442 |
| S 59-801[f] | H | U 1,30 | B, EB | Probe, EI, DCI(NH$_3$,i-Bu) FAB (glycerol, thio) | – | 443 |
| Salicylamide | M | U,B 1 | Q | Probe, EI | – | 444 |
| SCH 34343 | H | U 10,25,28,61 | BE | FAB (Xe, 6kV, +ve), HR | – | 445 |
| SKF-86466[f] | D,R | U 9,58 | Q QQQ | GC/MS (F-Si, DB-5), EI FAB | PFP | 446 |
| SKF 93944 | D | P 2,30,31,56 | Q | LC/MS (C$_{18}$), TSP | – | 17 |

Table 2. (cont.)

| Drug | In[a] | Metabolite[b] | | MS[c] | Technique[d] | Deriv.[e] | Ref |
|---|---|---|---|---|---|---|---|
| SKF 95018 | D | H | 4,14,41,57,83 | Q | LC/MS ($C_{18}$), TSP | - | 17 |
| SKF 102,081[f] | R | LZ | 2,3 | Q,BE | GC/MS (F-Si,D B-5), | TMS | 447 |
| " | I | A | 2,3,4,12,18 | Q | GC/MS (F-Si, DB-5), EI | Me,TMS | 448 |
| | | | 30 | BE | FAB (Xe), S ($^2$H) | $D_9$TMS | |
| Spirohydantoin mustard | M | L | 56 | BE | Probe, EI, FAB | - | 449 |
| Stiripentol | H | U | 4,19,25,30,89 | EB | GC/MS(F-Si,DB1),EI,S(D) | TMS,TMSD | 450 |
| Stobadine | R | LZ | 9,58, | EB | GC/MS(3% OV-17), EI(75) | - | 451 |
| Sulfadimethoxine | A,H,J | E | 31 | EB | FAB (Xe) | - | 342 |
| Sulfamethazine | P | B | 39,65 | B | Probe, EI, FAB | - | 452 |
| Sulphathiazole | R | U | 65 | B | Probe, EI | - | 453 |
| SX-PP 16 | R | A,U | 40 | EB | FAB (Xe) | Me | 454 |
| TA-1801[f] | R | U | 25 | B | GC/MS(OV1),EI(20),S($^2$H) | Me,TMS | 455 |
| " | A,D,R | U,F | 25,33,74 | B | GC/MS (3% OV-1), EI(20) | Me,EDMS Et,TMS | 456 |
| Tamoxifen | H | P | 1,9,22,57 | Q | GC/MS (F-Si, OV-101, SE-30), EI (70 eV) | TMS,Me HFB | 166 |
| " | R | H | 9,30 | EB | Probe, EI | - | 457 |
| " | H | A | 1,22 | Q | LC/MS (C-18), TSP | - | 458 |
| | | | | Q | GC/MS (SP-255 Cap), EI | TMS | |
| " | H | S | 1,14,22 | Q | LC/MS, TSP | - | 459 |
| Terodiline | R | LZ | 1(R),4(S) | B | GC/MS(F-Si,OV-1),EI,SIM | TMS | 460 |
| Thebaine | R | L,R | 17,20 | - | GC/MS (OV-1 Cap), EI | TMS | 461 |
| Thenyldiamine | F | - | 9,22 | Q | Probe, CI, DCI | - | 353 |
| Thiabendazole | M | U,F | 1,58 | EB | GC/MS (2% OV-1), EI | Me | 462 |
| 2-(2-thienyl)-allylamine | R | U | 14,16 | Q | GC/MS (F-Si, DB-5) CI ($CH_4$) | TMS | 463 |
| Thiopentone | H | U | 4,12 | EB | LC/MS ($C_{18}$), EI, CI | - | 464 |
| Thioridazine | H | U | 10 | BE | Probe, EI | - | 465 |
| Tiadenol | R | U | 10,28,30,55 | EB | Probe, EI | - | 466 |
| | | | | BE | FAB (Ar), B/E, $B^2$/E | | |
| Tiropamide | H | U | 29,56,59, | EB | GC/MS(F-Si,CP-Si118),EI | TMS | 467 |
| Tixocortol pivalate | H | U | 11,16,17,28 61,71,93 | Q | Probe,EI(70),CI,DCI($NH_3$) | - | 468 |
| Tracazolate | D,R | U | 2,4,52,59,65 74 | Q | GC/MS (OV-7, OV-1), EI | Me | 469 |
| Tranylcypromine | H | U,B | 1 | - | GC/MS | PFA | 470 |
| Trimoprostil | R | A | 12,15,18,99 | BE | FAB | - | 471 |
| Tripelennamine | R | U | 1,4,19,23,56 | Q | GC/MS(3% OV-11),CI($CH_4$) | TMS | 433 |
| " | F | - | 9,22 | Q | Probe, CI, DCI | - | 353 |
| Triprolidine | F | - | 2 | Q | Probe, EI, DCI, S($^{18}$O) | - | 472 |
| Tropatepine | R | A,U | 10 | Q | GC/MS (F-Si,CP-Si15),CI | - | 473 |
| Ufloxicin | D,J,R | U,F | 9,22,30 | EB | Probe, EI, FAB (Xe), CI | - | 474 |
| Urapidil | R | U | 23 | B | GC/MS (3% OV-1), EI | - | 293 |
| Valproic acid | H | T | 4 | EB | GC/MS(F-Si,DB-5),EI(70) | TMS | 475 |
| " | J | U | 4,12,15,18,30 37,52,55 | EB | GC/MS(F-Si,DB-5),EI(70) | TMS | 476 477 |
| " | H | S,U | 4,15,18,30, | EB | GC/MS(3% OV-1),EI(70eV) | TMS | 178 |
| " | R | U | 4,15,18,55 | EB | GC/MS(F-Si,DB1),EI,S(D) | TMS | 478 |
| Verapamil | H,R | LZ | 20 | EB | GC/MS (F-Si, DB-5) EI (70 eV) | TFA TBDMS | 479 |
| Wy-41770[f] | D,J M,R | U | 1,30 | Q,B | Probe, EI | - | 480 |
| Zearalenone | F | - | 74,75,89 | B | GC/MS, EI | TMS | 481 |
| Zolpidem | R | U | 2,22,78 | Q | GC/MS (SE-30Cap),EI(70) LC/MS/MS (C-18,$NH_4Ac$) | TMS | 482 |
| Zonisamide | H | U | 1,4,8,39 | QQQ | Probe, EI/CI (+/-) | - | 44 |

a) First column = Species: A = Rabbit, B = Baboon, C = Cat, D = Dog, E = Enzyme, F = Fungus, G = Goat, H = Human, I = Guinea-pig, J = Monkey, K = Hamster, L = Gerbil, M = Mouse, N = Horse, O = Cow, P = Pig, Q = Mini-pig, R = Rat, S = Sheep, T = Microbe.
Second column = medium [See footnote (a), Table 1]

b) 1 = Aromatic hydroxylation, 2 = Hydroxylation of methyl group (or ω-hydroxylation), 3 = ω-1 Hydroxylation, 4 = Other aliphatic hydroxylation, 5 = Benzylic hydroxylation, 6 = Hydroxylation of aromatic heterocyclic ring, 7 = Hydroxylation of aliphatic heterocyclic ring, 8 = Hydroxylation of nitrogen, 9 = Oxidation of nitrogen, 10 = Oxidation of sulphur, 11 = Oxidative C-C bond cleavage, 12 = Beta-oxidation, 13 = Allylic hydroxylation, 14 = Oxidative deamination, 15 = Elimination of hydrogen to give double bond, 16 = Reduction of double bond, 17 = Reduction of carbonyl group to OH, 18 = Oxidation of OH group to carbonyl, 19 = Methylation of OH, 20 = O-Debenzylation, 22 = N-Demethylation, 23 = N-Debenzylation, 24 = Oxidation of CHO to COOH, 25 = Ring opening, 26 = Ring contraction, 27 = Oxidation of NH$_2$ to NO$_2$, 28 = Oxidation of sulphur to SO$_2$, 29 = Hydroxylation at unspecified position, 30 = Glucuronide conjugation at OH, 31 = Glucuronide conjugation at nitrogen, 32 = Glucuronide conjugation at selenium, 33 = Glucuronide conjugation at a carboxylic acid, 34 = Sulphate conjugation at oxygen, 35 = Conjugation with bile acids, 36 = Conjugation with cholesterol, 37 = Conjugation with glycine, 38 = Conjugation with glucose, 39 = Acetylation at nitrogen, 40 = Conjugation with glutathione, 41 = Cleavage of amide bond, 42 = Elimination of HCN, 43 = Conjugation with N-acetylcysteine, 44 = Reduction of quinone, 45 = Formation of quinone methide, 46 = Conjugation with cysteine, 47 = Quinone formation, 48 = Covalent binding to protein, 49 = Aromatization, 50 = Cleavage with loss of an amino acid, 51 = N-De-acetylation, 52 = Formation of lactone ring, 53 = Formation of lactam ring, 54 = Conjugation with MESNA, 55 = Oxidation of CH$_2$OH to COOH, 56 = N-Dealkylation, 57 = O-De-alkylation, 58 = Methylation of nitrogen, 59 = N-Deethylation, 60 = Cyclisation, 61 = Methylation of sulphur, 62 = S-S Bond formation, 63, Formation of inorganic sulphate, 64 = Conjugation with carnitine, 65 = Loss of NH$_2$, 66 = Formation of N-formyl derivative, 67 = Ring opening to substituted acrylonitrile, 68 = Loss of sugar, 69 = Reduction of NO$_2$ to NH$_2$, 70 = Acetylation of oxygen, 71 = Alicyclic hydroxylation, 72 = Schiff base formation, 73 = O-De-acetylation, 74 = Ester hydrolysis, 75 = Decarboxylation, 76 = Epoxide formation, 77 Dimerization, 78 = Oxidation of CH$_3$ to COOH, 79 = Oxidation of dihydropyridine to pyridine, 80 = N-De-phenylation, 81 = Conversion of NO$_3$ to OH, 82 = Loss of NH-NH$_2$, 83 = Reduction of COOH to CH$_2$OH, 84 = Dehydration, 85 = Hydroxylation of epoxide to dihydrodiol, 86 = Sulphate conjugation with nitrogen, 87 = Formation of enol glucuronide, 88 = Formation of enol sulphate, 89 = Isomerization, 90 = Acetal cleavage, 91 = Epimerization, 92 = S-De-acetylation, 93 = S-De-alkylation, 94 = Unspecified conjugation, 95 = Conjugation with urea, 96 = Conjugation with glycerol, 97 = Peroxidation of heterocyclic ring, 98 = Chlorination of heterocyclic ring, 99 = Conjugation with taurine, 100 = Loss of NO$_2$, 101 = Loss of OH, 102 = Conversion of CH$_2$CH$_2$Cl to CH$_2$COCH$_3$, 104 = Methylation of selenium, 105 = Loss of phosphate ester, 106 = Addn. of =O to heterocyclic ring, 107 = Fluorine migration, 108 = Methylenedioxy ring cleavage, 109 = Loss of PhCOOH.

c) E = Electric sector, M = Magnetic sector, Q = Quadrupole, TOF = Time of flight

d) See footnote (d) Table 1. S = Paper containing work with stable isotopes. Isotopes are listed in parentheses.

e) See footnote (b) Table 1.

f) 4-Amino-5-Et-3-T = 4-Amino-5-ethyl-3-thiophenecarboxylic acid methyl ester, Bz-chinozolin = 3-{2-[2-(2-Benzimidazolyl)benzoxy]ethyl}chinozolin-4-one, CB-3717 = $N^{10}$-Propargyl-5,8-dideazafolic acid, 3,4-Di-amino-6-T = 3,4-Diamino-6-(2,5-dichlorophenyl)S-triazine maleate, 3,4-Di-Cl-BzOAc = 3,4-Dichlorobenzyloxyacetic acid, N-(Di-Me-Py)-Bz-sulphonamide = N-(2,6-Dimethyl-2-pyrimidinyl)-

benzene-sulphonamide, Di-Pr-NH$_2$-Et-I = 4-(2-Dipropylaminoethyl-2-(3H)-indoline, OH-Br-Bz-aminocyclohexanol = Trans-4'-(2-hydroxy-3,5-dibromobenzylamino)-cyclohexanol, LU-253 = 4-Oxa-5-exo-(N-methylcarbamoxy)tricyclo-[5,2,1,0$^{2,6}$-endo]dec-8-en-3-one, MOTP = (RS)-2-Methoxy-3-(octadecylcarbamoyloxy)propyl-2-(3-thiazolio)ethyl phosphate, ONO-802 = 16,16-Dimethyl-trans-delta-2-prostaglandin-E methyl ester, 2-S-Pr-glycine = 2-Mercapto-propionylglycine, Picolyl-di-Me-Bz = N-Picolyl-3,5-di-methylbenzamide, R-7000 = 7-Methoxy-2-nitronaptho[2,1-b]furan, S 59-801 = Alpha-[(dimethylamino)methyl]-2-(3-ethyl-5-methyl-4-isoxazolyl-1H-indole-3-methanol, SKF-86466 = 6-Chloro-2,3,4,5-tetrahydro-3-methyl-iH-3-benzazepine, SKF-102081 = 5-(2-Dodecylphenyl)-4,6-dithianoanedioic acid, TA-1801 = Ethyl-2-(4-chlorophenyl)-5-(2-furyl)oxazole-4-acetate, Wy-41770 = (5H-Dibenzo[ad]cyclohepten-5-ylidene)acetic acid, 450191-S = 5-[(2-Aminoacetamido)methyl]-1-[p-chloro-2-(o-chlorobenzoyl)phenyl]N,N-demethyl-1-H-triazole-3-carboxamide HCl.

trimetrexate[176], and the method of Murphy et al.[167] has been used to measure tamoxifen metabolites[483]. Although HPLC methods are available, GC/MS is still necessary for sensitive detection of busulfan[484]. Melphalan has been stabilised for analytical measurement by formation of a N-acetylcysteine derivative[485], and the sulfamic acid diester, NSC 329680, has been assayed as 1,7-diiodoheptane after reaction with sodium iodide[486]. Redox reactions have been detected during analysis of some heteroanthracycline antitumor antibiotics[487] using SIMS with Cs ions. A 1,5-hydrogen transfer during fragmentation of cyclophosphamide has been found to have a thermal component[488], and cyclisation of this drug during GC/MS without derivatization has been shown to parallel that of its 4-oxo-analogue[489]. Extensive degradation reactions in outdated maytansine formulations include ester hydrolysis[490], and P-N bond cleavage has been found for thiotepa in aqueous solution[491].

## 6. Antimicrobial Drugs

It is outside the scope of this review to cover the several hundred papers that have appeared on structural determination of new antibiotics but a few of the more significant ones can be cited. Among the most complex antibiotics to have been studied are the glycopeptides such as the kibdelins with molecular weights around 1800. These have been ionized by FAB from thioglycerol matrices containing organic acids[492]. The related actinoidin A$_2$, has also been ionized by FAB[493]. Slightly less complex are the polyether avilamycins (molecular weights around 1400) which give abundant ions at [M - 1]$^-$ from thioglycerol[494], maduramycin ([M + Na]$^+$ ion at 939)[495,496], lipiarmycin B which gives a [M + Na]$^+$ ion at m/z 1173 from thioglycerol[497], and portmicin which gives ions at m/z 828 and 851 ([M + Na]$^+$) by FD[498]. Other antibiotics with molecular weights in excess of 1000 ionized by FAB include actinomycins (ions around 1275)[499], iturins (1044-1072)[500], leucinostatins (1134-1234)[501], alamethicins (around 2000)[502], and izupeptins (1474)[503]. The polypeptide antifungal agent, verlamelin (MW = 885), on the other hand, has been analysed by conventional EI[504]. Also analysed by EI is the TMS derivative of mulundocandin which gave a cluster at m/z 1799 containing eleven TMS groups[505]. The compound gave no ions by conventional FAB but with 3-nitrobenzyl alcohol doped with either lithium iodide or sodium acetate, $[M + Li]^+$ and $[M + Na]^+$ ions respectively were observed[505]. Possibly the largest antibiotic yet analysed is the pentacyclic polypeptide nisin from Streptococcus lactis containing 34 amino acids. This has been ionized by FAB from thioglycerol containing water, acetonitrile and trifluoroacetic acid to give a $[M + thioglycerol]^+$ ion at m/z 3461.9[506]. The ion of highest abundance was at $[M - 216]^+$; this gave a monoisotopic mass of 3153.29 under high resolution conditions. The compound has also been examined by [252]Cf plasma desorption MS where a $[M + Na]^+$ ion at 3378.1 was formed[507]. Sodium removal gave a $[M + H]^+$ ion at 3353.3.

General mass spectrometric aspects of antibiotic analysis have been reviewed[508], and the EI fragmentation of dichloroacetamides[509], granaticin[510], griseochelin[511], quinine[512], and monobactam antibiotics[513] has been reported. The FAB spectra of paulomycins[514] and polyether antibiotics[515], and the FD spectra of anthracycline antibiotics have been studied[516,517]. Linked scanning has proved useful in deriving sugar sequence information[517] for these compounds and sequence ions are also present in the ammonia DCI spectra[518]. MS/MS studies of papulacandin antibiotics, combined with deuteroacetylation, has been used to identify galactose as the glycosidic sugar[519]. Anomalies observed in FAB spectra include dechlorination of several glycopeptide antibiotics[520], and the oxidation of fredericamycin A to the hydroquinone form in glycerol and DMF[521]. Separation and analysis of monobactams, using continuous-flow FAB with an aqueous methanol solvent containing glycerol, gives $[M + H]^+$ ions as the base peaks[522]. Similar ions are seen in CI spectra of polyether antibiotics after separation by SFC using low restrictor temperature, but more fragmentation occurs as the temperature is raised[523]. Beta-lactam antibiotics have been found to give informative thermospray spectra in both the positive and negative mode, and extensive fragmentation under MS/MS conditions[524]. Assays for dimetridazole[114], erythromycin[116,117], fludalanine[121], ipronidazole[114], isoniazid[130] and rimantidine[158,159] have been developed; details are in Table 1. Clinical aspects of the latter drug has also been extensively studied by GC/MS[525-528]. Quantification by GC/MS of artesunic acid, a metabolite of the antimalarial, dihydroquinghaosu, has utilised a pyrolysis product formed in the injector[111]. CI spectra obtained from a belt interface have enabled rapid screening for sulphonamide residues from tissues; multiple reaction monitoring was used for quantification[165]. Sulphonamides have also been confirmed in tissues by GC/MS[529].

Metabolic studies (Table 2) have been reported for amodiaquine[211], cefepime[242], ciprofloxicin[252], dapsone[263], furazolidone[314], isoniazid[332], ivermectin[333], metioprim[364], penicillamine[395], pivampicillin[412], pyrazinamide[426-428], sulphamethasine[452], sulphathiazole[453], and ufloxicin[474]. Pivaloylcarnitine has been identified by FAB as a metabolite of pivampicillin[412] and is the second example of this type of conjugation for drugs. The N-oxide metabolite of chloroquine has been found to decompose by a Cope rearrangement to an olefin during GC/MS analysis[244]. Degradation of several antibiotics has been studied by conventional mass spectrometry[530-534], and a thermospray interface has been used as a reactor for studying decomposition of beta-lactams[535]. In clinical studies of antimalarials, amidiaquine has been detected in human urine by GC/MS, 5 months after a single dose[536], and has been shown to form a reactive iminoquinone[537], and an oral solution of $[^2H_5]$mefloquine has been used as a standard to measure the bioavailability of tablets of the drug[538].

## 7. Drugs of abuse

**7.1 Drug Screening.**- The increased demand for screening for drugs of abuse, particularly in the United States, has produced a multitude of cheap methods. Many of these are based on radioimmunoassay and are not particularly reliable as cross-reactivity with other compounds is common at low concentrations. For example, phenylpropanolamine, cross-reacts in a test for amphetamines[539], ibuprofen cross-reacts with cannabinoids[539], and dextromorphan has given a false positive result in a test for phencyclidine[540]. Several other, similar interactions have been detected by GC/MS[541,542]. Most of these simple tests must be confirmed by other methods of which GC/MS is preferred[543]. Suitable methods for seven commonly abused drugs have been described by Mule and Casella[85]. Identification of drugs misused in sport has been discussed[544] and the detection of drugs in racehorses continues to receive much attention[545]. Rapid analysis is often necessary in this work and a LC/MS/MS method using an API source and capable of handling 60 samples an hour has been developed[545]. The authors point out that, as many of the drug metabolites present in urine and plasma are isomers, a preliminary chromatographic step is advantageous and sometimes necessary. Other areas where drug screening by mass spectrometry has been used is in drug identification in attempted suicide [546] and overdose cases[547], detection of drug residues in meat[548] and tissues[165], identification of drugs in cadavers[549], and identification of prescription drugs in herbal medications[550].

7.2 Cannabinoids.- Confirmation of cannabis abuse is usually achieved by the measurement of delta-9-tetrahydrocannabinol-11-oic acid (THC-11-oic acid) using GC/MS with a variety of derivatives and ionization methods[170,551,552],. The diuretic furosemide can interfere as its spectrum (methyl derivative) has common ions[553]. Although radioimmunoassay is used in the initial screening for the acid metabolite, results of current assays must be confirmed by GC/MS[554-557]. Evaluation of an ion trap in this context, as a cheap alternative to conventional mass spectrometers, has given a detection limit of 10 ng/ml of urine; the $[M - 1]^+$ ion was reported to be larger than $M^{+\cdot}$ but the relative abundances of fragment ions were found to vary with the concentration of the acid in the spectrometer[558]. The internal standard [$^2H_6$]delta-9-THC-11-oic acid has been synthesised and found to give less interference than the more common [$^2H_3$]-analogue in these assays[559], and the synthesis of several other deuterated THCs has been reported[560]. Use of a sensitive (0.02 ng/ml) assay for [$^2H_2$]THC in human plasma, with the [$^2H_7$]-analogue as the internal standard, has given a half-life of 4.1 ± 1.1 days[561]; the deuterated analogue was used to avoid interference from background levels of the drug. A similar study has been reported for [$^2H_2$]cannabinol where TBDMS derivatives gave superior detection limits to TMS derivatives[96]. Thermospray analysis of the synthetic cannabinoid, nabilone, with CID in a triple stage quadrupole has given low mass ions conveying considerable structural information[562]. Metabolism of this cannabinoid has been reported in several species[563]. GC/MS studies on the metabolism of methyl[358], propyl[425], and butyl[237] homologues of delta-8- and delta-9-tetrahydrocannabinol have revealed a greater tendency for the lower homologues to form acid metabolites as a function of their lower lipophilicity. Similar studies on the (+)-isomers of the naturally occurring (-)-isomers indicate a distinctly different metabolic pattern[265,266]. The oxidation of 7-alpha- but not 7-beta-hydroxy-delta-8-THC to the corresponding ketone, has been shown by $^{18}$O-labelling to involve incorporation of the oxygen atom from the alcohol[324]. Passive smoking has been shown by GC/MS to produce significant cannabinoid concentrations in urine[564], and THC has been found to be absorbed from plasma if stored in plastic containers[565].

7.3 Nicotine.- Nicotine metabolism has attracted modest attention. In addition to the major metabolite, cotinine[376], the drug forms an N-oxide[376] which has been quantified by GC/MS in urine as the derived oxazine following heating of the ionic compound with anisole[147]. The N-methyl-N'-nicotinium ion has been identified as a new metabolite in the guinea pig using thermospray MS from R-(+)-nicotine[379]. The alpha-carbon oxidation of (S)-nicotine has been shown by the use of deuterium labelling to involve loss of the C-5 hydrogen *trans* to the pyridine ring[380,566]. Other metabolic studies are listed in Table 2. A study by thermal desorption GC/MS has shown that volatile organic compounds were similar in the air from offices of smokers and non-smokers, but that the samples could be distinguished by the concentration of nicotine[567]. 3-Hydroxy-cotinine is a major nicotine metabolite in smokers[568] and cotinine is eliminated with a half-life of 12.2 hours[569].

7.4 Opiates.- Significant urinary concentrations of morphine have been detected after consumption of poppy seed[570-573] and various seed-containing products[574] indicating caution when interpreting positive forensic opiate analyses. Morphine and nor-morphine are metabolites of codeine[575] and nor-codeine[382] respectively, further confusing the conclusions as to what drug was consumed. Heroin consumption, however, may be detected by monitoring 6-acetylmorphine in urine[576]. Urinary sample clean-up of dihydrocodeine is aided by use of Sep-Pak $C_{18}$-cartridges[577]. Morphine and codeine have been detected in bovine hypothalami initiating some discussion of the reason for their presence[578]. Negative ion GC/MS has given improved sensitivity over EI for urinary opiate detection[579], and MIKES or linked-scanning at constant B/E have given high sensitivity for detection in hair samples[580]. Conventional GC/MS of hair samples, with detection in the low μg/g range, has allowed post mortem detection of opioid abuse[581]. An unusual daughter ion at m/z 87 in the ammonia CI spectra of 6-oxo-morphinans has enabled differentiation from the 8-oxo-isomer[582]. Metabolism of the opium alkaloid thebaine[461], and of the synthetic opiates, methadone[351] and dextromorphan[272] (Table 2), has been studied, and codeine phosphate shown to be esterified in tablets containing citric acid[583]. FAB spectra of natural opiates, the enkephalins, have been studied[584].

7.5 Phencyclidine.- Several studies on phencyclidine (PCP, "Angel dust") and other "designer heroins" have been reported (Tables 1,2). In addition, the EI-induced fragmentation of PCP[585] and $\underline{N}$-ethylphenylcyclohexylamine[586], and the pyrolytic fate of piperidinocyclohexanecarbonitrile, a PCP contaminant[587], have been studied.

7.6 Amphetamines.- Forensic detection of clandestine amphetamine synthesis relies on identification of minor reaction products[588,589] and impurities. ICP-MS has been used in this context to identify inorganic impurities in methamphetamine[590], and GC/MS has been used to identify by-products of the Leuckart synthesis of amphetamine after separation by HPLC[588]. The ability to detect a 0.25% change in $^{13}C$ content of methamphetamine[591] by MS has led to the suggestion that the "legal" drug be labelled with this isotope so that it may be differentiated from that synthesised illicitly. An LC/MS technique using a moving belt for resolving enantiomers of amphetamines as their $\underline{N}$-TFA-$\underline{l}$-prolyl derivatives on a chiral column has been described[592], and GC/MS has enabled the drugs to be detected in the skeletonised body of a 28 year old male user buried for five years[593]. Benzamphetamine[223], dimethylamphetamine[280], and methamphetamine[352] metabolism is reported in Table 2.

7.7 Cocaine.- Cocaine use is usually detected by measurement of the major metabolite, benzoylecgonine[89,90]; a compound that has been detected by GC/MS for as long as 48 hours after a single small dose[594]. However, it can also appear in urine as the result of consuming certain herbal teas[595]. Correlation between salivary and blood concentrations of cocaine has led to a noninvasive test following intravenous use[596]. Cocaine metabolite identification has been aided by ion cluster techniques using $[^2H_3]$analogues[259].

## 8 Cardiovascular Drugs

Drugs used to treat diseases of the cardiovascular system include beta-blockers, antiarrhythmics, anticoagulants, anti-angina agents and antihypertensives. Use of $^{12}C$ and $^{13}C$-pseudoracemates has given information on the pharmacokinetics and relative rates of metabolism of verapamil[479], disopramide[597], and warfarin[598,599] enantiomers; amiodarone has been shown to increase plasma concentrations of both forms of the latter drug[600]. Stable isotope labelling of mopidralazine has also shown retention of a $^{13}C$-methyl group from the dimethylpyrrole ring by some of its cyclised metabolites, but not by others[373], and has demonstrated stereoselectivity in debrisoquine[264] and verapamil[479] hydroxylation. Cibenzoline pharmacokinetics have been evaluated using a $^{15}N_2$-labelled analogue in patients with heart failure[601]. LAMMA has been used to measure verapamil in cardiac tissue[602], and GC/MS used to measure concentrations of this drug in fatalities[603], and to show decreased elimination after multiple dosing[604]. Piroximone, degrades both in basic and peroxide-containing solutions as shown by LC/MS/MS using a thermospray source[605]. In addition to the main drug groups discussed below, Table 1 lists assays for anagrelide[86], budralazine[94], clonidine[101], diltiazem[112], flecainide[120], moxonidine[143], rutoside[160], and verapamil[180], and Table 2 lists details of metabolism studies on bepridil[225,226], benzarone[220], benzazepine[221], detomidine[269], diltiazem[277], diltrizem[278,279], flecainide[308], hydralazine[323], mexiletine[367], molsidomine[372], oxaminozoline[387], penticainide[397], perhexiline[398,399], phenprocoumon[408], pinacidil[411], quinpirole[435], stobadine[451], and terodiline[460].

8.1 Beta Blockers.- The use of pseudoracemic mixtures, usually $^2H_2$-analogues, for differentiating diastereoisomers of beta-blockers has shown stereoselectivity in aromatic hydroxylation[419-421], and the use of $^{18}O_2$ has demonstrated sequential addition of hydroxy groups in the formation of dihydroxy metabolites[421]. Thermospray LC/MS stability of labetalol has been improved by lowering the temperature to 5 - 10° below that needed for complete vaporization[133]. Quantitative assays for nadolol (EI)[146] and tertatalol (CI)[169] have been reported (Table 1), but the former method has been criticized[606] for the use of the relatively unspecific ion at m/z 86: the use of TFA rather than TMS derivatives was suggested but the authors replied that they had difficulty preparing these in high purity[607]. The use of phosgene to prepare cyclic derivatives of beta-blockers has been suggested[608], and the incorporation of a methylene group from methanol and dichloromethane to give a cyclic oxazolidine derivative has accounted for an additional peak in some GC/MS studies[609]; deuterated solvents were used to confirm the source of the added group. A collection of 72 EI spectra of 22 beta-blockers and their metabolites has been published[610]. GC/MS finds wide applicability for study of these drugs: it has been shown to have a better detection limit than GLC with a nitrogen specific detector and has allowed beta-blockers to be detected for up to 48 hours in athletes[611]. GC/MS has also been used to study the pharmacokinetics of propranolol[612-615], bufuralol[616], and metoprolol[617]. Metabolism studies are summarised in Table 2.

8.2 Dihydropyridine calcium antagonists.- Metabolism of these antihypertensive drugs is dominated by oxidation to the pyridine form followed by hydrolysis of the ester groups. Studies on amlodipine[208,209], felodipine[303], and nilvadipine[381] are listed in Table 2. The $V_{max}$ of the (+)-isomer of nilvadipine has been measured at 0.43-0.54 times less than that of the (-)-form[618]. Quantification (Table 1) of the low therapeutic plasma levels is generally by negative CI with detection limits in the subnanogram/ml range with deuterated internal standards (felodipine[118], nilvadipine[148], and nitrendipine[149]). A $^{13}$C-standard has been used for isradipine[131] quantification. $^{13}$C-Analogues have also been used for bioavailability studies of nitrendipine[619]; pharmacokinetics usually exhibit wide variability between patients as shown by GC/MS[620]. Food has been shown to have little effect on nilvadipine uptake[621], and the drug shows triexponential kinetics[622]. EI Fragmentation of nimopidine[623] and isoxazolydihydropyridines[624], and negative ion spectra of 16 dihydropyridines[625] have been described.

8.3 Cardiac glycosides.- Positive FAB spectra of some cardiac glycosides give weak ions in the molecular ion region, but the abundance of these can be considerably enhanced if sodium chloride or potassium iodide is added to the glycerol matrix[626]. Sequence ions for the glycoside moiety are present in the negative FAB spectra[626]. Fragmentation is matrix dependent[627]; thioglycerol allows digoxin to be detected at 11 ng/ml in spiked human urine[627]. Laser desorption FT/ICR studies of 16 cardiac glycosides have produced abundant [M + K]$^+$ ions with some additional sugar sequence ions[628], and a study by $^{252}$Cf plasma desorption MS has revealed the presence of a new glycoside with a molecular weight of 1035 Daltons[629]. The metabolism of cinobufagin[250], gomphogenin[238], and gomphoside[319] has been studied (Table 2).

8.4 Angiotensin-converting Enzyme Inhibitors.- GC/MS methods for the assay of these antihypertensive drugs have been reviewed[630]. The drugs are usually stabilised for GC/MS studies by formation of N-ethylmaleimide derivatives[97,182] which have been shown to give a good response in electron-capture negative CI[631]. Lactam formation from compounds such as enalapril has been shown to be largely artefactual rather than metabolic as it occurs during sample preparation[291]. A GC/MS method used to evaluate a new enzymic assay for benazeprilat, another member of this group, has shown a 11-16% higher reading for the enzyme assay[632]. Captopril has been shown by

GC/MS to accumulate in the body in renal failure[633] and benazepril concentrations shown to be unaffected by the diuretic furosemide[634], a commonly co-prescribed drug.

## 9. Drugs affecting Central Function

### 9.1 Tricyclic antidepressants.

Several tricyclic antidepressants and phenothiazines have been studied by pulsed laser desorption with resonant two-photon ionization[635]. All gave molecular ions and fragmentation could be controlled by varying the laser wavelength and power density. With the exception of amitriptyline, soft ionization was possible at 222 or 245 nM with nitrogen-containing fragments predominating at the shorter wavelength. Both FAB and thermospray spectra of the glucuronide conjugate of 2-hydroxy-desipramine have been shown to give the aglycone ion at $m/z$ 283 although its abundance was higher in the thermospray spectrum[636]. GC/MS studies have given a half-life of 8 hours for 10-hydroxynortriptyline[637] and have shown the metabolite to be more selective than the parent drug on noradrenergic neurones[638]. Quantitative methods for the tricyclics clomipramine[100], desipramine[128], and quinupramin[157] are reported in Table 1, and metabolism studies for amitriptyline[205,206] and imipramine[329] reported in Table 2. Imipramine concentrations in blood have been found to vary by as much as 760% between pulmonary and peripheral venous sites[639], but little difference in concentration occurs with paracetamol.

### 9.2 Phenothiazines.

Metabolic studies on phenothiazines, such as chlorpromazine[250,416], promazine[416], and thioridazine[465], are listed in Table 2. FAB MS using xenon and a glycerol matrix has enabled a quaternary ammonium glucuronide of chlorpromazine to be identified in human urine[250]. It was noted that partial replacement of the aromatic chlorine atom by hydrogen occurred during analysis by FAB. FAB analysis without loss of resolution of phenothiazines separated by TLC has been achieved using threitol as the matrix[640,641]. This compound has matrix properties similar to glycerol but is solid at temperatures lower than 70°C. It melts, however, in the beam from the FAB gun. The resulting spectra give abundant [M + H]⁺ ions and fragments characteristic of the N-substituent. These are similar to CI spectra which have been interpreted in terms of the position of the added proton[642]. Clinical studies on phenothiazines involving use of stable isotopes have shown that carbamazepine induces its own metabolism[643], and that the epoxide-diol metabolic pathway is inhibited by the steroid danazol[644].

9.3 Benzodiazepines.- Several metabolic studies of benzodiazepines including bromazepam[227], camazepam[240], and oxazepam[389], and their related benzophenone analogues have been reported (Table 2); reduction of the benzophenones with sodium borohydride prevents cyclisation to the benzodiazepine[196]. N-Demethylation of temazepam, studied by deuterium labelling, shows C-hydroxylation as the rate-determining step[645]. The EI spectra of a large number of benzodiazepines and their metabolites have been published[646], and the electron-impact induced ring contraction of the heterocyclic ring in dihydro and tetrahydro-1,5-benzodiazepines has been investigated[647]. A comparison of methane CI spectra, recorded on quadrupole and magnetic instruments, has shown good agreement in the molecular ion region, but mass discrimination against low mass ions such as Cl⁻ in the quadrupole spectra[648]. The [$\underline{N}$-$^2H_3$]-analogue of lormetazepam has been reported to be a poor internal standard for quantitative studies because of a temperature-dependent isotope effect on the fragment ion peak ratios[135]. The [aromatic-$^2H_4$]-analogue was free of this phenomenon. On a somewhat controversial note, the benzodiazepine, N-desmethyldiazepam, previously thought to be a purely synthetic compound, has been identified by direct insertion EI MS in six human brains stored in paraffin in 1940, some 15 years before the compound was first synthesised. This raises the question of whether benzodiazepines occur naturally[649]. An interesting clinical observation, made by GC/MS, is that ethyl loflazepate metabolites can show half lives as long as 132 hours[650] and, therefore, must show considerable accumulation in the body.

9.4 Barbiturates.- Mass spectral data of three arylidene-thiobarbituric acids have been described[651], and stable isotope labelling has shown a higher metabolic clearance for S(+)- than for R(-)-hexobarbitone[652]. Unusual N-glucose conjugates of amobarbitone have been identified by GC/MS[210].

9.5 Valproic acid.- The anticonvulsant valproic acid (VPA) is extensively metabolised along pathways common to natural fatty acids (Table 2) and metabolism is inhibited by aspirin[653]. GC/MS is the preferred method of metabolite analysis. Mechanisms of metabolite formation have been investigated using deuterium labelling[478], and the [$^2H_4$]-analogue has been used as the intravenously administered reference compound in a bioavailability study of tablets of the drug; this showed almost complete absorption[654]. The carnitine conjugate of VPA has been included in a tandem mass spectrometric study of isomeric carnitines which showed that product ion abundances were characteristic of the isomer present[655]. Valproyl hydroxamate has been demonstrated to be an artefact formed in urine[656], and GC/MS has been used to demonstrate that "radiochemically pure" [1-$^{14}$C]-VPA is a mixture of four isomeric octanoic acids[657]. In the first of two GC/MS studies of clinical significance, no correlation was found between toxic effects and concentrations of the putative toxic metabolite, 2-propylpentene-4-oic acid[658]. In the second study, GC/MS measurements showed that neonatal blood of babies born to epileptic mothers on VPA therapy contains higher concentrations of monounsaturated metabolites than maternal blood[659]. Levels of two of these metabolites remained high for several weeks unlike the case of the cardioactive drug oxcarbazine[660] and its metabolites[661] which are eliminated rapidly after birth, even though the drug is obtained from the breast milk. In a death by valproic acid overdose, 750 μg/ml of the drug has been measured in blood[662], again by GC/MS.

9.6 Other Centrally Active Drugs.- Pharmacokinetic equivalence between unlabelled and [$^{2}$H$_{10}$]5,5-diphenylhydantoin has been demonstrated[663,664], and a method has been reported for estimating elimination times based on serum concentrations of a drug whose $K_m$ and $V_{max}$ are known[665,666]. Values obtained from a pulse-labelled study, where a dose of [$^{13}$C$_1$$^{15}$N$_2$]5,5-diphenylhydantoin was given on week four of treatment, were in good agreement with calculated values from data obtained in weeks 0 and 12. A binding study with several deuterated caffeines and phenobarbitone has shown multiple protein binding sites[667,668] for each drug, and dietary protein has been shown to affect theophylline clearance[669]

Other centrally active drugs that have been the subjects of metabolic studies are alpidem[202], cyproheptadine[262], fenetylline[304], indeloxazine[330,331], maprotiline[344], melperone[346], oxcarbazine[390], oxaprotiline[388], phenelzine[405,406], phendimetrazine[404], pheneturide[407], rulziracetam[442], stiripentol[450], tracazolate[469], tranylcypromine[470], tropatepine[473], zolpidem[482], and zonisamide[44]. Details are listed in Table 2. Quantitative assays (Table 1) have been reported for bromocriptine[93], clozapine[102], ethytoin[98], haloperidol[125], isofloxythepin[129], oxcarbazine[151], sertralin[163], and triazolam[175]. What is thought to be the first xenobiotic alpha-glucoside has been identified by FAB from indeloxazine in rats[331], and glucuronides of amitriptyline and nortriptyline[207] have been characterized by the same technique. The use of LC/MS/MS has enabled the N-hydroxymethyl metabolite of zolpidem to be identified[482]; the metabolite was found to decompose under LC/MS conditions by loss of formaldehyde, whereas under GC/MS conditions it was stable as the TMS derivative. Progabide shows E-Z isomerism during LC/MS analysis if polar solvents are used[670]. Fentons reagent has been used to model hepatic metabolism of xanthines with product detection by EI MS[671]. Phenelzine gives a single peak for GC/MS assay if reacted with pentafluorobenzaldehyde prior to PFP formation[672], or, alternatively, it may be reacted with acetyl acetone to give a cyclic derivative[673]. Molecular ions are weak or absent in EI spectra of many butyrophenones but both the $[M + H]^+$ and $[M - H]^-$ ions are abundant in the CI spectra[674]. In clinical studies using GC/MS, half lives of 8.29 hours and 1.54 - 3.15 min have been found for metapyramine[675] and tranylcypromine[676] respectively, steady state concentrations of tyramine have been demonstrated to occur after 4 - 5 days[677], and amiflamine metabolism has been shown to parallel that of debrisoquin[678].

## 10. Steroids

Metabolic studies on this pharmacologically diverse group of drugs are listed in Table 2. Included are budesonide[229-232], dexamethasone[270,271], fluoromethalone[311], mespirenone[350], methylprednisolone[361], norethindrone[383,384], and tixcortol[468]; hydroxylations, oxidations, and reductions dominate the metabolic profiles. $^{18}O$-Labelling has been used to demonstrate that cleavage of the acetal ring of budesonide proceeds by initial hydroxylation at C-22[229,232]. Metabolites were studied by LC/MS using a belt interface; this gave more abundant $[M + H]^+$ ions than previously observed by direct probe techniques[232]. However, considerable interference by plasticisers from the belt was reported. Sulphate conjugation sites of norethindrone have been investigated by GC/MS following acetylation of free hydroxyl groups prior to enzymic hydrolysis, and it was shown that a 17-beta-sulphate was formed by metabolites possessing the 3-alpha-5-alpha-configuration whereas the corresponding 5-beta-metabolites gave predominantly 3-alpha-sulphates[384]. Intact sulphates of anabolic steroids have been characterised at the 10 pg level by LC/MS/MS using ion spray in the negative ion mode[679]. Methanol thermospray MS of several steroids has been shown to give similar spectra to ammonia CI spectra[680], an increase in selectivity was found when ammonium acetate was replaced by pyridine. Penetration of dexamethasone into the human eye has been studied by GC/MS[681,682]. Quantitative studies of betamethasone[91], dexamethasone[91], pancuronium[153], prednisolone[91], and trenbolone[174] have been developed (Table 1), and good correlation has been observed between a GC/MS assay and a radioimmunoassay for norethisterone[683].

## 11. Drugs used to treat Pain and Inflammation

**11.1 Non-Steroid Antiinflammatory Agents.**- Metabolic studies on ebselen[288], etodolac[297,298], fenofibrate[305], flunixin[310], ketoprolac[337], ketotifen[338], meclomen[345], and phenylbutasone[409], are listed in Table 2. Ibuprofen has been found by GC/MS to be incorporated covalently into adipose tissue triglycerides[307,684]. Electron-impact induced fragmentation of antiinflammatory pyridazinones[685] and benzimidazoles[686] has been reported. Phenylbutasone decomposition has been studied[687], and the 1-O-acyl glucuronide of diflunisal has been shown by FAB MS to decompose into 9 isomers[688]. The GC/MS/MS method of Dawson et al.[92], (Table 1) has been used to measure the antiinflammatory drug, biphenylacetic acid, in synovial fluid at a level of 21 ng/ml[689]. Other quantitative methods have been published for benzofenac[88] and ibuprofen[126,127].

**11.2 Paracetamol and phenacetin.**- Toxicity of paracetamol and related compounds continues to be studied and has led to several papers on identification of glutathione conjugates. These can be derivatized with alkyl chloroformates for analysis by FAB[391]; reaction of these compounds with silylating reagents gives stable cyclic derivatives[690]. Both mono- and dimeric-paracetamol conjugates give abundant $[M + H]^+$ ions under FAB conditions; these can be induced to undergo collision-induced decomposition to give diagnostic fragment ions for MS/MS experiments[392,401]. Thermospray LC/MS/MS has been used to identify conjugates from bile injected directly into the instrument[400]. Dose-dependent clearance of phenacetin has been found using GC/MS[691].

**11.3 Local anaesthetics.**- The local anaesthetic, lidocaine, has been derivatized for GC/MS analysis in aqueous solution[692] and shown to be metabolised along several competing pathways[341]. Its bioavailability has been measured using stable isotopes[134], and the $[^2H_3]$-analogue shown to be pharmacokinetically equivalent to the parent drug[693]. Pulse labelling with $[^2H_3]$lignocaine and bupivacaine has shown equivalent first-order absorption following epidural administration[694]. Ketamine metabolism is listed in Table 2 and GC/MS studies have shown a half-life of 8-10 min[695].

11.4 Analgesics.- Metabolism of the analgesics aspirin[215], alfentanil[217], benoxinate[219], and propofol[418] have been studied mainly by GC/MS, and quantitative methods have been developed for alphaprodine[84], oxybuprocaine[152], phenazone[154], and tilidine[173]. GC/MS has been used in studies on pethidine[696] and tramadol[697] absorption, and on the partition of the former drug into tissues[698].

11.5 Antihistamines.- A considerable number of metabolic studies have been reported on antihistamines, many involving FAB and both GC/MS and LC/MS techniques. Complex and multiple metabolic pathways have been observed for most compounds. Drugs studied include bamipine[217], chlorpheniramine[247], chlorphenoxamine[248,249], doxylamine[282-287], methapyrilene[353-356], pyrilamine[285,431-433,434], roxatidine[440,441], tripelennamine[353], and triprolidine[472]. The isotope doublet technique has aided structural determination of ranitidine[437] and roxatidine[441] metabolites, with deuterium displacement also being used to determine hydroxylation sites for the latter drug. Glucuronide conjugates of doxylamine have been characterised by thermospray MS/MS; this ionization technique gave [M + H]$^+$ ions which were fragmented and analysed by MS/MS[285,287]. Glucuronides of this drug have also been identified by FAB[282]. Although chlorphenoxamine N-oxide was characterised, the compound was unstable under GC/MS conditions and decomposed into an olefin[248]. The DCI fragmentation of nine antihistamines and their metabolites has been reported[699], as well as an extensive collection of the EI spectra of alkanolamine-containing antihistamines[700].

12 Other miscellaneous studies

Miscellaneous metabolism studies are listed in Table 2. These include the first report of N-acetylation of primary and secondary aliphatic amines in man[284], and the first observation in mammals of glucose conjugation to a drug containing a non-acidic hydroxy group[239]. Metabolism of the cyclic undecapeptide, cyclosporin, (MW 1202) has been studied by a combination of GC/MS and FAB techniques with the identification of 13 compounds[260,261]. The synthetic prostaglandin derivatives alfaprostil[199], arbaprostil[213], enisoprost[294], rosaprostol[439], and trimoprostil[421], are metabolised predominantly by hydroxylation and beta-oxidation. The half-life of trimoprostil has been measured by negative ion GC/MS at 25-37 mins[701]. Electron-impact fragmentation of diuretic agents has been summarised[702], and a [NH$_3$NH$_4$]$^+$ adduct ion of thiazide diuretics has been found to decompose to give a fragment at m/z 303 in most spectra[703]. Water loss from the dihydroxy-monoterpene sobrenol occurs first from the tertiary position and then from the ring under electron-impact[704]. LC/MS, with linked scanning at constant B/E, has given the first analytical method for identification of the diagnostic tetrapeptide, tuftsin, in serum[705], and thermospray LC/MS has been applied to analysis of degradation products of metoclopramide[706]. Finally, use of stable isotopes has improved measurements of terodiline bioavailability[707], and the pharmacokinetics of oral doses of vigabatrin have been studied[708].

## References

1   J.DeGraeve, F.Berthou, and M.Prost, 'Chromatographic-Mass Spectrometric Methods. Technology and Applications in the Environmental, Pharmacological and Biochemical Fields', Masson, Paris, 1986.
2   'Ultratrace Analysis of Pharmaceuticals and Other Compounds of Interest', ed. S.Ahuja, J.Wiley, New York, 1986.
3   'Clarke's Isolation and Identification of Drugs in Pharmaceuticals, Body Fluids and Post-Mortem Material', ed. A.C.Moffat, Pharmaceutical Press, London, 1986.
4   S.P.Markey, J. Clin. Pharmacol., 1986, 26, 406.
5   G.K.Szabo and T.R.Browne, J. Clin. Pharmacol., 1986, 26, 400.
6   N.P.E.Vermeulen, W.Onkenhout, M.Van Der Graaff, B.J.Xu, and A.G.L.Burm, Chromatogr. Meth., 1987, 107.
7   A.L.Burlingame, D.Maltby, D.H.Russell, and P.T.Holland, Anal. Chem., 1988, 60, 294R.
8   T.A.Brettell and R.Saferstein, Anal. Chem., 1987, 59, 162R.
9   'Capillary Gas Chromatography-Mass Spectrometry in Medicine and Pharmacology', ed. H.Jaeger, A.Huthig Verlag, Heidelberg, 1987.
10  4th Symposium on LC/MS and GC/MS, Montreux, 1986, J. Chromatogr., 1987, 349.
11  6th International Symposium on Mass Spectrometry in the Life Sciences, Ghent, 1986, Biomed. Mass Spectrom., 1987, 14, No. 11.
12  'Drug Fate and Metabolism, Methods and Techniques' 5, ed. E.R.Garrett and J.L.Hirtz, Marcel Dekker, New York, 1985.
13  'Liquid Chromatography in Pharmaceutical Development, an Introduction', ed. I.W.Wainer, Aster Publishing Corp., Springfield, OR, 1985.
14  'New Methods in Drug Research' 1, ed. A.Makriyannis, J.R.Prous, Barcelona, 1985.
15  T.Nanba, N.Tsuneuchi, and M.Hattori, Pharm. Tech. Japan, 1986, 2, 31.
16  T.R.Covey, E.D.Lee, A.P.Bruins, and J.D.Henion, Anal. Chem., 1986, 58, 1451A.
17  T.J.A.Blake, J. Chromatogr., 1987, 394, 171.
18  M.L.Kimber, D.E.Games, and M.J.Whitehouse, Adv. Mass Spectrom., 1985, 1986, 583.
19  C.Lindberg and J.Paulson, J. Chromatogr., 1987, 394, 117.
20  J.R.Chapman and J.A.E.Pratt, J. Chromatogr., 1987, 394, 231.
21  C.E.Parker, R.W.Smith, S.J.Gaskell, and M.M.Bursey, Anal. Chem., 1986, 58, 1661.
22  J.R.Lloyd, M.L.Cotter, D.Ohori, and A.R.Oyler, Anal. Chem., 1987, 59, 2533.
23  M.Sakairi and H.Kambara, Anal. Chem., 1988, 60, 774.
24  A.P.Bruins, T.R.Covey, and J.D.Henion, Anal. Chem., 1987, 59, 2642.
25  A.J.Berry, D.E.Games, and J.R.Perkins, J. Chromatogr., 1986, 363, 147.
26  A.J.Berry, D.E.Games, and J.R.Perkins, Anal. Proc., 1986, 23, 451.
27  E.D.Lee, J.D.Henion, R.B.Cody, and J.A.Kinsinger, Anal. Chem., 1987, 59, 1309.
28  D.A.Laude, Jr., S.L.Pentoney, Jr., P.R.Griffiths, and C.L.Wilkins, Anal. Chem., 1987, 59, 2283.
29  R.E.Shomo, II, A.G.Marshall, and R.P.Lattimer, Int. J. Mass Spectrom. Ion Proc., 1986, 72, 209.

30  M.Karas, D.Bachmann, U.Bahr, and F.Hillenkamp, Int. J. Mass Spectrom. Ion Proc., 1987, 78, 53.
31  L.Van Vaeck, J.Claereboudt, S.De Nollin, W.Jacob, F.Adams, R.Gijbels, and W.Cautreels, Adv. Mass Spectrom., 1985, 1986, 1249.
32  J.Belanger and J.R.J.Pare, J. Pharmaceut. Biomed. Anal., 1986, 4, 415.
33  C.Fenselau and L.Yellet in 'Xenobiotic Conjugation Chemistry', ed. G.D.Paulson and J.Caldwell, ACS Symposium Ser., No. 299, Am. Chem. Soc, Washington, 1986.
34  F.M.Kaspersen and C.A.A.Van Boeckel, Xenobiotica, 1987, 17, 1451.
35  P.Rudewicz and K.M.Straub, Anal. Chem., 1986, 58, 2928.
36  K.M.Straub, P.Rudewicz and C.Garvie, Xenobiotica, 1987, 17, 413.
37  H.-M.Schiebel, O.Gries, R.Schuppel, and P.Schulze, J. Anal. Appl. Pyrolysis, 1987, 10, 267.
38  P.L.Jacobs, L.P.C.Delbressine, F.M.Kaspersen, and G.J.H.Schmeits, Biomed. Environ. Mass Spectrom., 1987, 14, 689.
39  H.Iwabuchi, A.Nakagawa, and K.-I.Nakamura, J. Chromatogr., 1987, 414, 139.
40  D.A.Kidwell, M.M.Ross, and R.J.Colton, Int. J. Mass Spectrom. Ion, Proc., 1987, 78, 315.
41  A.E.Ashcroft, J.R.Chapman, and J.S.Cottrell, J. Chromatogr., 1987, 394, 15.
42  M.S.Lee, R.A.Yost, and R.J.Perchalski, Annu. Rep. Med. Chem., 1986, 21, 313.
43  R.J.Perchalski, M.S.Lee, and R.A.Yost, J. Clin. Pharmacol., 1986, 26, 435.
44  M.S.Lee and R.A.Yost, Biomed. Environ. Mass Spectrom., 1988, 15, 193.
45  R.M.Facino, M.Carini, and P.Traldi, Chim. Oggi, 1985, 23.
46  S.A.Coran, M.Bambagiotti-Alberti, F.F.Vincieri, and G.Moneti, J. Pharmaceut. Biomed. Anal., 1987, 5, 509.
47  M.S.H.Hussain, P.York, A.Brown, and P.Timmins, J. Pharm. Pharmacol., 1987, 39, 127P.
48  M.C.Davies, J. Pharm. Pharmacol., 1987, 39, 122P.
49  M.C.Davies, S.S.Davis, and C.Washington, J. Pharm. Pharmacol., 1987, 39, 120P.
50  B.S.Middleditch and A.Zlatkis, J. Chromatogr. Sci., 1987, 25, 547.
51  D.L.Mount, J.W.Miles, and F.C.Churchill, J. Assoc. Off. Anal. Chem., 1986, 69., 624.
52  C.E.Wells, E.C.Juenge, and K.Wolnik, J. Pharm. Sci., 1986, 75, 724.
53  L.Castle, J.Gilbert, S.M.Jickells, and J.W.Gramshaw, J. Chromatogr., 1988, 437, 281.
54  T.Cairns, K.S.Chiu, E.G.Siegmund, B.Williamson, and J.C.Fischer, Biomed. Environ. Mass Spectrom., 1986, 13, 357.
55  B.Runesson and K.Gustavii, Acta Pharm. Suecica, 1986, 23, 151.
56  A.S.Salhab, J.Applewhite, M.W.Couch, R.T.Okita, and K.T.Shiverick, Drug Metab. Dispos., 1987, 15, 233.
57  T.-Y.Chou, P.Vouros, M.David, M.Saha, and R.W.Giese, Biomed. Environ. Mass Spectrom., 1987, 14, 23.
58  J.A.Thompson, A.M.Malkinson, M.D.Wand, S.L.Mastovich, E.W.Mead, K.M.Schullek, and W.G.Laudenschlager, Drug Metab. Dispos., 1987, 15, 833.
59  T.R.Browne, J. Clin. Pharmacol., 1986, 26, 485.
60  Symposium on New Methods of Investigation in Clinical Pharmacology, Pt. 1, Stable Isotopes and Mass Spectrometry, Therapie, 1987, 42, 417.
61  Proceedings of the Fifth Symposium on Frontiers of Pharmacology: Application of Stable Isotopes in Pharmacology, ed. T.R.Browne and T.A.Baillie, J. Clin. Pharmacol., 1986, 26(6).
62  'Synthesis and Applications of Isotopically Labeled Compounds', ed. R.R.Muccino, Elsevier, Amsterdam, 1986.
63  W.J.A.VandenHeuvel, J. Clin. Pharmacol., 1986, 26, 427.
64  T.A.Baillie and A.W.Rettenmeier, J. Clin. Pharmacol., 1986, 26, 448.
65  T.A.Baillie and A.W.Rettenmeier, J. Clin. Pharmacol., 1986, 26, 481.
66  W.J.A.VandenHeuvel, Xenobiotica, 1987, 17, 397.
67  W.Cautreels, H.Davi, and Y.Berger, Therapie, 1987, 42, 439.
68  J.L.Brazier, B.Ribon, J.B.Falconnet, Y.Cherrah, and Y.Benchekroun, Therapie, 1987, 42, 445.
69  T.R.Browne, D.J.Greenblatt, J.E.Evans, G.K.Szabo, B.A.Evans, and G.E.Schumacher, J. Clin. Pharmacol., 1986, 26, 463.
70  M.Eichelbaum, J. Clin. Pharmacol., 1986, 26, 469.
71  W.F.Trager, J. Clin. Pharmacol., 1986, 26, 443.

72  G.Pons, E.Rey, and G.Olive, Therapie, 1987, **42**, 457.
73  H.Andres and P.Voges, Therapie, 1987, **42**, 417.
74  A.Van Langenhove, J. Clin. Pharmacol., 1986, **26**, 383.
75  C.Julien-Larose, Therapie, 1987, **42**, 429.
76  W.J.A.VandenHeuvel, R.W.Walker and J.R.Carlin, Chromatogr. Methods, 1987, 139.
77  P.B.Farmer, Therapie, 1987, **42**, 451.
78  E.Gelpie, Adv. Mass Spectrom., 1985, 1986, 397.
79  A.Mignot and J.B.Fourtillan, Therapie, 1987, **42**, 423.
80  S.J.Gaskell and E.M.Finlay, Trends Anal. Chem., 1988, **7**, 202.
81  P.Levandoski, K.Straub, D.Shah, R.De Marinis, and C.de.Mey, Biomed. Environ. Mass Spectrom., 1986, **13**, 523.
82  K.H.Schellenberg, M.Linder, A.Groeppelin, and F.Erni, J. Chromatogr., 1987, **394**, 239.
83  W.A.Garland and M.P.Barbalas, J. Clin. Pharmacol., 1986, **26**, 412.
84  B.R.Kuhnert, W.T.Brashear, and C.D.Syracuse, J. Chromatogr., 1988, **426**, 392.
85  S.J.Mule and G.A.Casella, J. Anal. Toxicol., 1988, **12**, 102.
86  E.H.Kerns, J.W.Russell, and D.G.Gallo, J. Chromatogr., 1987, **416**, 357.
87  J.X.De Vries, I.Walter-Sack, and A.Ittensohn, J. Chromatogr., 1987, **417**, 420.
88  I.Koruna and M.Ryska, Cesk. Farm., 1985, **34**, 278. (Chem. Abs., 1986, **104**, 28281).
89  W.A.Joern, J. Anal. Toxicol., 1987, **11**, 110.
90  R.W.Taylor, N.C.Jain, and M.P.George, J. Anal. Toxicol., 1987, **11**, 233.
91  J.M.Midgley, D.G.Watson, T.Healey, and M.Noble, Biomed. Environ. Mass Spectrom., 1988, **15**, 479.
92  M.Dawson, C.M.McGee, P.M.Brooks, J.H.Vine, E.Lacey, and T.R.Watson J. Chromatogr., 1987, **420**, 129.
93  C.Julien-Larose, M.Guerret, D.Lavene, and J.R.Kiechel, Adv. Mass Spectrom., 1985, 1986, 1299.
94  Y.Fujimaki, H.Hakusui, and T.Takegoshi, J. Chromatogr., 1987, **421**, 367.
95  K.D.R.Setchell, M.B.Welsh, M.J.Klooster, W.F.Balistreri, and C.K.Lim, J. Chromatogr., 1987, **385**, 267.
96  E.Johansson, A.Ohlsson, J.-E.Lindgren, S.Agurell, H.Gillespie, and L.E. Hollister, Biomed. Environ. Mass Spectrom., 1987, **14**, 495.
97  T.Ito, Y.Matsuki, H.Kurihara, and T.Nambara, J. Chromatogr., 1987, **417**, 79.
98  N.Inotsume, A.Higashi, E.Kinoshita, T.Matsuoka, and M.Nakano, J. Chromatogr., 1986, **383**, 166.
99  G.Kaiser, R.Ackermann, W.Dieterle, and J.-P.Dubois, J. Chromatogr., 1987, **419**, 123.
100 A.Sioufi, F.Pommier, and J.P.Dubois, J. Chromatogr., 1988, **428**, 71.
101 N.Haring, Z.Salama, G.Reif, and H.Jaeger, Arzneim.-Forsch., 1988, **38**, 404.
102 U.Bondesson and L.H.Lindstrom, Psychopharmacology, 1988, **95**, 472.
103 C.G.Norbury, J. Chromatogr., 1987, **414**, 449.
104 H.G.Fouda, J.Lukaszewicz, and E.W.Luther, Biomed. Environ. Mass Spectrom., 1987, **14**, 599.
105 A.Nakagawa, Y.Matsushita, S.Muramatsu, Y.Tanishima, T.Hirota, W. Takasaki, Y.Kawahara, and H.Takahagi, Biomed. Chromatogr., 1988, **2**, 203.
106 K.Yamashita, K.Watanabe, R.Koga, S.Mizuguchi, Y.Hashimoto, T. Nakamura, and H.Umezawa, J. Chromatogr., 1988, **424**, 39.
107 A.K.Singh, U.Mishra, M.Ashraf, E.H.Abdennebi, K.Granley, D. Dombrouskis, D.Hewetson, and C.M.Stowe, J. Chromatogr., 1987, **404**, 223.
108 U.B.Ranalder, Drug Develop. Eval., 1986, **12**, 117.
109 T.R.Covy, D.Silvestre, M.K.Hoffman, and J.D.Henion, Biomed. Environ. Mass Spectrom., 1988, **15**, 45.
110 T.Irie, G.Idzu, Y.Hashimoto, M.Ishibashi, and H.Miyazaki, Yakugaku Zasshi., 1986, **106**, 900.
111 A.D.Theoharides, M.H.Smyth, R.W.Ashmore, J.M.Halverson, Z.M. Zhou, W.E.Ridder, and A.J.Lin, Anal. Chem., 1988, **60**, 115.
112 S.Nakamura, T.Suzuki, Y.Sugawara, S.Usuki, Y.Ito, T.Kume, M.Yoshikawa, H.Endo, M.Ohashi, and S.Harigaya, Arzneim.-Forsch., 1987, **37**, 1244.
113 K.Yamamoto, S.Kato, and H.Shimomura, J. Chromatogr., 1986, **362**, 274.

114 W.J.Morris, G.J.Nandrea, J.E.Roybal, R.K.Munns, W.Shimoda, and H.R. Skinner, Jr., J. Assoc. Off. Anal. Chem., 1987, 70, 630.
115 L.Kenne, B.Noren, and S.Stromberg, Acta Chem Scand., Ser. B, 1988, 42, 59.
116 K.Takatsuki, S.Suzuki, N.Sato, I.Ushizawa, and T.Shoji, J. Assoc. Off. Anal. Chem., 1987, 70, 708.
117 P.Ottoila and J.Taskinen, Biomed. Environ. Mass Spectrom., 1987, 14, 659.
118 M.Ahnoff, M.Ervik, and L.Johansson, J. Chromatogr., 1987, 394, 419.
119 S.A.Barker, L.C.Hsien, T.R.McDowell, and C.R.Short, Biomed. Environ. Mass Spectrom., 1987, 14, 161.
120 E.H.Taylor, E.E.Kennedy, and A.A.Pappas, J. Chromatogr., 1987, 416, 365.
121 G.K.Darland, R.Hajdu, H.Kroop, F.M.Kahan, R.W.Walker, and W.J.A. VandenHeuvel, Drug Metab. Dispos., 1986, 14, 668.
122 M.Jemal, E.Ivashkiv, D.Both, R.Koski, and A.I.Cohen, Biomed. Environ. Mass Spectrom., 1987, 14, 699.
123 H.Odagiri, S.Ichihara, E.Semura, M.Utoh, M.Tateishi, and I.Kuruma, J. Pharmacobio-Dyn., 1988, 11, 234.
124 N.Uchiyama, T.Takano, S.Saito, H.Morikawa, and M.Ohtawa, Chem. Pharm. Bull., 1986, 34, 3290.
125 N.Haring, Z.Salama, L.Todesko, and H.Jaeger, Arzneim.-Forsch., 1987, 37, 1402.
126 D.L.Theis, G.W.Halstead, and K.A.Halm, J. Chromatogr., 1986, 380, 77.
127 M.A.Young, L.Aarons, E.M.Davidson, and S.Toon, J. Pharm. Pharmacol., 1986, 38, 60P.
128 Y.Sasaki and S.Baba, J. Chromatogr., 1988, 426, 93.
129 M.Mogi, T.Ito, Y.Matsuki, Y.Kurata, and T.Nambara, J. Chromatogr., 1987, 399, 245.
130 G.Karlaganis, E.Peretti, and B.H.Lauterburg, J. Chromatogr., 1987, 420, 171.
131 C.Jean and R.Laplanche, J. Chromatogr., 1988, 428, 61.
132 E.J.Cone, D.Yousefnejad, W.D.Buchwald, and K.Kumor, J. Chromatogr., 1986, 383, 158.
133 M.S.Lant, J.Oxford, and L.E.Martin, J. Chromatogr., 1987, 294, 223.
134 G.Karlaganis and J.Bircher, Biomed. Environ. Mass Spectrom., 1987, 14, 513.
135 S.Takahashi, Biomed. Environ. Mass Spectrom., 1987, 14, 257.
136 B.D.Paul, J.M.Mitchell, L.D.Mell,Jr., R.Sroka and J.Irving, Clin. Chem. (Winston Salem, N.C.), 1987, 33, 971.
137 P.Francom, D.Andrenyak, H.-K.Lim, R.R.Bridges, R.L.Foltz, and R.T. Jones, J. Anal. Toxicol., 1988, 12, 1.
138 F.Hatch, K.McKellop, G.Hansen, and T.MacGregor, J. Pharm. Sci., 1986, 75, 886.
139 K.Hara, T.Nagata, and K.Kimura, Z. Rechtsmed, 1986, 96, 93 (Chem. Abs., 1986, 105, 110047).
140 C.Koppel, J.Tenczer, K.Eckart, and H.Schwarz, Adv. Mass Spectrom., 1985, 1986, 1441.
141 D.F.Woolner, D.Winter, T.J.Frendin, E.J.Begg, K.L.Lynn, and G.J.Wright, Br. J. Clin. Pharmacol., 1986, 22, 55.
142 V.Spiehler and R.Brown, J. Forensic Sci., 1987, 32, 906.
143 D.Trenk, F.Wagner, E.Jahnchen, and V.Planitz, J. Clin. Pharmacol., 1987, 27, 988.
144 S.P.Jindal, T.Lutz, and S.P.Bagchi, J. Chromatogr., 1987, 408, 356.
145 I.Irwin, J.W.Langston, and L.E.DeLanney, Life Sci., 1987, 40, 731.
146 M.Ribick, E.Ivashkiv, M.Jemal, and A.I.Cohen, J. Chromatogr., 1986, 381, 419.
147 P.Jacob,III, N.L.Benowitz, L.Yu, and A.T.Shulgin, Anal. Chem., 1986, 58, 2218.
148 Y.Tokuma, T.Fujiwara, and H.Noguchi, J. Pharm. Sci., 1987, 76, 310.
149 C.Fischer, B.Heuer, K.Heuck, and M.Eichelbaum, Biomed. Environ. Mass Spectrom., 1986, 13, 645.
150 K.Takatsuki, I.Ushizawa, and T.Shoji, J. Chromatogr., 1987, 391, 207.
151 G.E.von Unruh and W.D.Paar, Biomed. Environ. Mass Spectrom., 1986, 13, 651.
152 F.Kasuya, K.Igarashi and M.Fukui, Clin. Chem. (Winston Salem, N.C.), 1987, 33, 697.
153 K.P.Castagnoli, Y.Shinohara, T.Furuta, T.L.Nguyen, L.D.Gruenke, R.D.Miller, and N.Castagnoli, Jr., Biomed. Envirom. Mass Spectrom., 1986, 13, 327.
154 A.Nakagawa, K.Nakamura, K.Maeda, T.Kamataki, and R.Kato, Life. Sci., 1987, 41, 133.

155 Y.Benchekroun, B.Ribon, M.Desage, and J.L.Brazier, J. Chromatogr., 1987, 420, 287.
156 T.Marunaka, Y.Umeno, Y.Minami, E.Matsushima, M.Maniwa, K.Yoshida, and M. Nagamachi, J. Chromatogr., 1987, 420, 43.
157 S.Bouquet, J.Girault, U.H.Ly, M.C.Saux, M.A.Lefebvre, and J.B. Fourtillan, J. Chromatogr., 1986, 383, 393.
158 E.K.Fukuda, L.C.Rodriguez, N.Choma, N.Keigher, F.De Grazia, and W.A.Garland, Biomed. Environ. Mass Spectrom., 1987, 14, 549.
159 D.A.Herold, P.K.Anonick, M.Kinter, and F.G.Hayden, Clin. Chem. (Winston Salem, N.C.), 1987, 33, 1022.
160 Y.Sawai, K.Kohsaki, Y.Nishiyama, and K.Ando, Arzneim.-Forsch., 1987, 37, 729.
161 J.D.Ehrhardt, Adv. Mass Spectrom., 1985, 1986, 641.
162 H.L.Ung, J.Girault, M.A.Lefebvre, A.Mignot, and J.B.Fourtillan, Biomed. Environ. Mass Spectrom., 1987, 14, 289.
163 H.G.Fouda, R.A.Ronfeld, and D.J.Weilder, J. Chromatogr., 1987, 417, 197.
164 L.T.Friedhoff, J.Manning, P.T.Funke, E.Ivashkiv, J.Tu, W.Cooper, and D.A. Willard, Clin. Pharmacol. Ther., 1986, 40, 634.
165 E.M.H.Finlay, D.E.Games, J.R.Startin, and J.Gilbert, Biomed. Environ. Mass Spectrom., 1986, 13, 633.
166 C.Murphy, T.Fotsis, P.Pantzar, H.Adlercreutz, and F.Martin, J. Steroid Biochem., 1987, 26, 547.
167 C.Murphy, T.Fotsis, H.Adlercreutz, and F.Martin, J. Steroid Biochem., 1987, 28, 289.
168 L.-E.Edholm, C.Lindberg, J.Paulson, and A.Walhagen, J. Chromatogr., 1988, 424, 61.
169 C.Efthymiopoulos, S.Staveris, F.Weber, J.C.Koffel, and L.Jung, J. Chromatogr., 1987, 421, 360.
170 W.A.Joern, J. Anal. Toxicol., 1987, 11, 49.
171 B.D.Paul, L.D.Mell, Jr., J.M.Mitchell, R.M.McFinley, and J.Irving, J. Anal. Toxicol., 1987, 11, 1.
172 E.Bailey, P.B.Farmer, J.A.Peal, S.A.Hotchkiss, and J.Caldwell, J. Chromatogr., 1987, 416, 81.
173 J.Cordonnier, A.Wauters, and A.Heyndrickx, J. Anal. Toxicol., 1987, 11, 144.
174 S.-H.Hsu, R.H.Eckerlin, and J.D.Henion, J. Chromatogr., 1988, 424, 219.
175 G.Koves and J.Wells, J. Anal. Toxicol., 1986, 10, 241.
176 P.L.Stetson and W.D.Ensminger, J. Chromatogr., 1986, 383, 69.
177 L.M.R.Thienpont, P.G.Verhaeghe, and A.P.DeLeenheer, Biomed. Environ. Mass Spectrom., 1987, 14, 613.
178 T.Tatsuhara, H.Muro, Y.Matsuda, and Y.Imai, J. Chromatogr., 1987, 399, 183.
179 K.Singh, F.S.Abbott, and J.M.Orr, Res. Commun. Chem. Pathol. Pharmacol., 1987, 56, 211.
180 C.W.Jones, Clin. Chem. (Winston. Salem, N.C), 1987, 33, 1017.
181 W.Hammes and R.Weyhenmeyer, J. Chromatogr., 1987, 413, 264.
182 M.Jemal, E.Ivashkiv, D.Teitz, and A.I.Cohen, J. Chromatogr., 1988, 428, 81.
183 K.F.Blom, Org. Mass Spectrom., 1987, 22, 530.
184 J.M.Te Koppele, E.J.Van Der Mark, J.C.O.Boerrigter, J.Brussee, A.Van Der Gen, J.Van Der Greef, and G.J.Mulder, J. Pharmacol. Exp. Ther., 1986, 239, 898.
185 H.Tomisawa, H.Fukazawa, S.Ichihara, and M.Tateishi, Biochem. Pharmacol., 1986, 35, 2270.
186 E.C.Horning, K.Lertratanangkoon, and M.G.Horning, J. Chromatogr., 1987, 399, 321.
187 K.Lertratanangkoon and M.G.Horning, Drug Metab. Dispos., 1987, 15, 1.
188 K.Lertratanangkoon, E.C.Horning, and M.G.Horning, Drug Metab. Dispos., 1987, 15, 857.
189 D.A.Eastmond, M.T.Smith, L.O.Ruzo, and D.Ross, Mol. Pharmacol., 1986, 30, 674.
190 M.Heidmann, P.Fonrobert, M.Przybylski, K.L.Platt, A.Seidel, and F.Oesch, Biomed. Environ. Mass Spectrom., 1988, 15, 329.
191 A.R.Buckpitt, N.Castagnoli, Jr., S.D.Nelson, A.D.Jones, and L.S.Bahnson, Drug Metab. Dispos., 1987, 15, 491.
192 M.G.Horning, L.-S.Sheng, J.G.Nowlin, K.Lertratanangkoon, and E.C.Horning, J. Chromatogr., 1987, 399, 303.

193 A.B.Burke, P.Millburn, K.R.Huckle, and D.H.Hutson, Drug Metab. Dispos., 1987, 15, 581.
194 M.Sagi, Y.Nakagawa, M.Mizugaki, H.Yamanaka, M.Ishibashi, H.Takayama, and H.Miyazaki, Yakugaku Zasshi, 1988, 108, 325.
195 M.Sagi, Y.Nakagawa, M.Mizugaki, H.Yamanaka, M.Ishibashi, H.Takayama, and H.Miyazaki, Yakugaku Zasshi, 1988, 108, 350.
196 M.Koike, M.Mizobuchi, and S.Takahashi, J. Pharmacobio-Dyn., 1986, 9, 578.
197 M.Koike, R.Norikura, S.Futaguchi, T.Yamaguchi, K.Sugeno, K.Iwatani, Y.Ikenishi, and Y.Nakagawa, Drug Metab. Dispos., 1987, 15, 426.
198 P.S.Hong and K.K.Chan, Biomed. Environ. Mass Spectrom., 1987, 14, 167.
199 M.Kaykaty, G.Weiss, M.Barbalas, and P.Duke, Drug Metab. Dispos., 1987, 15, 303.
200 W.Meuldermans, J.Hendrickx, W.Lauwers, R.Hurkmans, E.Swysen, J. Thijssen, Ph.Timmerman, R.Woestenborghs, and J.Heykants, Drug Metab. Dispos., 1987, 15, 905.
201 K.Lavrijsen, J.van Houdt, W.Meuldermans, F.Knaeps, J.Hendrickx, W. Lauwers, R.Hurkmans, and J.Heykants, Xenobiotica, 1988, 18, 183.
202 P.Padovani, P.Guinebault, S.Vajta, A.Durand, J.Allen, J.P.Thenot, and P.L. Morselli, Eur. J. Drug Metab. Pharmacokin., 1987, 12, 295.
203 F.-J.Leinweber, A.J.Szuna, A.C.Loh, T.H.Williams, G.J.Sasso, I.Bekersky, E.Baggiolini, and J.Triscari, Xenobiotica, 1987, 17, 1405.
204 P.D.Dalrymple and P.J.Nicholls, Xenobiotica, 1988, 18, 75.
205 U.Breyer-Pfaff, A.Prox, H.Wachsmuth, and P. Yao, Drug Metab. Dispos., 1987, 15, 882.
206 A.Prox and U.Breyer-Pfaff, Drug Metab. Dispos., 1987, 15, 890.
207 B.Baier-Weber, A.Prox, H.Wachsmuth, and U.Breyer-Pfaff, Drug Metab. Dispos., 1988, 16, 490.
208 A.P.Beresford, D.McGibney, M.J.Humphrey, P.V.MacRae, and D.A.Stopher, Xenobiotica, 1988, 18, 245.
209 A.P.Beresford, P.V.MacRae, and D.A.Stopher, Xenobiotica, 1988, 18, 169.
210 W.H.Soine, P.J.Soine, B.W.Overton, and L.K.Garrettson, Drug Metab. Dispos., 1986, 14, 619.
211 J.L.Maggs, N.R.Kitteringham, A.M.Breckenridge, and B.K.Park, Biochem. Pharmacol., 1987, 36, 2061.
212 T.B.Felder, M.A.McLean, M.L.Vestal, K.Lu, D.Farquhar, S.S.Legha, R.Shah, and R.A.Newman, Drug Metab. Dispos., 1987, 15, 773.
213 B.A.Thornburgh, S.R.Shaw, G.E.Bronson, and A.J.Wickremasinha, Eur. J. Drug Metab. Pharmacokin., 1988, 13, 113.
214 G.Krumbiegel, J.Hallensleben, W.H.Mennicke, N.Rittmann, and H.J.Roth, Xenobiotica, 1987, 17, 981.
215 M.Grootveld and B.Halliwell, Biochem. Pharmacol., 1988, 37, 271.
216 M.J.Van Der Meer, H.K.L.Hundt, and F.O.Muller, J. Pharm. Pharmacol., 1986, 38, 781.
217 R.Neidlein and M.Kleiser, Arzneim.-Forsch., 1987, 37, 32.
218 M.Ryska, I.Koruna, and L.Polakova, Adv. Mass Spectrom., 1985, 1986, 1343.
219 F.Kasuya, K.Igarashi, and M.Fukui, J. Pharm. Sci., 1987, 76, 303.
220 S.G.Wood, B.A.John, L.F.Chasseaud, R.Bonn, H.Grote, K.Sandrock, A.Darragh, and R.F.Lambe, Xenobiotica, 1987, 17, 881.
221 K.Straub, M.Davis, and B.Hwang, Drug Metab. Dispos., 1988, 16, 359.
222 I.Koruna, M.Ryska, and E.Kleinerova, Adv. Mass Spectrom., 1985, 1986, 1341.
223 T.Inoue and S.Suzuki, Xenobiotica, 1986, 16, 691.
224 G.Christie, N.R.Kitteringham, and B.K.Park, Biochem. Pharmacol., 1987, 36, 3379.
225 P.L.Jacobs, G.J.H.Schmeits, H.Nieuwenhuyse, and J.Vink, Adv. Mass Spectrom., 1985, 1986, 1241.
226 W.N.Wu, J.F.Hills, S.Y.Chang, and K.T.Ng, Drug Metab. Dispos., 1988, 16, 69.
227 J.Fujii, N.Inotsume, and M.Nakano, Chem. Pharm. Bull., 1987, 35, 4338.
228 M.Hildebrand, M.Humpel, W.Krause, and U.Tauber, Eur. J. Drug Metab. Pharmacokin., 1987, 12, 31.
229 J.Paulson, C.Lindberg, and S.Edsbacker, Adv. Mass Spectrom., 1985, 1986, 613.
230 S.Edsbacker, P.Andersson, C.Lindberg, J.Paulson, A.Ryrfeldt, and A.Thalen, Drug Metab. Dispos., 1987, 15, 403.

231 S.Edsbacker, P.Andersson, C.Lindberg, A.Ryrfeldt, and A.Thalen, Drug Metab. Dispos., 1987, 15, 412.
232 C.Lindberg, J.Paulson, and S.Edsbacker, Biomed. Environ. Mass Spectrom., 1987, 14, 535.
233 M.Hassan and H.Ehrsson, Drug Metab. Dispos., 1987, 15, 399.
234 M.Hassan and H.Ehrsson, Eur. J. Drug Metab. Pharmacokin., 1987, 12, 71.
235 M.Hassan, Acta Pharm. Suecica, 1986, 24, 142.
236 D.H.Marchand, R.P.Remmel, and M.M.Abdel-Monem, Drug Metab. Dispos., 1988, 16, 85.
237 N.K.Brown and D.J.Harvey, Xenobiotica, 1988, 18, 417.
238 A.E.Mutlib, H.T.A.Cheung, and T.R.Watson, J. Steroid Biochem., 1988, 29, 135.
239 K.Nakano, H.Ando, Y.Sugawara, M.Ohashi, and S.Harigaya, Drug Metab. Dispos., 1986, 14, 740.
240 A.Nakamura, N.Tatewaki, A.Morino, and M.Sugiyama, Xenobiotica, 1986, 16, 1079.
241 D.R.Newell, D.L.Alison, A.H.Calvert, K.R.Harrap, M.Jarman, T.R.Jones, M. Manteuffel-Cymborowska, and P.O'Connor, Cancer Treatment Reports, 1986, 70, 971.
242 S.T.Forgue, P.Kari, and R.Barbhaiya, Drug Metab. Dispos., 1987, 15, 808.
243 T.Terada, C.Sakata, K.Ishabashi, T.Nakamura, R.Ishimura, T.Tsuchiya, and H.Noguchi, Xenobiotica, 1988, 18, 291.
244 C.Koppel, J.Tenczer, and K.Ibe, Arzneim.-Forsch., 1987, 37, 208.
245 M.J.Frearson, C.P.Offen, and K.Wilson, Adv. Mass Spectrom., 1985, 1986, 711.
246 C.Koppel, J.Tenczer, and A.Wagemann, Arzneim.-Forsch., 1986, 36, 1116.
247 R.C.Kammerer and M.A.Lampe, Biochem. Pharmacol., 1987, 36, 3445.
248 C.Koppel, J.Tenczer, I.Arndt, and K.Ibe, Arzneim.-Forsch., 1987, 37, 1062.
249 S.Goenechea, G.Rucker, H.Brzezinka, G.Hoffmann, M.Neugebauer, and T.Pyzik, Arzneim.-Forsch., 1987, 37, 854.
250 A.K.Chaudhary, J.W.Hubbard, G.McKay, and K.K.Midha, Drug Metab. Dispos., 1988, 16, 506.
251 S.Toma, S.Morishita, K.Kuronuma, Y.Mishima, Y.Hirai, and M.Kawakami, Xenobiotica, 1987, 17, 1195.
252 W.Gau, J.Kurz, U.Petersen, H.J.Ploschke, and C.Wuensche, Arzneim.-Forsch., 1986, 36, 1545.
253 W.Lauwers, L.Le Leune, and W.Meuldermans, Biomed. Environ. Mass Spectrom., 1988, 15, 323.
254 W.Meuldermans, A.Van Peer, J.Hendrickx, W.Lauwers, E.Swysen, M.Brokx, R. Woestenborghs, and J.Heykants, Drug Metab. Dispos., 1988, 16, 403.
255 W.Meuldermans, J.Hendrickx, W.Lauwers, R.Hurkmans, E.Mostmans, E.Swysen, J.Bracke, A.Knaeps, and J.Heykants, Drug Metab. Dispos., 1988, 16, 410.
256 G.C.Bolton. G.D.Allen, B.E.Davies, C.W.Filer, and D.J.Jeffery, Xenobiotica, 1986, 16, 853.
257 P.C.Ruenitz, R.F.Arrendale, G.D.George, C.B.Thompson, C.M.Mokler, and N.T. Nanavati, Cancer Res., 1987, 47, 4015.
258 J.H.Peters, G.R.Gordon, H.W.Nolen,III, M.Tracy, and D.W.Thomas, Biochem. Pharmacol., 1988, 37, 357.
259 S.P.Jindal and T.Lutz, J. Anal. Toxicol., 1986, 10, 150.
260 N.R.Hartman and I.Jardine, Biomed. Environ. Mass Spectrom., 1986, 13, 361.
261 N.R.Hartman and I.Jardine, Drug Metab. Dispos., 1987, 15, 661.
262 S.A.Chow and L.J.Fischer, Drug Metab. Dispos., 1987, 15, 740.
263 J.Uetrecht, N.Zahid, N.H.Shear, and W.D.Biggar, J. Pharmacol. Exp. Ther., 1988, 245, 274.
264 C.O.Meese, C.Fischer, and M.Eichelbaum, Biomed. Environ. Mass Spectrom., 1988, 15, 63.
265 D.J.Harvey, Biomed. Environ. Mass Spectrom., 1988, 15, 117.
266 D.J.Harvey and H.J.Marriage, Drug Metab. Dispos., 1987, 15, 914.
267 G.D.Paulson and V.J.Feil, Drug Metab. Dispos., 1987, 15, 671.
268 K.Ramakrishnan and J.Fisher, J. Med. Chem., 1986, 29, 1215.
269 J.S.Salonen, L.Vuorilehto, M.Eloranta, and A.Karjalainen, Eur. J. Drug Metab. Pharmacokin., 1988, 13, 59.
270 K.Minagawa, Y.Kasuya, S.Baba, G.Knapp, and J.P.Skelly, Steroids, 1986, 47, 175.

271  G.M.Rodchenkov, V.P.Uralets, V.A.Semenov, and P.A.Leclercq, J. High Resolut. Chromatogr., 1988, 11, 283.
272  C.Koppel, J.Tenczer, and K.Ibe, Arzneim.-Forsch., 1987, 37, 1304.
273  M.Sugiyama, T.Ando, Y.Okuyama, T.Sugimoto, and S.Chouka, Arzneim.-Forsch., 1986, 36, 1229.
274  R.C.Peffer, D.J.Abraham, M.A.Zemaitis, L.K.Wong, and J.D.Alvin, Drug Metab. Dispos., 1987, 15, 305.
275  A.H.Beckett and M.Stanojcik, J. Pharm. Pharmacol., 1987, 39, 409.
276  G.Blaich, E.Pfaff, and M.Metzler, Biochem. Pharmacol., 1987, 36, 3135.
277  E.LeBoeuf and O.Grech-Belanger, Drug Metab. Dispos., 1987, 15, 122.
278  Y.Sugawara, M.Ohashi, S.Nakamura, S.Usuki, T.Suzuki, Y.Ito, T.Kume, S. Harigaya, A.Nakao, M.Gaino and H.Inoue, J. Pharmacobio.-Dyn., 1988, 11, 211.
279  Y.Sugawara, S.Nakamura, S.Usuki, Y.Ito, T.Suzuki, M.Ohashi, and S. Harigaya, J. Pharmacobio.-Dyn., 1988, 11, 224.
280  T.Inoui and S.Suzuki, Xenobiotica., 1987, 17, 965.
281  B.A.Mico, J.E.Swagzdis, D.A.Federowicz, and K.Straub, J. Pharm. Sci., 1986, 75, 929.
282  J.O.Lay,Jr., W.A.Korfmacher, D.W.Miller, P.Siitonen, C.L.Holder, and A.B. Gosnell, Biomed. Environ. Mass Spectrom., 1986, 13, 627.
283  D.A.Ganes, K.W.Hindmarsh, and K.k.Midha, Xenobiotica, 1986, 16, 781.
284  D.A.Ganes and K.K.Midha, Xenobiotica, 1987, 17, 933.
285  W.A.Korfmacher, C.L.Holder, L.D.Betowski, and R.K.Mitchum, Biomed. Environ. Mass Spectrom., 1988, 15, 501.
286  C.L.Holder, H.C.Thompson, A.B.Gosnell, P.H.Siitonen, W.A.Korfmacher, C.E. Cerniglia, D.W.Miller, D.A.Casciano, and W.Slikker, J. Anal. Toxicol., 1987, 11, 113.
287  W.A.Korfmacher, C.L.Holder, L.D.Betowski, and R.K.Mitchum, J. Anal. Toxicol., 1987, 11, 182.
288  A.Muller, H.Gabriel, H.Sies, R.Terlinden, H.Fischer, and A.Romer, Biochem. Pharmacol., 1988, 37, 1103.
289  B.Monsarrat, M.Maftouh, G.Meunier, J.Bernadou, J.P.Armand, C.Paoletti, and B.Meunier, J. Pharmaceut. Biomed. Anal., 1987, 5, 341.
290  A.Gouyette, Biomed. Environ. Mass Spectrom., 1988, 15, 243.
291  O.H.Drummer, S.Kourtis, and D.Iakovidis, Arzneim.-Forsch., 1988, 38, 647.
292  O.H.Drummer and S.Kourtis, Arzneim.-Forsch., 1987, 37, 1225.
293  E.Benfenati, S.Caccia, and F.Della Vedova, J. Pharm. Pharmacol., 1987, 39, 312.
294  L.M.Allan, A.J.Hawkins, C.W.Vose, J.Firth, R.D.Brownsill, and J.A.Steiner, Xenobiotica, 1987, 17, 1233.
295  L.B.Piotrovsky, I.Y.Alexandrova, L.Soltes, M.Stefek, and T.Trnovec, Eur. J. Drug Metab. Pharmacokin., 1986, 11, 87.
296  H.S.Purba, J.L.Maggs, M.L'E.Orm, D.J.Back, and B.K.Park, Br. J. Clin. Pharmacol., 1987, 23, 447.
297  N.N.Singh, F.Jamali, F.M.Pasutto, R.T.Coutts, and A.S.Russell, J. Chromatogr., 1986, 382, 331.
298  E.S.Ferdinandi, D.Cochran, and R.Gedamke, Drug Metab. Dispos., 1987, 15, 921.
299  N.Haim, J.Nemec, J.Roman, and B.K.Sinha, Biochem. Pharmacol., 1987, 36, 527.
300  J.M.S.Van Maanen, J.de Vries, D.Pappie, E. van den Akker, V.M.La Fleur, J.Retel, J. van der Greef and H.M.Pinedo, Cancer Res., 1987, 47, 4658.
301  N.Haim, J.Nemec, J.Roman, and B.K.Sinha, Cancer Res., 1987, 47, 5835.
302  K.Hande, L.Anthony, R.Hamilton, R.Bennett, B.Sweetman, and R.Branch, Cancer Res., 1988, 48, 1829.
303  C.Baarnhielm, A.Backman, K.J.Hoffmann, and L.Weidolf, Drug Metab. Dispos., 1986, 14, 613.
304  G.Rucker, M.Neugebauer, and P.-G.Heiden, Arzneim.-Forsch., 1988, 38, 497.
305  J.Caldwell, M.Strolin-Benedetti, and A.Weil, Br. J. Clin. Pharmacol., 1986, 22, 219P.
306  A.Weil, J.Caldwell, and M.Strolin-Benedetti, Drug Metab. Dispos., 1988, 16, 302.
307  B.C.Sallustio, P.J.Meffin, and M.Thompson, J. Chromatogr., 1987, 422, 33,
308  C.Koppel, J.Tenczer, K.Eckart, and H.Schwarz, Adv. Mass Spectrom., 1985, 1986, 665.

309 I.Klebovich, L.Vereczkey, E.Toth, J.Tamas, M.Mak, G.Jalsovszky, and S.Holly, Xenobiotica, 1987, 17, 1247.
310 Ph.Jaussaud, D.Courtot, and J.L.Guyot, J. Chromatogr., 1987, 423, 123.
311 G.M.Rodchenkov, V.P.Uralets, and V.A.Semenov, J. Chromatogr., 1988, 426, 399.
312 D.J.Sweeny, S.Barnes, G.D.Heggie, and R.B.Diasio, Proc. Nat. Acad. Sci., 1987, 84, 5439.
313 D.J.Sweeny, S.Barnes, and R.B.Diasio, Cancer Res., 1988, 48, 2010.
314 L.H.M.Vroomen, J.P.Groten, K.Van Muiswinkel, A.Van Velduizen, and P.J.Van Bladeren, Chem-Biol. Inter., 1987, 64, 167.
315 A.J.Streeter, T.R.Krueger, and B.-A.Hoener, Pharmacology, 1988, 36, 283.
316 D.Mihailova, M.Velkov, and Z.Zhivkova, Eur. J. Drug Metab. Pharmacokin., 1987, 12, 25.
317 Y.Nishizawa, S.Abe, K.Yamada, T.Nakamura, I.Yamatsu, and K.Kinoshita, Xenobiotica, 1987, 17, 575.
318 Y.Ikeya, H.Taguchi, H.Mitsuhashi, H.Sasaki, T.Matsuzaki, M.Aburada, and E.Hosoya, Chem. Pharm. Bull., 1988, 36, 2061.
319 A.E.Mutlib, H.T.A.Cheung, and T.R.Watson, J. Steroid Biochem., 1987, 28, 65.
320 P.S.Callary, M.J.Egorin, L.A.Geelhaar, and M.S.B.Nayar, Cancer Res., 1986, 46, 4900.
321 M.J.Egorin, S.W.Snyder, A.S.Cohen, E.G.Zuhowski, B.Subramanyam, and P.S. Callary, Cancer Res., 1988, 48, 1712.
322 D.A.Ligtenstein, E.R.J.Wils, S.P.Kossen, and A.G.Hulst, J. Pharm. Pharmacol., 1987, 39, 17.
323 L.B.LaCagnin, H.D.Colby, and J.P.O'Donnell, Drug Metab. Dispos., 1986, 14, 549.
324 S.Narimatsu, K.Matsubara, T.Shimonishi, K.Watanabe, I.Yamamoto, and H. Yoshimura, Drug Metab. Dispos., 1988, 16, 156.
325 E.Bauer, J.McDougall, and B.D.Cameron, Xenobiotica, 1986, 16, 625.
326 S.Inui and S.Asada, Yakugaku-Zasshi., 1987, 107, 904.
327 H.Wikstrom, T.Elebring, G.Hallnemo, B.Andersson, K.Svensson, A.Carlsson, and H.Rollema, J. Med. Chem., 1988, 31, 1080.
328 I.Manz, K.Konig, H.O.Klein, U.Miemeyer, J.Pohl, P.Hilgard, N.Brock, and M.Przybylski, Adv. Mass Spectrom., 1985, 1986, 1521.
329 K.Iwasaki, T.Shiraga, K.Noda, K.Tada, and H.Noguchi, Xenobiotica, 1986, 16, 651.
330 H.Kamimura, Y.Enjoji, H.Sasaki, R.Kawai, H.Kaniwa, K.Niigata, and S.Kageyama, Xenobiotica, 1987, 17, 645.
331 H.Kamimura, R.Kawai, and H.Kudo, Xenobiotica, 1988, 18, 141.
332 E.Peretti, G.Karlaganis, and B.H.Lauterbach, J. Pharmacol. Exp. Ther., 1987, 243, 686.
333 S.-H.Lee, E.Sestokas, R.Taub, R.P.Buhs, M.Green, R.Sestokas, W.J.A. VandenHeuvel, B.H.Arison, and T.A.Jacob, Drug Metab. Dispos., 1986, 14, 590.
334 T.R.Woolf and J.D.Adams, Xenobiotica, 1987, 17, 839.
335 R.Sams and P.Pizzo, J. Anal. Toxicol., 1987, 11, 58.
336 W.Meuldermans, J.Hendrickx, R.Woestenborghs, A.Van Peer, W.Lauwers, J.De Cree, and J.Heykants, Arzneim.-Forsch., 1988, 38, 789.
337 E.J.Mroszczak, F.W.Lee, D.Combs, F.H.Sarnquist, B.-L.Huang, A.T.Wu, L.G.Tokes, M.L.Maddox, and D.K.Cho, Drug Metab. Dispos., 1987, 15, 618.
338 J.F.Le Bigot, J.M.Begue, J.R.Kiechel, and A.Guillouzo, Life Sci., 1987, 40, 883.
339 P.C.Canfell, N.Castagnoli,Jr., M.R.Fahey, P.J.Hennis, and R.D.Miller, Drug Metab. Dispos., 1986, 14, 703.
340 R.C.Kammerer and D.A.Schmitz, Xenobiotica, 1986, 16, 681.
341 R.T.Coutts, G.A.Torok-Both, L.V.Chu, Y.K.Tam, and F.M.Pasutto, J. Chromatogr., 1987, 421, 267.
342 D.M.Dulik and C. Fenselau, Drug Metab. Dispos., 1987, 15, 473.
343 R.Neidlein and H.-P.Deigner, Arzneim.-Forsch., 1988, 38, 260.
344 P.Beaumann, P.Bosshart, G.Gabris, M.Gastpar, L.Koeb, and B.Woggon, Arzneim.-Forsch., 1988, 38, 292.
345 Y.Matsuki, J.Dan, K.Fukuhara, T.Ito, and T.Nambara, Chem. Pharm. Bull., 1988, 36, 1431.
346 C.Koppel, J.Tenczer, and K.Ibe, J. Chromatogr., 1988, 427, 144.

347 D.M.Dulik, C.Fenselau, and J.Hilton, Biochem. Pharmacol., 1986, 35, 3405.
348 D.M.Dulik and C.Fenselau, Drug Metab. Dispos., 1987, 15, 195.
349 P.D.Wedlund, B.J.Sweetman, G.R.Wilkinson, and R.A.Branch, Drug Metab. Dispos., 1987, 15, 277.
350 M.Hilderbrand, W.Krause, G.Kuhne, and G.-A.Hoyer, Xenobiotica, 1987, 17, 623.
351 B.M.Gerardy, D.Kapusta, P.Dumont, and J.H.Poupaert, Drug Metab. Dispos., 1986, 14, 477.
352 T.Baba, H.Yamada, K.Oguri, and H.Yoshimura, Xenobiotica, 1987, 17, 1029.
353 C.E.Cerniglia, E.B.Hansen,Jr., K.J.Lambert, W.A.Korfmacher, and D.W. Miller, Xenobiotica, 1988, 18, 301.
354 S.S.Singer, W.Lijinsky, L.E.Kratz, N.Castagnoli,Jr., and J.E.Rose, Xenobiotica, 1987, 17, 1279.
355 R.C.Kammerer and D.A.Schmitz, Xenobiotica, 1987, 17, 1121.
356 R.C.Kammerer and D.A.Schmitz, Xenobiotica, 1986, 16, 671.
357 D.C.Mays, S.G.Hecht, S.E.Unger, C.M.Pacula, J.M.Climie, D.E.Sharp, and N. Gerber, Drug Metab. Dispos., 1987, 15, 318.
358 N.K.Brown and D.J.Harvey, Biomed. Environ. Mass Spectrom., 1988, 15, 389.
359 P.Kestell, A.P.Gledhill, M.D.Threadgill, and A.Gescher, Biochem. Pharmacol., 1986, 35, 2283.
360 M.D.Threadgill, D.B.Axworthy, T.A.Baillie, P.B.Farmer, K.C.Farrow, A.Gescher, P.Kestell, P.G.Pearson, and A.J.Shaw, J. Pharmacol. Exp. Ther., 1987, 242, 312.
361 G.M.Rodchenkov, V.P.Vralets, and V.A.Semenov, J. Chromatogr., 1987, 423, 15.
362 J.Martensson, T.Denneberg, and B.Kagedal, Eur. J. Clin. Pharmacol., 1986, 31, 119.
363 O.Beck, B.Jernstrom, M.Martinez, and D.B.Repke, Chem.-Biol. Inter., 1988, 65, 97.
364 P.Nielsen, N.Gyrd-Hansen, C.-E.Olsen, and W.Xia, Pharmacol. Toxicol., 1987, 61, 330.
365 K.-J.Hoffmann, O.Gyllenhaal, and J.Vessman, Biomed. Environ. Mass Spectrom, 1987, 14, 543.
366 H.U.Shetty and W.L.Nelson, J. Med. Chem., 1988, 31, 55.
367 O.Grech-Belanger, J.Turgeon, and M.Lalande, Res. Commun. Chem. Pathol. Pharmacol., 1987, 58, 53.
368 M.Nakaoka and H.Hakusui, Xenobiotica, 1987, 17, 1329.
369 P.R.Cid, E.G.Fernandes, and M.F.Brana, Eur. J. Drug Metab. Pharmacokin., 1986, 11, 255.
370 F.S.Chiccarelli, J.A.Morrison, D.B.Cosulich, N.A.Perkinson, D.N.Ridge, F.W.Sum, K.C.Murdock, D.L.Woodward, and E.T.Arnold, Cancer Res., 1986, 46, 4858.
371 J.E.Coutant, R.J.Barbuch, D.K.Santonin, and R.J.Cregge, Biomed. Mass Spectrom., 1987, 14, 325.
372 I.D.Wilson, K.V.Watson, J.Troke, H.P.A.Illing, and J.M.Fromson, Xenobiotica, 1986, 16, 1117.
373 A.Assandri, G.Tarzia, E.Bellasio, R.Ciabatti, G.Tuan, P.Ferrari, L.Zerilli, M.Lanfranchi, and G.Pelizzi, Xenobiotica, 1987, 17, 559.
374 M.J.Lynch, F.R.Mosher, W.R.Levesque, and T.J.Newby, Drug Metab. Dispos., 1987, 15, 253.
375 T.Kobayashi, H.Hohnoki, Y.Esumi, T.Ohtsuki, T.Washino, and S.Tanayama, Xenobiotica, 1988, 18, 49.
376 H.Nakayama, S.Fujihara, T.Nakashima, and Y.Kurogochi, Biochem. Pharmacol., 1987, 36, 4313.
377 G.A.Kyremateri, L.H.Taylor, J.D.DeBethizy and E.S.Vesell, Drug Metab. Dispos., 1988, 16, 125.
378 M.B.Mattammal, V.M.Lakshmi, T.V.Zenser, and B.B.Davis, J. Pharmacol. Exp. Ther., 1987, 242, 827.
379 W.F.Pool, A.A.Houdi, L.A.Damani, W.J.Layton, and P.A.Crooks, Drug Metab. Dispos., 1986, 14, 574.
380 L.A.Peterson and N.Castagnoli,Jr., J. Med. Chem., 1988, 31, 637.
381 S.Terashita, Y.Tokuma, T.Fujiwara, Y.Shiokawa, K.Okumura, and H.Noguchi, Xenobiotica, 1987, 17, 1415.

382 E.Bodd, A.S.Christophersen, and U.Fongen, Acta Pharmacol. Toxicol., 1986, 59, 252.
383 B.-L.Sahlberg, B.-M.Landgren, and M.Axelson, J. Steroid Biochem., 1987, 26, 609.
384 B.-L.Sahlberg, J. Steroid Biochem., 1987, 26, 481.
385 E.Houghton, M.C.Dumasia, P.Teal, M.S.Moss, and S.Sinkins, J. Chromatogr., 1986, 383, 1.
386 V.Dimov, K.Green, M.Bygdeman, and N.J.Christensen, Drug Metab. Dispos., 1986, 14, 494.
387 A.Guillouzo, L.Grislain, D.Ratanasavanh, M.T.Mocquard, J.-M.Begue, P.Du Vignaud, N.Bromet, P.Genissel, and B.Beau, Xenobiotica, 1988, 18, 757.
388 W.Dieterle, J.W.Faigle, H.-P.Kriemler, and T.Winkler, Xenobiotica, 1986, 16, 743.
389 S.F.Sisenwine, C.O.Tio, A.L.Liu, and J.F.Politowski, Drug Metab. Dispos., 1987, 15, 579.
390 H.Schutz, K.F.Feldmann, J.W.Faigle, H.-P.Kreimler, and T.Winkler, Xenobiotica, 1986, 16, 769.
391 T.A.Baillie, K.-J.Hoffmann, and D.B.Axworthy, Adv. Mass Spectrom., 1985, 1986, 1267.
392 J.O.Lay,Jr., D.W.Potter, and J.A.Hinson, Biomed. Environ. Mass Spectrom., 1987, 14, 517.
393 B.Spinar, I.Koruna, M.Ryska, and S.Smolik. Cesk. Farm., 1985, 34, 362.
394 K.H.Lehr, P.Damm, H.W.Fehlhaber, and P.Hajdu, Arzneim.-Forsch., 1987, 37, 1222.
395 R.H.Waring and S.C.Mitchell, Xenobiotica, 1988, 18, 235.
396 A.E.Pilkinton and R.H.Waring, Eur. J. Drug Metab. Pharmacokin., 1988, 13, 99.
397 H.Davi, A.Carayon, D.Berthet, W.Cautreels, R.Dommisse, and M.D.Faiez Zannad, Biomed. Environ. Mass Spectrom., 1986, 13, 559.
398 D.Decolin, A.M.Batt, J.M.Ziegler, and G.Siest, Biochem. Pharmacol., 1986, 35, 2301.
399 R.G.Cooper, S.A.Jenkins, D.A.Price Evans, and A.H.Price, Xenobiotica, 1988, 18, 389.
400 L.D.Betowski, W.A.Korfmacher, J.O.Lay,Jr., D.W.Potter, and J.A.Hinson, Biomed. Environ. Mass Spectrom., 1987, 14, 705.
401 P.E.Haroldsen, M.H.Reilly, H.Hughes, S.J.Gaskell, and C.J.Porter, Biomed. Environ. Mass Spectrom., 1988, 15, 615.
402 S.Ohta, H.Masumoto, K.Takeuchi, and M.Hirobe, Drug Metab. Dispos., 1987, 15, 583.
403 D.J.Gole, J.-L.Pirat, J.-M.Kamanka, and E.F.Domino, Drug Metab. Dispos., 1988, 16, 386.
404 A.Raisi and A.H.Beckett, Monatsh. Chem., 1986, 117, 1047.
405 B.C.Foster, R.T.Coutts, F.M.Pasutto, and A.Mozayani, Life Sci., 1988, 42, 285.
406 P.R.Ortiz de Montellano and M.D.Watanabe, Mol. Pharmacol., 1987, 31, 213.
407 J.Vachta, K.Valter, and Ph.Gold-Aubert, Eur. J. Drug Metab. Pharmacokin., 1986, 11, 195.
408 J.X.De Vries, Chromatographia, 1986, 22, 421.
409 S.Ichihara, H.Tomisawa, H.Fukazawa, M.Tateishi, R.Joly, and R.Heintz, Biochem. Pharmacol., 1986, 35, 3935.
410 O.Caputo, F.Viola, G.Grosa, F.Rocco, and G.Biglino, Eur. J. Drug Metab. Pharmakokin., 1986, 11, 91.
411 A.F.DeLong, S.W.Oldham, K.A.DeSante, G.Nell, and D.P.Henry, J. Pharm. Sci., 1988, 77, 153.
412 B.Melegh, J.Kerner, and L.L.Bieber, Biochem. Pharmacol., 1987, 36, 3405.
413 F.T.Delbeke, M.Debackere, J.A.A.Jonckheere, and P.A.De Leenheer, Biopharmaceut. Drug Dispos., 1986, 7, 389.
414 R.A.Budinsky, S.M.Roberts, E.A.Coates, L.Adams, and E.V.Hess, Drug Metab. Dispos., 1987, 15, 37.
415 J.Henion and T.Covy, Adv. Mass Spectrom., 1985, 1986, 629.
416 A.H.Beckett, G.E.Navas, and A.J.Hutt Xenobiotica, 1988, 18, 61.
417 B.Gober, K.Dressler, and P.Franke, Pharmazie, 1988, 43, 96.
418 P.J.Simons, I.D.Cockshott, E.J.Douglas, E.A.Gordon, K.Hopkins, and M.Rowland, Xenobiotica, 1988, 18, 429.

419 H.U.Shetty amd W.L.Nelson, J. Med. Chem., 1986, 29, 2004.
420 W.L.Nelson and H.U.Shetty, Drug Metab. Dispos., 1986, 14, 506.
421 H.A.Sasame, D.J.Liberato, and J.R.Gillette, Drug Metab. Dispos., 1987, 15, 349.
422 R.E.Talaat and W.L.Nelson, Drug Metab. Dispos., 1988, 16, 207.
423 R.E.Talaat and W.L.Nelson, Drug Metab. Dispos., 1988, 16, 212.
424 L.M.Gustavson and W.L.Nelson, Drug Metab. Dispos., 1988, 16, 217.
425 N.K.Brown and D.J.Harvey, Biomed. Environ. Mass Spectrom., 1988, 15, 403.
426 L.W.Whitehouse, B.A.Lodge, A.W.By, and B.H.Thomas, Biopharmaceut. Drug Dispos., 1987, 8, 307.
427 T.Yamamoto, Y.Moriwaki, S.Takahashi, T.Hada, and K.Higashino, Biochem. Pharmacol., 1987, 36, 2415.
428 E.Beretta, S.Botturi, P.Ferrari, G.Tuan, and L.F.Zerilli, J. Chromatogr., 1987, 416, 144.
429 A.Seago, M.H.Baker, J.Houghton, C.-S.Leung, and M.Jarman, Biochem. Pharmacol., 1987, 36, 573.
430 A.Seago, P.E.Goos, L.J.Griggs, M.Jarman, Biochem. Pharmacol., 1986, 35, 2911.
431 E.B.Hansen,Jr., C.E.Cerniglia, W.A.Korfmacher, D.W.Miller, and R.H.Heflich, Drug Metab. Dispos., 1987, 15, 97.
432 D.W.Kelly and W.Slikker,Jr., Drug Metab. Dispos., 1987, 15, 460.
433 S.Y.Yeh, Drug Metab. Dispos., 1987, 15, 466.
434 W.A.Korfmacher, C.L.Holder, A.B.Gosnell, and H.C.Thompson,Jr., J. Anal. Toxicol., 1986, 10, 142.
435 N.G.G.Whitaker and T.D.Lindstrom, Drug Metab. Dispos., 1987, 15, 107.
436 J.C.Maurizis, J.C.Madelmont, D.Parry, D.Dauzonne, R.Royer, and J.L.Chabard, Xenobiotica, 1986, 16, 635.
437 M.S.Lant, G.R.Manchee, L.E.Martin, and J.Oxford, Adv. Mass Spectrom., 1985, 1986, 615.
438 M.Koike, R.Norikura, K.Iwatani, K.Sugeno, S.Takahashi, and Y.Nakagawa, Xenobiotica, 1988, 18, 257.
439 U.Valcavi, E.Bosone, P.Farina, M.G.Castelli, C.Chiabrando, and E.Benfenati, Adv. Mass Spectrom., 1985, 1986, 707.
440 S.Iwamura, K.Shibata, Y.Kawabe, K.Tsukamoto, and E.Honma, J. Pharmacobio.-Dyn., 1987, 10, 229.
441 S.Honma, S.Iwamura, R.Kobayashi, Y.Kawabe, and K.Shibata, Drug Metab. Dispos., 1987, 15, 551.
442 A.Black and T.Chang, Eur. J. Drug Metab. Pharmacokin., 1987, 12, 135.
443 F.L.S.Tse, B.A.Orwig, J.M.Jaffe, and J.G.Dain, Xenobiotica, 1987, 17, 1259.
444 S.R.Howell, L.A.Kotkoskie, R.L.Dills, and C.D.Klaassen, J. Pharm. Sci., 1988, 77, 309.
445 H.Kim, B.Pramanik, A.Lapiguera, T.M.Chan, V.M.Girijavallabhan, S.Symchowicz, and C.Lin, Drug Metab. Dispos., 1988, 16, 325.
446 C.T.Gombar, K.Straub, P.Levandoski, L.Gutzait, J.Swagzdis, C.Garvie, G.Joseph, B.D.Potts, and B.A.Mico, Drug Metab. Dispos., 1986, 14, 540.
447 J.F.Newton, K.M.Straub, R.H.Dewey, C.D.Perchonock, T.B.Leonard, M.E.McCarthy, J.G.Gleason, and R.D.Eckardt, Drug Metab. Dispos., 1987, 15, 161.
448 J.F.Newton, K.M.Straub, G.Y.Kuo, C.D.Perchonock, M.E.McCarthy, J.G.Gleason, and R.K.Lynn, Drug Metab. Dispos., 1987, 15, 168.
449 R.F.Struck, M.C.Kirk, L.S.Rice, and W.J.Suling, J. Med. Chem., 1986, 29, 1319.
450 T.A.Moreland, J.Astoin, F.Lepage, F.Tombret, R.H.Levy, and T.A.Baillie, Drug Metab. Dispos., 1986, 14, 654.
451 M.Stefek, L.Benes, M.Jergelova, V.Scasnar, L.Turi-Nagy, and P.Kocis, Xenobiotica, 1987, 17, 1067.
452 G.D.Paulson and V.J.Feil, Drug Metab. Dispos., 1987, 15, 841.
453 P.A.Nelson, G.D.Paulson, and V.J.Feil, Xenobiotica, 1987, 17, 829.
454 A.Prox, J.Schmid, J.Nickl, and G.Engelhardt, Z. Naturforsch., Sect. C., 1987, 42, 465.
455 T.Kobayashi, J.Sugihari, and S.Harigaya, Drug Metab. Dispos., 1987, 15, 877.
456 T.Kobayashi, H.Ando, J.Sugihara, and S.Harigaya, Drug Metab. Dispos., 1987, 15, 262.
457 I.B.Parr, R.McCague, G.Leclercq, and S.Stoessel, Biochem. Pharmacol., 1987, 36, 1513.

458 E.A.Lien, E.Solheim, S.Kvinnsland, and P.M.Ueland, Cancer Res., 1988, 48, 2304.
459 E.A.Lien, P.M.Ueland, E.Solheim, and S.Kvinnsland, Clin. Chem. (Winston Salem, N.C.), 1987, 33, 1608.
460 B.Lindeke, O.Ericsson, A.Jonsson, B.Noren, S.Stromberg, and B.Vangbo, Xenobiotica, 1987, 17, 1269.
461 H.Kodaira and S.Spector, Proc. Nat. Acad. Sci., 1988, 85, 1267.
462 T.Tsuchiya, A.Tanaka, M.Fukuoka, M.Sato, and T.Yamaha, Chem. Pharm. Bull., 1987, 35, 2985.
463 W.P.Gordon, J.R.McCarthy, and S.Y.Chang, Biochem. Biophys. Res. Commun., 1987, 145, 575.
464 C.Bory, C.Chantin, R.Boulieu, J.Cotte, J.-C.Berthier, D.Fraisse, and M.-J.Bobenrieth, Compt. Rendus., 1986, 303, 7.
465 A.S.Papadopoulos and J.L.Crammer, Xenobiotica, 1986, 16, 1097.
466 R.Maffei Facino, M.Carini, O.Tofanetti, I.Casciarri, and E.Longoni, Arzneim.-Forsch., 1987, 37, 682.
467 R.Arigoni, R.Chiste, I.Setnikar, E.Benfenati, and R.Fanelli, Biomed. Environ. Mass Spectrom., 1988, 15, 205.
468 F.Chanoine, C.Grenot, N.Sellier, W.E.Barrett, R.M.Thompson, A.F.Fentiman,Jr., J.R.Nixon, R.Goyer, and J.L.Junien, Drug Metab. Dispos., 1987, 15, 868.
469 A.F.Heald, D.P.Dizio, K.M.Kirkland, P.Loftus, and J.O.Malbica, Drug Metab. Dispos., 1986, 14, 631.
470 G.B.Baker, D.R.Hampson, R.T.Coutts, R.G.Micetich, T.W.Hall, and T.S.Rao, J. Neural Transmission., 1986, 65, 233.
471 S.J.Kolis, E.J.Postma, T.H.Williams, and G.J.Sasso, Drug Metab. Dispos., 1986, 14, 465.
472 E.B.Hansen,Jr., R.H.Heflich, W.A.Korfmacher, D.W.Miller, and C.E.Cerniglia, J. Pharm. Sci., 1988, 77, 259.
473 P.Arnoux, M.Placidi, C.Aubert, and J.P.Cano, J. Chromatogr., 1986, 381, 75.
474 K.Sudo, O.Okazaki, M.Tsumura, and H.Tachizawa, Xenobiotica, 1986, 16, 725.
475 A.E.Rettie, A.W.Retteneeier, B.K.Beyer, T.A.Baillie, and M.R.Juchau, Clin. Pharmacol. Ther., 1986, 40, 172.
476 A.W.Rettenmeier, W.P.Gordon, K.S.Prickett, R.H.Levy and T.A.Baillie, Drug Metab. Dispos., 1986, 14, 454.
477 A.W.Rettenmeier, W.P.Gordon, K.S.Prickett, R.H.Levy, J.S.Lockard, K.E.Thummel, and T.A.Baillie, Drug Metab. Dispos., 1986, 14, 443.
478 W.A.Rettenmeier, W.P.Gordon, H.Barnes, and T.A.Baillie, Xenobiotica, 1987, 17, 1147.
479 W.L.Nelson, L.D.Olsen, D.B.Beitner, and R.J.Pallow,Jr., Drug Metab. Dispos., 1988, 16, 184.
480 R.T.Schillings and S.F.Sisenwine, Drug Metab. Dispos., 1986, 14, 405.
481 S.El-Sharkawy and Y.J.Abul-Hajj, Xenobiotica, 1988, 18, 365.
482 S.Vajta, J.P.Thenot, F.de Maack, G.Devant, and M.Lesieur, Biomed. Environ. Mass Spectrom., 1988, 15, 223.
483 C.Murphy, T.Fotsis, P.Pantzar, H.Adlercreutz, and F.Martin, J. Steroid Biochem., 1987, 28, 609.
484 W.D.Henner, E.A.Furlong, M.D.Flaherty, T.C.Shea, and W.P.Peters, J. Chromatogr., 1987, 416, 426.
485 H.Ehrsson, S.Eksborg, and A.Lindfors, J. Chromatogr., 1986, 380, 222.
486 J.I.Brodfuehrer and G.Powis, J. Chromatogr., 1984, 427, 247.
487 D.E.Main, W.Ens, R.Beavis, B.Schueler, K.G.Standing, J.B.Westmore, and C.-M.Wong, Biomed. Environ. Mass Spectrom, 1987, 14, 91.
488 E.Busker, E.Koberstein, U.Niemeyer, J.Engel, and M.Linscheid, Biomed. Environ. Mass Spectrom., 1988, 15, 163.
489 E.A.de Bruijn, A.T.van Oosterom, P.A.Leclercq, J.W.de Haan, L.J.M.van den Ven, and U.R.Tjaden, Biomed. Environ. Mass Spectrom., 1987, 14, 643.
490 J.A.Suchocki and A.T.Sneden, J. Pharm. Sci., 1987, 76, 738.
491 T.L.Pyatigorskaya, O.Y.Zhilkova, V.S.Shelkovsky, N.M.Arkhangelova, A.I. Grizodub and L.F.Sukhodub, Biomed. Environ. Mass Spectrom., 1987, 14, 143.
492 G.Folena-Wasserman, B.L.Poehland, E.W.K.Yeung, D.Staiger, L.B.Killmer, K. Snader, J.J.Dingerdissen, and P.W.Jeffs, J.Antibiot., 1986, 39, 1395.

493 J.J.Dingerdissen, R.D.Sitrin, P.A.DePhillips, A.J.Giovenella, S.F.Grappel, R.J.Mehta, Y.K.Oh, C.H.Pan, G.D.Roberts, M.C.Shearer, and L.J.Nisbet, J. Antibiot., 1987, 40, 165.
494 J.L.Mertz, J.S.Peloso, B.J.Barker, G.E.Babbitt, J.L.Occolowitz, V.L.Simson, and R.M.Kline, J. Antibiot., 1986, 39, 878.
495 T.T.Chang, H.-R.Tsou, and M.M.Siegel, Anal. Chem., 1987, 59, 614.
496 H.-R.Tsou, S.Rajan, T.T.Chang, R.R.Fiala, G.W.Stockton, and M.W.Bullock, J. Antibiot, 1987, 40, 94.
497 B.Cavalleri, A.Arnone, E.Di.Modugno, G.Nasini, and B.P.Goldstein, J. Antibiot., 1988, 41, 308.
498 Y.Kusakabe, N.Takahashi, Y.Iwagaya, and A.Seino, J. Antibiot., 1987, 40, 237.
498 J.W.Westley, C.-M.Liu, J.E.Blount, L.Todaro, L.H.Sello, and N.Troupe, J. Antibiot., 1986, 39, 1704.
500 F.Besson and G.Michel, J. Antibiot., 1987, 40, 437.
501 L.Radics, M.Kajtar-Peredy, C.G.Casinovi, C.Rossi, M.Ricci, and L.Tuttobello,J. Antibiot., 1987, 40, 714.
502 M.Przybylski, I.Manz, P.Fonroberts, I.Dietrich, and H.Bruckner, Adv. Mass Spectrom., 1985., 1986, 1519.
503 P.Spiri-Nakagawa, Y.Fukushi, K.Maebashi, N.Imamura, Y.Takanashi, Y.Tanaka, H.Tanaka, and S.Omura, J. Antibiot., 1986, 39, 1719.
504 G.L.Rowin, J.E.Miller, G.Albers-Schonberg, J.C.Onishi, D.Davis, and E.L. Dulaney, J. Antibiot., 1986, 39, 1772.
505 T.Mukhopadhyay, B.N.Ganguli, H.W.Fehlhaber, H.Kogler, and L.Vertesy, J. Antibiot., 1987, 40, 281.
506 M.Barber, G.J.Elliot, R.S.Bordoli, B.N.Green, and B.W.Bycroft, Experientia, 1988, 44, 266.
507 P.Roepstorff, P.F.Nielsen, I.Kamensky, A.G.Craig, and R.Self, Biomed. Environ. Mass Spectrom., 1988, 15, 305.
508 R.G.Smith, Drugs Pharm. Sci., 1986, 27, 141.
509 E.Arlandini, B.Giola, E.Pella, R.Tonani, and P.Traldi, Org. Mass Spectrom., 1986, 21, 747.
510 N.A.Klyuev, V.G.Zhilnikov, M.K.Kudinova, E.S.Sharoiko, N.V.Murenets, S.E.Esipov, M.V.Bibikova, I.A.Spiridnova, S.N.Vostrov, R.N.Elizarova, and L.P.Ivanitskaya, Antibiotic Med. Biotekhnol., 1987, 32, 668.
511 W.Schade, U.Grafe, and J.Schmidt, Biomed. Environ. Mass Spectrom., 1988, 15, 359.
512 H.Pohlmann, P.Franke, and S.Pfeifer, Pharmazie, 1987, 42, 827.
513 V.St.Georgiev, D.C.Coomber and G.B.Mullen, Org. Mass Spectrom., 1988, 23, 283.
514 A.D.Argoudelis, L.Baczynskyj, W.J.Haak, W.M.Knoll, S.A.Mizsak, and F.B.Shilliday, J. Antibiot., 1988, 41, 157.
515 M.M.Siegel and N.B.Colthup, Appl. Spectrosc., 1987, 41, 1227.
516 K.Hirayama, S.Akashi, T.Ando, I.Horino, Y.Etoh, H.Morioka, H.Shibai, and A.Murai, Shituryo Bunseki, 1987, 35, 31.
517 K.Hirayama, S.Akashi, T.Ando, I.Horino, Y.Etoh, H.Morioka, H.Shibai, and A.Murai, Biomed. Environ. Mass Spectrom., 1987, 14, 305.
518 C.Monneret and N.Sellier, Biomed. Environ. Mass Spectrom., 1986, 13, 319.
519 D.R.Mueller, B.M.Domon, W.Blum, F.Raschdorf, and W.J.Richter, Biomed. Environ. Mass Spectrom., 1988, 15, 441.
520 M.M.Siegel, W.J.McGahren, and G.A.Ellestad in: 'Mass Spectrometry in the Analysis of Large Molecules', J.Wiley, New York, 1986, p.207.
521 R.Misra, R.C.Pandey, B.D.Hilton, P.P.Roller, and J.V.Silverton, J. Antibiot., 1987, 40, 786,
522 A.E.Ashcroft, Org. Mass Spectrom., 1987, 22, 754.
523 H.T.Kalinoski, B.W.Wright, and R.D.Smith, Biomed. Environ. Mass Spectrom., 1988, 15, 239.
524 S.E.Unger and B.M.Warrack, Spectroscopy, 1986, 1, 33.
525 R.J.Wills, R.Belshe, D.Tomlinsin, F.De Grazia, A.Lin, S.Wells, J.Milazzo, and C.Berry, Clin. Pharmacol. Ther., 1987, 42, 449.
526 R.J.Wills, N.Choma, G.Buonpane, A.Lin, and N.Keigher, J. Pharm. Sci., 1987, 76, 886.
527 R.J.Wills, L.C.Rodriguez, N.Choma, and M.Oakes, J. Clin. Pharmacol., 1987, 27, 821.

528 E.V.Capparelli, R.C.Stevens, M.S.S.Chow, M.Izard, and R.J.Wills, Clin. Pharmacol. Ther., 1988, **43**, 536.
529 J.E.Matusik, C.G.Guyer, J.N.Geleta, and C.J.Barnes, J. Assoc. Off. Anal. Chem., 1987, **70**, 546.
530 G.Burton, C.R.Dobson, and J.R.Everett, J. Clin. Pharmacol., 1986, **38**, 758.
531 J.H.Beijnen, O.A.G.J.Van Der Houwen, H.Rosing, and W.J.M.Underberg, Chem. Pharm. Bull., 1986, **34**, 2900.
532 Y.Namiki, T.Tanabe, T.Kobayashi, J.Tanabe, Y.Okimura, S.Koda, and Y.Morimoto, J. Pharm. Sci., 1987, **76**., 208.
533 L.Vertesy, K.Heil, H.-W.Fehlhaber, and W.Ziegler, J. Antibiot., 1987, **40**, 388.
534 H.H.Tonnesen, A.-L.Grislingaas, S.O.Woo, and J.Karlsen, Int. J. Pharmacuet., 1988, **43**, 215.
535 M.M.Siegel, R.K.Isensee, and D.J.Beck, Anal. Chem., 1987, **59**, 989.
536 P.Winstanley, G.Edwards, M.Orme, and A.Breckenridge, Br. J. Clin. Pharmacol., 1987, **23**, 1.
537 J.L.Maggs, N.R.Kitteringham, A.M.Breckenridge, and B.K.Park, Biochem., Pharmacol., 1987, **36**, 2061.
538 S.Looareesuwan, N.J.White, D.A.Warrell, I.Forgo, U.G.Dubach, U.B.Ranalder, and D.E.Schwartz, Br. J. Clin. Pharmacol., 1987, **24**, 37.
539 M.Warner, Anal. Chem., 1987, **59**, 521A.
540 R.L.Boeckx, Clin. Chem. (Winston-Salem N.C.), 1987, **33**, 974.
541 J.J.Tasset, T.J.Schroeder, and A.J.Pesce, J. Anal. Toxicol., 1986, **10**, 258.
542 T.J.Schroeder, J.J.Tasset, E.J.Otten, and J.R.Hedges, J. Anal. Toxicol., 1986, **10**, 221.
543 M.A.Peat, Clin. Chem. (Winston-Salem, N.C.), 1988, 34 471.
544 D.A.Cowan, Anal. Proc., 1987, **24**, 69.
545 T.R.Covey, E.D.Lee and J.D.Henion, Anal. Chem., 1986, **58**, 2453.
546 M.Hayashida, M.Nihira, K.Hirakawa, T.Watanabe, T.Nagai, K.Sugahara, M.Tanaka, K.Suga, Y.Yamamoto, and T.Oksuka, Iyo Masu Kenkyukai Koenshu., 1985, **10**, 95.
547 E.M.Pare, J.R.Monforte, R.Gault, and H.Mirchandani, J. Anal. Toxicol., 1987, **11**, 272.
548 C.H.Van Peteghem, M.F.Lefevere, G.M.Van Haver, and A.P.De Leenheer, J. Agr. Food Chem., 1987, **35**, 228.
549 S.Pollak and W.Vycudilik, Arch. Kriminol., 1986, **178**, 25.
550 T.Cairns, E.G.Siegmund, and B.R.Rader, Pharmaceut. Res., 1987, **4**, 126.
551 D.Bourquin and R.Brenneisen, J. Chromatogr., 1987, **414**, 187.
552 L.Karlsson, J. Chromatogr., 1987, **417**, 309.
553 B.-I.Podowik, M.L.Smith, and R.O Pick, J. Anal. Toxicol., 1987, **11**, 215.
554 M.L.Abercrombie and J.S.Jewell, J. Anal. Toxicol., 1986, **10**, 178.
555 T.Vu Duc, J. Anal. Toxicol., 1987, **11**, 83.
556 R.E.Dubler, S.Thacker, and N.Wang, Clin. Chem. (Winston-Salem, N.C.)., 1987, **33**, 972.
557 S.J.Mule and S.J.Gross, Clin. Chem. (Winston-Salem, N.C.), 1987, **33**, 613.
558 H.H.McCurdy, L.J.Lewellen, L.S.Callahan, and P.S.Childs, J. Anal. Toxicol., 1986, **10**, 175.
559 M.A.ElSohly, D.F.Stanford and T.L.Little, Jr., J. Anal. Toxicol., 1988, 12, 54
560 A.R.Banijamali, N.A.-Taleb, C.J.Van der Schyf, A.Charalambous, and A. Makriyannis, J. Labelled. Comp. Radiopharm., 1988, **25**, 73.
561 E.Johansson, S.Agurell, L.-E.Hollister and M.M.Halldin, J. Pharm. Pharmacol., 1988, **40**, 374.
562 W.H.McFadden, D.A.Garteiz, and E.G.Siegmund, J. Chromatogr., 1987, **394**, 101.
563 H.R.Sullivan, G.K.Hanasono, W.M.Miller, and P.G.Wood, Xenobiotica, 1987, **17**, 459.
564 E.J.Cone, R.E.Johnson, W.D.Darwin, D.Yousefnejad, L.D.Mell, B.D.Paul and J.Mitchell, J. Anal. Toxicol., 1987, **11**, 89.
565 A.S.Christophersen, J. Anal. Toxicol., 1986, **10**, 129.
466 L.A.Peterson, A.Trevor, and N. Castagnoli, Jr.,J. Med. Chem., 1987, **30**,249.
567 C.W.Bayer and M.S.Balack, Biomed. Environ. Mass Spectrom., 1987, **14**, 363.
568 G.B.Neurath and F.G.Pein, J. Chromatogr., 1987, **415**, 400.
569 P.J.De Schepper, A.Van Hecken, P.Daenens, and J.M.Van Rossum, Eur. J. Clin. Pharmacol., 1987, **31**, 583.

570 B.C.Pettett, Jr., S.M.Dyszel, and L.V.Hood, Clin. Chem. (Winston-Salem, N.C.)., 1987, 33, 1251.
571 L.W.Hayes, W.Krasselt, and P.A.Mueggler, Clin. Chem. (Winston-Salem, N.C.), 1987, 33, 969.
572 L.W.Hayes, W.G.Krasselt, and P.A.Mueggler, Clin. Chem. (Winston-Salem, N.C.), 1987, 33, 806.
573 R.E.Struempler, J. Anal. Toxicol., 1987, 11, 97.
574 A.M.Zebelman, B.L.Troyer, G.L.Randall, and J.D.Batjer, J. Anal. Toxicol., 1987, 11, 131.
575 H.Quiding, P.Anderson, U.Bondesson, L.O.Boreus, and P.A.Hynning, Eur. J. Clin. Pharmacol., 1986, 30, 673.
576 G.Sticht, H.Kaeferstein, and M.Staak, Beitr. Gerichtl. Med., 1986, 44, 287.
577 G.R.Nakamura, W.J.Stall, and R.D.Meeks, J. Forensic Sci., 1987, 32, 535.
578 C.J.Weitz, L.I.Lowney, K.F.Faull, G.Feistner, ans A.Goldstein, Proc. Nat. Acad. Sci., 1986, 83, 9784.
579 M.Yashiki, F.West, and H. Brandenberger, Iyo Masu Kenkyukai Koenshu., 1985, 10, 99. (Chem. Abs., 1986, 104, 201691).
580 B.Pelli, P.Traldi, F.Tagliaro, G.Lubli, and M.Marigo, Biomed. Environ. Mass Spectrom., 1987, 14, 63.
581 H.Sachs and H.Brunner, Beitr. Gerichtl. Med., 1986, 44, 281.
582 B.Charles and J.-C.Tabet, Rapid Commun. Mass Spectrom., 1988 2, 86.
583 B.Silver and E.G.Sundholm, J. Pharm. Sci., 1987, 76, 53.
584 A.F.Casy, J. Pharm. Pharmacol., 1986, 38, 613.
585 C.C.Clark, J. Assoc. Off. Anal. Chem., 1986, 69, 814.
586 C.C.Clark, J. Forensic. Sci., 1987, 32, 917.
587 L.P.Lue, J.A.Scimeca, B.F.Thomas, and B.R.Martin, J. Anal. Toxicol., 1988, 12, 57.
588 M.Lambrechts, F.Tonnesen, and K.E.Rasmussen, J. Chromatogr., 1986, 369, 365.
589 A.C.Allen and W.O. Kiser, J. Forensic Sci., 1987, 32, 953.
590 S.-I.Suzuki, H.Tsuchihashi, K.Nakajima, A.Matsushita, and T.Nagao, J. Chromatogr., 1988, 437, 322.
591 I.A.Low, R.H.Liu, M.G.Legendre, E.G.Piotrowski, and R.L.Furner, Biomed. Environ. Mass Spectrom., 1986, 13, 531.
592 S.M.Hayes, R.H.Liu, W.-S.Tsang, M.G.Legendre, R.J.Berni, D.J.Pillion, S. Barnes, and M.H.Ho, J. Chromatogr., 1987, 398, 239.
593 T.Kojima, I.Okamoto, T.Miyazaki, F.Chikasue, M.Yashiki, and K.Nakamura, Forensic Sci. Int., 1986, 31, 93.
594 R.C.Baselt and R.Chang, J. Anal. Toxicol., 1987, 11, 81.
595 M.A.ElSohly, D.F.Stanford, and H.N.ElSohly, J. Anal. Toxicol., 1986, 10, 256.
596 L.K.Thompson, D.Yousefnejad, K.Kumor, M.Sherer, and E.J.Cone, J. Anal. Toxicol, 1987, 11, 36.
597 K.M.Giacomini, W.L.Nelson, R.A.Pershe, L.Valdivieso, K.Turner-Tamiyasu, and T.F.Blaschke, J. Pharmacokinet. Biopharmaceut., 1986, 14, 335.
598 L.D.Heimark, M.Gibaldi, W.F.Trager, R.A.O'Reilly, and D.A.Goulart, Clin. Pharmacol. Ther., 1987, 42, 388.
599 L.D.Heimark, S.Toon, M.Gibaldi, W.F.Trager, R.A.O'Reilly, and D.A.Goulart, Clin. Pharmacol. Ther., 1987, 42, 312.
600 R.A.O'Reilly, W.F.Trager, A.E.Rettie, and D.A.Goulart, Clin. Pharmacol. Ther., 1987, 42, 290.
601 J.W.Massarella, T.Silvestri, F.DeGrazia, B.Miwa, and D.Keefe, J. Clin. Pharmacol., 1987, 27, 187.
602 R.Mannhold and R.Kaufmann, Arch. Pharm. (Weinheim, Ger.), 1986, 319, 1028.
603 L.F.T.Chan, L.H.Chhuy, and R.J.Crowley, J. Anal. Toxicol., 1987, 11, 171.
604 P.Anderson, U.Bondesson, and U.De Faire, Eur. J. Clin. Pharmacol., 1986, 31, 155.
605 T.-M.Chen, J.E.Coutant, A.D.Sill and R.R.Fike, J. Chromatogr., 1987, 396, 382.
606 F.T.Delbeke and M.Debackere, J. Chromatogr., 1987, 416, 443.
607 A.I.Cohen, M.Jemal, E.Ivashkiv, and M.Ribick, J. Chromatogr., 1987, 416, 445.
608 O.Gyllenhaal and J.Vessman, J. Chromatogr., 1988, 435, 256.
609 V.Mok, L.V.Bui, and L.T.F.Chan, J. Chromatogr., 1987, 393, 335.
610 H.Maurer and K. Pfleger, J. Chromatogr., 1986, 382, 147.

611 F.T.Delbeke, M.Debackere, N.Desmet, and F.Maertens, J. Chromatogr., 1988, 426, 194.
612 L.S.Olanoff, T.Walle, T.D.Cowart, K.Walle, M.J.Oexmann, E.C.Conradi, Clin. Pharmacol. Ther., 1986, 40, 408.
613 T.C.Fagan, T.Walle, M.J.Oxemann, U.K.Walle, S.A.Bai, and T.E.Gaffney, Clin. Pharmacol. Ther., 1987, 41, 402.
614 T.Walle, U.K.Walle, L.S.Olanoff, and E.C.Conradi, Br. J. Clin. Pharmacol., 1986, 22, 317.
615 T.Walle, U.K.Walle, T.D.Cowart, E.C.Conradi, and T.E.Gaffney, J. Pharmacol. Exp. Ther., 1987, 241, 928.
616 T.H.Pringle, R.J.Francis, P.B.East, and R.G.Shanks., Br. J. Clin. Pharmacol., 1986, 22, 527.
617 S.Lindberg, P.Lundborg, C.G.Regardh, and B.Sandstrom, Eur. J. Clin. Pharmacol., 1987, 33, 363.
618 T.Niwa, Y.Tokuma, K.Nakagawa, H.Noguchi, Y.Yamazoe, and R.Kato, Res. Commun. Chem. Pathol. Pharmacol., 1988, 60, 161.
619 G.Mikus, C.Fischer, B.Heuer, C.Langen, and M.Eichelbaum, Br. J. Clin. Pharmacol., 1987, 24, 561.
620 M.R.Lawrence and F.B.Pipkin, Br. J. Clin. Pharmacol., 1987, 23, 683.
621 M.Terakawa, Y.Tokuma, A.Shishido, K.Yasuda, and H.Noguchi, J. Clin. Pharmacol., 1987, 27, 293.
622 M.Terakawa, Y.Tokuma, N.Kuwahara, A.Shishido, and H.Noguchi, J. Clin. Pharmacol., 1988, 28, 350.
623 Y.Takahashi, I.Hirako, H.Tamura, C.Shioyama, and H.Kondo, Iyakuhin Kenkyu, 1985, 16, 1112. (Chem. Abs., 1986, 104, 95355).
624 G.Knerr, J.I.McKenna, D.A.Quincy, and N.R.Natale, J. Heterocycl. Chem., 1987, 24, 1429.
625 J.D.Ehrhardt and J.M.Ziegler, Biomed. Environ. Mass Spectrom., 1988, 15, 525.
626 R.Isobe, T.Komori, F.Abe, and T.Yamauchi, Biomed. Environ. Mass Spectrom., 1986, 13, 585.
627 J.R.J.Pare, P.Lafontaine, J.Belanger, W.-W.Sy, N.Jordan, and J.C.K.Loo, J. Pharmaceut. Biomed. Anal., 1987, 5, 131.
628 R.E.Shomo, II, A.Chandrasekaran, A.G.Marshall, R.H.Reuning, and L.W.Robertson, Biomed. Environ. Mass Spectrom., 1988, 15, 295.
629 H.M.Yang, H.A.Lloyd, L.K.Pannell, H.M.Fales, R.D.MacFarlane, C.J.McNeal, and Y.Ito, Biomed. Environ. Mass Spectrom., 1986, 13, 439.
630 K.L.Duchin, D.N.McKinstry, A.I.Cohen, and B.H.Migdalof, Clin. Pharmacokinet., 1988, 14, 241.
631 M.Jemal, S.Black, and A.I.Cohen, Rapid. Commun. Mass Spectrom., 1987, 1, 129.
632 P.Graf, F.Frueh, and K.Schmid, J. Chromatogr., 1988, 425, 353.
633 O.H.Drummer, B.S.Workman, P.J.Miach, B.Jarrott, and W.J.Louis, Eur. J. Clin. Pharmacol., 1987, 32, 267.
634 I.De Lepeleire, A.Van Hecken, R.Verbesselt, G.Kaiser, A.Barner, I.Holmes, and P.J.De Schepper, Eur. J. Clin. Pharmacol., 1988, 34, 465.
635 R.Tembreull and D.M.Lubman, Anal. Chem., 1987, 59, 1082.
636 R.F.Suckow and T.B.Cooper, J. Chromatogr., 1988, 427, 287.
637 L.Bertilsson, C.Nordin, K.Otani, B.Resul, M.Scheinin, B.Siwers, and J. Sjoqvist, Clin. Pharmacol. Ther., 1986, 40, 261.
638 C.Nordin, L.Bertilsson, and B.Siwers, Clin. Pharmacol. Ther., 1987, 42, 10.
639 G.R.Jones and D.J.Pounder, J. Anal. Toxicol., 1987, 11, 186.
640 G.C.DiDonato and K.L.Busch, Anal. Chem., 1986, 58, 3231.
641 M.S.Stanley and K.L.Busch, Anal. Chim. Acta, 1987, 194, 199.
642 R.A.Flurer and K.L.Busch, Org. Mass Spectrom., 1988, 23, 118.
643 L.Bertilsson, T.Tomson, and G.Tybring, J. Clin. Pharmacol., 1986, 26, 459.
644 G.Kramer, M.Theilsohn, G.E.Von Unruh, and M.Eichelbaum, Therap. Drug Monitoring, 1986, 8, 387.
645 G.Maksay, Z.Tegyey, and L.Otuos, J. Chem. Soc., Perkin Trans. 2., 1988, 57.
646 H.Maurer and K.Pfleger, J. Chromatogr., 1987, 422, 85.
647 W.-G.Chai, G.-H.Wang, S.Jin, Z.-M.Lin, and P.Lau, Org. Mass Spectrom., 1987, 22, 660.

648 E.A.Stemmler, R.A.Hites, B.Arbogast, W.L.Budde, M.L.Deinzer, R.C.Dougherty, J.W.Eichelberger, R.L.Foltz, C.Grimm, E.P.Grimsrud, C.Sakashita, and L.J.Sears, Anal. Chem., 1988, 60, 781.
649 L.Sangameswaran, H.M.Fales, P.Friedrich and A.L.DeBlas, Proc. Natl. Acad. Sci. USA, 1986, 83, 9236.
650 M.Ito, K.Aizawa, I.Komiya, K.Miyagi, M.Fujita, and K.Umemura, Yakugaku Zasshi., 1986, 106, 703.
651 R.G.Sans and M.G.Chozas, Pharmazie., 1988, 43, 415.
652 M.Van der Graaff, N.P.E.Vermeulen, P.H.Hofman, and D.D.Breimer, Biochem. Pharmacol., 1987, 36, 1321.
653 F.S.Abbott, J.Kassam, J.M.Orr, and K.Farrell, Clin. Pharmacol. Ther., 1986, 40, 94.
654 F.Hoffmann, B.Ch.Jancik, and G.E.Von Unruh, Arzneim.-Forsch., 1986, 36, 1118.
655 S.J.Gaskell, C.Guenat, D.S.Millington, D.A.Maltby, and C.R.Roe, Anal. Chem., 1986, 58, 2801.
656 R.Libert, F.Van Hoof, A.Schanck, and E.Hoffmann, Biomed. Environ. Mass Spectrom., 1986, 13, 599.
657 R.G.Dickinson, B.T.Wood, R.M.Kluck, and W.D.Hooper, Therap. Drug Monitoring, 1986, 8, 462.
658 M.Paganini, G.Zaccara, F.Moroni, R.Campostrini, L.Bendoni, G.Arnetoli, and R.Zappoli, Eur. J. Clin. Pharmacol., 1987, 32, 219.
659 T.Kondo, K.Otani, T.Hirando, and S.Kaneko, Br. J. Clin. Pharmacol., 1987, 24, 401.
660 P.Anderson, U.Bondesson, I.Mattiasson, and B.W.Johansson, Eur. J. Clin. Pharmacol., 1987, 31, 625.
661 P.Bulau, W.D.Paar, and G.E.Von Unruh, Eur. J. Clin. Pharmacol., 1988, 34, 311.
662 R.J.Lokan and A.C.Dinan, J. Anal. Toxicol., 1988, 12, 35.
663 K.Manada, Y.Kasuya, and S.Baba, Drug Metab. Dispos., 1986, 14, 509.
664 J.H.Poupaert, P.Guiot, and P.Dumont, J. Labelled Comp. Radioparm., 1988, 25, 43.
665 T.R.Browne, D.J.Greenblatt, J.E.Evans, G.K.Szabo, B.A.Evans, and G.E. Schumacher, J. Clin. Pharmacol., 1987, 27, 318.
666 T.R.Browne, D.J.Greenblatt, J.E.Evans, B.A.Evans, G.K.Szabo, and G.E. Schumacher, J. Clin. Pharmacol., 1987, 27, 321.
667 Y.Cherrah, J.B.Falconnet, M.Desage, J.L.Brazier, R.Zini, and J.P.Tillement, Biomed. Environ. Mass Spectrom., 1987, 14, 653.
668 Y.Cherrah, R.Zini, J.B.Falconnet, M.Desage, J.P.Tillement, and J.L.Brazier, Biochem. Pharmacol., 1988, 37, 1311.
669 D.Juan, E.M.Worwag, D.A.Schoeller, A.N.Kotake, R.L.Hughes, and M.C. Frederiksen, Clin. Pharmacol. Ther., 1986, 40, 187.
670 B.Mompon, D.Loyaux, E.Kauffmann, and A.M.Krstulovic, J. Chromatogr., 1986, 363, 372.
671 S.Zbaida, R.Kariv, P.Fischer, and D.Gilhar, Xenobiotica, 1987, 17, 617.
672 A.Mozayani, R.T.Coutts, and T.J.Danielson, J. Chromatogr., 1987, 423, 131.
673 M.Lichtenwainer, M.McMullin, D.Hardy, and F.Rieders, J. Anal. Toxicol., 1988, 12, 98.
674 H.Hattori, O.Suzuki, and H.Brandenberger, J. Chromatogr., 1986, 382, 135.
675 A.M.Bougerolle, J.L.Chabard, G.Dordain, A.Frydman, J.Gaillot, J.J.Piron, J.Petit, and J.A.Berger, Eur. J. Drug Metab. Pharmacokin., 1986, 11, 113.
676 A.G.Mallinger, D.J.Edwards, J.M.Himmelhoch, S.Knopf, and J.Ehler, Clin. Pharmacol. Ther., 1986, 40, 444.
677 M.Grind, B.Siwers, C.Graffner, G.Alvan, L.L.Gustafsson, J.Helleday, J.E. Lindgren, S.Ogenstad and H.Selander, Clin. Pharmacol. Ther., 1986, 40, 155
678 G.Alvan, C.Graffner, M.Grind, L.L.Gustafsson, J.E.Lindgren, C.Nordin, S.Ross, H.Selander, and B.Siwers, Clin. Pharmacol. Ther., 1986, 40, 81.
679 L.O.G.Weidolf, E.D.Lee, and J.D.Henion, Biomed. Environ. Mass Spectrom., 1988, 15, 283.
680 P.R.Das, B.N.Pramanik, R.D.Malchow, and K.J.Ng, Biomed. Environ. Mass Spectrom., 1988, 15, 253.
681 J.M.Midgley, D.G.Watson, T.Healey, and M.Noble, J. Pharm. Pharmacol., 1987, 39, 51P.

682 D.Watson, M.J.Noble, G.N.Dutton, J.M.Midgley, and T.M.Healey, Arch. Opthalmol., 1988, 106, 686.
683 L.E.Golubovskya, E.N.Korenchuk, K.K.Pivnikskii, N.A.Kuzovkova, and N.D. Franchenko, Probl. Endokrinol., 1985, 31, 72.
684 K.Williams, R.Day, R.Knihinicki, and A.Duffield, Biochem. Pharmacol., 1986, 35, 3403.
685 S.N.Sawhney, S.Bhutani, and D.Vir, Org. Mass Spectrom., 1987, 22, 477.
686 M.He, H.Tang, and M.Zhang, Org. Mass Spectrom., 1988, 23, 145.
687 M.G.Quaglia, G.Carlucci, G.Cavicchio, and P.Mazzeo, J. Pharmaceut. Biomed. Anal., 1988, 6, 421.
688 J.Hansen-Moller, L.Dalgaard and S.Honore-Hansen, J. Chromatogr., 1987, 420, 99
689 M.Dawson, C.M.McGee, J.H.Vine, P.Nash, T.R.Watson, and P.M.Brooks, Eur. J. Clin. Pharmacol., 1988, 33, 639.
690 K.-J.Hoffmann and T.A.Baillie, Biomed. Environ. Mass Spectrom., 1988, 15, 637.
691 A.R.Boobis, S.Murray, C.J.Speirs, and D.S.Davies, Br. J. Clin. Pharmacol., 1987, 23, 643P.
692 R.T.Coutts, G.A.Torok-Both, Y.K.Tam, L.V.Chu, and F.M.Pasutto, Biomed. Environ. Mass Spectrom., 1987, 14, 173.
693 A.G.L.Burm, A.G.de Boer, J.W.van Kleef, N.P.E.Vermeulen, L.G.J.de Leede, J.Spierdijk, and D.D.Breimer, Biopharmaceut. Drug Dispos., 1988, 9, 85.
694 A.G.L.Burm, N.P.E.Vermeulen, J.W.van Kleefe, A.G.de Boer, J.Spierdijk, and D.D.Breimer, Clin. Pharmacokinet., 1987, 13, 191.
695 L.Y.Leung and T.A.Baillie, J. Med. Chem., 1986, 29, 2396.
696 J.Jacobsen, H.Flachs, J.O.Dich-Nielsen, J.Rosen, A.B.Larsen, and E.F.Hvidberg, Br. J. Aneasth., 1988, 60, 623.
697 W.Lintz, H.Barth, G.Osterloh, and E.Schmidt-Bothelt, Arzneim.-Forsch., 1986, 36, 1278.
698 J.L.Gabrielsson, P.Johansson, U.Bondesson, M.Karlsson, and L.K.Paalzow, J. Pharmacokinet. Biopharmaceut., 1986, 14, 381.
699 W.A.Korfmacher, C.L.Holder, C.E.Cerniglia, D.W.Miller, E.B.Hansen, K.L. Lambert, A.B.Gosnell, W.Slikker, L.G.Rushing, and H.C.Thompson, Spectrosc. Int. J., 1985, 4, 181.
700 H.Maurer and K.Pfleger, J. Chromatogr., 1988, 428, 43.
701 R.J.Wills, H.E.Gallo-Torres, R.Bertko, and B.H.Min, J. Pharmacol. Exp. Ther., 1987, 241, 433.
702 A.F.Casy, J. Pharmaceut. Biomed. Anal., 1987, 5, 247.
703 T.Cairns, E.G.Siegmund, J.J.Stamp, T.L.Barry, and G.Petzinger, Rapid Commun. Mass Spectrom., 1988, 2, 25.
704 A.Selva and P.Ventura, Adv. Mass Spectrom. 1985, 1986, 751.
705 J.O.Naim, D.M.Desiderio, J.Trimble, and J.R.Hinshaw, Anal. Biochem., 1987, 164, 221.
706 J.L.Gower, M.J.Redrup, I.O'Brien, and J.A.E.Pratt, Adv. Mass Spectrom, 1985, 1986, 589.
707 B.Hallen, O.Guilbaud, S.Stromberg, and B.Lindeke, Biopharmaceut. Drug Dispos., 1988, 9, 229.
708 K.D.Haegele and P.J.Schechter, Clin. Pharmacol. Ther., 1986, 40, 581.

# 10
# Metal-containing and Inorganic Compounds Investigated by Mass Spectrometry

BY J. CHARALAMBOUS AND K. W. P. WHITE

## 1 Introduction

The material reviewed in this report appeared in articles in the period July 86 - June 88, and concerned the mass spectral behaviour of metallic and inorganic compounds. The format employed in the last review[1] has largely been retained - only selected references are given in each section. During the period under review mass spectrometry has been used for the study of a wide range of metallic and inorganic compounds. The vast majority of applications have involved EI techniques, and have been concerned simply with compound characterization. An extensive review dealing with the applications of FAB spectrometry to organometallic chemistry and coordination chemistry,[2] and a book[3] relating to various aspects of the mass spectrometric behaviour of metal-containing and inorganic compounds have been published in the period under review.

## 2 Main-group organometallic compounds

EI studies of the compounds RLi (R = Me, $Pr^n$, $Pr^i$, $Bu^n$, $Bu^t$, $Bu^i$, or $Bu^{sec}$) have been carried out in the range 10-70 eV.[4] The spectra show mass peaks corresponding to ions with the general formulae $[R_{n-1}Li_n]^+$ and $[HR_{n-2}Li_n]^+$, and the value of n being dependent on the compound and ion type. The greater stability of the $[R_{n-1}Li_n]^+$ ions relative to the $[HR_{n-2}Li_n]^+$ ions is reflected in the considerably higher relative abundance of the former. Stable molecular ions are not detected for any species. This is a consequence of the high polarity of the Li-C bonds in alkyllithium compounds. A decrease in electron energy results in characteristic changes of the mass spectra. RLi (R = Me, $Pr^i$, $Bu^t$, or $Bu^{sec}$) are composed only of tetramers, whereas in the gas-phase RLi (R = $Pr^n$, $Bu^n$, or $Bu^i$) are mixtures of tetramers and hexamers.

The EI spectra of the carboranyl-mercury compounds $(9\text{-}C_2H_{11}B_{10})HgX$ (X = Cl, Br, or I), $(9\text{-}C_2H_{11}B_{10})_2Hg$, and $(C_2H_{11}B_{10}C\text{-})_2Hg$ have been studied.[5] The fragmentation patterns obtained suggest that the strengths of the Hg-B and Hg-C bonds increase on transition from o- to m- and p-isomers, and the Hg-B bond is stronger than the Hg-X bond. EI and CI techniques have both been used for the identification of the compounds $(Me_3Si)_3CSnR_2F$ (R = Me or Ph),[6] and of oligosilazanes, such as (1).[7]

The EI spectra of organoderivatives of 2,3-disubstituted 1,3-butadienes of types (2) and (3) (Met = Si, Ge, Sn, or Pb) have been reported.[8] These compounds show molecular ions and peaks for $[Me_3Met]^+$, $[Me_2Met]^{+\cdot}$, $[MeMet]^+$ and the base metal ion. An important fragmentation consists of the expulsion of $Me_4Met$ from the molecular ion.

Mass spectrometry has been used extensively for the characterization of a wide range of organotin compounds. In-beam EI, isobutane CI and FD mass spectra for several such compounds, e.g. $(Bu_3Sn)_2O$, $(Bu_3Sn)_2S$ and $Ph_3SnOH$, have been obtained.[9] For these compounds, primary cleavage of the Sn-C, Sn-O, and Sn-S bonds takes place prior to the other reactions so that the analysis of their mass spectra is relatively simple. In the EI spectra of the organotin derivatives of 3-indolylacetic acid (IAAH) and N-methyl-3-indolylacetic acid, the molecular ions are either of low abundance or completely absent.[10] In contrast, in the isobutene CI spectra, the molecular ion is observed along with high abundances of high mass ions. In all of the spectra, the most abundant ion corresponds to $[IAA - CO_2]^{+\cdot}$. EI studies have confirmed the dimetallic character of the piperazine bis(dithiocarbamato) compounds (4; R = Me, Bu, c-$C_6H_{11}$, or Ph).[11] The EI and isobutane CI spectra of several 1,3-dithian-2-yltin compounds (5; $R^1$ = Me, $Bu^n$, or Ph; $R^2$ = H, Ph, or PhCH=CH) are characterized by a molecular ion of low abundance, and by ions formed via primary fragmentation due to cleavage of the Sn-C bonds.[12] This is diagnostic of the fragmentation of tetraalkyltin compounds by EI. The base peak in each spectrum is due to cleavage of the C(2)-Sn bond with retention of the positive charge on the dithiane species. EI and CI techniques have been used for the characterization of several organostannanes, e.g. (6)[13] and (7).[14] The ammonia DCI spectrum of the C-glycopyranosyl compound (8; R = $PhCH_2$) has the highest m/z value corresponding to the ion $[M - Bu]^+$.[15] The identification and characterization of trace amounts of organotin compounds in aqueous media has been accomplished using EI and/or CI techniques.[16,17] Molecular ions are absent in both the EI and FAB spectra of a series of hexamethylphosphoramide (HMPA) adducts of

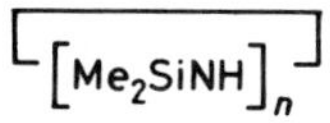
[Me2SiNH]n

(1)

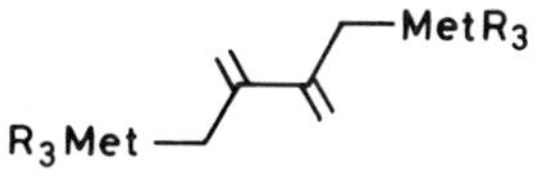
MetR3
R3Met

(2)

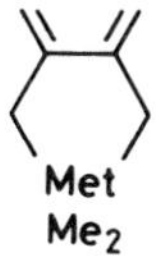
Met
Me2

(3)

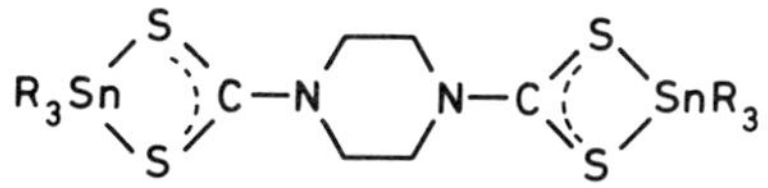
R3Sn
SnR3

(4)

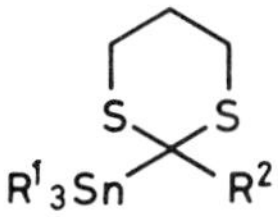
R¹3Sn
R²

(5)

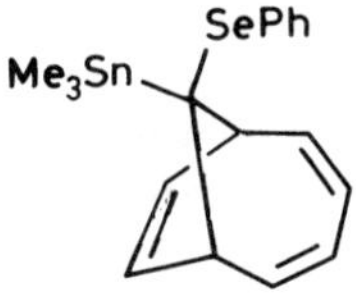
SePh
Me3Sn

(6)

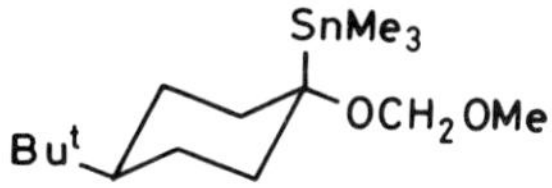
SnMe3
OCH2OMe
Bu^t

(7)

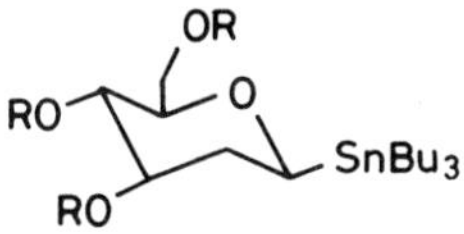
OR
RO
RO
SnBu3

(8)

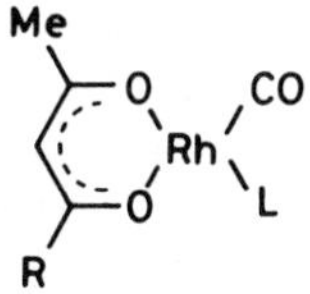
Me
Rh
CO
L
R

(9)

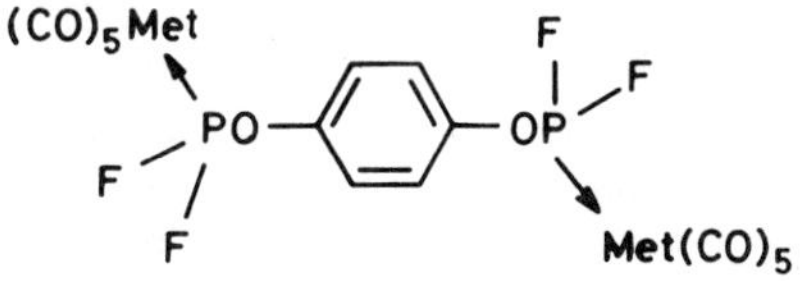
(CO)5Met
Met(CO)5

(10)

phenyltin(IV) halides [Ph$_3$SnX.HMPA (X = Cl, Br, or I); Ph$_2$SnX$_2$.HMPA (X = Br or I); Ph$_2$SnX$_2$.2HMPA (X = Br or I)] and phenyllead(IV) halides [Ph$_3$PbX.HMPA (X = Cl, Br, or I); Ph$_2$PbX$_2$.HMPA (X = Br or I) and Ph$_2$PbX$_2$.2HMPA (X = Cl, Br, or I)], but in the FAB spectra (glycerol/HMPA matrix) there is a much higher proportion of metal-containing ions which also have an HMPA molecule attached.[18] This is also the case even in a HMPA-free matrix such as 4-nitrophenyloctyl ether. The FAB spectra exhibit preferential loss of halide compared to phenyl, the reverse of that observed in the EI spectra. In a FAB spectrometric study of tris(4-chlorophenyl)- and tris(4-fluorophenyl)tin halides (X = Cl, Br, or I), the 4-nitrophenyl-**n**-octyl ether matrix employed was observed to coordinate weakly to the complex.[19] In contrast, for their monohexamethylphosphoramide adducts, little ligand displacement occurred. Fluoride and chloride migration to tin are observed in the FAB spectra for the free organometallics.

In the EI spectra of the tris(perfluoroorgano)bismuth compounds, BiR$_3$ (R = C$_2$F$_5$, **n**-C$_3$F$_7$, **n**-C$_4$F$_9$ and C$_6$F$_5$), no molecular ions are detectable, except in the case of the perfluorophenyl compound.[20] In all cases, abundant [BiR$_2$]$^+$ ions are present. Molecular ions are observed in the EI spectra of the compounds MeMet(SC$_6$H$_4$NH$_2$-4)$_2$ (Met = As or Bi).[21] The compound PhBi(SC$_6$H$_4$Cl-4)$_2$ shows mainly organic ions together with base metal ions and [PhBi]$^+$. Laser ionization TOF spectrometry has been sucessfully utilized for the detection of the compounds Me$_2$Met (Met = Se or Te).[22]

## 3 Transition-metal Organometallic Compounds

<u>Carbonyl and Related Compounds</u>.- Under EI conditions, the fragmentation pathways of the $\gamma$-oxoisocyanide complexes Met(CO)$_5$CNR$^1$R$^2$CH$_2$C(=O)R$^3$ (Met = Cr, Mo, or W; R$^1$-R$^3$ = Me; R$^1$, R$^2$ = H, R$^3$ = Me; R$^1$-R$^3$ = H) proceed via two main routes which involve loss of [MetCNH] or HCN respectively.[23] Low voltage EI mass spectra have been reported for the compounds (9) (R = CH$_3$, L = CO or Ph$_3$P; R = Ph, L = CO).[24] The fragmentation pathways were investigated using linked scan techniques, and were found to involve initial successive loss of the carbonyl groups which is then followed by cleavage of the other ligand bonds. This illustrates that the metal-CO bond is weaker than the others. The probe temperature was

shown to have a significant effect on the abundance of the
molecular ion. Abundant molecular ions were obtained below 200 °C.
At higher probe temperatures, the molecular ion was not observed
for the complex (9) (R = $CH_3$, L = CO). This was due to thermal
decomposition in the inlet system, and not related to the stability
of the complex itself after EI ionization. The EI mass spectra of
the compounds (10) (Met = Mo or W) and of their isomers derived
from 1,2- and 1,3-dihydroxybenzene show abundant molecular ions and
a sequence of ions formed by loss of up to ten CO groups.[25] The
mass spectra of mixtures of the complexes $Re(CO)_5X$ (X = Cl, Br, or
I) and ammonia show the molecular mass peaks $[Re(CO)_5X]^+$ and ions
arising by partial or complete decarbonylation $[Re(CO)_nX]^+$ (n = 3
or 4), $[ReX]^+$ and $[Re]^+$ as well as $[Re(CO)_3NH_3]^+$ and $[ReNH_3X]^+$.[26]
The formation of ammonia-containing ions proceeds by substitution
of the CO groups of the carbonyl-containing rhenium ions by an
ammonia molecule.

The CI spectrum of the compounds $(CO)_5MnMetPh_3$ (Met = Si, Ge,
or Sn) have been investigated using $CH_4$, $i$-$C_4H_{10}$, $NH_3$ and $H_2$ as
reactant gases.[27] The reactivity of the substrate towards the
reactant gas increases on going from the silicon to the tin
derivative, and for a given substrate methane is the most reactive
gas. The two most important processes of the protonated sample are
those which lead to the ions $[M - Ph]^+$ and $[MetPh_3]^+$, and are the
same as those observed in the condensed phase. The mass spectra
obtained with $CH_4$, $i$-$C_4H_{10}$, or $H_2$ are strongly influenced by the
sample temperature in the evaporation probe. An increase in this
temperature causes an increase in the relative abundance of the ion
$[M - Ph]^+$. The reversibility of this effect confirms that it is not
due to decomposition of the sample in the solid phase.

Electron ionization has been used for the characterization of
the complex [bis(dicyclohexylphosphino)ethane]platinum(0) and of
the related complex [bis(dicyclohexylphosphino)ethane](diphenyl-
acetylene)platinum(0), both of which show fairly abundant molecular
ions.[28] This technique has also been used for the characterization
of several phosphole complexes of type $(CO)_5WL$ (L = phosphole, e.g.
(11) or (12)),[29] of iridium(I) and iridium(III) pyrazolato
complexes such as (13) and (14),[30] and of various o-quinonedimethane complexes of type $(CO)_3FeL$ (L = o-quinonedimethane, e.g.
(15)).[31] For the characterization of the latter, methane CI has
also been employed. Several $RPCH_2PR$-bridged $Fe_2(CO)_6$ complexes,

(11), (12), (13), (14), (15), (16), (17), (18), (19), (20)

e.g. (16), have been examined with FD techniques,[32] and EI spectrometry has been used to investigate the gas-phase photochemistry of the complex $Mn_2(CO)_{10}$.[33]

<u>Complexes containing hydrocarbon ligands</u>.- Electron ionisation mass spectrometry has been used extensively for the characterization of various alkyl, aryl, and vinyl $\eta^1$ complexes, e.g (17),[34] (18; R = Et, Pr, or $(CH_2)_4CH_3$),[35] and (18; R = $CH=CH_2$ or $CH=CHCH_3$).[36] In compounds (18; R = alkyl) rearrangement of a $\beta$-hydrogen from the alkyl ligand onto the metal produces the rhenium hydride $[(\eta^5-C_5H_5)Re(NO)(Ph_3P)(H)]^+$ which subsequently fragments by loss of benzene.[35] This behaviour has also been observed in the spectrum of the hydride ($\eta^5-C_5H_5)Re(NO)(PPh_3)(H)$.[35] The positive-ion FAB mass spectra of the complexes (19) and (20) (Y = $Mn(CO)_3(Ph_2PCH_2CH_2PPh_2)$ or $Fe(CO)_2(\eta-C_5H_5)$) show features common to conventional EI spectra of similar complexes.[37] Thus, ready stepwise loss of the CO groups is followed by loss of CN or Ph groups. The base peaks for (19) and (20) (Y = $Mn(CO)_3(Ph_2PCH_2CH_2PPh_2)$) corresponds to the ion $[Mn(CN)(Ph_2PCH_2CH_2PPh_2)]^+$, formed by CN transfer to the metal, and are reminiscent of similar F-atom transfer reactions found for fluorocarbon complexes. In addition, cleavage of the cyanocarbon ligand affords the ion $[Mn(CO)_3(Ph_2PCH_2CH_2PPh_2)]^+$, which apparently loses the three CO groups simultaneously to give the ion $[Mn(Ph_2PCH_2CH_2PPh_2)]^+$. With both iron complexes, loss of $C_2(CN)_4$ from $[M - 2CO]^+$ is an important process, but this fragmentation is only found in the Mn-cyclobutenyl complex. Fragmentation of coordinated $Ph_2PCH_2CH_2PPh_2$ involves loss of $C_2H_4PPh_2$ which generates a diphenylphosphido ligand. The negative-ion spectra of (19) and (20) (Y = $Mn(CO)_3(Ph_2PCH_2CH_2PPh_2)$ are characterized by the ready loss of CO and $Ph_2PCH_2CH_2PPh_2$; ions of the type $[MnC_2PhC_2(CN)_4]^-$ and their fragmentation products are particularly stable, as is $[Mn(PPh_2)]^-$. The FAB spectrum of the $\sigma$-acetylide compounds such as (21) and (22) contain ions corresponding to the cation which fragment by loss of $PPh_3$ and of the aryldiazo moiety.[38] Similarly, the FAB spectrum of the complexes (23; Met = Fe or Ru, R = Ph, $L_2 = Ph_2PCH_2CH_2PPh_2$; Met = Ru, R = $C_6F_5$, L = $PPh_3$) show ions due to the cation which fragments by competitive loss of the two vinylidene substituents.[39] In these spectra, the ions at highest m/z value correspond to the cation with a molecule of the matrix attached.

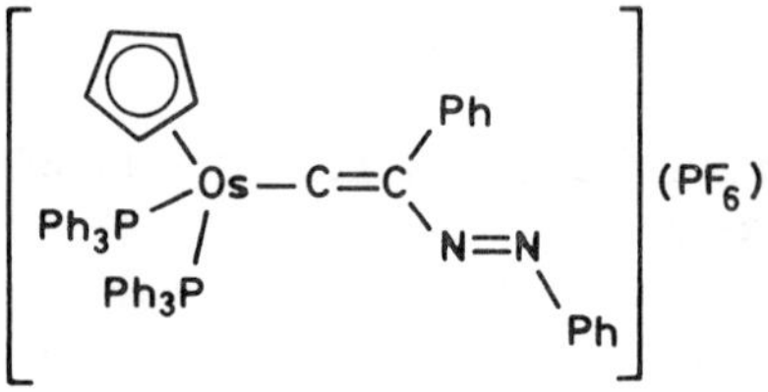

(21)

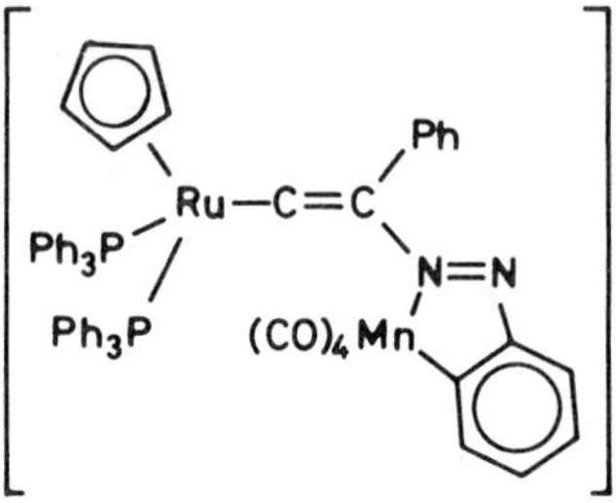

(22)

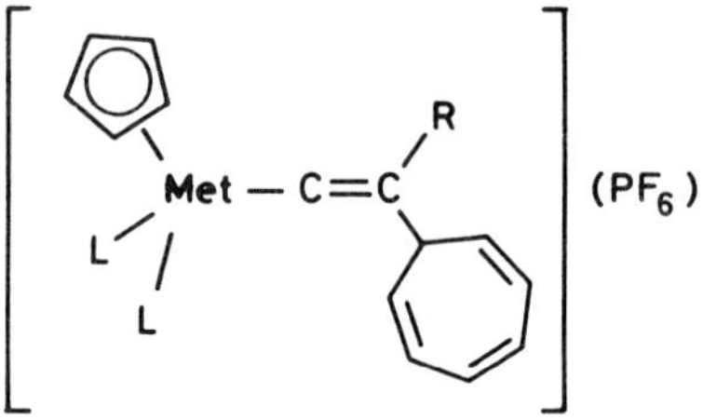

(23)

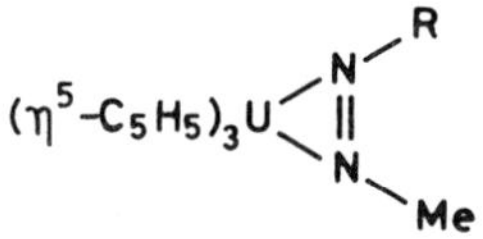

(24)

(25)

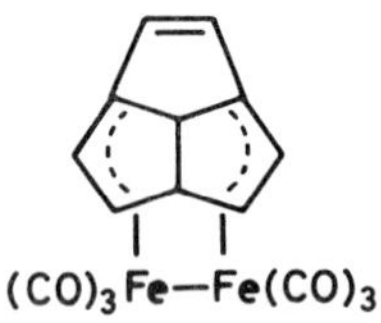

(26)

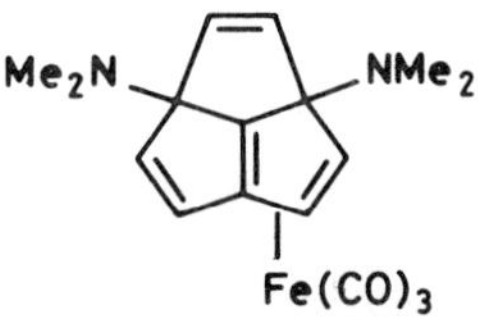

(27)

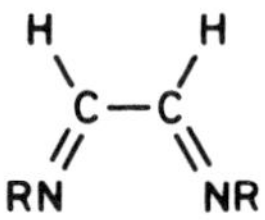

(28) R = Pr$^i$, Bu$^t$, or C$_6$H$_{11}$

Molecular ions have been observed in the EI spectra of the $\eta^2$-iminoalkyl complexes (24; R = Et, Bu$^n$, or c-C$_6$H$_{11}$).[40] For R = C$_6$H$_{11}$, linked scans and MIKE spectra showed that the primary fragmentation processes involve loss of the species C$_5$H$_5$˙, H˙, C$_6$H$_{11}$˙, C$_8$H$_{14}$N˙ and C$_8$H$_{13}$N. The last route involves a novel hydrogen rearrangement which has been accounted for in terms of the formation of tricyclopentadienyluranium(IV) hydride. This species has been detected occasionally, but never isolated in the condensed phase chemistry of tricyclopentadienyl(IV) compounds. EI mass spectrometry has been used to characterize the complexes (25; X = O or CH$_2$) and other related compounds.[41] This technique is ineffective in studying the carbene complex [Ir(PhC=NHMe)Cl(CO)(PPh$_3$)$_2$(O$_3$SCF$_3$)]O$_3$SCF$_3$, but its FAB spectrum shows a large peak corresponding to the cation.[42] Hydridotris(pyrazolyl)borate (Tp) complexes of the type TpMo(CO)$_2$($\eta^2$-COMe) and TpMo(CO)($\eta^2$-COMe)L (L = P(OMe)$_3$ or PEt$_3$) have been characterized by methane CI mass spectrometry.[43]

The EI spectra of $\eta^3$-allyl-$\eta^5$-cyclopentadienyltitanium(III) compounds ($\eta^5$-C$_5$H$_5$)$_2$TiR (R = allyl, 1-methylallyl, 2-methylallyl, 1,3-dimethylallyl, and 1,1-dimethylallyl) generally show molecular ions of low abundance which decompose via two routes involving loss of R· and (C$_5$H$_5$)$_2$Ti respectively.[44] The major metallic ions are formed by loss of C$_5$H$_5$ or C$_2$H$_2$ from (C$_5$H$_5$)$_2$Ti, and [TiC$_3$H$_3$]$^+$ is formed from [C$_5$H$_5$TiC$_3$H$_3$]$^+$ by elimination of C$_5$H$_5$. Several iron carbonyl complexes containing $\eta^3$ or $\eta^4$ bonded dihydroacepentalene ligands, e.g. (26) and (27) have been studied using EI or isobutane CI spectrometry.[45] Mass analysis of the 18-electron mixed isocyanide-1,4-diaza-1,3-butadiene (DAB) complexes [Cr(CO)$_3$(CNR)(DAB)] (R = Me, CHMe$_2$, CMe$_3$, or xylyl; DAB = 28) was carried out using SIMS, EI, PICI and NICI mass spectrometry.[46] With the exception of the PICI spectra, abundant molecular ions were not present. In all cases, well-defined, structurally informative fragment ions were identified which arise by loss of CO and/or the intact RNC and DAB ligands. Cleavage of the RNC ligands (dealkylation) occurs both in the SIMS and EI spectra to give fragment ions containing the species [Cr(CNH)]$^+$, whereas only in the NICI spectra are fragments observed in which the cyano moiety [Cr(CN)]$^-$ is present. High-abundance carbonyl-containing fragments are seen only in the case of the CI spectra and are most common in the NICI measurements. This difference arises because the CO

ligands are very effective in delocalizing the negative charge
(through Cr-to-CO π-backbonding) and, as a consequence, CO-
containing fragments are much more stable. The EI and methane CI
spectra of *trans*-μ-(2,3,5,6-tetramethylidenebicyclo[2.2.2]octane)-
bis(tricarbonyliron) complexes, e.g. (29)[47] and the FAB spectrum of
the rhodium complex (30) have been reported.[48] EI and CI techniques
have been used for the characterization of several complexes
containing $\eta^4$ or $\eta^4, \eta^2$ bonded silole ligands, e.g. (31) and (32).[49]

Prominent molecular ions have been observed in the EI spectra
of a wide range of metallocene metallocycles (33; E = S, R = H or
Me, Met = Ti;[50] E = Se, R = H, Met = Ti, Zr, Hf, Mo, or W;[51] E =
Se, R = Me, Met = Ti;[51] E = Te, R = H, Met = Ti[52]). The spectrum of
the analogous compound (33; E = O, R = H, Met = Ti)[53] shows a
molecular ion of low abundance, but in this case, ions due to the
dimeric compound [($C_5H_5$)$_2$Ti($O_2C_6H_4$-o)]$_2$ are also present. Linked
scan studies have been used to establish the fragmentation
behaviour of all the metallocycles. Prominent decomposition
reactions involve loss of intact cyclopentadienyl and $C_6H_4E_2$
ligands, but loss of species such as HE· and $H_2E$ are also observed.
Linked scan techniques have also demonstrated that the
fragmentation behaviour of the titanocene derivatives (34; X = Cl,
Br, or I) and (35; E = C, Si, or Ge) is highly influenced by the
cyclopentadienyl ring substituents and shows several interesting
rearrangements (e.g. Reactions 1 and 2).[54] A comparative study of
the EI and methane NICI mass spectra of a large number of
derivatives of dicyclopentadienyl titanium(IV) dichloride
($\eta^5$-RC$_5$H$_4$)$_2$TiCl$_2$ (R = e.g. Me, Et, Pr$^n$, or Bu$^n$) has shown that in
the EI spectra, most of the molecular ion peaks are not
discernible, whereas the NICI spectra exhibit large molecular ion
peaks and a series of structurally informative fragment ions.[44]
Several cyclopentadienyl complexes of vanadium(II),[55] e.g. ($C_5H_5$)$_2$V
and ($C_5H_5$)($C_5H_7$)VL (L = CO or PEt$_3$), and the complexes
($C_5H_5$)$_2$MetL(XCH=NPh) (Met = Nb or Ta; X = O or S; L = CO)[56] have
been charaterized by EI mass spectrometry. The EI spectra of the
alkylcyclopentadienyl carbonyl complexes R$_n$C$_5$H$_{5-n}$Met(CO)$_3$ (R = Me,
n = 0-5; R = Bu$^t$, n = 1; Met = Re[57] or Mn[58]) show molecular ions
which decompose by the initial loss of CO. In general, the spectra
of the rhenium complexes exhibit a much greater number of metal-
containing ions than their manganese analogues. This feature was
accounted for in terms of the greater strength of the Re-C$_5$H$_{5-n}$ and

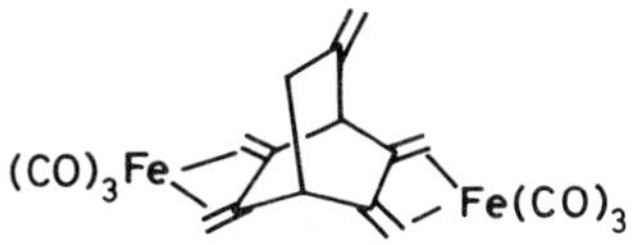

10: *Metal-containing and Inorganic Compounds Investigated by Mass Spectrometry* 333

(29)

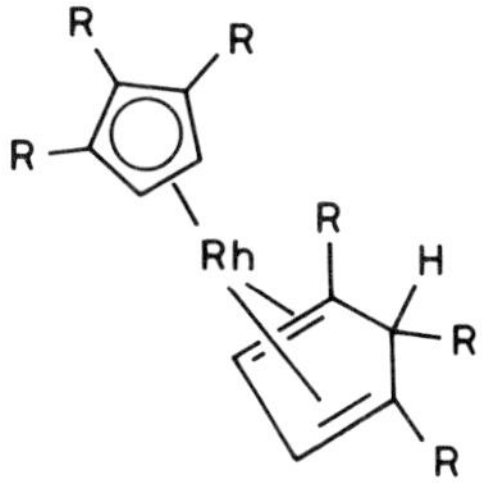

(30) R = CO$_2$Me

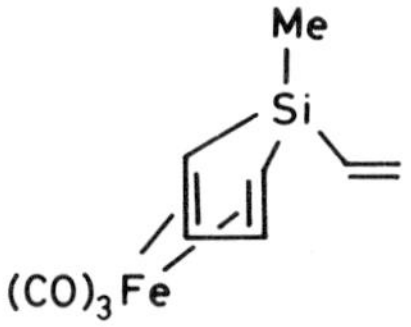

(31)

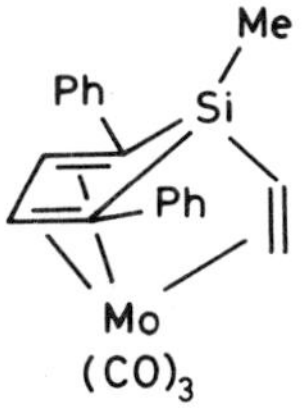

(32)

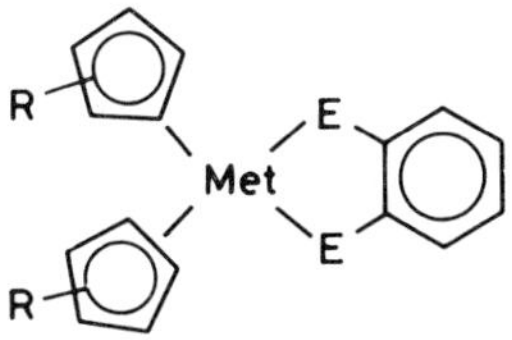

(33)

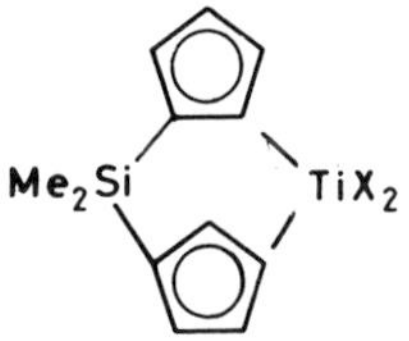

(34)

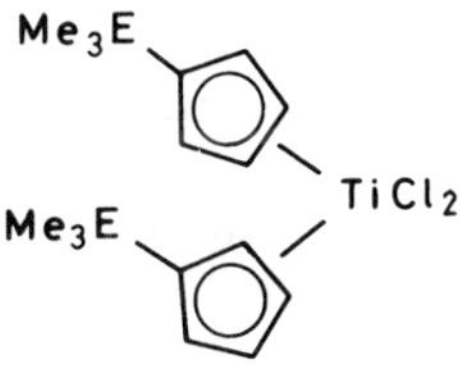

(35)

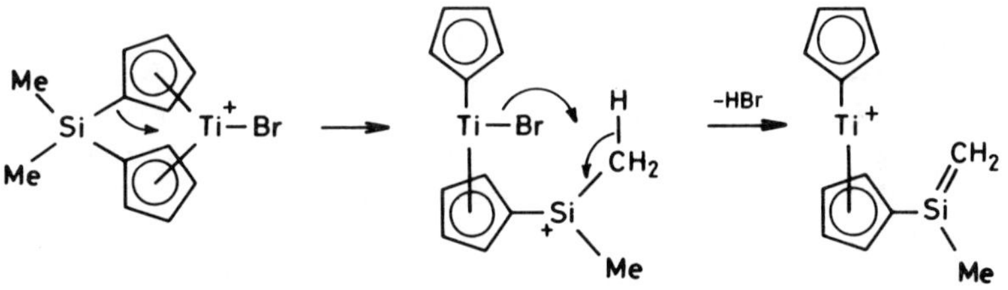

(1)

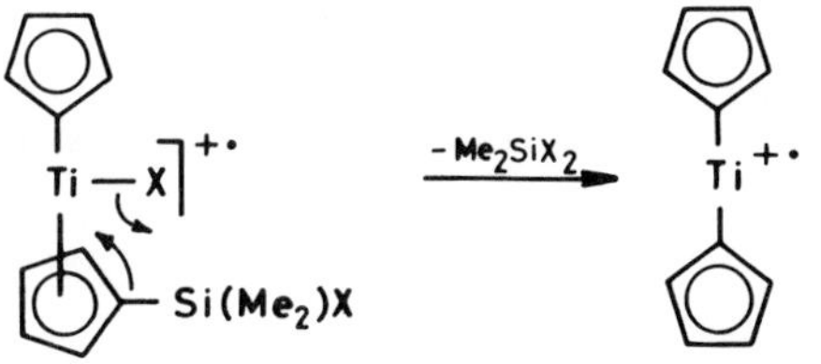

(2)

Re-CO bonds relative to the corresponding Mn-ligand bonds. Another interesting difference is the absence of doubly-charged metal containing ions and of ions arising through dehydrogenation processes in the spectra of the manganese complexes, though such ions are commonly observed in the spectra of the rhenium complexes. EI mass spectrometry has beeen used for the characterization of several manganese complexes such as [(CH$_3$C$_5$H$_4$)(CO)$_2$Mn]$_2$AsPh, (36) and (37).[59] In the EI spectra of the complexes (37; R = Bu$^t$, Ph, or c-C$_6$H$_{11}$), the ion of highest mass corresponds to [M/2]$^+$, but, in contrast, the NICI spectrum of (37; R = Bu$^t$) exhibits an ion corresponding to [M - CO]$^-$.

In the EI spectra of the complexes (38; X = Y = Cl, Br, NH$_2$, SMe, or OEt; X = NMe$_2$, Y = Cl or Br), the molecular ion of the dimer is of high abundance and only the halo complexes give rise to a prominent peak due to the monomer.[60] Fragmentation of the symmetrical complexes (X = Y) leads mainly to the ions [(Me$_5$C$_5$)CoX]$^+$, [(Me$_5$C$_5$)$_2$Co$_2$]$^+$ and [(Me$_5$C$_5$)Co - H]$^+$. When X = NMe$_2$, loss of NMe$_2$H also occurs and the ion [(Me$_5$C$_5$)CoNMe$_2$]$^+$ is also very prominent. Several rhodium,[61] and iron[62-66] metallocene complexes, e.g. (39),[61] (40),[62] (41; R = Me or Ph),[63] the ytterbium complex C$_5$H$_5$YbCl[67] and the actinide complexes (C$_5$H$_5$)$_3$MetMet´(C$_5$H$_5$)(CO)$_2$ (Met = Th or U; Met´ = Fe or Ru)[68] have been characterized by mass spectrometric techniques. The charge separation reactions (Coulomb explosion) of the ferrocene, cobaltocene, and nickelocene dications, generated by EI ionization in the gas phase have been reported.[69] These sandwich dications display a common behaviour in their unimolecular decomposition reactions. They are characterized by two distinct charge separation pathways. One corresponds to the reaction [MetC$_{10}$H$_{10}$]$^{2+}$ ----> [MetC$_5$H$_5$]$^+$ + [C$_5$H$_5$]$^+$, and the other to the generation of the ions [C$_{10}$H$_{10}$]$^{+\cdot}$ and [Met]$^+$ (Met = Fe, Co, or Ni). From the translational energy releases, associated with the charge separation processes, the interchange distances of the exploding dications in their transition structures were calculated. The results obtained demonstrated that, irrespective of the nature of the metal, the interchange distance for the formation of [MetC$_5$H$_5$]$^+$/[C$_5$H$_5$]$^+$ is 5.0 Å, and 6.2 Å for the generation of [C$_{10}$H$_{10}$]$^{+\cdot}$ and [Met]$^+$. FAB mass spectrometry has been used for the characterization of the complex bis[$\eta^5$-(trimethoxysilylpropyl)-cyclopentadienyl]dichlorotitanium(IV),[70] and of several neutral and cationic cyclopentadienyl sulphido complexes of molybdenum [e.g.

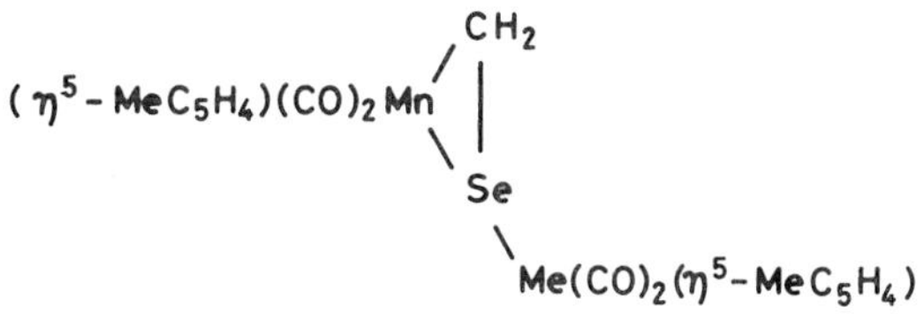

(36) $(\eta^5\text{-MeC}_5\text{H}_4)(\text{CO})_2\text{Mn}$—CH$_2$—Se—Mn(CO)$_2$($\eta^5$-MeC$_5$H$_4$)

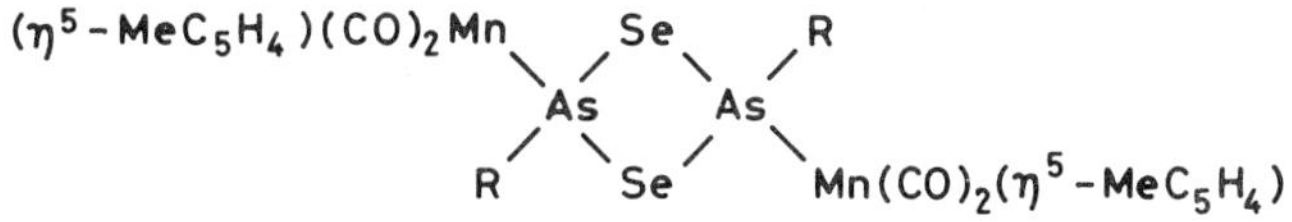

(37) $(\eta^5\text{-MeC}_5\text{H}_4)(\text{CO})_2\text{Mn}$, As, Se, R / R, Se, As, Mn(CO)$_2$($\eta^5$-MeC$_5$H$_4$)

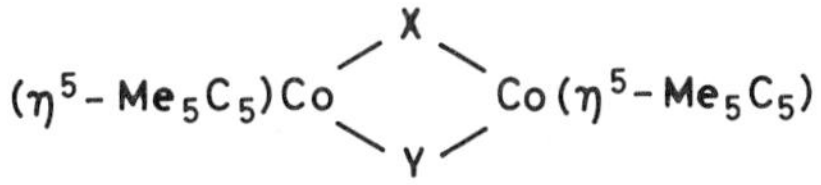

(38) $(\eta^5\text{-Me}_5\text{C}_5)\text{Co}$(X)(Y)Co($\eta^5$-Me$_5$C$_5$)

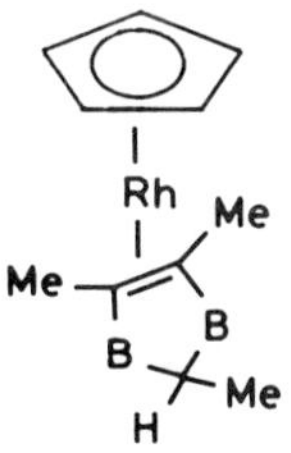

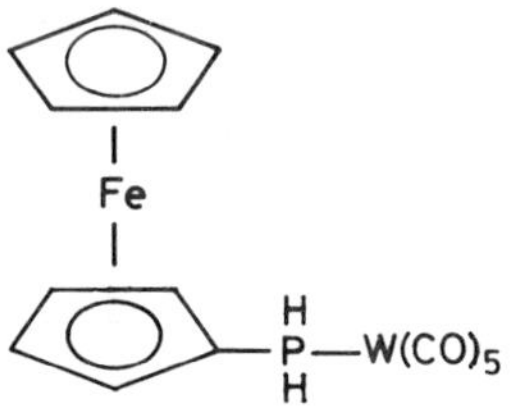

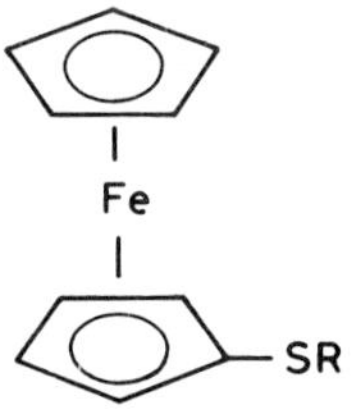

(39) (40) (41)

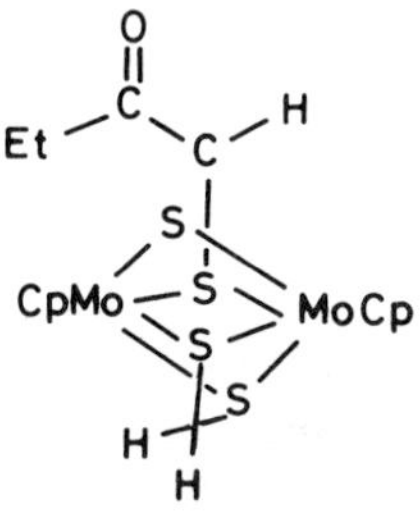

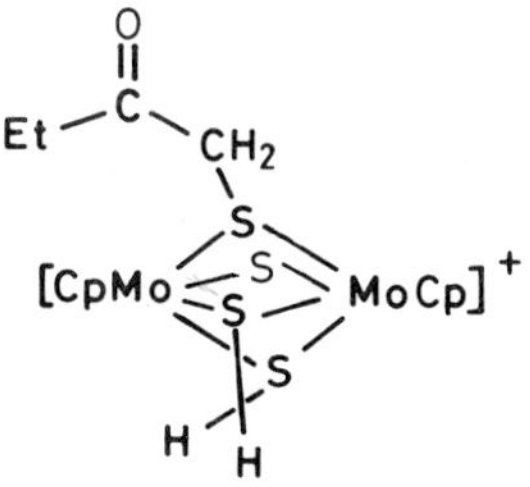

(42) (43)

(42) and (43)].[71]

In the hydrogen or methane CI spectra of the compounds $C_5H_5Fe(CO)_2X$ (X = Cl, Br, or I), the molecular ion is observed with a fairly high abundance.[72] In contrast, when isobutane is used as the reagent gas, the low recombination energy of this reactant system prevents an exchange reaction and only the protonated molecule is observed. The spectra of these compounds become simpler as the reagent gas is changed from $CH_4$ or $H_2$ to $Bu^tH$; with this last gas, almost all the total ion current is due to the protonated molecule and to the fragment ion $[(C_5H_5)Fe(CO)_2]^+$. When $H_2$ or $CH_4$ is used as the reagent gas, ions obtained by loss of some CO groups from the protonated molecule are observed. The differences in the $H_2$, $CH_4$, and $Bu^tH$ CI spectra of the compounds $C_5H_5Fe(CO)_2Met(Me)_3$ (Met = Si or Sn) are related to the protonating ability of the reagent gases. For example, when $H_2$ is used, a low relative abundance of the protonated molecule is observed, but a wide fragmentation occurs which, mainly, leads to the ion $[M - CH_3]^+$. For the compounds $(C_5H_5)Fe(CO)_2MetPh_3$ (Met = Ge or Sn), the relative abundances of the ions $[M]^{+\cdot}$ and $[M + H]^+$ were not established owing to the low intensity of the spectra, due to the low volatility of these aryl derivatives, and to the low relative abundance of the ions in the molecular ion region.

<u>Transition-metal cluster compounds</u>.- The EI method has been used for characterizing a wide range of polymetallic clusters such as $Ph_2Sn[Mn(CO)_4CNR]_2$ (R = Me, Et, $SiMe_3$, or $GeMe_3$),[73] (44; R = $Bu^t$ or Ph),[74] (45),[74] (46),[75] (47; Met = Cr or Mo),[75] (48),[76] (49),[77] and of various complexes derived from Roussin's red salt, e.g. $(\mu-RS)_2Fe_2(NO)_4$ (R = Me, Et, or $CH_2=CHCH_2$) and $(\mu_3-S)_2Fe_2(NO)_4$.[78] These compounds all exhibit molecular ions of low abundance. Well defined molecular ions have been observed in the FD spectra of the complexes (50) and $(C_5H_5)_2Co_2(\mu_3-S)_2Fe(NO)_2$.[78] Comparative studies of the spectra of the compounds (51; R = Me, Et, $Pr^i$, $Bu^t$, $NEt_2$, or $4\text{-}MeOC_6H_5$) have indicated that the positive-ion EI spectra exhibit the molecular ion, whereas the negative-ion EI spectra give ions of highest m/z corresponding to the ion $[M - CO]^-$.[79] In these spectra, the base peak corresponds to the ion $[M - CO]^-$ with the molecular ion being absent. In contrast, the molecular ion is observed in the isobutane PICI spectra. For these compounds and other metal clusters the argon NICI spectra have proven to be a very selective

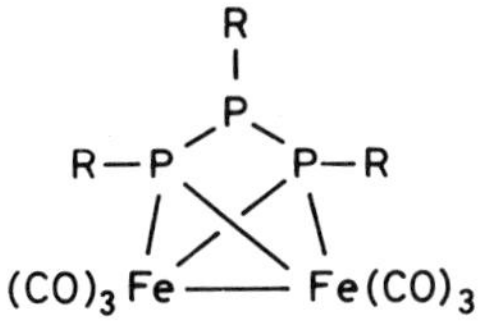

(44) 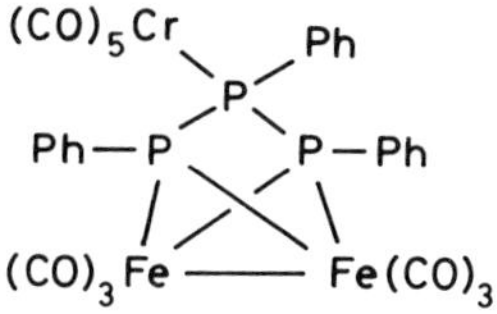 (45) 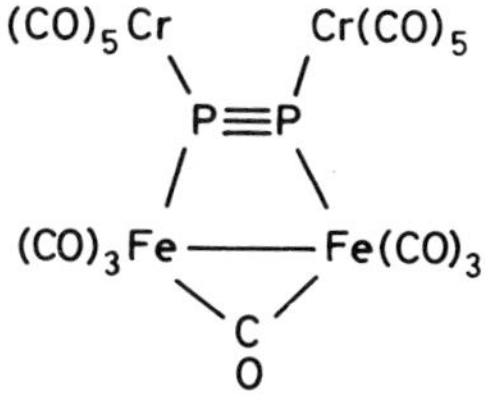 (46)

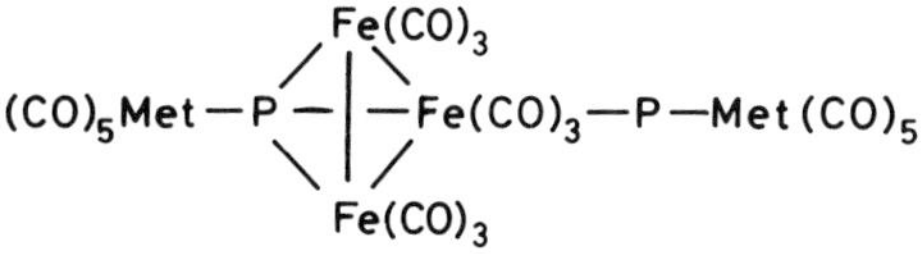

(47)

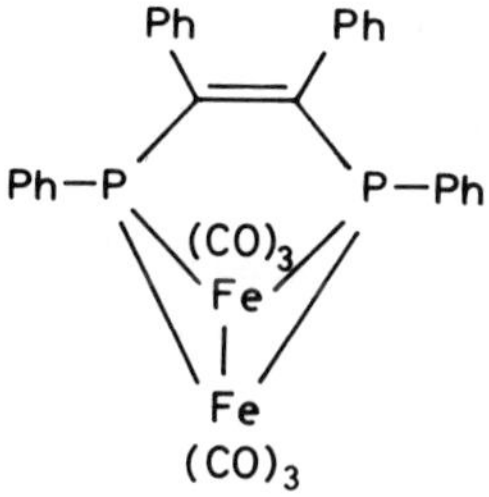

(48) 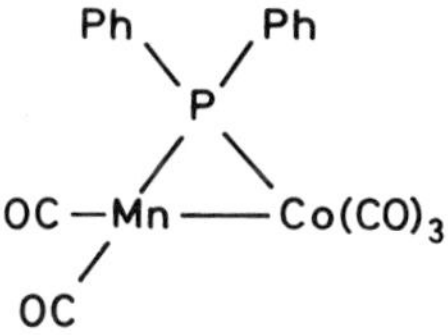 (49)

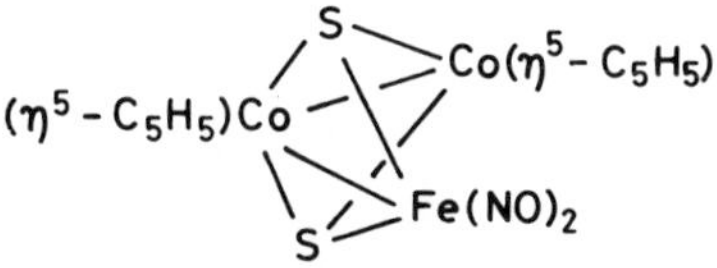

(50)

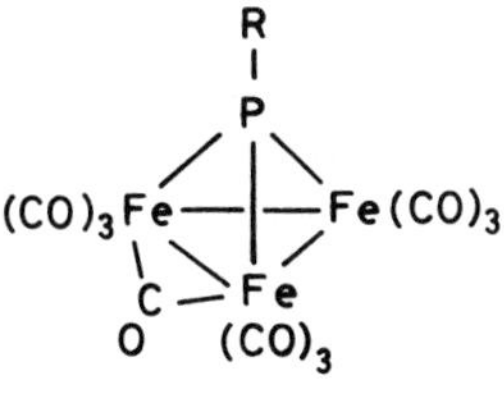

(51) 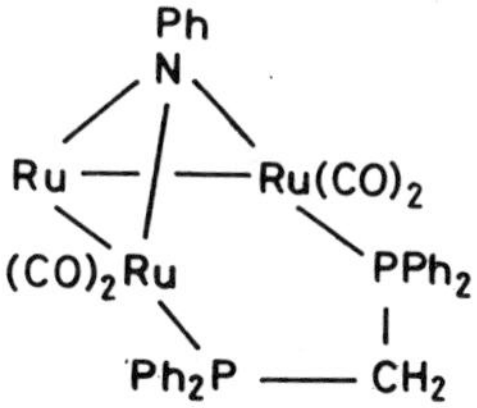 (52)

and sensitive analytical method. The FAB spectra of the compounds $Ru_3(\mu_3\text{-}NPh)_2(CO)_9$, $[Ru_3(\mu_3\text{-}NPh)_2(CO)_8]_2(Ph_2PCH_2CH_2PPh_2)$ and (52) show relatively large molecular ion peaks and several fragment ions arising by loss of the CO ligands.[80] FAB mass spectrometry has proved particularly useful in determining the composition of large charged clusters. The negative-ion spectra of several rhenium compounds, e.g. $[(Ph_3P)_2N][Re_4(CO)_{16}]$, $[Et_4N]_2[Re_8(CO)_{24}]$ and $[Et_4N]_3[Re_7C(CO)_{21}]$ show the molecular ions as well as ions corresponding to loss of $[(Ph_3P)_2N]$, $[Et_4N]$ and of approximately half of the carbonyl ligands.[81] By contrast, in the spectra of mixed metal cluster compounds such as $[Et_4N]_2[Re_7C(CO)_{21}Rh(CO)(PPh_3)]$, no molecular ions are observed and usually the peak of highest m/z is due to $[M - X]^-$ (X = counter ion).[82] However, in some cases the positive-ion spectra show the molecular ion as well as the ion $[M + X]^+$. A systematic study of a large number of cationic phosphine-containing clusters including polyhydrides such as $[Au_2Re_2(H)_6(PPh_3)_6]PF_6$ has indicated that FAB spectrometry is excellent in giving the correct molecular formula including the number of hydride ligands, and when combined with other techniques can provide reliable characterization of compounds of this type.[83] In general, the FAB spectra of these compounds with the use of 3-nitrobenzyl alcohol as the matrix gave well resolved peaks for the cluster cation and/or cluster cation plus anion pair. Comparison of observed and calculated isotopic ion distributions for these peaks reliably gave the correct molecular formula. The spectra also showed cluster fragments arising from loss of species such as $AuPPh_3$, $PPh_3$, Ph and H.

## 4 Chelate, Macrocyclic, and Other Complexes

**Neutral Chelates.**- The study of the mass spectra of metal $\beta$-diketonates has continued to attract much attention and some interesting results illustrating various factors which affect the spectra of such compounds and that could complicate their interpretation have been reported. An EI mass spectrometric study of molybdenyl acetylacetonate has demonstrated that the formation of both fragment and polymeric ions is strongly dependent on the instantaneous source pressure.[84] Scan-by-scan analysis showed that the relative abundances of some ions can change by factors as high as 15 over a period of 60 s, whereas other ions are remarkably transient, appearing in only a few scans during which the source pressure corresponds to that just required for their formation. The EI spectra of the complexes $UO_2(R^1COCHCOR^2)_2$ (e.g. $R^1 = R^2 =$ Me;[85, 86] $R^1 = R^2 = CF_3$;[86] $R^1 = R^2 = Bu^t$[86]) and Met($CF_3COCHCOBu^t$)$_3$.Q (Met = Er, Nd, or La; Q = 8-crown-6)[87] have been reported. The molecular ion is not observed in the EI spectra of the lanthanide complexes, but is shown by the FD spectrum of the neodymium complex. Several compounds of the type (53; $PR_3$ = $PMe_3$ or $PPh_2Me$; $R^1$, $R^2$ = e.g. Me, Ph, or NHMe), recorded at a probe temperature of ca. 100 °C, exhibit molecular ions and fragment ions which are related to the structure of the neutral molecules.[88] In contrast, their FD spectra exhibit only molecular ions which is indicative of low proton affinity. Linked scans at constant B/E have established that the primary fragmentation pathways in the EI spectra lead to the ions $[C_6Cl_5PR_3]^{+\cdot}$, $[M - C_6Cl_5PR_3]^{+\cdot}$, and $[C_6HCl_5]^{+\cdot}$. At a probe temperature of 140 °C, the spectrum of complex (53; $PR_3$ = $PPh_2Me$; $R^1 = R^2$ = Me) shows an abundant ion of composition $[M + Cl]^+$, whose formation involves a solid-state reaction of the complex in the probe to give the nickel(III) complex $[NiCl(C_6Cl_5)(acac)PPh_2Me]$, and the subsequent ionization of the latter. The SIMS spectrum of the chromium complex (54; Met = Cr, n = 3) shows the molecular ion, fragment ions arising mainly by loss of intact ligand radicals, and the ion $[M + Ag]^+$ which is due to the sample support used.[89] The pattern of the molecular and fragment ions is similar to that observed in the EI spectrum. In contrast, the SIMS spectrum of the analogous mercury(II) complex (54; Met = Hg, n = 2) is simpler than its EI spectrum showing a more prominent molecular ion, and fewer fragment ions. No silver attachment was observed, not even when the powdery samples were burnished onto silver surfaces.

The EI spectra of the complexes (55; Met = Tc; $R^1$ = Ph, $R^2$ = Me or Ph) exhibit molecular ions of low abundance, and various other metal-containing ions.[90] In contrast, in the spectrum of the complex $ReL_3$ (55; Met = Re; $R^1 = R^2$ = Ph), the only metal-containing ion present is $[Re(II)L]^{+\cdot}$. In the EI spectra of the complexes (56; Met = Ni or Cu; R = Me or Et), the molecular ion is prominent, especially in the case of the nickel complexes.[91] In all cases, the base peaks corresponds to an ion arising by loss of half of the ligand moiety from the molecular ion. EI spectrometry has

(53) (54) (55) (56) (57) (58) (59) (60)

been used to establish the dimetallic character of the nickel complex (57; Met = Ni), whereas in the case of the analogous copper complex this was accomplished using ammonia DCI spectrometry.[92]

The SIMS spectra of the complexes Met(ox)n (oxH = 58; R = H; Met = Cr, Ru, Ir, Rh, or Os, n = 3; Met = Pd or Pt, n = 2), VO(OH)(ox)$_2$, and MetO$_2$(ox)$_2$ (Met = Mo or W) have been studied.[89] Using etched or iodized silver foils, the compounds all exhibit abundant molecular ions, and ions corresponding to [M + Ag]$^+$. Fragment ions arising by loss of one or more intact ligand radicals are present in most of the spectra, and in certain cases, there are also ions arising by the loss of CO which is due to the assumed different bond strength between the metal and the chelating ligand at the O and N donor sites. Negatively-charged ions have been observed only for the rhodium and ruthenium complexes, and for the iridium complex only in the presence of [Cl]$^-$. In the ammonia or methane NICI spectra of the nickel dioximato complexes (59; R$^1$ = R$^2$ = H; R$^1$ = H, R$^2$ = Me; R$^1$ = R$^2$ = Me, R$^1$ = R$^2$ = Ph), the molecular ion is the base peak.[93] Fragmentation involves loss of O, OH and H$_2$O, and is more pronounced in the complexes of the hydrogen-substituted ligands. The PICI spectra exhibit [M + H]$^+$ as the major ion and [M]$^{+\cdot}$ as the second most abundant ion, both in the presence of ammonia or methane at 0.5 torr. EI mass spectrometry has indicated that the palladium(II)[94] and nickel(II)[95] benzamidino complexes ML$_2$ (LH = 60; R$^1$ = Ph, R$^2$ = aryl or allyl) are dimeric in the vapour state, whereas the corresponding platinum(II)[95] complexes are monomeric. In all of the complexes, the most favoured fragmentation routes involve loss of intact ligand radicals. In the EI spectra of the dihalide bridged iron(III) dithiocarbamato complexes [Fe(S$_2$CNR$_2$)$_2$X]$_2$($\mu$-Y$_2$) (R = e.g. Me, Et, or Pr$^i$; X = Cl, Br, or I; Y = Br or I), the ion of highest mass corresponds to [Fe(S$_2$CNR$_2$)X]$^{+\cdot}$.[96] Molecular ions have been observed in the EI spectra of several cobalt(III) diselenocarbamato complexes, e.g. Co(Se$_2$CNMe$_2$)$_3$.[97] In all cases, the ion [M - ligand]$^+$ is a major fragment ion, and the fragmentation patterns are similar to those of the corresponding dithiocarbamate complexes. The SIMS spectrum of the complex Ni(S$_2$CNEt$_2$)$_2$ exhibits the molecular ion, fragment ions arising by loss of S or the intact ligand, and the ions [M + Ag]$^+$ and [M + AgS]$^+$ due to the sample support used.[89]

<u>Anionic Complexes</u>.- The positive-ion FAB spectrum of the compound

$K_4[Fe(CN)_6]$ shows peaks due to the ions $\{K_5[Fe(CN)_n]\}^+$ (n = 2-6), $\{K_4[Fe(CN)_n]\}^+$ (n = 2, 4-5), $\{K_3[Fe(CN)_n]\}^+$ (n = 2-5), and $\{HK_4[Fe(CN)_6]\}^+$ indicating not only sequential loss of up to four CN ligands, but also a change of no less than four or five in the formal oxidation state on the metal centre in the anion of the compound.[98] This type of behaviour is also shown by several alkali metal salts of cyanometallates (Met = W(V), W(IV), Mo(IV), Cr(III), Fe(III), Fe(II), Mn(II), or Ni(II)). Sequential loss of CN ligands and wide changes in the oxidation state of the metal is also shown in the negative-ion FAB spectra of organic salts of cyanometallates e.g. $(^nBu_4N)_3[(W(CN)_8]$. The cyanometallates give rise to a wide range of fragment ions, in both their positive- and negative-ion FAB mass spectra, indicating the stability in the gas phase of ions possessing unusual oxidation states and coordination numbers. The negative-ion SIMS spectrum of $[NO][Co(ox)_2(NO_3)(py)]$ (oxH = 58; R = Me) has the molecular ion absent, but there is a large peak due to the ion $[M - H]^-$, and peaks arising by the loss of groups such as NO, $NO_3$, py, and Me.[99]

<u>Cationic Complexes</u>.- The positive-ion FAB mass spectra of a wide range of compounds of type $[(triphos)Met(\eta^3-E_2X)]Y$ (triphos = 1,1,1-tris(diphenylphosphinomethyl)ethane) (Met = Ni, E = X = P, Y = $BF_4$ or $PF_6$; Met = Co, E = P, X = S or Se; E = As, X = S, Se, or Te; Y = $BF_4$) with glycerol or thioglycerol as the matrix show peaks due to the cation $[(triphos)Met(\eta^3-E_2X)]^+$ and a characteristic fragmentation which involves the stepwise loss of the three atoms of the inorganic ring and the ejection of organic radicals from the triphos ligand.[100] In thioglycerol, adduct ions with sulphur atoms or sulphur organic radicals are also present. These adducts infer interactions of the matrix with (a) the atoms of the inorganic ring of the complex cations, and (b) the metal atom of the (triphos)Met moiety. The FAB mass spectra of complexes of type $[RuLL^1L^2]X_2$, $[RuL_2L^1]X_2$, and $[RuL_3]X_2$ (L, $L^1$, and $L^2$ include 2,2'-dipyridyl, 1,10-phenanthroline, their derivatives, and related compounds; X = e.g. $Cl^-$, $PF_6^-$, $CF_3SO_3^-$ and $CF_3CO_2^-$) have been investigated.[101] The compounds of type $[RuL_3]X_2$ show ions corresponding to $[RuL_3X]^+$, $[RuL_3]^+$, $[RuL_2]^+$, $[RuL]^+$, and $[L + H]^+$. An interesting observation is the presence of prominent doubly charged intact cations of type $[RuL_3]^{2+}$ in the spectra obtained, when 3-nitrobenzyl alcohol is employed as the matrix. This feature contrasts with the behaviour

of the corresponding iron, cobalt, and nickel complexes which show
no doubly charged ions of any kind in their positive-ion FAB
spectra with the same matrix. This indicates that the reduction
potential of the complex is not a major factor in the production of
doubly charged ions.

The positive-ion FAB spectra of the complexes
[Tc(MeO$_3$P)$_6$][BPh$_4$], [Tc(MeO$_2$PMe)$_6$][BPh$_4$] and the $^{252}$Cf plasma
desorption (PD) spectrum of the former have been reported.$^{102}$ The
FAB spectrum of [Tc(MeO$_3$P)$_6$][BPh$_4$] shows a peak due to the cation,
and ions arising by the successive loss of three trimethylphospite
ligands. The $^{252}$Cf PD mass spectrum is similar. The FAB spectrum of
[Tc(MeO$_2$PMe)$_6$][BPh$_4$] exhibits an ion due to the cation, and ions
arising by the subsequent loss of the dimethylmethylphosphite
ligands. The negative-ion FAB spectra of both compounds, and the PD
spectrum of [Tc(MeO$_3$P)$_6$][BPh$_4$] all show a strong signal due to the
tetraphenylborate anion. In the positive-ion spectra of several
compounds of type [Ph$_3$PtR][CF$_3$SO$_3$] (R = e.g. methyl, 2-propenyl,
4-nitrophenyl), the peak with highest m/z corresponds to the ion
[Ph$_3$PtR]$^+$.$^{103}$

FAB mass spectrometry has been used to study the complex
formation between tris(3,6-dioxaheptyl)amine and alkali metal
cations.$^{104}$ The close agreement of the results obtained with both
i.r. and n.m.r. data demonstrate the value of FAB spectrometry in
the study of complex formation. SIMS studies have shown that
complexes of the type [Ag(CNR)$_4$]X (R = Me, X = PF$_6$; R = c-C$_6$H$_{11}$,
But, X = ClO$_4$) undergo vacuum-promoted ligand loss.$^{105}$

<u>Carboxylate and Related Complexes</u>.- SIMS studies of aqueous
solutions of 1R,2R-cyclohexanediamine (1R,2R-dach) platinum(II)
complexes containing the dicarboxylate anions oxalate (oxal),
saccharate (sac), glutarate (glut), and malonate (1-mala) have
proved useful in determining the nature of the species which exist
in solution.$^{106}$ The spectrum of the oxalate complex for which X-ray
studies have indicated a square planar monomeric structure, showed
a very pronounced peak due to the ion [Pt(oxal)(1R,2R-dachH)]$^+$ and
peaks of very low abundance due to dimetallic species. In contrast,
the spectra of all the other complexes exhibited prominent peaks
due to the dimetallic ions [Pt$_2$(X)$_2$(1R,2R-dach)(1R,2R-dachH)]$^+$ (X =
glut or sac) and [Pt$_2$(X)(1R,2R-dach)(1R,2R-dach-H)]$^+$ (X = glut,
sac, or 1-mala) indicating the presence of dimetallic species in

solution. FAB mass spectrometry has been used to study the mass spectra of alkali metal, alkaline earth metal and transition metal complexes of EDTA.[107,108] In the case of EDTA containing Fe(III), both oxidation and reduction reactions were observed. The DCI-EC spectrum of [Ru$_2$(C$_4$H$_7$CO$_2$)$_4$Cl] exhibits the molecular ion, whereas in its FAB spectrum the peak of highest m/z value corresponds to [M - Cl]$^+$.[109]

**Macrocycles.**- A review which critically evaluates ten ionization methods for corrins and vitamin B$_{12}$ has been published.[110] FAB spectrometry has been used for the characterization of vitamin B$_{12}$ and of its $^{13}$C-labelled analogue,[111] of adenosylcobinamide hydroxide,[112] and of dicyanocobalt(III)-hepta(2-phenylethyl)cobyrinate.[113] The FAB spectrum of vitamin B$_{12}$ obtained using 3-nitrobenzyl alcohol as the matrix shows a more abundant [M + H]$^+$ ion than spectra obtained using other matrices.[114] Furthermore, there is no chemical degradation of the sample by the matrix observed when 3-nitrobenzyl alcohol is employed. This matrix has also proved useful for the recording of the positive- and negative-ion FAB spectrum of chlorophyll which exhibit the ion [M]$^{+\cdot}$ or [M]$^{-\cdot}$ respectively, together with a fragment ion arising by the loss of the phytyl moiety. The multiphoton ionization mass spectrum of chlorophyll is similar, exhibiting an abundant molecular ion and the ion [M - phytyl]$^{+\cdot}$.[115] FAB mass spectrometry has been employed for the characterization of the porphyrins CoPBr(py) and MnPCl (P$^{2-}$ = 5,10,15,20-tetrakis[4-(hexyloxycarbonyl)phenyl]porphyrin).[116] EI and hydrogen CI spectrometry techniques have been employed for the analysis of various porphyrin complexes present in geological samples.[117]

Comparative studies of the EI, FAB, and ammonia DCI spectra of hydrolytically labile dihalogenotitanium(IV) tetraphenylporphyrins (ttp) of formula TiX$_2$(tpp) (X = F, Cl, or Br) have been reported.[118] These complexes have weak axial metal-ligand bonds that are readily hydrolysed to the oxotitanium(IV) complex TiO(tpp). The EI mass spectrum of TiO(tpp) exhibits large peaks for the molecular ion. The other major peaks correspond to the loss of water, of one or more phenyl group, and to doubly charged ions. The mass spectra of the compounds TiX$_2$(tpp) (X = F, Cl, or Br) all exhibit peaks due to the hydrolysis product TiO(tpp) and its

fragments. The molecular ions are not observed except for a very
small peak due to [TiF$_2$(tpp)]$^{+\cdot}$, but, in contrast, the formation of
[TiX(tpp)]$^+$ (following loss of halide) is observed in every case.
The FAB spectra were obtained in various matrices. TiO(tpp) always
exhibits important peaks of [TiO(tpp)]$^{+\cdot}$ and [TiO(tpp) + H]$^+$ with
3-nitrobenzyl alcohol as the matrix. Addition products of the
matrix with TiO(tpp) are also observed. The only formed fragment is
[Ti(tpp)]$^+$. The spectrum of [TiF$_2$(tpp)]$^{+\cdot}$ again shows large peaks
corresponding to the hydrolytic product TiO(tpp), but the molecular
ion [TiF$_2$(tpp)]$^{+\cdot}$ is also well developed. 3-Nitrobenzyl alcohol is
the only matrix in which TiX$_2$(tpp) (X = Cl or Br) could be
dissolved, but they both give spectra typical of pure TiO(tpp). The
soft character of the FAB ionization technique is offset by the
hydrolysis or substitution reactions of the titanium complexes with
the matrix. In ammonia DCI-EC, the molecular ion is the most
abundant ion in the spectrum of TiF$_2$(tpp) with small peaks for
[TiF(tpp)]$^-$ and [TiO(tpp)]$^{-\cdot}$. The spectrum of TiCl$_2$(tpp) shows a
main peak for [TiCl(tpp)]$^-$ with a molecular ion more abundant than
the hydrolytic product [TiO(tpp)]$^{-\cdot}$. Unlike the EI spectrum, the
DCI spectrum shows no fragmentation at masses lower than that of
TiO(tpp). This confirms the soft ionization character of DCI. The
DCI-EC technique has also proved useful in characterizing the
complexes FeX(tpp) (X = F, Cl, or Br), all of which exhibit
abundant molecular ions.[109]

LD-FTMS has been shown to be an effective method for
characterizing various synthetic metalloporphyrins.[119] It was
established that molecular weight information can easily be
obtained and accurate masses may be determined for confirmation of
molecular formula. Although some structurally important
fragmentation occurs, it is not as extensive as with CI methods,
but compares favourably with FAB without the associated matrix
problems. The multiphoton ionization (MPI) spectra of the
tetraphenylporphine complexes Met(tpp) (Met = Co, Ni, or Cu) have
indicated that at 266 nm the complexes exhibit molecular ionization
without ligand predissociation.[120] The laser power dependences for
[Cu(tpp)]$^+$ and [Cu]$^+$ ion formation show that [Cu(tpp)]$^+$ is not
formed by an ion/molecule reaction with [Cu]$^+$. FAB studies of the
nickel thiocyanate complexes of O$_2$N$_3$-donor macrocycles NiL(NSC)$_2$ (L
= e.g. 61; R = H or Me) proved useful for structural
elucidation.[121] Although no molecular ions were observed for these

species under the conditions employed, prominent peaks corresponding to [NiL(NCS)]$^+$ and [NiL]$^+$ were evident in each spectrum. $^{252}$Cf plasma desorption mass spectrometry has been employed for the study of the barium and manganese macrocycles derived from (62) and (63), and has demonstrated the tetrametallic character of the complex [Mn$_4$(L)(ClO$_4$)$_4$] (LH$_4$ = 63).[122]

## 5 Miscellaneous Inorganic Compounds

EI mass spectrometry has been used to study the gas-phase composition of several lithium alkoxides (LiOR; e.g. R = Et, Pr$^i$, Bu$^n$, or Bu$^t$).[123] The most abundant ions shown by the spectra correspond to tetramers, hexamers, or larger aggregates, often with the loss of one or more H atoms. The large clusters are markedly stabilized by hydroxide ligands contributed via water contamination. Fragmentation pathways of the clusters have been investigated using high resolution mass spectrometry and by direct analysis of daughter ions. Similar studies have also been reported for lithium diisopropylamide and lithium 2,2,6,6-tetramethylpiperidine. The EI mass spectra of the cyclic tetramer of sulphur nitride,[124] and of a polymeric species of composition [SNBr$_{0.4}$]$_n$[125] have been reported. The photoionization mass spectrum of SBr$_2$ has been measured, and the data obtained utilized for the determination of the S-Br bond energy.[126] The ammonia DCI-EC spectra of various doubly- and triply-bridged polyazaheterophanes (e.g. 64 and 65; n = 6 or 8) show peaks due to the ions [M + H]$^+$, [M + nNH$_4$]$^+$ (n = 1 or 2) and various fragment ions, whereas, in contrast, the corresponding EI spectra do not reveal the molecular ion.[127] Negative-ion SIMS spectrometry has been used to investigate the anionic structure of a range of molten chloroaluminate salts (1-methyl-3-ethylimidazolium chloride mixed with AlCl$_3$ in various proportions).[128] The main features of the spectra are peaks due to [Cl]$^-$, [AlCl$_4$]$^-$, and various oxide-containing ions which appear essentially as surface contamination. Basic solutions exhibit only peaks for [Cl]$^-$ and [AlCl$_4$]$^-$, whereas very acidic solutions show the presence of additional peaks tentatively assigned to the ions [Al$_3$Cl$_{10}$]$^-$ and [Al$_2$Cl$_7$(Al(OH)$_3$)$_n$]$^-$ (n = 1, 2, or 3).

The EI mass spectra of niobium and tantalum oxyfluorides,[129] and of niobium sulphide trichloride (NbSCl$_3$)[130] have been investigated. The latter exhibits an abundant molecular ion, and

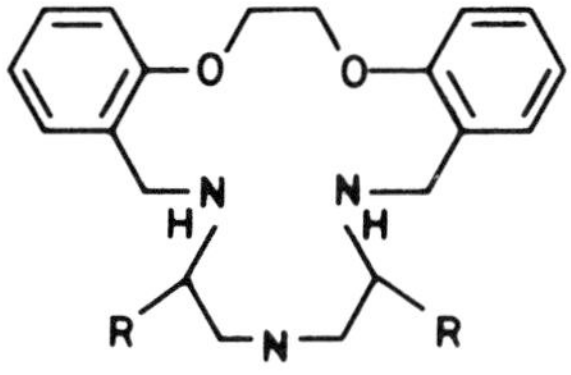

10: *Metal-containing and Inorganic Compounds Investigated by Mass Spectrometry*    347

(61)

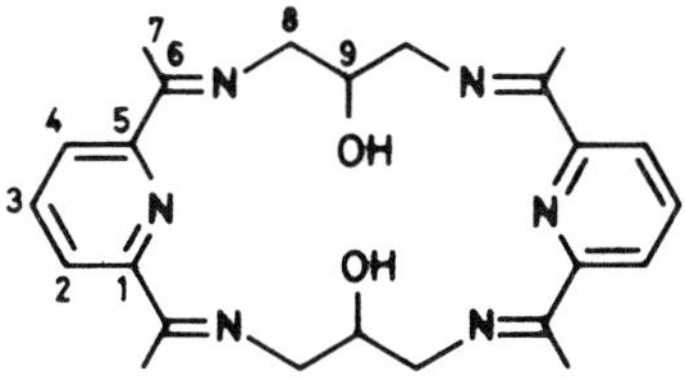

(62)

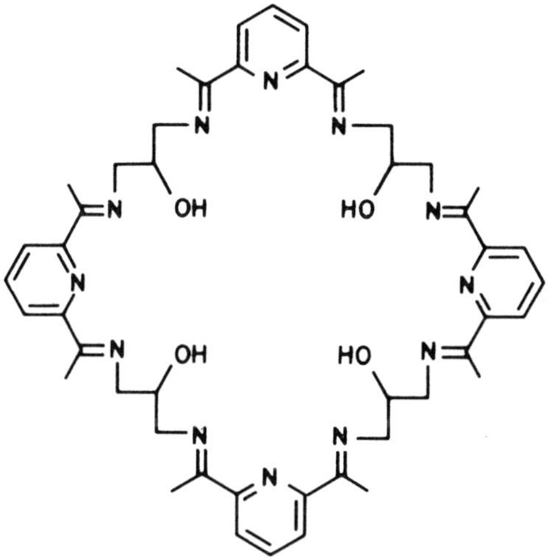

(63)

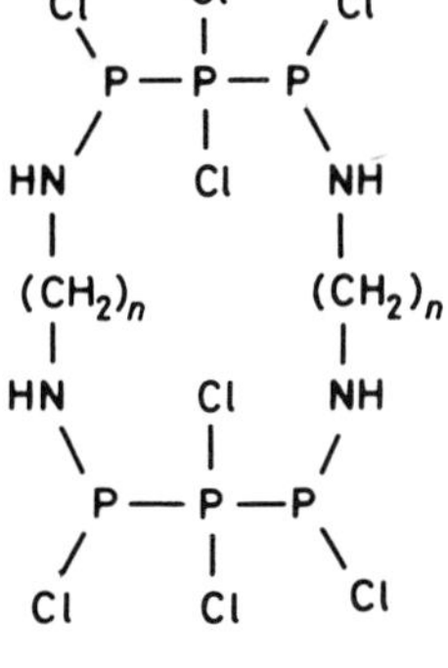

(64)

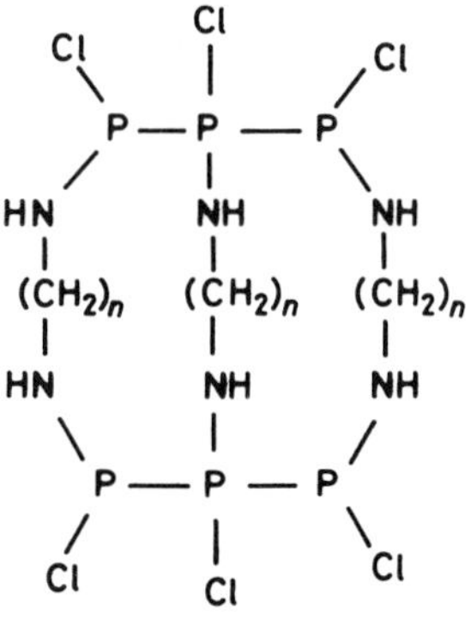

(65)

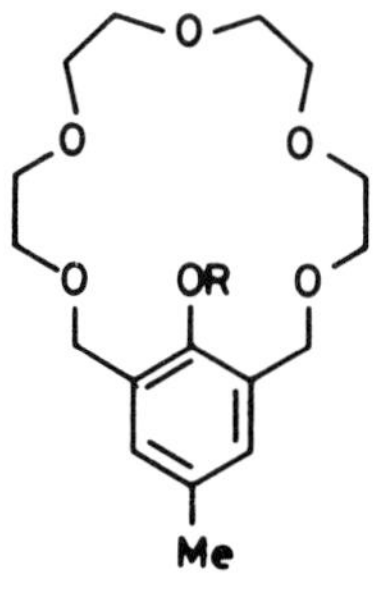

(66)

fragment ions such as $[NbSCl_n]^+$ (n = 1 or 2), $[NbCl]^+$, and $[NbS]^+$. The abundance of $[NbCl]^+$ is considerably higher than that of $[NbS]^+$. In the EI mass spectra of $Met(ClO_4)_4$ (Met = Zr or Hf), the predominant fragmentation pathway involves the loss of $ClO_4^·$ followed by elimination of $Cl_2O_7$ ($Cl_2O_6$), and eventually $ClO_3^·$ ($ClO_2^·$).[131] In contrast, the copper perchlorate molecular ion dissociates to $[Cu]^+$ by sequentially splitting off $ClO_4^·$ radicals.

The ambient temperature FAB spectra of a wide range of novel gaseous polyatomic binary and ternary oxides arising from various lanthanide (Ln) Schiff base complexes, simple salts and sesquioxides have been observed.[132] The new binary oxides detected, as singly positive ions, are $Ln_2O_3$, $Ln_3O_3$, $Ln_3O_4$, $Ln_4O_4$, $Ln_4O_5$, $Ln_4O_6$, $Ln_5O_6$, $Ln_5O_7$, $Ln_5O_8$, $Ln_6O_8$, $Ln_6O_9$, $Ln_7O_{10}$, $Ln_8O_{11}$, $Ln_8O_{12}$ and $Ln_9O_{13}$; the ternary gaseous oxides were $CeEuO_2$, $CeEu_2O_3$, and $Ce_2EuO_4$; $LaYbO_2$, $La_2YbO_4$ and $LaYb_2O_4$; $NdHoO_3$, $Nd_2HoO_4$, and $NdHo_2O_4$; $YTmO_3$; $Y_xTm_{3-x}O_4$, x = 1-2; $Y_xTm_{4-x}O_6$, x = 1-3; $Y_xTm_{5-x}O_7$, x = 1-4; $Y_xTm_{6-x}O_9$, x = 1-5. Some of these oxides show the cations in unusual oxidation states. The gadolinium-gallium ternary oxides, $GdGaO_2$, $GdGaO_3$, and $Gd_2GaO_4$ have also been detected. The gaseous polyatomic oxides are produced through a reductive condensation process involving primary species $[Ln]^+$ and $[LnO]^+$ formed when the rare earth compounds are struck by fast xenon atoms.

The EI, CI and thermal ionization mass spectra of alkali halides such as CsI, CsBr, CsCl, CsF, RbI, KI, NaI, LiI, LiF, and NaF have been reported.[133] The EI and methane PICI spectra show the base metal ions and metal halide ion clusters. The $CH_4/N_2O$ (4:1) electron-capture NICI spectra consist of halide ions and halide ion clusters. The molecular ions and quasi-molecular ions obtained are of very low abundance. The CI sensitivity is about an order of magnitude higher than the EI sensitivity. The thermal ionization spectra show only the metal cations. Perrhenate salts ($MetReO_4$; Met = $NH_4$ or alkali metal), supported on graphite, have been examined by low-flux positive- and negative-ion SIMS.[134] The relative abundance of $[ReO_4]^-$ with respect to $[ReO_3]^-$ in the spectra of the neat salts varies dramatically with the nature of the cation. $MetReO_4$ (Met = $NH_4$ or Li) shows the ion $[ReO_3]^-$ in high abundance, but, in contrast, in the spectra of $MetReO_4$ (Met = Na, K, or Rb) the ion $[ReO_3]^-$ has lower abundance. The positive-ion spectra also show large counterion effects, abundant clusters of the $[Met_2ReO_4]^+$ type only being observed for $MetReO_4$ (Met = Na, K, or Rb). The similarities between the spectra of 1:1 mixtures of MetReO$_4$/Met´Cl and Met´ReO$_4$/MetCl (Met and Met´ are alkali metal cations) suggest that intimate mixing occurs on the molecular scale when simple physical mixtures of the solid salts are subjected to ion beam bombardment.

## 6 Reactions of Gaseous Metal and Metal-containing Ions with Organic Compounds

The study of the reactions of metal and metal-containing ions with organic compounds continues to attract considerable attention. A relevant review on the subject[135] and a paper containing over 100 references concerning the reactions of transition metal ions with organic substrates have been published.[136] CI/FAB mass spectrometry has been used to study the reaction of cobalt/oxygen cluster ions with isobutane.[137] The reaction leads to novel cluster product ions, e.g. [Co$_x$O$_x$C$_4$H$_{10}$]$^+$ (x = 1-4), which fragment primarily by loss of C$_4$H$_{10}$ indicating that the hydrocarbon is weakly bound to the Co$_x$O$_x$ moiety. The reactivities of several transition metal ions with pentene have been compared using FTMS techniques, and it has been shown that Cu$^+$, like Fe$^+$, Co$^+$, and Ni$^+$, activates allylic C-C bonds.[138] Mn$^+$ has been found to be more reactive towards alkynes than Fe$^+$ which was previously considered to have the highest reactivity towards alkynes.[139] For the Fe$^+$/alkyne system, labelling studies have established that $\beta$-hydrogen transfer is involved in the major decomposition pathways of the [Fe(alkyne)]$^+$ ions.[136] Neutralization-reionization mass spectrometry has been used for the first time to generate and characterize in the gas phase neutral FeCH$_x$ (x = 0-3).[140] In contrast to [FeCH]$^+$ and FeCH which rearrange to [HFeC]$^+$ and HFeC respectively, neither the ionic [FeCH$_x$]$^+$ (x = 0, 2, or 3) nor its neutral analogue rearranges to the hydridometal complexes HFeCH$_x$ (x = 1 or 2). Studies of the gas-phase reactions of various linear nitriles with Fe$^+$,[141-143] Co$^+$,[141, 142] Ni$^+$,[141] and Cu$^+$[144] have been reported. Fragmentation products arising from the ion [Met(RCN)]$^+$ suggest "end-on" coordination of the nitrile to the metal (Met = Fe, Co, or Ni) and "side-on" co-ordination of the nitrile to copper. A study of the reaction of the macrocycles (L = 66; R = H or Me) with various acetylacetonato- and cyclopentadienyl-metal containing ions (Met = Fe, Co, Ni, Cr, Mn, Rh, Nd, Pr, Ga, In, or Tl) has shown that new ions containing metal-macrocycle bonds are formed, e.g. [acacCoL]+ and [C5H5RhL]+, and the reactivity of the ions [Met(acac)2]+ with respect to the macrocycles varies in the order Co > Fe > Cr > Pr, Ga, and that the methyl substituted macrocycle is more reactive than the hydrogen substituted analogue.[145]

## References

1. J.Charalambous in 'Mass Spectrometry', ed. M.E.Rose (Specialist Periodical Reports), The Royal Society of Chemistry, London, 1987, Vol 9, p. 373.
2. M.I.Bruce and M.J.Liddel, Appl. Organomet. Chem., 1987, 1, 191.
3. F.Adams, R.Gijbels, and R.Van Grieken, Eds., 'Inorganic Mass Spectrometry', John Wiley and Sons, New York, 1988.
4. D.Plavsic, D.Srzic, and L.Klasinc, J. Phys. Chem., 1986, 90, 2075.
5. V.I.Vasyukova, Yu.S.Nekrasov, V.Ts.Kampel', and V.I.Bregadze, Bull. Acad. Sci., USSR Div. Chem. Sci. (Eng. Transl.), 1987, 36, 517.
6. S.S.Al-Juaid, S.M.Dhaher, C.Eaborn, P.B.Hitchcock, and J.D.Smith, J. Organomet. Chem., 1987, 325, 117.
7. Y.Blum and R.M.Laine, Organometallics, 1986, 5, 2081.
8. R.B.Bates, J.J.White, and K.H.Schram, Org. Mass Spectrom., 1987, 22, 295.
9. S.Sugimoto, M.Arime, S.Kawabata, and Y.Ono, Nippon Kagaku Kaishi, 1986, 11, 1700.
10. K.C.Molloy, T.G.Purcell, M.F.Mahon, and E.Minshall, Applied Organometal. Chem., 1987, 1, 507.
11. K.C.Molloy and T.G.Purcell, J. Organomet. Chem., 1986, 303, 179.
12. J.Klaveness, F.Rise, and K.Undheim, J. Organomet. Chem., 1986, 303, 189.
13. M.J. Goldstein and J.P.Barren, Helv. Chim. Acta, 1986, 69, 548.
14. J.S.Sawyer, A.Kucerovy, T.L.Macdonald, and G.J.McGarvey, J. Am. Chem. Soc., 1988, 110, 842.
15. P.Lesimple, J.-M.Beau, and P.Sinay, Carbohydr. Res., 1987, 171, 289.
16. M.D.Muller, Anal. Chem., 1987, 59, 617.

17  M.A.Unger, W.G.MacIntyre, J.Greaves, and R.J.Hugget, Chemosphere, 1986, 15, 461.
18  J.M.Miller, H.Mondal, I.Wharf, and M.Onyszchuk, J. Organomet. Chem., 1986, 306, 193.
19  J.M.Miller and I.Wharf, Can. J. Spectrosc., 1987, 32, 1.
20  D.Naumann and W.Tyrra, J. Organomet. Chem., 1987, 334, 323.
21  Th.Klapotke, J. Organomet. Chem., 1987, 331, 299.
22  R.Lareiprete and M.Stuke, J. Crystal Growth, 1986, 77, 235.
23  V.E.Bar, F.Beck, and W.P.Fehlhammer, Chemiker-Zeitung, 1987, 331.
24  Z.Hua, Org. Mass Spectrom., 1987, 22, 761.
25  G.A.Bell, D.W.H.Rankin, and P.F.Reinisch, J. Chem. Soc., Dalton Trans., 1987, 3023.
26  V.A.Bogdanov, Yu.I.Savel'ev, R.N.Shchelokov, and V.A.Piven, Bull. Acad. Sci. USSR, Div. Chem. Sci. (Engl. Transl.), 1987, 36, 857.
27  D.Perugini, G.Innorta, S.Torroni, and A.Foffani, J. Organomet. Chem., 1986, 308, 167.
28  M.Hackett, J.A.Ibers, and G.M.Whitesides, J. Am. Chem. Soc., 1988, 110, 1436.
29  J.Svara, A.Marinetti, and F.Mathey, Organometallics, 1986, 5, 1161.
30  A.L.Bandini, G.Banditelli, G.Minghetti, U.Vettori, and P.Traldi, Inorg. Chim. Acta, 1988, 142, 101.
31  E.Bonfantini, J. L.Metral, and P.Vogel, Helv. Chim. Acta, 1987, 70, 1791.
32  D.Seyferth, T.G.Wood, and R.S.Henderson, J. Organomet. Chem., 1987, 336, 163.
33  D.A.Prinslow and V.Vaida, J. Am. Chem. Soc., 1987, 109, 5097.
34  J.O.Lay, Jr, N.T.Allison, W.Yongskulrote, and R.Ferede, Org. Mass Spectrom., 1986, 21, 371.
35  G.L.Crocco, T.R.Sharp, and J.A.Gladysz, Int. J. Mass Spectrom. Ion Processes, 1986, 73, 181.
36  G.S.Bodner, D.E.Smith, W.G.Hatton, P.C.Heah, S.Georgiou, A.L.Rheingold, S.J.Geib, J.P.Hutchinson, and J.A.Gladysz, J. Am. Chem. Soc., 1987, 109, 7688.
37  M.I.Bruce, D.N.Duffy, M.J.Liddel, M.R.Snow, and E.R.T.Tiekink, J. Organomet. Chem., 1987, 335, 365.
38  M.I.Bruce, M.G.Humphrey, and M.J.Liddell, J. Organomet. Chem., 1987, 321, 91.

39  M.I.Bruce, M.G.Humphrey, G.A.Koutsantonis, and M.J. Liddell, J. Organomet. Chem., 1987, 326, 247.
40  G.Paolucci, S.Daolio, and P.Traldi, J. Organomet. Chem., 1986, 309, 283.
41  P.Lei and P.Vogel, Organometallics, 1986, 5, 2500.
42  M.Barber, B.L.Booth, P.J.Bowers, and L.Tetler, J. Organomet. Chem., 1987, 328, C25.
43  M.D.Curtis, K.-B.Shiu, and W.M.Butler, J. Am. Chem. Soc., 1986, 108, 1550.
44  G.Fu, Y.Qian, Y.Xu, and S.Chen, J. Organomet. Chem., 1986, 314, 113.
45  H.Butenschon and A.de.Meijere, Tetrahedron, 1986, 42, 1721.
46  L.D.Detter and R.A.Walton, Polyhedron, 1986, 5, 1321.
47  R.Gabioud and P.Vogel, Helv. Chim. Acta, 1986, 69, 865.
48  M.I.Bruce, P.A.Humphrey, J.K.Walton, B.W.Skelton, and A.H.White, J. Organomet. Chem., 1987, 333, 393.
49  F.Carre, R.J.P.Corriu, C.Guerin, B.J.L.Henner, B.Kolani, and W.W.C.Wong Chi Man, J. Organomet. Chem., 1987, 328, 15.
50  H.Kopf and Th.Klapotke, Z. Naturforsch, 1986, 41b, 667.
51  H.Kopf and Th.Klapotke, J. Organomet. Chem., 1986, 310, 303.
52  H.Kopf and Th.Klapotke, J. Chem. Soc., Chem. Commun., 1986, 1192.
53  K.Mitteilung, H.Kopf, and Th.Klapotke, Monatshefte fur Chemie, 1986, 117, 1003.
54  J.Muller, F.Ludemann, and H.Kopf, J. Organomet. Chem., 1986, 303, 167.
55  R.M.Kowaleski, F.Basolo, W.C.Trogler, R.W.Gedridge, T.D.Newbound, and R.D.Ernst, J. Am. Chem. Soc., 1987, 109, 4860.
56  J.-F.Leboeuf, J.-C.Leblanc, and C.Moise, J. Organomet. Chem., 1987, 335, 331.
57  I.R.Lyatifov, G.I.Gulieva, E.I.Mysov, V.N.Babin, and R.B.Materikova, J. Organomet. Chem., 1986, 326, 83.
58  I.R.Lyatifov, G.I.Gulieva, E.I.Mysov, V.N.Babin, and R.B.Materikova, J. Organomet. Chem., 1986, 326, 89.
59  L-R.Frank, K.Evertz, L.Zsolnai, and G.Huttner, J. Organomet. Chem., 1987, 335, 179.
60  U.Koelle, B.Fuss, M.Belting, and E.Raabe, Organometallics, 1986, 5, 980.
61  K.Geilich and W.Siebert, Z. Naturforsch, 1986, 41b, 671.

62  J.Svara and F.Mathey, Organometallics, 1986, 5, 1159.
63  R.V.Honeychuck, M.O.Okoroafor, L. H.Shen, and
    C.H.Brubaker, Jr., Organometallics, 1986, 5, 182.
64  W.Siebert, D.Buchner, H.Pritzkow, H.Wadepohl, and
    F. W.Grevels, Chem. Ber., 1987, 120, 1511.
65  H.Ma, P.Weber, M.L.Ziegler, and R.D.Ernst, Organometallics,
    1987, 6, 854.
66  Y.Stenstrom, A.E.Koziol, G.J.Palenik, and W.M.Jones,
    Organometallics, 1987, 6, 2079.
67  S.J.Swamy and H.Schumann, J. Organomet. Chem., 1987, 334, 1.
68  R.S.Sternal and T.J.Marks, Organometallics, 1987, 6, 2621.
69  T.Drewello, C.B.Lebrilla, H.Schwarz, and T.Ast, J. Organomet.
    Chem., 1988, 339, 333.
70  B.L.Booth, G.C.Ofunne, C.Stacey, and P.J.T.Tait, J.
    Organomet. Chem., 1986,, 315, 143.
71  D.E.Coons, R.C.Haltiwanger, and M.R. DuBois, Organometallics,
    1987, 6, 2417.
72  D.Perugini, G.Innorta, S.Torroni, and A.Foffani, Inorg. Chim.
    Acta, 1987, 133, 243.
73  M.Moll, H.Behrens, P.Merbach, K. H.Trummer, G.Thiele, and
    K.Wittmann, Z. Naturforsch, 1986, 41b, 606.
74  J.Borm, K.Knoll, L.Zsolnai, and G.Huttner, Z. Naturforsch,
    1986, 41b, 532.
75  H.Lang, G.Huttner, L.Zsolnai, G.Mohr, B.Sigwarth, U.Weber,
    O.Orama, and I.Jibril, J. Organomet. Chem., 1986, 304, 157.
76  K.Knoll, G.Huttner, and L.Zsolnai, J. Organomet. Chem., 1986,
    307, 237.
77  H.Lang, G.Huttner, B.Sigwarth, I.Jibril, L.Zsolnai, and
    O.Orama, J. Organomet. Chem., 1986, 304, 137.
78  D.Seyferth, M.K.Gallagher, and M.Cowie, Organometallics,
    1986, 5, 539.
79  K.Knoll and G.Huttner, J. Organomet. Chem., 1987, 329, 369.
80  M.I.Bruce, M.G.Humphrey, O.B.Shawkataly, M.R.Snow, and
    E.R.T.Tiekink, J. Organomet. Chem., 1987, 336, 199.
81  C.M.T.Hayward and J.R.Shapley, Organometallics, 1988, 7, 448.
82  T.J.Henly, J.R.Shapley, A.L.Rheingold, and S.J.Geib,
    Organometallics, 1988, 7, 441.
83  P.D.Boyle, B.J.Johnson, B.D.Alexander, J.A.Casalnuovo,
    P.R.Gannon, S.M.Johnson, E.A.Larka, A.M.Mueting, and
    L.H.Pignolet, Inorg. Chem., 1987, 26, 1346.

84   G.C.DiDonato and K.L.Busch, Org. Mass Spectrom., 1986, 21, 571.
85   M.N.Bhattacharjee, M.K.Chaudhuri, M.Devi, R.N.D.Purkayastha, Z.Hiese, and D.T.Khathing, Int. J. Mass Spectrom. Ion Processes, 1986, 71, 109.
86   V.M.Adamov, B.N.Belyaev, S.O.Berezinskii, G.V.Sidorenko, D.N.Suglobov, and G.A.Firsanov, Radiokhimiya, 1986, 28, 172.
87   T.N.Martynova, L.D.Nikulina, and V.A.Logvinenko, J. Therm. Anal., 1987, 32, 533.
88   B.Corain, B.Longato, A.M.Maccioni, B.Pelli, P.Traldi, and F.R.Kreissl, Talanta, 1988, 35, 27.
89   B.Wenclawiak, A.Eicke, W.K.Sichtermann, and A.Benninghoven, Fresenius Z. Anal. Chem., 1987, 329, 447.
90   G.Bandoli, U.Mazzi, H.Spies, R.Munze, E.Ludwig, E.Ulhemann, and D.Scheller, Inorg. Chim. Acta, 1987, 132, 177.
91   S.Dilli, A.M.Maitra, and E.Patsalides, J. Chromatogr., 1986, 358, 359.
92   G.Cros and J.-P.Laurent, Inorg. Chim. Acta, 1988, 142, 113.
93   J.Charalambous, J.S.Morgan, L.Operti, G.A.Vaglio, and P.Volpe, Inorg. Chim. Acta, 1988, 144, 201.
94   J.Barker, N.Cameron, M.Kilner, M.M.Mahoud, and S.C.Wallwork, J. Chem. Soc., Dalton Trans., 1986, 1359.
95   J.Barker, M.Kilner, and R.O.Gould, J. Chem. Soc., Dalton Trans., 1987, 2687.
96   M.Lalia-Kantouri, G.A.Katsoulos, and F.D.Vakoulis, J. Therm. Anal., 1986, 31, 447.
97   A.M.Bond, R.Colton, D.R.Mann, and J.E.Moir, Aust. J. Chem., 1986, 39, 1385.
98   K.R.Jennings, T.J.Kemp, and B.Sieklucka, Inorg. Chim. Acta, 1988, 141, 163.
99   E.Miki, M.Tanaka, K.Mizumachi, and T.Ishimori, Bull. Chem. Soc. Jpn., 1986, 59, 3275.
100  G.Cetini, L.Operti, G.A.Vaglio, M.Peruzzini, and P.Stoppioni, Polyhedron, 1987, 6, 1491.
101  J.M.Miller, K.Balasanmugam, J.Nye, G.B.Deacon, and N.C.Thomas, Inorg. Chem., 1987, 26, 560.
102  D.W.Wester, D.H.White, F.W.Miller, R.T.Dean, J.A.Schreifels, and J.E.Hunt, Inorg. Chim. Acta, 1987, 131, 163.
103  M.H.Kowalski, T.R.Sharp, and P.J.Stang, Org. Mass Spectrom., 1987, 22, 642.

104  J.M.Miller, S.J.Brown, R.Theberge, and J.H.Clark, J. Chem. Soc., Dalton Trans., 1986, 2525.

105  L.D.Detter, S.J.Pachuta, R.G.Cooks, and R.A.Walton, Talanta, 1986, 33, 917.

106  M.Noji, K.Suzuki, T.Tashiro, M.Suzuki, K.-I.Harada, K.Masuda, and Y.Kidani, Chem. Pharm. Bull., 1987, 35, 221.

107  K.Isa, T.Kinoshita, H.Kido, R.Nakata, and K.Mizuta, Nippon Kagaku Kaishi, 1986, 11, 1657.

108  K.Isa, H.Torii, T.Kinoshita, R.Nakata, and K.Mizuta, Shitsuryo Bunseki, 1986, 34, 171.

109  E.Forest, P.Maldivi, J.-C.Marchon, and H.Virelizier, Spectros. Int. J., 1987, 5, 129.

110  H.M.Schiebel and H.-R.Schulten, Mass Spectrom. Rev., 1986, 5, 249.

111  K.Kurumaya, T.Sakamoto, Y.Okada, and M.Kajiwara, J. Chromatogr., 1988, 435, 235.

112  B.P.Hay and R.G.Finke, J. Am. Chem. Soc., 1987, 109, 8012.

113  R.Stepanek, B.Krautler, P.Schulthess, B.Lindemann, D.Ammann, and W. Simon, Analytica Chim. Acta, 1986, 182, 83.

114  M.Barber, D.Bell, M.Eckersley, M.Morris, and L.Tetler, Rapid Commun. Mass Spectrom., 1988, 2, 18.

115  J.Grotemeyer, U.Bosel, K.Walter, and E.W.Schlag, J. Am. Chem. Soc., 1986, 108, 4233.

116  D.Ammann, M.Huser, B.Krautler, B.Rusterholz, P.Schulthess, B.Lindermann, E.Halder, and W.Simon, Helv. Chim. Acta, 1986, 69, 849.

117  M.I.Chicarelli and J.R.Maxwell, Trends Anal. Chem., 1987, 6, 158.

118  E.Forest, J.Ulrich, J.C.Marchon, and H.Virelizier, Org. Mass Spectrom., 1987, 22, 45.

119  R.S.Brown and C.L.Wilkins, Anal. Chem., 1986, 58, 3196.

120  J.B.Morris and M.V.Johnston, Int. J. Mass Spectrom. Ion Processes, 1986, 73, 175.

121  K.R.Adam, A.J.Leong, L.F.Lindoy, B.J.McCool, A.Ekstrom, I.Liepa, P.A.Harding, K.Henrick, M.McPartlin, and P.A.Tasker, J. Chem. Soc., Dalton Trans., 1987, 2537.

122  S.Brooker, V.McKee, W.B.Shepard, and L.K.Pannell, J. Chem. Soc., Dalton Trans., 1987, 2555.

123  J.D.Kahn, A.Hagg, and P.v.R.Schleyer, J. Phys. Chem., 1988, 92, 212.

124 V.I.Spitsyn, I.D.Kolli, and E.M.Orlova, Russ. J. Inorg. Chem. (Engl. Trans.), 1986, 31, 163.
125 U.Demant and K.Dehnicke, Z. Naturforsch, 1986, 41b, 929.
126 R.Minkwitz, R.Lekies, H.W.Jochims, E.Ruhl, and H.Baumgartel, Z. Naturforsch, 1986, 41b, 784.
127 P.Castera, J. P.Faucher, M.Granier, and J.-F.Labarre, Phosphorus and Sulfur, 1987, 32, 37.
128 G.Franzen, B.P.Gilbert, G.Pelzer, and E.DePauw, Org. Mass Spectrom., 1986, 21, 443.
129 H.G.Nieder-Vahrenholz and H.Schafer, Z. Anorg. Allg. Chem., 1987, 544, 122.
130 Yu.V.Mironov, V.P.Fedin, P.P.Semyannikov, and V.E.Fedorov, Russ. J. Inorg. Chem. (Engl. Trans.), 1986, 31, 617.
131 V.A.Sipachev, Yu.S.Nekrasov, N.I.Tuseev, D.V.Gromov, and V.P.Babaeva, Polyhedron, 1988, 7, 345.
132 I.A.Kahwa, J.Selbin, T.C.Y.Hsieh, and R.A.Laine, Inorg. Chim. Acta, 1988, 141, 131.
133 F.Aladar Bencsath and F.H.Field, Anal. Chem., 1986, 58, 679.
134 O.W.Hand, S.M.Scheifers, and R.G.Cooks, Int. J. Mass Spectrom. Ion Processes, 1987, 78, 131.
135 J. Allison, Prog. Inorg. Chem., 1986, 34, 627.
136 C.Schulze, H.Schwarz, D.A.Peake, and M.L.Gross, J. Am. Chem. Soc., 1987, 109, 2368.
137 R.B.Freas and J.E.Campana, J. Am. Chem. Soc., 1986, 108, 4659.
138 D.A.Peake and M.L.Gross, J. Am. Chem. Soc., 1987, 109, 600.
139 C.Schulze and H.Schwarz, J. Am. Chem. Soc., 1988, 110, 67.
140 C.B.Lebrilla, T.Drewello, and H.Schwarz, Organometallics, 1987, 6, 2268.
141 C.B.Lebrilla, T.Drewello, and H.Schwarz, Int. J. Mass Spectrom. Ion Processes, 1987, 79, 287.
142 C.B.Lebrilla, T.Drewello, and H.Schwarz, J. Am. Chem. Soc., 1987, 109, 5639.
143 C.B.Lebrilla, C.Schulze, and H.Schwarz, J. Am. Chem. Soc., 1987, 109, 98.
144 C.B.Lebrilla, T.Drewello, and H.Schwarz, Organometallics, 1987, 6, 2450.
145 D.V.Zagorevskii, A.V.Bulatov, V.G.Kartsev, and I.K.Yakushchenko, Bull. Acad. Sci., USSR Div. Chem. Sci. (Engl. Transl.), 1987, 36, 714.

# 11
# High-temperature Mass Spectrometric Studies of Inorganic Systems

BY E. R. PLANTE AND J. W. HASTIE

## 1 Introduction

High temperature mass spectrometry (HTMS) has found wide use for the molecular identification and thermodynamic characterization of the vapor phase over high temperature inorganic solid and liquid systems. Initial studies, carried out during the period 1948-1953, are attributed to Ionov[1], Honig[2], and Chupka and Inghram[3]. In the interim, numerous reviews which include the basis of the method, its limitations, and surveys of studies carried out using the technique, have been published. The first review in 1960 by Inghram and Drowart[4] introduced a brief discussion of the method and tabulated the results obtained in the field at that time. As outlined below, interim results have been tabulated in many subsequent reviews and documentation provided for most of the work carried out in high temperature mass spectrometry. A second review by Drowart[5] covered new techniques introduced in the field during its first ten years in addition to providing a comprehensive bibliography of the numerous studies completed. Grimley[6] and Drowart and Goldfinger[7] provided reviews which emphasized the basis of the method, including calibration procedures used to obtain absolute vapor pressures and estimation methods used for total electron impact ionization cross sections. In 1969, Drowart[8] emphasized the relationship between the mass spectrometric measurements and the thermodynamic results obtained and discussed some of the remaining problem areas. In 1971, Hastie[9] reviewed the application of mass spectrometry to molten mixed halide systems. Stafford[10] stressed problem areas of HTMS while Gorokhov and Semenov[11] discussed the problems of unravelling the ionization efficiency curves formed from electron impact ionization of complex species and suggested that some of the complex positive ions underwent temperature dependent fragmentation. This effect could lead to erroneous results in thermodynamic data obtained from vapor pressures, especially with regard to second and third law methods, if not taken into account. Chatillon et al.[12] reviewed mass spectrometric techniques for the determination of solution activities in condensed phases using multiple Knudsen cells with one of the pure phases serving as a reference. The review by Gingerich[13] stressed vaporization processes for various classes of high temperature materials such as carbides, alloys, nitrides etc. and covered most of the literature through 1976. In 1984, Hastie[14] reviewed recent developments in mass spectrometric sampling of very complex systems, underscoring high pressure sampling mass spectrometry and kinetic processes, and indicated areas in the field where further work was needed. More recently, Murad[15] has reviewed instrumentation in modern mass spectrometry. The most recent review by Drowart[16] has covered the recent literature through 1984 and contains, in addition to the usual coverage of thermochemical studies, data on experimental and theoretical cross sections. The present review emphasizes recent work, covering mainly the literature from 1985 through 1987, and is restricted to areas which deal with high temperature thermodynamics and related studies. In addition to the various topical reviews, several books on high temperature mass spectrometry have also been published recently.[17,18]

## 2 Classical Method

The now classical approach to high temperature mass spectrometry combines the Knudsen effusion method for vapor pressure measurement with mass spectrometric identification and evaluation of the ion current of ionic species (usually singly charged positive ions) produced by electron impact. Prior to the use of mass spectrometry, the vapor species above slightly volatile materials, as determined by gravimetric Knudsen effusion measurements, were identified by indirect methods. For such measurements, the vapor pressure is related to the rate of vaporization by the equation:

$$P = \frac{\Delta m}{C\,a\,t} (2\pi R\,T/M)^{1/2} \qquad (1)$$

where $\Delta m/a\,t$, the rate of vaporization, is the weight loss in time $t$ through an orifice of area $a$. The Clausing factor, $C$ corrects for non-ideality of molecular effusion through a real orifice, $R$ is the Universal gas constant, $T$ the absolute temperature, and $M$ the molecular weight of the effusing species. When using a mass spectrometer to detect the vapor species, the vapor pressure of the i'th species is related to the ion current $I_i$ and the temperature by the expression:

$$P_i = k_i I_i T_i \tag{2}$$

where $P_i$ is the pressure, $k_i$ the mass spectrometer constant, $I_i$ the ion current derived from the molecular species i, and T the absolute temperature. The value of the mass spectrometer constant $k_i$, which is species dependent, may be determined in several ways. The more common calibration methods include:

1. Quantitative vaporization of a known amount of the substance under investigation.
2. Quantitative vaporization of a known amount of a standard substance (such as Ag).
3. Use of a standard or known pressure together with ionization cross section data.
4. Use of ion ratios and ionization cross section data for a reaction whose equilibrium constant is known.

For the vaporization of a known amount of material it may be assumed that the vaporization process yields identical pressures by the Knudsen gravimetric and mass spectrometer methods. This will strictly be true only if a single vapor species is formed. Equations (1) and (2) can be set equal and solved for the mass spectrometer constant which gives :

$$k_i = \frac{\Delta m \ (2\pi R)^{1/2}}{C \ a \ M^{1/2} \ I_i \ T^{1/2} \ t} \tag{3}$$

When the calibration is based on vaporization of a known mass of material, no cross sections are required for determination of the mass spectrometer constant. If there is more than a single isotopic species present, the sum may be made over all species or use may be made of the isotopic ratio of the species measured. For consistency, and to avoid introducing error, the same choice of ion currents needs to be made for equations (2) and (3). One can also use the species-dependent constant in equation (3) to define a mass spectrometer constant which includes only geometrical factors. This would be defined as:

$$k = \frac{k_i}{\sigma_i \gamma_i} = \frac{k_j}{\sigma_j \gamma_j}. \tag{4}$$

The species-independent mass spectrometer constant k can now be used to define the species-dependent constant for species j in terms of ionization cross section, $\sigma$, and multiplier sensitivity factors, $\gamma$.

When a standard substance such as silver is used to determine the mass spectrometer constant, several species-dependent factors need to be taken into account. One of the most frequently used expressions is equation (5) which has also been discussed in prior reviews:

$$P_u = I_u T_u \frac{P_s}{I_s T_s} \frac{\sigma_s}{\sigma_u} \frac{\gamma_s}{\gamma_u} \frac{\Delta E_s}{\Delta E_u} \; . \tag{5}$$

The subscripted s terms and u terms represent the pressure, ion current, temperature, maximum ionization cross section, multiplier efficiency, and energy difference between the ionization energy and appearance potential of the standard and the unknown respectively. Several approximations have been included in the derivation of this equation and are frequently made in its use. The E terms serve the purpose of correcting the maximum cross section at $E_{max}$ to the ionizing voltage used. As an approximation one can write the equation:

$$\sigma_e = \sigma_{max} (E - AP)/(E_{max} - AP) \tag{6}$$

where $\sigma_e$ is the cross section at the ionizing voltage actually used, AP the appearance potential, $\sigma_{max}$ the maximum cross section, and $E_{max}$ the ionizing energy at the maximum cross section. One may note that in equation (5) it has been assumed that the factor $(E_{max} - AP)$ is constant and can be ignored. For some molecules, this may not be a good approximation and application of this correction factor may not be appropriate. The factor $\gamma$, is used to correct for the difference in the average number of electrons formed at the first collision of the ions with the conversion dynode of the multiplier. This factor may be measured or, if not, is frequently assumed to be inversely proportional to the square root of the molecular weight of the impinging ion. Equation (5) is most frequently used for studies involving magnetic mass spectrometers and may need to be modified for quadrupole mass spectrometers, depending on the type of detector and tuning being used. For example, quadrupole spectrometers used in our laboratory and tuned for a mass range of 1 - 200u (u is universal mass unit) have shown significant mass discrimination and have been calibrated with a standard gas mixture containing He, Ne, Ar, $N_2$, Kr and Xe. In this case, the ion current is set equal to:

$$I_i = k_m f_i \sigma_i A_i \tau_i \tag{7}$$

where $k_m$ is the machine constant, $f_i$ the fraction of species i present in the gas mixture, $\sigma_i$ the cross section, $A_i$ the isotopic abundance, and $\tau_i$ the transmission coefficient. The transmission coefficient is then expressed as the ratio of $\tau_i/\tau_{ref}$ where $\tau_{ref}$ is a reference transmission. The transmission coefficient defined in this way also includes the multiplier efficiency besides differences in transmission through the mass filter. Our experience shows that this factor may depend strongly on the multiplier being used. For example, replacement of the multiplier with a new or different type completely changes the transmission function. The mass discrimination is less important when the spectrometer is tuned for a broader mass range, such as 1-800u, or if the detector is equipped with a conversion dynode. It seems clear that some electron multipliers may not have proper characteristics to be used for quantitative high temperature mass spectrometry because the output current is not a linear function of the input current. Unfortunately, when this happens it is frequently assumed that the mass spectrometer, rather than the detector, is not suitable for high temperature mass spectrometry.

## 3 Electron Impact Ionization Cross Sections

It is customary in high temperature mass spectrometry to use electron energies from a few volts above the appearance potential to as high as 100 eV. Many of the earlier studies were done using 70 eV ionizing electrons but lower ionizing energies have become more common in later studies in order to (hopefully) decrease the effects of fragmentation processes and ionization of hydrocarbon background vapors. The ions formed will generally be a combination of singly charged ions, formed from neutral parent species, and a number of multiply charged ions and fragment ions, depending on the ionizing energy used.

Three basic assumptions are used or must be considered in the analysis of the mass spectral ion data. These are: (1) the ion intensity used results from the ionization of only a single type of parent molecule; (2) the cross section of the molecular precursor is the sum of the cross sections of the atomic components of the precursor as suggested by Otvos and Stevenson[19]; and (3) the pressure is proportional to the ion current times the absolute temperature [Eqn. (2)].

Each of these basic assumptions may be subject to some degree of uncertainty. An implicit assumption is that the identification of an ion should make sense from a chemical or chemical thermodynamic point of view. For example, ionization of the vapor over NaCl yields mainly $Na^+$, $Na_2Cl^+$, $NaCl^+$, and $Na_3Cl_2^+$. From the thermodynamic data for NaCl and other non-mass spectrometric vaporization data, it is quite certain that the Cl-containing ions are formed from different molecular precursors but it is less certain that the $Na^+$ is formed only by fragmentation of NaCl, which is a common assumption. Study of the appearance potential and ionization efficiency curves may give information as to which ions represent a parent vapor molecule and which are fragments formed by dissociative ionization of vapor molecules. For example, in the NaCl case, the appearance potential of $Na^+$ will be significantly higher than its ionization potential showing that it is formed as a fragment. Finally, a plot of $\ln I^+T$ vs $1/T$ for the $Na^+$ and $NaCl^+$ signals will yield slopes which are the same, within experimental error, demonstrating that $Na^+$ is formed largely from NaCl.

The second assumption, that the total ionization cross section is the sum of the atomic ionization cross sections, is generally believed to be correct within a factor of two. However, the application of this assumption has been shown to be incorrect in some specific cases. For example, White et al.[20] concluded from microbalance and mass spectrometric measurements that the ratio of $\sigma\gamma$ of $Li/Li_2O$ is 1.6 while additivity would suggest a value less than 0.5. In addition, application of the additivity rule requires that the assignment of fragments to specific precursors can be made unambiguously. Thus, fragmentation of different molecular species to yield the same ion can make it difficult to check on the additivity rule. For example, in the $NaF-ZrF_4$ system, Sidorov et al.[21] showed that the molecule $NaZrF_5(g)$ fragments into $Na^+$ and $ZrF_3^+$ but both of these ions are also formed by other species in the equilibrium vapor, while the intensity of $NaZrF_5^+$ is very small. For application of the additivity rule, however, one will be required to sum the intensity of all fragments of the parent molecule. Therefore it is necessary to determine what fraction of the $Na^+$ and $ZrF_3^+$ ion intensity are due to other possible vapor species.

The third assumption essentially requires the same information as needed in assumptions (1) and (2) but, in addition, requires that the ionization process is independent of temperature. In studies of temperature dependent fragmentation processes Efimova and Gorokhov[22] suggested that the total ionization cross section is temperature independent but that the partial cross sections for individual ions are temperature dependent in certain cases. Relative ionization cross section measurements can be obtained by HTMS using several methods. These include: (1) the quantitative vaporization of known amounts of two elements or compounds and comparison of the integrated ion currents; (2) vaporization of a compound or azeotrope which evaporates congruently; (3) vaporization using double cell techniques; and (4) modulation and phase angle measurements of effusing vapor species. Method (1) is the most commonly applied. The reference substance may be a gas such as Ar (e.g., see Hastie[14]) or a condensed phase material such as Ag (e.g., see Sheldon and Gilles[23]). The composition of congruently vaporizing compounds or azeotropes may be used to determine the relative pressures of the vapor species if the vaporization process is not too complicated. Double cell methods described by Milne[24] require a sufficiently large orifice between the two cells to permit viscous flow between the chambers which limits the applicable temperature range. Berkowitz[25] devised a constant-flow double cell technique using a porous metal diaphragm between the two cells which increased the applicable temperature range. The double cell techniques were designed to determine the relative ionization cross sections of monomer and dimer alkali halide species as a check on the additivity rule. Beam modulation techniques have been applied by Fite and Brackmann[26] to determine the relative ionization cross section of $H/H_2$. Pottie[27], Hastie[14] and Drowart[16] have tabulated literature data for ionization cross sections pertinent to HTMS, but such data are still relatively sparse.

Sidorov[28] and co-workers have pioneered the isothermal evaporation method as a procedure for determining relative ionization cross sections and applied it to the two-component metal fluoride systems which have complex vapor compositions. This method is similar in concept to the double cell methods but the variation in ion current as a function of time is due to the change in the concentration of the condensed and vapor phase components during the vaporization process. The data obtained consist of ion current, time, effusion cell parameters and weight of the sample for a total evaporation experiment, or the weight-loss at some interim time during the experiment. Analysis of the data allows one to determine relative ionization cross sections and contributions to a single ion intensity from two, or more, fragmentation processes as well as the partial pressures of vapor molecules. When applied to solutions, the Gibbs-Duhem equation can be used to relate the activities of components as a function of composition in the solution phase.

## 4 Recent Literature Data

The present review covers literature mainly from the period 1985 through 1987. Literature citations were obtained primarily by using a computer search of Chemical Abstracts based on authors known to be active in HTMS, in addition to key phrases including mass spectrometry, high temperature, vaporization, and thermodynamics. Some citations were also obtained in selected fields from literature reviews not currently in print. Rather than discussing each citation, the subject data of each paper has been categorized into tables which lists the material studied, the principle thermodynamic property determined, and the reference. The various categories selected are as follows and the Tables are collected together in the Appendix.

4.1 Ion-Molecule Measurements.-The study of ion-molecule equilibria is a rapidly expanding area of high temperature mass spectrometry, with unique applications. This area has recently been reviewed by Sidorov, Zhuravleva and Sorokin[29]. Measurements may involve both negative and positive ions generated thermally in the Knudsen cell as well as neutral molecular species whose concentrations are determined by normal electron bombardment methods. Typical thermodynamic data obtained in these studies include electron affinities, heats of formation of negative and positive ions, activities of components in solution, and the demonstration of compatibility of thermodynamic data by providing key data for closure of thermodynamic cycles. Data reported during the literature survey period are given in Table 1.

**4.2 Gaseous Oxide Measurements.**-Table 2 lists data in which the major objective has been to obtain thermodynamic data for gaseous oxide species. Most of the recent studies have involved all-gaseous reactions which are advantageous because the equilibria are independent of activities in the condensed phase. A disadvantage is the necessity of having accurate thermodynamic data for vapor species other than the one being determined. Thus, in Table 2, all data have been obtained by equilibrium measurements of all-gaseous reactions except for the data given by Kleinschmidt[52], Banchorndhevakul[55], and Marushkin[56].

**4.3 Activity of Oxide Systems.**-Thermodynamic activity studies of binary and higher order oxide systems are summarized in Table 3. Measurement techniques used include the method of Belton and Fruehan[77] or Neckel and Wagner[78], which allows for the determination of activities from the evaluation of relative ion intensities at fixed temperature and as a function of composition. This technique has found widespread use for activity determinations in mixed alkali oxide silicate systems, alloy systems and oxide systems in which the components have partial pressures which produce ionization currents whose ratios can be determined. Another approach is the use of double or multiple cell techniques such as that pioneered by Chatillon[12] in which the ratio of ion current at reduced activity to that of the pure material contained in a different cell in its reference state is determined. Some activity data are still being obtained by determination of the partial pressures of the components at reduced activity while activities of non-volatile components are calculated by use of a Gibbs-Duhem integration.

**4.4 Simple Halide Systems.**-There is still considerable interest in simple halide systems as summarized in Table 4. Many studies address the tendency of halides to form polymeric vapor species. Much of the recent work has involved classical Knudsen-effusion mass spectrometric vaporization measurements, while a few studies have also used isomolecular exchange reactions to determine heats of formation of a particular vapor species with reference to already established values of other species involved in the reaction. Techniques for introducing fluorinating agents to produce molecular species of interest were devised by Korobov[95] and Bondarenko et al.[96] to provide Pt-fluoride species. Measurements by DeMercurio et al.[91] used angular distribution mass spectrometry to investigate fragmentation in the Na-I and K-I systems as well as to obtain thermodynamic data for polymer species.

**4.5 Mixed Halide Systems.**—Studies of mixed halide systems are summarized in Table 5. Data have been obtained on activities in the condensed phase, or thermodynamic data for the formation of complex mixed halide species such as the mixed species of $NaI-PbI_2$ discussed by Hilpert[110]. Included with the mixed halide salts for this review are the decomposition studies of the hexafluoroxenates by Kiselev[116].

**4.6 Alloy Systems.**—Recent studies on alloy systems are summarized in Table 6. Most interest in alloy systems is centered on the determination of activities, obtained usually by the methods of Belton and Fruehan[77] or Neckel and Wagner[78]. Besides activity data it is also possible to determine phase boundary data as a function of temperature. Activity data are sometimes expressed in terms of excess thermodynamic functions. Some of the intermetallic systems have been studied at sufficiently low temperatures so that intermetalic compounds coexist and thermodynamic data such as heats or Gibbs energy changes of intermetallic phases can be determined, as in the studies of Schmidt[133] and Storms[134].

**4.7 Oxy-salts.**—Table 7 lists studies reported for oxy-salts. The introduction of permanent gases or low temperature vapors into the Knudsen cell to react with high temperature vapors or salts was necessary to produce the desired vapor species for some of the studies; e.g., Gorokhov[136], Milushin[137,138], Farber[141], Ahlrichs[150], Binnewies[151,152], and Shinmel[153]. Other data were obtained using conventional vaporization methods. Brittain et al.[159] studied the kinetics of vaporization of $ZnSO_4$, noting the failure to obtain equilibrium between $SO_2$, $SO_3$, and $O_2$.

**4.8 Group VB and VIB.**—Studies of group VB and VIB compounds listed in Table 8 include determination of heats of formation of As-Se vapors and Na and K vapors formed with Sb. Activities have been determined in various phases containing Cu-As-S or Cu-Sb-S phases. Nakamura et al.[165] have measured vapor pressures and heats of vaporization of GeS, SnS and CdS using a transpiration mass-spectrometer system based on that developed by Hastie and Bonnell (see ref. 14).

## Appendix: Tables

Table 1 Survey of ion-molecule reaction studies

| Material or Species | Property* | Reference |
|---|---|---|
| $AlF_4^-$ | $\Delta H_f^o$ | Nikitin 30 |
| $AuF_3$, $Au_2F_6$, $Au_3F_9$, $AuF_4^-$, $MnF_3$ | $\Delta H_r^o$ | Chilingarov 31 |
| $CoF_4^-$ | $\Delta H_f^o$ | Sidorov 32 |
| $CsBO_2(c)$, $Cs_2BO_2^+$, $Cs(BO_2)_2^-$ | $\Delta H_f^o$ | Sidorova 33 |
| $DyI_3$, $HoI_3$, $CsI$, $NaI$ | ion-molecule equil., activities | Kaposi 85a |
| $FeF_3^-$, $FeF_4^-$, $Fe_2F_5^-$, $Fe_2F_6^-$, $Fe_2F_7^-$ | $\Delta H_f^o$, EA | Sidorov 35 |
| $FeF_3^-$, $CuF_3^-$ | $\Delta H_f^o$ | Kuznetsov 36 |
| $GaF_4^-$, $Ga_2F_7^-$, $GaAlF_7^-$ | $\Delta H_f^o$ | Zhuravleva 37 |
| $LaCl_4^-$, $La_2Cl_7^-$ | $\Delta H_f^o$ | Butman 38 |
| $PO_2^-$, $PO_3^-$, $NaPO_2(g)$ | $\Delta H_f^o$ | Rudnyi 39 |
| Mo-U-Fe-F-O-e$^-$ | $\Delta H$ of 13 reactions | Borshchevskii 40 |
| $NaCrO_4^-$ (g) | $\Delta H_f^o$, $D_o^o(Na^+\text{-}NaCrO_4^-)$ | Vovk 41 |
| $PtF_6^-$, $RuF_4^-$, $RuF_5^-$ | EA | Korobov 42 |
| $VF_3$, $VF_4^-$, $VF_5^-$ | $\Delta H_f^o$, EA | Igolkina 43, 44 |

*Terminology for this and other tables is:

| | |
|---|---|
| $\Delta H_f^o$, $\Delta S_f^o$ | are standard enthalpy and entropy of formation, respectively |
| $\Delta H_v$ | enthalpy of vaporization, |
| $D_{o,at}^o$ | dissociation energy to atoms at 0 K |
| $\Delta G^E$, $\Delta H^E$, $\Delta S^E$ | are excess Gibbs, enthalpy and entropy functions, respectively |
| $D_o^o$ | atomization energy |
| EA | electron affinity, |
| VP | vapor pressure |
| ADMS | angular distribution mass spectrometry |

Table 2 Survey of oxide vaporization studies

| Material or Species | Property | Reference |
|---|---|---|
| $Al_2O(g)$, $Al_2O_2(g)$ | $\Delta H_f^o$ | Milushin 45 |
| $BO_2(g)$ | EA | Semenikhin 46 |
| $EuNbO_3(g)$, $EuNbO_2(g)$, $EuNb_2O_6(g)$, $NbO(g)$, $NbO_2(g)$ | $D_{o,at}^o$, $\Delta H_f^o$ | Balducci 47 |
| $EuTiO_3$ | $D_{o,at}^o$ | Balducci 48 |
| $Eu\text{-}X\text{-}O_n$, n= 1-4, X=Ti, Zr, Hf, V, Nb, Ta, Cr, Mo, W | $\Delta H_f^o$ of ternary oxide gases | Balducci 49 |
| $Li_4O(g)$, $Li_5O(g)$ | $\Delta H_f^o$ | Wu 50 |
| $Nb_2O_4(g)$, $Nb_2O_5(g)$, $Nb_4O_9(g)$, $Nb_4O_{10}(g)$ | $\Delta H_f^o$ | Balducci 51 |
| $PaO_2(g)$, $PaO(g)$ | VP, $\Delta H_f^o$ | Kleinschmidt 52 |
| $Ti_2O_3$, $Ti_2O_4$, $TiO_2$ | $D_{o,at}^o$, AP | Balducci 48, 53 |
| $V_2O_3(c)$, $V_2O_3(c)\text{-}VO(c)$ | $\Delta H_v$, V, VO, $VO_2$ | Banchorndhevakul 54 |
| (0.001<O/V<0.145)(ss) | VP(V, VO), $\Delta S^E$ | Banchorndhevakul 55 |
| $W_2O_6$, $W_3O_9$, $W_3O_8$, $W_4O_{12}$ | $\Delta H_s^o$, $\Delta H_f^o$ | Marushkin 56 |

Table 3 Survey of activity determinations in oxide systems

| Material or Species | Property | Reference |
|---|---|---|
| $B_2O_3$-$SiO_2$(g,l) | P vs comp, $\Delta G^E$, $\Delta H^E$ | Shults 57 |
| $Cu_2O$-$K_2O$-$SiO_2$(l) | activity $SiO_2$ | Kowalska 58 |
| $GeO_2$- $SiO_2$(g,l) | activities | Shul'ts 59 |
| $K_2O$-$Na_2O$-$SiO_2$(g,l) | $\Delta G^E$, activities | Chastel 60 |
| $KAlSi_3O_8$-$NaAlSi_3O_8$(g,l) | activities | Frazer 61 |
| (K-Na)$AlSi_4O_{10}$ (K-Na)$AlSi_5O_{12}$ | activities | Frazer 62 |
| $Na_2O$-$SiO_2$(g,l) | activity | Shul'ts 63 |
| $Na_2O$-$B_2O_3$-$SiO_2$(g,l) | VP, $\Delta H_v$, activity | Asano 64 |
| FeO-$SiO_2$, FeO-$SiO_2$-CaO, $Al_2O_3$-$SiO_2$ | activity of FeO or $SiO_2$ | Dhima 65 |
| $Na_2O$-$Cs_2O$-$B_2O_3$-$SiO_2$(g,l) | Na and Cs borate pressures, $\Delta H_v$ | Asano 66 |
| 0.47<O/Ti<1.40 | activities, Ti, TiO | Granier 67 |
| PbO-$B_2O_3$(l) | activity | Semenikhin 68 |
| PbO-ZnO, ZnO-$B_2O_3$ | activity | Semenikhin 69 |
| PbO-ZnO-$B_2O_3$ | activity | Semenikhin 70 |
| $Sc_2O_3$-$HfO_2$(ss) | relative P vs comp. | Semenov 71 |
| $Sc_2O_3$(c), $Sc_2O_3$-$ZrO_2$(ss) | VP ScO, activities | Belov 72, 73 |
| $UO_{2-x}$/$UO_2$ | U activity, VP as fn of T and x | Storms 74 |
| $UO_2$(c), $(U_{1-y}Nb_y)O_{2+x}$ | VP vs composition | Matsui 75 |
| $ZrO_2$-$HfO_2$(ss), $HfO_2$-$Y_2O_3$(ss), $ZrO_2$-$Y_2O_3$(ss) | activity, $\Delta G^E$ | Belov 76 |

Table 4 Survey of simple halide studies

| Material or Species | Property | Reference |
|---|---|---|
| $AlF_3$ | mass spectrum | Menz 79 |
| $Br_2$, Br/C | desorption | Sigurdsson 80 |
| $CaI_2$, $(CaI_2)_2$ | VP, $\Delta H_s^o$, $\Delta S_s^o$ | Saha 81 |
| $CdCl_2$, $CdBr_2$, $CdI_2$ | VP, $\Delta H_s^o$, $\Delta S_s^o$ | Skudlarski 82 |
| CsI | mass discr., $\sigma$ | Gorokhov 83 |
| $(CsI)_i$, i=3,4 | $\Delta S_s^o$, $\Delta H_s^o$ | Hilpert 84 |
| $FeI_3$, $(FeI_2)_i$, i= 1,2,3 | $\Delta H_s^o$, $\Delta S_s^o$, VP | Hilpert 85 |
| $FeI_2(s)$, $(FeI_2)_{1,2}(g)$ | $\Delta H_s^o$, VP | Mucklejohn 86 |
| FeF, $FeF_2$, MgF | $\Delta H_o^o(f)$, isomolecular | Gorokhov 87 |
| FeF, $FeF_2$, MgF, NiF | $\Delta H_f^o$ | Khodeyev 88 |
| $HgI_2$, $Hg_2I_2(c)$ | $\Delta H_s^o$, VP | Piechotka 89 |
| $HoI_3$, $Ho_2I_6$ | $\Delta H_s^o$, $\Delta H_f^o$ | Kaposi 90 |
| $(KI)_n$, n=1-4 | ADMS, $\Delta H_s^o$ | DeMercurio 91 |
| $MoF_4(c)$, $MoF_5$ | $\Delta H_d^o$ | Malkerova 92 |
| $(NaI)_n$, n=1-4 | ADMS, $\Delta H_s^o$ | DeMercurio 91 |
| $PbF_2(g)$, $Pb_2F_4(g)$ | $\Delta H_s^o$, VP | Korenev 93 |
| $(PbI_2)_i$, i=1,2 | $\Delta H_s^o$, $\Delta S_s^o$, VP | Hilpert 94 |
| $PtF_6(g)$ | $\Delta H_f^o$ | Korobov 95 |
| $PtF_4 + PtF_6 = Pt_2F_{10}$ | $\Delta H_r^o$ | Bondarenko 96 |
| $ScI_3$, $Sc_2I_6$ | $\Delta H_s^o$, VP | Dettinggmeijer 97 |
| $(SnI_2)_i$, i=1,2 | $\Delta H_s^o$, $\Delta S_s^o$, VP | Hilpert 94 |
| $TbF_4(c)$ | $\Delta H_d^o$, $VP(F_2)$ | Nikulin 98 |
| $ThI_4$, $ThOI_2(c)$ | VP | Flesch 99 |
| UBr, $UBr_2$, $UBr_3$, $UBr_4$, $UBr_5$ | $\Delta H_f^o$ | Lau 100 |
| $UF_5$ | $\Delta H_f^o$ | Bondarenko 101 |
| $WF_4(c)$, $WF_6$ | $\Delta H_d^o$ | Malkerova 92 |
| $ZnCl_2$ | VP, $\Delta H_s^o$, $\Delta S_s^o$ | Skudlarski 102 |
| $ZnI_2$, $(ZnI_2)_2$ | $\Delta H_s^o$, $\Delta S_s^o$, VP | Hilpert 103 |
| $ZrCl(c)$, $ZrCl_4(g)$ | $\Delta H_d^o$ | Makarov 104 |

Table 5 Survey of mixed halide studies

| Material or Species | Property | Reference |
|---|---|---|
| CsI-NaI | activities, $\Delta H^E$, $\Delta S^E$ | Kaposi 105 |
| KCl-KBr | VP, $\Delta G^E$ | Miller 106 |
| (LiF, NaF, KF, RbF)- $BeF_2$ | P, $\Delta G_{mix}$ vs comp. | Korenev 107 |
| $SnX_2$ + $NaX$<br>X = Cl, Br, I | $\Delta H_r^o$, complex ions | Mucklejohn 108 |
| $DyI_3(NaI)_{1,2}$, $HoI_3(CsI)_{1,2}$ | $\Delta H_f^o$ | Gavrilin 109 |
| $NaFeI_3$, $Na_2FeI_4$, $NaPbI_3$ | VP, $\Delta H_r^o$ | Hilpert 110 |
| TlCl-TlI | $\Delta G_m^E$, | Kapala 111, 112, 113 |
| NaI-CsI | activities | Kaposi 114, 115 |
| $M_2XeF_6$, M = Na, K, Rb, Cs | decomposition | Kiselev 116 |
| $NaSnI_3$, $ScSnI_3$, $ScSnI_5$ | K, $\Delta H_r^o$, $\Delta H_d^\sigma$ | Miller 117 |
| NaCl-NaBr(s) | $\Delta G_m^E$ | Miller 118 |

Table 6 Survey of alloy studies

| Material or Species | Property | Reference |
|---|---|---|
| Al-Ni | phase boundary | Hilpert 119 |
| As-Pd, As-Pd-B | activity, Pd, As | Storms 120 |
| B-Ni, B-Ni-C | activity, Ni | Storms 121 |
| B-Pd, B-Pd-C | activity, Pd, | Storms 122 |
| Co-Fe(s) | activities, $\Delta H^E$, $\Delta S^E$ | Tomiska 123 |
| Co-Fe(l) | activities, $\Delta H^E$, $\Delta S^E$ | Tomiska 124 |
| Co-Ti(l) | activities | Ueda 125 |
| Cu-Si (l) | activity, $\Delta G^E$ | Bergman 126 |
| $Cr_3Si$, $Cr_5Si_3$, $CrSi$, $CrSi_2$ | activity, $\Delta H_f^o$ | Myers 127 |
| Fe-Ni(l) | $\Delta G^E$, $\Delta H^E$ | Tomiska 128 |
| Fe-Ni(c) | $\Delta G^E$, $\Delta H^E$ | Tomiska 129 |
| Fe-Ta(l) | activity | Ichise 130 |
| Ni-Pd (l) | activity, $\Delta \bar{H}$ | Oishi 131 |
| Ni-Te, $NiTe_{0.634}$ | VP, Te, $Te_2$, $\Delta H_s^o$ | Viswanthan 132 |
| $TaAl_3(c)$, $Ta_2Al_3(c)$,<br>$Ta_2Al(c)$, $Ta_4Al(c)$ | $\Delta H_f^o$ | Schmidt 133 |
| $VSi_2(c)$, $V_6Si_5(c)$,<br>$V_5Si_3(c)$, $V_3Si(c)$ | $\Delta G_f^o$, $\Delta H_f^o$ | Storms 134 |
| $V_3Si(c)$, $V_5Si_3(c)$, | activities, $\Delta H_f^o$ | Myers 135 |

Table 7 Survey of oxy-salt studies

| Material or Species | Property | Reference |
|---|---|---|
| AlOH(g), AlClO(g) | $\Delta H_f^o$ | Gorokhov 136 |
| AlOH(g) | $\Delta H_f^o$ | Milushin 137 |
| AlClO(g) | $\Delta H_f^o$ | Milushin 138 |
| Ca, Sr, Ba(ReO$_4$)$_2$ | $\Delta H_f^o$, $\Delta S_f^o$ | Nikolaev 139 |
| CaPO$_2$(g), SrPO$_2$(g), BaPO$_2$(g) | $\Delta H_f^o$ | Lopatin 140 |
| CaOH(g), Ca(OH)$_2$(g), KCaO(g) | $\Delta H_f^o$ | Farber 141 |
| CaP$_4$O$_{11}$(c) | vapor phase identity | Golubchenko 142 |
| In$_2$MoO$_4$(g) | $\Delta H_f$ | Kaposi 143 |
| KAlO(g), KSiO(g) | $\Delta H_f^o$ | Farber 144 |
| KBO$_2$(c,l) | $\Delta H_v^o$ | Farber 145 |
| LiBO(g), Li(g), BO(g) | $\Delta H_r^o$ | Neubert 146 |
| Li,Na,K,Rb,--CsSbO$_3$(c) | VP all but Li | Semenov 147 |
| Mg(PO$_3$)$_2$(c), Mg$_2$P$_2$O$_7$ (c) | VP, $\Delta H_f^o$ | Lopatin 148 |
| NaPO$_2$, PO$_2^-$, PO$_3^-$ | $\Delta H_f^o$ | Rudnyi 149 |
| OPF(g) | $\Delta H_f^o$ | Ahlrichs 150 |
| POBr(g) | $\Delta H_f^o$ | Binnewies 151 |
| SPF(g) | $\Delta H_f^o$ | Binnewies 152 |
| Si$_2$OF$_6$(g) | $\Delta H_f^o$, $\Delta S_f^o$ | Shinmel 153 |
| Sr$_3$(PO$_4$)$_2$ | decomposition | Lopatin 154 |
| ThOI$_2$(c), ThI$_4$ | diss'n. press. | Flesch 155 |
| TlClO$_4$(g) | VP, $\Delta H_f^o$ | Nikolaev 156 |
| Tl$_2$CrO$_4$(g) | VP, $\Delta H_s^o$ | Kuligina 157 |
| UO$_2$F$_2$(g), UOF$_4$(g), UF$_5$(g) | $\Delta H_f^o$, $P_{total}$ | Lau 158 |
| ZnSO$_4$(c), Zn$_2$SO$_5$(c) | kinetics of decomp. | Brittain 159 |

Table 8 Survey of Group VB and VIB studies

| Material or Species | Property | Reference |
|---|---|---|
| As-Se, $As_4Se_4(g)$, $As_4Se_3(g)$, $As_2Se_2(g)$, AsSe(g) | $\Delta H_f^o$ | Steblevskii 160 |
| $As-Cu_2S-Cu$ (l) | activities | Hino 161 |
| Cu-As-S, Cu-Sb-S | activities | Dziewidek 162 |
| Cu-Sb, Cu-Sb-S | activities | Hino 163 |
| Cu-Sb-S | activities | Dziewidek 164 |
| GeS(g), SnS(g), CdS(g) | VP, $\Delta H_v^o$ | Nakamura 165 |
| InAs-GaAs | activities | Chatillon 166 |
| InGaS(g) | $\Delta H_f^o$, isomolecular | Mukdeeprom 167 |
| NaSb(g), $K_2Sb_4(g)$, $K_4Sb_4(g)$, $Na_6Sb_4(g)$ | VP, $\Delta H_r^o$ | Scheuring 168, 169 |
| SbS (g) | $\Delta H_f^o$ | Hino 170 |
| Se(l), Tl doped | kinetic study $\Delta H_v^o$, $\Delta H_v^*$ | Huang 171 |
| $Tl_3AsS_4$ | VP ($TlAsS_2$), $\Delta H_f^o$ | Hirayama 172 |
| Te(c), $FeTe_{0.9}$(c)-Fe(c) | VP $Te_2$, Te | Saha 173 |

## References

1. N. I. Ionov, <u>Dokl. Akad. Nauk SSSR</u>, 1948, <u>59</u>, 467.
2. R. E. Honig, <u>J. Chem. Phys.</u>, 1954, <u>21</u>, 371 (1954)
3. W. A Chupka and M. G. Inghram, <u>J. Chem. Phys.</u>, 1953, <u>21</u>, 371.
4. M. G. Inghram and J. Drowart, in 'Proceedings of an International Symposium on High Temperature Technology', Asilomar Calif., 1959. New York, McGraw-Hill 1960, p. 219.
5. J. Drowart, in 'Condensation and Evaporation of Solids', ed. E. Rutner, P. Goldfinger and J. P. Hirth, Gordon and Breach, New York, 1964, p. 255.
6. R. T. Grimley, in 'Characterization of High Temperature Vapors', ed. J. L. Margrave, Wiley, New York, 1967, p. 195.
7. J. Drowart, and P. Goldfinger, <u>Angew. Chem., Intl. Ed.</u>, 1967, <u>6</u>, 581.
8. J. Drowart, in 'Mass Spectrometry', Proceedings of the International School on Mass Spectrometry, ed. J. Marsel, Ljubjana, Yugoslavia, 1969, p. 187.
9. J. W. Hastie, in 'Advances in Molten Salt Chemistry', ed. J. Braunstein, G. Mamantov, and G. P. Smith, Plenum Press, New York, 1971, Vol. 1. p. 225.
10. F. E. Stafford, <u>High Temperatures-High Pressures</u>, 1971, <u>3</u>, 213.
11. L. N. Gorokhov and G. A. Semenov, <u>Adv. Mass Spectrometry</u>, 1971, <u>5</u>, 349.
12. C. Chatillon, A. Pattoret, and J. Drowart, <u>High Temperature-High Pressure</u>, 1975, <u>7</u>, 119.
13. K. A. Gingerich, in 'Current Topics in Materials Science', ed. E. Kaldis, North-Holland Publishing Co.,1980, p. 345.
14. J. W. Hastie, <u>Pure Appl. Chem.</u>, 1984, <u>56</u>, 1583.
15. E. Murad, in 'Spectrometric Techniques', ed. George A. Vanasse., Academic Press Inc., New York, 1985, vol. 4, p.181.
16. J. Drowart, in 'Advances in Mass Spectrometry 1985', ed. J. F. J. Todd, John Wiley & Sons, New York, 1986, p. 195.
17. L. N. Gorokhov, 'Mass Spectrometry in Inorganic Chemistry', Znanie, Moscow, 1984
18. L. N. Sidorov, M. V. Korobov,and L. V. Zhuravleva, 'Mass Spectrometric Thermodynamic Investigations.', Moscow University, 1985
19. J. W. Otvos and D. P. Stevenson, <u>J. Amer. Chem. Soc.</u>, 1956, <u>78</u>, 546.
20. D. White, K. S. Seshadri, D. F. Dever, D. E. Mann and M. I. Linevsky, <u>J. Chem. Phys.</u>, 1963, <u>39</u>, 2463.
21. L. N. Sidorov, P. A. Akishin, V. I. Belousov, and V. V. Shol'ts, <u>Russian J. Phys. Chem.</u>, 1964, <u>38</u>, 641.
22. A. G. Efimova and L. N. Gorokhov, <u>Teplofiz. Vysok. Temp.</u>, 1978, <u>16</u>, 1195.
23. R. I. Sheldon and P. W. Gilles, in 'Characterization of High Temperature Vapors and Gases', ed. J. W. Hastie, NBS-SP-561, US Gov't Printing Office, 1979, p231.
24. T. I. Milne, <u>J. Chem Phys.</u>, 1958, <u>28</u>, 717.
25. J. Berkowitz, H. A. Tasman, and W. A. Chupka, <u>J. Chem. Phys.</u>, 1962, <u>36</u>, 2170.
26. W. L. Fite and R. T. Brackmann, <u>Phys. Rev.</u>, 1958, <u>112</u>, 1141.
27. R. F. Pottie, <u>J. Chem. Phys.</u>, 1966, <u>44</u>, 916.
28. L. N. Sidorov and V. B. Shol'ts, <u>Int. J. Mass Spectrom. Ion Phys.</u>, 1972, <u>8</u>, 437.
29. L. N. Sidorov, L. V. Zhuravleva, and I. D. Sorokin, <u>Mass Spect. Rev.</u>, 1986, <u>5</u>, 73.
30. M. I. Nikitin, N. A. Igolkina, E. V. Skokan, I. D. Sorokin, and L. N. Sidorov, <u>Zh. Fiz. Khim.</u>, 1986, <u>60</u>, 62.
31. N. S. Chilingarov, M. V. Korobov, S. V. Rudometkin, A. S. Alikhanyan, and L. N. Sidorov, <u>Int. J. Mass Spectrom. Ion Processes.</u>, 1986, <u>69</u>, 175.
32. L. N. Sidorov, V. V. Nikulin, N. S. Chilingarov, N. S. Korobov, <u>Zh. Fiz. Khim.</u>, 1987, <u>61</u>, 1078.
33. I. V. Sidorova and L. N. Gorokhov, <u>Teplofiz. Vys. Temp.</u>, 1987, <u>25</u>, 1100.
34. O. Kaposi, L. Lelik, M. V. Korobov, N. S. Chilingorov, E. N. Garilov, L. N. Sidorov, and J. D. Sorokin, in 'Advances in Mass Spectrometry 1985', Part B, J. Wiley and Sons, New York, p. 1003.

35  L. N. Sidorov, A. Ya. Borshchevskii, O. V. Boltalina, I. D. Sorokin, and E. V. Skokan, Int. J. Mass Spectrom. Ion Processes, 1986, 73, 1.
36  S. V. Kuznetsov M. V. Korobov, L. N. Savinova, and L. N. Sidorov, Zh. Fiz. Khim., 1986, 60, 1285.
37  L. V. Zhuravleva, M. I. Nikitin, I. D. Sorokin, and L. N Sidorov, Int J. Mass Spectrom. Ion Processes, 1985, 65, 253.
38  M. F. Butman, L. S. Kudin, G. G. Burdukovskaya, K. S. Krasnov, and N. V Bozhko, Zh. Fiz. Khim., 1987, 61, 2880.
39  E. B. Rudnyi, O. M. Vovk, L. N. Sidorov, I. D. Sorokin, and A. S. Alikhanyan, Teplofiz. Vys. Temp., 1986, 24, 62.
40  A. Ya. Borshchevskii, L. N. Sidorov, O. V Boltalina, Dokl. Akad. Nauk SSSR. (Phys. Chem.), 1985, 285, 377.
41  O. M. Vovk, E. B. Rudnyi, L. N. Siderov, Vestn. Mosk. Univ. Ser. 2: Khim., 1987, 28, 221.
42  M. V. Korobov, S. V. Kuznetsov, S. V. Chilingarov, L. N. Sidorov, Dokl. Akad. Nauk SSSR. (Phys. Chem.), 1987, 295, 131.
43  N. A. Igolkina, M. I. Nikitin, O. V. Boltalina, and L. N. Sidorov, High Temp. Sci., 1986, 21, 111.
44  N. A. Igolkina, M. I. Nikitin, L. N. Sidorov, and O. V. Boltalina, High Temp. Sci., 1987, 23 89.
45  M. I. Milushin, A. M. Emel'yanov, L. N. Gorokhov, Teplofiz. Vys. Temp., 1987, 25, 52 .
46  V. I. Semenikhin, I. I. Minaeva, I. D. Sorokin, M. I. Nikitin, E. B. Rudnyi, and L N Sidorov, Teplofiz. Vys. Temp., 1987, 25, 666.
47  G. Balducci, G. Gigli, and M. Guido, Ber. Bunsen-Ges. Phys. Chem., 1987, 91, 635.
48  G. Balducci, G. Gigli, and M. Guido, J. Chem. Phys., 1985, 83, 1909.
49  G. Balducci, G. De Maria, G. Gigli, and M. Guido, High Temp. Sci., 1986, 22, 145.
50  C. H. Wu, Chem. Phys. Lett., 1987, 139, 357.
51  G. Balducci, G. Gigli, and M. Guido, J. Chem. Phys., 1986, 85, 5955.
52  P. D. Kleinschmidt, and J. W. Ward, J. Less-Common Met., 1986, 121, 61.
53  G. Balducci, G. Gigli, and M. Guido, J. Chem. Phys., 1985, 83, 1913.
54  W. Banchorndhevakul, T. Matsui, and K. Naito, Thermochim. Acta., 1985, 88, 301.
55  W. Banchorndhevakul, T. Matsui, and K. Naito, J. Nucl. Sci. Technol., 1986, 23, 873.
56  K. N. Marushkin, A. S. Alikhanyan, J. H. Greenberg, V. B. Lazarev, V. A. Malyusov, O. N. Rozanova, B. T. Melekh, and V. I. Gorgoraki, J. Chem. Thermodyn., 1985, 17, 245.
57  M. M. Shul'ts, G. G. Ivanov, and V. L. Stolyarova, Fiz. Khim. Stekla, 1986, 12, 285.
58  M. Kowalska, K. Skudlarski, and J. Botor, J. Chem. Thermodyn., 1986, 18, 997.
59  M. M. Shul'ts, V. L. Stolyarova, and G. G. Ivanov, Fiz. Khim. Stekla, 1987, 13, 830.
60  R. Chastel, C. Bergman, J Rogez, and J. C. Mathieu, Chem. Geol., 1987, 62, 19.
61  D. G. Fraser, and Y. Bottinga, Geochim. Cosmochim. Acta, 1985, 49, 1377.
62  D. G. Fraser, W. Rammensee, and A. Hardwick, Geochim. Cosmochim. Acta, 1985, 49, 349.
63  M. M. Shul'ts, V. L. Stolyarova, and G. G. Ivanov, Fiz. Khim. Stekla, 1987, 13, 168.
64  M. Asano and Y. Yasue, J. Nucl. Mater., 1986, 138, 65.
65  A. Dhima, B. Stafa, and M. Allibert, High Temp. Sci., 1986, 21, 143.
66  M. Asano, T. Kou, and Y. Yasue, J. Non-Cryst. Solids 1987, 92, 245.
67  B. Granier, C. Chatillon, and M. A. Allibert, High Temp. Sci., 1987, 23, 115.
68  V. I. Semenikhin, I. I. Minaeva, I. D. Sorokin, M. I. Nikitin, E. B. Rudnyi, and L N Sidorov, Fiz. Khim. Stekla, 1987, 13, 542.
69  V. I. Semenikhin, I. D. Sorokin, L. F. Yurkov, and L. N. Sidorov, Fiz. Khim. Stekla, 1987, 13, 667.

70  V. I. Semenikhin, I. D. Sorokin, L. F. Yurkov, and L. N. Sidorov, Fiz. Khim. Stekla, 1987, 13, 672.
71  G. A. Semenov, L. A. Kuligina, G. A. Tetrerin, E. M. Menchuk, and T. M. Shkol'nikova, Ukr. Khim. Zh., 1986, 52, 1123.
72  A. N. Belov, G. A. Semenov, G. A. Teterin, and T. M. Shkol'nikova, Zh. Fiz. Khim., 1987, 61, 893.
73  A. N. Belov, G. A. Semenov, G. A. Teterin, E. Menchuk, and T. M. Shkol'-nikova, Zh. Fiz. Khim., 1987, 61, 899.
74  E. K. Storms, J. Nucl. Mater., 1985, 132, 231.
75  T. Matsui and K. Naito, J. Nucl. Mater., 1985, 136, 69.
76  A. N. Belov and G. A. Semenov, Zh. Fiz. Khim., 1985, 59 589.
77  G. R. Belton and R. J. Fruehan, J. Phys. Chem., 1967, 71, 1403.
78  A. Neckel and S. Wagner, Ber. Bunsenges. Phys. Chem., 1969, 73, 210.
79  D. Menz, L. Kolditz, K. Heide, C. Schmidt, Ch. Kunert, Ch. Mensing, H. G. v. Schnering, and W. Hoenle, Z. Anorg. Allg. Chem., 1987, 551, 231.
80  A. Sigurdsson and L. Holmlid, J. Appl. Phys., 1987, 61, 2849.
81  B. Saha, K. Hilpert, and L. Bencivenni, in 'Advances in Mass Spectrometry 1985', Part B, J. Wiley and Sons, New York, NY, p. 1001.
82  K. Skudlarski, J. Dudek, J. Kapala, J. Chem. Thermodyn., 1987, 19, 857.
83  L. N. Gorokhov and N. E. Khandamirova, in 'Advances in Mass Spectrometry 1985', Part B, J. Wiley and Sons, New York, NY, p. 1031.
84  K. Hilpert, and L. Bencivenni, Surf. Sci., 1985, 156, 436.
85  K. Hilpert, R. Viswanathan, K. A. Gingerich, H. Gerads, and D. Kobertz, J. Chem. Thermodyn., 1985, 17, 423.
86  S. A. Mucklejohn, N. W. O'Brien, and T. R. Brumleve, J. Phys. Chem., 1985, 89, 2409.
87  L. N. Gorokhov, M. Yu. Ryzhov, and Yu. S. Khodeev, Zh. Fiz. Khim., 1985, 59, 2939.
88  Yu S. Khodeyev and M. Yu. Ryzhov, in 'Advances in Mass Spectrometry 1985', Part B, J. Wiley and Sons, New York, NY, p. 1027.
89  M. Piechotka, and E. Kaldis, J. Less-Common Met., 115, 315 (1986).
90  O. Kaposi, Z. Ajtony, A. Popovic, and J. Marsel, J. Less-Common Met., 1986, 123, 199.
91  T. A. DeMercurio and R. T. Grimley, in 'Proceedings of the Symposium on High Temperature Materials-Chemistry III, The Electrochemical Society' ed. A. Munir and D. Cubicciotti, 1986, Vol. 86-2, p. 59.
92  I. P. Malkerova, A. S. Alikhanyan, V. D. Butskii, V. S. Pervov, and V. I. Gorgoraki, Zh. Neorg. Khim., 1985, 30, 2761.
93  Yu. M. Korenev, A. N. Rykov, S. V. Kuznetsov, A. I. Boltalin, and A. V. Novoselova, Zh. Neorg. Khim., 1986, 31, 1832.
94  K. Hilpert, L. Bencivenni, and B. Saha, Ber. Bunsenges. Phys. Chem., 1985, 89, 1292.
95  M. V. Korobov, V. V. Nikulin, N. S. Chilingarov, and L. N. Sidorov, J. Chem. Thermodyn., 1986, 18, 235.
96  A. A. Bondarenko, M. V. Korobov, O. L. Sharova, A. A. Ryzhkov, and L. N. Sidorov, Zh. Neorg. Khim., 1987, 32, 25.
97  J. H. Dettingmeijer, H. R. Dielis, B. J. De Maagt and P. A. M. Vermeulen, J. Less-Common Met., 1985, 107, 11.
98  V. V. Nikulin, S. A. Goryachenkov, M. V. Korobov, Yu. M Kiselev, and L. N. Sidorov, Zh. Neorg. Khim., 1985, 30, 2530.
99  R. M. Flesch, O. Knacke, and E. Munstermann, Z. Anorg. Allg. Chem., 1986, 535, 123.
100 K. H. Lau and D. L. Hildenbrand, J. Chem. Phys., 1987, 86, 2949.
101 A. A. Bondarenko, M. V. Korobov, L. N. Sidorov, and N. M. Karasev, Zh. Fiz. Khim., 1987, 61, 2593.
102 K. Skudlarski, J. Dudek, and J. Kapala, J. Chem. Thermodyn., 1987, 19, 151.
103 K.Hilpert, L. Bencivenni, and B. Saha, J. Chem. Phys., 1985, 83, 5227.
104 A. V. Makarov, V. V. Ganin, S. I. Troyanov, and O. T. Nikitin, Vest. Mosk. Univ. Ser. 2: Khim., 1985, 40, 219.
105 O. Kaposi, L. Bencze, and L. V Zhuravleva, J. Chem. Thermodyn., 1986, 18, 635.

| | |
|---|---|
| 106 | M. Miller, and K. Skudlarski, Ber. Bunsenges. Phys. Chem., 1985, 89, 916. |
| 107 | Yu. M. Korenev, A. N. Rykov, and A. V. Novoselova, Vestn. Mosk. Univ., Ser 2 Khim., 1986, 27, 115. |
| 108 | S. A. Mucklejohn and N. W. O'Brien, in: 'Advances in Mass Spectrometry 1985', Part B, J. Wiley and Sons, New York, NY, p. 999 |
| 109 | E. N. Gavrilin, N. S. Chilingarov, E. V. Skokan, I. D. Sorokin, O. Kaposi, and L. N. Sidorov, Russian J. Phys Chem., 1987, 61, 265. |
| 110 | K. Hilpert, H. Gerads, D. Kobertz, and M. Miller, Ber. Bunsenges. Phys. Chem., 1987, 91, 200. |
| 111 | J. Kapala and K.Skudlarski, in 'Advances in Mass Spectrometry 1985', Part B, J. Wiley and Sons, New York, NY, p. 1017. |
| 112 | J. Kapala, K. Skudlarski, J. Chem Thermodyn., 1987, 19, 27. |
| 113 | J. Kapala and K. Skudlarski, Int. J. Mass Spectrom. Ion Processes., 1987, 77, 13. |
| 114 | O. Kaposi, L. Bencze, and L. V Zhuravleva, J. Chem. Thermodyn., 1986, 18, 635. |
| 115 | O. Kaposi, L. Bencze, Acta Chim. Hung., 1987, 124, 431. |
| 116 | Yu. M. Kiselev, N. E. Fadeeva, A. I. Popov, M. V. Korobov, V. V. Nikulin, and V. I. Spitsyn, Dokl. Akad. Nauk SSSR, (Chem.) 1987, 295, 378. |
| 117 | M. Miller, and K. Hilpert, Ber. Bunsenges. Phys. Chem., 1987, 91, 642. |
| 118 | M. Miller and K. Skudlarski, J. Chem. Thermodyn., 1987, 19, 565. |
| 119 | K. Hilpert, D. Kobertz, V. Venogopal, M. Miller, H. Gerads, F. J. Bremer, and H. Nickel, Z. Naturforsch., A: Phys. Sci., 1987, 42, 1327. |
| 120 | E. K. Storms and E. G. Szklarz, J. Less-Common Met., 1987, 136, 61. |
| 121 | E. K. Storms and E. G. Szklarz, J. Less-Common Met., 1987, 135, 229. |
| 122 | E. K. Storms and E. G. Szklarz, J. Less-Common Met., 1987, 135, 217. |
| 123 | J. Tomiska, Z. Metallkd., 1986, 77, 97. |
| 124 | J. Tomiska and A. Neckel, Z. Metallkd., 1986, 77, 649. |
| 125 | Y. Ueda, T. Nishi, T. Oishi, and K. Ono, Nippon Kinzoku Gakkaishi, 1986, 50, 1081. |
| 126 | C. Bergman, R. Chastel, and J. C. Mathieu, J. Chem. Thermodyn., 1986, 18, 835. |
| 127 | C. E. Myers, G. A. Murray, R. J. Kematick, and M. A. Frisch, 'Proceedings of the Symposium on High Temperature Materials-Chemistry III, The Electrochemical Society' ed. A. Munir and D. Cubicciotti, 1986, Vol. 86-2, p. 47. |
| 128 | J. Tomiska, Z. Metallkd., 1985, 76, 532. |
| 129 | J. Tomiska and A. Neckel, Ber. Bunsenges. Phys. Chem., 1985, 89, 1104. |
| 130 | E. Ichise and K. Horikawa, Ganstie, 1985, 11, 18., CA:104(26)228824y. |
| 131 | T. Oishi, S. Nishi and K. Ono, Trans. Jpn. Inst. Met., 1986, 27, 288. |
| 132 | R. Viswanthan, M. S. Baba, D. Darwin, A. Raj, R. Balasubramanian, B. Saha, and C. K. Mathews, J. Nucl. Mater., 1987, 149, 302. |
| 133 | S. R. Schmidt and H. F. Franzen, J. Less-Common Met., 1986, 116, 73. |
| 134 | E. K. Storms and C. E. Myers, High Temp. Sci., 1985, 20, 87. |
| 135 | C. E. Myers and R. J. Kemateck, J. Electrochem. Soc., 1987, 134, 720. |
| 136 | L. N. Gorokhov, M. I. Milushin, A. M. Emelyanov, and D. V. Chekhovskoi, in 'Advances in Mass Spectrometry 1985', Part B, J. Wiley and Sons, New York, NY, p. 1029. |
| 137 | M. I. Milushin, A. M. Emel'yanov, and L. N. Gorokhov, Teplofiz. Vys. Temp., 1986, 24, 806. |
| 138 | M. I. Milushin, A. M. Emel'yanov, and L. N. Gorokhov, Teplofiz. Vys. Temp., 1986, 24, 468. |
| 139 | E. N. Nikolaev, V. V. Ovchinnikov, G. A. Semenov, and A. M. Starodubtsev, Vestn. Leningr. Univ., Fiz., Khim., 1985,42, CA:104(6)40839c. |
| 140 | S. I. Lopatin and G. A Semenov, Dokl. Akad. Nauk SSSR (Phys. Chem), 1986, 287, 380. |
| 141 | M. Farber, R. D. Srivastava, J. W. Moyer, and J. D. Leeper, J. Chem. Soc., Faraday Trans. 1, 1987, 83, 3229. |
| 142 | S. V. Golubchenko, A. V. Poloznikov, R. G. Aziev, V. V. Men'shikov, L. V. Potemkin, and T. A. Oralov, Vestnik Moscow Univ.,Khim., 1986, 41, 79. |
| 143 | O. Kaposi, L. Lelik, G. A. Semenov, and E. N. Nikolaev, Acta Chim. Hung., 1985, 120, 79. |

144  M. Farber, R. D. Srivastava, J. W. Moyer, and J. D. Leeper, High Temp. Sci., 1986, 21, 17.
145  M. Farber, R. D. Srivastava, J. W. Moyer, and J. D. Leeper, J. Chem. Soc., Faraday Trans. 1., 1985, 81, 913.
146  A. Neubert, J. Chem Phys., 1985, 82, 939.
147  G. A. Semenov, and L. N. Smirnova, Dokl. Akad. Nauk SSSR. (Phys. Chem.) 1985, 284, 175.
148  S. I. Lopatin, G. A. Semenov, and Yu. L. A. Kutuzova, Izv. Akad. Nauk SSSR. Neorg. Mater., 1986, 22, 1506.
149  E. B. Rudnyi, O. M. Vovk, L. N. Sidorov, I. D. Sorokin, and A. S. Alikhanyan, Teplofiz. Vys. Temp., 1986, 24, 62.
150  R. Ahlrichs, R. Becherer, M. Binnewies, H. Borrmann, M. Lakenbrink, S. Schunck, and H. Schnockel, J. Am. Chem. Soc., 1986, 108 7905.
151  M. Binnewies, M. Lakenbrink, and H. Schnockel, High Temp. Sci., 1986, 22, 83.
152  M. Binnewies and H. Borrmann, Anorg. Allg. Chem., 1987, 552, 147.
153  M. Shinmel, T. Imai, T. Yokokawa, and C. R. Masson, J. Chem. Thermodyn., 1986, 18, 241.
154  S. I. Lopatin and G. A. Semenov, Izv. Akad. Nauk SSSR. Neorg. Mater., 1987, 23, 1200.
155  R. M. Flesch, O. Knacke, and E. Munstermann, Z. Anorg. Allg. Chem., 1986, 535, 123.
156  E. N. Nikolaev, G. A. Semenov, and O. V Romanova, Zh. Neorg. Khim., 1986, 31, 1339.
157  L. A. Kuligina, G. A. Semenov, and V. L. Stolyarova, Zh. Obshch. Khim., 1986, 56, 1678.
158  K. Lau, R. D. Brittain, and D. L. Hildenbrand, J. Phys. Chem., 1985, 89, 4369.
159  R. D. Brittain, K. H. Lau, D. R. Knittel, and D. L. Hildenbrand, J. Phys. Chem., 1986, 90, 2259.
160  A. V. Steblevskii, A. S. Alikhanyan, V. I. Gorgoraki and A. S Pashinkin, Zh. Neorg. Khim., 1986, 31, 834.
161  M. Hino and J. M. Toguri, Met. Trans., 1986, 17B, 755.
162  L. Dziewidek and J. Botor, in 'Advances in Mass Spectrometry 1985', Part B, J. Wiley and Sons, New York, NY, p. 1025.
163  M. Hino and J. M. Toguri, Metal. Trans., 1987, 18B, 189.
164  L. Dziewidek, J. Botor, and J. Norwisz, Arch. Hutn., 1986, 31, 491.
165  S. Nakamura and A. Fuwa, J. Japan. Inst. Metals, 1987, 51, 124.
166  C. Chatillon and M. Tmar, in 'Advances in Mass Spectrometry 1985', Part B, J. Wiley and Sons, New York, NY, p. 1013.
167  P. Mukdeeprom and J. G. Edwards, Thermochim. Acta, 1987, 112, 141.
168  T. Scheuring and K. G. Weil, Ber. Bunsenges. Phys. Chem., 1985, 89, 811.
169  T. Scheuring and K. G. Weil, Surf. Sci., 1985, 156, 457.
170  M. Hino, M. Nagamori, and J. M. Toguri, Met. Trans., 1986, 17B, 913.
171  J. Yun-Kuang Huang, P. W. Gilles, and J. E. Bennett, High Temp. Sci., 1986, 21, 169.
172  C. Hirayama, R. D. Straw, and Z. Kun, Thermochim. Acta, 1987, 111, 127.
173  B. Saha, R. Viswanathan, M. Saibaba, D. D. A. Raj, R. Balasubramanian, D. Karunasasar, and C. K. Mathews, J. Nucl. Mat., 1985, 130, 316.

# Subject Index

*ab initio* Calculations 3, 13-14, 15-16, 22, 35, 36, 37, 38-39, 46, 47, 48, 49, 50, 51, 52, 53, 54, 55, 59, 62, 65, 66, 67-68, 146, 155-157, 160
Abused drugs 194, 209, 234, 277, 278, 279, 282, 283, 284, 285, 286, 287, 288, 292-295, 296, 301
Accelerator mass spectrometry 79, 91, 101-102
Accurate mass measurement 77, 80, 83, 92, 97, 100, 102, 123, 126, 132, 136, 184, 206, 226, 244
Activation (critical) energy 40-41, 67
Alcohol cluster ions 44, 160
Alcohols and phenols 8, 41, 48, 51, 52, 53, 55, 56, 64, 65, 66, 67, 69, 134, 148, 149, 151, 155, 160, 188, 189-190, 200, 202, 203, 206, 208, 209, 224, 228, 275, 281
Aldehydes 12, 42, 56, 59, 63, 134, 151, 155, 190
Alkali metals 23
Alkaloids 148, 151, 153, 197-198, 209, 226, 228, 256, 277, 278, 283, 286, 288, 291, 294, 300
Alloys 366, 371
Amides 66, 146, 148, 149, 151, 291
Amine cluster ions 18, 44
Amines 3, 4, 8, 37, 38, 51, 52, 62, 64, 86, 131, 146-148, 151, 153, 183, 191-192, 196-198, 203, 205, 206, 208, 211, 259, 275, 303, 343
Amine salts 223
Amino acids 48, 77, 80, 124, 148, 151, 153, 183, 198-199, 208, 224
Amphetamines 209, 277, 278, 282, 284, 285, 292, 295
Anaesthetics 127, 278, 285, 302
Analgesics 277, 278, 279, 282, 286, 287, 292, 294-295, 299, 302, 303
Angular distribution 4, 6, 12, 127, 365, 370
Antibiotics 88, 131, 153, 198, 202, 224, 228, 242, 256, 258, 274, 275, 277, 278, 279, 282, 283, 284, 285, 286, 287, 288, 290-292
Anticancer drugs 276, 277, 278, 279, 281-290
Anticoagulants 295-296
Anticonvulsants 146, 148, 277, 278, 279, 280, 282, 283, 286, 288
Antidepressants 233, 277, 278, 279, 282, 285, 298
Antihistamines 282, 283, 284, 285, 287, 288, 303
Anti-inflammatory drugs 277, 278, 284, 285, 287, 302
Antimalarial drugs 277, 283, 291-292
Appearance energy 19, 36, 58, 67, 360, 361, 362
Archaeological studies 102
Artificial intelligence 120, 132
Atmospheric pressure ionization 87, 94, 136, 183, 191, 204, 223, 237, 238, 258-259, 274, 287, 292
Autodetachment 11, 12, 23-24
Autoionization 1, 3, 6, 8, 9, 11, 14, 22
Autoneutralization 12, 81-82
Avoided crossings 55

Barbiturates 226, 274, 279, 282, 299, 300
Barium 79
Bile acids 194-195, 289
Boron compounds 19, 23, 137, 146, 153, 160, 188, 323-324, 331, 334, 342, 343, 367, 368, 369, 371, 372

Cannabinoids 148, 209, 277, 279, 283, 285, 287, 292, 293-294
Carbohydrates 88, 137, 148, 151, 153, 183, 195-196, 210, 211, 228, 238, 258, 291, 303
Carbon dioxide cluster ions 21, 23, 146
Carboxylic acids 42, 51, 56-58, 60-61, 62, 65, 80, 88, 128, 146, 148, 149-151, 153, 182, 188, 191, 192, 193, 198, 199, 200, 201, 203, 208, 228, 275, 281, 299-300
Cardiovascular drugs 277, 278, 279, 280, 282, 283, 284, 286, 287, 288, 295-298, 300
Carnitines 153, 224, 289, 292, 299-300
Catecholamines 146, 197, 233, 242
Centrally acting drugs 277, 278, 279, 282, 283, 284, 285, 286, 287, 288, 298-301
$^{252}$Cf plasma desorption 83-84, 98-100, 104, 105, 125, 153-157, 258, 268-269, 282, 286, 291, 297, 343, 346
Charge exchange 15, 22, 44, 45, 94, 191, 227
Charge reversal 13, 65, 155
Charge stripping 15, 41, 54, 102
Chemical ionization 43, 88, 96, 121, 130, 133-134, 183, 190, 191, 195, 196, 197-198, 201, 205, 207-208, 209, 211, 225, 226, 227, 228, 235, 236, 243, 257, 266, 274, 275, 276, 277, 278, 279, 280, 282, 283, 284, 285, 286, 287, 288, 291, 292, 294, 298, 299, 301, 303, 324, 327, 331-332, 336, 341, 344, 345, 348, 349
Chemometrics 91, 184
Cluster ions 8, 9, 17-19, 21, 22, 23, 35, 44, 94, 232, 348, 349, 370
  Negatively charged 17, 18, 23, 146, 153, 160
Cobalt 86
Collision-induced decomposition 13, 19, 40-42, 43, 46, 48, 58, 60, 61, 62, 67, 69, 91, 94, 95, 96, 99, 122, 134-135, 149, 153, 190, 191, 193, 195, 201-202, 204, 206, 207, 211, 230, 240-241, 243, 257, 260, 261, 263, 265, 267, 268, 286, 293, 302
Collision-induced dissociative ionization 44-49, 65
Computer instrumentation 119-120
Configuration interaction 4, 14, 38
Contamination 275-276, 296, 297, 300, 346, 361
Copper 23, 85
Cosmochemistry 20, 22, 51, 102
Coulomb explosion 18, 20, 36, 334
Coumarins 148
CZE/MS 88, 222-223

DADI 346
Data acquisition 118, 119-127, 129
Data processing 118, 119, 120, 123, 127-138, 184-185
Depth profiling 122-123, 137
Derivatization 77, 183, 187-190, 193, 197, 206, 228, 257, 259, 275, 277-280, 282-288, 295, 296, 301, 302
  Boronates 188
  Fluorinated derivatives 192, 193, 194, 196, 197, 198, 200, 201, 207, 209, 277-279, 282-288, 296, 301
  Methoximes 188, 193, 194, 277-279, 282, 284-286
  Silylations 187-188, 192, 193, 194, 195, 196, 197, 198-199, 210, 254, 259, 277-280, 282-288, 291, 293, 301, 302
Detectors 103-105, 360
  Electron multipliers 360-361
  Focal plane 102, 103-104, 119, 124, 138, 241-242
  Microchannel electron multipliers 105, 122, 123
  Post-acceleration 104, 241-242
  Pulse-counting 104-105, 122
  Scintillation 105
  Time-array 104, 125-126

## Subject Index

Diketonate complexes 338-341, 349-350
Direct liquid introduction 105, 229, 233, 236-238, 276-280
Distonic ions 43, 46, 49, 51-52, 53, 56-63, 64, 66
Double-bond location 69, 134, 188-191, 192
Doubly charged clusters 18
Doubly charged ions 13-17, 41, 54-55, 86, 334, 342-343
Drugs 129, 148, 183, 199, 208, 224, 225, 228, 233, 234, 236, 238, 242, 244, 273-304
Dyes 149, 151, 224, 234

Electron affinity 23, 146, 148, 155, 157, 364, 367, 368
Electron capture 11-13, 16, 17, 18, 41, 145-148, 203, 206, 207, 208, 236, 279, 297, 344, 345, 346, 348
Electron capture induced decomposition 16
Electron ionization 4, 8-13, 17-18, 83, 99, 136, 357
Electron momentum spectroscopy 4-5, 8
Electrospray 88, 222-223, 238, 258-259
Enantiomer differentiation 43, 276, 295, 297, 299
Environmental analysis 47, 94, 105, 106, 127-128, 129, 130, 131, 133, 148, 182, 183, 184, 187, 191, 202-209, 234, 242-243
Esters 46, 49-50, 54, 56, 58, 62, 146, 148, 149, 151, 157, 190-191, 228, 275, 290
Ethers 42, 63, 64, 134, 148, 149, 151, 157, 190, 209, 211, 290, 291
Expert systems 103, 120-121, 132, 133
Explosives 87

Factor analysis 118, 135, 136-137
Fast atom bombardment 81, 83, 88, 104, 105, 134-135, 151-153, 182, 196, 198, 201, 204, 224, 229, 232, 235-236, 237-238, 240, 241, 242, 254, 256, 257-258, 259-261, 263-268, 273, 274-275, 278, 281, 282, 283, 284, 285, 286, 287, 288, 290, 291, 292, 294, 297, 298, 300, 302, 303, 324-326, 329, 331, 332, 334-336, 338, 341-343, 344-346, 348, 349
Fatty acids 146, 153, 188-189, 192, 200, 201, 210
Field desorption 155, 257, 260, 282, 285, 286, 290, 291, 324, 329, 336, 339
Field ionization 11, 136, 232, 287
Flavonoids 146, 148
Flavours and odours 131, 184
Flowing afterglow 155, 157
Flow tubes 36
Fluorescence 3, 6, 13-14, 21-22
Fluorescence excitation spectroscopy
Food and agriculture 136, 183, 199, 201-202, 211, 234, 235, 237, 242, 243, 293, 297
Forensic science 106, 130, 183, 208-209, 273, 292, 294-295
Fourier-transform methods 43, 80, 82, 84, 89, 97-99, 101, 126, 132, 183-184, 190, 225, 226, 243-244, 257, 274, 297, 345, 349

Gas-phase acidities/basicities 57-58, 155, 157, 232, 274
GC/FTIR/MS 76, 106, 183-185, 202
GC/MS 65, 76, 87-88, 95-96, 99, 121, 126, 127-131, 132, 134, 135, 181-211, 225, 227, 233, 253, 257, 259, 273
 of drugs 129-130, 183, 199, 274, 276, 277-280, 281-290, 291-292, 293-294, 295-299, 300, 301-304
Geochemistry and fuel 102, 106, 119, 125, 127, 128-129, 134, 136, 138, 151, 182, 183, 186, 192, 209-211, 344
Gibberellins 146
Glow discharge 80, 86, 138
Glutathiones 281, 289, 302
Glycopeptides 290, 291
Glycosides 146, 148, 151, 153, 197, 233, 234, 242, 274, 281, 289, 297, 298, 299, 300, 303, 324

Halides 2, 3, 5, 8, 9, 11, 12-13, 14, 15, 16-17, 21, 22, 42-43, 47, 127-128, 129, 130, 146, 148, 157, 160, 163, 191, 202, 203, 204, 205-208, 211, 243, 281, 290
Heats of formation 36-39, 43, 50, 51, 56, 364, 365, 366, 367, 368, 370, 371, 372, 373
High-pressure mass spectrometry 43-44, 155
High-temperature mass spectrometry 357-373
Hybrid mass spectrometers 89, 90, 91, 94-95, 122, 125, 190, 244, 279, 285, 286, 288
Hydrocarbons 2, 7, 8, 9-10, 12, 14, 17, 20, 37, 40, 41, 43, 68-69, 87-88, 131, 134, 145-146, 148, 151, 153, 155, 157, 163, 182, 188, 190, 203, 204, 208, 209, 210, 225, 228, 236, 244, 268, 281, 349
Hydrogen-bridged radical cations 52-53, 65

ICP/MS 80, 84-86, 93-94, 105, 138, 182, 295
Inert gas clusters 18, 20
Infrared spectroscopy of ions 19-20, 21
Inorganic compounds 78, 84-85, 86, 100, 102, 118, 122-124, 134, 137-138, 295, 323, 346-350, 357-373
   of actinides 9, 367, 368, 369, 370, 372
   of alkali metals 23, 231, 346, 348-349, 362-363, 365-366, 367, 368, 369, 370, 371, 372, 373
   of alkaline earth elements 23, 79, 369, 370, 371, 372
   of aluminium 137, 346, 367, 368, 369, 370, 371, 372
   of antimony 366, 372, 373
   of arsenic 366, 371, 373
   of cadmium 366, 370, 373
   of chromium 23, 367, 368, 371, 372
   of cobalt 23, 349, 367, 371
   of copper 348, 366, 367, 369, 371, 373
   of gallium 348, 367, 373
   of germanium 22, 366, 369, 373
   of gold 367
   of hafnium 348, 368, 369
   of halides 370, 371
   of indium 372, 373
   of inert gases 366
   of iron 23, 190, 367, 369, 370, 371, 373
   of lead 366, 369, 370, 371
   of manganese 137, 367
   of mercury 370
   of molybdenum 367, 368, 370, 372
   of nickel 23, 370, 371
   of niobium 346-348, 368, 369
   of palladium 24, 371
   of platinum 261, 365, 367, 370
   of rare earth elements 79, 348, 367, 368, 370, 371
   of rhenium 23, 348-349, 372
   of ruthenium 367
   of scandium 369, 370, 371,
   of selenium 366, 373
   of silicon 365, 369, 371, 372
   of tantalum 346, 368, 371
   of tellurium 371, 373
   of thallium 371, 372, 373
   of tin 366, 370, 371, 373
   of titanium 3, 368, 371
   of tungsten 368, 370
   of vanadium 367, 368, 371
   of yttrium 369
   of zinc 366, 369, 370, 372

## Subject Index

of zirconium  348, 362-363, 368, 369, 370
Instrument control  119-127
Internal standards  129, 185, 193, 195, 196, 197, 200, 201, 207, 209, 233, 243, 276, 277-281, 292, 293, 297, 299, 359, 363
Ion cyclotron resonance  43, 57-58, 89, 97-99, 155, 160, 184, 257, 297
Ion-induced dipole complexes  63-67, 68
Ionization cross-sections  1-3, 9-10, 11-12, 14, 18, 357-358, 359-360, 361-364
Ionization energy  3-4, 18, 36, 37, 46, 50, 51, 52, 54, 55, 360, 361, 362
Ion kinetic energy spectroscopy  12
Ion microprobe mass analysis  123
Ion mobility spectrometry  87, 127, 225, 226-227
Ion/molecule reactions  42-44, 95, 97, 121, 155-163, 228, 231-232, 238, 349-350, 364, 367
    in clusters  19, 42
Ionophores  228, 290, 339, 345-346, 349-350
Ion trapping  84, 95-99, 106, 121, 130, 183, 202, 207, 209, 225, 244, 293
Iridoids  146
Iron  85, 86
Isolated states  16, 23
Isotope studies  18, 44, 58, 63, 67-68, 79, 83, 85, 93, 101-102, 103, 104-105, 120, 121, 124, 128, 129, 132, 135-136, 138, 155, 184, 185, 188, 193, 194-195, 196, 197, 199, 200, 201, 202, 203, 205, 206, 207, 209, 234, 258, 261, 276, 277-281, 282, 288, 291, 293, 294, 295-296, 297, 298, 299, 300, 301, 302, 303, 304, 338, 344, 349, 359

KERDs  40
Keto/enol tautomers  51, 55, 65
Ketones  7, 42, 56, 59, 60, 61, 65, 134, 146, 148, 149, 151, 155, 157
K$^+$IDS  88-89, 155
Kinetic energy release  16-17, 39, 40, 42, 55, 63, 66, 122, 334
Knudsen cell studies  357-373

Ladder-switching  8
LAMMA  78, 80, 123-124, 138
Laser-induced fluorescence  6, 79, 107
Laser ionization methods  7-8, 12, 17, 19, 20-22, 23, 76-81, 87-88, 97, 99, 100, 107, 127, 134, 136, 153, 155-157, 182, 190, 202-203, 225, 227, 258, 274, 296, 297, 298, 326, 345
LC/MS  87, 88, 181, 194, 198, 205, 229-239, 242, 254, 257, 258-259, 266, 273-274, 275, 276-280, 282, 283, 284, 285, 286, 287, 288, 292-293, 295, 296, 300-301, 302, 303, 304
Lead  138
Leukotrienes  148, 151, 187-188
Library searching  118, 126, 127, 128, 130-131, 132, 133, 184, 199, 201
Linked scanning  41, 92, 190, 288, 291, 294, 304, 326, 331, 332, 339
Lipids  146, 151, 153, 192, 199, 201, 224, 228, 235, 244, 275, 302
Liquid matrix  80, 88, 100, 182, 224, 260, 267, 274, 275, 290, 291, 297, 298, 326, 338, 342-343, 344, 345
Lithium  85

Macrocycles  344-346, 349-350
MAGIC  229, 236
Magic numbers  18
Magnetic sector technology  76, 82, 87, 89-92, 95, 103, 138, 230
Maximum entropy model  126
Medical applications  91, 102, 105, 124, 128, 182, 183, 192, 199-201, 242, 256-257, 276, 293, 298, 299, 300, 301
Metabolism  85, 106, 128, 151, 183, 194-195, 197, 198, 199-201, 205, 207, 208, 234, 235, 236, 242, 256, 268, 273-276, 280-304
Metallic cluster ions  17, 18, 19, 21, 23

Metastable ions  40-41, 48, 49, 50, 56, 58, 64, 66, 67, 68, 122, 240, 257, 259, 260, 263, 267, 278, 279
Microorganisms  84, 91, 127, 136, 189, 192, 199, 201, 211, 240, 254, 256
Microwave-induced plasma ion source  85-86
MIKES  190, 204, 257, 261, 294, 331
Molecular beams  21, 23, 77, 124, 146, 227, 236
Monte Carlo calculations  138
Moving belt  225, 229, 234-236, 279, 292, 295, 301
MS/MS  41, 50, 69, 75-76, 90, 91-92, 94-95, 96, 97, 101, 105, 121-122, 126, 133, 134-135, 138, 183, 184, 192-193, 195, 198, 201-202, 206, 207, 211, 226, 229, 230, 234, 236, 239-244, 256-257, 260, 261, 263, 266, 268, 273, 274, 275, 277, 278, 279, 282, 284, 285, 286, 287, 288, 291, 292-293, 296, 300-301, 302, 303
Multichannel quantum defect theory  3
Multi-photon ionization  6-8, 18, 19, 76-77, 78, 83, 94, 190, 225, 227, 298, 344, 345
Multiply charged ions  13-17, 54-55, 361

Negative-ion chemical ionization  148-151, 190, 191, 192, 193, 194, 195, 197-198, 201, 203, 204, 205, 206, 207-209, 227-228, 236-237, 243, 277, 278, 279, 282, 284, 287, 288, 297, 299, 301, 331-332, 334, 336-338, 341, 348
Negative ions  9-10, 11-13, 17, 18, 19, 20, 23-24, 47-48, 87, 88, 104, 119, 145-163, 182, 203, 206, 232, 242, 260-261, 266-268, 269, 274, 276, 283, 286, 291, 294, 297, 303, 329, 336, 338, 341, 342, 343, 346, 348, 364
Neutralization/reionization  11, 35, 44-49, 51, 155, 349
Nitrates  148, 203
Nitriles and isonitriles  3, 22, 23-24, 43, 51, 146, 148, 155-157, 329, 342, 344, 349
Nitrites  67-68
Nitro-compounds  3, 38, 94, 146, 148, 151, 182, 191, 203-204, 206, 224, 235
Nitrosamines  148, 182, 208
Nuclear industry  93
Nucleosides & nucleotides  135, 148, 151, 153, 232, 244, 253-269, 344

Organometallic compounds  21, 130, 146, 149, 151, 163, 182, 224, 323-346
 of actinides  331, 334, 339
 of alkali metals  69, 323, 342, 343, 344
 of alkaline earth elements  344, 346
 of antimony  326
 of arsenic  205, 334, 342
 of bismuth  326
 of chromium  163, 326, 331-332, 336, 339, 341, 342, 349-350
 of cobalt  163, 334, 336, 341, 342, 343, 344, 345, 349-350
 of copper  339-341, 345, 349
 of gallium  349-350
 of germanium  163, 324, 327, 332, 336
 of gold  338
 of indium  349-350
 of iridium  327, 331, 341
 of iron  8, 163, 188, 327-329, 331, 332, 334, 336-338, 341, 342, 343, 344, 345, 349-350
 of lead  234, 324, 326
 of manganese  163, 327, 329, 332-334, 336, 342, 344, 346, 349-350
 of mercury  323-324, 339
 of molybdenum  326, 327, 331, 332, 334-336, 338-339, 341, 342
 of nickel  334, 339-341, 342, 343, 345-346, 349-350
 of niobium  332
 of osmium  329, 341
 of palladium  341
 of platinum  327, 341, 343
 of rare earth elements  332, 334, 339, 349-350

of rhenium 327, 329, 332-334, 338, 339
of rhodium 326-327, 332, 334, 338, 341, 349-350
of ruthenium 329, 334, 338, 341, 342-343, 344
of selenium 326, 332, 334, 341, 342
of silver 343
of tantalum 332
of technetium 339, 343
of tellurium 326, 332, 342
of thallium 349-350
of tin 182, 204-205, 324-326, 327, 336
of titanium 331, 332, 334, 344-345
of tungsten 326, 327, 332, 341, 342
of vanadium 332, 341
of zirconium 332

PAHs 128, 191, 202-203, 204, 205, 209, 227-228
Pattern recognition 127, 128, 132, 133-134, 137, 204
Penning ionization 11, 18-19, 22, 83
PEPICO 39-40, 50, 67
PEPICO-PDS 40, 67
Peptides 77, 80-81, 83, 88, 99, 100, 101, 104, 125, 134-135, 148, 151, 153, 155, 183, 198, 199, 210, 211, 223, 224, 230, 233, 234, 237, 239-240, 241, 244, 257, 281, 290, 291, 294, 303, 304
Pesticides 106, 128, 149, 151, 153, 183, 188, 202, 205-208, 226, 228, 234, 236-237, 243
Phase space theory 4
Pheromones 200-201
Phosphorus compounds 23, 55, 86, 146, 149, 151, 153, 196, 204, 207, 208, 209, 224, 228, 236-237, 260-261, 263-264, 290, 294, 327-329, 332, 334, 336-338, 339, 342, 343, 346, 367, 372
Photodissociation 6, 20-22, 40, 43, 96, 99, 244
Photoelectron-photoelectron coincidence 13
Photoelectron-photoion-photoion coincidence 14-15
Photoelectron spectroscopy 3-4, 6-8, 10, 12, 14, 22-24, 146, 148, 155, 157
Photoion-fluorescence photon coincidence 13-14, 16
Photoionization 1-8, 10, 13, 14, 15, 17, 18, 42, 79, 202-203, 210, 346
Photon burst mass spectrometry 79
Photon-induced decomposition 96, 99
PIPICO 14, 15, 16, 17
Plasma desorption mass spectrometry See $^{252}$Cf plasma desorption
Plasticisers 202, 204, 235, 275 301
Polymers 80, 84, 88, 127, 151, 153, 208, 210-211, 228, 231
Porphyrins 128, 131, 151, 153, 155, 210, 344-345
Predissociation 3, 17, 22
Principal components analysis 118, 128-129, 134, 135, 138
Probability-based matching 131, 184
Process control 118, 126-127
Propensity rules 3, 6, 13
Prostaglandins 96, 146, 148, 187, 188, 192-193, 233, 303
Proton affinity 37, 43, 52, 231, 232, 339
Pulse labelling 276, 300, 302
Purines & pyrimidines 88, 148, 256, 258, 266, 274, 281
Pyrolysis mass spectrometry 50-51, 91, 118, 127, 134, 136-137, 183, 205, 210-211, 275

QET See RRKM/QET
Quadrupole 79, 82, 86, 89, 90, 92-97, 98, 119, 120, 127, 138, 187, 202, 269, 360-361

Relative sensitivity coefficients 86, 138
Residual gas analysis 93

Resonance ionization mass spectrometry  6-7, 17, 19, 76-79, 104-105
Reverse library searching  127-128
Robotics  121
RRKM/QET  40, 66
Rydberg states  2-3, 6, 9, 11
Secondary ion mass spectrometry  12, 81-83, 90, 94, 99, 100, 101, 102, 122-123, 137, 138, 153, 223-224, 257, 269, 275, 285, 290, 331, 339, 341, 342, 343, 346, 348-349
Selected ion monitoring  129-130, 183, 185, 187, 188, 191, 193, 194, 195, 196, 197, 200-202, 203, 205, 206, 208, 233, 257, 259, 266, 276, 277-280, 287
Selected reaction monitoring  128, 193, 201-202, 206, 207, 210, 236, 243, 276, 292
Semiconductors  78
Sequencing
  Oligonucleotides  135, 151, 153, 263-265, 267-269
  Peptides  134-135, 151, 198, 211, 234, 239-240, 241, 244
  Sugars  291, 297
SFC/MS  98, 181, 224-229, 274, 291
Silicon compounds  3, 9, 14, 22, 23, 145, 146, 149, 151, 160-163, 324, 327, 332, 334, 336
Silver  359, 360, 363
Simplex algorithm  120
Smoking  203, 293-294
Spark source mass spectrometry  80, 86, 119, 124, 138
Spin-orbit interaction  3, 22
Sputtered neutral mass spectrometry  82-83
Sputtering  81-83
Steroids  40, 88, 128, 146, 148, 151, 183, 187, 193-195, 210, 224, 228, 233, 234, 274, 277, 279, 282, 283, 284, 285, 286, 288, 298, 301-302
  Fatty acid esters  148, 195
  Glycosides  151
  Sulphates  148, 151, 224, 301
Strontium  20
Sulphur compounds  3, 5, 12, 15, 18, 23, 48, 81, 146, 148-149, 151, 157, 203, 205, 209, 210, 234, 242, 244, 274, 275, 281, 290, 292, 299, 324, 326, 331, 332, 334-336, 339, 341, 342, 346-348, 366, 372, 373
Supersonic expansion  18
Surface analysis  78, 80, 81-83, 84, 88-89, 100, 102, 122-123, 137, 139
Surface-induced decomposition  41, 95, 101
Surfactants  153, 204, 205, 228, 238, 268
Synchrotron radiation  1, 3, 4, 13, 22

Tandem mass spectrometry  See MS/MS
Terpenoids  43, 128-129, 146, 148, 151, 209, 210, 304
Thermal analysis/mass spectrometry  106, 119, 124-125, 137
Thermal ionization  87, 103, 348
Thermospray  131, 194, 198, 225, 227, 229-234, 242, 253-254, 258-259, 260, 266, 273-274, 275, 276-280, 282, 283, 284, 285, 286, 287, 288, 291, 292, 293, 294, 296, 298, 301, 302, 303, 304
Thromboxanes  148, 193
Time-resolved ion-momentum spectrometry  92, 138
Titanium  86
TLC/MS  87, 224, 275, 298
Tobacco  203
TOF mass spectrometry  77, 78, 80, 81, 88-89, 92, 94, 98, 99-101, 104, 125-126, 182, 202-203, 225, 227, 257, 259, 269, 282, 326
Toxins  130, 148, 187, 197, 201, 207, 208-209, 228, 229, 234, 237, 242, 243, 256-257
Trajectory calculations  138
Transition state theory  24

Translational energy (EQ) spectrometer 15-16
Transpiration mass spectrometer 366
Triple quadrupole 81, 94, 95, 97, 120, 121-122, 133, 191, 192-193, 195, 206, 207, 226, 229, 230, 238, 239-240, 242, 243, 277, 278, 282, 284, 286, 287, 288, 293
Triply charged clusters 18

Uranium 103

Vanadium 86,
Variance diagram method 137
Vitamins 228, 229, 235, 344

Water clusters 19, 44, 84, 160

Ylid ions 38, 46, 47, 55, 64, 66

Zinc 85

# Author Index

*In this index the number given in parenthesis is the Chapter number of the citation and this is followed by the reference number or numbers of the relevant citations within that Chapter.*

Aamot, E. (6) 342
Aarons, L. (9) 127
Aarts, J.F.M. (1) 104, 296
Abbey, L.E. (2) 127
Abbott, B.J. (6) 268
Abbott, F.S. (5) 83; (9) 179, 653
Abbt-Braun, G. (6) 432
Abdel-Monem, M.M. (9) 236
Abdennebi, E.H. (9) 107
Abe, F. (5) 345, 355; (9) 626
Abe, S. (9) 317
Abe, Y. (2) 60
Abelt, C.J. (2) 93
Abercrombie, M.L. (6) 408; (9) 554
Aberth, W. (3) 195
Abian, J. (7) 117
Abouaf, R. (1) 139
Abraham, D.J. (9) 274
Abraham, N.G. (5) 200
Abramson, D. (5) 236; (7) 149
Abul-Hajj, Y.J. (9) 481
Aburada, M. (9) 318
Achiba, Y. (1) 74, 82, 83, 94
Ackerman, D.M. (5) 383
Ackerman, K.E. (6) 252
Ackermann, R. (9) 99
Ackland, M.J. (5) 324
Ackman, R.G. (4) 109
Actis, L.A. (5) 402
Adam, G. (5) 88, 92, 93
Adam, K.R. (10) 121
Adam, M.-Y. (1) 37, 38, 171
Adamov, V.M. (10) 86
Adams, F. (4) 150, 205; (9) 31

Adams, G.W. (5) 115
Adams, J. (2) 226, 227; (5) 284, 299, 307; (7) 250
Adams, J.D. (9) 334
Adams, L. (9) 414
Adams, N.G. (5) 484
Adams, V.H. (4) 75
Adinolfi, M. (5) 351
Adler, B. (4) 130
Adlercreutz, H. (9) 483
Adlof, R.O. (6) 300
Affrossman, S. (5) 433
Afghan, B.K. (5) 141; (6) 371
Agranat, I. (1) 238
Agurell, S. (9) 96, 561
Ahel, M. (6) 354
Ahlrichs, R. (11) 150
Ahmad, V.U. (5) 339, 340
Ahmed, A.A. (5) 337, 363
Ahmed, A.W. (6) 151
Ahnoff, M. (9) 118
Ahrens, A.F. (5) 527
Ahuja, S. (7) 84
Aihara, R. (3) 186
Airoldi, L. (6) 397
Aizawa, K. (9) 650
Ajami, M. (3) 91, 176
Ajtony, Z. (11) 90
Akashi, S. (7) 258; (9) 516, 517
Åkeby, H. (1) 272
Akishin, P.A. (11) 21
Akita, S. (3) 45
Akkök, S. (2) 162, 209
Alajajian, S.H. (5) 57
Alauddin, M.M. (2) 15; (5) 106
Albaiges, J. (6) 340, 356; (7) 93
Albersheim, P. (6) 248

Albers-Schonberg, G. (9) 504
Alborn, H. (7) 152
Albrecht, P. (6) 419
Alcock, L.R. (6) 73
Alder, L. (4) 165; (8) 29
Aldercreutz, H. (9) 166, 167
Aldinger, S. (4) 27
Aldous, K.M. (6) 331
Alexander, A.J. (7) 109; (8) 19
Alexander, B.D. (10) 83
Alexander, L. (6) 365, 366
Alexander, L.R. (3) 92
Alexandrova, I.Y. (9) 295
Alford-Stevens, A.L. (4) 82-84
Alikhanyan, A.S. (11) 31, 39, 56, 92, 149, 160
Alison, D.L. (9) 241
Al-Juaid, S.S. (10) 6
Allan, L.M. (9) 294
Allan, M. (1) 134-138; (4) 10
Allemann, M. (3) 143, 148, 151; (6) 82
Allen, A.C. (9) 589
Allen, G.D. (9) 256
Allen, J. (9) 202
Allen, R.H. (6) 277
Allenger, V.M. (6) 110
Allibert, M.A. (11) 65, 67
Allison, C.E. (4) 33
Allison, J. (3) 90, 139; (4) 60; (5) 125, 471, 554; (6) 70, 249, 360; (10) 135
Allison, N.T. (10) 34
Allmaier G.M. (5) 423

# Author Index

Allois, J. (6) 84
Alonso, J.M. (7) 160
Alvan, G. (9) 677, 678
Alvin, J.D. (9) 274
Amano, T. (1) 240, 244, 248, 249, 251, 254, 262
Amer, W.M.M. (5) 367
Ames, F. (3) 38
Amico, V. (5) 214
Ammann, D. (10) 113, 116
Amoureus, M. (6) 374
Amrani, B.E. (5) 429
Amster, I.J. (3) 149, 157; (7) 168
Amy, J.W. (3) 129
Ana, R.G. (9) 651
Anderegg, R.J. (4) 86; (5) 251
Anders, V. (5) 420
Andersen, T. (1) 336; (5) 492
Anderson, G.B. (5) 149
Anderson, L.W. (6) 278
Anderson, P. (9) 575, 604, 660
Anderson, S.L. (1) 76, 77, 235-237
Andersson, B. (9) 327
Andersson, P. (9) 230, 231
Andisio, G. (6) 449, 450
Ando, H. (9) 239, 456
Ando, K. (9) 160
Ando, T. (7) 258; (9) 273, 516, 517
Andrade, J.G. (5) 488
Andrenyak, D. (6) 412; (9) 137
Andres, H. (9) 73
Andrés, J.L. (5) 525
Andresen, B.D. (6) 345; (7) 7
Androrati, S.A. (5) 80
Andrzejewski, D. (5) 147; (7) 224
Anklam, E. (2) 84
Anonick, P.K. (9) 159
Anthony, L. (9) 302
Antolovic, D. (5) 66
Antonov, V.S. (3) 5
Anvia, F. (5) 486
Aouchiche, H. (1) 140
Aparicio, X. (6) 340
Apeloig, Y. (2) 32, 152
Appelhans, A.D. (1) 128-132; (3) 49
Applewhite, J. (9) 56
Appling, J.R. (1) 76
Arbogast, B.C. (5) 117, 137; (9) 648
Arey, J. (5) 255; (6) 336
Argoudelis, A.D. (9) 514

Aries, R.E. (4) 184
Arif, S. (5) 339
Arigoni, R. (9) 467
Arime, M. (10) 9
Arison, B.H. (9) 333
Arkhangelova, N.M. (9) 491
Arlandini, E. (5) 338, 397; (9) 509
Armand, J.P. (9) 289
Armentrout, P.B. (1) 204
Armstrong, D.P. (3) 36
Arndt, I. (9) 248
Arnetoli, G. (9) 658
Arnold, E.T. (9) 370
Arnold, F. (5) 259
Arnold, L.D. (5) 385
Arnone, A. (9) 497
Arnoux, P. (9) 473
Aronson, E.A. (4) 78
Arpino, P.J. (7) 35, 57, 73, 81
Arratoon, C. (6) 250
Arrendale, R.F. (7) 218; (9) 257
Artigas, F. (7) 116
Aruev, N.N. (4) 218
Asada, S. (9) 326
Asada, T. (3) 107; (7) 167
Asano, M. (11) 64, 66
Asfandiarov, N.L. (5) 70, 75, 80, 81
Ashcroft, A.E. (3) 193; (5) 391; (7) 157; (9) 41, 522
Ashfold, M.N.R. (1) 77
Ashford-Stevens, A.L. (6) 393
Ashley, D.L. (6) 366
Ashmore, R.W. (9) 111
Ashraf, H. (6) 287
Ashraf, M. (6) 390; (9) 107
Asmus, K.D. (2) 84
Aso, M. (6) 237
Assandri, A. (9) 373
Ast, T. (1) 176; (2) 54, 56, 57; (10) 69
Astoin, J. (9) 450
Ataka, M. (3) 186
Atalay, A. (6) 145
Athnasios, A.K. (5) 110
Atlas, E. (5) 256; (6) 328
Attina, M. (2) 81
Attygalle, A.B. (6) 157
Aubagnac, J.L. (5) 429
Aubert, C. (9) 473
Audier, H.E. (2) 192, 193, 200, 205, 208
Audisio, G. (5) 113; (6)

123
Aurelle, H. (5) 67, 304, 373
Austin, W.E. (3) 115
Avato, P. (6) 123
Avery, M.J. (6) 257
Axelson, J.E. (5) 83
Axelson, M. (9) 383
Axworthy, D.B. (9) 360, 391
Ayers, C.R. (5) 189; (6) 216
Aziev, R.G. (11) 142
Azpino, P. (5) 35
Azria, R. (1) 139, 220-222
Azuma, K. (3) 186

Baarnheilm, C. (9) 303
Baba, M.S. (11) 132
Baba, S. (6) 286; (9) 128, 270, 663
Baba, T. (6) 119; (9) 352
Babadjainian, A. (5) 334
Babaeva, V.P. (10) 131
Babbitt, G.E. (5) 357; (9) 494
Babcock, L.M. (5) 551, 565
Babin, V.N. (10) 57, 58
Bablievski, F.V. (4) 104
Bach, G. (5) 415
Bachmann, D. (3) 43; (9) 30
Back, D.J. (9) 296
Backman, A. (9) 303
Bacon, C.W. (7) 218
Baczynskyj, L. (9) 514
Badrinathan, C. (1) 174; (4) 8
Baer, T. (1) 34, 78, 79; (2) 42-44, 155, 222
Bagchi, S.P. (9) 144
Baggiolini, E. (9) 203
Bagus, P.S. (1) 54
Bahasadri, A. (3) 119
Bahnson, L.S. (9) 191
Bahr, U. (3) 43; (9) 30
Bai, S.A. (9) 613
Baidakov, E.L. (4) 218
Baier-Weber, B. (9) 207
Bailey, E. (6) 394, 395; (9) 172
Baillie, T.A. (2) 8; (5) 24; (7) 86; (9) 64, 65, 360, 391, 450, 475-478, 690, 695
Baker, G.B. (9) 470
Baker, M.H. (9) 429
Baker, R.M. (6) 202
Baker, T.R. (7) 141

Bakulev, V.A. (5) 74
Balack, M.S. (9) 567
Balansara, G. (5) 334
Balasanmugam, K. (5) 413, 420, 449; (10) 101
Balasubramanian, R. (11) 132, 173
Balazy, M. (5) 179, 184, 200; (6) 212
Balducci, G. (11) 47–49, 51, 53
Baldwin, M.A. (2) 121, 136; (7) 103
Baleriola-Lucas, C. (6) 267
Balistreri, W.F. (9) 95
Ball, H.A. (6) 353
Ballantine, D. (5) 209
Ballantine, J.A. (5) 243
Ballard, M.J. (7) 123
Ballini, J.-P. (5) 466, 467
Ballistreri, A. (5) 322
Ballou, C.E. (5) 380
Ballschmitter, K. (6) 181, 370
Balraadjsing, W. (5) 221, 222; (6) 265
Baltzer, P. (1) 172
Bambagiotti, A.M. (5) 177; (6) 174
Bambagiotti-Alberti, M. (7) 210; (9) 46
Banchornhevakul, W. (11) 54, 55
Bandini, A.L. (10) 30
Banditelli, G. (10) 30
Bandoli, G. (10) 90
Bandy, A.R. (6) 329, 330
Banichevich, A. (1) 171
Banijamali, A.R. (9) 560
Banna, M.S. (1) 38
Bar, V.E. (10) 23
Barbalas, M.P. (4) 101; (5) 230; (9) 83, 199
Barber, B.J. (5) 357
Barber, M. (9) 506; (10) 42, 114
Barber, R.C. (5) 10
Barbhaiya, R. (9) 242
Barbuch, R.J. (7) 201; (9) 371
Barcelo, D. (5) 275; (7) 93, 148
Baril, M. (3) 99, 100
Barker, B.J. (9) 494
Barker, C. (4) 5
Barker, J. (10) 94, 95
Barker, S.A. (9) 119
Barkofsky, D.F. (3) 79
Barlow, S.E. (2) 99; (5) 503, 507, 517, 553

Barner, A. (9) 634
Barnes, C.J. (9) 529
Barnes, H. (9) 478
Barnes, S. (9) 312, 313, 592
Barnhart, E.R. (3) 92; (6) 366
Barnhart, M. (6) 268
Baro, N. (5) 340
Baro, S. (5) 340
Baronavski, A.P. (1) 199, 201
Barone, G. (5) 351
Barren, J.P. (10) 13
Barrett, W.E. (9) 468
Barron, D. (5) 365, 367
Barry, A.J. (7) 30
Barry, T.L. (9) 703
Barth, H. (9) 697
Bartmess, J.E. (2) 11; (5) 288, 481
Barton, J.D. (5) 243
Basden, B. (1) 10, 11
Baselt, R.C. (9) 594
Basolo, F. (10) 55
Bassekier, J.J. (5) 219
Bateman, R.H. (3) 94; (7) 106
Bates, R.B. (10) 8
Batey, J.H. (3) 116
Batjer, J.D. (9) 574
Batt, A.H. (9) 398
Batten, C.F. (5) 260
Baudon, J. (1) 140
Bauer, E. (9) 325
Bauer, R. (5) 342
Baugh, P.J. (6) 111
Baumann, N. (4) 119, 157
Baumgärtel, H. (1) 226, 227; (10) 126
Baumhoer, G. (7) 128
Bawagan, A.O. (1) 50, 52–57
Bayer, C.W. (6) 108, 332; (9) 567
Bayer, F.L. (6) 29
Baykut, G. (2) 92; (5) 452
Bayona, J.M. (6) 356
Beal, E.J. (6) 192
Beau, B. (9) 387
Beau, J.-M. (10) 15
Beaudet, S. (2) 62
Beaufrere, B. (6) 299
Beaugrand, C. (3) 72
Beaumann, P. (9) 344
Beavis, R. (4) 56; (5) 434; (9) 487
Becherer, R. (11) 150
Beck, D.J. (9) 535
Beck, F. (10) 23
Beck, O. (9) 363

Becker, A. (3) 38
Becker, C.H. (3) 25, 26
Becker, G. (7) 241
Becker, S. (3) 37; (5) 458
Beckett, A.H. (9) 275, 404, 416
Beckmann, K. (1) 121, 225
Beer, B.R. (4) 31
Begg, E.J. (9) 141
Begley, P. (5) 202; (6) 389, 403
Begue, J.-M. (9) 338, 387
Behbehani, A.L. (4) 106
Behr, F.E. (5) 421
Behrens, H. (10) 73
Behrens, J., jun. (4) 53
Beijnen, J.H. (9) 531
Beitner, D.B. (9) 479
Bekersky, I. (9) 203
Beland, F.A. (5) 254; (6) 189
Belanger, J. (5) 290; (9) 32, 627
Bell, D. (10) 114
Bell, G.A. (10) 25
Bellander, T. (6) 399
Bellar, T.A. (4) 82, 84
Bellasio, E. (9) 373
Belov, A.N. (11) 72, 73, 76
Belser, W.T. (6) 365
Belshe, R. (9) 525
Belting, M. (10) 60
Belton, G.R. (11) 77
Belyaev, B.N. (10) 86
Benchekroun, Y. (9) 68, 155
Bencivenni, L. (11) 81, 84, 94, 103
Bencsath, F.A. (7) 99; (8) 35; (10) 133
Bencze, L. (11) 105, 114, 115
Bendoni, L. (9) 658
Benes, L. (9) 451
Benfenati, E. (9) 293, 439, 467
Benigni, A. (6) 204
Benner, B.A. (6) 323
Bennett, J.E. (11) 171
Bennett, R. (9) 302
Benninghoven, A. (3) 171, 172; (5) 20, 420; (10) 89
Benoit, F.M. (6) 343
Benowitz, N.L. (9) 147
Bensimon, M. (2) 82
Bentz, B.L. (3) 73, 123
Benz, A. (1) 119, 120
Beousov, V.I. (11) 21
Berecz, I. (4) 70

# Author Index

Beresford, A.P. (9) 208, 209
Beretta, E. (9) 428
Berezinskii, S.O. (10) 86
Berg, B.E. (7) 26
Berger, C. (3) 99
Berger, J.A. (9) 675
Berger, R.A. (6) 252
Berger, Y. (9) 67
Bergman, C. (11) 60, 126
Bergtold, D.S. (8) 17
Berkeley, R.C.W. (4) 180
Berkowitz, J. (11) 25
Berman, S.S. (5) 119
Bernadou, J. (9) 289
Berni, R.J. (9) 592
Bernius, M.T. (4) 202, 203
Bernstein, R.B. (2) 72
Berr, F. (6) 235
Berry, A.J. (7) 31, 32, 36, 47, 51, 52; (9) 25, 26
Berry, C. (9) 525
Berthelot, J. (5) 218, 219
Berthet, D. (9) 397
Berthier, J.-C. (9) 464
Berthou, F. (5) 35; (6) 243; (9) 1
Bertilsson, L. (9) 637, 638, 643
Bertko, R. (9) 701
Bertram, L.K. (3) 126
Bertrán, J. (5) 525
Bertrand, M.J. (5) 395; (6) 138-141, 149-151
Besnard, M.J. (1) 150, 151, 154, 157, 160, 165
Besson, F. (9) 500
Beswick, J.A. (1) 5
Betowski, L.D. (5) 269; (7) 119, 123, 126, 127; (9) 285, 287, 400
Beug-Deep, M.U.D. (3) 64
Beugnies, D. (2) 71
Beuhler, R.J. (3) 65
Beyer, B.K. (9) 475
Beylot, M. (6) 299
Beynon, J.H. (1) 141, 146, 177, 178, 187, 188, 190-193, 276; (2) 48, 49, 52, 53, 117, 198
Bhaduri, A.P. (5) 160
Bhakuni, D.S. (5) 79, 82, 207, 225
Bhatt, B.D. (6) 414
Bhattacharjee, M.N. (10) 85
Bhavanandan, V.P. (5) 375
Bhudiori, A.P. (5) 237

Bhutani, S. (9) 685
Bianchi, D. (4) 7
Bianchi, G. (6) 123
Bibikova, M.V. (9) 510
Bidault, F. (2) 182
Bieber, L.L. (9) 412
Biemann, K. (3) 107; (4) 156; (5) 389; (6) 268; (7) 180, 181, 183, 186, 192, 193
Bier, M.E. (2) 228; (3) 129
Bierbaum, V.M. (2) 99; (5) 489, 503, 507, 517, 518
Bieri, R.H. (5) 123; (7) 147
Biersack, J.P. (4) 219
Biggar, W.D. (9) 263
Biglino, G. (9) 410
Billen, J.P.J. (6) 157
Biller, J.E. (4) 156
Binghuan, L. (3) 84
Binnewies, M. (11) 150-152
Bircher, J. (9) 134
Bird, I. (6) 394
Bischofberger, P. (3) 143
Bishop, P. (3) 133
Bisling, P. (1) 225-227
Björk, G.R. (8) 10
Bjørnholm, T. (2) 201
Bjostad, L.B. (6) 301
Black, A. (9) 442
Black, M.S. (6) 108, 332
Black, R.M. (5) 202; (6) 403, 404
Black, S. (5) 267; (9) 631
Blackstock, W.P. (7) 135
Blackwell, I.G. (4) 71
Blaich, G. (9) 276
Blair, I.A. (6) 202
Blair, J.T. (1) 271
Blake, G.A. (1) 243
Blake, T.J.A. (9) 17
Blanchette, M.C. (1) 115; (2) 41, 124, 133
Blaschke, T.F. (9) 597
Blaser, O. (5) 138
Blaszkewicz, M. (7) 128
Blau, P.A. (8) 27
Blazso, M. (4) 176; (6) 430, 439, 441
Bley, W.G. (3) 117
Blilie, A.L. (7) 26
Blom, C.E. (1) 252
Blom, K.F. (4) 166-169; (6) 94; (9) 183
Blomkvist, G. (6) 62
Blonski, C. (5) 467
Blount, J.E. (9) 499

Blum, W. (6) 423; (7) 64; (9) 519
Blum, Y. (10) 7
Blumenthal, T. (5) 151, 266
Bobbie, B.A. (6) 364; (7) 230
Bobenrieth, M.-J. (9) 464
Bodd, E. (9) 382
Bodenhausen, G. (2) 83
Bodeur, S. (1) 16
Bodner, G.S. (10) 36
Boerboom, A.J.H. (3) 191; (4) 210, 211
Börlin, K. (1) 35
Boerrigter, J.C.O. (9) 184
Boesl, U. (3) 12-16, 18
Bogatova, N.G. (5) 102
Bogdanov, V.A. (10) 26
Bohatka, S. (4) 70
Bohme, D.K. (2) 90; (5) 265
Bolbach, G. (4) 56; (5) 434
Boldrini, G.P. (5) 101
Bolgar, M.S. (7) 107
Boltalin, A.I. (11) 93
Boltalina, O.V. (11) 35, 40, 43, 44
Bolton, G.C. (9) 256
Bombick, D.D. (3) 90; (5) 471
Bond, A.M. (10) 97
Bondarenko, A.A. (11) 96, 101
Bondesson, U. (9) 102, 575, 604, 660, 698
Bone, W.M. (5) 306
Bonfantini, E. (10) 31
Bonham, R.W. (3) 39
Boniface, C. (4) 113; (5) 387; (6) 89
Bonn, R. (9) 220
Bonnefond, C. (6) 84
Boobis, A.R. (9) 691
Boon, J.J. (4) 176, 183; (6) 430, 434, 436, 438, 444
Booth, B.L. (10) 42, 70
Bordas-Nagy, J. (1) 116-118; (2) 119, 124, 132
Borden, W.T. (2) 218
Bordet, J.C. (5) 195
Bordoli, R.S. (9) 506
Borel, C. (5) 327
Boreus, L.O. (9) 575
Borgeat, P. (6) 138-141
Borm, J. (10) 74
Bornhop, D.J. (7) 23
Borrmann, H. (11) 150,

152
Borsdorf, R. (5) 415
Borshchevskii, A.Ya. (11) 35, 40
Bory, C. (9) 464
Bosel, U. (10) 115
Bosone, E. (9) 439
Bosshart, P. (9) 344
Both, D. (4) 103; (9) 122
Both, G. (1) 261
Botor, J. (11) 58, 162, 164
Bott, P.A. (3) 106; (4) 31
Bottcher, C. (4) 222
Bottinga, Y. (11) 61
Botturi, S. (9) 428
Bouchoux, G. (2) 4, 50, 51, 63-65, 71, 74, 154, 181-184, 199; (5) 150, 515; (6) 173
Bougerolle, A.M. (9) 675
Boukef, K. (5) 334
Boulengeur, P. (7) 160
Boulieu, R. (9) 464
Boulton, A.A. (6) 260
Bouma, W.J. (2) 34, 36, 37, 164, 180
Bouquet, S. (9) 157
Bourquin, D. (9) 551
Boutton, T.W. (6) 51
Bowater, I.C. (6) 79
Bowen, R.D. (2) 203; (5) 253
Bowers, M.T. (1) 231, 232, 270; (2) 45
Bowers, P.J. (10) 42
Bowers, W.D. (3) 147
Bowie, J.H. (2) 69, 196; (5) 2, 103-105, 115, 142, 143, 146, 151-157, 162, 169-173, 235, 266, 267, 278, 518, 531, 543-547
Boyd, R.K. (1) 152, 155, 177, 178, 193, 194; (2) 52, 123; (3) 106; (4) 31; (5) 61, 139; (6) 178, 352, 382
Boyle, P.D. (10) 83
Bozhko, N.V. (11) 38
Bracewell, J.M. (4) 185
Bracke, J. (9) 255
Brackmann, R.T. (11) 26
Bradford, D.C. (3) 130
Brajter-Toth, A. (5) 231
Brakstad, F. (6) 342
Bramer-Weger, E. (5) 470
Brana, M.F. (9) 369
Branch, R.A. (9) 302, 349
Brand, H.R. (4) 18
Brand, W.A. (4) 28

Brandenberger, H. (5) 263; (9) 579, 674
Branscomb, C.J. (5) 128; (6) 387
Brashear, W.T. (9) 84
Brassell, S.C. (6) 418
Brauman, J.I. (1) 337; (5) 87, 144, 145, 483, 493, 511, 519, 537, 552
Brazier, J.L. (9) 68, 155, 667, 668
Breaux, E.J. (5) 417
Brechany, E.Y. (6) 160-162, 164
Bréchignac, C. (1) 215
Breckenridge, A.M. (9) 211, 536, 537
Breckenridge, W.C. (6) 222
Bregadze, V.I. (10) 5
Breimer, D.D. (9) 652, 693, 694
Bremer, F.J. (11) 119
Bremser, W. (4) 126, 127
Brennan, P.J. (5) 462
Brenneisen, R. (9) 551
Brent, D.A. (3) 194; (5) 190
Brenton, A.G. (1) 141, 146, 177, 178, 184, 185, 188, 190-194; (2) 48, 49, 52, 53, 116
Brettell, T.A. (9) 8
Brevard, H. (5) 175
Breyer-Pfaff, U. (9) 205-207
Bricker, D.L. (5) 560; (7) 189
Brickhouse, M.D. (5) 485
Bridges, R.R. (6) 412; (9) 137
Briehl, H. (2) 156
Briggs, D. (5) 435
Brinkhuis, H. (6) 435
Brinkman, U.A.T. (5) 275; (6) 374; (7) 148
Brion, C.E. (1) 49-55, 57
Brissey, G.M. (3) 206
Brisson, A. (2) 62
Brittain, R.D. (11) 158, 159
Brock, N. (9) 328
Brodbelt, J.S. (3) 134, 137; (6) 77
Brodfuehrer, J.I. (9) 486
Broeckx, R.L. (9) 540
Broekaert, J.A.C. (3) 75
Broido, E. (2) 189
Brokx, M. (9) 254
Bromet, N. (9) 387
Brondz, I. (4) 181; (6) 52, 53

Bronson, G.E. (9) 213
Brooker, S. (10) 122
Brooks, C.J.W. (6) 153
Brooks, P.M. (5) 166; (9) 92, 689
Brossel, J. (3) 2
Brotherton, S.A. (5) 118; (6) 188
Brown, A. (3) 48; (9) 47
Brown, C.E. (5) 448
Brown, L.C. (6) 401
Brown, N.K. (9) 237, 358, 425
Brown, R. (9) 142
Brown, R.D. (6) 398
Brown, R.S. (3) 193; (4) 63; (5) 456; (6) 81; (10) 119
Brown, S.J. (10) 104
Brown, W.L. (1) 274
Brown-Booth, L. (6) 300
Browne, T.R. (9) 5, 59, 69, 665, 666
Browner, R.F. (7) 143
Brownsill, R.D. (9) 294
Broyer, M. (1) 215
Brubaker, C.H., jun. (10) 63
Bruce, M.I. (5) 424, 426; (10) 2, 37-39, 48, 80
Bruckner, H. (9) 502
Brueckner, C. (5) 77
Bruin, G.J.M. (7) 5
Bruins, A.P. (5) 42, 221, 222, 270, 271; (6) 3, 265; (7) 85, 162, 163; (8) 33; (9) 16, 24
Brumleve, T.R. (11) 86
Brumley, W.C. (5) 147; (6) 311; (7) 235
Brunée, C. (3) 93
Brunner, H. (9) 581
Brussee, J. (9) 184
Brutschy, B. (1) 225-227
Bruynseraede, Y. (4) 74
Bryan, S.R. (4) 36, 37
Bryant, M.S. (5) 217
Brzezinka, H. (9) 249
Buchanan, B.B. (7) 181
Buchanan, M.V. (3) 142; (5) 121
Buchau, S. (2) 102
Buchner, D. (10) 64
Buchner, J.D. (6) 318
Buchwald, W.D. (9) 132
Buckley, T.J. (2) 85
Buckpitt, A.R. (9) 191
Budde, W.L. (4) 83, 84, 106, 107; (5) 117; (6) 317, 393; (9) 648
Buddle, E.L. (5) 270
Budinsky, R.A. (9) 414

Budzikiewicz, H. (2) 213; (5) 107, 262, 317, 319; (6) 172
Buell, G. (6) 270
Bütfering, L. (7) 104, 105; (8) 34
Buettner, D. (7) 13
Buhs, R.P. (9) 333
Bui, L.V. (9) 609
Bulatov, A.V. (10) 145
Bulau, P. (9) 661
Bullock, M..W. (9) 496
Buncel, E. (5) 509
Bundy, G.L. (5) 182; (6) 208
Bunn, T.L. (2) 42, 43, 45
Buonpane, G. (9) 526
Burdukovskaya, G.G. (11) 38
Burgers, P.C. (2) 30, 131, 147, 148, 151, 165, 167
Burke, A.B. (9) 193
Burkes, L.J. (7) 60, 70
Burleigh, D. (6) 270
Burlingame, A.L. (1) 2; (2) 8; (5) 24, 25, 442; (6) 9; (7) 86, 95, 243; (8) 4; (9) 7
Burm, A.G.L. (5) 109; (9) 6, 693, 694
Burmistrov, E.A. (5) 56
Burns, M.S. (3) 185
Burrows, J.C. (5) 330
Bursey, M.M. (2) 59, 61; (5) 234, 295; (7) 110, 111, 190; (9) 21
Burton, G. (9) 530
Busch, K.L. (7) 8–11, 166; (9) 640-642; (10) 84
Buschek, J. (2) 144
Buser, H.-R. (5) 134, 138; (6) 357, 367, 368
Bush, E.D. (5) 190
Bushaw, B.A. (3) 32-34
Busker, E. (9) 488
Bustamente, S.W. (5) 502
Butcher, V. (1) 314
Butenschon, H. (10) 45
Butler, J.L. (2) 155
Butler, W.M. (10) 43
Butman, M.F. (11) 38
Butskii, V.D. (11) 92
By, A.W. (9) 426
Bycroft, B.W. (9) 506
Bygdeman, M. (9) 386
Byrd, G.D. (5) 446; (6) 323

Caballol, R. (2) 153
Cacace, F. (2) 81
Caccia, S. (9) 293
Cadet, J. (8) 24
Cadez, I.M. (1) 103
Cafolla, A.A. (1) 312
Cagiotti, M.R. (5) 342
Cagnac, B. (3) 2
Cahvzac, Ph. (1) 215
Cairns, T. (6) 240, 310; (7) 87, 236; (9) 54, 550, 703
Caixach, J. (5) 296; (6) 344
Calaway, W.F. (3) 28
Calcamese, S. (5) 214
Calder, A.G. (6) 281
Caldwell, J. (9) 172, 305, 306
Callahan, J.H. (5) 294
Callahan, L.S. (6) 409; (9) 558
Callary, P.S. (9) 320, 321
Calvert, A.H. (9) 241
Cambi, R. (1) 42
Cameron, A.E. (3) 165
Cameron, N. (10) 94
Campana, J.E. (2) 20; (3) 166; (5) 294, 298, 310; (6) 363; (7) 229; (10) 137
Campbell, J.-A. (5) 133; (6) 326
Campostrini, R. (9) 658
Canas, B.J. (6) 311; (7) 235
Canfell, P.C. (9) 339
Cannon, B.D. (3) 32-34
Cano, J.P. (9) 473
Capiau, C. (6) 269
Capparelli, E.V. (9) 528
Cappellini, L. (6) 397
Caprioli, R.M. (6) 44; (7) 83, 156
Caputo, O. (9) 410
Carasea, A. (7) 228
Carayon, A. (9) 397
Carbonell, E. (5) 525
Carini, M. (5) 338; (7) 260; (9) 45, 466
Carlberg, G.E. (5) 131; (6) 191
Carlin, J.R. (5) 430; (9) 76
Carlsen, L. (2) 139, 219, 221
Carlson, T.A. (1) 9, 19
Carlsson, A. (9) 327
Carlstedt, I. (6) 118
Carlucci, G. (9) 687
Carminati, C. (6) 204
Carpenter, J.E. (1) 271

Carr, R. (1) 159
Carr, S.A. (5) 378; (6) 269
Carre, F. (10) 49
Carreau, S. (6) 241
Carrell, C.B. (6) 268
Carrington, N.J. (5) 216
Carron, J. (5) 141; (6) 371
Carruthers, R.A. (6) 245
Carter, G.T. (4) 119
Carter, J.F. (5) 243
Carter, J.G. (1) 126
Caruso, J.A. (3) 71
Casallni, A. (6) 431
Casalnuovo, J.A. (10) 83
Casanovas, A.M. (6) 443
Casciano, D.A. (9) 286
Casciarri, I. (9) 466
Casella, G.A. (9) 85
Casinovi, C.G. (9) 501
Cass, G.R. (6) 339
Cassani, G. (6) 155
Casson, A. (6) 111
Castagnoli, K.P. (9) 153
Castagnoli, N., jun. (9) 153, 191, 339, 354, 380, 566
Castelli, M.G. (9) 439
Castera, P. (10) 127
Castle, L. (6) 313; (9) 53
Castleman, A.W., jun. (1) 198; (2) 220; (5) 124
Castro, M.E. (3) 158
Casy, A.F. (9) 584, 702
Catalan, J. (5) 486
Cauletti, C. (1) 37, 38
Cautreels, W. (9) 31, 67, 397
Cavalieri, E.L. (8) 21
Cavalleri, B. (9) 497
Cavicchio, G. (9) 687
Cech, D. (8) 29
Ceciarelli, N. (5) 324
Cederbaum, L.S. (1) 46, 48, 196, 197
Ceja, P. (3) 129
Celii, F.G. (1) 264, 285, 290, 291
Celikkaya, A. (4) 105
Cerniglia, C.E. (9) 286, 353, 472, 431, 699
Cerny, R.L. (5) 308, 407, 412, 421; (7) 178, 249, 257; (8) 21
Cerveau, G. (5) 429
Cetini, G. (10) 100
Chabard, J.L. (9) 436, 675
Chae, W.-G. (5) 232; (7) 255; (9) 647

Chain, C. (6) 115
Chait, B.T. (3) 63, 157, 198, 205; (4) 58
Chakravarty, T. (4) 177, 196, 199
Chalasinski, G. (5) 521
Chan, C. (6) 349
Chan, G.L.-Y. (5) 83
Chan, K.K. (9) 198
Chan, L.F.T. (9) 603, 609
Chan, T.M. (9) 445
Chandrasekaran, A. (9) 628
Chang, C. (5) 232
Chang, C.J. (7) 255
Chang, K.-T. (6) 291
Chang, R. (9) 594
Chang, S. (5) 184; (6) 212
Chang, S.Y. (9) 226, 463
Chang, T. (9) 442
Chang, T.-C. (1) 89
Chang, T.T. (7) 261; (9) 495, 496
Chanoine, F. (9) 468
Chantin, C. (9) 464
Chapman, J.R. (4) 16; (5) 9, 391; (7) 53, 130; (9) 20, 41
Chapman, T.E. (6) 226
Chapuis, J.-C. (5) 335
Charalambous, A. (9) 560
Charalambous, J. (10) 1, 93
Charalambous, P. (3) 74
Charles, B. (9) 582
Charles, M.J. (5) 129; (6) 386
Chase, S. (6) 348
Chasseaud, L.F. (9) 220
Chassot, P.H. (4) 10
Chastel, R. (11) 60, 126
Chatfield, D.A. (3) 91, 176
Chatham, A.T. (6) 398
Chatillon, C. (11) 12, 67, 166
Chattopadhyaya, J. (5) 405
Chaudhary, A.K. (9) 250
Chaudhary, T. (3) 205
Chaudhuri, M.K. (10) 85
Chaukiyal, D.C. (5) 91
Chaves das Neves, H.J. (6) 251, 279, 280
Chawla, A.S. (5) 289
Chekhovskoi, D.V. (11) 136
Chen, C.-X. (5) 379
Chen, D. (5) 468
Chen, E.C.M. (5) 260
Chen, H.-N. (4) 211

Chen, L. (3) 161; (4) 61
Chen, N. (5) 379
Chen, N.-Y. (5) 379
Chen, S. (5) 230, 280; (10) 44
Chen, T.-M. (9) 605
Chen, Y. (5) 241
Chen, Y.-Z. (5) 379
Cheney, J.C. (1) 149
Chenili, R. (5) 334
Cherrah, Y. (9) 68, 667, 668
Cheshrovsky, O. (5) 54, 55
Chess, E.K. (7) 37, 46
Chessebeur, M. (6) 238
Chester, S.N. (6) 323
Chester, T.L. (7) 72
Cheung, H.T.A. (9) 238, 319
Chhuy, L.H. (9) 603
Chiabrando, C. (6) 204; (9) 439
Chiasera, G. (4) 15
Chicarelli, M.I. (10) 117
Chiccarelli, F.S. (9) 370
Chikasue, F. (9) 593
Childs, P.S. (6) 409; (9) 558
Childs, R.A. (6) 245
Chilingarov, N.S. (11) 31, 32, 34, 42, 95, 109
Chiste, R. (9) 467
Chittin, B.G. (5) 135; (6) 178, 179
Chiu, K.S. (9) 54
Cho, D. (7) 159
Cho, D.K. (9) 337
Choay, J. (5) 378
Choi, J.C. (1) 294, 295
Choma, N. (9) 158, 526, 527
Chorev, M. (7) 185
Chou, D.K.H. (5) 315
Chou, T.-Y. (9) 57
Chouka, S. (9) 273
Chow, M.S.S. (9) 528
Chow, S.A. (9) 262
Chowdary, D. (8) 25
Chowdhury, A.K. (5) 494, 506, 564
Chowdhury, S. (5) 480, 496-498, 538
Chozas, M.G. (9) 651
Christakopoulos, A. (6) 350
Christensen, N.J. (9) 386
Christie, G. (9) 224
Christie, W.H. (3) 27, 29
Christie, W.W. (6) 160-164
Christodoulides, A.A. (1)

123, 125, 126; (5) 59
Christophersen, A.S. (9) 382, 565
Christophorou, L.G. (1) 122-126; (5) 59
Chronister, E.L. (1) 91
Chtaib, M. (3) 118
Chu, L.V. (9) 341, 692
Chuit, C. (5) 429
Chulia, A.J. (5) 358
Chupka, W.A. (1) 25, 26; (11) 3, 25
Churchill, F.C. (9) 51
Chutjian, A. (5) 57
Ciabatti, R. (9) 373
Ciabattoni, G. (6) 206
Cid, P.R. (9) 369
Claereboudt, J. (9) 31
Claramunt, R.M. (5) 486
Clark, C.C. (9) 585, 586
Clark, J. (3) 74
Clark, J.H. (10) 104
Clark, T. (5) 488
Clarke, R.J. (6) 404
Clary, D.C. (1) 338
Clay, K.L. (6) 303
Clayton, B.E. (6) 258
Clayton, P.T. (5) 348
Clement, R.E. (6) 6, 31; (7) 230
Clementi, S. (4) 95
Climie, J.M. (9) 357
Coad, P. (5) 447
Coad, R.A. (5) 447
Coates, E.A. (9) 414
Coates, M.L. (5) 455
Coates, P.M. (6) 285
Cochran, D. (9) 298
Cochran, R.L. (7) 89
Cocker, J. (6) 401
Cockshott, I.D. (9) 418
Cody, R.B. (3) 149, 152, 153, 159; (4) 62; (5) 448; (7) 48, 239; (9) 27
Coggiola, M.J. (1) 277; (5) 7
Cohen, A. (6) 264
Cohen, A.I. (4) 103; (5) 267; (9) 122, 146, 182, 607, 630, 631
Cohen, A.S. (9) 321
Cohen, P. (7) 243
Colby, H.D. (9) 323
Cole, C.R. (7) 197
Cole, R.B. (7) 196
Cole, W.J. (6) 153
Coles, A.D. (3) 193
Coline, E.S. (6) 50
Colon, R.P. (5) 209
Colthup, N.B. (9) 515
Colton, R. (10) 97

Colton, R.J. (9) 40
Combs, D. (9) 337
Comer, J. (1) 312
Comisarow, M.B. (3) 142
Comita, P.B. (5) 87, 493
Compton, R.N. (1) 62
Comtet, G. (1) 172
Conceicao, J. (5) 54, 55
Cone, E.J. (9) 132, 564, 596
Connan, J. (6) 419
Connolly, G. (6) 358
Connolly, J.D. (5) 90
Conrad, T. (6) 354
Conradi, E.C. (9) 612, 614, 615
Conzemius, R.J. (3) 200
Cook, J.P.D. (1) 58
Cook, K.D. (5) 28
Cooks, R.G. (2) 54–58, 228; (3) 46, 102, 103, 127–129, 134, 136, 137, 179; (4) 139; (5) 232, 439; (6) 16, 77; (7) 165, 171, 172, 177, 188, 255; (10) 105, 134
Coomber, D.C. (9) 513
Coons, D.E. (10) 71
Cooper, D. (1) 153
Cooper, D.A. (6) 63
Cooper, R. (5) 401; (7) 259
Cooper, R.G. (9) 399
Cooper, T.B. (9) 636
Cooper, W. (9) 164
Coppella, S.J. (4) 69
Corain, B. (10) 88
Coran, L. (5) 177
Coran, S.A. (6) 174; (7) 210; (9) 46
Corbett, M. (6) 274
Corda, L. (5) 178
Cordonnier, J. (9) 173
Corneliussen, P.E. (6) 311; (7) 235
Cornish, T.J. (1) 78, 79
Cornwall, M. (5) 220
Cornwell, K.L. (7) 181
Corriu, R.J.P. (5) 429; (10) 49
Corsaro, M.M. (5) 351
Cortner, J.A. (6) 285
Cosby, P.C. (1) 277
Costa, M.L. (1) 314
Costantantini, A. (6) 242
Costello, C.E. (3) 107; (7) 245
Cosulich, D.B. (9) 370
Cotte, J. (9) 464
Cottee, F.H. (4) 71
Cotter, M.L. (5) 301; (9) 22

Cotter, R.J. (3) 6, 61, 174, 175; (5) 311, 320, 420; (6) 358
Cottrell, C.E. (4) 61
Cottrell, J.S. (3) 192, 194; (5) 391; (9) 41
Couch, M.W. (9) 56
Couderc, F. (5) 67
Courtot, D. (9) 310
Cousin, J. (7) 35, 57, 73
Coutant, J.E. (7) 201; (9) 371, 605
Coutts, R.T. (9) 297, 341, 405, 470, 672, 692
Covey, T.R. (5) 227, 271; (7) 85, 124, 146, 162, 163; (8) 33; (9) 16, 24, 109, 415, 545
Cowan, D.A. (9) 544
Cowart, T.D. (9) 612
Cowie, M. (10) 78
Cox, D.M. (1) 200
Cox, X.B.'(4) 36
Cozzi, E. (6) 204
Cozzolino, F. (5) 175
Cozzolino, T. (4) 133
Craig, A.G. (3) 62, 168; (9) 507
Craig, J.C. (6) 333, 396
Crain, P.F. (8) 5, 6, 7
Cramers, C.A. (5) 62, 233; (6) 97, 127, 261, 283
Crammer, J.L. (9) 465
Craycraft, M. (1) 332; (5) 55
Creelman, R.A. (5) 168
Cregge, R.J. (7) 201; (9) 371
Cremer, D. (5) 48
Cremonesi, P. (8) 21
Cretney, W.J. (6) 98
Crews, C. (6) 60
Crews, H.M. (3) 202
Crews, T. (5) 230
Crislain, L. (9) 387
Crocco, G.L. (10) 35
Crofton, M.W. (1) 245, 247, 250, 253, 255
Crooks, P.A. (9) 379
Cros, G. (10) 92
Crosa, J.H. (5) 402
Cross, K.P. (4) 136
Crouch, R.L. (4) 38
Crow, F.W. (5) 360; (7) 96
Crowley, R.J. (9) 603
Crummett, W. (6) 59
Cryer, D.R. (6) 285
Curl, C.L. (5) 343
Curl, R.F. (1) 207, 275; (5) 53

Currie, G.J. (2) 69; (5) 104, 115, 154, 155, 162, 278
Currie, L.A. (4) 144
Curtis, A.J. (6) 358
Curtis, J.M. (1) 193, 194; (2) 48, 49, 52
Curtis, M.D. (10) 43
Curtiss, L.A. (2) 27, 28; (5) 524
Cutie, S.S. (5) 165

Dabrowski, J. (6) 246
Dähling, P. (5) 470
Daenens, P. (9) 569
D'Agastino, P.A. (6) 405, 406
Dagaut, J. (2) 183; (6) 154
DaGue, B. (7) 156
Dahinden, R. (7) 125
Dahl, B. (4) 93; (6) 424
Dahl, D.A. (1) 131, 132; (3) 49
Dain, J.G. (9) 443
Daiser, S.M. (4) 40
Dalgaard, L. (9) 688
Dalrymple, P.D. (9) 204
Damani, L.A. (9) 379
Damen, R. (3) 188
Damewood, J.R. (5) 548
Damha, M.J. (5) 444
Damm, P. (9) 394
Dammann, V. (4) 42
Damrauer, R. (5) 545, 546, 549, 550, 553
Damste, J.S.S. (6) 351, 418
Dan, J. (9) 345
Daniel, R.M. (4) 178
Danieli, B. (5) 397
Danielson, J.W. (6) 99
Danielson, T.J. (9) 672
Danilova, N.A. (5) 89
Danis, P.O. (1) 307
Dannenberg, J.J. (2) 25
Daolio, S. (3) 53; (10) 40
Darland, G.K. (9) 121
Darragh, A. (9) 220
Darvill, A.G. (6) 248
Darwin, D. (11) 132
Darwin, W.D. (9) 564
Das, P.R. (5) 249; (9) 680
Dass, C. (5) 392, 398
Datskos, P.G. (1) 124
D'Auria, M.V. (5) 349
Dauzonne, D. (9) 436
Daves, C.D. (6) 372
Davi, H. (9) 67, 397

David, M. (9) 57
Davidson, E.M. (9) 127
Davidson, E.R. (1) 51,
  54, 55, 68; (5) 66
Davies, B.E. (9) 256
Davies, D.S. (9) 691
Davies, J.G. (5) 396
Davies, J.P. (5) 243
Davies, M.C. (9) 48, 49
Davis, B.A. (6) 256, 260
Davis, B.B. (9) 378
Davis, D. (9) 504
Davis, D.R. (8) 7
Davis, I.L. (5) 383
Davis, M. (9) 221
Davis, S.S. (9) 49
Davoli, E. (6) 397
Dawson, M. (5) 166; (9) 92, 689
Dawson, P.H. (3) 110–113
Day, R. (9) 684
Day, R.A. (3) 58
Dazzo, F.B. (6) 196
Deacon, G.B. (10) 101
Deakyne, C.A. (2) 85, 86
Dean, J.R. (3) 202
Dean, R.T. (10) 102
Debackere, M. (9) 413, 606, 611
DeBethizy, J.D. (9) 377
De Bièvre, P. (3) 188
DeBlas, A.L. (9) 649
de Boer, A.G. (9) 693, 694
De Boer, F. (5) 221
Debono, M. (6) 268
de Bruijn, D.P. (1) 112
de Bruijn, E.A. (9) 489
Decarroz, C. (8) 24
De Castiglione, R. (7) 242
Declerck, D.M. (4) 92; (6) 289
Decolin, D. (9) 398
Decosta, B. (6) 159
De Cree, J. (9) 336
De Faire, U. (9) 604
DeFrees, D.J. (1) 333, 334; (2) 29; (5) 490, 491
de Galan, L. (3) 70
De Graeve, J. (5) 35; (9) 1
De Grazia, F. (5) 230; (9) 158, 525, 601
Degrève, F. (3) 187; (4) 39
de Haan, J.W. (9) 489
De Haseth, J.A. (7) 80
Dehennin, L. (6) 225
Dehmer, J.L. (1) 4, 12, 28, 61, 63–67, 70, 72, 75

Dehmer, P.M. (1) 25, 26, 61, 63–67, 72, 75
Dehnicke, K. (10) 125
Deigner, H.-P. (9) 343
Deinzer, M.L. (5) 117, 133, 137, 414; (7) 256; (8) 12; (9) 648
De Jong, A.P.J.M. (5) 62, 68, 86, 233; (6) 261, 283
De Jong, G.J. (5) 275; (7) 148
DeJong, H.J. (4) 147
Dekok, A. (6) 374
de Koning, L.J. (1) 268; (2) 84, 94; (5) 512–514; (7) 238
de Koster, C.G. (2) 96
DeKrey, M.J. (3) 127, 128
Delacretaz, G. (1) 215
Delaney, M.F. (4) 98
Delaney, T.E. (7) 60
de Lange, F. (6) 420
deLange, W. (1) 142
DeLanney, L.E. (9) 145
DeLarge, W. (5) 522
Delbeke, F.T. (9) 413, 606, 611
Del Bene, J.E. (2) 88
Delbressine, L.P.C. (9) 38
de Leede, L.G.J. (9) 693
De Leenheer, A.P. (4) 92; (6) 37, 50, 289, 316; (9) 177, 413, 548
de Leer, E.W.B. (6) 351
de Leeuw, J.W. (4) 176, 183; (6) 351, 418, 420, 421, 430, 433, 435, 436, 444
De Lepeleire, I. (9) 634
Delgass, W.N. (7) 177
Dell, A. (5) 380
Della-Negra, S. (3) 197; (5) 434
Della Vedova, F. (9) 293
Delmore, J.E. (1) 128–132; (3) 49
DeLong, A.F. (9) 411
DeLuca, M.J. (1) 328, 330
Deluca, S.J. (4) 197; (7) 54
Delwiche, J. (1) 22, 147, 163, 167
De Maack, F. (5) 164; (9) 482
De Maagt, B.J. (11) 97
Demant, U. (10) 125
De Maria, G. (11) 49
De Marinis, R. (9) 81
de Meijere, A. (10) 45

DeMercurio, T.A. (11) 91
de Mey, C. (9) 81
Demirev, P. (3) 174; (5) 311, 461
Demirigian, J.C. (3) 207; (6) 21
Denisenko, S.N. (5) 75
Denneberg, T. (9) 362
Dennis, R.J. (7) 135
De Nollin, S. (9) 31
DePauw, E. (10) 128
DePhillips, P.A. (9) 493
Deprun, C. (3) 197; (5) 434
DePuy, C.H. (2) 99; (5) 489, 503, 507, 517, 518, 531, 545–547, 549, 553
Derendyev, B.G. (4) 120
Derrick, P.J. (2) 8, 47, 202; (4) 33; (5) 24; (7) 86, 194
Deruaz, D. (3) 72
Desage, M. (9) 155, 667, 668
Desai, D.E. (5) 260
DeSante, K.A. (9) 411
De Schepper, P.J. (9) 569, 634
Desiderio, D.M. (5) 392, 398; (7) 191; (9) 705
Desmet, N. (9) 611
de Souza, G.G.B. (1) 30, 32
Deterding, L.J. (2) 226; (5) 284; (7) 250
De Tommasi, N. (5) 333
De Torres, M. (6) 344
Detter, L.D. (5) 281; (10) 46, 105
Dettingmeijer, J.H. (11) 97
Deutsch, H. (1) 99
Devant, G. (9) 482
Dever, D.F. (11) 20
Devi, M. (10) 85
de Vries, J. (9) 300
De Vries, J.X. (9) 87, 408
DeWald, J. (4) 24
Dewar, M.S. (2) 24
Dewey, R.H. (9) 447
De Wit, J.S.M. (5) 228
De Wolf, M. (4) 217
de Zeeuw, M.A. (6) 420
Dhaher, S.M. (10) 6
Dheandhanoo, S. (2) 90; (3) 120, 121; (4) 137
Dhima, A. (11) 65
Dias, J.R. (4) 225
Diasio, R.B. (9) 312, 313
Dich-Nielsen, J.O. (9)

# Author Index

696
Dickinson, R.G. (9) 657
DiDonato, G.C. (7) 9,
 166; (9) 640; (10) 84
Diehl, J.W. (3) 212; (6)
 78
Dielis, H.R. (11) 97
Dielmann, G. (7) 159, 184
Dieter, K.M. (2) 24
Dieterle, W. (9) 99, 388
Dietrich, I. (9) 502
Dietz, T.M. (8) 23
Dietze, H.J. (5) 458
Dilli, S. (10) 91
Dillow, G.W. (5) 120,
 279, 283
Dills, R.L. (9) 444
di Modugno, E. (9) 497
Dimov, V. (9) 386
Dinan, A.C. (9) 662
Dinelli, B.M. (1) 245
Dingerdissen, J.J. (9)
 492, 493
Dino, J.J. (5) 411; (7)
 253
Dios, Z. (4) 70
Dipple, A. (8) 13
Dirksen, M.L. (8) 18
Dizdaroglu, M. (8) 15-18
Dizio, D.P. (9) 469
Djazi, F. (2) 51, 63, 65,
 182
Djerassi, C. (2) 49, 197
Djurić, N.Lj. (1) 103
Djuric, Z. (5) 254; (6)
 189
Dobson, B.R. (1) 70
Dobson, C.R. (9) 530
Dobson, R.L. (6) 322
Dobychin, S.L. (5) 440
Dodd, J.A. (5) 511, 537
Dodonov, A.F. (4) 173
Doehl, J. (7) 26
Doherty, S.J. (7) 8
Dolnikowski, G.G. (3)
 140; (4) 29; (5) 261
Dolson, D.A. (1) 229
Dombrouskis, D. (9) 107
Domcke, W. (1) 23, 33
Domino, E.F. (9) 403
Dommisse, R. (9) 397
Dommröse, A.-M. (2)
 169-171
Domon, B. (7) 125, 245;
 (9) 519
Donais, A. (6) 150
Donau, R. (4) 165; (8) 29
Donnelly, J.R. (5) 140,
 422; (6) 185, 369; (7)
 223, 233
Donnison, A.M. (4) 178
Donohue, D.L. (3) 27

Doose, G.M. (6) 310; (7)
 236
Dordain, G. (9) 675
Dorland, L. (5) 68
Dorn, S.B. (5) 438, 443
Dorsaz, D.A.C. (5) 328
Dotan, I. (2) 99
Dougherty, R.C. (5) 117;
 (9) 648
Douglas, E.J. (9) 418
Dousset, P. (5) 466, 467
Dowd, P. (5) 453
Downing, G.V. (5) 273
Downs, J.J. (6) 318
Dowsett, M.G. (4) 44, 216
Drablos, F. (4) 116
Dratz, E.A. (6) 195
Drawert, F. (6) 83
Drayer, N.M. (6) 226
Dressler, K. (9) 417
Dressler, R. (1) 134-136;
 (4) 10
Drewello, T. (1) 181; (2)
 84, 137, 139, 172, 216;
 (10) 69, 140-142, 144
Driedger, A.R. (6) 329
Droege, J.B.M. (4) 179
Drosdowsky, M.A. (6) 241
Drover, D.R. (6) 405
Drowart, J. (11) 4, 5, 7,
 8, 12, 16
Drummer, O.H. (9) 291,
 292, 633
D'Silva, A.P. (6) 322
Du, P. (2) 218
Dubach, U.G. (9) 538
Dube, G. (5) 415
Dubler, R.E. (9) 556
Dubois, J.-P. (9) 99, 100
Dubois, M.A. (5) 342
DuBois, M.R. (10) 71
Duchin, K.L. (9) 630
Duckworth, H.E. (5) 10
Duddeck, H. (5) 377
Dudek, J. (11) 82, 102
Duffield, A. (9) 684
Duffield, A.M. (6) 388
Duffin, K.L. (7) 8
Duffy, D.N. (5) 426; (10)
 37
Duffy, M.L. (6) 125, 126
Dugat, D. (5) 419
Dujardin, G. (1) 150,
 151, 154, 157, 160,
 165, 166
Duke, P. (9) 199
Dulak, J.G. (6) 318
Dulaney, E.L. (9) 504
Duliban, J. (4) 129
Dulik, D.M. (9) 342, 347,
 348
Dumasia, M.C. (6) 39,

228; (9) 385
Dumont, P. (9) 351, 664
Dunbar, R.C. (1) 277
Dunn, W.J. (4) 140, 141;
 (6) 341
Dunnebier, G. (3) 57
Dunyuan, X. (6) 112
Dupont, L.M. (4) 183
Duran, M. (5) 525
Durand, A. (9) 202
Durani, S. (5) 159
Durden, D.A. (6) 256
Dutsch, S. (6) 82
Dutton, G.N. (9) 682
Du Vignaud, P. (9) 387
Dybowski, C. (4) 166
Dyckes, D.F. (5) 238; (7)
 134
Dyke, J.M. (1) 314
Dyszel, S.M. (9) 570
Dzhemilev, U.M. (5) 97
Dziewidek, L. (11) 162,
 164

Eaborn, C. (10) 6
Eagles, J. (5) 344, 371
East, P.B. (9) 616
Eastmond, D.A. (9) 189
Ebata, T. (1) 71
Ebdon, L. (3) 202
Ebel, S. (2) 221
Eberhardt, W. (1) 159
Eccles, A.J. (3) 48
Echt, O. (1) 214
Eckardt, R.D. (9) 447
Eckart, K. (7) 185, 241;
 (9) 140, 308
Eckenrode, B.A. (3) 109
Eckerlin, R.H. (9) 174
Eckersley, M. (10) 114
Edholm, L.-E. (9) 168
Edmonds, C.G. (8) 5-7, 10
Edsbacker, S. (7) 136;
 (9) 229-232
Edwards, D.J. (9) 676
Edwards, G. (9) 536
Edwards, J.G. (11) 167
Efimova, A.G. (11) 22
Efthymiopoulos, C. (9)
 169
Eganhouse, R.P. (6) 355
Egestad, B. (5) 336; (6)
 236
Egge, H. (5) 312
Eggers, D.F., jun. (3)
 165
Eglinton, G. (4) 88; (6)
 422, 423
Egorin, M.J. (9) 320, 321
Egsgaard, H. (2) 139,
 219, 221

Ehler, J. (9) 676
Ehrenberg, L. (5) 240
Ehrhardt, J.D. (9) 161, 625
Ehrsson, H. (9) 233, 234, 485
Eiceman, G.A. (6) 31
Eichelbaum, M. (9) 70, 149, 264, 619, 644
Eichelberger, J.W. (4) 82-84; (5) 117; (6) 317, 393; (9) 648
Eichinger, P.C.H. (2) 196; (5) 104, 151-153, 162, 172, 173, 277
Eicke, A. (10) 89
Eijkel, G.B. (6) 438
Einhorn, J. (6) 171
Ekiel, I. (5) 314
Eksborg, S. (9) 485
Ekstrom, A. (10) 121
Eland, J.H.D. (1) 147-149, 162, 163, 167-170, 172
Elder, J.F., jun. (6) 103
Elebring, T. (9) 327
Elgamal, M.H.A. (5) 88
Elgerero, J. (5) 486
Elguindi, K. (5) 265
Elizarova, R.N. (9) 510
Ellenberger, S.R. (5) 402
Eller, T.D. (5) 194; (6) 202, 203
Ellestad, G.A. (9) 520
Elliot, G.J. (9) 506
Ellis, A.R. (1) 314
Ellis, H.B., jun. (1) 333, 334; (5) 490, 491
Ellison, G.B. (1) 316, 318-321, 325, 333, 334; (5) 45, 96, 473, 490, 491
Elmore, D. (3) 183
Eloranta, M. (9) 269
El-Sayed, M.A. (1) 91, 95; (4) 81
El-Sharkawy, S. (9) 481
ElSohly, H.N. (9) 595
ElSohly, M.A. (6) 410; (9) 559, 595
Elson, C.M. (7) 68
Elzinga, H. (6) 292
Emary, W.B. (3) 46
Emel'yanov, A.M. (11) 45, 136-138
Emery, S.L. (4) 140, 141; (6) 341
Emken, E.A. (6) 300
Endo, H. (9) 112
Engel, J. (9) 488
Engelbach, R.J. (5) 254; (6) 189

Engelhardt, G. (9) 454
Engelke, F. (3) 23
Engelmann, R., jun. (3) 35
Englert, G.G. (1) 315
Enjoji, Y. (9) 330
Enke, C.G. (3) 108, 109, 140, 199; (4) 29, 35, 60, 135, 136, 209
Ens, W. (4) 56; (8) 14; (9) 487
Ensminger, W.D. (8) 28; (9) 176
Eping, J. (6) 400
Epstein, P.S. (6) 348
Er, H. (5) 163
Erickson, B.J. (4) 161; (6) 272
Ericsson, O. (9) 460
Erke, C.G. (6) 70
Ermakov, A.I. (5) 76, 78, 81
Erni, F. (7) 114; (9) 82
Ernst, R.D. (10) 55, 65
Ernstberger, B. (1) 228
Ers, W. (5) 434
Ervik, M. (9) 118
Esbensen, K. (6) 342
Esipov, S.E. (9) 510
Esnaud, H. (6) 262
Esteban, N.V. (7) 120
Esterbauer, H. (6) 210, 211
Esumi, Y. (9) 375
Etoh, Y. (7) 258; (9) 516, 517
Etzold, G. (7) 12, 13
Eustache, H. (6) 84
Evans, B.A. (9) 69, 665, 666
Evans, D.H. (5) 124
Evans, J.E. (5) 315; (7) 139; (9) 69, 69, 665, 666
Evans, J.V. (6) 144
Evans, M.J. (6) 138-141
Evans, R. (4) 184
Evans, S. (3) 192, 193
Everett, J.R. (9) 530
Evershed, R.P. (4) 88; (5) 30; (6) 1, 38, 120-122, 157, 239, 422
Everskere, R.P. (5) 210
Evertz, K. (10) 59
Ewald, H. (3) 101
Ewald, M. (6) 356
Ewig, C.S. (5) 482
Eyler, J.R. (1) 201-203; (2) 92; (5) 452

Facchin, B. (3) 53

Facino, R.M. (7) 260; (9) 45
Fadeeva, N.E. (11) 116
Fagan, T.C. (9) 613
Fahey, M.R. (9) 339
Faiez Zannad, M.D. (9) 397
Faigle, J.W. (9) 388, 390
Fairbank, W.M., jun. (3) 31, 35
Faix, O. (6) 437
Falardeau, P. (5) 187; (6) 207
Falconnet, J.B. (9) 68, 667, 668
Fales, H.M. (5) 460; (8) 26; (9) 629, 649
Falick, A.M. (3) 195; (5) 442; (7) 243
Fal'ko, V.S. (5) 74
Fan, T. (7) 156
Fanelli, R. (6) 204, 397; (9) 467
Farber, M. (11) 141, 144, 145
Farbrot, A. (7) 26
Farcasiu, D. (2) 73
Farina, P. (9) 439
Farmer, P.B. (6) 394, 395; (9) 77, 172, 360
Farncombe, M.J. (4) 163
Farquhar, D. (9) 212
Farrar, J.M. (1) 267
Farrar, T.C. (4) 66
Farrell, K. (9) 653
Farrow, K.C. (9) 360
Fassel, V.A. (3) 66; (6) 322
Fasser, A.C. (5) 126
Fassett, J.D. (3) 9
Fatiadi, A.J. (5) 446
Fatima, I. (5) 340
Faucher, J.-P. (10) 127
Fauler, J. (5) 181
Faull, K.F. (9) 653
Faure, R. (5) 334
Favre-Bonvin, J. (5) 361
Fawcett, T. (3) 214
Federowicz, D.A. (9) 281
Fedin, V.P. (10) 130
Fedorov, V.E. (10) 130
Fehlhaber, H.-W. (5) 377; (9) 394, 505, 533
Fehlhammer, W.P. (10) 23
Feigerle, C.S. (1) 70, 324
Feil, V.J. (9) 267, 452, 453
Feistner, G. (9) 578
Feizi, T. (6) 245
Feld, H. (3) 171, 172
Felder, T.B. (9) 212

## Author Index

Feldman, K.F. (9) 390
Fellas, R. (5) 175
Fellenberg, A.J. (6) 177
Feller, D. (1) 55
Feng, H. (5) 372
Feng, R. (2) 121, 130, 136, 140, 145
Fenselau, C. (5) 311, 320; (6) 358; (9) 33, 342, 347, 348
Fentiman, A.R., jun. (9) 468
Fenwick, G.R. (5) 330, 343, 344
Fenzlaff, H.-P. (5) 73
Fenzlaff, M. (1) 127
Ferdinadi, E.S. (9) 298
Ferede, R. (10) 34
Ferguson, E.E. (2) 99, 220, 223
Fergusson, G.J. (7) 108
Fernandes, A.A. (6) 148
Fernandes, E.G. (9) 369
Fernandez, M.T. (2) 78, 79
Ferrante, G. (5) 314
Ferrari, P. (9) 373, 428
Ferrett, T.A. (1) 18, 20, 70
Ferretti, A. (6) 214
Fetterolf, D.D. (6) 445
Fiala, R.R. (9) 496
Fic, G. (4) 130
Fiedler, R. (3) 114
Field, F.H. (3) 63, 157, 198, 205; (4) 58; (8) 35; (10) 133
Field, F.W. (7) 99
Fies, W.J., jun. (4) 20
Fifer, E.K. (5) 254; (6) 189
Figueras, A. (5) 296; (6) 344
Fike, R.R. (9) 605
Filer, C.W. (9) 256
Filges, U. (2) 210, 211
Filippi, A. (5) 322
Fillaux, J. (2) 183
Filley, J. (5) 489
Filpus-Luyckx, P.E. (3) 64
Findeis, A.F. (5) 293
Fink, J.K. (6) 440
Finke, R.G. (10) 112
Finlay, E.M.H. (7) 211; (9) 80, 165
Fiola, J.W. (7) 9
Firsanov, G.A. (10) 86
Firth, J. (9) 294
Fischer, C. (9) 149, 264, 619
Fischer, H. (9) 288

Fischer, J.C. (9) 54
Fischer, L.J. (9) 262
Fischer, P. (9) 671
Fischer, S. (5) 188
Fish, W. (5) 402
Fished, D.L. (5) 464
Fisher, J.J. (2) 70; (9) 268
Fite, W.L. (11) 26
Fitzgerald, G.A. (6) 206
Fitzpatrick, F.A. (5) 182; (6) 208
Flachs, H. (9) 696
Flaherty, M.D. (9) 484
Flament, J.P. (1) 175
Flammang, R. (2) 71, 74, 156, 157, 198
Flanagan, M. (6) 114
Flanagan, V.P. (6) 214
Fleishman, S.H. (5) 64
Flesch, G.D. (3) 66
Flesch, R.M. (11) 99, 155
Fletcher, R.A. (4) 144
Flock, H.H. (6) 243
Florencio, H. (2) 139
Flory, D.A. (5) 269; (7) 127
Floyd, A.J. (6) 380
Flurer, R.A. (9) 642
Foffani, A. (10) 27, 72
Fokkens, R.H. (2) 84; (7) 187, 238
Folena-Wasserman, G. (9) 492
Foltz, R.L. (5) 117, 208; (6) 412; (9) 137, 648
Fong, K.-L. (5) 383
Fongen, U. (9) 382
Fonroberts, P. (9) 190, 502
Forbes, R.A. (4) 142, 143; (5) 563
Ford, W.K. (1) 159
Forest, E. (5) 416; (10) 109, 118
Forgo, I. (9) 538
Forgue, S.T. (9) 242
Forino, R. (7) 242
Fornarini, S. (2) 91
Forte, L. (2) 90
Fortier, G. (6) 275, 276
Foster, B.C. (9) 405
Foster, R.F. (5) 144, 145
Fotsis, T. (9) 166, 167, 483
Fouda, H.G. (9) 104, 163
Fought, E.R. (7) 7
Foulger, B.E. (5) 202; (6) 389, 403
Fournie, J.J. (7) 247
Fournier, J. (1) 148, 170, 172

Fournier, P.G. (1) 140, 148, 149, 156, 170-172
Fourtillan, J.B. (9) 79, 157, 162
Fowler, B.R. (6) 98
Fox, A. (2) 90; (5) 265; (6) 447
Fox, H. (4) 44, 216
Fradin, A. (5) 84; (6) 215
Fraisse, D. (4) 4; (5) 158, 296, 387; (9) 464
Franchenko, N.D. (9) 683
Francis, P.L. (6) 263
Francis, R.J. (9) 616
Francois, M. (3) 119
Francom, P. (6) 412; (9) 137
Frank, L.-R. (10) 59
Franke, P. (5) 90; (7) 13; (9) 417, 512
Franskin-Hubin, M.-J. (1) 163
Franz, R. (8) 22
Franzen, G. (10) 128
Franzen, H.F. (11) 133
Fraser, D.G. (11) 61, 62
Fraser-Monteiro, M.L. (2) 44
Frearson, M.J. (9) 245
Freas, R.B. (5) 298; (10) 137
Frederiksen, M.C. (9) 669
Freeman, C.G. (2) 87
Freeman, J.P. (5) 254; (6) 189
Freeman, P.K. (5) 133
Freeman, R.R. (1) 274
Frei, R.W. (5) 275; (7) 148
Freiser, B.S. (3) 159; (4) 142, 143; (5) 287
Freiser-Monteiro, L. (2) 155
Freiser-Monteiro, M.L. (2) 155
French, C.L. (1) 51, 54
French, J.B. (6) 73
Frendin, T.J. (9) 141
Frenking, G. (2) 23
Frenzel, H. (4) 40, 41
Fricke, F.L. (3) 71
Friedberg, S.J. (5) 303
Friedhoff, L.T. (9) 164
Friedman, L. (3) 65
Friedrich, P. (9) 649
Friend, C.M. (4) 12
Frimmel, F.H. (6) 432
Frisch, M.A. (11) 127
Froelicher, S.W. (5) 287
Fromson, J.M. (9) 372
Frost, L. (1) 52, 53

Frueh, F. (9) 632
Fruehan, R.J. (11) 77
Frydman, A. (9) 675
Fu, G. (5) 241, 280; (10) 44
Fuciarelli, A.F. (8) 19
Fuerst, O. (5) 198, 199; (7) 202, 251
Fujihara, S. (6) 255; (9) 376
Fujii, A. (1) 71
Fujii, I. (5) 394
Fujii, J. (9) 227
Fujii, M. (1) 84
Fujikawa, H. (1) 158
Fujimaki, Y. (9) 94
Fujimoto, Y. (5) 352, 356
Fujita, M. (9) 650
Fujita, Y. (3) 97
Fujiwara, H. (3) 211; (5) 417; (6) 80
Fujiwara, T. (9) 148, 381
Fukazawa, H. (9) 185, 409
Fukirda, D.S. (6) 268
Fukuda, E.K. (9) 158
Fukuhara, K. (9) 345
Fukui, M. (9) 152, 219
Fukuoka, M. (9) 462
Fukushi, Y. (9) 503
Fulara, J. (1) 285
Fulton, J.L. (7) 42
Funke, P.T. (9) 164
Furlei, I.I. (5) 56, 89, 95, 97, 98, 102
Furlong, E.A. (9) 484
Furlong, J.J.P. (3) 63, 157
Furner, R.L. (9) 591
Furst, O. (6) 198, 199, 201
Furst, P. (6) 376
Furuta, T. (6) 224, 286; (9) 153
Fuss, B. (10) 60
Futaguchi, S. (9) 197
Futrell, J.H. (4) 174
Fuwa, A. (11) 165

Gaarenstroom, S.W. (4) 153
Gabioud, R. (10) 47
Gabriel, H. (9) 288
Gabrielli, R. (2) 91
Gabrielsson, J.L. (9) 698
Gabris, G. (9) 344
Gäumann, T. (2) 83, 159, 160
Gaffney, T.E. (9) 613, 615
Gafner, F. (5) 335
Gage, D.A. (5) 168

Gaillot, J. (9) 675
Gaino, M. (9) 278
Gal, I. (4) 70
Galantino, M. (7) 242
Gale, R.W. (7) 42
Galensa, R. (5) 362
Galin, F.Z. (5) 56
Gallagher, M.K. (10) 78
Gallegos, E.J. (3) 86; (5) 302; (6) 74
Galletti, G.C. (5) 371
Gallo, D.G. (9) 86
Gallo, R. (6) 450
Gallo-Torres, H.E. (9) 701
Games, D.E. (7) 30-32, 36, 47, 51, 52, 77, 82, 158, 211; (9) 18, 25, 26, 165
Gamian, A. (6) 246
Gandhe, B.R. (6) 439
Ganes, D.A. (9) 283, 284
Ganguli, B.N. (9) 505
Ganin, V.V. (11) 104
Gannon, P.R. (10) 83
Gao, X. (8) 30
Garcia, J. (5) 358
Gard, G.L. (5) 272
Garg, H.S. (5) 207
Garg, V.N. (6) 414
Garilov, E.N. (11) 34
Garland, W.A. (4) 101; (5) 230; (9) 83, 158
Garozzo, D. (5) 322
Garrettson, L.K. (9) 210
Garrigues, P. (6) 356
Garteiz, D.A. (7) 98; (9) 562
Garvey, J.F. (2) 72, 114
Garvie, C. (7) 206; (9) 36, 446
Garza, C. (6) 300
Gasa, S. (5) 313
Gaskell, S. (5) 346
Gaskell, S.J. (4) 96, 97; (5) 350; (6) 92, 227; (7) 111, 113, 196, 208, 209; (9) 21, 80, 401, 655
Gastpar, M. (9) 344
Gates, B. (4) 166
Gates, C. (5) 126
Gau, W. (9) 252
Gaudin, D. (4) 158; (5) 409
Gault, R. (9) 547
Gaumann, T. (3) 155
Gavrilin, E.N. (11) 109
Gearhart, H.L. (4) 26
Gedamke, R. (9) 298
Gedridge, R.W. (10) 55
Geelhaar, L.A. (9) 320

Geeraert, E. (6) 102
Geerdink, R.B. (5) 275; (6) 374; (7) 148
Geib, S.J. (10) 36, 82
Geiger, J.F. (3) 56
Geilich, K. (10) 61
Geladi, P. (4) 164
Geleta, J.N. (9) 529
Gelius, U. (1) 172
Gellene, G.I. (2) 107-110, 114, 115
Gellert, U. (6) 83
Gelpi, E. (6) 24; (7) 116, 117; (9) 78
Genissel, P. (9) 387
Geno, P.W. (3) 170
Genuit, W. (4) 176; (6) 430
George, G.D. (5) 175; (9) 257
George, M.P. (6) 411; (9) 90
Georgiou, S. (10) 36
Gerads, H. (11) 85, 110, 119
Gerard, P. (1) 9, 19
Gerardy, B.M. (9) 351
Gerber, N. (9) 357
Gerbier, L. (5) 429
Gergely, R.J. (6) 352
Gerhard, R. (1) 127
Gerhardt, P. (5) 51
Gerke, G.K. (3) 34
Gerken, M. (7) 228
Germershausen, J.I. (5) 404
Gershey, R.M. (6) 352
Gerwick, W.H. (5) 209
Gescher, A. (9) 359, 360
Geusic, M.E. (1) 274
Ghaderi, S. (3) 149, 150
Giacomini, K.M. (9) 597
Giam, C.S. (3) 156; (8) 20
Gibaldi, M. (9) 598, 599
Giblin, D.E. (7) 178
Gibson, B.W. (5) 293, 442; (7) 243
Gielsdorf, W. (5) 112
Giese, R.W. (9) 57
Giger, W. (6) 354
Gigli, G. (11) 47-49, 51, 53
Gijbels, R. (4) 217; (9) 31
Gilbar, D. (9) 671
Gilbert, B.P. (10) 128
Gilbert, J. (6) 57, 60, 313, 314; (7) 211; (9) 53, 165
Gilbert, T.D. (6) 428
Giles, B. (6) 348

Gilhaus, M. (5) 279
Gill, J.B. (3) 92
Gill, J.P. (4) 88; (6) 422
Gill, P.M.W. (1) 180; (2) 38, 175, 176, 178, 179
Gilles, P.W. (11) 23, 171
Gillespie, H. (9) 96
Gillette, J.R. (9) 421
Gilmour, C.C. (6) 347
Gilon, C. (7) 185
Gingerich, K.A. (11) 13, 85
Giola, B. (5) 338, 397; (7) 242; (9) 509
Giordani, A.B. (5) 285, 286
Giovenella, A.J. (9) 493
Girault, J. (9) 157, 162
Girijavallabhan, V.M. (9) 445
Giss, G.N. (3) 206
Given, P.S. (5) 110
Gladysz, J.A. (10) 35, 36
Gleason, J.G. (9) 447, 448
Glebov, L.S. (6) 152
Gledhill, A.P. (9) 359
Gleispach, H. (5) 191, 193; (6) 205, 210, 211, 213, 217
Glish, B.G. (6) 252
Glish, G.L. (3) 124-126; (4) 59; (5) 63, 121, 247
Glombitza, K.W. (5) 331
Goad, L.J. (5) 210; (6) 38, 120, 122, 239
Goates, S.R. (7) 43
Gober, B. (9) 417
Goenechea, S. (9) 249
Goeringer, D.E. (3) 27, 29
Goff, M. (3) 173
Gold-Aubert, Ph. (9) 407
Goldfinger, P. (11) 7
Goldstein, A. (9) 578
Goldstein, B.P. (9) 497
Goldstein, M.J. (10) 13
Gole, D.J. (9) 403
Golubchenko, S.V. (11) 142
Golubovskya, L.E. (9) 683
Gomaa, K.S. (6) 193
Gombar, C.T. (9) 446
Gonne, K.P. (7) 202
Goodings, J.M. (5) 264, 264
Goodley, P.C. (3) 203
Goodman, S.D. (4) 62
Goos, P.E. (9) 430
Gordon, B.M. (6) 103

Gordon, E.A. (9) 418
Gordon, G.R. (9) 258
Gordon, J. (5) 324
Gordon, J.S. (3) 70
Gordon, W.P. (9) 463, 476-478
Gorf, S.M. (6) 395
Gorgoraki, V.I. (11) 56, 92, 160
Gorokhov, L.N. (11) 11, 17, 22, 33, 45, 83, 87, 136-138
Goryachenkov, S.A. (11) 98
Gosnell, A.B. (9) 282, 286, 434, 699
Goti, A. (6) 242
Goto, J. (6) 231
Goto, M. (7) 161
Gotoh, T. (4) 160
Goulart, D.A. (9) 598-600
Gould, R.O. (10) 95
Gould, V. (4) 23
Goulden, P.D. (5) 141; (6) 371
Gouyette, A. (9) 290
Gower, J.L. (9) 706
Goyer, R. (9) 468
Graber, P. (6) 270
Grabowski, J.J. (5) 516
Grace, L.I. (3) 198; (4) 58
Graefe, V. (5) 85
Graf, P. (9) 632
Grafe, U. (9) 511
Graffner, C. (9) 677, 678
Grange, A.H. (3) 79
Granier, B. (11) 67
Granier, M. (10) 127
Granley, K. (9) 107
Grappel, S.F. (9) 493
Grasserbauer, M. (4) 151
Grassi, M. (5) 113
Graul, S.T. (2) 21; (5) 500, 505
Gray, A.L. (3) 66, 67
Gray, H.A. (6) 339
Gray, N.A.B. (4) 17
Grayson, M.A. (6) 12
Greaves, J. (5) 123; (7) 147; (10) 17
Grech-Belanger, O. (9) 277, 367
Green, B.N. (5) 378; (9) 506
Green, K. (9) 386
Green, M. (9) 333
Green, V. (3) 92; (6) 365
Greenberg, J.H. (11) 56
Greenblatt, D.J. (9) 69, 665, 666
Greene, S.L. (5) 204; (6) 306

Gregor, I.K. (5) 120, 279, 283, 557, 559
Gregorski, K.S. (4) 200
Greibrokk, T. (7) 26
Gremaud, M. (4) 10
Grenot, C. (9) 468
Grese, R.P. (3) 144; (7) 169
Greter, J. (6) 290
Grevels, F.-W. (10) 64
Gries, O. (9) 37
Griest, W.H. (6) 415
Griffin, L.L. (2) 189
Griffis, D.P. (4) 36, 37
Griffiths, I.W. (1) 276
Griffiths, P.R. (3) 154; (7) 28, 49; (9) 28
Griffiths, W.J. (1) 143-145, 182, 183; (2) 116, 117; (5) 475-477, 523
Griggs, L.J. (9) 430
Grimalt, J. (6) 340
Grimley, R.T. (11) 6, 91
Grimm, C. (5) 117; (9) 648
Grimm, F.A. (1) 9, 19
Grimshaw, J.W. (6) 313; (9) 53
Grimsrud, E.P. (5) 117, 250, 480; (6) 76, 326; (9) 648
Grind, M. (9) 677, 678
Grislingaas, A.-L. (9) 534
Grisogono, A.M. (1) 52, 53
Gritsenko, I.V. (4) 131
Grix, R. (3) 179
Grizodub, A.I. (9) 491
Grob, K., jun. (6) 100, 101
Grob, R.L. (6) 320
Grobe, I. (6) 437
Groebel, W. (6) 376
Groepellin, A. (7) 114; (9) 82
Groiss, H. (3) 189; (4) 21
Grolimund, K. (7) 64
Gromov, D.V. (10) 131
Gronert, S. (5) 553
Grootveld, M. (9) 215
Grosa, G. (9) 410
Gross, A.F. (5) 110
Gross, M.J. (2) 226, 227
Gross, M.L. (1) 268; (2) 94, 95; (3) 144; (5) 36, 40, 284, 299, 305, 307-309, 318, 347, 360, 407, 412, 414, 421; (6)

33, 170; (7) 169, 178, 187, 200, 204, 244, 246, 248-250, 252, 256, 257; (8) 12, 21; (10) 136, 138
Gross, S.J. (9) 557
Grossman, P. (3) 143; (6) 82
Grote, H. (9) 220
Grotemeyer, J. (3) 8, 12-19; (10) 115
Groten, J.P. (9) 314
Grotjahn, L. (5) 362, 407, 412; (7) 257
Grover, J.R. (1) 223, 224
Gruebele, M. (1) 243, 246, 256-260
Gruen, D.M. (3) 28
Gruenke, L.D. (6) 333, 396; (9) 153
Grützmacher, H.-Fr. (2) 169-171, 185, 210-213
Gu, Y. (4) 146
Guaita, C. (5) 322
Guarna, A. (6) 242
Guckert, J.B. (6) 156, 194
Guenat, C.R. (5) 411; (7) 196, 253; (9) 655
Guerin, C. (10) 49
Guerret, M. (9) 93
Guest, M.F. (1) 41
Guette, C. (5) 218, 219
Guevara, P. (5) 209
Guevremont, R. (4) 19; (6) 184, 352; (7) 234
Guezzenec, J. (6) 154
Guglielmetti, G. (6) 155
Guichardant, M. (5) 164, 195
Guido, D.M. (5) 182; (6) 208
Guido, M. (11) 47-49, 51, 53
Guiffrida, M. (5) 322
Guilbaud, O. (9) 707
Guilhaus, M. (1) 146
Guillouzo, A. (9) 338, 387
Guinebault, P. (9) 202
Guiot, P. (9) 664
Guiotto, A. (5) 212
Gujer, R. (5) 371
Gulick, W.M. (5) 118; (6) 188
Gulieva, G.I. (10) 57, 58
Gunstone, F.D. (6) 160
Gunzinger, J. (5) 326
Guoying, S. (6) 418
Gupta, M.P. (5) 327, 364
Gupta, R. (8) 5
Gurg, W.Y. (5) 504

Gurka, D.F. (3) 208; (6) 19, 319
Gurprasad, N.P. (5) 242; (6) 392
Gustafson, O. (1) 310
Gustafsson, J.A. (7) 138
Gustafsson, L.L. (9) 677, 678
Gustavii, K. (9) 55
Gustavson, L.M. (9) 424
Gutteridge, C.S. (4) 178, 180, 182, 184
Gutzait, L. (9) 446
Guyer, C.G. (9) 529
Guyon, P.M. (1) 34
Guyot, J.L. (9) 310
Gyllenhaal, O. (9) 365, 608
Gyrd-Hansen, N. (9) 364

Ha, T. (2) 194
Haak, W.J. (9) 514
Haan, E. (5) 233; (6) 283
Habfast, K. (4) 28
Hachey, D.L. (6) 51, 300
Hackett, M. (10) 28
Hada, T. (9) 427
Hadad, C.M. (5) 548
Haddad, G.N. (1) 59
Hadley, J.S. (5) 84; (6) 215
Haegele, K.D. (9) 708
Haegg, C. (4) 212-214
Häkansson, P. (3) 62, 196
Haering, N. (5) 112, 132, 223; (9) 101, 125
Haessner, R. (5) 415
Hagervall, T.G. (8) 10
Hagg, A. (10) 123
Hagman, A. (6) 104, 350 ·
Hahn, J.H. (3) 23
Haim, N. (9) 299, 301
Hajdu, R. (9) 121, 394
Hakusui, H. (9) 94, 368
Halder, E. (10) 116
Hales, D.A. (1) 204
Halim, H. (2) 23
Halket, J.M. (6) 427, 448
Hall, D. (3) 74
Hall, E. (3) 74
Hall, E.R. (5) 185
Hall, H.K. (5) 297
Hall, T.W. (9) 470
Halldin, M.M. (9) 561
Hallen, B. (9) 707
Hallensleben, J. (9) 214
Halliwell, B. (9) 215
Hallnemo, G. (9) 327
Halm, K.A. (9) 126
Halstead, G.W. (9) 126
Haltiwanger, R.C. (10) 71

Halushka, P.V. (6) 202
Halverson, J.M. (9) 111
Hamasur, B. (6) 350
Hamdan, M. (1) 184, 185, 276
Hamill, R.L. (6) 268
Hamilton, R. (9) 302
Hamm, C.W. (4) 155
Hammerum, S. (2) 3, 201, 202, 207, 208
Hammes, W. (9) 181
Hammong, M.J. (4) 23
Hampson, D.R. (9) 470
Hampton, L. (6) 365
Han, C.-C. (5) 483, 511, 519
Hanasono, G.K. (9) 563
Hand, O.W. (3) 103; (7) 171; (10) 134
Handa, S. (5) 437
Hande, K. (9) 302
Handy, N.C. (5) 502
Hanley, L. (1) 235-237
Hanratty, M.A. (1) 290, 291
Hansen, E.B., jun. (9) 353, 431, 472, 699
Hansen, G. (9) 138
Hansen-Moller, J. (9) 688
Hanson, J.R. (5) 324
Hansson, G.C. (6) 118
Hara, K. (9) 139
Harada, K. (4) 115; (10) 106
Haraguchi, K. (6) 375
Harborne, J.B. (5) 359, 368, 369
Hard, O.W. (5) 439
Harding, L.B. (1) 325
Harding, P.A. (10) 121
Hardis, J.E. (1) 12, 70
Hardwick, A. (11) 62
Hardwick, J.L. (1) 294, 295
Hardy, D. (9) 673
Hardy, D.R. (6) 192
Hardy, M. (5) 214
Harigaya, S. (9) 112, 239, 278, 279, 455, 456
Haroldsen, P.E. (5) 99; (6) 303; (7) 209; (9) 401
Harper, A.M. (4) 174
Harrap, K.R. (9) 241
Harris, F.M. (1) 143, 144, 145, 182, 183, 187-189, 276; (2) 80, 116, 117, 173; (5) 475-477, 523
Harris, J.H. (2) 116
Harrison, A.G. (2) 123, 150; (5) 116, 176, 474

Harrison, D.E. (4) 221
Harrison, W.W. (3) 73
Hart, K.J. (4) 135
Hart, K.M. (6) 116, 117
Hartley, R.D. (5) 371
Hartman, N.R. (6) 271; (9) 260, 261
Hartung, E. (4) 54
Hartwick, R.A. (6) 105, 106
Harvan, D.J. (3) 106; (4) 31, 155
Harvey, D.J. (5) 41; (6) 169; (9) 237, 265, 266, 358, 425
Hasegawa, T. (6) 254
Hashimoto, H. (6) 254
Hashimoto, Y. (9) 106, 110
Hashizume, T. (8) 5, 6
Haslam, E. (5) 371
Hass, J.R. (2) 222; (3) 106; (4) 31
Hassan, M. (9) 233, 234, 235
Hassanein, A.M. (4) 220
Hasselbring, L. (4) 166
Hastie, J.W. (11) 14
Hatakeda, K. (4) 160
Hatano, H. (7) 50
Hatch, F. (9) 138
Hatton, W.G. (10) 36
Hattori, H. (9) 674
Hattori, M. (5) 291; (9) 15
Hattori, T. (7) 18
Hau, C.-C. (5) 537
Hauer, C.R. (7) 179
Hauser, F.M. (5) 402
Havlas, Z. (2) 159, 160
Hawkins, A.J. (9) 294
Hawthorne, S.B. (6) 113; (7) 55, 74
Hay, B.P. (10) 112
Hayashi, A. (7) 18
Hayashi, T. (6) 259, 282
Hayashida, M. (9) 546
Hayden, F.G. (9) 159
Hayer, M. (6) 292
Hayes, L.W. (9) 571, 572
Hayes, R.N. (4) 143; (5) 104, 157, 169, 171, 173, 531, 543, 544, 547
Hayes, S.M. (9) 592
Hayhurst, A.N. (5) 52
Hayward, C.M.T. (5) 428; (10) 81
Hayward, M.J. (2) 55
Hazlett, R.N. (6) 192
He, M. (9) 686
Heah, P.C. (10) 36
Heal, J.W. (4) 44

Heald, A.F. (9) 469
Healey, T. (9) 91, 681, 682
Hearmon, R.A. (4) 163
Hearn, M.J. (5) 435
Heath, J.R. (1) 207, 275; (5) 53
Hecht, S.G. (9) 357
Heckles, K. (3) 204
Hedges, J.I. (6) 436
Hedges, J.R. (9) 542
Hedin, A. (3) 196
Hedin, C.M. (5) 140; (6) 369
Heerma, W. (2) 147
Heflich, R.H. (9) 431, 472
Heggie, G.D. (9) 312
Hehre, W.J. (2) 7
Heide, K. (4) 54; (11) 79
Heiden, P.-G. (9) 304
Heidmann, M. (9) 190
Heidmann, W.A. (6) 377, 383
Heil, K. (9) 533
Heilemann, J. (5) 366
Heimann, P.A. (1) 18
Heimark, L.D. (9) 598, 599
Hein, R.E. (5) 448
Heinis, T. (5) 480, 498
Heinrich, N. (2) 23, 143, 152, 214-216
Heintz, R. (9) 409
Helleday, J. (9) 677
Heller, D.N. (5) 311, 320, 420
Heller, S.R. (4) 3, 107
Heller, T. (3) 171, 172
Helleur, R.J. (6) 446
Hellner, L. (1) 150, 151, 154, 157, 160, 165
Helm, H. (1) 277
Hemling, M.E. (6) 269
Hempfling, R. (4) 189
Henchman, M. (5) 527
Henderson, R.S. (10) 32
Hendrickx, F. (3) 188
Hendrickx, J. (9) 200, 201, 254, 255, 336
Hendriks, H. (5) 221, 222; (6) 265
Heni, M. (5) 71, 72
Henion, J.D. (3) 153; (5) 211, 227, 270, 271; (7) 23, 48, 85, 124, 146, 162, 163; (8) 33; (9) 16, 24, 27, 109, 174, 415, 545, 679
Henke, W. (3) 23
Henly, T.J. (10) 82
Hennequin, J.-F. (4) 80

Henner, B.J.L. (10) 49
Henner, W.D. (9) 484
Hennis, P.J. (9) 339
Henrick, K. (10) 121
Henry, D.P. (9) 411
Her, G.-R. (5) 261; (7) 142
Hercules, D.M. (3) 156; (4) 47; (5) 420, 449, 454, 457
Herd, C.R. (5) 551
Herion, J. (4) 43
Herlihy, W.C. (6) 268
Herman, Z. (1) 184, 190
Herold, D.A. (5) 189; (6) 142, 216; (9) 159
Herrmann, K. (5) 362
Hertz, H.S. (6) 264
Hess, E.V. (9) 414
Hess, J. (4) 101
Hessels, J.K.C. (6) 436
Heuck, K. (9) 149
Heuer, B. (9) 149, 619
Heumann, K.G. (6) 93
Heusler, H. (6) 400
Hewetson, D.W. (4) 110; (6) 308, 390; (9) 107
Heyes, M.P. (5) 226
Heykants, J. (9) 200, 201, 254, 255, 336
Hierl, P.M. (5) 527
Hiese, Z. (10) 85
Higashi, A. (9) 98
Higashino, K. (9) 427
Higgins, D. (2) 26
Higgins, J. (7) 35
Higuchi, R. (5) 341, 400
Hilberath, Th. (3) 37
Hildebrand, M. (9) 228, 350
Hildenbrand, D.L. (11) 100, 158, 159
Hilgard, P. (9) 328
Hilgers, H. (7) 71
Hilgers, W. (4) 43
Hilker, D.R. (6) 331
Hill, G.R. (4) 175
Hill, H.H., jun. (6) 31; (7) 50
Hill, J.C. (3) 169; (4) 57
Hill, N. (2) 159
Hillenkamp, F. (3) 43, 44; (9) 30
Hillier, I.H. (1) 41
Hills, J.F. (9) 226
Hilpert, K. (11) 81, 84, 85, 94, 103, 110, 117, 119
Hilpert, L.R. (5) 122; (6) 338

Hiltmann, A. (5) 377
Hilton, B.D. (9) 521
Hilton, J. (9) 347
Himmelhoch, J.M. (9) 676
Hindmarsh, K.W. (9) 283
Hines, J.W. (8) 27
Hino, M. (11) 161, 163, 170
Hinshaw, J.R. (9) 705
Hinshaw, J.V. (6) 27
Hinson, J.A. (7) 119, 207; (9) 392, 400
Hipp, D.E. (2) 110
Hippe, Z. (4) 122, 129, 130
Hirai, Y. (9) 251
Hirakawa, K. (9) 546
Hirako, I. (9) 623
Hirando, T. (9) 659
Hirano, C. (6) 175
Hiraoka, K. (1) 233; (2) 66; (5) 533, 535, 540
Hirata, P. (6) 391
Hirata, Y. (7) 75, 76
Hirayama, C. (11) 172
Hirayama, K. (7) 258; (9) 516, 517
Hirobe, M. (9) 402
Hirose, H. (7) 155
Hirota, E. (1) 249
Hirota, N. (6) 224
Hirota, T. (9) 105
Hitchcock, A. (1) 164, 167
Hitchcock, P.B. (10) 6
Hites, R.A. (5) 117, 129, 135, 248; (6) 186, 187, 325, 362, 386; (9) 648
Hitzman, C.J. (4) 38
Ho, C.-H. (6) 415
Ho, J. (1) 331
Ho, M.-H. (1) 15, 27, 29; (9) 592
Ho, W.-C. (1) 252, 263
Hoehn, M. (7) 137, 140, 184
Hoener, B.-A. (9) 315
Hoenle, W. (11) 79
Hoffman, M.K. (9) 109
Hoffmann, E. (9) 656
Hoffmann, F. (9) 654
Hoffmann, G. (9) 249
Hoffmann, J.A. (6) 268
Hoffmann, K.-J. (9) 303, 365, 391, 690
Hofman, P.H. (9) 652
Hohenester, E. (5) 191; (6) 213
Hohne, B. (4) 133
Hohnoki, H. (9) 375
Holder, C.L. (5) 128; (6) 387; (7) 126; (9) 282, 285–287, 434, 699
Holick, M.F. (6) 159
Holland, D.M.P. (1) 70
Holland, G.E. (3) 200
Holland, J.F. (3) 108, 109; (4) 60, 90, 209; (6) 49, 70; (7) 79
Holland, P.T. (1) 2; (2) 8; (5) 25; (6) 9; (7) 95; (8) 4; (9) 7
Hollingsworth, R.J. (6) 196
Hollister, L.-E. (9) 96, 561
Holloway, P.J. (6) 435
Holly, S. (9) 309
Holman, R.T. (6) 160–162
Holman, R.W. (2) 95
Holmes, I. (9) 634
Holmes, J.L. (1) 115–118; (2) 11, 30, 39, 41, 119, 120, 122, 124, 132–135, 144, 148, 149, 151, 165–168; (5) 155
Holmlid, L. (11) 80
Holzer, G.U. (7) 54
Homann, K.H. (5) 51
Honda, S. (6) 45
Hondo, T. (7) 44
Honeychuck, R.V. (10) 63
Hong, P.S. (9) 198
Honig, R.E. (3) 123; (11) 2
Honigs, D.E. (6) 67
Honma, E. (9) 440, 441
Honore-Hansen, S. (9) 688
Honovich, J.P. (3) 175; (5) 311
Hood, L.V. (9) 570
Hooper, W.D. (9) 657
Hop, C.E.C.A. (1) 115; (2) 41, 119, 122, 124, 132, 134, 135, 149, 151, 165
Hopkins, K. (9) 418
Hoppilliard, Y. (2) 50, 51, 63, 154, 184; (5) 150, 515; (6) 173
Horiba, M. (6) 95
Horie, K. (6) 134
Horie, N. (7) 8
Horiike, M. (6) 175
Horikawa, K. (11) 130
Horino, I. (7) 258; (9) 516, 517
Horitsu, H. (5) 192
Horlick, G. (4) 207
Horning, E.C. (6) 11; (9) 186, 188, 192
Horning, M.G. (9) 186–188, 192
Hoschi, P. (5) 352

Hoshi, M. (5) 356
Hosoya, E. (9) 318
Hostettmann, K. (5) 326–328, 335, 364
Hotchkiss, S.A. (9) 172
Houdi, A.A. (9) 379
Houghton, E. (6) 39, 228; (9) 385
Houghton, J. (9) 429, 429
Houk, R.S. (3) 66, 200, 201
Hounsell, E.F. (6) 245
Houriet, R. (2) 63, 65, 71, 74, 82, 83
Howell, S.R. (9) 444
Hoyer, G.-A. (9) 350
Hrovat, D.A. (2) 218
Hruska, F.E. (8) 14
Hsiao, K.-J. (6) 291
Hsieh, T.C.Y. (10) 132
Hsien, L.C. (9) 119
Hsu, B.H. (5) 439
Hu, Z.-H. (4) 211
Hua, Z. (10) 24
Huang, B.-L. (9) 337
Huang, E.C. (7) 56
Huang, L.-Q. (3) 200, 201
Huang, S.-K. (6) 170
Huang, S.D. (3) 82
Huang, Y. (4) 143
Huang, Z.H. (6) 166, 167
Huazhang, L. (3) 84
Hubbard, H.L. (6) 202
Hubbard, J.W. (9) 250
Hubin-Franskin, M.-J. (1) 22, 147, 167
Huckle, K.R. (9) 193
Hudgins, D.M. (2) 102, 104
Hudson, C.E. (2) 186–189, 195
Hudson, M.-J. (6) 168
Huestis, D.L. (5) 7
Huff, S.M. (4) 73
Hugget, R.J. (10) 17
Hughes, H. (7) 209; (9) 401
Hughes, R.L. (9) 669
Hui, E.T. (1) 224
Huizing, H.J. (5) 221, 222; (6) 265
Hulst, A.G. (6) 407; (9) 322
Humfeld, S. (6) 302
Hummel, S.V. (5) 274; (6) 384; (7) 222
Humpel, M. (9) 228
Humphrey, M.G. (10) 38, 39, 80
Humphrey, M.J. (9) 208
Humphrey, P.A. (10) 48
Hundt, H.K.L. (9) 216

# Author Index

Huneck, S. (5) 90
Huneke, J.C. (3) 54
Hung, P.D. (5) 92
Hung, S.-H. (6) 291
Hunt, D.F. (3) 158, 160; (5) 285, 286; (7) 179, 182, 240
Hunt, J.B. (7) 107
Hunt, J.E. (5) 464; (10) 102
Hunt, S.L. (5) 421
Hunter, J.A. (2) 67
Hunter, R.L. (3) 147
Hunter, S.R. (1) 126
Hunter, S.W. (5) 462
Hur, D.B. (7) 231
Hurkmans, R. (9) 200, 201, 255
Hus, S.-H. (9) 174
Huser, M. (10) 116
Hussain, M.S.H. (9) 47
Hussein, N. (6) 238
Hutchinson, D.W. (7) 157
Hutchinson, J.P. (10) 36
Hutson, D.H. (9) 193
Hutt, A.J. (9) 416
Huttner, G. (5) 282; (10) 59, 74-77, 79
Hutzinger, O. (6) 59, 451
Hvidberg, E.F. (9) 696
Hwang, B. (9) 221
Hynning, P.A. (9) 575
Hyver, K.J. (6) 129, 130

Iakovidis, D. (9) 291
Ibe, K. (9) 244, 248, 272, 346
Ibers, J.A. (10) 28
Ibrahim, R.K. (5) 365
Ichihara, S. (9) 123, 185, 409
Ichise, E. (11) 130
Idaka, E. (5) 192
Ido, Y. (3) 45
Idzu, G. (9) 110
Igarashi, K. (9) 152, 219
Igeta, H. (6) 75
Igolkina, N.A. (11) 30, 43, 44
Igual, J. (2) 153
Iida, T. (6) 231
Ijames, C.F. (3) 41
Ikegawa, S. (6) 232
Ikekawa, N. (5) 352, 356
Ikenishi, Y. (9) 197
Ikeya, Y. (9) 318
Illenberger, E. (1) 127; (5) 71-73
Illies, A.J. (1) 234
Illing, H.P.A. (9) 372
Imai, T. (11) 153

Imai, Y. (9) 178
Imaizumi, S. (6) 119
Inamura, N. (9) 503
Inghram, M.G. (11) 3, 4
Innis, D.P. (7) 72
Innorta, G. (10) 27, 72
Inotsume, N. (9) 98, 227
Inoue, H. (9) 278
Inoue, T. (9) 223, 280
Inui, S. (9) 326
Ioffe, B.V. (4) 171
Ionov, N.I. (11) 1
Iorizzi, M. (5) 354
Ireland, J.C. (4) 106
Irie, T. (6) 137; (9) 110
Irion, M.P. (2) 220
Irving, J. (9) 136, 171
Irwin, I. (9) 145
Irzel, G.H. (6) 440
Isa, K. (10) 107, 108
Isabelle, L.M. (6) 85, 117
Isenhour, T.L. (4) 124; (6) 90
Isensee, R.K. (9) 535
Isern-Flecha, I. (5) 232; (7) 255
Ishabashi, K. (5) 258; (9) 243
Ishibashi, B. (9) 194
Ishibashi, M. (6) 131-137, 143; (9) 110, 195
Ishihara, M. (3) 107; (7) 167
Ishii, D. (7) 161
Ishikawa, K. (4) 154, 160
Ishikawa, Y. (5) 313
Ishimori, T. (10) 99
Ishimura, R. (5) 258; (9) 243
Iskandarova, V.N. (5) 56
Isobe, R. (5) 345, 394, 400; (9) 626
Isono, K. (8) 9
Issachar, D. (4) 91
Itakura, Y. (5) 353
Ito, K. (1) 163
Ito, M. (1) 71; (9) 650
Ito, T. (3) 69; (9) 97, 129, 345
Ito, Y. (9) 112, 278, 279, 629
Itoh, M. (6) 135, 136
Ittensohn, A. (9) 87
Ivanitskaya, L.P. (9) 510
Ivanov, G.G. (11) 57, 59, 63
Ivashkiv, E. (4) 103; (9) 122, 146, 164, 182, 607
Iwabuchi, H. (7) 15; (9) 39

Iwagaya, Y. (9) 498
Iwamura, S. (9) 440, 441
Iwasaki, K. (9) 329
Iwata, S. (1) 311
Iwatani, K. (9) 197, 438
Izard, M. (9) 528

Jackman, L.M. (5) 383
Jackson, B.J. (7) 56
Jackson, J.F. (6) 58
Jacob, P. (9) 147
Jacob, T.A. (9) 333
Jacob, W. (9) 31
Jacobs, D.C. (1) 73, 80, 81
Jacobs, D.L. (3) 169; (4) 57
Jacobs, P.L. (9) 38, 225
Jacobsen, J. (9) 696
Jacobsen, K.A. (5) 465; (8) 26
Jacobson, C.-E. (6) 290
Jacobsson, S. (6) 104, 350
Jacolot, F. (6) 243
Jacquot, A. (6) 84
Jaeger, H. (5) 22, 34, 112, 132, 244; (9) 101, 125
Jaffe, J.M. (9) 443
Jaffe, K. (5) 220
Jagod, M.-F. (1) 247, 250
Jahnchen, E. (9) 143
Jahnke, P. (7) 159
Jain, N.C. (6) 411; (9) 90
Jain, R. (5) 159
Jain, T. (5) 383
Jakab, E. (4) 195; (6) 441
Jakas, M.M. (4) 221
Jakobs, C. (5) 68, 86
Jakubowski, N. (3) 76-78; (4) 206
Jalsovszky, G. (9) 309
Jamali, F. (9) 297
Jamieson, W.D. (4) 19; (6) 184, 352; (7) 234
Jancik, B.Ch. (9) 654
Jandke, J. (6) 273
Jandon, P. (6) 173
Janghorbani, M. (3) 68
Jankowski, K. (4) 158; (5) 409, 410; (8) 2
Janoschek, R. (5) 44
Janssen, G. (6) 158
Jaquen, D. (4) 30
Jardine, I. (4) 161; (5) 462; (6) 271, 272; (9) 260, 261
Jarman, M. (9) 241, 429,

430
Jarrold, M.F. (1) 274
Jarrott, B. (9) 633
Jarvis, B.B. (5) 204; (6) 306
Jaudon, P. (2) 50, 51; (5) 150
Jaussaud, Ph. (9) 310
Jean, C. (9) 131
Jede, R. (3) 57
Jefferies, T.M. (6) 380
Jeffery, D.J. (9) 256
Jeffery, P.D. (5) 202; (6) 403
Jeffs, P.W. (9) 492
Jemal, M. (4) 103; (9) 122, 146, 182, 607, 631
Jemaz, M. (5) 267
Jena, P. (2) 17
Jenkins, S.A. (9) 399
Jennings, K.R. (2) 78, 79; (4) 163; (5) 253, 427; (6) 10; (10) 98
Jennings, W. (6) 26
Jensen, N.J. (5) 36, 40, 305, 309, 318; (6) 33; (7) 204, 244, 246, 248, 252
Jensen, S. (5) 240
Jergelova, M. (9) 451
Jernstrom, B. (9) 363
Jetten, J. (7) 137
Jewell, J.S. (6) 408; (9) 554
Jiang, H. (8) 18
Jiang, S.-J. (3) 201
Jiang, X.-Y. (2) 57; (5) 232; (7) 255
Jibril, I. (10) 75, 77
Jickells, S.M. (6) 313; (9) 53
Jin, S. (9) 647
Jin, Z.K. (1) 77
Jindal, S.P. (9) 144, 259
Jocelyn Paré, J.R. (8) 2
Jochims, H.W. (10) 126
Joern, W.A. (9) 89, 170
Johansson, B.W. (9) 660
Johansson, E. (9) 96, 561
Johansson, L. (9) 118
Johansson, P. (9) 698
Johlman, C.L. (4) 63; (5) 161; (6) 81
John, B.A. (9) 220
Johnathan, P. (1) 177, 178, 184, 190
Johns, R.B. (5) 149; (6) 66
Johnsen, S. (4) 85
Johnson, B.J. (10) 83
Johnson, C.A.F. (2) 67
Johnson, D.M. (7) 110
Johnson, D.W. (6) 177
Johnson, M.A. (1) 328, 329, 330
Johnson, N.J. (8) 28
Johnson, R.E. (9) 564
Johnson, R.L. (3) 190
Johnson, R.S. (7) 180, 181, 186
Johnson, S.B. (6) 162
Johnson, S.M. (10) 83
Johnson, T.C. (7) 181
Johnston, M. (3) 145, 146
Johnston, M.V. (1) 88-90; (10) 120
Johnstone, R.A.W. (3) 204
Joly, J.P. (4) 7
Joly, R. (9) 409
Jonathan, P. (1) 141; (2) 53
Jonckheere, J.A.A. (9) 413
Jones, A.C. (6) 111
Jones, A.D. (5) 300; (6) 266; (9) 191
Jones, C.S. (6) 76
Jones, C.W. (9) 180
Jones, D.S. (7) 106
Jones, G. (4) 67
Jones, G.R. (9) 639
Jones, H.R.N. (5) 52
Jones, M.D. (6) 229
Jones, M.T. (5) 561
Jones, M.W. (6) 111
Jones, P.R. (4) 199
Jones, R.A. (8) 30
Jones, R.M.L. (5) 396
Jones, R.T. (6) 412; (9) 137
Jones, T.R. (9) 241
Jones, W.M. (10) 66
Jonsson, A. (9) 460
Jordan, C.A. (2) 127
Jordan, K.D. (5) 64
Jordan, N. (9) 627
Jordan, R.M. (3) 10
Jordi, P.E. (7) 64
Jordon, K.C. (3) 390
Jørgensen, B. (3) 28
Jorgenson, J.W. (5) 228
Jortay, C. (2) 71
Joseph, G. (9) 446
Jouvet, C. (1) 85
Juan, D. (9) 669
Juchau, M.R. (9) 475
Juenge, E.C. (9) 52
Julien-Larose, C. (9) 75, 93
Jung, L. (9) 169
Jungalwala, F.B. (5) 315
Junge, T. (4) 111
Jungen, M. (1) 35
Junien, J.L. (9) 468
Junk, G.A. (6) 257
Jurenitsch, J. (5) 323
Just, G. (5) 419

Kababia, S. (1) 238
Kabalka, G.W. (5) 253
Kaddouri, A. (1) 172
Kaeferstein, H. (9) 576
Kaever, V. (5) 181
Kagedal, B. (9) 362
Kageshima, H. (5) 65
Kageyama, S. (9) 330
Kahan, F.M. (9) 121
Kahn, J.D. (10) 123
Kahwa, I.A. (10) 132
Kaiser, C. (5) 383
Kaiser, G. (9) 99, 634
Kaiser, M.A. (6) 28
Kaiser, U. (3) 54
Kaiya, H. (5) 192
Kajdas, C. (5) 100
Kajiura, H. (7) 17
Kajiwara, M. (10) 111
Kajtar-Peredy, M. (9) 501
Kakehi, K. (6) 45
Kakinuma, H. (7) 16
Kalcher, J. (5) 44, 47
Kaldis, E. (11) 89
Kaldor, A. (1) 200
Kalhan, S.C. (6) 148
Kalinoski, H.T. (7) 33, 46, 58, 59, 65, 69, 220; (9) 523
Kallos, G.J. (5) 165
Kamada, S. (6) 282
Kamanka, J.-M. (9) 403
Kamar, A. (2) 68
Kamataki, T. (9) 154
Kambara, H. (3) 83; (7) 164; (8) 36; (9) 23
Kamensky, I. (3) 168; (9) 507
Kamikihara, T. (6) 176
Kamimura, H. (6) 254; (9) 330, 331
Kammei, Y. (3) 107, 186
Kammerer, R.C. (9) 247, 340, 355, 356
Kampel, V.Ts. (10) 5
Kanamaru, H. (6) 95
Kaneda, K. (6) 119
Kaneko, S. (9) 659
Kanematsu, K. (5) 394
Kaneti, J. (5) 488
Kaniwa, H. (9) 330
Kannan, N. (6) 373
Kantasubrata, J. (5) 387
Kanter, E.P. (1) 261
Kapala, J. (11) 82, 102, 111-113
Kapetanovic, I.M. (6) 284

Kapil, R.S. (5) 159
Kapoor, V.K. (5) 289
Kaposi, O. (11) 34, 90, 105, 109, 114, 115, 143
Kapusta, D. (9) 351
Karajalainen, A. (9) 269
Karas, M. (3) 43, 44; (9) 30
Karasek, F.W. (6) 6, 59, 182
Karasev, N.M. (11) 101
Karellas, N.S. (5) 264
Kargacin, M.E. (4) 201
Karger, B.L. (7) 141
Kari, P. (9) 242
Kariv, R. (9) 671
Karlaganis, G. (9) 130, 134, 332
Karlsen, J. (9) 534
Karlson, H. (6) 118
Karlsson, L. (1) 35; (9) 552
Karlsson, M. (9) 698
Karni, M. (2) 32
Karpas, Z. (2) 86, 89
Karran, N. (5) 276
Kartsev, V.G. (10) 145
Karunasasar, D. (11) 173
Kaslter, A. (3) 2
Kaspersen, F.M. (9) 34, 38
Kass, S.R. (5) 549
Kassam, J. (9) 653
Kassel, D.B. (5) 125; (6) 249, 295, 360
Kaster, L. (5) 86
Kasuya, F. (9) 152, 219
Kasuya, Y. (6) 224, 286; (9) 270
Kath, G.S. (4) 25
Kato, R. (9) 154, 618
Kato, S. (9) 113
Kato, Y. (7) 14
Katori, M. (6) 143
Katsoulos, G.A. (10) 96
Katsumata, S. (1) 74
Kauffmann, E. (9) 670
Kaufman, D.G. (5) 411; (7) 253
Kaufmann, R. (9) 602
Kaushik, V.K. (6) 414
Kautiainen, A. (6) 395
Kawabata, S. (10) 9
Kawabe, Y. (9) 440, 441
Kawaguchi, K. (1) 244, 249
Kawahara, Y. (9) 105
Kawai, R. (9) 330, 331
Kawakami, M. (9) 251
Kawano, Y. (5) 400
Kawasaki, T. (5) 91, 370
Kayganich, K.A. (5) 125; (6) 360
Kaykaty, M. (9) 199
Kazaryan, A.G. (5) 74
Kebarle, P. (2) 101; (5) 292, 480, 496-498, 538; (7) 109; (8) 19
Keefe, D. (9) 601
Keesee, R.G. (5) 124
Keigher, N. (9) 158, 526
Keinert, B. (4) 54
Kelleher, F.M. (5) 375
Kelleman, K. (5) 402
Keller, F. (1) 7, 14
Keller, R.A. (3) 35
Kellerhals, H. (3) 143, 148, 151; (6) 82
Kelley, P.E. (3) 130-132, 135, 136, 138
Kelly, D.W. (9) 432
Kemateck, R.J. (11) 127, 135
Kemp, T.J. (5) 427; (10) 98
Kemp, T.R. (3) 105
Kendall, R.A. (5) 521
Kenne, L. (9) 115
Kennedy, E.E. (9) 120
Kennedy, R.A. (1) 229
Kenttamaa, H.I. (7) 188
Kenyon, G.L. (5) 442; (7) 243
Keough, T. (5) 420; (7) 70
Kerner, J. (9) 412
Kerns, E.H. (9) 86
Kerr, C.R. (5) 83
Kessler, H. (7) 241
Kestell, P. (9) 359, 360
Kesuya, Y. (9) 663
Ketterle, W. (2) 111
Khan, M.R. (4) 199
Khandamirova, N.E. (11) 83
Khanna, S.N. (2) 17
Khathing, D.T. (10) 85
Khodeyev, Yu.S. (11) 87, 88
Khoobehi, B. (7) 43
Khundkar, L.R. (1) 93
Khvostenko, O.G. (5) 70, 75, 80, 81
Khvostenko, V.I. (5) 70, 75, 80, 81, 95
Kiang, P.H. (6) 320
Kidani, Y. (10) 106
Kido, H. (10) 107
Kidwell, D.A. (9) 40
Kiechel, J.R. (9) 93, 338
Kiermeier, A. (1) 87, 228
Kilcoyne, L.D. (1) 59
Kilius, L.R. (3) 181
Killmer, L.B. (9) 492
Kilner, M. (10) 94, 95
Kilpatrick, G. (7) 101
Kim, H. (9) 445
Kim, H.-S. (1) 270
Kim, H.-Y. (5) 186
Kimber, M.L. (9) 18
Kimura, K. (1) 60, 74, 82-84, 94, 230; (8) 9; (9) 139
Kindvall, G. (1) 309
King, M.A. (1) 288, 289
King, R.M. (4) 216
Kingham, D.R. (3) 173
Kingma, A.W. (6) 253
Kingston, E.E. (2) 198; (5) 243
Kingston, R.G. (1) 146, 187, 188
Kinoshita, E. (9) 98
Kinoshita, K. (9) 317
Kinoshita, T. (10) 107, 108
Kinsel, G.R. (1) 90
Kinsinger, J.A. (3) 149, 152, 153; (4) 62; (5) 448; (7) 48; (9) 27
Kinter, M. (2) 59; (6) 142; (9) 159
Kiplinger, J.P. (5) 234, 288, 295, 481; (7) 190
Kirby, S.P. (2) 12
Kirchner, N.J. (1) 231, 232
Kirk, K.L. (8) 26
Kirk, M.C. (9) 449
Kirkland, K.M. (9) 469
Kirle, K.L. (5) 465
Kiselev, Yu.M. (11) 98, 116
Kiser, W.O. (9) 589
Kishimoto, Y. (5) 311
Kiss, L. (4) 70
Kitsopoulos, T. (1) 340; (5) 58
Kitteringham, N.R. (9) 211, 224, 537
Klaassen, C.D. (9) 444
Klapotke, Th. (10) 21, 50-53
Klapstein, D. (1) 286, 287
Klasinc, L. (10) 4
Klass, G. (5) 543
Klaveness, J. (10) 12
Klebovich, I. (9) 309
Klee, M.S. (6) 28
Klein, H.O. (9) 328
Klein, P.D. (6) 51, 300
Klein, R.A. (6) 165
Kleinerova, E. (5) 224; (9) 222
Kleinschmidt, P.D. (11)

52
Kleiser, M. (9) 217
Kleopfer, R.D. (7) 228, 232
Klesper, E. (7) 39
Klewer, M. (3) 4
Kleywegt, G.J. (4) 121
Kline, R.M. (5) 357; (9) 494
Klinpott, E.-S. (5) 366
Klooster, M.J. (9) 95
Kluch, R.M. (9) 657
Kluft, E. (2) 196
Kluge, H.-J. (3) 37, 38
Klyuev, N.A. (9) 510
Knacke, O. (11) 99, 155
Knaeps, A. (9) 255
Knaeps, F. (9) 201
Knapp, D.R. (5) 194; (6) 202, 203
Knapp, G. (9) 270
Knerr, G. (9) 624
Knight, J.S. (2) 87
Knihinicki, R. (9) 684
Knittel, D.R. (11) 159
Knoll, J.E. (6) 96
Knoll, K. (5) 282; (10) 74, 76, 79
Knoll, W.M. (9) 514
Knop, G. (5) 259
Knopf, S. (9) 676
Knoppel, H. (5) 127
Knorr, F.J. (3) 176, 177
Knyazov, I.N. (3) 5
Kobayashi, R. (9) 441
Kobayashi, T. (9) 375, 455, 456, 532
Koberstein, E. (9) 488
Kobertz, D. (11) 85, 110, 119
Koch, C.A. (4) 24
Koch, T.H. (8) 23
Koch, W. (1) 181; (2) 6, 23, 143, 172
Kocis, P. (9) 451
Koda, S. (9) 532
Kodaira, H. (9) 461
Kodama, S. (6) 296
Koeb, L. (9) 344
Köfel, P. (3) 143, 148, 151
Koehler, M.G. (4) 140
Koelle, U. (10) 60
König, W.A. (5) 408
Koepke, S.R. (8) 13
Köster, H. (5) 408
Koffel, J.C. (9) 169
Koga, R. (9) 106
Koga, Y. (5) 316
Kogler, H. (9) 505
Kohsaki, K. (9) 160
Koike, M. (9) 196, 197, 438
Koizumi, K. (5) 376
Kojima, T. (9) 593
Kok, R.M. (5) 62, 68, 86, 233; (6) 261, 283
Kok, W.T. (7) 5
Kolaitis, L. (3) 80–82; (8) 32
Kolani, B. (10) 49
Kolasinski, K.W. (1) 81
Kolditz, L. (11) 79
Kolis, S.J. (9) 471
Kolli, I.D. (10) 124
Kolsaker, P. (5) 131; (6) 191
Kolset, K. (4) 85; (6) 342
Komai, Y. (4) 14
Komiya, I. (9) 650
Komori, T. (5) 341, 345, 353, 400; (9) 626
Kondo, H. (9) 623
Kondo, N. (7) 161
Kondo, T. (9) 659
Kondow, T. (1) 218, 219; (5) 65
Kondrat, R.W. (2) 206
Konig, K. (9) 328
Koolstra, W. (6) 292
Koot, W. (1) 112
Kopf, H. (10) 50–54
Kopnarski, M. (3) 56
Koppel, C. (9) 140, 244, 246, 248, 272, 308, 346
Kopperaal, D.W. (5) 27
Kordau, T. (5) 534
Korenchuk, E.N. (9) 683
Korenev, Yu.M. (11) 93, 107
Korfmacher, W.A. (5) 128, 252, 254, 255; (6) 189, 190, 336, 337, 387; (7) 119, 126; (9) 282, 285–287, 353, 400, 431, 434, 472, 699
Korobov, M.V. (11) 18, 31, 32, 34, 36, 42, 95, 96, 98, 101, 116
Koruna, I. (5) 108, 224; (9) 88, 218, 222, 393
Kos, A.J. (5) 488
Koshevoi, O.G. (4) 208
Koski, R. (4) 103; (9) 122
Kossen, S.P. (9) 322
Kossman, H. (1) 38
Kostiainen, R. (5) 205; (6) 307
Kostner, G.M. (6) 210, 211
Kostyanovskii, R.G. (5) 75
Kotake, A.N. (9) 669
Kotkoskie, L.A. (9) 444
Kott, M.G. (8) 28
Kou, T. (11) 66
Kourtis, S. (9) 291, 292
Koutsantonis, G.A. (10) 39
Kovacic, P. (5) 448
Kovacs, R.M. (4) 70
Koves, G. (9) 175
Kowaleski, R.M. (10) 55
Kowalska, M. (11) 58
Kowalski, B.R. (4) 201
Kowalski, M.H. (10) 103
Koyano, I. (2) 77
Koziol, A.E. (10) 66
Koziol, J. (4) 122, 129
Kraan, G.P.B. (6) 226
Kraft, R. (7) 12, 13
Kraka, E. (5) 48
Kramell, R. (5) 77
Kramer, G. (9) 644
Krane, J. (6) 342
Kraska, S. (6) 391
Krasnov, K.S. (11) 38
Krasselt, W.G. (9) 571, 572
Kratz, L.E. (9) 354
Krause, M.O. (1) 9, 19
Krause, W. (9) 228, 350
Krautler, B. (10) 113, 116
Kreimler, H.-P. (9) 390
Kreiner, W.A. (1) 247
Kreise, D. (1) 214
Kreissl, F.R. (10) 88
Kremer, J.H. (4) 192
Kresbach, G.M. (7) 141
Kreuger, C. (4) 165
Krieger, L.A. (5) 250
Krieger, M.S. (6) 113
Kriemler, H.-P. (9) 388
Kriep, B. (5) 312
Krishnamurthy, T. (5) 201, 203, 204; (6) 305, 306; (7) 212, 221
Kristo, M.J. (3) 140, 199; (4) 29, 35
Krönert, U. (3) 37
Krolik, S. (6) 131-133; (7) 217
Krolik, S.T. (7) 106
Kroop, H. (9) 121
Kroto, H.W. (1) 207; (5) 53
Krstulovic, A.M. (9) 670
Krueger, T.R. (9) 315
Kruger, C. (6) 376
Kruggel, F. (5) 377
Krumbiegel, G. (9) 214
Kubiak, F.M. (6) 144
Kubota, E. (3) 107

# Author Index

Kucerovy, A. (10) 14
Kuche, P.G. (5) 199
Kuchino, Y. (8) 8
Kuchitsu, K. (1) 218; (5) 534
Kuck, D. (2) 201
Kudin, L.S. (11) 38
Kudinova, M.K. (9) 510
Kudo, H. (9) 331
Kühlewind, H. (1) 87
Kuehn, A. (5) 73
Kuehnle, R. (4) 111
Kuhara, T. (6) 46
Kuhl, P.G. (6) 198
Kuhn, R. (1) 283, 284, 287–289
Kuhne, G. (9) 350
Kuhnert, B.R. (9) 84
Kujiwara, T. (5) 245
Kukovetz, W.R. (5) 193
Kuksis, A. (5) 19; (6) 34
Kuligina, L.A. (11) 71, 157
Kulik, W. (2) 147
Kumar, P. (5) 82
Kume, T. (9) 112, 278
Kumor, K. (9) 132, 596
Kun, Z. (11) 172
Kunakova, R.V. (5) 97
Kunert, Ch. (11) 79
Kung, C.-Y. (1) 229
Kunihiro, F. (3) 107; (7) 167
Kunish, K. (5) 377
Kuo, C.-H. (1) 270, 306
Kuo, G.Y. (9) 448
Kupferberg, H.J. (6) 284
Kuppermann, A. (2) 114
Kurata, Y. (9) 129
Kurepa, M.V. (1) 103
Kurihara, H. (9) 97
Kurogochi, Y. (6) 255; (9) 376
Kuroki, H. (6) 375
Kuronuma, K. (9) 251
Kurosawa, T. (6) 232, 233
Kurth, H. (5) 331
Kuruma, I. (9) 123
Kuruyama, K. (10) 111
Kurz, J. (9) 252
Kusakabe, Y. (9) 498
Kusunose, E. (6) 119
Kusunose, M. (6) 119
Kutschabsky, L. (5) 92
Kutschera, W. (3) 184
Kutuzova, Yu.L.A. (11) 148
Kuwahara, N. (9) 622
Kuwahara, Y. (4) 118; (6) 176
Kuznetsov, S.V. (11) 36, 42, 93

Kuzovkova, N.A. (9) 683
Kvinnsland, S. (9) 458, 459
Kyrematen, G.A. (9) 377

Laasasenaho, K. (7) 131
Labarre, J.-F. (10) 127
La Barre, S. (5) 349
Labastie, P. (1) 215
LaBelle, R.D. (3) 35
Lablanquie, P. (1) 32, 147, 149, 162–164, 167
Labows, J.N., jun. (4) 27
LaCagnin, L.B. (9) 323
Lacey, E. (5) 166; (9) 92
Lacey, M.P. (7) 70
Lacey, R.F. (4) 162
Ladd, D.L. (5) 383
La Fleur, V.M. (9) 300
Lafontaine, P. (9) 627
Lafortune, F. (4) 56; (5) 444; (8) 14
Laframboise, S. (4) 79
Lagarde, M. (5) 164, 195
Lagna, W.M. (3) 87
Laine, R.M. (10) 7, 132
Laird, S. (7) 118
Lakenbrink, M. (11) 150, 151
Lakshmi, V.M. (9) 378
Lalande, M. (9) 367
Lalevic, M. (6) 329, 330
Lalia-Kantouri, M. (10) 96
Lamb, J.H. (6) 394
Lambe, R.F. (9) 220
Lambert, K.L. (9) 353, 699
Lambert, W.E. (6) 50
Lambrechts, M. (9) 588
Lammert, S.A. (7) 97, 226
Lampe, M.A. (9) 247
Lancaster, M. (7) 106
Landgren, B.-M. (9) 383
Lane, K.R. (5) 555, 556
Lane, S.J. (7) 38, 135
Lanfranchi, M. (9) 373
Lang, H. (10) 75, 77
Lange, C. (5) 114, 218, 219
Langen, C. (9) 619
Langenbeck, U. (6) 147
Langer, G. (4) 70
Langford, M.L. (2) 173
Langley, G.J. (7) 103
Langston, J.W. (9) 145
Lanouette, M. (5) 242; (6) 392
Lant, M.S. (7) 90, 115; (9) 133, 437
Lanzetta, R. (5) 351

Lapeza, C.R. (6) 365
Lapiguera, A. (9) 445
Laplanche, R. (9) 131
Laramée, J.A. (5) 137
Larciprete, R. (10) 22
Larice, J.L. (5) 387
Larka, E.A. (10) 83
Larkin, R.J. (1) 36
Larsen, A.B. (9) 696
Larson, J.W. (5) 536, 542
Larsson, L. (6) 194, 304
Larsson, M. (1) 173, 266, 309, 310
Larter, S.R. (4) 174
Laschic, N.S. (6) 293
Later, D.W. (7) 23
Lattimer, R.P. (9) 29
Lau, B.P.-Y. (5) 148; (6) 309, 312; (7) 216, 217
Lau, K.H. (11) 100, 158, 159
Lau, P. (9) 647
Laubli, T. (6) 100, 101
Laude, D.A., jun. (3) 154; (4) 63; (6) 81; (7) 49; (9) 28
Laudenschlager, W.G. (9) 58
Laukien, F.H. (3) 143; (5) 563
Laurent, J.-P. (10) 92
Lauterburg, B.H. (9) 130, 332
Lauwers, W. (9) 200, 201, 253–255, 336
Lavene, D. (9) 93
Lavrijsen, K. (9) 201
Lawrence, A.H. (4) 79; (5) 257
Lawrence, J. (5) 141; (6) 371
Lawrence, M.R. (9) 620
Lawson, A.M. (5) 348; (6) 245
Lawson, J.A. (6) 206
Lay, J.O., jun. (7) 119, 207; (9) 282, 392, 400; (10) 34
Layrez, J. (5) 486
Layton, W.J. (9) 379
Lazarev, V.B. (11) 56
Lea, A.G.H. (5) 371
Leach, S. (1) 308, 315
Leary, J.J. (4) 90; (6) 49; (7) 79
Lebedev, A.T. (5) 74
Lebedev, K.S. (4) 131
LeBel, G.L. (6) 343, 379; (7) 216
Le Beyec, Y. (3) 197; (5) 434
Le Bigot, J.F. (9) 338

Leblanc, J.-C. (10) 56
LeBoef, E. (9) 277
Leboeuf, J.-F. (10) 56
Lebrilla, C.B. (2) 84, 137, 172; (10) 69, 140–144
Leck, J.H. (3) 115
Leclerc, J.C. (4) 108
Leclercq, G. (9) 457
Leclercq, P.A. (9) 271, 489
Ledbetter, M.C. (3) 65
Lederer, K. (6) 440
Lee, A.R. (1) 141
Lee, C.R. (6) 262
Lee, E.D. (3) 153; (5) 211, 227; (7) 23, 48, 85, 124, 146; (9) 16, 27, 545, 679
Lee, F.W. (9) 337
Lee, M.L. (7) 21, 22, 56
Lee, M.S. (5) 231; (7) 198, 199, 203; (9) 42–44
Lee, R.E. (5) 287, 508
Lee, S.-A. (3) 35
Lee, S.-H. (9) 333
Lee, S.K. (1) 254
Lee, T.J. (5) 502
Lee, Y.T. (1) 36, 313; (5) 502
Leem, F. (5) 418
Leeper, J.D. (11) 141, 144, 145
Lefebvre, M.A. (9) 157, 162
Lefebvre-Brion, H. (1) 3, 7, 8, 13, 14, 25, 26
Lefevere, M.F. (4) 92; (6) 37, 50, 289, 316; (9) 548
Legendre, M.G. (6) 413; (9) 591, 592
Leger, D. (5) 141; (6) 371
Legha, S.S. (9) 212
Lehmann, A. (5) 415
Lehr, K.H. (9) 394
Lei, P. (10) 41
Leinweber, F.-J. (9) 203
Leis, H.-J. (5) 191; (6) 205, 210, 211, 213
Leisin, O. (1) 121, 225
Leiter, K. (1) 214
Lekies, R. (10) 126
Le Leune, L. (9) 253
Leleyer, M. (5) 432
Lelik, L. (11) 34, 143
Lentz, D. (5) 72
Leonard, T.B. (9) 447
Leone, A.M. (6) 263
Leong, A.J. (10) 121

Leopold, D.G. (1) 317, 322, 326, 327, 331; (5) 487, 495
Lepage, F. (9) 450
Leroi, G.E. (1) 15, 27–29
LeRouzo, H. (1) 161
Leroy, G. (2) 31
Leroy, Y. (7) 160
Lertratanangkoon, K. (9) 186–188, 192
Lesieur, M. (5) 164; (9) 482
Lesimple, P. (10) 15
Letokhov, V.S. (3) 5
Leung, C.-S. (9) 429
Leung, L.Y. (9) 695
Levandoski, P. (9) 81, 446
Levesque, W.R. (9) 374
Levin, I. (2) 225
Levin, R.D. (2) 11
Levinger, N.E. (1) 265
Levsen, K. (5) 39; (7) 215
Levy, R.H. (9) 450, 476, 477
Lewellen, L.J. (6) 409; (9) 558
Lewin, E.E. (6) 330
Lewis, D.E. (5) 142, 143
Lewis, I.A.S. (7) 101
Lewis-Enright, D.P. (6) 395
Leyh, B. (1) 8, 21, 22
Leyh-Nihant, B. (2) 224
Li, H. (5) 379
Li, L. (3) 20
Lias, S.G. (2) 10
Liberato, D.J. (5) 346; (7) 102, 120; (9) 421
Libert, R. (9) 656
Lichtenwainer, M. (9) 673
Lida, T. (6) 259
Liddell, M.J. (5) 424, 426; (10) 2, 37–39
Lidgard, R.O. (5) 174; (6) 388
Lie, B. (2) 172
Lie, R.H. (9) 591
Liebich, H.M. (6) 47
Liebl, H. (3) 98
Liebman, J.F. (2) 11, 88
Liedós, A. (5) 525
Liedtke, R.J. (2) 197
Lie Ken Jie, M.S.F. (6) 160
Lien, E.A. (9) 458, 459
Liepa, I. (10) 121
Liesch, J.M. (4) 25; (6) 23
Lifshitz, C. (1) 1, 109, 111, 238; (2) 1, 138, 225
Lignori, A. (5) 386
Ligocki, M.P. (6) 117
Ligon, W.V. (5) 438, 441, 443; (6) 30
Ligtenstein, D.A. (9) 322
Liguori, A. (8) 31
Lijinsky, W. (9) 354
Lim, C.K. (9) 95
Lim, H.-K. (6) 412; (9) 137
Limin, Q. (6) 112
Lin, A. (9) 525, 526
Lin, A.J. (9) 111
Lin, C. (9) 445
Lin, T. (2) 187
Lin, Z.-M. (9) 647
Lindanes, E. (7) 26
Lindberg, C. (7) 100, 136; (9) 19, 168, 229–232
Lindberg, S. (9) 617
Lindeke, B. (9) 460, 707
Lindemann, B. (10) 113, 116
Linder, M. (7) 114; (9) 82
Lindfors, A. (9) 485
Lindgren, J.-E. (9) 96, 677, 678
Lindle, D.W. (1) 18, 20
Lindner, B. (4) 193
Lindoy, L.F. (10) 121
Lindstrom, G. (5) 138
Lindstrom, L.H. (9) 102
Lindstrom, T.D. (9) 435
Lineberger, W.C. (1) 265, 317, 322–324, 326, 327, 331, 335–337, 339; (5) 487, 492, 495
Linevsky, M.I. (11) 20
Ling, Y.-C. (4) 202, 203
Linscheid, M. (7) 128; (9) 488
Linskers, H.F. (6) 58
Linton, R.W. (4) 36, 37
Lintz, W. (9) 697
Lipman, R. (8) 25
Lipsky, S.R. (6) 125, 126
Liska, T. (7) 106
Litherland, A.E. (3) 180, 181
Little, T.L. (6) 410; (9) 559
Littlejohn, D.P. (3) 150
Liu, A.C. (4) 12
Liu, A.L. (9) 389
Liu, B.N. (6) 166
Liu, C.-M. (9) 499
Liu, D.-J. (1) 252, 263
Liu, R.H. (6) 413; (9) 592

# Author Index

Liu, S.H. (1) 18
Liu, X.D. (4) 150, 205
Liu, Y. (1) 207; (5) 53
Lizon, E. (1) 164
Lloyd, D. (4) 67, 68
Lloyd, H.A. (9) 629
Lloyd, J. (7) 159
Lloyd, J.R. (5) 301; (9) 22
Lochner, A. (6) 297
Locht, R. (1) 195; (4) 11
Lockard, J.S. (9) 478
Locquet, J.-P. (4) 74
Lodge, B.A. (9) 426
Lodi, L. (5) 101
Loftus, P. (9) 469
Logvinenko, V.A. (10) 87
Loh, A.C. (9) 203
Loh, S.K. (1) 204
Loh, S.Y. (4) 65
Lohninger, H. (4) 87; (6) 88
Loiffler, S. (5) 51
Lokan, R.J. (9) 662
Lomakii, G.S. (5) 80
Long, F.A. (1) 109
Longato, B. (10) 88
Longevialle, P. (2) 154
Longoni, E. (9) 466
Loo, J.A. (3) 63, 157
Loo, J.C.K. (9) 627
Looareesuwan, S. (9) 538
Looker, J.H. (5) 360
Looper, J.R. (6) 79
Lopatin, S.I. (11) 140, 148, 154
Lopez-Avila, Y. (6) 114, 391
Lorant, A. (6) 84
Lorenčak, P. (2) 156, 157
Lorent, V. (1) 140
Lorenzi, R. (5) 324
Lorquet, J.C. (2) 224
Los, J. (3) 4
Lossing, F.P. (2) 41, 133, 134
Loucks, D.H. (3) 203
Louis, W.J. (9) 633
Louris, J.N. (3) 134, 136, 137; (6) 77
Low, I.A. (6) 413; (9) 591
Lowney, L.I. (9) 578
Loyaux, D. (9) 670
Lu, K. (9) 212
Lubli, G. (9) 580
Lubman, D.M. (3) 7, 10, 11, 30, 80-82; (7) 61-63; (8) 32; (9) 635
Lubman, E.W. (3) 20
Lucatorto, T.B. (3) 9
Lucchese, R.R. (1) 10, 11

Ludemann, F. (10) 54
Ludwig, E. (10) 90
Lue, L.P. (9) 587
Luger, P. (5) 332
Lugrin, A. (1) 164
Lukaszewicz, J. (9) 104
Luke, M.A. (6) 310; (7) 236
Lukin, V.G. (5) 74
Lundborg, P. (9) 617
Lurie, I.S. (6) 63
Luthe, H. (6) 147
Luther, E.W. (9) 104
Lutz, T. (9) 144, 259
Ly, U.H. (9) 157
Lyapina, N.K. (5) 95, 98
Lyatifov, I.R. (10) 57, 58
Lycke, W. (3) 188
Lykke, K.R. (1) 335-337, 339; (5) 492
Lynch, M.J. (9) 374
Lynn, D.G. (5) 220
Lynn, K.L. (9) 141
Lynn, R.K. (9) 448
Lyo, I.W. (1) 159
Lyon, P.A. (5) 421
Lys, I. (3) 174

Ma, C.Y. (6) 415
Ma, H. (10) 65
Maas, W.P.M. (2) 216
Mabud, Md.A. (2) 54-57; (3) 127, 128
McAdoo, D.J. (2) 5, 186-190, 195
MacAllister, J.M.R. (5) 433
McBay, E.H. (3) 126; (6) 415
MacBride, J.A.H. (6) 366
McCague, R. (9) 457
McCarthy, I.E. (1) 52
McCarthy, J.R. (9) 463
McCarthy, M.E. (9) 447, 448
Maccioni, A.M. (5) 178; (6) 449; (10) 88
McCloskey, J.A. (8) 1, 5-7, 9, 10
McCluer, R.H. (7) 139
McConnell, J.R. (5) 297
McCool, B.J. (10) 121
McCord, B.R. (4) 49
McCorkle, D.L. (1) 123, 125; (5) 59
McCormick, T.J. (5) 425; (7) 107
MacCross, M. (5) 404
McCulla, W.H. (3) 36
McCurdy, H.H. (6) 409;
(9) 558
McCurry, J.D. (6) 372
McDaniel, M.L. (5) 215
McDonald, R.N. (5) 494, 504, 506, 561
Macdonald, T.L. (10) 14
McDowall, M.A. (7) 158
McDowell, T.R. (9) 119
McElvany, S.W. (1) 199, 201; (5) 554
McErlane, K.M. (5) 83
McEwan, M.J. (2) 87
McEwen, C.N. (2) 16; (5) 29
McFadden, W.H. (7) 97, 98, 226; (9) 562
Macfarlane, R.D. (3) 59, 169, 170; (4) 57; (5) 420, 459, 462, 463, 468; (9) 629
Macfie, H.J.H. (4) 182
McFinley, R.M. (9) 171
McGahren, W.J. (7) 261; (9) 520
McGarvey, G.J. (10) 14
McGee, C.M. (5) 166; (9) 92, 689
McGhee, W.D. (5) 506
McGibney, D. (9) 208
McGift, J.C. (5) 200
McGill, I.J.M. (2) 67
MacGregor, T. (9) 138
McIlrath, T.J. (1) 274
MacIntyre, W.G. (10) 17
McIver, R.T. (3) 147
McKay, G. (9) 250
McKee, M.L. (2) 40
McKee, V. (10) 122
McKeel, W.J. (4) 25
McKellop, K. (9) 138
McKenna, J.I. (9) 624
MacKenzie, S.L. (6) 275, 276
McKinnon, H.S. (5) 139; (6) 178
McKinstry, D.N. (9) 630
McKown, H.S. (3) 27, 124, 125; (4) 59
McLafferty, F.W. (1) 113; (2) 9, 121, 130, 136, 140; (3) 63, 149, 157; (4) 65, 117; (6) 17, 86; (7) 168, 174
McLaughlin, F.A. (6) 98
McLean, A.D. (1) 333, 334; (2) 29; (5) 490, 491; (6) 110
McLean, M.A. (9) 212
McLean, M.M. (5) 269; (7) 127
McLennan, W.H. (4) 148, 149, 194, 198

McLeod, J.K. (2) 164
McLeod, L. (2) 37
McLeod, R.A. (5) 208
Maclouf, J. (5) 184; (6) 212
McLuckey, S.A. (3) 124–126; (4) 59; (5) 63, 247
McMahon, D.H. (6) 32
McMahon, T.B. (2) 70, 101; (5) 536, 542
McMillan, J.G. (2) 106
McMullin, M. (9) 673
McNeal, C.J. (5) 14, 420, 462, 463, 468; (9) 629
MacNeil, J.D. (5) 126
McPartlin, M. (10) 121
MacRae, P.V. (9) 208, 209
Maddock, J. (6) 229
Maddox, M.L. (9) 337
Madelmont, J.C. (9) 436
Madge, J.A. (6) 179
Madhusudanan, K.P. (5) 79, 82, 158–160, 206, 207, 225, 237
Madigan, M.J. (5) 348
Madix, R.J. (1) 73, 81
Madson, M.A. (6) 145
Maebashi, K. (9) 503
Maeda, K. (9) 154
Märk, T.D. (1) 97–101, 198, 205, 206, 208–214, 216, 217; (2) 19
Maertens, F. (9) 611
Maffei Facino, R. (5) 338; (9) 466
Maftouh, M. (9) 289
Maggio, J.E. (5) 377
Maggio, V.L. (3) 92; (6) 366
Maggs, J.L. (9) 211, 296, 537
Magnolato, D. (5) 371
Mahara, R. (6) 232, 233
Mahato, S.B. (5) 325, 329, 332
Mahle, N.H. (5) 37
Mahon, M.F. (10) 10
Mahoud, M.M. (10) 94
Maibach, H.I. (6) 333, 396
Maier, J.P. (1) 264, 278–293, 298, 300, 302, 307
Mailahn, W. (6) 324
Main, D.E. (4) 56; (5) 434; (6) 202; (9) 487
Maitra, A.M. (10) 91
Mak, M. (9) 309
Makarov, A.V. (11) 104
Maker, P.D. (3) 3
Makern, H. (6) 236

Makern, S. (6) 236
Maketa, A. (5) 313
Makriyannis, A. (9) 560
Maksay, G. (9) 645
Maksoud, S.A. (5) 367
Malakhov, K.V. (5) 440
Malatesta, V. (7) 242
Malbica, J.O. (9) 469
Malchow, R.D. (9) 680
Maldivi, P. (10) 109
Male, V.L. (6) 239
Malinovich, Y. (1) 150, 154, 160
Malinowski, E.R. (4) 145
Malkerova, I.P. (11) 92
Malkinson, A.M. (9) 58
Mallard, W.G. (2) 11
Malle, E. (5) 191; (6) 210, 211, 213
Mallet, A.I. (6) 258
Mallet, N.N. (6) 151
Mallinger, A.G. (9) 676
Mallis, L.M. (7) 254
Maltby, D. (1) 2; (2) 8; (6) 9; (7) 95, 197; (8) 4; (9) 7, 655
Malutzki, R. (1) 38
Malyusov, V.A. (11) 56
Mamer, O.A. (6) 293, 294
Mamyrin, B.A. (4) 218
Manada, K. (9) 663
Manaresi, P. (5) 322
Manchee, G.R. (9) 437
Mangoni, L. (5) 351
Maniwa, M. (9) 156
Mann, D.E. (11) 20
Mann, D.R. (10) 97
Mannervik, S. (1) 266
Mannhold, R. (9) 602
Manning, J. (9) 164
Manteuffel-Cymborowska, M. (9) 241
Manuel, G. (5) 550
Manz, I. (9) 328, 502
Mao, F.M. (3) 115
Maquestiau, A. (2) 71, 74, 156, 157, 198
Maquin, F. (2) 159, 160, 174
Marcell, P.D. (6) 277
March, R.E. (1) 156; (2) 68, 105, 106, 118, 123
Marchand, D.H. (9) 236
Marchon, J.-C. (5) 416; (10) 109, 118
Marigo, M. (9) 580
Mariotte, A.M. (5) 358
Maripuu, R. (1) 35
Maris, F.A. (5) 275; (7) 148
Markey, S.P. (5) 226; (7) 175; (9) 4

Markham, K.R. (5) 367
Markides, K.E. (7) 21, 22, 56
Marks, J. (1) 337; (5) 493
Marks, T.J. (10) 68
Marmet, P. (1) 105–108; (5) 49, 50
Marquis, F. (5) 189; (6) 216
Marr, G.V. (1) 70
Marriage, H.J. (9) 266
Marschall, H.-U. (5) 336; (6) 236
Marsel, J. (11) 90
Marsh, J.B. (6) 285
Marshall, A.G. (3) 161–163; (4) 61; (9) 29, 628
Martell, A.E. (5) 468
Martensson, J. (9) 362
Martin, B.R. (9) 587
Martin, C. (6) 299
Martin, D.J. (4) 22
Martin, F. (9) 166, 167, 483
Martin, G.G. (4) 125
Martin, L.E. (7) 90, 115; (9) 133, 437
Martin, M. (5) 181; (6) 295
Martin, S.A. (3) 107; (5) 389; (7) 186
Martin, S.J. (5) 37
Martin, W.B. (1) 96; (3) 47
Martineau, A. (5) 187; (6) 207
Martines, R.A. (5) 303
Martinetti, A. (10) 29
Martinez, M. (9) 363
Martinez, R.I. (3) 120–122; (4) 137–139
Marting, G.J. (4) 125
Martinsen, D.P. (4) 107
Marty, J.-C. (6) 340, 426
Martynova, T.N. (10) 87
Marunaka, T. (9) 156
Marushkin, K.N. (11) 56
Marx, K.-H. (5) 181
Marzec, A. (4) 191
Marzouki, Z.M.H. (6) 193
Mascherpa, A. (6) 431
Masfurer, J. (5) 200
Mason, R.P. (5) 384
Mason, R.S. (2) 78–80
Massarella, J.W. (9) 601
Massey, R.C. (3) 202
Masson, C.R. (11) 153
Massot, R. (4) 108
Massy-Westropp, R.A. (5) 115

Mastovich, S.L. (9) 58
Masuda, K. (10) 106
Masuda, Y. (6) 375
Masumoto, H. (9) 402
Masumoto, K. (4) 112
Masuoka, T. (1) 158
Masur, M. (2) 185, 212, 213
Materikova, R.B. (10) 57, 58
Matern, H. (5) 336
Matern, S. (5) 336
Mathews, C.K. (11) 132, 173
Mathews, W.R. (5) 182; (6) 208; (7) 181
Mathey, F. (10) 29, 62
Mathian, B. (6) 221
Mathiasson, L. (6) 399
Mathieu, J.C. (11) 60, 126
Mathur, D. (1) 174, 185, 187-189; (2) 173; (4) 8
Matjiuk, V.M. (3) 5
Matousek, G.S. (6) 148
Matsubara, K. (9) 324
Matsubara, T. (7) 18
Matsuda, A. (8) 8
Matsuda, F. (6) 259
Matsuda, H. (3) 96, 97, 178; (4) 211
Matsuda, Y. (9) 178
Matsui, E. (5) 437
Matsui, T. (5) 352, 356; (11) 54, 55, 75
Matsuki, Y. (9) 97, 129, 345
Matsumoto, I. (6) 46
Matsumoto, K. (7) 75, 76, 153-155
Matsuo, T. (3) 97, 178, 191; (5) 399; (6) 296
Matsuoka, T. (9) 98
Matsushima, E. (9) 156
Matsushima, T. (6) 285
Matsushita, A. (9) 590
Matsushita, Y. (9) 105
Matsuzaki, T. (9) 318
Mattammal, M.B. (9) 378
Matthiesen, U. (6) 335
Mattiasson, I. (9) 660
Mattly, D. (5) 25
Matusik, J.E. (9) 529
Mauer, T. (6) 348
Maul, J.L. (4) 40, 41
Maume, G. (6) 238
Mauney, T. (4) 217
Maurer, H. (9) 610, 646, 700
Maurer, J. (5) 323
Maurizis, J.C. (9) 436
Mavrodiev, V.K. (5) 56, 89, 97, 102
Mawhinney, T.P. (6) 145
Maxwell, J.R. (10) 117
May, R.G. (5) 385
May, R.J. (6) 30
Maybaum, J. (8) 28
Mayer, B. (5) 191, 193; (6) 205, 211, 213
Mayernik, J.A. (8) 20
Maynard, A.T. (5) 234
Mays, D.C. (9) 357
Mazur, M. (4) 129
Mazurek, M.A. (6) 339
Mazurov, V.A. (5) 80
Mazzeo, P. (9) 687
Mazzi, U. (10) 90
Mead, E.W. (9) 58
Mead, R.D. (1) 337
Means, J.C. (6) 347
Medhurst, L.J. (1) 18
Meeks, R.D. (9) 577
Meemken, H.-A. (6) 376
Meese, C.O. (5) 198, 199; (6) 198-201; (7) 202, 251; (9) 264
Meffin, P.J. (9) 307
Mehran, M.F. (6) 26
Mehrotra, K.N. (5) 79
Mehrotra, R. (5) 206
Mehta, R.J. (9) 493
Meier, D. (6) 437
Meili, J. (4) 34
Melby, S.J. (5) 516
Melegh, B. (9) 412
Melekh, B.T. (11) 56
Melendres, C.A. (5) 524
Mell, L.D. (9) 136, 171, 564
Mellon, F.A. (7) 130
Menchuk, E. (11) 73
Mennicke, W.H. (9) 214
Men'shikov, V.V. (11) 142
Mensing, Ch. (11) 79
Menz, D. (11) 79
Meot-Ner, M. (2) 85, 86, 88, 89, 98, 100; (5) 478, 479, 528-530, 532, 541
Merbach, P. (10) 73
Mercer, J.G. (6) 121
Mercer, R.S. (2) 123, 150; (5) 474
Meredez, M. (5) 486
Merian, E. (6) 59
Mermet, J.M. (3) 72
Mertz, J.L. (5) 357; (9) 494
Metcalf, G.S. (4) 175
Metral, C.J. (3) 58
Metral, J.-L. (10) 31
Metz, R.B. (1) 340; (2) 85; (5) 58
Metzger, J. (4) 4, 113; (5) 387; (6) 89
Metzler, M. (9) 276
Meuldermans, W. (9) 200, 201, 253-255, 336
Meunier, B. (9) 289
Meunier, G. (9) 289
Meuzelaar, H.L.C. (4) 73, 148, 149, 174, 175, 177, 192, 194-196, 198, 199
Mevel, V. (6) 84
Mezzena, C. (4) 15
Miach, P.J. (9) 633
Micetich, R.G. (9) 470
Michaelis, K. (5) 244
Michaelis, N. (5) 112
Michalik, P. (7) 217
Michel, G. (9) 500
Michiels, E. (4) 150, 217
Michinton, A. (1) 58
Mico, B.A. (9) 281, 446
Middleditch, B.S. (6) 297; (9) 50
Midgley, J.M. (9) 91, 681, 682
Midha, K.K. (9) 250, 283, 284
Miemeyer, U. (9) 328
Miersch, O. (5) 94
Mifatkov, M.S. (5) 89
Migdalof, B.H. (9) 630
Mignot, A. (9) 79, 162
Mihailova, D. (9) 316
Mihara, S. (4) 115
Mikaya, A.I. (6) 152
Miki, E. (10) 99
Mikus, G. (9) 619
Milazzo, J. (9) 525
Milborrow, B.N. (5) 216
Miles, C.J. (7) 214
Miles, J.W. (9) 51
Milkman, I.W. (1) 294, 295, 306
Millburn, P. (9) 193
Miller, A.E.S. (1) 317; (5) 495
Miller, C.E. (6) 67
Miller, C.M. (3) 35
Miller, D.J. (6) 113; (7) 55, 74
Miller, D.W. (9) 282, 286, 353, 431, 472, 699
Miller, F.W. (10) 102
Miller, J.C. (1) 62
Miller, J.E. (9) 504
Miller, J.M. (5) 413; (10) 18, 19, 101, 104
Miller, M. (11) 106, 110, 117-119
Miller, M.K. (4) 45, 46
Miller, R.D. (9) 153, 339

Miller, T.A. (1) 229
Miller, T.M. (1) 322, 323, 327; (5) 487
Miller, W.M. (9) 563
Millet, A. (2) 205
Millie, P. (1) 16, 39, 148, 172
Millington, D.S. (7) 60, 197; (9) 655
Millowar, B.V. (5) 174
Mills, D.A. (6) 398
Milman, B.L. (5) 43
Milne, G.W.A. (4) 107
Milne, T.I. (11) 24
Milton, D.J. (2) 80; (3) 193
Milushin, M.I. (11) 45, 136-138
Milvihill, M.J. (6) 378
Min, B.H. (9) 701
Minaeva, I.I. (11) 46, 68
Minagawa, K. (9) 270
Minale, L. (5) 354
Minami, R. (5) 313
Minami, Y. (9) 156
Minatogawa, Y. (6) 259, 282
Minghetti, G. (10) 30
Minkwitz, R. (10) 126
Minor, J.L. (6) 247
Minshall, E. (10) 10
Minze, R. (10) 90
Mio, T. (6) 296
Mirale, L. (5) 349
Mirchandani, H. (9) 547
Mirocha, C.J. (4) 110; (6) 308
Mironov, Yu.V. (10) 130
Misaizu, F. (5) 534
Misev, L. (1) 284
Mishima, Y. (9) 251
Mishra, U. (9) 107
Misra, R. (9) 521
Missler, S.R. (6) 363
Mitchell, A.L. (2) 76
Mitchell, D. (5) 253
Mitchell, J.M. (9) 136, 171, 564
Mitchell, S.C. (9) 395
Mitchum, R.K. (5) 140, 422; (6) 185, 369; (7) 126, 223, 233; (9) 285, 287
Mitsuhashi, H. (9) 318
Mitsuke, K. (1) 218; (5) 534
Mittal, S. (5) 159
Mitteilung, K. (10) 53
Miura, H. (6) 231
Miwa, B. (9) 601
Miyagi, K. (9) 650
Miyahara, K. (5) 91, 370

Miyazaki, H. (6) 131-137, 143; (9) 110, 194, 195
Miyazaki, K. (6) 237
Miyazaki, T. (9) 593
Miyazawa, T. (8) 8
Mizise, S. (5) 535, 540
Mizobuchi, M. (9) 196
Mizsak, S.A. (9) 514
Mizugaki, M. (9) 194, 195
Mizuguchi, S. (9) 106
Mizumachi, K. (10) 99
Mizuno, S. (6) 119
Mizuta, K. (10) 107, 108
Mocquard, M.T. (9) 387
Möhringer, C. (5) 408
Moeller, L. (7) 138
Moffat, J.B. (5) 488
Mogi, M. (9) 129
Mohr, G. (10) 75
Mohtasham, J. (5) 272
Moini, M. (1) 202, 203; (2) 92
Moir, J.E. (10) 97
Moise, C. (10) 56
Mok, V. (9) 609
Mokler, C.M. (9) 257
Moldoveanu, S. (4) 128
Moll, M. (10) 73
Molloy, K.C. (10) 10, 11
Molnar, J. (4) 70
Momigny, J. (1) 195; (4) 11
Mommers, A.A. (1) 115, 116; (2) 122, 124; (4) 6
Mompon, B. (9) 670
Monch, W. (6) 335
Mondal, H. (10) 18
Mondando, G. (5) 322
Moneti, G. (6) 242; (7) 210; (9) 46
Monforte, J.R. (9) 547
Monneret, C. (9) 518
Monsarrat, B. (9) 289
Monteiro, C. (2) 64
Montreuil, J. (7) 160
Moore, J.M. (6) 63
Moore, W.T. (7) 156
Morales, A. (5) 292
Moran, S. (1) 318, 319, 321, 333, 334; (5) 96, 490, 491
Moran, T.F. (2) 125-128
Moreland, T.A. (9) 450
Morelli, J.J. (5) 457
Moreti, G. (5) 384
Morey, D. (7) 228
Morgan, E.D. (6) 157; (7) 77
Morgan, H.W. (4) 178
Morgan, J.S. (10) 93
Morgan, S.L. (6) 447

Morgan, T.J. (1) 171
Morgner, H. (1) 119-121, 225
Mori, K. (7) 16
Mori, Y. (5) 355
Morii, H. (5) 316
Morikawa, H. (9) 124
Morimoto, Y. (9) 532
Morin, M. (5) 268
Morin, N. (4) 30
Morin, P. (1) 30-32, 39, 147, 164, 167
Morino, A. (9) 240
Morioka, H. (7) 258; (9) 516, 517
Morise, K. (2) 66
Morishita, S. (9) 251
Morita, M. (7) 18
Moriwaki, Y. (9) 427
Morizur, J.-P. (2) 64
Mornex, R. (6) 299
Morokuma, K. (5) 526
Moroni, F. (9) 658
Morris, A. (1) 314
Morris, J.B. (1) 88; (10) 120
Morris, M. (10) 114
Morris, W.J. (9) 114
Morrison, G.H. (4) 202, 203
Morrison, J.A. (9) 370
Morrison, J.D. (1) 269
Morrow, J.C. (2) 44
Morselli, P.L. (9) 202
Morton, T.H. (2) 204, 206
Moschel, R.C. (8) 13
Moseley, J.T. (1) 294, 295, 306
Mosely, J.T. (5) 472
Moser, R. (6) 205, 210
Mosher, F.R. (9) 374
Moshev, B.G. (3) 5
Moss, M.S. (6) 228; (9) 385
Mossoba, M.M. (6) 311; (7) 235
Mostmans, E. (9) 255
Mount, D.L. (9) 51
Mourgues, P. (2) 205
Mousselmal, M. (1) 171
Mowrer, J. (5) 240
Moyer, J.W. (11) 141, 144, 145
Moylan, C.R. (5) 537
Mozayani, A. (9) 672
Mroszczak, E.J. (9) 337
Msonthi, J.D. (5) 326, 335
Mucklejohn, S.A. (11) 86, 108
Mueggler, P.A. (9) 571, 572

Muehlradt, P.F. (5) 312
Mueller, D. (5) 377
Mueller, D.R. (9) 519
Mueller, E. (5) 332
Mueller, P. (4) 111
Mueller, R. (4) 187
Müller-Fiedler, R. (1) 56
Mueller-Harvey, I. (5) 371
Münster, H. (2) 213; (5) 317, 319
Muething, J. (5) 312
Mueting, A.M. (10) 83
Muir, A.U. (5) 396
Muissler, S.R. (7) 229
Mukdeeprom, P. (11) 167
Mukherjee, P.K. (1) 52
Mukhopadhyay, T. (9) 505
Mulder, G.J. (9) 184
Mule, S.J. (9) 85, 557
Mullen, G.B. (9) 513
Muller, A. (9) 288
Muller, D. (7) 125
Muller, F.O. (9) 216
Muller, J. (10) 54
Muller, L. (6) 219
Muller, L.L. (5) 450, 451
Muller, M.D. (6) 183, 346, 357; (10) 16
Mullin, A.S. (1) 265
Mullins, M.G. (6) 267
Munns, R.K. (9) 114
Munslow, W.D. (5) 140; (6) 369; (7) 233
Munson, B. (4) 166; (6) 416
Munson, T.O. (6) 445
Munstermann, E. (11) 99, 155
Murad, E. (11) 15
Murai, A. (7) 258; (9) 516, 517
Murakani, T. (6) 359
Murali Krishna, C. (8) 24
Muramatsu, S. (9) 105
Muramatsu, T. (8) 8
Murayama, K. (5) 197; (6) 209
Murdock, K.C. (9) 370
Murenets, N.V. (9) 510
Muro, H. (9) 178
Murphy, C. (9) 166, 167, 483
Murphy, R.C. (5) 84, 99, 179, 180, 184, 200; (6) 212, 215, 303
Murphy, S.J. (7) 77
Murray, G.A. (11) 127
Murray, K.K. (1) 265, 322, 335, 339; (5) 487
Murray, S. (6) 230; (9) 691

Murthy, K.R. (6) 414
Murthy, U.S. (5) 158
Mushbrush, G.W. (6) 192
Mustiet, F.A.J. (6) 253
Mutlib, A.E. (9) 238, 319
Myers, C.E. (11) 127, 134, 135
Myher, J.J. (6) 34
Mylchreest, I.C. (7) 30–32, 36, 47, 51, 52, 77
Mysov, E.I. (10) 57, 58

Nadasdi, L. (5) 442
Nagai, T. (9) 546
Nagamachi, M. (9) 156
Nagamori, M. (11) 170
Nagano, Y. (1) 82, 83
Nagao, T. (5) 355; (9) 590
Nagaoka, K. (5) 390
Nagata, T. (9) 139
Nagel, D.L. (8) 21
Nagel, G.T. (6) 226
Nahoul, K. (6) 225
Naikwadi, K.P. (6) 182
Naim, J.O. (9) 705
Nair, A.G.R. (5) 361
Naito, K. (11) 54, 55, 75
Naito, S. (6) 119
Najmus-Saquib, Q. (5) 339
Nakagawa, A. (7) 15; (9) 39, 105, 154
Nakagawa, K. (9) 618
Nakagawa, Y. (9) 194, 195, 197, 438
Nakajima, K. (9) 590
Nakajima, T. (6) 254
Nakamura, A. (9) 240
Nakamura, G.R. (9) 577
Nakamura, K. (7) 15; (9) 39, 154, 593
Nakamura, S. (2) 66; (4) 115; (9) 112, 278, 279; (11) 165
Nakamura, T. (5) 258; (9) 106, 243, 317
Nakanaga, T. (1) 248, 262
Nakanishi, K. (8) 25
Nakano, K. (9) 239
Nakano, M. (9) 98, 227
Nakao, A. (9) 278
Nakaoka, M. (9) 368
Nakas, J. (6) 254
Nakashima, T. (5) 390; (6) 255; (9) 376
Nakata, H. (6) 135, 136
Nakata, R. (10) 107, 108
Nakatsuka, I. (6) 95
Nakayama, F. (6) 237
Nakayama, H. (9) 376

Namba, T. (5) 291
Nambara, T. (6) 231; (9) 97, 129, 345
Namboodiri, K. (5) 510
Namiki, Y. (9) 532
Nanavati, N.T. (9) 257
Nanba, T. (9) 15
Nanbu, S. (1) 311
Nandrea, G.J. (9) 114
Narimatsu, S. (9) 324
Naruse, H. (6) 259, 282
Nash, P. (9) 689
Nasini, G. (9) 497
Natale, N.R. (9) 624
Natali, A. (6) 242
Naumann, D. (10) 20
Navas, G.E. (9) 416
Nayak, D. (3) 119
Nayar, M.S.B. (9) 320
Naylor, R.D. (2) 12
Naylor, S. (5) 293, 384
Naze, L. (4) 80
Neckel, A. (11) 78, 124, 129
Needham, L.L. (6) 365
Neelima, (5) 160, 237
Negri, R.E. (4) 215
Neiderhoot, W.J. (5) 422
Neidhart, B. (7) 128
Neidlein, R. (9) 217, 343
Neihof, R.A. (5) 310
Nekhoroshev, S.A. (4) 120
Nekrasov, Yu.S. (10) 5, 131
Nell, G. (9) 411
Nelsen, S.F. (2) 98
Nelson, C.C. (8) 9
Nelson, D. (5) 418
Nelson, H.H. (1) 201
Nelson, P.A. (9) 453
Nelson, P.R. (2) 125–127
Nelson, R.J. (7) 141
Nelson, S.D. (6) 220; (9) 191
Nelson, W.L. (9) 366, 419, 420, 422–424, 479, 597
Nemec, J. (9) 299, 301
Nenner, I. (1) 5, 6, 16, 17, 22, 30–32, 39, 147, 149, 162–164, 167
Nesbet, R.K. (4) 223
Netting, A.G. (5) 174
Neubert, A. (11) 146
Neudert, R. (4) 126, 127
Neudorfl, P. (5) 257
Neugebauer, M. (9) 249, 304
Neumann, H.G. (8) 22
Neumark, D.M. (1) 335, 336, 340; (5) 58, 492
Neurath, G.B. (9) 568

Neusser, H.J. (1) 86, 87, 228
Newbound, T.D. (10) 55
Newby, T.J. (9) 374
Newell, D.R. (9) 241
Newman, N. (3) 26
Newman, R.A. (9) 212
Newton, J.F. (9) 447, 448
Ng, K.J. (6) 264; (9) 226, 680
Nguyen, A.C. (5) 442
Nguyen, N.T. (2) 194; (3) 88, 89; (7) 2, 6
Nguyen, T.L. (9) 153
Nibbering, N.M.M. (1) 142, 268; (2) 84, 94, 196, 216; (5) 499, 501, 512-514, 522, 544; (7) 187, 238
Nicholls, P.J. (9) 204
Nichols, P.D. (6) 156
Nickel, H. (11) 119
Nickl, J. (9) 454
Nicod, B. (2) 182
Nicol, G. (5) 497
Nicolaysen, L.C. (3) 92
Niederhut, W.N. (6) 363; (7) 223, 229
Nieder-Vahrenholz, H.G. (10) 129
Niehuis, E. (3) 171, 172
Nielsen, P. (9) 364, 507
Niemeyer, U. (9) 488
Niessen, W.M.A. (5) 32; (7) 91, 92, 150, 151
Nieuwenhuyse, H. (9) 225
Nihei, Y. (3) 186
Nihira, M. (9) 546
Niigata, K. (9) 330
Nijssen, L.M. (7) 137
Nikitin, M.I. (11) 30, 37, 43, 44, 46, 68
Nikitin, O.T. (11) 104
Nikolaev, E.N. (11) 139, 143, 156
Nikulin, V.V. (11) 32, 95, 98, 116
Nikulina, L.D. (10) 87
Nimlos, M.R. (1) 320, 325; (5) 45
Nimpf, J. (6) 210
Nip, M. (4) 176; (6) 430, 435
Nisbet, L.J. (9) 493
Nishi, M. (5) 91
Nishi, N. (1) 94
Nishi, S. (11) 131
Nishi, T. (11) 125
Nishibe, S. (5) 213
Nishihara, M. (5) 316
Nishimura, S. (8) 8
Nishimura, Y. (1) 297-305

Nishimuta, T. (7) 16
Nishina, Y. (2) 18
Nishio, H. (6) 296
Nishioka, I. (5) 372
Nishiuama, Y. (9) 160
Nishiyama, I. (5) 352, 356
Nishizawa, Y. (7) 18; (9) 317
Nitz, S. (6) 83
Niu, B. (1) 330
Niwa, T. (4) 89; (6) 48; (9) 618
Niwa, Y. (4) 154, 160
Nixon, J.R. (9) 468
Njoroge, F.G. (6) 148
Nobes, R.H. (1) 179, 180; (2) 37, 176, 177
Noble, M. (9) 91, 681, 682
Noda, K. (9) 329
Noda, N. (5) 370
Nöel, M. (3) 100
Noguchi, H. (5) 245; (9) 148, 243, 329, 381, 618, 621, 622
Noguecra Ramos, P. (6) 280
Noiton, D. (6) 267
Noji, M. (10) 106
Nolen, H.W., III (9) 258
Nonaka, G.I. (5) 372
Norbury, C.G. (9) 103
Nordenson, S. (6) 342
Nordgren, I. (6) 350
Nordin, C. (9) 637, 638, 678
Noren, B. (9) 115, 460
Noren, K. (6) 381
Norikura, R. (9) 197, 438
Norin, H. (6) 350
Normand, D. (5) 502
Norris, J.R. (4) 178, 180
Norstrom, R.J. (4) 99; (6) 378
Norwisz, J. (11) 164
Noto, T. (6) 254
Noturianni, L.J. (6) 380
Nouts, N. (2) 51
Novelli, L. (4) 95
Novic, M. (4) 132
Novoselova, A.V. (11) 93, 107
Nowicki, H.G. (6) 327
Nowlin, J.G. (9) 192
Nowotny, P. (6) 244
Noy, T. (6) 97
Numaziri, Y. (7) 14
Nunn, A.D. (5) 425
Nunn, N.J. (5) 140; (6) 369
Nye, J. (10) 101

Nygren, M. (5) 138
Nyquist, R.A. (5) 37

Oakes, J.M. (1) 316; (5) 473
Oakes, M. (9) 527
Obase, H. (1) 297, 298, 300-302, 304
Oberrauch, E. (4) 95
Obreterar, T. (5) 387
O'Brien, I. (9) 706
O'Brien, N.W. (11) 86, 108
O'Brien, R.J. (3) 79
O'Brien, S.C. (1) 207, 275; (5) 53
Occolowitz, J.L. (5) 357; (6) 268; (9) 494
Ochsner, M. (1) 264, 283, 287
O'Connor, P. (9) 241
O'Connor, T. (4) 13
Odagiri, H. (9) 123
Odham, G. (5) 163; (6) 194, 304
Odom, R.W. (3) 24
O'Donnell, J.P. (9) 323
Oechsner, H. (3) 55, 56
Oeerlage, M.J.M. (3) 4
Oehme, M. (5) 127; (6) 107
Oesch, F. (9) 190
Oexmann, M.J. (9) 612
Offen, C.P. (9) 245
Offord, R.E. (5) 396
Ofunne, G.C. (10) 70
Ogenstad, S. (9) 677
Ogilvie, K. (5) 444
Ogura, K. (5) 437
Oguri, K. (9) 352
Oh, Y.K. (9) 493
O'Hair, R.A.J. (5) 278
O'Halloran, M.A. (1) 66
Ohashi, M. (9) 112, 239, 278, 279
Ohazaki, O. (9) 474
Ohga, M. (5) 316
Ohlendorf, R. (7) 128
Ohlsson, A. (9) 96
Ohnishi, S. (2) 19
Ohori, D. (9) 22
Ohsawa, A. (6) 75
Ohshima, T. (4) 109
Ohta, K. (5) 526
Ohta, S. (9) 402
Ohtawa, M. (9) 124
Ohtsuki, T. (9) 375
Oishi, T. (11) 125, 131
Oka, T. (1) 242, 245, 247, 250, 252, 253, 255, 263

Okabe, M. (5) 370
Okada, S. (2) 60
Okada, Y. (10) 111
Okamoto, I. (9) 593
O'Keefe, P.W. (6) 331
Okimura, Y. (9) 532
Okita, R.T. (9) 56
Okoroafor, M.O. (10) 63
Oksukaa, T. (9) 546
Okumura, K. (9) 381
Okumura, M. (5) 502
Okuyama, F. (5) 470
Okuyama, Y. (9) 273
Olanoff, L.S. (9) 612, 614, 615
Oldham, R.G. (4) 107
Oldham, S.W. (9) 411
Olivares, J.A. (3) 88, 89; (7) 2, 6
Olive, G. (9) 72
Ollet, D.G. (6) 157
Olsen, C.-E. (9) 364
Olsen, I. (6) 52, 53
Olsen, L.D. (9) 479
Olson, E.S. (3) 212; (6) 78
Olsson, B. (1) 173, 309
Olthoff, J.K. (3) 174, 175; (5) 311
O'Malley, R.M. (1) 96; (3) 47
Omura, S. (9) 503
O'Neill, R.A. (6) 248
Onishi, J.C. (9) 504
Onkenhout, W. (5) 109; (9) 6
Ono, A. (8) 11
Ono, K. (11) 125, 131
Ono, M. (5) 370
Ono, Y. (10) 9
Onuska, F. (5) 141; (6) 31, 180, 371
Onyszchik, M. (10) 18
Operti, L. (10) 93, 100
Opsal, R.B. (3) 85; (6) 71, 72
Oralov, T.A. (11) 142
Orama, O. (10) 75, 77
O'Reilly, R.A. (9) 598–600
Orlando, R. (6) 416
Orlando, T.M. (1) 76
Orlova, E.M. (10) 124
Orm, M.L'E. (9) 296
Orme, M. (9) 536
Orning, L. (5) 321
Orr, J.M. (9) 179, 653
Ortiz, J.V. (5) 46
Ortiz de Montellano, P.R. (9) 406
Orwig, B.A. (9) 443
Osawa, T. (3) 69

Osman, R. (5) 510
Osterloh, G. (9) 697
Osterman-Golkar, S. (6) 395
Otani, K. (9) 637, 659
Otson, R. (6) 115, 349
Otten, E.J. (9) 542
Otto, A. (7) 12
Ottoila, P. (9) 117
Otuos, L. (9) 645
Otvos, J.W. (11) 19
Ouwerkerk, C.E.D. (3) 191; (4) 210
Ovchinnikov, R.V. (11) 139
Overton, B.W. (9) 210
Owari, M. (3) 186
Owens, G.D. (7) 60, 70
Owens, K.G. (1) 92
Oxborrow, G.S. (6) 99
Oxemann, M.J. (9) 613
Oxford, J. (7) 90, 115; (9) 133, 437
Oyler, A.R. (9) 22
Ozturk, F. (2) 92

Paal, A. (4) 70
Paalzow, L.K. (9) 698
Paar, W.D. (9) 151, 661
Pace-Asciak, C.R. (5) 183
Pacey, N. (4) 185
Pachuta, S.J. (10) 105
Pacula, C.M. (9) 357
Padias, A.B. (5) 297
Padieu, P. (6) 238
Padovani, P. (9) 202
Paganini, M. (9) 658
Page, B.D. (6) 309
Page, J.A. (7) 227
Page, R.H. (1) 36
Pagura, C. (3) 53
Paik, Y.H. (5) 453
Paillasseur, J.-L. (6) 426
Pal, B.C. (5) 325, 329
Palenik, G.J. (10) 66
Pallix, J.B. (3) 25, 26
Pallow, R.J., jun. (9) 479
Palmer, P.T. (4) 135
Pan, C.H. (9) 493
Panatella, J.E. (5) 417
Pancirov, R.G. (2) 73
Pandey, R.C. (9) 521
Pang, H.M. (7) 61–63
Pankow, J.F. (6) 85, 116, 117
Pannell, L.K. (5) 460, 465; (8) 26; (9) 629; (10) 122
Pant, P. (5) 225

Pantzar, P. (9) 166, 483
Paoletti, C. (9) 289
Paolucci, G. (10) 40
Papadopoulos, A.S. (9) 465
Papadopoulos, V. (6) 241
Pappas, A.A. (9) 120
Pappie, D. (9) 300
Pare, E.M. (9) 547
Pare, J.R.J. (5) 290; (9) 32, 627
Parent, D.C. (3) 35
Parenteau, L. (1) 220, 221, 222
Pares, P. (6) 356
Park, B.K. (9) 211, 224, 296, 537
Parker, C.D. (5) 454
Parker, C.E. (5) 228; (6) 227; (7) 110, 111, 113; (9) 21
Parker, E.H.C. (4) 44, 216
Parker, I. (6) 314
Parker, J.E. (2) 67
Parks, C.D. (6) 447
Parmentier, G. (6) 158
Parr, A.C. (1) 4, 12, 28, 70
Parr, I.B. (9) 457
Parr, V.C. (7) 106
Parrilli, M. (5) 351
Parry, D. (9) 436
Pashinkin, A.S. (11) 160
Pasper, J. (5) 112
Pasutto, F.M. (9) 297, 341, 692
Pasztor, A.J. (5) 37
Patarasakulchai, N. (5) 418
Patricot, M.C. (6) 221
Patrono, C. (6) 206
Patsalides, E. (10) 91
Patterson, D.G., jun. (3) 92; (6) 365, 366
Patterson, J.R. (5) 126
Pattoret, A. (11) 12
Paul, B.D. (9) 136, 171, 564
Paulson, B.P. (2) 128
Paulson, G.D. (9) 267, 452, 453
Paulson, J. (7) 100, 136; (9) 19, 168, 229, 230, 232
Paulson, J.F. (5) 520, 527
Paulson, S.E. (1) 334; (5) 491
Paulus, H. (3) 56
Paumgartner, G. (6) 223, 234, 235

Pavlath, A.E. (4) 200
Pawlak, J. (8) 25
Payne, P.A. (4) 96
Pazzagli, M. (6) 242
Peake, D.A. (6) 170; (10) 136, 138
Peal, J.A. (6) 394; (9) 172
Pearson, P.G. (9) 360
Peat, M.A. (9) 543
Pechine, J.-M. (5) 150; (6) 173
Pedley, J.B. (2) 12
Peeters, D. (2) 31
Peffer, R.C. (9) 274
Pein, F.G. (9) 568
Pelissolo, F.J.C. (4) 125
Pelizzi, G. (9) 373
Pell, R.J. (4) 26
Pella, E. (9) 509
Pellerite, M.J. (5) 144
Pelletier, O. (6) 222, 250
Pelli, B. (5) 212, 338; (9) 580; (10) 88
Pellin, M.J. (3) 28
Peloso, J.S. (5) 357; (9) 494
Pelzer, G. (10) 128
Pentoney, S.L., jun. (3) 154; (7) 49; (9) 28
Pepe, C. (6) 154
Perchalski, R.J. (7) 198, 203; (9) 42, 43
Perchonock, C.D. (9) 447, 448
Pereira, W.E. (6) 321
Peres, T. (1) 238; (2) 225
Peretti, E. (9) 130, 332
Perfetti, G.A. (6) 311; (7) 235
Perkins, D.D. (7) 143
Perkins, J.R. (7) 30-32, 36, 51, 52; (9) 25, 26
Perkinson, N.A. (9) 370
Perone, S.P. (4) 142, 143
Perri, E. (8) 31
Perseo, G. (5) 397; (7) 242
Pershe, R.A. (9) 597
Perugini, D. (10) 27, 72
Peruzzini, M. (10) 100
Pervov, V.S. (11) 92
Pesce, A.J. (9) 541
Pesch, R. (7) 159
Peterman, P.H. (4) 99
Peters, J.H. (9) 258
Peters, W.P. (9) 484
Petersen, U. (9) 252
Peterson, D.M. (5) 303
Peterson, J.R. (1) 277

Peterson, L.A. (9) 380, 566
Pethrick, R.A. (5) 433
Petit, B.R. (6) 91
Petit, J. (9) 675
Petitjean, M. (4) 4
Petiton, M. (5) 378
Petre, J. (6) 269
Petrosyan, V.S. (5) 74
Petrucciani, T. (5) 177; (6) 174
Pettersen, R.C. (6) 247
Pettersson, L.G.M. (1) 272
Pettett, B.C., jun. (9) 570
Pettiette, C.L. (1) 332; (5) 54, 55
Petzinger, G. (9) 703
Peyerimhoff, S.D. (1) 170, 171
Pfaendler, P. (2) 83
Pfaff, E. (9) 276
Pfeifer, S. (9) 512
Pfleger, K. (9) 610, 646, 700
Pfleiderer, W. (5) 232; (7) 255
Philip, R.P. (6) 65, 428
Phillipou, (6) 219
Phillips, L.R. (5) 306
Phillips, R.J. (6) 129
Phillipson, D.W. (8) 5, 7
Piancastelli, M.N. (1) 18, 20, 37, 38
Picart, D. (6) 243
Piccinelli, A. (6) 204
Pichiarella, L. (1) 123
Pick, R.O. (9) 553
Piechotka, M. (11) 89
Pietra, F. (4) 15
Pignolet, L.H. (10) 83
Pilate, F. (5) 322
Pilet, P.E. (5) 167
Pilkinton, A.E. (9) 396
Pillion, D.J. (9) 592
Pinchin, W.H. (6) 73
Pinedo, H.M. (9) 300
Pinkse, F.A. (7) 238
Pinkston, D. (4) 91
Pinkston, J.D. (7) 20, 60, 70
Pinnaduwage, L.A. (1) 126
Piotrovsky, L.B. (9) 295
Piotrowski, E.G. (6) 413; (9) 591
Pipkin, F.B. (9) 620
Pirat, J.-L. (9) 403
Piron, J.J. (9) 675
Pitoizet, N. (6) 238
Pitts, J.N. (5) 255; (6) 336

Pitzer, E.W. (4) 114
Piven, V.A. (10) 26
Pivnikskii, K.K. (9) 683
Pizza, C. (5) 333
Pizzo, P. (9) 335
Placidi, M. (9) 473
Planitz, V. (9) 143
Platt, K.L. (9) 190
Plattner, R.D. (5) 239; (7) 218, 219
Plavsic, D. (10) 4
Pleasance, S. (7) 30-32, 36, 51, 52, 158
Pleshakov, M.G. (5) 76
Pleshkova, A.P. (5) 76
Plessis, P. (1) 105-108; (5) 49, 50
Pleva, M.A. (6) 5
Ploschke, H.J. (9) 252
Plummer, E.W. (1) 159
Plummer, J.A. (6) 267
Poblet, J.M. (2) 153
Poc, M. (5) 404
Pochan, S. (5) 396
Podda, G. (5) 178
Podell, E.R. (6) 277
Podojil, M. (6) 124
Podowik, B.-I. (9) 553
Poehland, B.L. (9) 492
Pogolotti, A.L. (8) 11
Pohl, J. (9) 328
Pohlmann, H. (9) 512
Polak, M. (1) 243, 246, 256-260
Polakova, L. (9) 218
Polen, V.R. (5) 100
Poliakoff, E.D. (1) 15, 27-29
Politowski, J.F. (9) 389
Pollak, S. (9) 549
Pollinger, P. (3) 52
Poll-The, B. (5) 86
Poloznikov, A.V. (11) 142
Polyakova, A.A. (5) 95
Pommier, F. (9) 100
Poneter, L. (5) 393
Pons, G. (9) 72
Pool, W.F. (9) 379
Pople, J.A. (5) 7, 27, 28
Popli, S.P. (5) 206
Popov, A.I. (11) 116
Popovic, A. (11) 90
Poppe, A. (5) 262
Poppe, H. (7) 150, 151
Poreti, M. (3) 155
Porter, C. (4) 32; (7) 170
Porter, C.J. (2) 123; (7) 209, 217; (9) 401
Porter, J.K. (7) 218
Porter, Q.N. (5) 149
Porter, R.F. (2) 102-104,

# Author Index

107, 112–114
Porubek, D.J. (6) 220
Posey, L.A. (1) 328, 329
Postma, E.J. (9) 471
Postma, R. (2) 39, 41, 143, 165–168
Potapov, V.K. (3) 5
Potemkin, L.V. (11) 142
Potter, D.W. (7) 119, 207; (9) 392, 400
Pottie, R.F. (11) 27
Potts, B.D. (9) 446
Poulos, A. (6) 177
Poulter, L. (7) 243
Pounder, D.J. (9) 639
Poupaert, J.H. (9) 351, 664
Pourezaei, K. (3) 119
Poussel, E. (3) 72
Pouwels, A.D. (6) 438
Powell, R.G. (5) 239
Powis, G. (9) 486
Powis, I. (2) 14
Pradelles, P. (5) 184; (6) 212
Pramanik, B.N. (5) 249; (9) 445, 680
Prasad, D. (4) 69
Prasad, G. (5) 79
Pratap, R. (5) 79
Pratt, J.A.E. (7) 130; (9) 20, 706
Pratt, S.T. (1) 61, 63–67, 72, 75
Preac-Mursic, V. (5) 188
Preiss, A. (5) 77, 92
Prescott, M.C. (6) 38, 122
Presser, R. (3) 209
Preuss, M. (4) 42
Price, A.H. (9) 399
Price, K.R. (5) 330, 343, 344
Price, S.D. (1) 148, 149, 172
Price Evans, D.A. (9) 399
Prickett, K.S. (9) 476, 477
Pringle, T.H. (9) 616
Prini, A. (4) 79
Prinslow, D.A. (10) 33
Pritchard, D.G. (6) 447
Pritzkow, H. (10) 64
Prokai, L. (6) 442
Proliac, A. (5) 374
Prome, D. (5) 304, 373
Prome, J.-C. (5) 35, 67, 304, 373
Proot, M. (6) 128
Prost, M. (5) 35; (9) 1
Prouse, B. (4) 9
Provost, L.R. (6) 405,
406
Prox, A. (9) 205–207, 454
Przybylski, M. (9) 190, 328, 502
Pullen, B.P. (1) 9, 19
Purba, H.S. (9) 296
Purcell, T.G. (10) 10, 11
Purkayastha, R.N.D. (10) 85
Pusset, J. (5) 349
Putteman, M.L. (5) 185
Puzo, G. (7) 247
Pyatigorskaya, T.L. (9) 491
Pyle, S.M. (7) 123
Pyzik, T. (9) 249

Qian, Y. (5) 280; (10) 44
Qin, X.-Z. (2) 163
Quaglia, M.G. (9) 687
Quattlebaum, J.C. (3) 39
Quian, J.H, (5) 486
Quiding, H. (9) 575
Quilliam, M.A. (6) 352
Quincy, D.A. (9) 624
Quinn, F.M. (1) 70
Quinones, L. (3) 64

Raabe, E. (10) 60
Rabinowitz, J.R. (5) 510
Rader, B.R. (9) 550
Radermacher, L. (4) 48
Radics, L. (9) 501
Radloff, C. (2) 194
Radom, L. (1) 179, 180; (2) 7, 34–38, 164, 175–180
Radom, T.H. (5) 100
Raftery, M.J. (5) 104, 146, 156, 170, 235
Raglione, T.V. (6) 105, 106
Rahbee, A. (4) 64
Rain, U. (5) 323
Raisi, A. (9) 404
Raj, A. (11) 132
Raj, D.D.A. (11) 173
Rajakyla, E. (7) 131
Rajan, S. (9) 496
Raksit, A.B. (2) 102–104, 113
Raksit, W. (2) 112
Raleigh, J.A. (8) 19
Ramakrishnan, K. (9) 268
Ramdahl, T. (5) 131; (6) 191
Rammensee, W. (11) 62
Ramsey, E.D. (7) 158
Ramstein, R. (7) 64
Ranalder, U.B. (9) 108,
538
Randall, G.L. (9) 574
Rankin, D.W.H. (10) 25
Rao, B.K. (2) 17
Rao, T.S. (9) 470
Rapin, J. (2) 83; (3) 155
Rapp, D. (1) 110
Rapp, U. (7) 140, 184
Rappe, C. (5) 138
Rapson, C.A. (4) 128
Raschdorf, F. (7) 125; (9) 519
Raseev, G. (1) 8, 14, 21
Rasmussen, K.E. (9) 588
Ratanasavanh, D. (9) 387
Ratcliffe, A.H. (5) 324
Ratnayake, W.M.N. (4) 109
Raushel, F.M. (7) 254
Ray, D. (1) 265
Raynaud, J. (5) 374
Raznikov, V.V. (4) 172, 173
Raznikova, M.O. (4) 172
Read, A.E. (5) 524
Read, R.W. (5) 202; (6) 403, 404
Rebbert, R.E. (6) 323
Reddish, T. (1) 312
Redrup, M.J. (9) 706
Reents, W.D. (2) 93
Rees, H.H. (6) 38, 121
Reeves, V.B. (6) 214
Regardh, C.G. (9) 617
Reggiani, G. (6) 59
Rehfuss, B.D. (1) 247, 250, 255
Reichmann, K.C. (1) 200
Reid, C.J. (1) 189
Reif, G. (9) 101
Reilly, J.P. (1) 68, 69, 92; (3) 42, 85; (6) 71, 72
Reilly, M.H. (7) 209; (9) 401
Reinauer, H. (6) 302
Reiner, E.J. (2) 123; (6) 364
Reinhard, M. (6) 353
Reinhold, V.N. (5) 11, 378; (7) 142
Reinisch, P.F. (10) 25
Reissinger, M. (6) 59
Rejwan, M. (2) 225
Remmel, R.P. (9) 236
Rempel, D.L. (3) 144; (7) 169
Remuzzi, G. (6) 204
Renner, M. (4) 9
Rennie, M.J. (4) 50, 51
Repke, D.B. (9) 363
Resch, K. (5) 181
Resul, B. (9) 637

Retel, J. (9) 300
Rettenmeier, A.W. (9) 64, 65, 475–478
Rettie, A.E. (9) 475, 600
Reuning, R.H. (9) 628
Revol, A. (6) 221
Rey, E. (9) 72
Reye, C. (5) 429
Reymard, P. (5) 167
Reynolds, W.E. (3) 131
Rezanka, T. (6) 124
Rheingold, A.L. (10) 36, 82
Rhodes, G. (5) 286
Ribick, M. (9) 146, 607
Ribon, B. (9) 68, 155
Ricca, T.L. (3) 161, 162
Ricci, M. (9) 501
Riccio, R. (5) 349, 354
Rice, D.W. (6) 345
Rice, L.S. (9) 449
Richards, A.M. (2) 43
Richards, J.M. (4) 73, 195, 196
Richardson, C.H. (3) 173
Richli, U. (5) 371
Richter, E. (6) 400
Richter, W.J. (6) 423; (7) 125; (9) 519
Ridder, W.E. (9) 111
Ridge, D.N. (9) 370
Rieders, F. (9) 673
Riesz, P. (8) 24
Riis, E. (3) 35
Rijmerse, E. (6) 374
Rijpstra, W.I.C. (6) 420
Rimke, H. (3) 38
Ringo, N.T. (6) 318
Rinsma, W.J. (4) 179
Riou, J.P. (6) 299
Ripley, S. (5) 242; (6) 392
Riscado, A.M.V. (6) 251
Rise, F. (10) 12
Rittmann, N. (9) 214
Rivera, J. (5) 296; (6) 344
Riviere, M. (7) 247
Rizzo, A. (5) 205; (6) 307
Roach, J.A.G. (7) 224
Robbins, J.H. (8) 18
Roberts, C.R. (5) 503
Roberts, G.D. (5) 383; (9) 493
Roberts, L.J. (5) 196
Roberts, S.M. (9) 414
Robertson, G.W. (4) 185
Robertson, L.W. (9) 628
Robien, W. (5) 323
Robinett, R.S.R. (6) 145
Robins, R.H. (7) 96

Robinson, D. (3) 209
Robinson, J.N. (2) 75
Robinson, P.R. (6) 229
Robson, J.N. (6) 417
Rocca, F. (9) 410
Rocchiccioli, F. (6) 146
Rodchenkov, G.M. (9) 271, 311, 361
Rodighiero, P. (5) 212
Rodler, M. (2) 161
Rodriguez, L.C. (9) 158, 527
Roe, C.R. (9) 655
Roehl, J.E. (4) 77
Röllgen, F.W. (5) 470; (7) 104, 105; (8) 34
Roelofs, W.L. (6) 301
Roepstorff, P. (9) 507
Roesner, P. (4) 111
Rösslein, M. (1) 285, 290–293
Rogan, E.G. (8) 21
Rogez, J. (11) 60
Rohle, R. (6) 218
Rokushika, S. (7) 50
Rolando, C. (4) 30; (5) 268
Rollema, H. (9) 327
Roller, P.P. (9) 521
Rolli, E. (2) 63, 65, 71, 74
Rollins, K. (6) 227, 401; (7) 113
Roman, J. (9) 299, 301
Romanoska, E. (6) 246
Romanova, O.V. (11) 156
Romer, A. (9) 288
Rondahl, L. (6) 350
Ronfeld, R.A. (9) 163
Rose, J.E. (9) 354
Rose, K. (5) 396; (6) 270
Rose, M.E. (5) 31; (7) 1, 94
Rosen, J.D. (5) 18; (7) 225; (9) 696
Rosen, M.E. (6) 117
Rosenberg, C. (6) 402
Rosenfeld, J.M. (5) 208
Rosi, M. (1) 52
Rosing, H. (9) 531
Rosman, K.J.R. (3) 188
Ross, D. (9) 189
Ross, M.M. (1) 199; (5) 294, 310; (9) 40
Ross, R.M. (5) 189; (6) 142, 216
Ross, S. (9) 678
Rossi, C. (9) 501
Rossini, A. (6) 123, 450
Rostad, C.E. (6) 321
Rotard, W. (6) 324
Roth, H.D. (2) 93

Roth, H.J. (9) 214
Rovira, X. (6) 443
Rowin, G.L. (9) 504
Rowland, M. (9) 418
Rowland, S.J. (6) 417
Roy, D. (1) 39
Roy, P. (1) 39, 167
Roybal, J.E. (9) 114
Royds, R.N. (1) 168
Royer, R. (9) 436
Rozanova, O.N. (11) 56
Rozeboom, M.D. (2) 95; (5) 288
Ruatta, S.A. (1) 235
Rubin, I.B. (5) 121
Rubio, F. (5) 230
Rubiro, F.M. (5) 397
Rucker, G. (9) 249, 304
Rucklidge, J.C. (3) 181
Ruderauer, F.G. (5) 20
Rudewicz, P. (7) 121, 133, 195, 206; (9) 35, 36
Rudnyi, E.B. (11) 39, 41, 46, 68, 149
Rudometkin, S.V. (11) 31
Rüdenauer, F.G. (3) 51, 52
Rueff, J. (6) 280
Rühl, E. (1) 225, 226, 227
Ruenitz, P.C. (9) 257
Ruhl, E. (10) 126
Rumack, D.T. (2) 98
Rumpf, B.A. (4) 33
Rumpf, P.J.A. (2) 47
Runesson, B. (9) 55
Ruotsalainen, H. (5) 415
Rushing, L.G. (5) 128, 252, 254, 255; (6) 189, 190, 336, 337, 387; (9) 699
Russell, A.S. (9) 297
Russell, D.H. (1) 2, 273; (2) 8; (3) 158; (5) 25, 560; (6) 9; (7) 95, 173, 189, 254; (8) 4; (9) 7
Russell, W.J. (9) 86
Ruster, W. (3) 38
Rusterholz, B. (10) 116
Ruttink, P.J.A. (2) 30, 39, 41, 46, 138, 165–168, 217
Ruzo, L.O. (9) 189
Ryan, J.J. (7) 216
Ryan, P.A. (4) 16
Rykov, A.N. (11) 93, 107
Rynkowski, J.M. (5) 433
Ryrfeldt, A. (9) 230, 231
Ryska, E. (5) 224
Ryska, M. (5) 108; (9)

88, 218, 222, 393
Ryzhkov, A.A. (11) 96
Ryzhov, M.Yu. (11) 87, 88

Sachs, H. (9) 581
Sackmann, M. (6) 235
Sadana, K.L. (8) 14
Safe, S. (5) 139; (6) 59, 178, 179
Saferstein, R. (9) 8
Sagi, M. (9) 194, 195
Saha, B. (11) 81, 94, 103, 132, 173
Saha, M. (9) 57
Sahlberg, B.-L. (9) 383, 384
Sahoo, S. (5) 419
Sahu, N.P. (5) 332
Saibaba, M. (11) 173
Saiki, Y. (5) 355
St. Georgiev, V. (9) 513
Saito, M. (7) 44
Saito, S. (9) 124
Saito, T. (5) 197; (6) 209
Saiz-Jiminez, C. (6) 433, 436
Sakairi, M. (3) 83; (7) 164; (8) 36; (9) 23
Sakamoto, T. (10) 111
Sakashita, C. (5) 117; (9) 648
Sakata, C. (5) 258; (9) 243
Sakkers, P.J.D. (7) 131
Sakurai, T. (3) 178, 191
Sakushima, A. (5) 213
Salama, F. (1) 170
Salama, Z. (9) 101, 125
Salara, Z. (5) 132
Salari, H. (6) 140
Saleh, N.A.M. (5) 337, 363, 367
Salehpour, M. (3) 62
Salem, N. (5) 186
Salerno, R. (6) 242
Salhab, A.S. (9) 56
Salimgareeva, I.M. (5) 102
Saliot, A. (6) 154, 340, 426
Salisbury, C.D. (5) 126
Sallans, L. (5) 555
Sallustio, B.C. (9) 307
Salomon, K.E. (5) 552
Salonen, J.S. (9) 269
Salvatori, T. (4) 95
Sams, R. (9) 335
Samson, J.A.R. (1) 59
Samukawa, K. (4) 14
Sana, M. (2) 31

Sanche, L. (1) 220-222
Sanders, E.F. (5) 417
Sanders, J.K.M. (5) 384
Sanders, S.W. (5) 112
Sanders-Loehr, J. (5) 402
Sanderson, N.E. (3) 74
Sandra, P. (6) 128
Sandrock, K. (9) 220
Sandstrom, A. (5) 405
Sandstrom, B. (9) 617
Sangameswaran, L. (9) 649
Santi, D.V. (8) 11
Santikarn, S. (7) 142
Santonin, D.K. (7) 201; (9) 371
Sarasa, J.P. (2) 153
Sarnquist, F.H. (9) 337
Sarrasin, B. (6) 149-151
Sarver, E.W. (4) 197; (5) 201, 203, 204; (6) 305, 306; (7) 212, 221
Sasaki, H. (9) 318, 330
Sasaki, Y. (9) 128
Sasame, H.A. (9) 421
Sasso, G.J. (9) 203, 471
Sati, O.P. (5) 91
Sato, K. (1) 74, 84, 94; (3) 107; (7) 167
Sato, M. (5) 376; (9) 462
Sato, N. (1) 311; (9) 116
Satoh, H. (3) 186
Satsangi, N. (5) 303
Satzger, R.D. (3) 71
Saudubray, J.-M. (5) 86
Saugy, M. (5) 167
Sauter, A.D. (6) 318
Saux, M.C. (9) 157
Savage, C.M. (5) 3
Savagnac, A. (5) 67, 304, 373
Savel'ev, Yu.I. (10) 26
Savinova, L.N. (11) 36
Savolainen, H. (6) 402
Savory, J. (5) 189; (6) 142, 216
Savoy, L.A. (5) 396
Sawai, Y. (9) 160
Sawhney, S.N. (9) 685
Sawyer, D.T. (5) 161
Sawyer, J.S. (10) 14
Saxon, R.P. (5) 7
Saykally, R.J. (1) 239, 243, 246, 256-260
Scasnar, V. (9) 451
Schade, W. (5) 85; (9) 511
Schaefer, H.F. (5) 502
Schafer, H. (10) 129
Schaffer, R. (6) 264
Schaftenaar, G. (2) 138
Schall, W. (6) 295
Schanck, A. (9) 656

Schantz, M.M. (6) 323
Scharff, C. (4) 4
Schaufelberger, D. (5) 364
Schechter, P.J. (9) 708
Scheier, P. (1) 99, 208-213, 216, 217
Scheifers, S.M. (10) 134
Scheinin, M. (9) 637
Schellenberg, K.H. (6) 364; (7) 114; (9) 82
Scheller, D. (10) 90
Schenk, P.A. (4) 176; (6) 418, 420, 421, 430, 435
Scherkenbeck, J. (5) 377
Scheuler, B. (5) 434
Scheuring, T. (11) 168, 169
Schey, K. (3) 179
Schiebel, H.M. (5) 469; (9) 37; (10) 110
Schilling, A.B. (5) 185
Schilling, M.L. (2) 93
Schillings, R.T. (9) 480
Schirmer, J. (1) 46, 48, 196, 197
Schlag, E.W. (1) 228; (3) 8, 12-19; (10) 115
Schleyer, P.von R. (2) 7, 33; (5) 488; (10) 123
Schmeits, G.J.H. (9) 38, 225
Schmelzeisen-Redeker, G. (7) 104, 105; (8) 34
Schmid, J. (9) 454
Schmid, J.C. (6) 419
Schmid, K. (9) 632
Schmid, P. (6) 183
Schmidbauer, N. (6) 107
Schmidt, C. (11) 79
Schmidt, E.R. (5) 423
Schmidt, J. (2) 23, 152; (5) 77, 85, 90, 93; (9) 511
Schmidt, M. (6) 400
Schmidt, S.R. (11) 133
Schmidt, V. (1) 38
Schmidt-Bothelt, E. (9) 697
Schmit, J.-P. (2) 62; (3) 118
Schmitz, B. (6) 165
Schmitz, D.A. (9) 340, 355, 356
Schmitz, F.P. (7) 39, 71
Schneider, E. (7) 215
Schneider, G. (5) 77
Schneider, W.P. (5) 182; (6) 208
Schnockel, H. (11) 150, 151
Schoeller, D.A. (4) 100;

(9) 669
Schoenmakers, P.J. (7) 24, 25, 41
Scholler, R. (6) 225
Scholze, C. (4) 41
Schonfelder, A. (6) 298
Schooley, D.L. (6) 144
Schram, K.H. (5) 406; (8) 3; (10) 8
Schreifels, J.A. (4) 13; (10) 102
Schrijver, J. (7) 140
Schroeder, E. (7) 159
Schroeder, H.F. (7) 213
Schroeder, T.J. (9) 541, 542
Schruber, K. (5) 77
Schubert, R. (6) 197; (7) 205
Schühle, U. (3) 25
Schueler, B. (3) 24; (4) 56; (9) 487
Schullek, K.M. (9) 58
Schulten, H.-R. (4) 186–191; (5) 469; (6) 427, 432, 448; (8) 22; (10) 110
Schulthess, P. (10) 113, 116
Schulz, C. (3) 37
Schulz, C.P. (1) 265
Schulz, J. (4) 102
Schulze, C. (10) 136, 139, 143
Schulze, P. (9) 37
Schumacher, G.E. (9) 69, 665, 666
Schumann, H. (10) 67
Schunk, S. (11) 150
Schuppel, R. (9) 37
Schutz, H. (9) 390
Schuyl, P.J.W. (6) 351
Schwanebeck, W. (3) 213
Schwarting, G.A. (5) 315
Schwartz, D.E. (9) 538
Schwartz, H.E. (7) 40
Schwartz, J.C. (3) 102; (6) 16; (7) 165, 172
Schwartzman, M.L. (5) 200
Schwarz, H. (1) 114, 181; (2) 6, 10, 84, 137, 139, 141–143, 146, 152, 162, 172, 191, 209, 214, 216; (7) 185, 241; (9) 140, 308; (10) 69, 136, 139–144
Schwarz, W. (5) 100
Schweer, H. (5) 198, 199; (6) 197–201; (7) 202, 205, 251
Schweikert, E.A. (3) 64
Schweitzer, E.L. (3) 28

Schweitzer, G.K. (3) 36
Scimeca, J.A. (9) 587
Scoble, H.A. (3) 107; (4) 156; (7) 183
Scott, D.R. (4) 140, 141; (6) 341
Scott, R. (6) 114
Scott, S.L. (5) 498
Scribe, P. (6) 154
Scrimgeour, C.M. (4) 50, 51; (6) 274
Scrivens, J.H. (4) 163
Scriver, C.R. (6) 293
Scudder, P. (6) 245
Seago, A. (9) 429, 430
Sears, L.J. (5) 117; (6) 326; (9) 648
Sears, T.J. (1) 241
Sedgewick, J.B. (2) 125, 127, 128
Seetharaman, T.R. (5) 361
Segall, H.J. (5) 300; (6) 266
Segar, K.R. (1) 90
Sehram, K.H. (5) 403
Seibert, G. (5) 377
Seibert, K. (5) 196
Seid, R.C. (5) 306
Seidel, A. (9) 190
Seifert, K. (3) 57
Seiler, N. (6) 42
Seino, A. (9) 498
Sek, D. (6) 439
Sekiya, H. (1) 299, 301, 302
Sekreta, E. (1) 68, 69, 92
Selander, H. (9) 677, 678
Selbin, J. (10) 132
Self, R. (5) 359, 369, 371; (9) 507
Selgren, S.F. (2) 108–110, 115
Selim, M.I. (7) 66
Selinger, A. (2) 220
Sellier, N. (9) 468, 518
Sello, L.H. (9) 499
Selva, A. (9) 704
Sembdner, S. (5) 77, 94
Semenikhin, V.I. (11) 46, 68–70
Semenov, G.A. (11) 11, 71, 72, 73, 76, 139, 140, 143, 147, 148, 154, 156, 157
Semenov, V.A. (9) 271, 311, 361
Semura, E. (9) 123
Semyannikov, P.P. (10) 130
Sepsy, K. (4) 70
Serio, M. (6) 242

Serpentie, S. (6) 221
Servais, C. (4) 11
Seshadri, K.S. (11) 20
Sestokas, E. (9) 333
Sestokas, R. (9) 333
Setchell, K.D.R. (9) 95
Setnikar, I. (9) 467
Settlage, J.A. (5) 34, 223, 244
Settle, F.A., jun. (6) 5
Sevenhans, W. (4) 74
Severini, F. (6) 449, 450
Seyberth, H.W. (5) 198, 199; (6) 197–201; (7) 202, 205, 251
Seydel, U. (4) 193
Seyferth, D. (10) 32, 78
Seymour, M.P. (6) 380
Sgamellotti, A. (1) 52, 196, 197
Shabanowitz, J. (3) 158, 160; (5) 285, 286; (7) 179, 182, 240
Shackleton, C. (5) 346
Shackleton, C.H.L. (6) 36
Shadid, J.B. (2) 96
Shafran, Yu.T. (5) 74
Shah, D. (9) 81
Shah, R. (9) 212
Shaik, S.S. (5) 509
Shalaby, M. (7) 129
Shanks, R.G. (9) 616
Shao, J.-D. (2) 44
Shapley, J.R. (5) 428; (10) 81, 82
Sharkey, A.G. (5) 457
Sharman, M. (6) 314
Sharoiko, E.S. (9) 510
Sharova, O.L. (11) 96
Sharp, D.E. (9) 357
Sharp, P. (6) 177
Sharp, T.R. (10) 35, 103
Shaul, G.M. (7) 123
Shaw, A.J. (9) 360
Shaw, D.L. (6) 318
Shaw, R.W. (3) 29
Shaw, S.R. (9) 213
Shawkataly, O.B. (10) 80
Shchelokov, R.N. (10) 26
Shea, T.C. (9) 484
Shear, N.H. (9) 263
Shearer, M.C. (9) 493
Sheets, R.M. (5) 272
Sheldon, J.C. (2) 69; (5) 142, 143, 146, 157, 277, 531, 544–547
Sheldon, R.I. (11) 23
Shelkovsky, V.S. (9) 491
Shen, H. (5) 241
Shen, J. (4) 72
Shen, L.-H, (10) 63
Shen, M.H. (1) 267

Shen, Y.R. (1) 36
Sheng, L.-S. (9) 192
Shepard, W.B. (10) 122
Sherer, M. (9) 596
Sherma, J. (5) 38
Sherman, W.R. (6) 252
Sherry, J. (5) 141; (6) 371
Shetty, H.U. (9) 366, 419, 420
Shibai, H. (7) 258; (9) 516, 517
Shibasaki, H. (6) 286
Shibata, K. (9) 440, 441
Shiel, M.M. (7) 194
Shields, C.G. (2) 126, 128
Shilliday, F.B. (9) 514
Shima, N. (5) 65
Shimamura, M. (6) 259, 282
Shimoda, W. (9) 114
Shimomura, H. (9) 113
Shimonishi, T. (9) 324
Shindo, N. (5) 197; (6) 209
Shiner, V.J. (5) 66
Shinmel, M. (11) 153
Shinohara, H. (1) 94, 230
Shinohara, Y. (9) 153
Shiokawa, Y. (9) 381
Shioyama, C. (9) 623
Shiraga, T. (9) 329
Shire, D. (5) 466, 467
Shirley, D.A. (1) 18, 20, 313
Shiromaru, H. (1) 230
Shishido, A. (9) 621, 622
Shiu, K.-B. (10) 43
Shiverick, K.T. (9) 56
Shizukuishi, K. (7) 14
Shkol'nikova, T.M. (11) 71-73
Shmakov, V.S. (5) 95, 98
Shoda, T. (2) 66
Shohet, J.L. (3) 150
Shoji, T. (6) 315; (9) 116, 150
Shol'ts, V.V. (11) 21, 28
Shomo, R.E., (1) 131; (9) 29, 628
Short, C.R. (9) 119
Short, R.T. (1) 266
Shortland, D. (5) 348
Shubert, R. (5) 223
Shulgin, A.T. (9) 147
Shul'ts, E.E. (5) 97
Shul'ts, M.M. (11) 57, 59, 63
Shuskov, G.V. (5) 75
Shute, L.A. (4) 180
Sichtermann, W.K. (5) 420; (10) 89
Sicre, M.-A. (6) 340, 426
Sidky, E.Y. (4) 66
Sidorenko, G.V. (10) 86
Sidorov, L.N. (11) 18, 21, 28-32, 34-37, 39-44, 46, 68-70, 95, 96, 98, 101, 109, 149
Sidorova, I.V. (11) 33
Siebert, W. (10) 61, 64
Sieck, L.W. (5) 529, 530, 539
Siegel, M.M. (4) 119, 157; (7) 261; (9) 495, 515, 520, 535
Siegmund, E.G. (6) 240, 310; (7) 87, 98, 236; (9) 54, 550, 562, 703
Sieklucka, B. (5) 427; (10) 98
Siekmann, A. (6) 298
Siekmann, L. (6) 218, 288, 298
Sies, H. (9) 288
Siest, G. (9) 398
Sigray, P. (1) 266, 309, 310
Sigurdsson, A. (11) 80
Sigwarth, B. (10) 75, 77
Siitonen, P.H. (6) 387; (9) 282, 286
Sill, A.D. (9) 605
Silman, R.E. (6) 263
Silver, B. (9) 583
Silverton, J.V. (9) 521
Silvestre, D. (9) 109
Silvestri, T. (9) 601
Sim, P.G. (5) 61; (6) 352, 382; (7) 68
Simmleit, N. (4) 187, 188, 190, 191
Simon, J.D. (4) 81
Simon, M. (1) 32, 164; (4) 99; (6) 378
Simon, W. (10) 113, 116
Simona, M.G. (6) 270
Simoneit, B.R.T. (6) 339
Simons, D.S. (3) 40; (5) 446
Simons, J. (5) 521
Simons, J.K. (7) 43
Simons, P.J. (9) 418
Simonsick, W.J. (6) 186, 187, 325
Simson, V.L. (5) 357; (9) 494
Sin, C.H. (7) 61, 62, 63
Sinay, P. (5) 378; (10) 15
Sindona, G. (5) 386; (8) 31
Singer, S.S. (9) 354
Singh, A.K. (6) 287, 390; (9) 107
Singh, C. (5) 206
Singh, G. (5) 79
Singh, K. (9) 179
Singh, N.N. (9) 297
Sinha, B.K. (9) 299, 301
Sinkins, S. (6) 228; (9) 385
Sinninghe, J.S. (6) 420, 421
Siouffi, A.M. (5) 387; (9) 100
Sipachev, V.A. (10) 131
Sisenwine, S.F. (9) 389, 480
Sitrin, R.D. (9) 493
Siwers, B. (9) 637, 638, 677, 678
Sjoberg, E.J. (3) 104
Sjoqvist, J. (9) 637
Sjovall, J. (6) 236, 381
Sjovall, S. (5) 336
Skachilova, S.Ya. (5) 76
Skarping, G. (6) 399
Skelly, J.P. (9) 270
Skelton, B.W. (10) 48
Skinner, H.R. (9) 114
Skokan, E.V. (11) 30, 35, 109
Skowronski, R.P. (3) 59
Skudlarski, K. (11) 58, 82, 102, 106, 111-113, 118
Skyiepal, M. (2) 189
Slikker, W., jun. (9) 286, 432, 699
Slomp, A. (4) 15
Slowikowski, D.L. (5) 406
Small, G. (4) 134
Smalley, R.E. (1) 207, 275, 332; (5) 53-55
Smirnova, L.N. (11) 147
Smith, A. (6) 281
Smith, B.J. (5) 189; (6) 216
Smith, C.G. (5) 37
Smith, C.S. (6) 447
Smith, D. (5) 484
Smith, D.C. (4) 23
Smith, D.E. (10) 36
Smith, D.H. (3) 27, 29
Smith, D.L. (4) 220; (6) 15; (7) 176; (8) 7
Smith, G.P. (2) 67
Smith, J.D. (10) 6
Smith, J.F. (7) 101
Smith, J.L. (4) 25
Smith, J.S. (7) 231
Smith, K. (4) 51; (5) 243
Smith, L.M. (4) 99
Smith, M.L. (9) 553

Smith, M.P. (4) 5
Smith, M.T. (9) 189
Smith, P.B. (5) 37, 165
Smith, R.D. (3) 88, 89;
  (7) 2, 3, 4, 6, 33, 34,
  37, 42, 45-47, 58, 59,
  65, 67, 69, 220; (9)
  523
Smith, R.G. (9) 508
Smith, R.M. (6) 331
Smith, R.W. (6) 227; (7)
  110, 111, 113; (9) 21
Smolik, S. (9) 393
Smyth, M.H. (9) 111
Snader, K. (9) 492
Sneden, A.T. (9) 490
Snell, R.P. (6) 99
Snow, M.R. (5) 426; (10)
  37, 80
Snyder, A.P. (4) 192
Snyder, L.R. (7) 84
Snyder, S.W. (9) 321
Sobolewski, A.L. (1) 23,
  24, 33
Söler, F. (5) 409, 410
Soine, P.J. (9) 210
Soine, W.H. (9) 210
Sokoloski, E.A. (5) 460,
  465; (8) 26
Solanas, A.M. (6) 356
Soler, F. (4) 158
Solgadi, D. (1) 85, 175
Solheim, E. (9) 458, 459
Solomon, J.S. (4) 152
Solsten, R.T. (3) 211
Soltes, L. (9) 295
Somms, J.R. (7) 70
Sonesson, A. (6) 304
Sonnek, D. (1) 266
Sorkau, E. (4) 130
Sorokin, A.A. (5) 76, 78,
  81
Sorokin, I.D. (11) 29,
  30, 34, 35, 37, 39, 46,
  68, 69, 70, 109, 149
Southwell-Keely, P.T. (5)
  418
Southworth, S.H. (1) 4,
  12, 70
Sovocool, G.W. (5) 140,
  422; (6) 185, 369; (7)
  223, 233
Sozzi, G. (2) 192, 193,
  200, 205
Speck, A.J. (7) 140
Spector, S. (9) 461
Speirs, C.J. (9) 691
Spencer, L. (5) 161
Speranza, M. (2) 91
Sphon, J.A. (5) 147; (6)
  311; (7) 235
Spiehler, V. (9) 142

Spierdijk, J. (9) 693,
  694
Spies, H. (10) 90
Spies, R.B. (6) 345
Spinar, B. (9) 393
Spiridnova, I.A. (9) 510
Spiri-Nakagawa, P. (9)
  503
Spiteller, G. (6) 273
Spitsyn, V.I. (10) 124;
  (11) 116
Spitznagel, G.W. (5) 488
Spooner, N. (6) 122
Sprafke, A. (2) 185
Sprott, G.D. (5) 314
Spyrou, S.M. (1) 126
Squires, R.R. (2) 21; (5)
  287, 485, 500, 505,
  508, 555, 556, 558, 562
Srinivasa, R. (5) 133
Srivastava, R.D. (11)
  141, 144, 145
Sroka, R. (9) 136
Srzic, D. (10) 4
Staak, M. (9) 576
Stabler, S.P. (6) 277
Stace, A.J. (1) 102
Stacey, C. (10) 70
Stachowiak, K. (5) 238;
  (7) 134
Stafa, B. (11) 65
Stafford, F.E. (11) 10
Stafford, G.C., jun. (3)
  129-132, 135, 136
Stahl, D. (1) 181; (2)
  174
Staiger, D. (5) 383; (9)
  492
Stalick, W.M. (6) 192
Stall, W.J. (9) 577
Stalling, D.L. (4) 99
Stamatovic, A. (1) 205,
  212, 213, 216, 217
Stamp, J.J. (9) 703
Standing, K.G. (4) 56;
  (5) 434, 444; (8) 14;
  (9) 487
Stanford, D.F. (9) 559,
  595
Stanford, D.I. (6) 410
Stang, P.J. (10) 103
Stanley, M.S. (7) 8, 10,
  11; (9) 641
Stanney, K. (2) 76
Stanojcik, M. (9) 275
Stapelkamp, W. (4) 133
Staringer, W. (3) 189;
  (4) 21
Stark, D. (1) 170
Starodubtsev, A.M. (11)
  139
Startin, J.R. (6) 60,

  314; (7) 211; (9) 165
Statheropoulos, M. (4)
  224
Stauffer, D.B. (4) 65,
  117; (6) 86
Staveris, S. (9) 169
Steblevskii, A.V. (11)
  160
Stedman, G. (5) 243
Stefanidis, S. (6) 149,
  150
Stefanov, K. (6) 163, 164
Stefek, M. (9) 295, 451
Steffenrud, S. (6)
  138-141
Steifer, S. (3) 206
Steiger, W. (3) 51, 52
Steimle, T.C. (1) 306
Stein, H. (5) 377
Stein, J. (5) 317
Steiner, J.A. (9) 294
Steiner, P.A. (2) 125,
  126
Steinnes, E. (6) 342
Stellard, F. (6) 223,
  234, 235
Stemmler, E.A. (5) 117,
  135, 248; (6) 362; (9)
  648
Stene, D.O. (5) 184; (6)
  212
Stenhagen, G. (7) 152
Stenmark, K.R. (5) 180
Stenstrom, Y. (10) 66
Stepanek, R. (10) 113
Stephanou, E. (6) 353
Stephans, R.J. (6) 195
Stephens, D.R. (3) 132
Stepnowski, R.M. (3) 139
Sternal, R.S, (10) 68
Stetson, P.L. (8) 28; (9)
  176
Stetter, K.O. (8) 5, 6
Stevens, R.C. (9) 528
Stevens Miller, A.E. (1)
  324
Stevenson, D.P. (11) 19
Steward, E.M. (4) 114
Stiachan, M.G. (5) 149
Sticht, G. (9) 576
Stillwell, W.G. (5) 217
Stocke, D. (5) 127
Stockton, G.W. (9) 496
Stoessel, S. (9) 457
Stoesser, R. (8) 29
Stolk, H. (4) 194
Stoll, N. (4) 54
Stolyarova, V.L. (11) 57,
  59, 63, 157
Stopher, D.A. (9) 208,
  209
Stoppioni, P. (10) 100

Storms, E.K. (11) 74, 120–122, 134
Stotter, P.L. (5) 303
Stovell, P. (6) 263
Stowe, C.M. (9) 107
Strack, D. (5) 366
Straub, K. (9) 81, 221, 281, 446
Straub, K.M. (7) 121, 195, 206; (9) 35, 36, 447, 448
Straw, R.D. (11) 172
Strayer, M.R. (4) 222
Streeter, A.J. (9) 315
Streit, G.E. (5) 565
Strife, R.J. (3) 138
Stringer, M.B. (2) 196; (5) 104, 154, 155, 162
Stroh, J. (7) 88
Strokov, I.I. (4) 131
Strolin-Benedetti, M. (9) 305, 306
Stromberg, S. (9) 115, 460, 707
Strong, J.M. (6) 278
Strubinger, J.R. (7) 66
Struck, R.F. (9) 449
Struempler, R.E. (9) 573
Struewer, D. (4) 206
Studnicka, H. (3) 52
Stuewer, D. (3) 76–78
Stuke, M. (10) 22
Stults, J.T. (3) 108; (4) 209; (5) 168; (7) 186
Su, A.C.L. (5) 520
Su, K. (5) 60
Su, T. (5) 520
Subhan, M. (5) 470
Subramaniam, R. (8) 11
Subramanyam, B. (9) 321
Suchocki, J.A. (9) 490
Suckow, R.F. (9) 636
Sudo, K. (9) 474
Sülzle, D. (2) 141, 146
Suga, K. (9) 546
Sugano, S. (2) 18
Sugawara, Y. (9) 112, 239, 278, 279
Sugeno, K. (9) 197, 438
Sugihara, J. (9) 455, 456
Sugimoto, S. (10) 9
Sugimoto, T. (9) 273
Sugiyama, M. (9) 240, 273
Suglobov, D.N. (10) 86
Suguhara, K. (9) 546
Sui, K.W.M. (5) 119
Sukhodub, L.F. (9) 491
Suling, W.J. (9) 449
Sullivan, H.R. (9) 563
Sullivan, S.A. (5) 518
Sum, F.W. (9) 370
Sumino, K. (6) 296

Summers, W.R. (3) 64
Sun, Y. (5) 468
Sundholm, E.G. (9) 583
Sundqvist, B.U.R. (3) 62, 196
Sunner, J. (5) 292
Suon, K.N. (4) 4
Sutonen, P.H. (5) 128
Suttara, V. (5) 339
Suzer, S. (4) 105
Suzuki, K. (10) 106
Suzuki, M. (10) 106
Suzuki, O. (5) 263; (9) 674
Suzuki, S. (2) 77; (9) 116, 223, 280, 590
Suzuki, T. (4) 118; (6) 176; (9) 112, 278, 279
Svara, J. (10) 29, 62
Svec, H.J. (3) 66
Svensson, K. (9) 327
Svensson, S. (1) 172
Svensson, W.A. (1) 9
Swackhamer, D.L. (5) 129; (6) 386
Swagzdis, J. (9) 281, 446
Swamy, S.J. (10) 67
Swanson, B.J. (8) 23
Swanto, F. (6) 302
Sweeley, C.C. (4) 90, 91; (6) 49, 295; (7) 79
Sweeney, C.W. (5) 69
Sweeny, D.J. (9) 312, 313
Sweetman, B.J. (9) 302, 349
Swere, M. (6) 370
Swerev, M. (6) 181
Swysen, E. (9) 200, 254, 255
Sy, W.-W. (9) 627
Syka, J.E.P. (3) 128, 129, 131, 135, 136
Symchowicz, S. (9) 445
Syracuse, C.D. (9) 84
Szabo, G.K. (9) 5, 69, 69, 665, 666
Szabo, I. (4) 70, 212, 213, 214
Szaflarski, D.M. (1) 91, 95; (4) 81
Szekely, G. (4) 70
Szekely, T. (4) 52, 176; (6) 430
Szerman-Joly, E. (6) 241
Szklarz, E.G. (11) 120–122
Szulejko, J.E. (2) 70
Szuna, A.J. (9) 203


Taagepera, M. (5) 486
Tabet, J.-C. (2) 61, 97;
 (3) 155; (6) 171; (9) 582
Tachizawa, H. (9) 474
Tada, H. (5) 534
Tada, K. (9) 329
Taft, R.W. (5) 486
Taggart, R.L. (6) 330
Taghavini, E. (5) 101
Taghizadeh, K. (4) 177, 192
Tagliaro, F. (9) 580
Taguchi, H. (9) 318
Taguchi, V. (7) 230
Taguchi, V.Y. (6) 364
Taha, A.M. (6) 193
Tait, P.J.T. (10) 70
Takahagi, H. (9) 105
Takahashi, N. (9) 498
Takahashi, S. (5) 229; (9) 135, 196, 427, 438
Takahashi, Y. (9) 623
Takai, R. (6) 95
Takanashi, Y. (9) 503
Takano, T. (9) 124
Takasaki, W. (9) 105
Takatsuki, K. (6) 315; (9) 116, 150
Takayama, H. (9) 194, 195
Takegoshi, T. (9) 94
Takeshita, H. (6) 232, 233
Takeuchi, G. (5) 176
Takeuchi, K. (9) 402
Takeuchi, T. (7) 161
Takimoto, H. (2) 66
Talaat, R.E. (9) 422, 423
Taleb, N.A. (9) 560
Tam, Y.K. (9) 341, 692
Tamas, J. (9) 309
Tamiri, T. (6) 385
Tamura, H. (9) 623
Tanabe, J. (9) 532
Tanabe, S. (6) 373
Tanabe, T. (9) 532
Tanaka, A. (9) 462
Tanaka, H. (9) 503
Tanaka, K. (1) 251; (3) 45
Tanaka, M. (9) 546; (10) 99
Tanaka, T. (5) 311
Tanaka, Y. (9) 503
Tanayama, S. (9) 375
Tang, H. (9) 686
Tang, X. (4) 56
Taniguchi, S. (2) 60
Tanishima, Y. (9) 105
Tarabe, S. (5) 276
Tarantelli, F. (1) 196, 197
Tarzia, G. (9) 373
Tas, A.C. (4) 179; (7)

137, 140
Tashiro, T. (10) 106
Tasker, P.A. (10) 121
Taskinen, J. (9) 117
Tasman, H.A. (11) 25
Tasset, J.J. (9) 541, 542
Tateishi, M. (9) 123, 185, 409
Tatewaki, N. (9) 240
Tatsuhara, T. (9) 178
Tatsukawa, R. (5) 276; (6) 373
Tatsuzawa, S. (3) 186
Taub, R. (9) 333
Tauber, U. (9) 228
Taylor, C.E. (3) 66
Taylor, E.H. (9) 120
Taylor, G.W. (7) 118
Taylor, J.H. (6) 391
Taylor, J.W. (4) 49, 215
Taylor, K.J. (5) 55
Taylor, L.C.E. (3) 190, 194
Taylor, L.H. (9) 377
Taylor, R.W. (6) 411; (9) 90
Taylor, W.S. (5) 551
Teale, P. (6) 39, 228; (6) 385
Tecklenburg, R.E., jun. (1) 273
Tedder, J.M. (2) 75, 76
Tegelaar, E.W. (6) 435
Tegyey, Z. (9) 645
Teillet, D. (1) 139
Teitz, D. (9) 182
Te Koppele, J.M. (9) 184
Telnaes, N. (4) 93; (6) 424
Tembreull, R. (3) 11; (9) 635
Temkin, A. (5) 3
Tenaschuk, D. (6) 275, 276
Tenczer, J. (9) 140, 244, 246, 248, 272, 308, 346
Tenhosaari, A. (4) 170; (6) 87
Téoule, R.E.A. (4) 158; (5) 409
Terada, T. (5) 258; (9) 243
Terakawa, M. (9) 621, 622
Terashita, S. (9) 381
Terhune, R.W. (3) 3
Terjeson, R.J. (5) 272
Terlinden, R. (9) 288
Terlouw, J.K. (1) 114; (2) 10, 39, 41, 122, 124, 129, 132, 138, 142, 143, 146, 147, 148, 149, 151, 162, 165-168
Ternan, M. (6) 110
Terry, K. (6) 180
Terwilliger, D.T. (4) 106
Tessaro, S.V. (5) 126
Teterin, G.A. (11) 71-73
Tetler, L. (10) 42, 114
Tews, E.C. (4) 142, 143
Thacker, S. (9) 556
Thalen, A. (9) 230, 231
Thebault, P. (5) 395
Theberge, R. (10) 104
Theilsohn, M. (9) 644
Theis, D.L. (9) 126
Thenot, J.P. (9) 202, 482
Theoharides, A.D. (9) 111
Thiel, W. (2) 26
Thiele, G. (10) 73
Thienpont, L.M.R. (6) 37; (9) 177
Thijssen, J. (9) 200
Thiry, P. (1) 164
Thoma, H. (6) 451
Thomas, A.D. (5) 450, 451
Thomas, B.F. (9) 587
Thomas, B.H. (9) 426
Thomas, D.W. (6) 195; (9) 258
Thomas, M.R. (6) 300
Thomas, N.C. (10) 101
Thomas, R. (6) 80
Thompson, C.B. (9) 257
Thompson, H.C., jun. (9) 286, 434, 699
Thompson, J.A. (9) 58
Thompson, L.K. (9) 596
Thompson, M. (9) 307
Thompson, R.M. (9) 468
Thomson, C. (2) 26
Thornburgh, B.A. (9) 213
Thorne, G.C. (4) 96, 97; (6) 92
Thorne, N.A. (3) 187; (4) 39
Thornton, D.C. (6) 329
Thrane, K. (6) 342
Threadgill, M.D. (9) 359, 360
Thummel, K.E. (9) 477
Tibbels, S.R. (8) 21
Tiekink, E.R.T. (5) 426; (10) 37, 80
Till, F. (4) 52
Tillement, J.P. (9) 667, 668
Timmerman, Ph. (9) 200
Timmins, A. (4) 109
Timmins, P. (9) 47
Tindeur, Y. (8) 13
Ting, B.T.G. (3) 68
Tio, C.O. (9) 389
Tippett, P.A. (6) 258
Tittel, F.K. (1) 207; (5) 53
Titus, R. (3) 208; (6) 19, 319
Tjaden, U.R. (9) 489
Tjandra, H. (5) 418
Tmar, M. (11) 166
Todaro, L. (9) 499
Todd, D.E. (4) 23
Todd, J.F.J. (3) 131, 136; (5) 381, 382; (6) 18; (7) 47
Todesko, L. (5) 132; (9) 125
Toelg, G. (3) 78
Tofanetti, O. (9) 466
Toguri, J.M. (11) 161, 163, 170
Tohma, M. (6) 232, 233
Tokes, L.G. (9) 337
Tokimitsu, Y, (5) 341
Tokuma, Y. (5) 245; (9) 148, 381, 618, 621, 622
Tolstikov, G.A. (5) 56, 70, 89, 95, 97, 98
Tolun, E. (5) 381, 382, 398
Tom, A. (6) 438
Toma, S. (9) 251
Tomalia, P.A. (5) 297
Tomas-Barberan, F.A. (5) 359, 368, 369
Tomasello, P. (1) 45, 48, 53
Tomasz, M. (8) 25
Tombret, F. (9) 450
Tomer, K.B. (5) 228, 308, 309, 318, 347, 360, 411, 412, 414, 421; (7) 178, 187, 200, 204, 244, 246, 248, 249, 253, 256, 261; (8) 12, 21
Tomingas, T. (6) 335
Tomisawa, H. (9) 185, 409
Tomiska, J. (11) 123, 124, 128, 129
Tomiyasu, I. (6) 119
Tomkins, F.S. (1) 66
Tomlinsin, D. (9) 525
Tomson, T. (9) 643
Tonani, R. (9) 509
Tondeur, Y. (6) 363; (7) 223, 229, 233
Tong, H.Y. (5) 116
Tonini, C. (6) 155
Tonnesen, F. (9) 588
Tonnesen, H.H. (9) 534
Toon, S. (9) 127, 599
Tordeur, Y. (5) 422
Torgerson, D.F. (3) 59
Torii, H. (10) 108

Tornqvist, M. (5) 240; (6) 395
Torok-Both, G.A. (9) 341, 692
Torroni, S. (10) 27, 72
Tortajada, J. (2) 64, 181-184
Toth, E. (9) 309
Traas, L. (3) 188
Tracy, M. (9) 258
Traeger, J.C. (2) 195; (4) 6
Trager, W.F. (9) 71, 598-600
Trainor, T.M. (5) 130, 136; (6) 361
Traldi, P. (5) 113, 177, 178, 212, 338; (6) 174, 449; (9) 45, 509, 580; (10) 30, 40, 88
Tramer, A. (1) 85
Trapa, V.J. (1) 336; (5) 492
Trarrontaro, W. (5) 220
Trauner, M.C. (2) 126
Trautmann, N. (3) 38
Travis, J.C. (3) 9
Travniczek, C. (3) 189; (4) 21
Treher, E.N. (5) 425
Trehy, M.L. (7) 214
Trenk, D. (9) 143
Trevor, A. (9) 566
Trimble, J. (9) 705
Triscari, J. (9) 203
Trnovec, T. (9) 295
Trogler, W.C. (10) 55
Troke, J. (9) 372
Trombini, C. (5) 101
Tronc, M. (1) 133, 134
Troskosky, J.A. (6) 105, 106
Troupe, N. (9) 499
Troyanov, S.I. (11) 104
Troyer, B.L. (9) 574
Trummer, K.-H. (10) 73
Tsaconas, C. (6) 238
Tsai, P. (5) 380
Tsang, W.-S. (9) 592
Tse, F.L.S. (9) 443
Tsou, H.-R. (9) 495, 496
Tsuchihashi, H. (9) 590
Tsuchiya, T. (5) 258; (9) 243, 462
Tsuge, S. (7) 75, 76, 153-155
Tsuji, M. (1) 297-305
Tsukada, M. (5) 65
Tsukamoto, K. (9) 440
Tsukioka, T. (6) 359
Tsumura, M. (9) 474
Tsuneuchi, N. (9) 15

Tsuneyuki, S. (5) 65
Tu, J. (9) 164
Tuan, G. (9) 373, 428
Tuckina, M.Ya. (5) 440
Tümmler, R. (5) 100
Tuinman, A.A. (4) 159
Tumas, W. (5) 144, 145, 552
Tunlid, A. (5) 163; (6) 194
Tureček, F. (2) 13, 158-160; (5) 111
Turgeon, J. (9) 367
Turi-Nagy, L. (9) 451
Turk, J. (5) 215
Turner, L.K. (4) 203
Turner-Tamiyasu, K. (9) 597
Tuseev, N.I. (10) 131
Tutcher, B. (1) 77
Tuttle, J.H. (6) 347
Tuttobello, L. (9) 501
Tyrra, W. (10) 20
Tzeilhin, M. (5) 304

Ubukata, M. (8) 9
Uccella, N. (5) 386; (8) 31
Uchiyama, N. (9) 124
Udseth, H.R. (3) 89; (7) 3, 4, 6, 33, 34, 37, 45, 46, 65, 69, 220
Ueda, T. (8) 8
Ueda, Y. (11) 125
Ueland, P.M. (9) 458, 459
Uetrecht, J. (9) 263
Uhrig, M.S. (6) 103
Ulendeeva, A.D. (5) 98
Ulhemann, E. (10) 90
Ulrich, J. (4) 158; (5) 409, 416; (10) 118
Um, I.-H (5) 509
Umani-Ronchi, A. (5) 101
Umemura, K. (9) 650
Umeno, Y. (9) 156
Umezawa, H. (9) 106
Underberg, W.J.M. (9) 531
Undheim, K. (10) 12
Ung, H.L. (9) 162
Unger, M.A. (10) 17
Unger, S.E. (5) 33, 401, 425; (7) 107, 122, 259; (9) 357, 524
Upadhyaya, A.K. (5) 79
Uralets, V.P. (9) 271, 311
Ushizawa, I. (6) 315; (9) 116, 150
Usuki, S. (9) 112, 278, 279
Utamura, T. (5) 376

Utoh, M. (9) 123
Uunk, L.G.M. (7) 25
Uy, O.M. (5) 311, 320


Vachon, L. (5) 251
Vachta, J. (9) 407
Vager, Z. (1) 261
Vaglio, G.A. (10) 93, 100
Vaida, V. (10) 33
Vajta, S. (9) 202, 482
Vakoulis, F.D. (10) 96
Valcavi, U. (9) 439
Valdivieso, L. (9) 597
Vallis, L. (4) 182
Valter, K. (9) 407
van Baar, B.L.M. (2) 142, 143, 148, 162, 167
Van Berkel, W.W. (5) 513
Van Bladeren, P.J. (9) 314
Van Boeckel, C.A.A. (9) 34
van Breeman, R.B. (6) 358
van Dalen, A.C.K. (6) 418, 420, 421
van den Akker, E. (9) 300
Van Den Berg, G.A. (6) 253
van den Berg, J.A. (3) 48
Van den Heuvel, W.J.A. (5) 273, 430; (6) 23, 56; (9) 63, 66, 76, 121, 333
van den Ven, L.J.M. (9) 489
Van Der Biesen, J.J.H. (5) 502
Van Der Gen, A. (9) 184
Van der Graaf, M. (5) 109; (9) 6, 652
Van der Greef, J. (4) 179; (7) 137, 140; (9) 184, 300
van der Hart, W.J. (1) 268; (2) 94
Van Der Houwen, O.A.G.J. (9) 531
Van Der Lijn, P. (6) 264
Van Der Mark, E.J. (9) 184
Van Der Meer, M.J. (9) 216
van der Plas, P.S.C. (3) 70
Van der Schyf, C.J. (9) 560
Van Der Wel, H. (5) 544
van der Wiel, M.J. (3) 4
Van der Zande, W.J. (1) 112
van Doren, J.M. (2) 99;

(5) 503, 507, 517
van Duijneveldt, F.B. (2) 39
Van Espen, P. (4) 150, 205
Vangbo, B. (9) 460
van Graas, G. (6) 429
van Haver, G.M. (6) 316; (9) 548
Van Hecken, A. (9) 569, 634
Van Heijenoort, J. (5) 377
Van Heijenoort, Y. (5) 377
van Helden, S.P. (2) 166
Van Hoof, F. (9) 656
van Houdt, J. (9) 201
van Houte, J.J. (2) 96
van Kleefe, J.W. (9) 693, 694
van Kuijk, F.J.M. (6) 195
van Laar, J. (6) 407
Van Langenhove, A. (9) 74
van Lenthe, J.H. (2) 30, 166
Van Lier, J.E. (8) 24
Van Lonkhuyzen, H. (1) 296
Van Loon, J.C. (6) 73
Van Maanen, J.M.S. (9) 300
Van Muiswinkel, K. (9) 314
van Oosterom, A.T. (9) 489
Van Peer, A. (9) 254, 336
Van Peteghem, C.H. (6) 316; (9) 548
Van Rossum, J.M. (9) 569
van Smeerdijk, D.G. (6) 434
van Thuijl, J. (2) 96; (4) 147
Van't Klooster, H.A. (4) 121, 179
Van Vaeck, L. (9) 31
Van Velduizen, A. (9) 314
Van Wazer, J.P. (5) 482
Varhegyi, G. (4) 52
Varma, R.S. (5) 253
Varmuza, K. (3) 189; (4) 21, 87; (6) 88
Vasconcelos, A.M.P. (6) 279, 280
Vasyukova, V.I. (10) 5
Vaughan, G. (5) 174, 216
Vaughan, M.A. (4) 207
Vederas, J.C. (5) 385
Vékey, K. (1) 191, 192
Velkov, M. (9) 316
Venkatasubramanian, V.S.

(5) 10
Venogopal, V. (11) 119
Venton, D.L. (5) 185
Ventura, F. (5) 296; (6) 344
Ventura, P. (9) 704
Verbesselt, R. (9) 634
Verdine, G.L. (8) 25
Verdun, F.R. (3) 162
Vereczkey, L. (9) 309
Verhaege, B.J. (4) 92; (6) 289
Verharghe, P.G. (9) 177
Verhoeven, F.C.C.J.G. (7) 24
Verhoeven, M. (2) 146
Verhulst, A. (6) 158
Verma, S. (2) 57
Vermeulen, N.P.E. (5) 109; (9) 6, 652, 693, 694
Vermeulen, P.A.M. (11) 97
Vernin, G. (4) 4, 113; (6) 89
Verpeau, J.N. (5) 268
Vertesy, L. (9) 505, 533
Vesell, E.S. (9) 377
Vessman, J. (9) 365, 608
Vestal, C.H. (7) 108
Vestal, M.L. (5) 238, 269; (7) 127, 134; (9) 212
Vettori, U. (10) 30
Viari, A. (5) 466, 467
Viaris de Lesegno, P. (4) 80
Vickermann, J.C. (3) 48; (5) 431
Vidal, E. (5) 334
Vierhapper, H. (6) 244
Vieth, W. (3) 76, 77; (4) 204, 206
Viggiano, R.A. (5) 520, 527
Vigny, P. (5) 466, 467
Vijayakumar, C.T. (6) 440
Villeneuve, S. (2) 131
Vincent, M.A. (1) 41
Vincenti, M. (2) 58; (6) 155
Vincieri, F.F. (5) 177; (6) 174; (7) 210; (9) 46
Vincze, A. (7) 11
Vine, J.H. (5) 166; (6) 267; (9) 92, 689
Vink, J. (9) 225
Vinson, G.P. (7) 118
Vinson, L.K. (2) 25
Viola, F. (9) 410
Vir, D. (9) 685
Virelizier, H. (4) 158;

(5) 409, 416; (6) 171; (10) 109, 118
Vischer, A. (5) 188
Visentini, J. (6) 406
Viste, A. (4) 9
Viswanadham, S.K. (3) 156; (5) 420, 449, 457
Viswanathan, K.S. (1) 68, 69
Viswanathan, R. (11) 85, 132, 173
Voelkel, N.F. (5) 184; (6) 212
Vogel, P. (10) 31, 41, 47
Vogelbacher, U.J. (2) 156
Voges, P. (9) 73
Vogt, N.B. (6) 342
Voigt, D. (5) 88
Voirin, B. (5) 361
Volk, K.J. (5) 231
Volpe, P. (10) 93
von Ardenne, H. (5) 100
Vonnahme, T.L. (6) 369; (7) 233
Vonnature, T.L. (5) 140
von Niessen, W. (1) 40–48, 52, 53, 58
von Schnering, H.G. (11) 79
Von Trebra, R.J. (8) 23
Von Unruh, G.E. (9) 151, 644, 654, 661
Voorhees, K.J. (4) 197; (7) 54
Vorbrodt, H.-M. (5) 93
Vorob'ev, A.S. (5) 89, 98
Voronin, V.G. (5) 78, 81
Voronira, T.N. (5) 80
Vosburger, O. (3) 150
Vose, C.W. (9) 294
Vostrov, S.N. (9) 510
Vouros, P. (5) 130, 136; (6) 159; (7) 141; (9) 57
Vovk, O.M. (11) 39, 41, 149
Voyksner, R.D. (5) 190, 246; (7) 112, 132, 226; (8) 27
Vralets, V.P. (9) 361
Vrbanac, J.J. (4) 91; (5) 194; (6) 202, 203
Vroomen, L.H.M. (9) 314
Vu Duc, T. (9) 555
Vuorilehto, L. (9) 269
Vuos, P. (6) 361
Vycudilik, W. (9) 549

Wachsmuth, H. (9) 205, 207
Wachter, E.A. (4) 66

## Author Index

Waddell, D.S. (5) 61, 139; (6) 178, 382
Wade, A.P. (4) 135
Wadepohl, H. (10) 64
Wadman, S.K. (5) 62, 68, 233; (6) 261, 283
Wagemann, A. (9) 246
Wagner, F. (9) 143
Wagner, H. (4) 126; (5) 342
Wagner, J.R. (8) 24
Wagner, M. (6) 348
Wagner, S. (11) 78
Wait, R. (6) 168
Wakimoto, T. (5) 276; (6) 373
Wald, B. (5) 362
Waldhausi, W. (6) 244
Waldrich, H. (3) 101
Walhagen, A. (9) 168
Walker, R.W. (5) 273, 430; (9) 76, 121
Wall, R. (7) 27
Walle, K. (9) 612
Walle, T. (9) 612-615
Walle, U.K. (9) 613-615
Wallwork, S.C. (10) 94
Walter, K. (3) 12-16, 18; (10) 115
Walter-Sack, I. (9) 87
Walters, E.A. (1) 223, 224
Walther, H. (2) 111
Walton, J.K. (10) 48
Walton, R.A. (5) 281; (10) 46, 105
Wanczek, K.-P. (3) 148, 151
Wand, M.D. (9) 58
Wang, B.H. (3) 63, 157
Wang, C.P. (4) 124; (6) 90
Wang, D. (5) 558, 562
Wang, G.-H. (3) 195; (9) 647
Wang, L. (1) 313
Wang, M. (3) 163
Wang, N. (9) 556
Wang, T.-C.L. (3) 161
Wannberg, B. (1) 172
Warburton, J. (3) 95
Ward, J.W. (11) 52
Waring, R.H. (9) 395, 396
Warner, C.D. (8) 21
Warner, M. (6) 2; (7) 19; (9) 539
Warner, W.D. (2) 95
Warrack, B.M. (5) 33; (7) 122; (9) 524
Warrell, D.A. (9) 538
Washida, N. (1) 230
Washington, C. (9) 49
Washino, T. (9) 375
Wasserman, M.B. (5) 204; (6) 305
Wasterdahl, G. (6) 304
Watanabe, K. (6) 131-134, 143, 231; (9) 106, 324
Watanabe, M.D. (9) 406
Watanabe, S. (7) 161
Watanabe, T. (9) 546
Watson, C.H. (1) 201; (5) 452
Watson, D. (6) 230; (7) 118; (9) 91, 681, 682
Watson, J.T. (3) 109, 140; (4) 1, 29, 60, 209; (5) 125, 261; (6) 70, 360; (7) 186
Watson, K.V. (9) 372
Watson, T.R. (5) 166; (9) 92, 238, 319, 680
Watt, P.W. (6) 274
Waugh, A.R. (3) 173
Wauters, A. (9) 173
Weaver, A. (1) 340; (5) 58
Webb, R.L. (5) 383
Weber, D. (5) 148; (6) 309, 312
Weber, F. (9) 169
Weber, J.J. (4) 147
Weber, P. (10) 65
Weber, P.C. (5) 188
Weber, U. (10) 75
Weber, W.P. (5) 550
Weber-Grabau, M. (3) 138; (4) 28
Wedlund, P.D. (9) 349
Weeks, S.J. (6) 322
Wehner, T.A. (5) 273
Wei, H. (5) 60
Weidolf, L.O.G. (5) 211, 270; (9) 303, 679
Weigold, E. (1) 52, 53, 58
Weil, A. (9) 305, 306
Weil, D.A. (5) 450
Weil, K.G. (2) 220; (11) 168, 169
Weilder, D.J. (9) 163
Weinhold, F. (1) 271; (5) 524
Weinstein, H. (5) 510
Weintraub, S.T. (5) 303
Weiske, T. (1) 181; (2) 142, 172, 191, 209, 216
Weisner, S. (6) 297
Weiss, G. (9) 199
Weiss, M. (5) 176
Weisshaar, J.C. (1) 271
Weitz, C.J. (9) 578
Welch, M.J. (6) 264
Welkie, D.G. (3) 50
Weller, R.R. (3) 156; (8) 20
Wells, C.E. (9) 52
Wells, J. (9) 175
Wells, R.J. (6) 388
Wells, S. (9) 525
Welsh, M.B. (9) 95
Wenclawiak, B. (10) 89
Wentrup, C. (2) 156, 157
Wentworth, W.E. (5) 260
Wenz, H.W. (3) 213
Werdellorn, D.F. (5) 196
Werner, H.W. (5) 20
Werz, R. (3) 188
Wesdemiotis, C. (1) 113; (2) 9, 121, 130, 136, 140, 145
West, F. (9) 579
West, J.B. (1) 70
Westcott, J.Y. (5) 180, 184; (6) 212
Wester, D.W. (10) 102
Wester, R.C. (6) 333, 396
Westerdahl, G. (5) 163; (6) 194
Westley, J.W. (9) 499
Westmore, J.B. (2) 15; (4) 56; (5) 106, 444; (8) 14; (9) 487
Wetmore, R.W. (1) 152, 155
Wetzel, D.M. (5) 493
Wetzel, P. (5) 377
Weyhenmeyer, R. (9) 181
Wharf, I. (10) 18, 19
Whitaker, N.G.G. (9) 435
Whitaker, T.J. (3) 32, 33, 34
White, A.H. (10) 48
White, A.R. (6) 248
White, D. (11) 20
White, D.C. (6) 156, 194
White, D.H. (10) 102
White, E. (5) 446; (6) 264
White, E.L. (2) 61
White, F. (6) 14
White, F.A. (4) 2; (5) 12
White, J.J. (10) 8
White, M.G. (1) 15, 27, 29, 76, 223
White, N.J. (9) 538
White, P.Y. (5) 266
Whitehouse, L.W. (9) 426
Whitehouse, M.J. (9) 18
Whitesides, G.M. (10) 28
Whitley, T.A. (1) 9
Whitmore, T.N. (4) 67, 68
Whitney, J. (7) 204
Whitten, J.L. (1) 237
Wickham-Jones, C.T. (1) 321

Wickramayake, P.P. (5) 119
Wickremasinha, A.J. (9) 213
Wieboldt, R.C. (7) 23
Wigfield, Y.Y. (5) 242
Wightman, R.H. (8) 2
Wijenberg, J.H.O.J. (2) 30
Wikkerink, B. (5) 68
Wikstrom, H. (9) 327
Wilante, C. (2) 31
Wilby, A.H. (3) 204
Wilder, C. (5) 238; (7) 134
Wilhartitz, P. (4) 151
Wilk, Z.A. (4) 47
Wilkie, C.A. (5) 448
Wilkins, C.L. (3) 41, 154, 206, 210; (4) 63, 123; (5) 161, 447, 450, 455, 456, 564; (6) 20, 79, 81; (7) 49; (9) 28; (10) 119
Wilkinson, G.R. (9) 349
Wilkinson, R. (5) 141; (6) 371
Wilkinson, R.J. (6) 180
Willard, D.A. (9) 164
Williams, C.A. (5) 368
Williams, D. (4) 32; (7) 170
Williams, D.H. (2) 203; (5) 293, 384, 388, 393; (7) 243
Williams, D.T. (6) 379; (7) 216
Williams, E.R. (2) 130; (3) 63, 157; (4) 65
Williams, F. (2) 163
Williams, J.G. (3) 67
Williams, K. (9) 684
Williams, T.D. (5) 251
Williams, T.H. (9) 203, 471
Williams, T.N. (4) 67
Williams, W.K. (7) 143
Williamson, B. (9) 54
Wills, M.R. (5) 189; (6) 142, 216
Wills, R.J. (9) 525, 526–528, 701
Wils, E.R.J. (6) 407; (9) 322
Wilson, H.K. (6) 401
Wilson, I.D. (9) 372
Wilson, K. (9) 245
Wilson, W.E. (4) 155
Windig, W. (4) 73, 174, 175, 177, 192, 194–196, 198
Wingfield, P.T. (6) 270

Wingfield, Y.Y. (6) 392
Winkler, P.C. (7) 143
Winkler, T. (9) 388, 390
Winkoun, D. (1) 157, 165, 166, 175
Winniczek, J.W. (1) 267
Winstanley, P. (9) 536
Winston, S. (7) 179
Winter, C.D. (5) 421
Winter, C.K. (5) 300; (6) 266
Winter, D. (9) 141
Winter, H.W. (2) 156
Wise, M.B. (4) 142; (5) 121; (7) 237
Wise, S.A. (6) 323
Wishnok, J.S. (5) 217
Wittmaack, K. (5) 436
Wittmann, K. (10) 73
Wölfli, W. (3) 182
Wöllnik, H. (3) 97, 167, 179
Wöste, L. (1) 215
Woestenborghs, R. (9) 200, 254, 336
Woggon, B. (9) 344
Wold, S. (4) 164
Wolf, B.A. (5) 215
Wolf, R. (2) 171
Wolfe, S. (5) 509
Wolnik, K. (9) 52
Wolstenholme, J. (4) 42
Wolter, A. (5) 408
Wolthers, B.G. (6) 226, 292
Wong, C.-M. (4) 18; (9) 487
Wong, L.K. (9) 274
Wong, M.W. (1) 179, 180; (2) 176–178
Wong, W.W. (6) 51
Wong Chi Man, W.W.C. (10) 49
Woo, S.O. (9) 534
Wood, B.T. (9) 657
Wood, G. (6) 14
Wood, G.M. (4) 2; (5) 12
Wood, J.S. (5) 273
Wood, K.V. (3) 46
Wood, P.G. (9) 563
Wood, R. (6) 114
Wood, S.G. (9) 220
Wood, T.G. (10) 32
Woodward, D.L. (9) 370
Woolf, T.R. (9) 334
Woolfit, A.R. (7) 157
Woolner, D.F. (9) 141
Workman, B.S. (9) 633
Wort, F.S. (1) 162, 168
Worthy, W. (2) 22
Worwag, E.M. (9) 669
Wratten, S.J. (3) 211;

(6) 80
Wray, V. (5) 362, 366
Wright, B. (3) 193
Wright, B.W. (7) 34, 37, 59, 65, 69, 220; (9) 523
Wright, G.J. (9) 141
Wright, L.A. (6) 222
Wright, L.G. (7) 165
Wrobel, B. (5) 94
Wronka, J. (2) 70; (5) 563; (7) 141
Wu, A.T. (9) 337
Wu, C.H. (11) 50
Wu, W.N. (9) 226
Wu, Y. (5) 241
Wuensche, C. (9) 252
Wuethrich, C. (5) 138
Wyatt, D.M. (6) 109
Wynalda, M.-A. (5) 182; (6) 208
Wysocki, V.H. (2) 228; (7) 188
Wyttenbach, M. (1) 292
Wyttenbach, T. (1) 284, 287, 291, 307


Xia, W. (9) 364
Xu, B.J. (5) 109; (9) 6
Xu, X. (5) 241
Xu, Y. (2) 127; (5) 280; (10) 44


Yagi, Y. (5) 376
Yakota, K. (6) 134
Yakovlev, A.V. (4) 218
Yakubovskaya, L.N. (5) 80
Yakushchenko, I.K. (10) 145
Yamabe, S. (2) 60; (5) 540
Yamada, H. (4) 115; (9) 352
Yamada, K. (9) 317
Yamada, T. (5) 352, 356
Yamaguchi, K. (6) 75
Yamaguchi, S. (1) 301, 303, 305
Yamaguchi, T. (9) 197
Yamaha, T. (9) 462
Yamaizumi, Z. (8) 8
Yamamoto, I. (9) 324
Yamamoto, K. (9) 113
Yamamoto, S. (6) 134; (7) 16
Yamamoto, T. (4) 112; (9) 427
Yamamoto, Y. (9) 546
Yamanaka, H. (9) 194, 195
Yamashita, K. (6)

# Author Index

Yamatsu, I. (9) 106, 133–135, 143; (9) 317
Yamauchi, T. (5) 345, 355; (9) 626
Yamauchi, Y. (7) 44
Yamazoe, Y. (9) 618
Yanagisawa, J. (6) 237
Yanez, M. (2) 81
Yang, B. (1) 77
Yang, C.L.C. (5) 447
Yang, C.R. (5) 379
Yang, J.M. (3) 115
Yang, M. (3) 42
Yang, S.H. (1) 332; (5) 54, 55
Yang, Y.M. (5) 460; (9) 629
Yaniger, S.I. (5) 448
Yano, I. (6) 119
Yao, P. (9) 205
Yaozu, C. (6) 112
Yashiki, M. (9) 579, 593
Yasuda, K. (9) 621
Yasue, Y. (11) 64, 66
Yasuhara, A. (6) 334
Yates, B.F. (2) 34–36, 164, 180
Yates, J.R. (3) 158, 160; (7) 179, 182, 240
Yatsenko, A.E. (6) 152
Yavorskii, A.S. (5) 80
Yeh, L.I. (5) 502
Yeh, S.F. (6) 291
Yeh, S.Y. (9) 433
Yellet, L. (9) 33
Yeoh, B.L. (5) 324
Yergey, A.L. (7) 102, 120
Yergey, J.A. (5) 186
Yeung, E.W.K. (9) 492
Yinon, J. (5) 246
Yokogama, S. (6) 296
Yokoi, K. (5) 390
Yokokawa, T. (11) 153
Yokoyama, S. (8) 8
Yonekawa, W.D. (6) 284
Yonekawa, Y. (4) 118; (6) 176
Yongskulrote, W. (10) 34
Yonker, C.R. (3) 88; (7) 2, 67

Yoo, H.-S. (1) 230
York, P. (9) 47
York, W.S. (6) 248
Yoshida, K. (9) 156
Yoshida, T. (3) 45
Yoshida, Y. (3) 45
Yoshikawa, M. (9) 112
Yoshimura, H. (9) 324, 352
Yoshitake, A. (6) 95
Yost, R.A. (5) 231, 274; (6) 184, 384; (7) 198, 199, 203, 214, 222, 234; (9) 42–44
Young, A.B. (2) 68, 105, 118, 123
Young, C.E. (3) 28
Young, D.C. (6) 159
Young, I.M. (6) 263
Young, J.P. (3) 29
Young, M.A. (9) 127
Yousefnejad, D. (9) 132, 564, 596
Yu, L. (9) 147
Yu, Q.T. (6) 166, 167
Yudkoff, M. (6) 285
Yun-Kuang Huang, J. (11) 171
Yurkov, L.F. (11) 69, 70

Zabriskie, N.A. (7) 43
Zaccara, G. (9) 658
Zagorevskii, D.V. (10) 145
Zaharevitz, D.W. (6) 278
Zahid, N. (9) 263
Zaikin, V.G. (6) 152
Zajfman, D. (1) 261
Zappey, H. (7) 187
Zappoli, R. (9) 658
Zare, R.N. (1) 73, 80, 81; (3) 23
Zaretskii, Z.V.I. (2) 49, 49
Zaugg, S.D. (7) 54
Zbaida, S. (9) 671
Zebelman, A.M. (9) 574
Zech, W. (4) 189
Zechman, J.M. (4) 27

Zeevaart, J.A.D. (5) 168
Zelei, B. (6) 439
Zelenov, V.V. (4) 173
Zemaitis, M.A. (9) 274
Zenkevich, I.G. (4) 171
Zenser, T.V. (9) 378
Zerilli, L.F. (9) 373, 428
Zewail, A.H. (1) 93
Zhang, J.Y. (6) 166, 167
Zhang, M. (9) 686
Zhang, Q. (1) 207; (5) 53
Zhang-Liang, Z. (5) 333
Zhao, F.Z. (5) 379
Zhaolin, L. (6) 112
Zhilkova, O.Y. (9) 491
Zhilnikov, V.G. (9) 510
Zhivkova, Z. (9) 316
Zhon, C. (5) 241
Zhou, Z.M. (9) 111
Zhu, N.-Z. (3) 158
Zhuravleva, L.V. (11) 18, 29, 37, 105, 114
Ziegler, J.M. (9) 398, 625
Ziegler, M.L. (10) 65
Ziegler, W. (9) 533
Zielinska, B. (5) 255; (6) 336
Zingaro, R.A. (5) 459
Zini, R. (9) 667, 668
Zitrin, S. (6) 385
Zlatkis, A. (6) 297; (9) 50
Zoeller, J.H. (5) 459
Zoepfl, H.J. (7) 12
Zorn, J. (7) 61
Zsolnai, L. (10) 59, 74–77
Zuev, A.P. (5) 76
Zuhowski, E.G. (9) 321
Zulah, I.M. (5) 185
Zumberge, J.E. (4) 94; (6) 425
Zummack, W. (2) 162
Zunbo, M. (6) 152
Zupan, J. (4) 132
Zykov, B.G. (5) 75, 80